ANNUAL REVIEW OF ECOLOGY AND SYSTEMATICS

VOLUME 22, 1991

RICHARD F. JOHNSTON, *Editor*
University of Kansas

PETER W. FRANK, *Associate Editor*
University of Oregon

FRANCES C. JAMES, *Associate Editor*
Florida State University

ANNUAL REVIEWS INC. 4139 EL CAMINO WAY P.O. BOX 10139 PALO ALTO, CALIFORNIA 94303–0897

ANNUAL REVIEWS INC.
Palo Alto, California, USA

International Standard Serial Number: 0066–4162
International Standard Book Number: 0–8243–1422-0
Library of Congress Catalog Card Number: 71-135616

⊗ The paper used in this publication meets the minimum requirements of American National Standard for Information Sciences—Permanence of Paper for Printed Library Materials, ANZI Z39.48-1984.

Typesetting by Kachina Typesetting Inc., Tempe, Arizona; John Olson, President; Janis Hoffman, Typesetting Coordinator; and by the Annual Reviews Inc. Editorial Staff

PRINTED AND BOUND IN THE UNITED STATES OF AMERICA

ANNUAL REVIEW OF ECOLOGY AND SYSTEMATICS

PREFACE

Volume 22 of the *Annual Review of Ecology and Systematics* (ARES) is the first to appear without the name of Charles D. Michener on it. Dr. Michener, one of the two original associate editors, retired as of January 1, 1991. The series will continue on its way much as before, of course, but in a fundamental way it will not be quite the same. This is because "Mich," as he is known to his friends and associates, was its editorial father.

In the 1960s, Dr. J. Murray Luck, a Stanford biochemistry professor and founder of Annual Reviews Inc., determined that an *Annual Review of Ecology and Systematics* would be an appropriate addition to the family of some 14 annual reviews that had been successfully launched since the initial appearance of the *Annual Review of Biochemistry* in 1932. In his search for an editor for this new series, Dr. Luck spoke to Dr. Michener who said he thought that such a new series was a good idea and could fill an important niche in the review literature. At the same time, he made it clear that he would not be interested in the editorship. As a result of their further discussions, and with the ultimate approval of the Board of Directors of Annual Reviews Inc., I became the editor, with Mich and Dr. Peter W. Frank, as associate editors. All of us, including the additional 5 members of the first editorial committee, came together in mid-year of 1968 to plan the first volume.

The intervening 23 years have seen a gratifying acceptance of the *Annual Review of Ecology and Systematics,* validating Mich's perception that the time was right for such a series, as well as his insights into the contributions and contents for each volume to date. His telling relationship to the ongoing production of ARES was always evident at the annual meetings of the editorial committee, when his commentary on the proposed list of topics and authors was broad-ranging, weighted with the right detail, and focussed on issues rather than personality. I considered Mich to be the professional soul of the overall editorial process, as well as of our annual meetings.

We shall miss his participation in the future production of ARES, but we have the past 21 volumes as an historic record of his indispensable contribution to this series. Volumes yet to come will also be a part of his legacy to the community of systematics and ecology. For the present, which can be neither history nor legacy, I think of it as belonging to Mich, to whom Volume 22 is affectionately dedicated.

Richard F. Johnston

Annual Review of Ecology and Systematics
Volume 22, 1991

CONTENTS

(*continued*)

INDEXES

RELATED ARTICLES FROM OTHER *ANNUAL REVIEWS*

From the *Annual Review of Entomology*, Volume 36 (1991)

Evolution of Oviposition Behavior and Host Preference in Lepidoptera, J. N. Thompson and O. Pellmyr

Biosystematics of the Chewing Lice of Pocket Gophers, R. A. Hellenthal and R. D. Price

Sampling and Analysis of Insect Populations, E. Kuno

Transmission of Retroviruses by Arthropods, L. D. Foil and C. J. Issel

Behavioral Ecology of Pheromone-Mediated Communication in Moths and Its Importance in the Use of Pheromone Traps, J. N. McNeil

Lyme Borreliosis: Relation of Its Causative Agent to Its Vectors and Hosts in North America and Europe, R. S. Lane, J. Piesman, and W. Burgdorfer

From the *Annual Review of Genetics*, Volume 25 (1991)

The Inheritance of Acquired Characteristics, O. C. Landman

Gene Transfer Between Distantly Related Bacteria, Julian Davies

Genetic Mechanisms for Adapting to a Changing Environment, D. A. Powers

Ten Unorthodox Perspectives on Evolution Prompted by Comparative Population Genetic Findings on Mitochondrial DNA, J. C. Avis

Spontaneous Mutation, J. Drake

From the *Annual Review of Phytopathology*, Volume 29 (1991)

Research Relating to the Recent Outbreak of Citrus Canker in Florida, R. E. Stall and E. L. Civerolo

Integration of Molecular Data with Systematics of Plant Parasitic Nematodes, B. C. Hyman and T. O. Powers

Maintaining Genetic Diversity in Breeding for Resistance in Forest Trees, G. Namkoong

Factors Affecting the Efficacy of Natural Enemies of Nematodes, R. M. Sayre and D. E. Walter

From the *Annual Review of Plant Physiology and Plant Molecular Biology*, Volume 42 (1991)

The Roles of Heat Shock Proteins in Plants, E. Vierling

The Role of Homeotic Genes in Flower Development and Evolution, E. S. Coen

Functional Aspects of the Lichen Symbiosis, R. Honegger

The Self-Incompatibility Genes of Brassica: *Expression and Use in Genetic Ablation of Flora. Tissues*, J. B. Nasrallah, T. Nishiol, M. E. Nasrallah

ANNUAL REVIEWS INC. is a nonprofit scientific publisher established to promote the advancement of the sciences. Beginning in 1932 with the *Annual Review of Biochemistry,* the Company has pursued as its principal function the publication of high quality, reasonably priced *Annual Review* volumes. The volumes are organized by Editors and Editorial Committees who invite qualified authors to contribute critical articles reviewing significant developments within each major discipline. The Editor-in-Chief invites those interested in serving as future Editorial Committee members to communicate directly with him. Annual Reviews Inc. is administered by a Board of Directors, whose members serve without compensation.

Annu. Rev. Ecol. Syst. 1991. 22:1–18

EVOLUTIONARY RATES: STRESS AND SPECIES BOUNDARIES

P. A. Parsons

Department of Zoology, University of Adelaide, Adelaide, South Australia 5000, Australia

KEY WORDS: metabolic rate, phenotypic plasticity, paleontology, marginal populations, environmental probe

---

INTRODUCTION

From the molecular to the biogeographic levels of organization, stressful environmental conditions underlie much evolutionary change. An understanding of evolutionary rates therefore depends upon a fusion of ecology and genetics especially under stressful conditions (59). Following Hoffmann & Parsons (30), this article defines stress as an environmental factor—usually a physical one—that causes a potentially injurious change in a system. Defined in this way, stress has a major impact on many evolutionary and ecological processes since it will cause irreversible change irrespective of the density of organisms. At the limits of resistance, stress can impose high selection intensities with resultant rapid population shifts (4). Even though fundamental changes in the natural order of systems are likely at these times, much of the literature of evolutionary biology focusses upon more benign situations (30). However, extreme stress periods are unpredictable in occurrence and tend to be of short duration so that observations covering many generations are needed for their doumentation.

There are some situations where severe stress can be regarded as an "environmental probe"; analysis of such situations leads to generalizations difficult to perceive under more optimal conditions. For example, biological systems can be described in terms of energetic costs, from which suggestive associations between habitat, life history characteristics, and stress resistance

0066-4162/91/1120-0001$02.00

have emerged (28–30). Indeed, in many organisms phenotypic and genotypic variability tend to be high (38, 59) under severe environmental stress (provided that the stress is not so severe as to cause death). Because much relevant data derives from experiments not primarily concerned with stress and variability, it is not surprising that there are exceptions to this generalization. In many agricultural plants, heritability has been found to change with stress levels, but without consistent trends. To some extent this may reflect the complication of phenotypic plasticity which tends to be high in plant species and populations experiencing non-extreme environmental fluctuations (30). Given the need for additional work, especially extrapolations to field conditions, there is now sufficient evidence to accept the association between stress and variability as a working hypothesis (in populations not previously subjected to natural selection for the stress in question). There are supportive data for the association on recombination, temperature-dependent catalytic properties of enzyme systems, and morphology assessed by fluctuating asymmetry (36, 60, 62).

At the metabolic level, stressful conditions normally require the continuous expenditure of excess energy which ultimately becomes incompatible with survival. Rates of energy metabolism have been measured in numerous species including many living under demonstrably stressful conditions. For example, in species exposed to xeric desert conditions, a lower-than-expected specific metabolic rate has been observed in Peromyscus (44) and subterranean mole-rats (40), and in invertebrates including the desert harvester ant, *Pogonomyrex rugosus,* and carabid and tenebrionid beetle species (39, 88). More generally, low metabolic rate is a common feature of species living in environments with decreased food availability and increased high temperature or desiccation stress (30). Those few studies addressing the association between stress resistance and metabolic rate at the intraspecific level such as in the mole-rat, *Spalax ehrenbergi* (50) and in beef cattle (22) give parallel conclusions. In particular, desiccation-resistant selection lines of *Drosophila melanogaster* have decreased metabolic rate, fecundity (a fitness measure), and behavioral activity compared with unselected control lines (28, 29). Similarly, lines selected for tolerance to acrolein, which is an unsaturated aldehyde and an atmospheric pollutant, have reduced metabolic rate and spontaneous locomotor activity (74).

In summary, an organism can increase its tolerance to a wide range of stresses by reducing its metabolic energy requirement, so that individuals genetically tolerant to stress should have a reduced metabolic rate. This is associated with reduced fitness measured under nonstressful conditions.

Species boundaries typically have environmental stress levels exceeding those of more central areas and, from the above considerations, the likelihood of higher genotypic and phenotypic variability. However, stress causes a drain in metabolic energy resources rendering organisms susceptible to further stressful environmental perturbations which would reduce the possibility of adaptation to more extreme conditions. If genetic responses to a stress do occur, populations would lose their ability to respond further because a correlated response to selection is likely to be diminished available metabolic energy expressed as reduced fitness. Simplistically, therefore, species boundaries should be characterized by high levels of environmentally induced variability, but the concomitant metabolic costs may preclude range expansions into more stressful habitats. On this argument, rapid evolutionary change may not be the expectation for populations in ecologically marginal habitats. The interaction between the enhancement of variability by environmental stress and its metabolic cost is the major theme of this article.

STRESS: AN ORGANISMIC APPROACH

New reactions of organisms are unlikely to be totally adaptive (21) since each genotype is characterized by its specific "norm of reaction," which includes adaptive modifications in response to different environments (42, 71). Progressive evolution then consists of the continuous acquisition of new reaction norms, especially in response to substantial environmental perturbations; these include climatic stress which is demonstrably important in the determination of physiological races within species, and the location of species borders at both the individual species and ecosystem levels (11, 16, 30, 42, 51, 61, 71).

Ecologists such as Andrewartha & Birch (1) emphasize quantitative traits (especially those related to physical features of the environment) important in determining the distribution and abundance of organisms, based upon the assumption that the organism is the unit of selection. Indeed, in *D. melanogaster,* such ecological phenotypes tend to vary clinally in a manner determined by *a priori* considerations of geographic variation in climate (78). Periods of short-term high and low temperature stresses are the most effective in determining climatic races. This is leading to a slowly increasing emphasis upon genetic variation for resistance to extreme stresses in natural populations (55, 56, 59).

Other ecologists, such as Lack (37), emphasize interactions between organisms in the form of competition. Andrewartha & Birch (1) studied insect populations which often fluctuate in response to physical factors, while Lack (37) worked with birds where population sizes tend to be less variable. These divergent views appear partly to be a reflection of the groups of organisms being studied. This continuing debate has resulted in a dichotomy between an emphasis upon survival under climatic extremes vs traits associated with competitive ability (70). Indeed, approaches to the evolutionary ecology of

Drosophila range from those in which environmental stress is almost ignored and the focus is upon competition (48) to those in which environmental stress predominates (55, 59). However, at species boundaries, the extreme end of the stress gradient in terms of resistance is approached so that the persistence of an environmental stress will result in nonreversible damage and mortality irrespective of the density or frequency of organisms. This means that competitive interactions would tend to be unimportant. Even so, competition may interact with stress by influencing levels of stress resistance so that a narrowing of tolerance limits occurs (64), but such effects would not be long term especially under the environmental perturbations expected at species boundaries (30).

STRESS: METABOLIC COST

Organisms unable to respond to an environmental stress enter a stressed state measurable at the biochemical, morphological, physiological, or behavioral levels. Some methods of assessment are limited because of low sensitivity, a need for long exposures, inconsistencies, and difficulties in extrapolating across species (34). However, a general early warning signal of the stressed state is likely to be an increase in respiration, since repairing damage caused by stress may require increased energy expenditure. This means that the energy available for growth and reproduction will be reduced compared with maintenance requirements (52, 53). Stressful conditions including salinity, heavy metals, oxygen deprivation, extremes of heat and cold, and desiccation all pose costs for the maintenance of normal protoplasmic homeostasis and the normal functioning of membranes and enzymes.

Monitoring changes in the concentrations of energy carriers provides a widely applicable biochemical approach for stress assessment (34). One measure is the adenylate energy charge (AEC) = ATP + ½ ADP/ATP + ADP + AMP, where ATP, ADP, and AMP are measured amounts of adenosine triphosphate, diphosphate, and monophosphate, respectively. The AEC is a measure of the amount of metabolically available energy stored in the adenine nucleotide pool and can be used to express the deviation of an organism from a steady metabolic state. Under stress, the AEC is reduced due to a drain of metabolic energy resources, and as a consequence the vulnerability of the organism to further environmental stresses is increased. Quantitatively, AEC is 0.8–0.9 in optimal and nonstressed organisms, 0.5–0.7 when conditions are limiting or nonoptimal (but upon return to normal conditions the characteristics of this state are resumed), and 0.5 when conditions are severely stressful and a return to normal conditions is associated with some unrestored fitness losses.

The AEC approach indicates that an organism may increase its resistance to stress by a reduction in metabolic rate and predicts that (*a*) organisms genetically resistant to environmental stress should have a lowered metabolic rate, and (*b*) resistance to different stresses should be positively correlated because genes that decrease metabolic rate are likely to increase resistance to a number of stresses. Evidence for prediction (*a*) has already been cited. In desiccation resistance selection experiments in *D. melanogaster* (28, 29), resistances to a number of generalized environmental stresses were genetically associated; these included starvation, toxic levels of ethanol and acetic acid, irradiation with extremely high doses of Co60-γ rays, and high temperatures in accord with prediction (*b*) (Table 1). Genotypes with low rates of metabolism may then be favored under a wide range of stresses. Exceptions would include biochemical changes largely specific for a particular kind of stress; for example, increased resistance to insecticides may involve mechanisms specific to a particular class of insecticides (54), and heavy metal resistance in grasses is usually specific to a particular metal (79).

The correlated responses in Table 1 suggest that a substantial portion of the genetic variation in resistance to different environmental stresses has a common basis. This is confirmed by correlations between the desiccation, ethanol, and starvation resistances of isofemale lines set up from the F_2s of a cross between one of the selected and one of the control lines (29), as well as correlations between 26 isofemale strains set up from a natural population (Table 2). In lines of *D. melanogaster* selected for postponed senescence, which showed increased resistance to starvation, desiccation, and ethanol (73), an association between stress resistance and lipid content was found, but this did not occur in the lines in Table 1. However, this does not preclude qualitative rather than quantitative changes, especially as molecular studies in plants and animals have identified specific components of membrane lipids that may contribute to stress resistance (24). Investigations of this nature need to be carried out on populations under stress at species boundaries.

A mammalian example providing a model as the limits of the distribution of a species is approached comes from cattle selected for growth rate under moderate to highly stressful tropical conditions when growth is normally suppressed because of ecto- and endoparasites, high temperatures, poor nutrition, and eye diseases (22). The selected line was more heat tolerant, had lower maintenance requirements and greater resistance to infection compared with the controls. As a consequence, this line had a higher growth rate in the presence of these stresses, which was achieved through increases in resistance to environmental stresses associated with a fall in metabolic rate. In other words, while increased resistance to stress appears readily achievable under natural and artificial conditions, at the ultimate limit, the fall in metabolic rate is restrictive in determining the habitat range of a species. Similarly, the oxygen consumption of the desert harvester ant, *Pogonomyrex rugosus*, of the

Table 1 A comparison of lines of *D. melanogaster* selected for desiccation resistance with control lines.

	Metabolic rate[a] ($O_2mg^{-1}hr^{-1}$)	Activity[b] Low humidity	Activity[b] High humidity	Fecundity[c]	Starvation[d]	Ethanol[e]	Acetic Acid[e]	Heat[f]	Radiation[g]
Selected	2.1	10.4	6.0	131	98	63	23	0.30	0.36
Control	2.6	12.6	9.0	162	79	28	12	0.55	0.75
Significance Level	< 0.01	< 0.05	< 0.05	< 0.01	< 0.001	< 0.01	< 0.001	< 0.01	< 0.01

[a] $O_2mg^{-1}hr^{-1}$ based upon 20 flies
[b] Number of flies (out of 15) that moved in a one-minute interval
[c] Total eggs laid in 3 successive 24 hour periods at 3–5 days of age
[d] LT50 in hours for 20 flies in a humid atmosphere
[e] LT50 in hours for 20 flies
[f] Proportion of 20 flies knocked down in vial after 1.5 hr at 37°C
[g] Proportion of 20 flies dead 22 hours after exposure to Co60γ irradiation of dose 1.2 kGy

Table 2 Correlations between 26 isofemale strains of *D. melanogaster* for resistance to three stresses (defined in table 1).

	Ethanol	Desiccation	Starvation 1[a]	Starvation 2[a]
Ethanol				
Desiccation	0.65[d]			
Starvation 1[a]	0.41[b]	0.53[c]		
Starvation 2[a]	0.44[b]	0.53[c]	0.98[c]	

[a] Two experiments were carried out on starvation
Significance levels: [b] $P < 0.05$, [c] $P < 0.01$, [d] $P < 0.001$

Mojave Desert is only 39% of that predicted from the allometric relationship of standard body weight and mass (39).

Based upon the premise that external environmental factors such as temperature are primary in shaping biogeographic ranges, a continent-wide analysis of avian (passerine) distributions showed that the winter distribution and abundance patterns of several avian species are directly related to their physiological demands (67). An association of temperature extremes and physiological mechanisms was found at species borders, whereby birds are limited to regions where they do not raise their metabolic rates beyond about 2.5 times the basal level in order to remain warm. This means that on a broad scale, the winter ranges of a large number of passerines are limited by the energy expenditure needed to compensate for colder ambient temperatures. Similar arguments for metabolic limits to distribution have been presented in rodents (6). More generally, metabolic rates and costs at species boundaries need detailed investigation, especially as the increased cost of maintenance under stress can be substantial; for example, in the crustacean, *Cancer irroratus,* maintenance costs were doubled when larvae were exposed to copper and cadmium (35), and sea otters, *Enhydra lutris,* nearly doubled their metabolic rate following exposure to crude oil from a damaged tanker (17). Looked at in another way, species have a characteristic sustained metabolic rate covering normal activities which falls between 1.5 and 7 times the resting metabolic rate; when these limits are consistently exceeded, the existence of the species is threatened (63).

MARGINAL POPULATIONS: GENOTYPES AND PHENOTYPES

The proposition that physiological constraints restrict range extensions into increasingly harsh environments leads to the various hypotheses and/or explanations in the literature for species boundaries (2, 30, 42). The main ones have been recently summarized (30):

1. Low genetic variation in stress response traits: (*a*) resulting from directional selection and physiological limits, (*b*) as a consequence of population structure, or (*c*) expressed under stressful conditions,
2. Restrictions due to a requirement for simultaneous genetic change in several characters simultaneously,
3. Negative genetic correlations between stressful and favorable conditions, or among fitness traits under stressful conditions,
4. Gene flow from central populations that may swamp genotypes favored in marginal populations,
5. Small population sizes that may lead to lowered fitness via inbreeding,
6. Lack of response to directional selection especially under circumstances of heterozygote advantage under stress, and
7. Dispersal ability that may be restricted to limits in genetic variation.

In general, information to distinguish between these alternatives is insufficient, although some data indicate that genetic variation at species boundaries may be limiting. One example concerns coastal Australian populations of *D. melanogaster,* where populations from Melbourne and Darwin have higher desiccation resistance than those from Brisbane and Townsville, as predicted from climatic considerations (56, 78). Populations from the more stressful environments have reduced genetic variation, presumably because climatic selection is more intense in these localities. Such studies are needed on less cosmopolitan species than *D. melanogaster* to enable easier identification of climatically marginal populations. Genetic variability assessments are more frequently based upon variation at electrophoretic loci; however, natural selection primarily acts upon ecologically important traits at the phenotypic level. The diffuse debate on the causes of genetic variability levels in central vs. marginal populations (8, 58, 77) reflects a heavy dependence upon electrophoretic data. It seems more appropriate to commence at the phenotypic level before reducing the argument to the genetic level. It is, however, encouraging that analyses of electrophoretic data on a world-wide basis emphasize climatic selection involving extremes (49, 59) so indicating convergence between the alternative approaches.

Marginal population studies should therefore place more emphasis upon genetic variation in quantitative phenotypic traits important in determining the location of species boundaries. These include resistance to climatic stresses and the metabolic and general fitness costs associated with extremes. More generally, in understanding evolutionary rates, environmental stress resistance and metabolic rates are important primary phenotypic traits; studies commencing at the genetic level, especially those involving electrophoretic variants (26), are likely to be incomplete.

METABOLIC RATE, STRESS, AND EVOLUTIONARY CHANGE

Metabolic Rate

In Britain, the diversity of butterflies and nocturnal moths is highly correlated with sunshine and temperature during the growing season (83). This correlation suggests that species numbers are strongly influenced by the amount of energy available during this season and supports the species-energy hypothesis of Wright (87), whereby species diversity is causally regulated by solar thermal energy. While conclusions based upon multiple regressions must be regarded as provisional, the finding of parallel effects upon diversity in ectothermic butterflies and heavy bodied, endodermic, nocturnal moths appears convincing.

The measurement of energy intake and utilization by animals is a major area of study within ecological physiology. Energy availability has often been found to impose constraints on animals, so that low energy environments are restrictive. Indeed it can be argued (84) that high metabolic rates enable organisms to achieve a greater degree of specialization in a given direction along a pathway of adaptation than do low metabolic rates. This means that high metabolic rates may increase the scope for adaptation, the opportunity for selection, and ultimately speciation.

Some physiological processes and behaviors are so energetically expensive that they require nearly all of the energy intake of an animal. For example, the energetic cost of maintaining a temperature excess during sustained endothermy in insects is very substantial (47). Energetically expensive behaviors include vocalization by frogs, associated with mating (66), and web-building in spiders (80). Indeed, arguments have been presented for an association between metabolic rate and fitness based upon reproductive success in various organisms (20, 66). The association does, however, break down at extremely high metabolic rates, as shown by various "shaker" mutants in *D. melanogaster* which have a high metabolic rate and reduced longevity (82). For example, hyperkinetic, Hk^{1p}, males tend to switch rapidly from one behavior to another, and they have reduced mating speed caused by the frequent loss of females during courtship (10). Extremes of high and low metabolic rates are therefore associated with reduced fitness, as expected (21, 42). In spite of this qualification, an association between metabolic rate and fitness is a working hypothesis in this article, while it also emphasizes the need for extensive investigations at the intraspecific level. The working hypothesis is supported, for example, by a generalization of calling energetics in frogs, whereby calling is primarily supported by aerobically generated ATPs (69) thus providing a link with the AEC approach for stress assessment.

Stress

As already discussed, extrinsic changes such as in climatic conditions also appear important for much evolutionary change. This implies a certain level of stress, as was recognized over 60 years ago (21): "Any environmental heterogeneity which requires special adaptations, which are either irreconcilable or difficult to reconcile, will exert upon the cohesive power of the species a certain stress."

In recent years, there have been extensive studies on speciation in the true fruitfly, *Rhagoletis pomonella* (12). About 150 years ago in North America, a sympatric shift occurred from its native host, hawthorn, to the introduced domestic apple. There is now differential host utilization by hawthorn and apple flies resulting in a system of positive assortative mating which helps to maintain genetic variation between the host races (19). Ambient temperature may form one suggested environmental factor that differentially affects apple and hawthorn flies. Because *R. pomonella* is univoltine with a life expectancy of only 3–6 weeks in the field, flies must eclose as adults at times that closely match the fruit phenologies of their host plants. Since apples generally ripen and abscise 3–4 weeks ahead of hawthorns the temporal windows for infesting the fruits differ. In the utilization of apple fruits from hawthorns, a period of substantial climatic stress for selection for development under a new temperature regime is therefore implied.

In the cactophilic *Drosophila mulleri* species complex, *Opuntia*-breeding is considered to be the ancestral state, and in a shift to columnar cactus breeding, fitness tends to decline (68). This switch appears most likely to be accomplished in isolation when there is neither gene flow from *Opuntia*-breeding populations, nor competition from other species occupying the columnar habitat. The force behind this shift could have been an increase in temperature and aridity causing a major decrease in the density of Opuntias and a concomitant increase in columnars. This shift could happen under desert conditions, so that speciation appears to be associated with climatic stress which occurs during the resource shift from one host plant to another, thus providing a parallel with the *R. pomonella* situation.

This is not the place to enter into discussions of modes of speciation (13, 15, 81) except to comment that environmental stress may have been rather underemphasized as a factor. In this connection, it is important to note that ecological systems appear to be the most diverse when subjected to a certain degree of stress (46). Indeed the greatest number of species is usually seen at intermediate levels of disturbance (65). Clearly, the stress level cannot normally go much beyond the resistance limits of the most sensitive species, since species extinctions would follow. On the other hand, stress could lead to population bottlenecks of very few individuals. In some circumstances additive genetic variance is increased at this time through the disruption of interrelationships between quantitative characters, and this is a situation favorable for character change and perhaps speciation via founder events (14).

Evolutionary Change

In the determination of evolutionary rates, a case can therefore be made for two important interacting continuums: (*a*) the availability of metabolic energy beyond maintenance needs, and (*b*) the intensity of stress. At species boundaries, metabolic energy requirements become restrictive, since energy availability is likely to be reduced to maintenance levels only. Yet these are the regions where a consequence of stress is likely to be high genetic variability which could, in principle, underlie rapid phenotypic shifts. Conversely, in the benign habitats of more central regions of the distribution of a species, there would be available metabolic energy permitting adaptive change; however, these are the habitats where the genetic variability of traits important in determining distribution and abundance may be low. On these arguments, evolutionary change and speciation may be most likely in populations from habitats between the extremes, where there is a certain degree of environmental stress associated with sufficient metabolic energy to accommodate some adaptive change. In parallel with these rather theoretical arguments (62a), it is fascinating that in observations on the evolution of Ordovician trilobites, rapid phyletic evolution appears sustainable under narrowly fluctuating, slowly changing physical environments, while stasis may prevail under widely fluctuating, rapidly changing, physical environments (75, 76) where stress levels are presumably maximized.

PHENOTYPIC PLASTICITY

Genetic considerations predominate in many discussions of evolutionary change. However, the analysis of environmental effects upon the phenotype and their evolutionary consequences is now attracting attention (30, 86). Phenotypic plasticity can be defined as the amount by which the expression of individual characters of a genotype are changed by different environments (7), and it encompasses an enormous diversity of variability especially in plants grown in stressful environments. Animals tend to have less morphological plasticity than plants, but many physiological processes influencing stress resistance show a high degree of plasticity.

Ectotherms often undergo changes in metabolic rate in response to temperature, and this can be modified by acclimation. Following acclimation at warm temperatures, the respiration rate of individuals at intermediate temperatures is slowed, while the reverse occurs following acclimation at cold temperatures (9, 72). In *D. melanogaster,* high temperature increases metabolic rate. However, the metabolic rate of females grown at 25°C and 30°C was lower than that of flies grown at 15°C when tested in the 20–30°C range (31). In this way organisms can maintain a similar metabolic rate over a range of temperatures, so minimizing the direct effects of extremes. Metabolic acclimation similar to that of *D. melanogaster* has been recorded in *D. immigrans* and *D. hydei* but not in *D. willistoni* (32, 33).

These species differences appear relatable to habitat differences. Organisms living in variable environments should exhibit an increased ability to counter the effects of environmental change thereby minimizing variation in metabolic rate and other physiological processes. In this regard, *D. willistoni,* a largely tropical species, is more restricted in its distribution across diverse habitats than are the other three species, which occur mainly in temperate regions where temperature fluctuations are larger. These observations therefore suggest that the ability of organisms to undergo acclimation is under genetic control.

More generally, the differences in plasticity levels between species and populations can be interpreted in terms of the adjustment by natural selection of plasticity levels in response to the degrees of environmental variability experienced by a population. This suggests that plasticity levels may be influenced by selection for stress resistance. For example, increased stress resistance could reduce the plasticity level if the same mechanism controls the plastic response and genetic variation in stress resistance. While a number of plants from differing habitats fit this explanation (30), direct evidence that stress resistance and acclimation responses are associated comes from the *D. melanogaster* lines selected for increased desiccation resistance (Table 1). Following the observation that a nonlethal period of desiccation increases subsequent desiccation resistance in *D. melanogaster,* the data in Table 3 show that acclimation to desiccation was markedly effective in the control lines, but in the selected lines little change occurred (25). The response in the unselected lines developed within 2 hr of exposure to low humidity and persisted for more than 28 hr after the prior stress period. Hence increased resistance to the stress of desiccation reduces the plastic response to a nonlethal period of desiccation stress. In addition, prior exposure to a heat shock causes a smaller increase in desiccation resistance (25).

The potential importance of desiccation and high temperature resistance in the ecology of *Drosophila* populations suggests that acclimatory responses can influence the fitness of flies in the wild. In particular, a prior nonlethal period of desiccation/high temperature stress could have a marked effect on desiccation resistance, and such periods may be encountered by flies at certain times of the year. While there are no data on the desiccation stress experienced by flies in the wild, fecundity data indicate that stressful conditions are often experienced (5).

Organisms that rarely experience stressful conditions seem to show poorer

Table 3 Desiccation resistance of selected and unselected lines of *Drosophila melanogaster* with and without a prior acclimation period. Lines were selected for increased desiccation resistance as described in (28). Flies were acclimated by placing them at 0% humidity for 9 hr before allowing them to recover for 9 hr (after 25).

	Desiccation Resistance[a]	
	Non-acclimated	Acclimated
Selected lines		
line 1	23.6	24.8
line 2	22.1	23.2
line 3	24.2	25.0
Control lines		
line 1	16.2	19.9
line 2	17.7	21.4
line 3	15.0	19.2

[a] Mean time taken for 50% of the females to die (in hours), based on six replicate groups of 20 females.

acclimatory ability than those from more variable environments. Undoubtedly, maintaining a plastic phenotype is likely to entail a metabolic cost that leads to selection against genotypes with a high level of plasticity when variable environments do not occur (7, 23). Hence, it is to be expected that the reduced metabolic rate associated with the acquisition of increased stress resistance (as found experimentally in *D. melanogaster*) will be associated with a lower level of acclimation (25). This will be the expected situation as species borders are approached.

High phenotypic plasticity is likely under five conditions (30): (*a*) stress periods are frequent, (*b*) individuals experience substantial environmental variability, (*c*) the absolute magnitude of environmental change is not excessive, (*d*) environmental cues show reliability, (*e*) the metabolic cost of plasticity should not be excessive. At the end of the stress continuum at species boundaries, the absolute magnitude of environmental change would place a premium on selection to adapt to these changes with associated metabolic costs, and this would preclude the simultaneous development of a high level of plasticity. Plasticity is therefore likely to be maximized under conditions of some environmental stress when simultaneously there is sufficient available metabolic energy to permit plasticity to develop. High plasticity is therefore most likely to occur in habitats similar to those where the interaction between stress intensity and available metabolic energy appears permissive of appreciable evolutionary change.

PHENOTYPIC VS GENOTYPIC ASSESSMENTS

In spite of the emphasis upon metabolic rate for understanding evolutionary rates, there are few studies of its variation in natural populations. The gene-to-physiology approach (3) has been mainly used, for example, in elegant studies on the impact of phosphoglucose isomerase (PGI) variation on flight capacity, survivorship, and mating success in butterfly populations (85). While this approach can reveal pathways from gene to phenotype, it cannot tell us the number of genes nor explore the interactions between genes with respect to a particular physiological trait, so it is largely restricted to single loci where much of the genetic variation is directly translatable into phenotypic variation. An example of the difficulty in this approach comes from alcohol dehydrogenase variation which only explains a small proportion of ethanol tolerance levels in *D. melanogaster* (43, 45) so that the approach is very limited in this situation.

Comparative physiology has usually focussed upon contrasts between populations, species, and higher taxa. The analysis of variation within populations, which requires the quantitative analysis of physiological traits, has been largely overlooked and is necessary for exploring relationships between energetics and fitness. Physiological traits based upon the organism as the unit of selection should form the basic data. In genetically well-known organisms, genetic activity can be localized to chromosomes and regions of chromosomes, especially if extreme strains from a population, or those derived by selection procedures, are used. These analyses should reveal involvements of genetic markers, such as electrophoretic loci, with physiological traits in a more direct manner than from using variation at electrophoretic loci as the primary assessment. This physiology-to-gene approach is theoretically fully informative and is based upon the methods and techniques of quantitative genetics as discussed elsewhere (27, 58) with reference to ecologically important traits.

This article concerns the study and interaction between two categories of quantitative traits—metabolic rate and stress resistance. Simultaneous intraspecific-level studies are needed on these traits throughout the distribution of a species. Isofemale strains (the progeny of a set of single inseminated females taken at random from the wild) provide an effective way of investigating variation within and between populations. This is because isofemale strains tend to maintain their own unique characteristics derived from the founder female for a number of laboratory generations (57), although meaningful heritability estimates require maintenence at population sizes exceeding 50 and testing within five generations of establishment (27). The pattern of variation within and between isofemale strains suggests that loci of large effect contribute to most variation in morphological and some environmental stress traits (18, 41). Therefore the analysis of quantitative traits to the gene level is feasible. Even without such sophisticated analyses, isofemale strains enable quick inferences to be made at the organismic level for any species that can be cultured in the laboratory, and derived from environments ranging from optimal to extreme, and for traits ranging from molecular and physiological to ecological and behavioral. Studies of pairs of traits based upon many isofemale strains may be informative concerning correlations, and hence on possible shared developmental and physiological processes as illustrated in Table 2 for a number of stress traits. Combinations of environmental stress traits, fitness traits, and metabolic rate could be studied in this way and would assist in the reduction of the interspecific level conclusions of the ecological physiologist to the intraspecific level. This reduction is a necessary step for an understanding of evolutionary rates at the phenotypic level in populations derived from the differing habitats within the distribution of a species.

CONCLUSIONS

1. Evolutionary rates are discussed in terms of two interacting continuums: (*a*) the availability of metabolic energy, and (*b*) the intensity of environmental stress.

2. Species margins are often relatable to climatic extremes, where metabolic costs may be sufficient to preclude major range expansions even though variability may be high.

3. Various genetical explanations on restrictions to range expansions are secondary to phenotypic approaches.

4. Simultaneous quantitative genetic analyses are needed on metabolic rate and stress resistance, emphasizing relationships with fitness throughout the distribution of a species.

5. In agreement with observations and theory on species diversity, maximum evolutionary rates may occur in moderately stressed regions of the distribution of a species where variability would be higher than in benign regions, and the metabolic costs of stress are not unduly restrictive. Phenotypic plasticity is expected to be the highest under similar conditions. Observations in the geological record are consistent with those conclusions.

6. The level of physical stress, especially climatic, is therefore emerging as a tentative connecting environmental link between paleontology, physiology, ecology, and evolutionary genetics.

ACKNOWLEDGMENT

I am most grateful to Peter R. Sheldon for alerting me to his work on evolutionary patterns of Ordovician trilobites, and for some helpful correspondence.

Literature Cited

1. Andrewartha, H. G., Birch, L. C. 1954. *The Distribution and Abundance of Animals*. Chicago: Univ. Chicago Press
2. Antonovics, J. 1975. The nature of limits to natural selection. *Ann. Missouri Bot. Gard.* 63:224–47
3. Arnold, S. J. 1987. Genetic correlations and the evolution of physiology. In *New Directions in Ecological Physiology*, ed. M. E. Feder, A. F. Bennett, W. W. Burggren, R. B. Huey, pp. 189–215. Cambridge: Cambridge Univ. Press
4. Boag, P. T., Grant, P. R. 1981. Intense natural selection in a population of Darwin's finches (Geospizinae) in the Galapagos. *Science* 214:82–85
5. Bouletreau, J. 1978. Ovarian activity and reproductive potential in a natural population of *Drosophila melanogaster*. *Oecologia* 33:319–42
6. Bozinovic, F., Rosenmann, M. 1989. Maximum metabolic rate of rodents: physiological and ecological consequences on distributional limits. *Func. Ecol.* 3:173–81
7. Bradshaw, A. D. 1985. Evolutionary significance of phenotypic plasticity in plants. *Adv. Genet.* 13:115–55
8. Brussard, P. F. 1984. Geographic patterns and environmental gradients: the central marginal model in *Drosophila* revisited. *Annu. Rev. Ecol. Syst.* 15:25–64
9. Bullock, T. H. 1955. Compensation for temperature in the metabolism and activity of poikilotherms. *Biol. Rev.* 30:311–42
10. Burnet, B., Connolly, K., Mallinson, M. 1974. Activity and sexual behavior of neurological mutants in *Drosophila melanogaster*. *Behav. Genet.* 4:227–35
11. Busby, J. R. 1986. *Bioclimate Prediction System*. Canberra: Bur. Fauna Flora
12. Bush, G. L., Howard, D. J. 1986. Allopatric and non-allopatric speciation: Assumptions and evidence. In *Evolutionary Processes and Theory*, ed. S. Karlin, E. Nevo, pp. 411–38. New York. Academic
13. Carson, H. L. 1987. The genetic system, the deme, and the origin of species. *Annu. Rev. Genet.* 21:405–23
14. Carson, H. L. 1990. Increased genetic variance after a population bottleneck. *Trends Ecol. Evol.* 5:228–30
15. Carson, H. L., Templeton, A. R. 1984. Genetic revolutions in relation to speciation phenomena: The founding of new populations. *Annu. Rev. Ecol. Syst.* 15:97–131
16. Caughley, G., Short, J., Grigg, G. C., Nix, H. 1987. Kangaroos and climate: an analysis of distribution. *J. Anim. Ecology* 56:751–61
17. Davis, R. W., Williams, T. M., Thomas, J. A., Kastelein, R. A., Cornell, L. H. 1988. The effects of oil contamination and cleaning on sea otters *(Enhydra lutris)*. II. Metabolism, thermoregulation, and behavior. *Can. J. Zool.* 66:2782–90
18. Deery, B. J., Parsons, P. A. 1972. Ether resistance in *Drosophila melanogaster*. *Theoret. Appl. Genet.* 42:208–14
19. Feder, J. L., Chilcote, C. A., Bush, G. L. 1988. Genetic differentiation between sympatric host races of the apple maggot fly *Rhagoletis pomonella*. *Nature* 336:616–64
20. Feder, M. E. 1987. The analysis of physiological diversity: the prospects for pattern documentation and general questions in ecological physiology. In *New Directions in Physiological Ecology*, ed. M. E. Feder, A. F. Bennett, W. W. Burggren, R. G. Huey, pp. 38–75. Cambridge: Cambridge Univ. Press
21. Fisher, R. A. 1930. *The Genetical Theory of Natural Selection*. Oxford: Clarendon
22. Frisch, J. E. 1981. Changes occurring in cattle as a consequence of selection for growth rate in a stressful environment. *J. Agric. Sci. Camb.* 96:23–38
23. Heslop-Harrison, J. 1964. Forty years of genecology. *Adv. Ecol. Res.* 2:159–247
24. Hochachka, P. W., Somero, G. N. 1984. *Biochemical Adaptation*. Princeton: Princeton Univ. Press
25. Hoffmann, A. A. 1990. Acclimation for desiccation resistance in *Drosophila melanogaster* and the association between acclimation responses and genetic variation. *J. Insect Physiol.* 36:885–91
26. Hoffmann, A. A., Nielsen, K. M., Parsons, P. A. 1984. Spatial variation of biochemical and ecological phenotypes in *Drosophila*-electrophoretic and quantitative variation. *Dev. Genetics* 4:439–50
27. Hoffmann, A. A., Parsons, P. A. 1988. The analysis of quantitative variation in natural populations with isofemale strains. *Génét. Sél. Evol.* 20:87–98
28. Hoffmann, A. A., Parsons, P. A. 1989. An integrated approach to environmental stress tolerance and life-history variation: Desiccation tolerance in *Drosophila*. *Biol. J. Linn. Soc.* 37:117–37
29. Hoffmann, A. A., Parsons, P. A. 1989. Selection for desiccation resistance in

Drosophila melanogaster: Additive genetic control and correlated responses for other stresses. *Genetics* 122:837–45
30. Hoffmann, A. A., Parsons, P. A. 1991. *Evolutionary Genetics and Environmental Stress*. Oxford: Oxford Univ. Press
31. Hunter, A. S. 1964. Effects of temperature on *Drosophila*-I. Respiration of *D. melanogaster* grown at different temperatures. *Comp. Biochem. Physiol.* 11: 411–17
32. Hunter, A. S. 1966. Effects of temperature on *Drosophila*-III. Respiration of *D. willistoni* and *D. hydei* grown at different temperatures. *Comp. Biochem. Physiol.* 19:171–77
33. Hunter, A. S. 1968. Effects of temperature on *Drosophila*-IV. Adaptation of *D. immigrans*. *Comp. Biochem. Physiol.* 24:327–33
34. Ivanovici, A. M., Wiebe, R. J. 1981. Towards a working 'definition' of 'stress': a review and critique. In *Stress Effects on Natural Ecosystems,* ed. G. W. Barrett, R. Rosenberg, pp. 13–27. New York: Wiley
35. Johns, D. M., Miller, D. C. 1982. The use of bioenergetics to investigate the mechanisms of pollutant toxicity in crustacean larvae. In *Physiological Mechanisms of Marine Pollutant Toxicity,* ed. W. B. Vernberg, A. Calabrese, F. P. Thurberg, F. J. Vernberg, pp. 261–88. New York: Academic
36. Karvountzi, E., Goulielmos, G., Kalpaxis, D. L., Alahiotis, S. N. 1989. Adaptation of *Drosophila* enzymes to temperature-VI. Acclimation studies using the malate dehydrogenase (MDH) and lactate dehydrogenase (LDH) systems. *J. Therm. Biol.* 14:55–61
37. Lack, D. 1966. *Population Studies of Birds*. Oxford: Oxford Univ. Press
38. Langridge, J., Griffing, B. 1959. A study of high temperature lesions in *Arabidopsis thaliana*. *Aust. J. Biol. Sci.* 12:117–35
39. Lighton, J. R. B., Bartholomew, G. A. 1988. Standard energy metabolism of a desert harvester ant, *Pogonomyrex rugosus:* effects of temperature, body mass, group size, and humidity. *Proc. Natl. Acad. Sci. USA* 85:4765–69
40. Lovegrove, B. G. 1986. The metabolism of social subterranean rodents: adaptation to aridity. *Oecologia* 69:551–55
41. Matheson, A. C., Parsons, P. A. 1973. The genetics of resistance to long term exposure to CO_2 in *Drosophila melanogaster,* an environmental stress leading to anoxia. *Theoret. Appl. Genet.* 42:261–68
42. Mayr, E. 1963. *Animal Species and Evolution*. Harvard: Belknap
43. McKenzie, J. A., Parsons, P. A. 1974. Microdifferentiation in a natural population of *Drosophila melanogaster* to alcohol in the environment. *Genetics* 77:385–94
44. McNab, B. K., Morrison, P. 1963. Body temperature and metabolism in subspecies of *Peromyscus* from arid and mesic environments. *Ecol. Monogr.* 33:63–82
45. Merçot, H., Massad, L. 1989. ADH activity and ethanol tolerance in third chromosome substitution lines in *Drosophila melanogaster*. *Heredity* 62:35–44
46. Moore, P. D. 1983. Ecological diversity and stress. *Nature* 306:17
47. Morgan, K. R. 1987. Temperature regulation, energy metabolism and mate-searching in rain beetles (*Pleocoma* spp.), winter-active, endodermic scarabs (Coleoptera). *J. Exp. Biol.* 128:107–22
48. Mueller, L. D. 1988. Evolution of competitive ability in *Drosophila* by density-dependent natural selection. *Proc. Natl. Acad. Sci. USA* 85:4383–6
49. Nevo, E., Beiles, A., Ben-Shlomo, R. 1984. The evolutionary significance of genetic diversity: ecological, demographic and life-history correlates. In *Evolutionary Dynamics of Genetic Diversity,* ed. G. S. Mani, pp. 13–213. Berlin: Springer Verlag
50. Nevo, E., Shkolnik, A. 1974. Adaptive metabolic variation of chromosome forms in mole rats *Spalax*. *Experientia* 30:724–26
51. Nix, H. A. 1981. The environment of Terra Australis. In *Ecological Biogeography of Australia,* ed. A. Keast, pp. 103–13. The Hague: Junk
52. Odum, E. P. 1983. *Basic Ecology*. Philadelphia: Saunders College Publ.
53. Odum, E. P. 1985. Trends expected in stressed ecosystems. *BioScience* 35: 419–22
54. Oppenoorth, F. J. 1985. Biochemistry and genetics of insecticide resistance. In *Comprehensive Insect Physiology, Biochemistry and Pharmacology* 12, ed. G. A. Kerkut, I. L. Gilbert, pp. 771–3. Oxford: Pergamon
55. Parsons, P. A. 1974. Genetics of resistance to environmental stresses in *Drosophila* populations. *Annu. Rev. Genet.* 7:234–65
56. Parsons, P. A. 1980. Adaptive strategies in natural populations of *Drosophila:* Ethanol tolerance, desiccation resistance, and development times in climati-

cally optimal and extreme environments. *Theor. Appl. Genet.* 57:257–66
57. Parsons, P. A. 1980. Isofemale strains and evolutionary strategies in natural populations. *Evol. Biol.* 13:175–217
58. Parsons, P. A. 1983. *The Evolutionary Biology of Colonizing Species*. Cambridge: Cambridge Univ. Press
59. Parsons, P. A. 1987. Evolutionary rates under environmental stress. *Evol. Biol.* 21:311–47
60. Parsons, P. A. 1988. Evolutionary rates: effects of stress upon recombination. *Biol. J. Linn. Soc.* 35:49–68
61. Parsons, P. A. 1989. Environmental stresses and conservation of natural populations. *Annu. Rev. Ecol. Syst.* 20:29–49
62. Parsons, P. A. 1990. Fluctuating asymmetry: an epigenetic measure of stress. *Biol. Rev.* 65:131–45
62a. Parsons, P. A. 1991. *Stress and evolution. Nature*. In press
63. Peterson, C. C., Nagy, K. A., Diamond, J. 1990. Sustained metabolic scope. *Proc. Natl. Acad. Sci. USA* 87:2324–28
64. Peterson, C. H., Black, R. 1988. Density-dependent mortality caused by physical stress interacting with biotic history. *Am. Nat.* 131:257–70
65. Petraitis, P., Latham, R. E., Neisenbaum, R. A. 1989. The maintenance of species diversity by disturbance. *Q. Rev. Biol.* 64:393–418
66. Pough, F. H. 1989. Organismal performance and Darwinian fitness: approaches and interpretations. *Physiol. Zool.* 62:199–236
67. Root, T. 1988. Environmental factors associated with avian distributional limits. *J. Biogeogr.* 15:489–505
68. Ruiz, A., Heed, W. B. 1988. Host-plant specificity in the cactophilic *Drosophila mulleri* species complex. *J. Anim. Ecol.* 57:237–49
69. Ryan, M. J. 1988. Energy, calling, and selection. *Am. Zool.* 28:885–98
70. Salt, G. W. 1984. *Ecology and Evolutionary Biology*. Chicago: Univ. Chicago Press
71. Schmalhausen, I. I. 1949. *Factors of Evolution*. Philadelphia: Blakiston
72. Schmidt-Nielsen, K. 1984. *Scaling: Why is Animal Size so Important?* Cambridge: Cambridge Univ. Press
73. Service, P. M. 1987. Physiological mechanisms of increased stress resistance in *Drosophila melanogaster* selected for postponed senescence. *Physiol. Zool.* 60:321–26
74. Sierra, L. M., Comendador, M. A., Aguirrezabalaga, I. 1989. Mechanisms of resistance to acrolein in *Drosophila melanogaster. Génét. Sél. Evol.* 21: 427–36
75. Sheldon, P. R. 1987. Parallel gradualistic evolution of Ordovician trilobites. *Nature* 330:561–63
76. Sheldon, P. R. 1990. Shaking up evolutionary patterns. *Nature* 345:772
77. Soulé, M. 1973. The epistasis cycle: a theory of marginal populations. *Annu. Rev. Ecol. Syst.* 4:165–87
78. Stanley, S. M., Parsons, P. A. 1982. The response of the cosmopolitan species, *Drosophila melanogaster,* to ecological gradients. *Proc. Ecol. Soc. Aust.* 11:121–30
79. Symeonidis, L., McNeilly, T., Bradshaw, A. D. 1985. Differential tolerance of three cultivars of *Agrostis capillaris* L. to cadmium, copper, lead, nickel and zinc. *New Phytol.* 101:309–15
80. Tanaka, K. 1989. Energetic cost of web construction and its effect on web relocation in the web-building spider *Agelena limbata. Oecologia* 81:459–64
81. Templeton, A. R. 1980. The theory of speciation *via* the founder principle. *Genetics* 94:1011–38
82. Trout, W. E., Kaplan, W. D. 1970. A relation between longevity, metabolic rate, and activity in shaker mutants of *Drosophila melanogaster. Exp. Gerontol.* 5:83–92
83. Turner, J. R. G., Gatehouse, C. M., Corey, C. A. 1987. Does solar energy control organic diversity? Butterflies, moths and the British climate. *Oikos* 48:195–205
84. Vermeij, G. J. 1987. *Evolution and Escalation: an Ecological History of Life*. Princeton: Princeton Univ. Press
85. Watt, W. B., Carter, P. A., Donohue, K. 1986. Females' choice of 'good genotypes' as mates is promoted by an insect mating system. *Science* 223: 1187–90
86. West-Eberhard, M. J. 1989. Phenotypic plasticity and the origins of diversity. *Annu. Rev. Ecol. Syst.* 20:249–78
87. Wright, D. H. 1983. Species-energy theory: an extension of species-area theory. *Oikos* 41:496–506
88. Zachariassen, K. E., Andersen, J., Maloiy, G. M. O., Kamau, J. M. Z. 1987. Transpiratory water loss and metabolism of beetles from arid area in East Africa. *Comp. Biochem. Physiol.* 86A:403–8

Annu. Rev. Ecol. Syst. 1991. 22:19–36

ECOLOGY OF PARAPATRIC DISTRIBUTIONS

C. M. Bull

School of Biological Sciences, Flinders University, GPO Box 2100, Adelaide, South Australia, Australia

KEY WORDS: parapatry, distribution boundaries

INTRODUCTION

Parapatry describes a distributional pattern in which pairs of taxa have separate but contiguous distributions, abutting along common boundaries. Smith first proposed the word "parapatry" (81, 122) to differentiate situations where the ranges of two taxa are in contact, both from sympatry where ranges overlap, and from allopatry where ranges are separate. The condition of parapatry was known much earlier. For instance, Darwin (38) wrote that the location of a species border was often determined by the border of a related species.

More recently numerous examples of parapatry, from a wide range of taxonomic groups and geographic regions, have been described (53, 61, 64, 71, 82). Parapatry might previously have been more common since continuous patches of habitat where species ranges could have been in contact have now been fragmented by clearance. It may still be more common than is realized, because parapatry is difficult to detect. Detection usually involves taxonomic separation of closely related, morphologically similar taxa, and collection on a fine geographic scale (71). Parapatry has been most frequently reported in birds (40, 46), perhaps because their high visibility makes fine-scale surveys relatively easy. Even among birds, detailed collection may reveal parapatry where sympatry has previously been reported (51).

Initial discussion of parapatry revolved around speciation mechanisms, and

0066-4162/91/1120–0019$02.00

whether taxa with contiguous distributions actually were in contact and had the potential to interbreed. Smith (123) argued that parapatry was a special case of allopatry where genic interchange was possible without sympatry. All other cases of allopatry, where individuals of two taxa never meet, he defined as dichopatry (a term that has gained little acceptance). Others (14, 26, 70, 80) claimed parapatry to be a special case of sympatry, because contact usually implied some small range overlap.

Parapatry has now become a legitimate separate category of distribution, differentiated from sympatry by the extent of range overlap. For instance Futuyma & Mayer (48) defined two populations as parapatric if they "occupied separate but adjoining areas, such that only a small fraction of individuals in each encounters the other." Key (71) said parapatry involved two populations that "occupied contiguous territories that overlapped only very narrowly in relation to both the length of the overlap zone, and the vagility of the individuals." He suggested the overlap should be no more than a small multiple of the dispersal range of individuals. These definitions recognize some overlap between parapatric taxa. There is no sharp demarcation, but rather a continuum between cases of allopatry, parapatry, and sympatry.

Parapatry is still invoked in discussions of the speciation process. An implicit, though rarely stated, assumption has been that the terms allopatry, sympatry, and (by implication) parapatry refer to closely related species (112), although the parapatric pair are not necessarily the most closely related in the species complex (63, 95). Haffer (52, 53) viewed parapatry as the result of secondary contact of differentiating taxa, reproductively isolated but incompletely speciated because they are not yet ecologically compatible. In this context parapatry differs specifically from sympatry in that parapatric taxa have not previously co-existed and do not currently co-exist. Some mechanism prevents overlap of their ranges. Parapatry does not refer to the state where species have overlapping ranges in which they occupy different, mutually exclusive, habitats. This distributional relationship, allotopy (112), implies a form of ecological compatibility of the species, by use of different niches. Parapatry often results from the first meeting between taxa that have been isolated in geographical refugia and have spread so their ranges now abut. Overlap with habitat segregation may result from that contact, but parapatry is the state (which may be stable for a long time) before there is any overlap.

When two taxa first make contact, parapatry can go through two temporal phases: a non-equilibrium invasion phase in which the range of one expands as it replaces the other, and an equilibrium phase in which a stable boundary is maintained by some balance of the fitness of each taxon. Moving boundaries have been directly observed as one species invades and takes over areas previously occupied by the other (42, 65, 103, 118). This can result in local extinction of the invaded species. Moving boundaries of hybridizing taxa have been inferred from relict mitochondrial DNA in locations where the nuclear genome has been replaced (77).

Whether or not Smith intended parapatry to include hybridizing taxa has been variously interpreted (81, 123). Nevertheless, parapatry is now taken to imply both hybridizing and nonhybridizing contact and, in the evolutionary context, to cover the range of differentiation of taxa from partial to complete reproductive isolation. Key (71) divided parapatry into two categories: hybridization parapatry where the contacting taxa form a narrow hybrid zone; and ecological parapatry, where they may have a narrow overlap zone, but without any hybridization. In fact, parapatric boundaries form a continuum of cases from those with no hybrids to those with many hybrids in the overlap. At the latter end of the continuum a genetic mechanism, negative heterosis, has been proposed as the major factor preventing range overlap of the parental taxa (13, 61). Key (71) defined hybridization parapatry as "parapatry in which the restriction of interpenetration results primarily from the populations mating more or less freely with each other, but either leaving no fertile progeny, or leaving progeny of reduced fertility." Narrow hybrid zones, or tension zones, have been intensively investigated and reviewed (13, 61, 71). Many of 170 cases coincided with environmental ecotones (13, 61), implying a role for ecological factors in preventing overlap. Ecological factors should be more important in those parapatric boundaries where hybridization is more rare. This review concentrates on parapatry where there is little or no hybridization. Haffer (53) called taxa in this relationship paraspecies and saw them as further differentiated in the speciation process than those with hybridization parapatry. They are referred to as species for the rest of this review.

Key (71) defined *ecological parapatry* as "parapatry in which the restriction of interpenetration results primarily from a sharp ecological interface between the habitats of the two populations, sometimes reinforced by competition between them." Haffer (53) further subdivided this relationship into: (*a*) *ecological parapatry*, as defined above, between species not necessarily related, which have distinctly different habitat requirements and occupy widely different ecological zones; and (*b*) *competitive parapatry* where "geographic exclusion in a uniform habitat zone results from competition between populations of two species with nearly identical ecological requirements." The two definitions include hypotheses for how parapatry is maintained. These mechanisms are inferred in many examples of parapatry but are rarely supported either by experimental tests or by adequate data on spatial and temporal dynamics (53).

The study of parapatry has wide implications. For studies of speciation, parapatric taxa may represent a rich source of information about the final stage of differentiation before full species status is achieved (136). For community ecologists, clues to ecological processes permitting coexistence may be derived from parapatry where pairs of species cannot coexist. In biogeography, parapatry provides patterns where the spread of a species could be prevented by another species rather than by physical features. Parapatric boundaries are also an ecological phenomenon in their own right. They are widespread among taxa and geographic regions, and a general theory explaining their structure and maintenance needs to be developed.

MAINTENANCE OF ECOLOGICAL PARAPATRY

The central ecological question about parapatry concerns the mechanisms that prevent range overlap of the contacting species. Five major mechanisms have been suggested. Some are related, and at some boundaries combinations of mechanisms may be involved.

Ecotonal Change

This is the mechanism implicit in Key's (71) original definition of ecological parapatry occurring primarily at a sharp ecological interface. Each species is better adapted for conditions on its own side of the boundary and avoids conditions for which it is less suited across the boundary. An extreme form, in which conditions are intolerable for each species across the boundary, does not require close evolutionary or ecological relationships between the contacting species. Parapatric boundaries also occur on less extreme gradients of elevation (59, 131), climate (74), soil structure (84), water speed (92, 127), and vegetation (126, 127).

Small environmental changes may prevent range expansion of single species. For instance the edge of the range of an established species may coincide with a specific altitude (90, 101) or climatic condition (50, 116, 135), and colonizing species can expand their ranges only into climatically suitable areas (41, 85). The range edge may represent the limit of conditions the species can tolerate physiologically (78, 117) or behaviorally (4), but the relationship between climate and population variables is rarely simple (115). More usually a combination of physical and biological variables influences distribution. A single factor is rarely important (17, 30), except where there is an abrupt distribution edge (17). Where boundaries coincide with climatic changes, other factors correlated with climate may have a more direct influence on an individual's chance of surviving and reproducing (31). Distribution limits can be further complicated by dispersal causing labile boundaries (129).

In relation to parapatry many studies infer that an environmental gradient alone, without any biological interaction, can limit the ranges of both species at the same place. For instance, Ford (45) suggested that cases of parapatry between species of Australian quail thrushes *(Cinclosoma)* at points of sharp habitat change were maintained by adaptations of each species to different habitats. Other studies comment on the coincidence of the parapatric boundary with an environmental change (65, 76, 108) or suggest that parapatry is maintained because each species selects the habitat on its side of the boundary (68, 127).

However, it is unlikely that a small or gradual environmental cline will alone maintain a parapatric boundary. Key (71) originally proposed that the effect of a sharp ecological interface was sometimes reinforced by competition. Some form of biological interaction is probably required to prevent expansion into the less favorable habitat, except at extreme ecotones. The parapatry between two chipmunk species *(Eutamias),* at an ecotone, probably resulted from precise habitat choice, which had evolved in each species to avoid habitats where adverse interspecific interactions would occur (127). That is, the boundary is now maintained by the ecotone, although interactions played a role in the past.

An ecotone also represents a resource gradient. Slade & Robertson (121) suggested that along a gradient of increasing resources for one species and diminishing resources for the other, the resources of each may become too sparse in the center, so neither can spread, and a boundary, or even a gap, between the two species is formed. A more likely outcome (35) is that one or both species would switch from specialist to generalist resource use, and if there were no interaction, they would overlap.

Interspecific Competition

Interspecific competition may prevent species from invading each other's ranges across parapatric boundaries (82). Haffer (52, 53) considered that many parapatric species pairs with nearly identical ecological requirements were ecologically incompatible. He thought competition and mutual exclusion should maintain parapatry even in regions of uniform habitat. However, few competition models predict this result without some change in fitness of at least one species across the boundary.

Mayr (80) suggested interspecific competition combined with a gradual climatic cline could maintain parapatric boundaries. Key (71) disagreed because normal climatic fluctuations would allow periodic expansions of the ranges of each species. Then, with a delicate balance of competition, the colonists would only be eliminated slowly, allowing wide overlap rather than sharp separation along gradual clines. For this reason Key (71) considered competition less important than ecotones in maintaining parapatry.

Nevertheless, competition is often inferred to explain parapatry (44, 134). This view is supported by evidence of interspecific territoriality in the contact zone of some parapatric bird species (42, 79). Also, the experimental removal of one species from plots in the parapatric overlap zone led to improved performance, in salamanders (54), or increased range of habitats, in rodents (86), of the other species. Each of two chipmunk species *(Eutamias)* is more successful in interspecific agonistic interactions over food on its own side of a parapatric boundary (16). Similarly, the ant *Wasmannia* recruits to baits more efficiently and aggressively than the species it replaces parapatrically (34).

Theoretical models show that competition along environmental gradients can lead to parapatry. MacArthur (83) modelled one species competitively replacing another along a resource cline. Modifications of this model (121), to include additional resource requirements for the cost of defending territories interspecifically, produced a gap between species on a resource replacement cline. Here neither species could persist, even with adequate resources for a single species. In this model, parapatric contact results from dispersal into the gap. From optimal foraging theory Cody (35) predicted that two potentially competing species along a resource replacement cline would abut parapatrically over areas where either species could exist alone, if they were *K*-selected, but would extensively overlap if *r*-selected.

The included niche model (84) combines a more abrupt environmental change with competition, to predict parapatric boundaries at the physiological limit of the competitively dominant species. The competitively inferior species, with wider physiological tolerance, is restricted to a subset of its potential range where the dominant species cannot persist. This model explains parapatric boundaries, for instance in salamanders (69), gophers (84), and chipmunks (32, 58, 119).

Terborgh (131) compared the relative importance of competition and ecotones as determinants of distributional limits of bird species on altitudinal transects in Peru. Using different slopes, where ecotonal changes occurred at different altitudes, and where congeners were variably present, he deduced that direct or diffuse competition accounted for about two thirds of all distributional limits. The general emphasis on competition and changing competitive fitness along environmental gradients in explaining parapatry may reflect its real importance. Alternatively, it may reflect a bias toward competition among ecologists (55, 105). Other ecological processes, discussed below, may also explain parapatry.

Predation

Predators can limit species distributions. Parapatry could result if a competitively dominant species is more susceptible to predation, and if there is an ecotone or cline where the predator becomes less effective (55, 100). Alternatively, where the combined density of two prey species increases the size of the predator population, the more susceptible prey may be eliminated by the increased pressure. For instance, one species of leafhopper is replacing another parapatrically in California. While less susceptible to an egg parasitoid, its presence enhances the parasitoid population, to the detriment of the other species (118). This "apparent competition" (66) is analogous to an interaction through infectious disease, discussed in the next section.

Parasites and Disease

A parasite and host might coevolve to form a unit, which, through cross-infection, reduces the fitness of naive hosts of another species (10, 47, 105). In Cornell's (37) parasite model for parapatry, each host species has unique parasite and vector species. A vector can transmit its parasite to invading individuals of the other host species, to their detriment if they lack previous experience of, or evolved resistance to, the parasite. The numerical superiority of the resident population and its parasites will prevent invaders becoming established, leading to parapatry without an environmental gradient. Parapatry could also develop if a gradient limits the spread of one host species, when only that species has a parasite-vector system inimical to the other. Parasite-induced advantage has been an explanation for parapatric invasions (104, 105), for instance of the white-tailed deer replacing other cervids (5).

Freeland (47) suggested that species could not coexist without effective barriers to interspecific parasite transmission. He proposed that species related ecologically or phylogenetically would be most susceptible to cross-infection, and hence to interactions via their parasites. Thus, species divergence should reduce parasite-induced interaction, and parasite-induced parapatry. This reflects Haffer's (53) contention that parapatric species are those that have not yet diverged sufficiently for coexistence, although Haffer was contemplating interspecific competition as the mechanism.

In another model, "apparent competition" (66), one of two species that share parasites or infectious diseases can exclude the other because combined host density allows the level of infestation to become higher than on a single host population, and too high for one species to tolerate. That species is then excluded, even when it can persist alone with the disease.

Although these models can explain parapatry, and appear to be important in some cases of invasion parapatry (5, 104, 105), they have not yet been used to explain cases of stable parapatric boundaries.

Reproductive Interference

In narrow hybrid zones, reduced fitness of hybrids diminishes the vigor of colonization attempts across the boundary (11, 61). By analogy, interspecific pairing can reduce reproductive potential of colonizing females and generate parapatry, even where no, or very few, hybrids are produced. Anderson (6) proposed that completely intersterile species may still interbreed where their ranges come in contact, reducing fitness through wastage of reproductive potential. Simulations (6) showed that random pairing alone produced species replacement clines 62% as steep as those from interspecific competition, implying comparable potential of the two processes to maintain parapatric boundaries. More reproductive potential was lost by the numerically inferior species, so a resident established population would resist low density invasion (6). This model was used to explain the parapatric boundary between two species of flycatcher in Europe, and the incursion of one through numerical supremacy onto islands within the range of the other (3).

A similar proposal, the "satyr effect" (110, 111), has males mating indiscriminately, reducing reproductive potential of heterospecific females through the production of sterile eggs, or the blocking of, or physical damage to, genitalia. Simulations of a linear series of demes showed satyrization alone could generate stable parapatry if dispersal between demes was low. Increased interdeme migration led to one species excluding the other (110, 111). The satyr effect was used to explain the parapatric replacement of one tick species by another in Africa (111), and the stable parapatry of two mosquito species in the Bahamas (110, 125).

Another form of reproductive interference is the jamming of reproductive signals, such that one species cannot transmit as effectively to conspecifics in the presence of the other. Where the effect is symmetrical, a resident species will resist invasion from a low density colonizer, leading to parapatry, as proposed in acoustically signalling frogs (98), and in chemically signalling ticks (8). These examples are discussed in the next section.

TWO CASE STUDIES

There are few cases where a parapatric boundary has been closely mapped, and where alternative hypotheses for its maintenance have been experimentally tested. Two such cases come from South Australia.

One involves two morphologically similar frog species, *Ranidella riparia* and *R. signifera,* which were once regarded as the same species (133). Subsequent allozyme electrophoresis showed genetic differentiation five times greater than between other sibling *Ranidella* species (95). *Ranidella riparia,* endemic to the Flinders Ranges of South Australia, has its southern limit at a stable parapatric boundary with the more widespread *R. signifera* (92). No hybrids are found in the narrow overlap zone where *R. riparia* breeds in swift, rocky, west-flowing creeks, and *R. signifera* in slower, more muddy, east-flowing creeks. Habitat suitable for each species is available beyond the boundary, but neither species extends further into the range of the other (92).

Laboratory experiments showed that *R. signifera* tadpoles were displaced more by flowing water (91), and more often chose sheltered habitats, especially in the presence of *R. riparia* (96). In field cages in still creeks, or sheltered cages in flowing creeks, tadpoles of both species survived and grew equally well; but *R. signifera* had reduced success when mixed with *R. riparia* in flowing water (93). Even in still water tadpoles of *R. riparia* caused a feeding shift in those of *R. signifera* (97). These results suggest *R. signifera* is prevented from spreading into the swift flowing creeks beyond its range because its tadpoles are adapted to calmer water, and are competitively inferior to *R. riparia*.

Reproductive interference may prevent *R. riparia* from extending into calmer water, a habitat it successfully inhabits outside the range of *R. signifera* (92). Males of *R. riparia* have a relatively complex vocal repertoire (99). Yet their advertisement calls average 24 dB lower than *R. signifera* and may be inaudible to conspecific females in the dense continuous chorus of *R. signifera* (98). Experimentally transplanted *R. riparia* males moved away more rapidly from areas where *R. signifera* males were calling (94). Acoustic jamming may prevent successful colonization by *R. riparia* of creeks where *R. signifera* is established and calling (98).

The second example concerns three tick species that have parapatric distributions in South Australia (124). All infest the same major host, the sleepy lizard *Trachydosaurus rugosus,* whose distribution is continuous across the tick boundaries. The boundary between *Aponomma hydrosauri* and *Amblyomma limbatum* near Mt. Mary has remained stable for over 20 years (C. M. Bull, in preparation). On three transects the boundary center oscillated by less than 800 m over eight years, with no unidirectional trend. The boundary is close to a vegetational ecotone (24, 102, 124). To the north, ground conditions may be too arid for *Ap. hydrosauri,* which is less tolerant of desiccation (25) and seeks less stressful conditions (73, 124). However, transplanted *Am. limbatum* survive as well on either side of the boundary in the litter microhabitats where they wait for hosts (24).

The mechanisms preventing *Am. limbatum* from spreading south across the boundary remain unknown. Interspecific competition appears an unlikely explanation because there are many underused hosts, and because neither species was affected by experimental coinfestation at or above maximum field levels (22). Predators, mainly ants, attack detached ticks, but the ant distributions do not coincide with the tick boundary (23, 33). There is no evidence that lizards from across the boundary are less suitable hosts for attachment and engorgement by ticks (24).

Reproductive interference could be involved in maintaining the boundary. Females initiate mating on the hosts by emitting an excitant pheromone soon after attachment (7). Attached conspecific males will not respond when females of the other species are also attached (8), perhaps because of signal jamming. However, only 13% of lizards close to the boundary carried *Ap. hydrosauri* females, so this would be an ineffectual barrier to colonization by *Am. limbatum* (C. M. Bull, in preparation).

Tick dispersal is slow (20). Over eight years very few *Ap. hydrosauri* colonized north of the boundary, while more *Am. limbatum* colonized south. At least one successful colonization, 1 km south of the boundary, persisted from 1985 to 1988. On the three transects studied, there were consistently low densities of hosts, and low infestation levels per host, in areas immediately adjacent to the boundary (C. M. Bull, in preparation). A role which this population trough may play in maintaining the boundary is explored below.

MODELS OF PARAPATRY

In some cases, such as the reptile ticks, conventional models cannot adequately explain the maintenance of stable parapatry. Other cases of nonhybridizing parapatry have seldom been studied in this detail, so there are few empirical data for generalizations to be developed. Where explanations for parapatry have been sought there was sometimes disagreement (58, 127), often no conclusion (49, 67, 68, 72), and rarely experimental testing of proposed mechanisms. Low dispersal rates, possibly with a population trough, may be key elements in the maintenance of parapatry. Troughs or gaps between parapatric species were often mentioned in early descriptions of parapatry (15, 37, 121) and were detected at the reptile-tick boundary. Low dispersal and the presence of troughs may influence parapatry in two possible ways.

Density Dependent Advantage

The first mechanism is derived by analogy with narrow hybrid zones. Many of these coincide with environmental ecotones (9, 19, 107), but in others, ecological change is gradual relative to the abrupt hybrid zone (128) or undetectable (89). Many hybrid zones are maintained by negative heterosis (61), with a balance of selection against hybrids counteracting dispersal of parental individuals into the zone (11). Low dispersal rate relative to the zone width means that very small, maybe undetectable, levels of selection against hybrids are sufficient to maintain stable zones (12). Thus, in some examples, evidence for reduced hybrid fitness was not found (21). Hybrid zones may be located at density troughs, regions of local decrease of population, with no requirement for an environmental gradient (13, 60, 61). With low dispersal the zone becomes trapped at the trough because neither species can overcome the numerical disadvantage implicit in the negative heterosis mechanism (11). Simulation models show how this may persist for thousands of generations (87). The rarer species always suffer greater proportional losses from interspecific crosses. A hybrid zone in *Podisma* follows such a trough (88).

Stable parapatry without hybridization may be maintained in analogous ways. This requires density dependent interactions which depress each species more when it is less numerous. Interactions through reproductive interference, and through parasites and disease, could act in this way as already discussed. With these mechanisms, both a slowly dispersing species and a species colonizing from a population trough will be numerically disadvantaged and less able to penetrate the range of the resident species. The outcome of the species contact will depend upon the strength and symmetry of the interaction and the dispersal rates. Where dispersal is slow, weak interactions will be sufficient to maintain parapatry. Increased dispersal will require stronger interspecific interactions. An asymmetrical advantage to one species may cause the boundary to move and to stabilize only where an environmental cline alters the relative advantages. The greater the asymmetry the further the boundary will move along a cline, in extreme cases leading to the exclusion of one species.

The Deme Model

An alternative model, where interactions do not need to be density dependent, is derived from models of the dynamics of local and regional populations. Migration is a major, but often ignored, component of population dynamics (129, 130). Models in which migration influences the spatial dynamics of populations (43, 56, 57, 113, 114) view regional populations as groups of local populations occupying isolated patches of variably suitable habitat (2). The regional population in these models is only maintained when dispersal from local populations is sufficient to recolonize local extinctions. Empirical evidence supports these models. Local populations on small islands have measurable extinction rates that are higher for smaller or more recently established populations (109, 132); local populations of herbivorous insects frequently establish or go extinct on individual plants (1). Many local populations rely entirely on repeated migratory recruitment for their establishment and maintenance (36), a phenomenon called the rescue effect (18).

A characteristic of the edge of a species range is that there are fewer suitable patches where local populations can establish (17, 138). In peripheral populations of birds, death rates exceed birth rates, and local populations are only sustained by continual migration (137). Carter & Prince (27) developed a model from epidemic dynamics to show that a population could not persist in a region with a high ratio of the rate of patch extinction to the rate of dispersal to reoccupy vacant patches. Their model predicts the abrupt species boundaries on gentle climatic or altitudinal clines that they found for the British prickly lettuce *Lactuca* (28, 29, 106). Small drops in fitness from the presence of another species could have an equivalent effect.

A model for parapatry is illustrated in Figure 1. Assume a species *A* is spreading its range from a source population onto a landscape of "dunes" of ecological suitability such that the distance between adjacent ridges is further than the dispersal range of individuals. Local populations can establish and persist for some time on the ridges. In the troughs there is normally no effective reproduction, and local populations are maintained by migration from the ridges. However, in good years, the trough population can reproduce

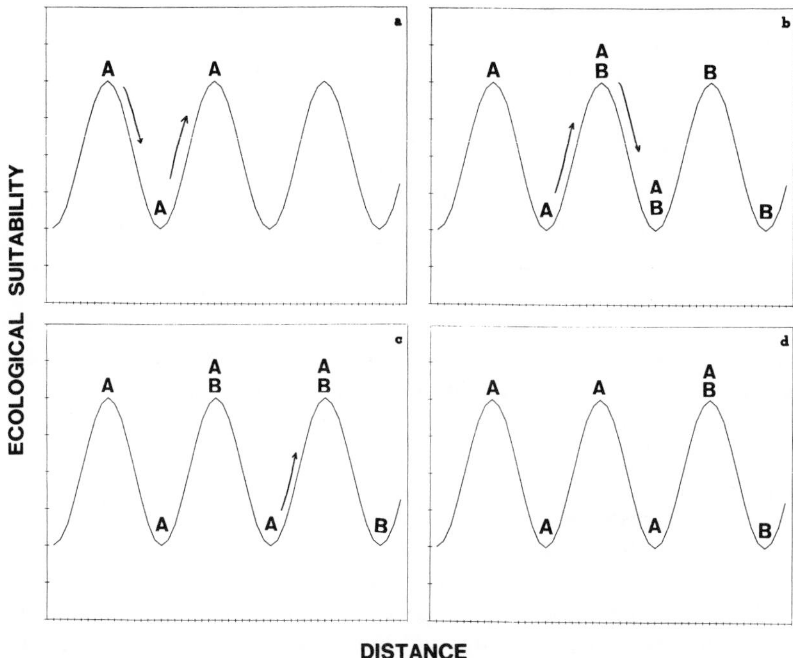

Figure 1 A model for parapatry with slowly dispersing species in a patchy environment.

sufficiently to generate dispersers itself, and the next unoccupied ridge can be attained. Once established, local populations persist on the ridges as long as there is occasional interpopulation migration via the troughs. Ephemeral presence in the troughs is maintained by dispersal counteracting extinction, less effectively if dispersal is lower.

Now suppose there is secondary contact with another species, *B*. Assume they interact so weakly that local populations of neither are influenced on the ridges. In the troughs, however, small interactions may be sufficient to overbalance the equilibrium between dispersal and extinction. The lower the dispersal the smaller the interaction needed to exclude one species. Then either (*a*) the ranges of both species are held at that trough in which contact was first made, because neither can now build up numbers sufficient to colonize and permanently maintain the next ridge; or (*b*) one species, *B,* is held, but *A* is better adapted, or less affected by interactions, and can expand onto the next ridge. There *A* coexists with *B,* then spreads to the next trough, excludes B from there, so *B* becomes isolated on the previous ridge. That local population of *B,* now not supported by migration, will eventually go extinct. Meanwhile *A* expands further into the range of *B,* replacing it in this manner, until some environmental change along a gradient reduces its adaptive or competitive advantage in the troughs. The spatial and temporal pattern shown at the reptile tick boundary near Mt. Mary is that predicted by this model (C. M. Bull, in preparation).

CONCLUSION

This view of parapatry emphasizes dispersal as a parameter of prime importance in influencing interspecific interactions. Relatively mobile species will have labile distribution edges and will frequently mix and interact in communities (129). They will be able to colonize and exploit patches within the distribution of other species (120). Only sharp ecotonal changes or strong ecological interactions will lead to parapatry in those species. In contrast, less mobile species like wingless orthopterans (61, 67, 71), amphibians (19, 64, 69, 75, 92), and ants (34, 49, 103) will often form abrupt parapatric boundaries or hybrid zones (62). Even within a taxonomic group, birds tend to disperse less in the tropics (39), and to form more parapatric boundaries there (52, 131) than in temperate forests (90). Parapatry is a phenomenon where interactions with influences beyond the local population may have major implications. Cases of parapatry, previously dismissed as simple examples of competitive exclusion, or of the effect of an ecotone, may prove to be fertile testing ground in a new and richer perspective of past and present processes important in ecological communities.

ACKNOWLEDGMENTS

I have been supported by research funds from the Australian Research Council and the Flinders University Research Budget. Many have contributed ideas and data, including Ross Andrews, Dale Burzacott, Neil Chilton, Francois Odendaal, and Trevor Petney. G. R. Brooks, N. B. Chilton, G. M. Hewitt, K. H. L. Key, and M. J. Littlejohn kindly commented on a draft manuscript.

Literature Cited

1. Addicott, J. F. 1978. The population dynamics of aphids on fireweed: a comparison of local populations and metapopulations. *Can. J. Zool.* 56:2554–64
2. Addicott, J. F., Aho, J. M., Antolin, M. F., Padilla, D. K., Richardson, J. S., et al. 1987. Ecological neighbourhoods: scaling environmental patterns. *Oikos* 49:340–46
3. Alerstam, T., Ebenman, B., Sylven, M., Tamm, S., Ulfstrand, S. 1978. Hybridization as an agent of competition between two bird allospecies: *Ficedula albicollis* and *F. hypoleuca* on the island of Gotland in the Baltic. *Oikos* 31:326–31
4. Alkon, P. U., Saltz, D. 1988. Foraging time and the northern range limits of the Indian crested porcupine *(Hystrix indica* Kerr). *J. Biogeogr.* 15:403–08
5. Anderson, R. C. 1972. The ecological relationships of meningeal worm and native cervids in North America. *J. Wildl. Dis.* 8:304–10
6. Anderson, R. F. V. 1977. Ethological isolation and competition of allospecies in secondary contact. *Am. Nat.* 111: 939–49

7. Andrews, R. H., Bull, C. M. 1980. Mating behaviour in the Australian reptile tick *Aponomma hydrosauri*. *Anim. Behav*. 28:1280–86
8. Andrews, R. H., Petney, T. N., Bull, C. M. 1982. Reproductive interference between three parapatric species of reptile tick. *Oecologia* 52:281–86
9. Baker, R. J., Davies, S. K., Bradley, R. D., Hamilton, M. J., van der Bussche, R. A. 1989. Ribosomal-DNA, mitochondrial DNA, chromosomal, and allozymic studies on a contact zone in the pocket gopher, *Geomys*. *Evol*. 43:63–75
10. Barbehenn, K. R. 1969. Host-parasite relationships and species diversity in mammals: an hypothesis. *Biotropica* 1:29–35
11. Barton, N. H. 1979. The dynamics of hybrid zones. *Heredity* 43:341–59
12. Barton, N. H., Hewitt, G. M. 1982. A measurement of dispersal in the grasshopper *Podisma pedestris* (Orthoptera: Acrididae). *Heredity* 48:237–49
13. Barton, N. H., Hewitt, G. M. 1985. Analysis of hybrid zones. *Annu. Rev. Ecol. Syst*. 16:113–48
14. Bigelow, R. S. 1965. Hybrid zones and reproductive isolation. *Evolution* 19: 449–58
15. Brewer, R. 1963. Ecological and reproductive relationships of black-capped and Carolina chickadees. *Auk* 80:9–47
16. Brown, J. H. 1971. Mechanisms of competitive exclusion between two species of chipmunks. *Ecology* 52:305–11
17. Brown, J. H. 1984. On the relationship between the abundance and distribution of species. *Am. Nat*. 124:255–79
18. Brown, J. H., Kodric-Brown, A. 1977. Turnover rates in insular biogeography: effects of immigration on extinction. *Ecology* 58:445–49
19. Bull, C. M. 1978. The position and stability of a hybrid zone between the Western Australian frogs *Ranidella insignifera* and *R. pseudinsignifera*. *Aust. J. Zool*. 26:305–32
20. Bull, C. M. 1978. Dispersal of the Australian reptile tick *Aponomma hydrosauri* by host movement. *Aust. J. Zool*. 26:689–97
21. Bull, C. M. 1979. A narrow hybrid zone between two Western Australian frog species *Ranidella insignifera* and *R. pseudinsignifera:* the fitness of hybrids. *Heredity* 42:381–89
22. Bull, C. M., Burzacott, D., Sharrad, R. D. 1989. No competition between two tick species at their parapatric boundary. *Oecologia* 79:558–562
23. Bull, C. M., Chilton, N. B., Sharrad, R. D. 1988. The risk of predation for two reptile tick species near their parapatric boundary. *Exp. Appl. Acarol*. 5:93–99
24. Bull, C. M., Sharrad, R. D., Petney, T. N. 1981. Parapatric boundaries between Australian reptile ticks. *Proc. Ecol. Soc. Aust*. 11:95–107
25. Bull, C. M., Smyth, M. 1973. The distribution of three species of reptile ticks, *Aponomma hydrosauri* (Denny), *Amblyomma albolimbatum* Neumann, and *Amb. limbatum* Neumann. II. Water balance of nymphs and adults in relation to distribution. *Aust. J. Zool*. 21:103–10
26. Cain, A. J. 1953. Geography, ecology and coexistence in relation to the biological definition of the species. *Evolution* 7:76–83
27. Carter, R. N., Prince, S. D. 1981. Epidemic models to explain biogeographic distribution limits. *Nature* 293:644–45
28. Carter, R. N., Prince, S. D. 1985. The geographical distribution of prickly lettuce *(Lactuca serriola)*. I. A general survey of its habitats and performance in Britain. *J. Ecol*. 73:27–38
29. Carter, R. N., Prince, S. D. 1985. The geographical distribution of prickly lettuce *(Lactuca serriola)*. II. Characteristics of population near its distributional limit in Britain. *J. Ecol*. 73:39–48
30. Caughley, G., Grice, D., Barker, R., Brown, B. 1988. The edge of the range. *J. Anim. Ecol*. 57:771–85
31. Caughley, G., Short, J., Grigg, G. C., Nix, H. 1987. Kangaroos and climate: an analysis of distribution. *J. Anim. Ecol*. 56:751–61
32. Chappell, M. A. 1978. Behavioral factors in the altitudinal zonation of chipmunks *(Eutamias)*. *Ecology* 59:565–79
33. Chilton, N. B. 1989. Life cycle adaptations and their implications in the distribution of two parapatric species of tick. PhD thesis. Flinders Univ., South Aust.
34. Clark, D. B., Guayasamin, C., Pazmino, O., Donosa, C., de Villacis, Y. P. 1982. The tramp ant *Wasmannia auropunctata:* autecology and effects on ant diversity and distribution on Santa Cruz Island, Galapagos. *Biotropica* 14:196–207
35. Cody, M. L. 1974. Optimization in ecology. *Science* 183:1156–64
36. Connor, E. F., Faeth, S. H., Simberloff, D. 1983. Leafminers on oak: the role of immigration and in situ reproductive recruitment. *Ecology* 64:191–204
37. Cornell, H. 1974. Parasitism and dis-

tributional gaps between allopatric species. *Am. Nat.* 108:880–03
38. Darwin, C. 1869. *On the Origin of Species by Means of Natural Selection*. London: Murray. 596 pp. 6th ed.
39. Diamond, J. 1973. Distributional ecology of New Guinea birds. *Science* 179:759–69
40. Dixon, K. L. 1989. Contact zones of avian congeners on the southern Great Plains. *Condor* 91:15–22
41. Easteal, S., Van Beurden, E. K., Floyd, R. B., Sabath, M. D. 1985. Continuing geographical spread of *Bufo marinus* in Australia: range expansion between 1974 and 1980. *J. Herpetol.* 19:185–88
42. Emlen, S. T., Rising, J. D., Thompson, W. L. 1975. A behavioral and morphological study of sympatry in the indigo and lazuli buntings of the Great Plains. *Wilson Bull.* 87:145–202
43. Fahrig, L., Merriam, G. 1985. Habitat patch connectivity and population survival. *Ecology* 66:1762–68
44. Findley, J. S. 1954. Competition as a possible limiting factor in the distribution of *Microtus*. *Ecology* 35:418–20
45. Ford, J. 1983. Evolutionary and ecological relationships between quail thrushes. *Emu* 83:152–72
46. Ford, J. 1987. Hybrid zones in Australian birds. *Emu* 87:158–78
47. Freeland, W. J. 1983. Parasites and the coexistence of animal host species. *Am. Nat.* 121:223–36
48. Futuyma, D. J., Mayer, G. C. 1980. Non-allopatric speciation in animals. *Syst. Zool.* 29:254–71
49. Greenslade, P. J. M. 1974. Distribution of two forms of the meat ant, *Iridomyrmex purpureus* (Hymenoptera: Formicidae), in parts of South Australia. *Aust. J. Zool.* 22:489–504
50. Greenslade, P. J. M. 1987. Environment and competition as determinants of local geographical distribution of five meat ants, *Iridomyrmex purpureus* and allied species (Hymenoptera: Formicidae). *Aust. J. Zool.* 35:259–73
51. Haffer, J. 1967. Some allopatric species pairs of birds in north-western Colombia. *Auk* 84:343–65
52. Haffer, J. 1969. Speciation in Amazonian forest birds. *Science* 165:131–37
53. Haffer, J. 1986. Superspecies and species limits in vertebrates. *Z. Zool. Syst. Evol.* 24:169–90
54. Hairston, N. G. 1980. The experimental test of an analysis of field distributions: competition in terrestrial salamanders. *Ecology* 61:817–26
55. Hairston, N. G. 1980. Species packing in the salamander genus *Desmognathus:* what are the interspecific interactions involved. *Am. Nat.* 115:354–66
56. Hanski, I. 1982. Dynamics of regional distribution: the core and satellite species hypothesis. *Oikos* 38:210–21
57. Hanski, I. 1985. Single species spatial dynamics may contribute to long-term rarity and commonness. *Ecology* 66:335–43
58. Heller, H. C. 1971. Altitudinal zonation of chipmunks *(Eutamias):* interspecific aggression. *Ecology* 52:312–19
59. Heller, H. C., Gates, D. M. 1971. Altitudinal zonation of chipmunks *(Eutamias):* energy budgets. *Ecology* 52: 424–33
60. Hewitt, G. M. 1975. A sex-chromosome hybrid zone in the grasshopper *Podisma pedestris* (Orthoptera: Acrididae). *Heredity* 35:375–87
61. Hewitt, G. M. 1988. Hybrid zones—natural laboratories for evolutionary studies. *Trends Ecol. Evol.* 3:158–67
62. Hewitt, G. M. 1990. Divergence and speciation as viewed from an insect hybrid zone. *Can. J. Zool.* 68:1701–15
63. Hillis, D. M. 1985. Evolutionary genetics of the Andean lizard genus *Pholidobolus* (Sauria: Gymnophthalmidae): phylogeny, biogeography, and a comparison of tree construction techniques. *Syst. Zool.* 34:109–26
64. Hillis, D. M., Frost, J. S., Wright, D. A. 1983. Phylogeny and biogeography of the *Rana pipiens* complex: a biochemical evaluation. *Syst. Zool.* 32:132–43
65. Hillis, D. M., Simmons, J. E. 1986. Dynamic change of a zone of parapatry between two species of *Pholidobolus* (Sauria: Gymnophthalmidae). *J. Herpetol.* 20:85–86
66. Holt, R. D., Pickering, J. 1985. Infectious disease and species coexistence: a model of Lotka-Volterra form. *Am. Nat.* 126:196–211
67. Howard, D. J., Harrison, R. G. 1984. Habitat segregation in ground crickets: experimental studies of adult survival, reproductive success, and oviposition preferences. *Ecology* 65:61–68
68. Howard, D. J., Harrison, R. G. 1984. Habitat segregation in ground crickets: the role of interspecific competition and habitat selection. *Ecology* 65:69–76
69. Jaeger, R. G. 1970. Potential extinction through competition between two species of terrestrial salamanders. *Evolution* 24:632–42
70. Key, K. H. L. 1968. The concept of stasipatric speciation. *Syst. Zool.* 17:14–22
71. Key, K. H. L. 1982. Species, parapatry,

and the morabine grasshoppers. *Syst. Zool.* 30:425–58
72. Key, K. H. L., Balderson, J. 1972. Distributional relations of two species of *Psednura* (Orthoptera: Pyrgomorphidae) in the Evans Head area of New South Wales. *Aust. J. Zool.* 20:411–22
73. Klomp, N. I., Bull, C. M. 1987. Responses to environmental cues by unfed larvae of the Australian reptile ticks *Aponomma hydrosauri* and *Amblyomma limbatum*. *J. Parasitol.* 73:462–466
74. Kohlmann, B., Nix, H., Shaw, D. D. 1988. Environmental predictions and distributional limits of chromosomal taxa in the Australian grasshopper *Caledia captiva* (F.). *Oecologia* 75:483–93
75. Littlejohn, M. J. 1976. The *Litoria ewingi* complex (Anura: Hylidae) in south-eastern Australia. IV. Variation in mating call structure across a narrow hybrid zone between *L. ewingi* and *L. paraewingi*. *Aust. J. Zool.* 24:283–93
76. Lougheed, S. C., Lougheed, A. J., Rae, M., Handford, P. 1989. Analysis of a dialect boundary in chaco vegetation in the rufous-collared sparrow. *Condor* 91:1002–05
77. Marchant, A. D. 1988. Apparent introgression of mitochondrial DNA across a narrow hybrid zone in the *Caledia captiva* species complex. *Heredity* 60:39–46
78. Marshall, J. K. 1968. Factors limiting the survival of *Corynephorus canescens* (L.) Beauv. in Great Britain at the northern edge of its distribution. *Oikos* 19:206–16
79. Martens, J. 1982. Circular distributional overlap and speciation in the chiffchaff *(Phylloscopus collybita)*. The *lorenzii* problem. *Z. Zool. Syst. Evolut.* 20:82–100
80. Mayr, E. 1963. *Animal Species and Evolution*. Cambridge: Harvard Univ. Press. 797 pp.
81. Mayr, E. 1978. Origin and history of some terms in systematic and evolutionary biology. *Syst. Zool.* 27:83–88
82. Mayr, E. 1978. Review of "Modes of Speciation" by MJD White. *Syst. Zool.* 27:478–82
83. MacArthur, R. H. 1972. *Geographical Ecology,* New York: Harper & Row. 269 pp.
84. Miller, R. S. 1967. Pattern and process in competition. *Adv. Ecol. Res.* 4:1–74
85. Murray, M. D., Nix, H. A. 1987. Southern limits of distribution and abundance of the biting midge *Culicoides brevitarsis* Kieffer (Diptera: Ceratopogonidae) in south-eastern Australia: an application of the GROWEST model. *Aust. J. Zool.* 35:575–85
86. Neet, C. R., Hausser, J. 1990. Habitat selection in zones of parapatric contact between the common shrew *Sorex araneus* and Millet's shrew *S. coronatus*. *J. Anim. Ecol.* 59:235–50
87. Nichols, R. A. 1989. The fragmentation of tension zones in sparsely populated areas. *Am. Nat.* 134:969–77
88. Nichols, R. A., Hewitt, G. M. 1986. Population structure and the shape of a chromosomal cline between two races of *Podisma pedestris* (Orthoptera: Acrididae). *Biol. J. Linn. Soc.* 29:301–16
89. Nichols, R. A., Hewitt, G. M. 1988. Genetical and ecological differentiation across a hybrid zone. *Ecol. Entomol.* 13:39–49
90. Noon, B. R. 1981. The distribution of an avian guild along a temperate elevational gradient: the importance and expression of competition. *Ecol. Monogr.* 51:105–24
91. Odendaal, F. J., Bull, C. M. 1980. Influence of water speed on tadpoles of *Ranidella signifera* and *R. riparia* (Anura:Leptodactylidae). *Aust. J. Zool.* 28:79–82
92. Odendaal, F. J., Bull, C. M. 1982. A parapatric boundary between *Ranidella signifera* and *R. riparia* (Anura: Leptodactylidae) in South Australia. *Aust. J. Zool.* 30:49–57
93. Odendaal, F. J., Bull, C. M. 1983. Water movements, tadpole competition and limits to the distribution of the frogs *Ranidella riparia* and *R. signifera*. *Oecologia* 57:361–67
94. Odendaal, F. J., Bull, C. M. 1986. Evidence for interactions over calling sites between males of the frogs *Ranidella signifera* and *R. riparia*. *J. Herpetol.* 20:256–59
95. Odendaal, F. J., Bull, C. M., Adams, M. 1983. Genetic divergence between two morphologically similar *Ranidella* species (Anura: Leptodactylidae). *Copeia* 1983:275–79
96. Odendaal, F. J., Bull, C. M., Nias, R. C. 1982. Habitat selection in tadpoles of *Ranidella signifera* and *R. riparia* (Anura: Leptodactylidae). *Oecologia* 52: 411–14
97. Odendaal, F. J., Bull, C. M., Richards, S. J. 1984. Interactions during feeding between tadpoles of *Ranidella signifera* and *R. riparia* (Anura: Leptodactylidae). *J. Herpetol.* 18:489–92
98. Odendaal, F. J., Bull, C. M., Telford, S. R. 1986. Influence of the acoustic environment on the distribution of the

frog species. *Ranidella riparia*. *Anim. Behav*. 34:1836–43
99. Odendaal, F. J., Telford, S. R., Bull, C. M. 1983. The vocabulary of calls of *Ranidella riparia* (Anura: Leptodactylidae). *Copeia* 1983:534–37
100. Paine, R. T. 1971. A short-term experimental investigation of resource partitioning in a New Zealand rocky intertidal habitat. *Ecology* 52:1096–106
101. Patterson, B. D., Meserve, P. L., Lang, B. K. 1989. Distribution and abundance of small mammals along an elevational transect in temperate rainforests of Chile. *J. Mammal*. 70:67–78
102. Petney, T. N., Bull, C. M. 1984. Microhabitat selection by two reptile ticks at their parapatric boundary. *Aust. J. Ecol*. 9:233–239
103. Porter, S. D., van Eimeren, B., Gilbert, L. E. 1988. Invasion of red imported fire ants (Hymenoptera: Formicidae): microgeography of competitive replacement. *Ann. Entomol. Soc. Am*. 81:913–18
104. Price, P. W., Westoby, M., Rice, B. 1988. Parasite mediated competition: some predictions and tests. *Am. Nat*. 131:544–55
105. Price, P. W., Westoby, M., Rice, B., Atsatt, P. R., Fritz, R. S., et al. 1986. Parasite mediation in ecological interactions. *Annu. Rev. Ecol. Syst*. 17:487–505
106. Prince, S. D., Carter, R. N. 1985. The geographical distribution of prickly lettuce *(Lactuca serriola)*. III. Its performance in transplant sites beyond its distributional limit in Britain. *J. Ecol*. 73:49–64
107. Rand, D. M., Harrison, R. G. 1989. Ecological genetics of a mosaic hybrid zone: mitochondrial differentiation of crickets by soil type. *Evolution* 43:432–49
108. Reimchen, T. E., Stinson, E. M., Nelson, J. S. 1985. Multivariate differentiation and allopatric populations of threespine sticklebacks in the Sangan River watershed, Queen Charlotte Islands. *Can. J. Zool*. 63:2944–51
109. Rey, J. R., Strong, D. R. 1983. Immigration and extinction of salt marsh arthropods on islands: an experimental study. *Oikos* 41:396–401
110. Ribeiro, J. M. C. 1988. Can satyrs control pests and vectors. *J. Med. Entomol*. 25:431–40
111. Ribeiro, J. M. C., Spielman, A. 1986. The satyr effect: a model predicting parapatry and species extinction. *Am. Nat*. 128:513–28
112. Rivas, L. R. 1964. A re-interpretation of the concepts "sympatric" and "allopatric" with proposal of the additional terms "syntopic" and "allotopic". *Syst. Zool*. 13:42–43
113. Roff, D. A. 1974. Spatial heterogeneity and the persistence of populations. *Oecologia* 15:245–58
114. Roff, D. A. 1974. The analysis of a population model demonstrating the importance of dispersal in a heterogeneous environment. *Oecologia* 15: 259–75
115. Rogers, D. J., Randolph, S. E. 1986. Distribution and abundance of tsetse flies (*Glossina* spp). *J. Anim. Ecol*. 55:1007–25
116. Root, T. 1988. Environmental factors associated with avian distributional boundaries. *J. Biogeogr*. 15:489–505
117. Root, T. 1988. Energy constraints on avian distributions and abundances. *Ecology* 69:330–39
118. Settle, W. H., Wilson, L. T. 1990. Invasion by the variegated leafhopper and biotic interactions: parasitism, competition, and apparent competition. *Ecology* 71:1461–70
119. Sheppard, D. H. 1971. Competition between two chipmunk species *(Eutamias)*. *Ecology* 52:320–29
120. Shorrocks, B., Rosewell, J., Edwards, K., Atkinson, W. 1984. Interspecific competition is not a major organising force in many insect communities. *Nature* 310:310–12
121. Slade, N. A., Robertson, P. B. 1977. Comments on competitively induced disjunct allopatry. *Occas. Pap. Mus. Nat. Hist. Kansas*. 65:1–8
122. Smith, H. M. 1955. The perspective of species. *Turtox News* 33:74–77
123. Smith, H. M. 1965. More evolutionary terms. *Syst. Zool*. 14:57–58
124. Smyth, M. 1973. The distribution of three species of reptile ticks, *Aponomma hydrosauri* (Denny), *Amblyomma albolimbatum* Neumann, and *Amb. limbatum Neumann*. I. Distribution and hosts. *Aust. J. Zool*. 21:91–101
125. Spielman, A., Feinsod, F. M. 1979. Differential distribution of peridomestic *Aedes* mosquitoes on Grand Bahama Island. *Trans. R. Soc. Trop. Med. Hyg*. 73:381–84
126. Stangl, F. B. 1986. Aspects of a contact zone between two chromosomal races of *Peromyscus leucopus* (Rodentia: Cricetidae). *J. Mammal*. 67:465–73
127. States, J. B. 1976. Local adaptations in chipmunk *Eutamias amoenus* populations and evolutionary potential at

species borders. *Ecol. Monogr.* 46:221–56

128. Szymura, J. M., Barton, N. H. 1986. Genetic analysis of a hybrid zone between the fire bellied toads, *Bombina bombina* and *B. variegata,* near Cracow in southern Poland. *Evolution* 40:1141–59
129. Taylor, L. R. 1986. Synoptic dynamics, migration and the Rothamstead insect survey. *J. Anim. Ecol.* 55:1–38
130. Taylor, L. R., Taylor, R. A. J. 1977. Aggregations, migration and population mechanics. *Nature* 265:415–21
131. Terborgh, J. 1985. The role of ecotones in the distribution of Andean birds. *Ecology* 66:1237–46
132. Toft, C. A., Schoener, T. W. 1983. Abundance and diversity of orb spiders on 106 Bahamanian islands: biogeography at an intermediate level. *Oikos* 41:411–26
133. Tyler, M. J., Roberts, J. D. 1973. Noteworthy range extensions for some South Australian frogs. *S. Aust. Nat.* 48:20–21
134. Vaughan, T. A. 1967. Two parapatric species of pocket gophers. *Evolution* 21:148–58
135. Walker, P. A. 1990. Modelling wildlife distributions using a geographic information system: kangaroos in relation to climate. *J. Biogeogr.* 17:279–89
136. White, M. J. D. 1978. *Modes of Speciation.* San Francisco: Freeman. 455 pp.
137. Wiens, J. A., Rotenberry, J. T. 1981. Censusing and evaluation of avian habitat occupancy. *Stud. Avian Biol.* 6:522–32
138. Williams, P. H. 1988. Habitat use by bumble bees (*Bombus* spp.). *Ecol. Entomol.* 13:223–37

Annu. Rev. Ecol. Syst. 1991. 22:37–63

MATE CHOICE IN PLANTS: An Anatomical to Population Perspective

Diane L. Marshall and Michael W. Folsom

Department of Biology, University of New Mexico, Albuquerque, New Mexico 87131

KEY WORDS: mating, seed paternity, sexual selection, pollination

---

INTRODUCTION

Whenever the potential fathers of seeds differ in quality and whenever pollen is available in excess of the amount necessary to sire seeds, plants have both the opportunity and the selective pressure to mate nonrandomly (181). Plant mating may be nonrandom at several genetic, structural, and temporal levels. Genetically, mates may be sorted on the basis of relatedness to the seed parent (121, 137), complementarity of maternal and paternal genotypes (174), and the characters of pollen and pollen donors. Structurally, the physiological decisions that regulate mating may occur among the pollen grains and ovules within individual flowers and fruits (159, 181), among the fruits along branches (103), or across entire plants. Temporally, processes that produce nonrandom mating may occur both before and after pollen arrives on stigmas, during all of the steps from pollen germination through seed maturation, and under varying environmental and physiological conditions across seasons.

Nonrandom mating, which occurs whenever the paternity of seeds is different from that which would result from random use of the pollen available, can occur by mechanisms under the control of pollen donors and pollen tubes, maternal tissues, and embryos (134, 159, 181). All have clear fitness interests in the mating process: Pollen donors and maternal plants can improve fitness by increasing the number and quality of offspring, and embryos must garner sufficient maternal resources to survive to maturity, germinate, and grow to reproductive size. The interests of pollen donors have been relatively

0066-4162/91/1120-0037$02.00

uncontroversial. However, the interests of the maternal plant in producing the best possible offspring through choice of the sires of her seeds have been more difficult to demonstrate. The interests of the maternal plant are thought to be under less selection than the interests of the pollen donor and have been more controversial in the literature (30, 94, 134).

There are several reasons why the existence and importance of mate choice in plants remain controversial. First, demonstration of nonrandom mating requires the use of a marker for seed paternity (98, 100, 101, 157). Second, once nonrandom mating is found, it is difficult to determine whether maternal plants, pollen donors, or embryos influence the outcome of mating (94). While differences in pollen tube growth can be tested in vitro in the absence of maternal tissue, maternal sorting cannot be demonstrated in the absence of pollen and pollen tubes. Third, many of the processes that may be under maternal control occur in styles and ovaries where they are difficult to observe. Thus, it is not simple to ascertain whether sorting occurs before or after fertilization and hence whether the haploid genotype of the pollen tube or the diploid genotype of the embryo is being assessed. Finally, lack of basic information about patterns of and variation in fertilization and embryo development limits our understanding of the level of information available to maternal plants during nonrandom mating.

In spite of inherent difficulties in testing for mate choice in plants, unless offspring fitness is random with respect to seed paternity, it would be more surprising to find that plants have no ability to sort among the pollen available than to find that pollen is used nonrandomly. In fact, we know that plants do not simply mate at random. Physiological self-incompatibility and other mechanisms that regulate mating with self and with close relatives are well known (121, 137). However, while self-incompatibility can produce nonrandom mating, sorting among compatible mates is more analogous to mate choice in animals. But, the existence and operation of mechanisms that sort among compatible donors in plants are less thorougly tested. The infrequent documentation of sorting among compatible mates and the lack of information about its mechanisms may be a major source of the controversy surrounding mate choice in plants. Therefore, we concentrate here on nonrandom mating among compatible mates.

The possibility of mate choice in plants was reviewed extensively in 1983 by Willson & Burley (181). We concentrate primarily on the considerable body of evidence that has accumulated since then; we focus more intently on structural evidence for the mechanisms that sort among mates; and we consider primarily angiosperms, whereas Willson & Burley emphasize comparison of angiosperms and gymnosperms.

Our goal is to consider the following major issues. (*a*) Is there evidence, aside from avoidance of mating with self, that mating is nonrandom? (*b*) What are the mechanisms by which mate choice can occur, and what is the physiological and morphological evidence that they operate? (*c*) Can mechanisms that sort among mates be distinguished from pollen competition, embryo competition, or embryo choice? (*d*) What are the functional and evolutionary relationships between choice among compatible mates and discrimination among self- and non-self pollen? And (*e*) is there evidence that these mechanisms evolved in the context of selection for mate choice; does mate choice affect maternal fitness?

ASSUMPTIONS AND DEFINITIONS

Mate choice can be defined as the differential reception and use of pollen by plants acting as seed parents (159, 181). The differential use of pollen must be due to mechanisms under the control of the maternal plant and should act to sort among the sires of seeds rather than among the seeds themselves. Nonrandom mating may alter both the identity and the number of mates that sire seeds on a maternal plant (77, 102). There are difficulties in applying this definition (30, 94, 134) that will emerge as we discuss possible mechanisms of mate choice.

Mate choice cannot occur unless plants receive more pollen than is necessary for full seed set (181). The assumption that excess pollen is available underlies much of the work we discuss. Evidence that seed set is not limited by pollen receipt comes from data showing that hand pollinations do not increase fruit and seed set, that fruit and seed abortion occur in response to stress (85, 158), that fertilizer addition increases fruit and seed set (181), and that stigmas routinely obtain more pollen grains than are necessary to fertilize seeds (153). While plant reproduction is sometimes limited by the amount of pollen received (e.g. 16), limitation of seed set by resource availability appears to be frequent.

Likewise, speculations about the presence of mate choice assume that there is sufficient phenotypic and genotypic variation in pollen donors and in the pollen received by maternal plants to make choice possible. On theoretical grounds, it has been argued that variation in fitness will be depleted by selection (49) and that there will be no basis for mate choice in most populations (30). However, it is also possible to predict that a balance between mutation and selection, variation in environments in time and space, and trade-offs in the components of fitness may maintain selectable variation from generation to generation (29, 136). Empirical studies show that some variation in pollen germination and pollen tube growth is expressed (see 13), and selection for a variety of characters is possible in the stylar environment (see 123).

Even if pollen is variable, selection on mate identity and number can only occur if plants receive pollen from a variety of donors. Data from paternity analysis of field collected seeds of wild radish *(Raphanus sativus)* (38, 43, 44) and other plants (23, 111) show that plants' seeds are sired by more than one pollen donor, so several types of pollen must have been available. Less direct evidence, from surveys of plant mating systems, indicates that mating with more than one pollen donor may be relatively common (144). More to the point, multiple paternity of seeds within fruits has been demonstrated in a few natural populations (14, 38, 43, 44, 46, 139). Experimental pollinations of wild radish suggest that this multiple paternity is due to simultaneous deposition of mixed pollen loads or pollen carryover (99). Therefore, evidence of multiple paternity in plants is good evidence that plants have diverse pollen available on single stigmas.

While multiple paternity has been demonstrated in only a few species, studies of a variety of species indicate that pollen carryover, and hence the deposition of mixed pollen loads, may be relatively common in insect-pollinated plants (45, 57, 60, 88, 143, 164, 165, 172). Thus, the opportunity for maternal influence on the sires of seeds may be widespread.

EVIDENCE FOR NONRANDOM MATING AMONG COMPATIBLE MATES

The most direct evidence that nonrandom mating is possible comes from studies in which the pool of available pollen is known and the resulting paternity of seeds is determined by a genetic marker. Most such studies involve control over the pollen load applied to stigmas and can thus be used only as evidence for postpollination sorting among mates. Some of these studies manipulate the pollen load on stigmas by using pollen from different donors on different flowers (e.g. 10, 11, 17). These can test only for sorting among fruits, so any sorting that is detectable will likely involve postzygotic processes. Nonetheless, differences in maturation of fruits sired by different donors are known (10, 11, 17, 101).

More convincing data come from studies in which pollen from two or more donors is mixed and applied to single stigmas. In these cases, information about both kinds of pollen is likely to be available to the maternal plant in a small amount of space and time. However, relatively few studies have been done that use pollen from multiple compatible donors followed by paternity analysis. Mixtures of two or more kinds of non-self pollen have been applied to a variety of cultivated species: corn (124, 125), onion (37), lima beans (9), alfalfa (6), cultivated *Phlox* (87) and zucchini (131). In all of these cases, mating was nonrandom.

The available data support the possibility that, when pollen of several types is available, noncultivated species also mate nonrandomly. At least six experiments with wild radish, *Raphanus sativus,* suggest that mating is nonrandom (97, 98, 100–102; D. L. Marshall, D. M. Oliveras, M. W. Folsom, O. S. Fuller, unpublished). The differences in number of seeds sired by various pollen donors were often large, with the best donor siring more than twice as many seeds as the worst. Pollen donor effects were largest when the number of kinds of pollen mixed was highest (D. L. Marshall, submitted). The closely related *Raphanus raphanistrum* (155, 184) and *Brassica campestris* (Starck, unpublished) also have nonrandom seed paternity when pollen from multiple donors is applied. All of these experiments used greenhouse plants and thus may not reflect the capabilities of field plants. However, more recent experiments with *R. sativus* show that hand pollinated field plants also mate nonrandomly (D. L. Marshall & O. S. Fuller, unpublished).

Documentation of nonrandom mating is available for only a few species of noncrucifers, but all of these studies were done on field plants. In *Campsis radicans* (14), seed paternity is also nonrandom after mixed pollinations, but in contrast to *Raphanus sativus,* the effects of pollen donors are stronger across fruits in single-donor crosses than within fruits after mixed pollination. Mating is also nonrandom in *Erythronium grandiflorum* (35) and *Hibiscus moscheutos* (A. A. Snow & T. P. Spira, submitted). Among conifers, nonrandom seed paternity has been reported in *Pinus radiata* (113), *Picea albies* (32), and *Pseudotsuga menziezii* (4).

While nonrandom mating may also affect mate number per fruit, testing for this process requires assessment of seed paternity within fruits. While a variety of studies compare the results of mixed and single pollination (7, 12, 86, 145, 169), very few actually compare fruits that are known to be singly and multiply sired (98, 100, 101). Those studies, all with wild radish, reveal that the weight of fruits and seeds increases as the number of fathers per fruit increases. These results suggest that wild radish allocates resources where mate number is highest. A companion study (102), shows that, in spite of this effect on fruit filling, the mean number of mates per fruit is lower than that expected by chance. Taken together, these results suggest that wild radish selects an intermediate mate number.

The evidence presented above can only be interpreted as support for postpollination nonrandom mating. Studies that also include prepollination events are much harder to conduct. These require knowledge of the pollen production of available mates, the amount and kind of pollen that reaches stigmas, and the resulting paternity of seeds. A few studies encompass all of those events but do not distinguish among all of the parts of the mating process. For example, in white spruce orchards, both pollen production and seed paternity were known. Pollen production was correlated with, but did not completely explain, seed paternity (150). Likewise, analyses of *Pinus sylvestris* orchards (116) and populations of *Ipomoea sp.* (149), wild radish (38), and *Asclepias sp.* (23) found that seed paternity was not random and that the patterns could be explained partly, but not entirely, by prepollination events.

The data presented in this section show that nonrandom mating is possible and probably frequent in plants. Taken alone, they do not distinguish whether the maternal plant or the pollen and pollen donor determine these patterns nor whether particular mechanisms of nonrandom mating are important.

POTENTIAL MECHANISMS OF MATE CHOICE IN PLANTS

Prepollination Mechanisms

Although they are more difficult to demonstrate than postpollination mechanisms that sort among mates, there may be a variety of means by which plants acting as seed parents can influence opportunities to mate by affecting the nature of the pollen load deposited on their stigmas. Patterns of floral morphology and phenology that alter the probabilities of deposition of self and outcross pollen are well known (see 137). However, opportunities for plants to affect the precise identity of pollen on stigmas are probably limited, although intraspecific differences in morphology of pollinia and stigmas of *Asclepias syriaca* (114) may provide relatively specific mating relationships. Even so, opportunities to affect the diversity rather than the identity of pollen deposited may be important because plants that receive a wide array of pollen may be able to promote competition among pollen grains and may have the chance to choose among potential mates (51, 52).

Inflorescence and floral characters can influence pollinator movement rather precisely (see 170, 185), but they have been considered primarily as features that increase rates of pollen removal (61, 90). While changes in inflorescence structure can have a greater effect on the rate of pollen removal than on the amount of pollen deposited on stigmas (8, 36, 82, 133, 157), this is not always the case (147). Level of reward can also affect the amount of pollen deposited as well as the amount of pollen removed (50, 163). In fact, in *Ipomopsis aggregata,* the intensity of selection on characters related to pollen receipt was similar to that on characters related to pollen donation (24). More important, it may be a mistake to assume that pollen receipt is adequate if there are just enough pollen grains to fertilize all of the seeds (e.g. 8). The minimum amount of pollen may be inadequate to allow pollen competition and may not include pollen from a diversity of donors.

Plants can increase the potential for mate choice by increasing the number of mating opportunities, that is, multi-ovuled flowers may increase the diversity of mates within fruits. Increases in the number of flowers and the length of time over which they are presented also increase the diversity of mates across fruits (159, 181, 183). Indirect data on the relationship between flower number and visitation pattern suggest that plants with small inflorescences attract few pollinators, each of which visits a high proportion of the flowers, while large inflorescences attract many pollinators, each of which pollinates a smaller proportion of the flowers (57, 82). More direct data from studies of dye movement and genetic diversity of seeds in *Amianthium muscaetoxicum* indicate that progeny diversity is affected by both the size of the seed parent and the timing of flower production (126).

Within flowers, characters that affect the amount of pollen carryover, e.g. the nature of the reward (50) or floral morphology (166, 172), alter the opportunities of plants to mate multiply or to compare mates within flowers (51). Stigma characters such as size and receptivity (52) may also affect the size of the pollen load received and used. For example, annual relatives of the peanut, *Arachis sp.*, have easily pollinated stigmas that can accommodate about 15 pollen grains, while perennials have small stigmas, guarded by a ring of hairs, and can only accept one to three pollen grains (93). Flowers in the genus have two to three ovules. Thus, the opportunity for competition among pollen grains and choice of mates within flowers is much reduced in the perennial species. In *Clintonea borealis,* restriction of the period during which pollen tubes can grow may cause pollen grains to accumulate on the stigma and increase the opportunities for mate choice (52), while in *Amianthium muscaetoxicum,* relatively prolonged stigma receptivity may be favorable to pollen competition (127).

In most of the examples cited above, it is difficult to distinguish the degree to which inflorescence and floral characters may affect pollen donation and pollen receipt. Both are likely to be important. From the perspective of the seed parent, what is missing are studies in which prepollination characters are related to both the amount and the nature of pollen deposition. That is, what are the paternities of seeds of plants with different floral and inflorescence characters, and how is fitness affected relative to the effects on pollen donation?

Postpollination Mechanisms—Prezygotic Events

Once pollen arrives on stigmas, maternal plants have considerably more opportunity for mate choice. Prezygotic sorting among potential mates can be inferred from studies showing that pollen germination (117, 129), pollen tube growth (42), and/or fertilization (109) differ among maternal plants or are affected by the condition of the maternal tissue. The possibilities for maternal control range from relatively precise physiological mechanisms through more general structural characteristics of the female tissue. Although hundreds of excellent studies have considered various aspects of angiosperm reproduction, few reports have dealt specifically with how any portion of this process could affect mate choice in compatible crosses. Here we sample relevant results from a diverse array of species to compose a picture of mechanisms that could be involved in mate choice in plants. Because observation of these mechanisms usually requires sacrifice of pollinated flowers, the connections between observations of any mechanism and nonrandom seed paternity are always indirect.

POLLEN GERMINATION Once on the stigma the pollen grain is in an environment that can stimulate, delay, or stop germination. Relationships between the stigma and pollen grain in compatible crosses are complex (see 39, 66–68). However, a few key events must occur. For a pollen grain to germinate it must adhere to the stigmatic surface, hydrate, become metabolically active, and produce a pollen tube (66).

The first interaction between stigmatic tissue and pollen grains occurs during adhesion. The ease with which pollen grains attach to a stigma depends both on the structure of the pollen grain and the stigma's surface. It is possible that these surfaces have evolved toward tight adhesion of pollen of the same species and little adhesion of pollen grains of other species (68). Which of the several possible mechanisms of pollen adhesion operates depends on whether the stigma is "wet" or "dry" (72). While "wet" stigmas tend to retain whatever is transferred to them by a nonspecific adhesion, "dry" stigmas appear more selective in nature (68).

Following adhesion, the pollen grain must take up water. Although relative humidity and the physiological state of the pollen affect this process, most hydration occurs via the stigma (66). This represents an opportunity for direct maternal effects on pollen germination.

In some cases, adhesion and hydration may operate together to determine which pollen grains germinate. For example, *Primula sieboldii* has both distylous flowers and dimorphic pollen grains. The thrum flowers produce larger pollen and smaller stigmatic papillae than the pin flowers (142). This may provide a reciprocal fit of pollen and papillae across the floral morphs. In *P. vulgaris,* pollen grains of thrum flowers hydrate faster, and their germination is much more sensitive to state of hydration than are pollen grains of pin flowers (152). If pollen germination is similar in *P. sieboldii,* then the smaller pollen grains of pin flowers may germinate more easily because their hydration requirements are not as strict as those of the large pollen grains of thrum flowers. The smaller stigmatic papillae on thrum flowers may not be capable of hydrating self-pollen grains. While this example represents limitation of self-pollination by the structure of the stigma, differences in pollen grain size among compatible donors might also interact with stigmatic papillae to produce differences in hydration.

The maternal tissue may also control pollen germination in *Leucaena*, where there are both a critical number of pollen grains per stigma for pollen germination and a critical number of fertilized ovules per pod for fruit maturation. The stigmatic cells contain a proteinaceous inhibitor which may block germination until sufficient pollen grains are present (56). Germination is pH dependent, but groups of pollen grains can raise the pH of stigmatic extracts to levels that allow pollen germination (54).

Maternal tissue may delay as well as block pollen germination. Delays in pollen germination up to 2 hr were routinely seen in some populations of *Talinum mengesii* (117). Reciprocal crosses showed that this delay was under maternal control, but the mechanism was not determined.

A variety of maternally produced substances are known to affect the rate of pollen germination. Calcium is often necessary for pollen germination (19) and may be the basis for cases in which the proportion of pollen germinating increases as the number of pollen grains per stigma increases. The content of free calcium in pollen is very low, but most plant tissues contain far higher levels of calcium than do pollen grains. So, it is likely that pollen receives the necessary calcium for germination from the stigmatic tissue. Either maternal denial of supplemental calcium and/or production of calmodulin or calcium-binding agents could control levels of free calcium and the rate of pollen germination. Other possibilities for maternal control of pollen germination include sucrose content of nectar (81) and availability of oxygen (53).

Some studies include pollen germination inhibitors obtained from extracts of stigmatic tissue and of other plant parts (74, 75, 122). Although these reports are interesting from a physiological standpoint, their relevance to events surrounding normal pollen germination is not clear. Breakdown of stigmatic cells during the process of germination has been reported in *Petunia* (65), but not in most plants. Therefore, most pollen grains probably would not be exposed to any of the extracted compounds tested.

POLLEN TUBE GROWTH After germination the pollen tube must grow through the stigma, enter the style, obtain nutrients, and grow to the ovule to effect fertilization (see 66, 83). Thus, maternal plants may affect pollen tube growth by regulating the supply of nutrients, through chemical signals, or by structural features that limit and direct the path of pollen tube growth.

Maternal tissue affects mating by controlling the amount and quality of the nutritive matrix in which pollen tubes grow since growing pollen tubes quickly become parasitic on maternal tissue. For example, in *Petunia* pollen tube growth was associated with decreases in starch content of contiguous stylar cells (65). And, in peach, pollen tubes grow through the stigma and style but stop for several days at the obturator, a small mass of tissue that grows out from the placenta (64). Continued growth of the pollen tube depends on breakdown of reserves in the cells of the obturator and production of a nutritive secretion (5). This control over pollen tube growth might decrease competition by allowing all pollen tubes to catch up at the base of the ovary, or it might increase competition for ovules in the last phase of pollen tube growth.

The nutritive tissue of the maternal plant can also determine where in the gynoecium pollen tubes grow. Nearly all styles, whether they are open or closed, possess some form of a mucilaginous matrix through which the pollen tubes grow on their way to the ovary (see 83). Based on histochemical staining, this matrix may contain come combination of carbohydrates, lipids, and proteins (Table 5, 83), and because calcium is necessary for normal pollen tube growth (19, 135), it is probably also present. In *Phaseolus acutifolius* secretory cells cover only a portion of the stylar canal, giving the transmitting tissue a limited "carrying capacity" for pollen tubes (92). Similarly, in *Zea mays,* a progressive decrease occurs in the surface area of the transmitting tissue that is accompanied by abscission of the stigma as pollen tubes grow into the ovary wall, thus limiting further pollination (71). Gradients in the content of the nutritive matrix of the transmitting tissue are also possible. In *Gasteria verrucosa* pollen tubes grow in a fluid filled pathway which has a higher concentration of glucose/fructose in the stylar region than that in the ovule containing locule (179). These mechanisms may control the speed and number of pollen tubes in the style; however, it is not known whether they can determine which pollen tubes enter the ovary.

Early observations that pollen tubes in culture grew toward ovary tissue (21) led to speculation that the style might affect both the speed and the direction of pollen tube growth through the presence of gradients of chemotropic agents in the style. While many tissues and compounds were investigated, calcium appeared to be the primary agent of these responses (106), which is not surprising given the general requirement of calcium for pollen tube growth (19) and the widespread presence of calcium in plant tissues. However, calcium does not direct pollen tube growth in all species, notably not in *Lilium leucanthum* (175).

Recently, interest in chemotropic control of pollen tube growth has waned for several reasons (see 69). Gradients of compounds that would produce continuous chemotropic control have been difficult to find (105). Pollen inserted into the center of styles has been found to produce pollen tubes that grow toward both the stigma and the ovary (63). Nonetheless, some intriguing data remain that are not easily dismissed. Experimental pollinations of *Nicotiana alata* suggest that some of the cases cited to disprove chemotropy may need to be reexamined. For *N. alata,* when pollen tube growth was initiated at the center of an unpollinated style, pollen tubes were equally likely to grow toward the stigma as toward the ovary. However, when this experiment was repeated in a style whose stigma had been pollinated, growth toward the stigma was suppressed (115). Thus directionality of pollen tube growth in the style might be induced by pollination. And, for *Crotalaria retusa,* extracts of the style and ovary increase both the mean and the variance of pollen tube growth rate (96). Thus, some pollen tubes increased in speed more than others, suggesting that maternal tissues could produce chemicals with selective effects on pollen tube growth.

If chemical control of pollen tube growth is unimportant, the maternal tissue may still affect pollen tube growth by structural features of the stigma and style. Since pollen tubes grow only from the tip, small changes in the structure of the gynoecium may suffice to redirect pollen tubes (69).

One of the first structures encountered by a growing pollen tube is the cuticle, which may serve as a physical barrier to the entrance of pollen tubes into the stigma and may respond to changes in the hydration of stigmatic cells to regulate the flow of water and metabolites to germinating pollen grains (67, 70). For example, in *Crocus,* penetration of the cuticle was accomplished by production of a cutinase enzyme (70). The pollen grain produced an inactive form of the enzyme, and a compound in the stigmatic secretion transformed the precursor into an active enzyme. This mechanism might affect competition among pollen grains if the release of the stigmatic secretion depends on the number of pollen grains on the stigma.

Variations in cuticle and stigma structure have been implicated in the determination of self-fertility in *Vicia faba* (91). Highly self-fertile lines of this plant had stigmas composed of small papillae cells covered by a thin cuticle, while those with lower levels of selfing had stigmas with large papillae cells covered by a thick cuticle. The implication is that increases in the thickness of the cuticle increase the difficulty of pollen grain hydration. Whether these differences in cuticle thickness also affect compatible crosses is not known.

The importance of structural control over the path of pollen tube growth is emphasized in a recent study by Sanders & Lord (140). When tiny latex beads were applied directly to the stylar transmitting tissue of *Hemerocallis flava, Raphanus raphanistrum,* and *Vicia faba,* the beads moved through the transmitting tissue at the same rate and in the same direction as pollen tubes. Some caution is warranted in interpreting these results since pollen tubes are anchored on the stigma and the beads are not. Still, these data suggest the possibility of strong maternal control over the growth of pollen tubes in the style.

FERTILIZATION The final opportunity for interaction of the female tissue and the microgametophyte is during fertilization. In ovaries which contain more than one ovule, time course studies of fertilization (73, 109) and analysis of mature seed distributions in gynoecia (76, 131) suggest that it is not always the first ovule in the pollen tube's path that is fertilized. In fact, the first pollen tubes in *Raphanus* ovaries sometimes grow past all available ovules, fertilizing nothing (M. W. Folsom & D. L. Marshall, unpublished). Thus, the physical order of ovules and speed of pollen tube growth are not the sole determinants of fertilization.

Other possibilities are structural or physiological features that test the qualities of the male genome or increase the amount of competition among pollen tubes for fertilization. These characteristics may include structures such as obturators or changes in the spatial relationships between ovules and transmitting tissue. For example, it has long been held that obturators physically direct growth of the pollen tube to the micropyle (40, 95) as in *Raphanus raphanistrum* where the obturator directs growth of the pollen tube out of the transmitting tissue (i.e. the septum) and toward the micropyle (73). In *Mangifera indica* a structure called the ponticulus or bridge begins development after pollination. It originates at the base of the style and may act as an extension of the transmitting tissue that provides the shortest route for the growing pollen tube between the stylar tissue and the ovule (177).

Postzygotic Mechanisms: Seed and Fruit Abortion

Nonrandom seed abortion is relatively difficult to demonstrate. Simple differences in the final paternity of seeds may be due either to prefertilization events or to differential seed abortion. However, a few data sets indicate that nonrandom seed abortion is possible (see 85). In *Cryptantha flava,* there is always some abortion of developing seeds since the four ovules can produce only one or two seeds. Experimental, random removal of developing seeds produced lower quality offspring than natural abortion, suggesting that abortion does sort among offspring (25). When wild radish plants were given reduced water, such that plants were induced to abort seeds, seed paternity was different than in related plants which had been given adequate water (101). In this case, seed abortion resulted in nonrandom seed paternity largely because pollen donors fertilized seeds in particular parts of the fruit and because seed abortion was more frequent in the stylar end of the fruit.

Even if nonrandom seed abortion can be demonstrated, it is difficult to interpret because, by the time seeds are filling, there are several genetically distinct entities with potentially different fitness interests (see 58, 107, 132, 176). The maternal plant's fitness might be maximized by abortion of some seeds and optimal allocation of resources among the rest; however, the female gametophyte, endosperm, and embryo, in that order, may have highest fitness by garnering additional resources for a particular embryo to which they are more closely related than to other embryos. Likewise, there may be conflicts among embryos as any particular embryo may have the highest fitness by not being the one aborted (168). So, when seed abortion is nonrandom, it may be the result of sorting under the control of the maternal plant, conflicts between the maternal plant and the embryos and associated tissues, or competition among embryos.

Distinguishing among these mechanisms requires careful examination of the structural and physiological bases of embryo abortion (see 58). Assessment of the processes leading to differential survival of embryos must account for the nature and sequence of changes occurring in maternal tissue, embryos, and endosperm. If abortion of developing seeds is due to maternal choice, then structural changes in the maternal tissues of the seed (nucellus and integuments) may precede changes in embryo and endosperm development. In contrast, if differential survival of embryos is due to competition among seeds for available resources, then variation in rates of embryo and endosperm development should occur before or in the absence of changes in the maternal tissues.

From the few studies that provide sufficient description of all tissues involved in seed abortion, there is some evidence for both of these mechanisms. For example, changes in the nucellus and integuments that precede changes in the embryo or endosperm have been described in a few species. In many species of Rosaceae two ovules are initiated, but one aborts after anthesis. The earliest evidence of abortion in these species is deposition of lignin and callose in the integuments, followed later by separation of the integuments from the nucellus, and finally deterioration of the nucellus (18, 128, 167). Disorganization of the embryo and endosperm follows the changes in maternal tissues. Transport of nutrients to the developing embryo may be blocked by deposition of lignin, callose, or cutin in the integuments (20, 78, 128). In these examples, a fixed proportion of developing seeds is typically aborted, so there must be a maternal component to abortion. During abortion of ovules from interspecific crosses among *Datura* species, tumors derived from maternal tissue form and arrest growth of hybrid embryos (141). Culturing of normal selfed embryos in vitro in the presence of tumor extract caused death of these embryos. Thus, this may be a mechanism that prevents interspecific hybridization. The mechanism was never seen in "compatible" crosses.

In contrast to the examples of maternal control, there are several cases where there are known or assumed differences in vigor of embryos, and the first evidence of abortion is a decreased growth rate of the endosperm and embryo, followed at a later stage by changes in the maternal tissue. For example, in numerous studies of interspecific and intergeneric hybrids, the first evidence of abortion is retardation of endosperm development. Embryo growth is affected at a later stage, but the maternal tissues continue to develop normally for a considerable period of time (22). Abortions occurring in low yield alfalfa strains (33) and embryo lethal mutants of *Lilium* (27) follow the same sequence of events. In *Lilium,* competition with higher quality seeds further decreases embryo and endosperm growth rate and results in increased seed abortion (28). While all of these differences among embryos result from fairly gross genetic abnormalities, these studies suggest that finer scale genetic differences such as those occurring among half-sibs within fruits might also lead to differences in endosperm and embryo development and competitive ability.

Conflicts among the embryos within fruits may also determine patterns of seed abortion (168). For example, the embryos that produce the strongest hormonal signals are likely to get the most resources (112), a form of exploitation competition. Uma Shaanker et al (168) suggest that more direct interference competition among embryos is also possible because water extracts of *Dalbergia sisoo* embryos can suppress the growth of other embryos in culture (55). However, extracted compounds may include combinations that do not occur within the plant, and in the plant, each embryo is completely surrounded by maternal tissue. Thus, inhibitors cannot move among embryos without passing through maternal tissue.

Kress (84) predicted that competition among embryos will be most severe when the seeds within fruits have several fathers and so suggested examination of data on the pattern of variation in seed size within singly and multiply sired fruits. In wild radish, the variance in seed weight is indistinguishable in singly and multiply sired fruits (97). And, contrary to the prediction, multiply sired fruits get more resources than do singly sired fruits (97, 98, 100).

Ideally, we need data on the patterns of hormone production by embryos of different genotypes when those embryos are surrounded by full and half siblings. Less direct data on patterns of final seed size, as influenced by the paternity of neighboring seeds, could be informative. Here, neighboring seeds need to be identified as those that obtain resources from the same vascular bundle.

Even if it can be determined that the maternal plant has a large influence on patterns of seed abortion, it is still necessary to consider whether nonrandom abortion is due to the identity of the sire of the seed or to the combination of maternal and paternal genes in the embryo and/or endosperm. Whether sorting is among mates or offspring will depend on the mechanism of sorting and the timing and nature of gene expression. For example, in wild radish, nonrandom seed abortion occurs because pollen donors sire seeds in nonrandom locations within fruits, and seed abortion depends on position within the fruit (101). Thus, the process selects for pollen donors that can fertilize seeds in specific locations within fruits but is not dependent on the expression of offspring genes. Even after gene expression begins, it is possible that differences in early embryo success are largely due to the paternal genome if selection has produced embryo growth genes whose expression depends on the parent of origin (59). Alternatively, abortion may occur because the embryo is simply defective (178), but this is clearly not always the cause of abortion since it is possible to "rescue" and culture aborting embryos (118). Maternally controlled abortion may be a response to the expression of the diploid genotype of the developing seed. To sort among these possibilities we need information about when gene expression in general and expression of the paternal genome in particular begins in plant zygotes.

Short of seed abortion, seeds may be provisioned differently so that final seed size varies with paternity (1, 3, 10, 17, 97, 100, 108, 109, but see 47). In these cases, there is little doubt that the diploid genome of the seed was expressed, so it is difficult to attribute seed weight differences to maternal response to paternal genes, the embryo genotype, or to competition among embryos.

Differential fruit abortion can also alter the paternity of seeds if fruit abortion is based on the identity or number of pollen donors that sired the seeds within fruits (see 85). Nonrandom fruit abortion due to the identity of compatible donors has been demonstrated in only a few plants (10, 11, 17, 97). Less direct evidence comes from cases in which natural, and therefore potentially selective, abortion of fruits produces higher quality offspring than random fruit abortion (161, 162). Cases in which fruit filling or fruit abortion are correlated with the number of donors per fruit are even less common. In wild radish, multiply sired fruits are filled to greater weights than are singly sired fruits (97, 98, 100), and in some cases, multiply sired fruits are less likely to abort (97).

Fruit abortion may also alter the composition of a plant's mates if abortion of fruits along branches depends on the prior pollination history of a branch or a plant. The result may be an increase in diversity of mates if fruits sired by unusual pollen donors are more likely to be filled, or a decrease in diversity of mates if prior pollinations of inflorescences reduce the number of susceptible mates. The first process may occur in wild radish, where fruits sired by rare pollen donors can be selectively filled (103). The second occurs in *Campsis radicans,* where the number of acceptable pollen donors is reduced on inflorescences with several developing fruits (11).

Since fruit abortion occurs late, relative to the other potential mechanisms of maternal control, it is especially difficult to determine the basis of any selection. By the time fruits can be "compared" along branches, information about both the paternal and the embryo genotype is surely available. Thus, it is difficult to ascertain whether mate choice or embryo choice is involved. In cases where fruit production is correlated with seed number (160), it is probably not the specific embryo genotype that is being selected. However, in many cases, fruit abortion may be a point at which choice can truly be based on indications of the quality of the developing offspring.

CONFOUNDING EFFECTS

Consideration of mate choice in plants has been hampered by the difficulty of dissociating it from other mechanisms of nonrandom mating. That is, both pre- and postzygotic sorting by the maternal plant are confounded with the possibility of competition among pollen donors and pollen tubes for access to ovules. Postzygotic sorting can involve sorting among offspring as well as among mates, and it may also be confounded with competition among embryos for resources. Proposals for determining which of these processes operate range from closer consideration of the mechanisms, through better data analysis, to changes in experimental procedures. In deducing which sorting processes occur, it is important to consider what is physiologically and structurally possible, to carefully observe mating, and to consider what information is available at various stages of pollination, fertilization, and seed and fruit development.

The literature cited above suggests that the maternal plant has enormous opportunity to control fertilization and seed development. All pollen tube growth, fertilization, and embryo development take place inside maternal tissue. Any of these processes can be stopped if the maternal plant withholds resources. Pollen-pollen interactions are possible on the stigma (D. L. Marshall, M. W. Folsom unpublished), and interactions among pollen tubes may occur in the style (35); however, interactions among embryos must be entirely mediated by maternal tissue.

N. M. Waser & M. V. Price (submitted) point out that prezygotic sorting is easier to interpret than postzygotic sorting as fewer genetic entities are involved. Slowing of pollen tube growth (173), for example, is more likely to be in the interests of the maternal plant than the paternal plant.

Lyons et al (94) discuss statistical procedures at length. They emphasize the importance of measuring both the effects of parents on mating and the interactions between parents and they consider which kinds of designs are likely to be most informative. To their discussion, we add that it is important to evaluate the power of experiments in which negative results are obtained.

Finally, both Lyons et al (94) and Marshall & Ellstrand (97) have suggested that adjusting maternal condition may be a critical part of experiments to test for the existence of mate choice. If changing maternal but not paternal condition alters mating, then the maternal tissue must be involved. This might be done by chemically turning off the incompatibility system (94). However, the interpretation of such experiments will be difficult until we know more about the interaction between incompatibility and sorting among compatible mates.

In wild radish, maternal stress resulted in a change in mating patterns (97). Similarly maternal plant age in wild radish (D. L. Marshall & D. M. Oliveras, in preparation) and flower age in *Brassica campestris* (L. A. Starck, personal communication) alter mating patterns and suggest that the maternal tissue plays an important role.

MATE CHOICE AND SELF-INCOMPATIBILITY

We have concentrated primarily on the sorting among mates that may occur in compatible crosses; however, many plants also sort among close relatives and nonrelatives (see 121, 137), and several species discriminate against mates that are too distantly related (171). It is increasingly clear that the functional and evolutionary relationships among these mechanisms need to be understood.

Both sorting among compatible mates, as reviewed here, and sorting among relatives can occur at all stages from prepollination mechanisms through seed and fruit development. Examples of mechanisms that limit self-pollination range from temporal and spatial separation of anthers and stigmas, through physiological self-incompatibility (see 121, 137), to differential pollen tube growth of self- and outcross pollen (34, 41, A. A. Snow, T. P. Spira, submitted) and postfertilization failures of selfed seeds or fruits (151). Thus, at virtually every stage of mating, sorting among compatible mates might occur by mechanisms related to self-incompatibility.

Does this mean that all nonrandom mating is based on relatedness? Currently, we cannot answer that question. Recent data suggest that incompatibility is more complicated than we once suspected. For example, in *Raphanus* (89) and *Brassica* (186), a gametophytic incompatibility system may underlie the sporophytic incompatibility system typical of these genera. And, both cryptic self-incompatibility and cryptic self-fertility, which can be detected only when self-pollen competes against outcrossed pollen, may be frequent (e.g., 15, 26).

In addition to specific incompatibilities due to shared S-(self-incompatibility) alleles, degree of relationship also may affect sorting among pollen donors between pollination and fertilization. For example, in *Nicotiana alata*, the effect of the degree of sequence similarity of S-alleles of pollen donor and seed parent on pollen tube growth is being explored (see 62). In wild radish, degree of relatedness had a small but significant effect on pollen tube growth (120). Thus, much of what is seen in comparing purportedly compatible donors could possibly reflect different degrees of relatedness.

On the other hand, subtle effects of inbreeding and outbreeding cannot explain all aspects of mating, at least in wild radish. The same pollen donor may perform best on maternal plants from several populations (D. L. Marshall, submitted). Recent crosses in which all maternal and paternal plants had different S-alleles revealed that mating was still nonrandom (D. L. Marshall & M. W. Folsom, unpublished).

Experiments are needed that distinguish among and evaluate the importance of all the mechanisms of mate sorting in plants. These experiments need to be repeated in plants with and without physiological self-incompatibility. We need to discover the relative magnitudes of effects attributable to sorting among relatives and nonrelatives, sorting among pollen donors of similar relatedness to the parent plant, and of sorting among moderately similar and very dissimilar pollen donors.

These experiments may not reveal the evolutionary relationships among the various processes that sort among mates. That is, if the mechanisms of sorting among compatible mates differ from those that produce incompatibility, the processes may be evolutionarily distinct. However, if the mechanisms are similar, then it is interesting to ask which function came first and if selection for one kind of sorting (e.g. among compatible mates) has altered the mechanisms of other kinds of sorting (e.g. incompatibility).

NONRANDOM MATING AND OFFSPRING FITNESS

While the data reviewed above may reveal the presence and potential mechanisms of nonrandom mating, these data cannot tell us whether those mechanisms evolved in the context of selection for increased choice among mates. That is, because these processes exist does not mean that they occur because the plants that possessed them sorted more effectively among mates. One way to address whether mate choice is, and perhaps was, the function of these mechanisms is to consider their effect on maternal and offspring fitness.

The fitness of the maternal plant can be improved by nonrandom mating if mating is based on good genes or if nonrandom mating improves the ability of offspring to compete for fertilizations. Evidence that offspring characteristics may be affected comes, indirectly, from studies showing that gene expression occurs during pollen development, that selection for particular characters is possible in the style, and that pollen donor identity affects offspring growth. Stronger evidence comes from studies showing that nonrandom mating or the opportunity for nonrandom mating alters progeny success.

First, many genes are expressed during pollen tube growth, and these overlap extensively with the genes detectable during seedling growth (see 123). Thus, it is possible for selection in the style to affect genetically based characters. Second, paternal effects on offspring characteristics such as seed size and seedling growth are known (1, 3, 10, 79, 104, 119). Third, when plants are provided with greater opportunities for nonrandom mating by increasing the amount of pollen on the stigma (182; earlier refs. in 13) or by allowing for selective fruit (161, 162) or seed (25, 138) abortion, the size or growth rate of the resulting offspring can be improved relative to those produced under reduced opportunity for selection. However, this is not uniformly the case. Increasing the size of the pollen load in *Raphanus raphanistrum* (154) and *Epilobium* (A. A. Snow, submitted) did not improve offspring growth.

The most direct evidence for effects of mating on offspring quality comes from the small number of cases in which both mating patterns and offspring growth are known. In wild radish, the pollen donor whose seeds were least susceptible to fruit and seed abortion produced offspring that grew to the greatest weight in the greenhouse (104). In *Campsis radicans,* the pollen donors whose fruits are least susceptible to abortion produce offspring which germinate (10) and grow (Bertin, submitted) at greater rates than other offspring.

Evidence for the effects of mate number on offspring success is limited. Increasing the number of mates may increase the genetic diversity of offspring. While mixtures of genotypes outyield pure stands in some cases (2, 130), studies that compare the relationship between fitness and genetic diversity among sibships give mixed results (80, 110, 148). For wild radish, where mating patterns act to increase genetic diversity of offspring through selective allocation of resources to multiply sired fruits (97, 98, 100), comparisons of the fitness of groups of full-sib and half-sib progeny give equivocal results (79).

There is even less evidence to test whether nonrandom mating improves the mating success of progeny. This requires information about the heritability of the mating characters themselves. While some kinds of floral characters such as flower color (e.g. 156) are known to be genetically based, there are relatively few studies that document selection on these during mating (but see 24).

Evidence for the heritability of and potential for selection on characters related to postpollination mating success is also scarce. While selection on corn was able to produce progeny that performed better as pollen donors (124), and selection based on pollen load size may improve some aspects of pollen donor success in zucchini (146), selection on pollen tube growth in *R. raphanistrum* did not improve the mating performance of progeny (155), perhaps because pollen performance may be affected by the environment of the pollen parent (184).

CONCLUSIONS AND CONTINUING PROBLEMS

Based on studies that use genetic markers, sorting among compatible donors clearly occurs. However, in spite of a wealth of information on the possibility of mechanisms that sort among mates, there is little evidence to prove that specific maternal mechanisms produce this sorting. Studies designed to separate incompatibility from other forms of mate sorting and studies designed to powerfully test for both maternal and paternal effects on mating are needed. In addition, more information on variation among plants and among crosses in mechanisms that operate at the physiological level is necessary.

Given the difficulty in distinguishing mate choice in plants, should we bother to use the term at all? Since nonrandom mating does occur in plants and since possibilities for maternal control over these mechanisms exist, if we do not use the term mate choice we will have to invent another. Given the points of similarity in some aspects of plant and animal mating (180), using different terminology might create more problems than it resolves. There is certainly complexity in plant mate choice since the options for mating may include self-pollen, highly outcrossed pollen, and a variety of compatible donors. However, all of these possibilities are perhaps best seen as a continuum of categories of mates. Some plants may distinguish primarily among self- and outcross pollen, others may distinguish among all possible mates, and some plants show little of any of these processes (48).

There is some resistance to the application of the term *mate choice* to plants due to a misunderstanding that the term implies conscious choice in animals, which it does not (180), and to a perceived passivity of plants relative to animals. In fact, maternal mechanisms effecting mate choice may grade from mainly physiological (active) to mostly structural (passive). Active mechanisms may require signaling between the pollen grain or tube and the maternal plant, and these mechanisms could be relatively specific. On the other hand, structural mechanisms established during floral development may be more passive. These may set up both barriers and controls that are insensitive to paternal genotypes. However, they may serve as tests of the capabilities of the paternal genotype since only genotypes that surmount the difficulties encountered during germination, pollen tube growth, and location of the ovule's micropyle achieve fertilization.

Mate choice in plants has also been controversial because of the suggestion that it could result in sexual selection (30, 31, 94, 159, 181). Among the types of mate sorting that may occur, sorting among compatible mates most clearly creates this opportunity. Other kinds of sorting that are due to similarity or dissimilarity among maternal and pollen donor genotypes cannot produce directional selection, but consistent sorting among compatible mates might do so. Thus, that portion of plant mate choice that produces differences in the mating success of pollen donors due to information about the heritable characteristics of the pollen donors (and not due to relatedness or the diploid genotype of the progeny) can produce sexual selection. Of course, we do not know what fraction of the nonrandom mating this represents for any species.

In conclusion, the possibility of mate choice in plants is rich in mechanisms to consider and opportunities to affect maternal, paternal, and offspring fitness. While its importance is by no means proven, nonrandom mating certainly occurs and maternal control is clearly possible. The lack of specific evidence seems more an opportunity for future research than a reason to dismiss the issue.

ACKNOWLEDGMENTS

We thank Ann Evans and Diana Oliveras for comments on an earlier version of the manuscript, and Pam Diggle for suggesting relevant references on seed abortion. Support for this work was provided by NSF grants BSR-8818522 and BSR-8958233.

Literature Cited

1. Andersson, S. 1990. Paternal effects on seed size in a population of *Crepis tectorum* (Asteraceae). *Oikos* 59:3–8
2. Antonovics, J., Ellstrand, N. C. 1984. Experimental studies of the evolutionary significance of sexual reproduction. I. A test of the frequency-dependent selection hypothesis. *Evolution* 38:103–15
3. Antonovics, J., Schmitt, J. 1986. Paternal and maternal effects on propagule size in *Anthoxanthum odoratum*. *Oecologia* 69:277–82
4. Apsit, V. J., Nakamura, R. R., Wheeler, N. C. 1989. Differential male reproductive success in douglas-fir. *Theor. Appl. Genet.* 77:681–84
5. Arbeloa, A., Herrero, M. 1987. The significance of the obturator in the control of pollen tube entry into the ovary in peach *(Prunus persica)*. *Ann. Bot.* 60:681–85
6. Barnes, D. K., Cleveland, R. W. 1963. Genetic evidence for nonrandom fertilization in alfalfa as influenced by differential pollen tube growth. *Crop Sci.* 3:295–97
7. Bawa, K. S., Webb, C. J. 1984. Flower, fruit, and seed abortion in tropical forest trees: implications for the evolution of paternal and maternal reproductive patterns. *Am. J. Bot.* 71:736–51
8. Bell, G. 1985. On the function of flowers. *Proc. R. Soc. Lond. B* 224:223–65
9. Bemis, W. P. 1959. Selective fertilization in lima beans. *Genetics* 44:555–62
10. Bertin, R. I. 1982. Paternity and fruit production in trumpet creeper *(Campsis radicans)*. *Am. Nat.* 119:694–709
11. Bertin, R. I. 1985. Nonrandom fruit production in *Campsis radicans:* between-year consistency and effects of prior pollination. *Am. Nat.* 126:750–59
12. Bertin, R. I. 1986. Consequences of mixed pollinations in *Campsis radicans*. *Oecologia* 70:1–5
13. Bertin, R. I. 1988. Paternity in plants. In *Plant Reproductive Strategies,* ed. J. Lovett Doust, L. Lovett Doust, pp. 30–59. New York: Oxford Univ. Press
14. Bertin, R. I. 1990. Paternal success following mixed pollinations of *Campsis radicans*. *Am. Midl. Nat.* 124:153–63
15. Bertin, R. I., Barnes, C., Guttman, S. I. 1989. Self sterility and cryptic self-fertility in *Campsis radicans* (Bignoniaceae). *Bot. Gaz.* 150:397–403
16. Bierzechudek, P. 1981. Pollinator limitation of plant reproductive effort. *Am. Nat.* 117:838–40
17. Bookman, S. S. 1984. Evidence for selective fruit production in *Asclepias*. *Evolution* 38:72–86
18. Bradbury, D. 1929. A comparative study of the developing and aborting fruits of *Prunus serasus*. *Am. J. Bot.* 16:525–42
19. Brewbaker, J. L., Kwak, B. H. 1963. The essential role of calcium ion in pollen germination and pollen tube growth. *Am. J. Bot.* 50:859–65
20. Briggs, C. L., Westoby, M., Selkirk, P. M., Oldfield, R. J. 1987. Embryology of early abortion due to limited maternal resources in *Pisum sativum* L. *Ann. Bot.* 59:611–19
21. Brink, R. A. 1924. The physiology of pollen. IV. Chemotropism; effects on growth of grouping grains; formation and function of callose plugs; summary and conclusions. *Am. J. Bot.* 11:417–36

22. Brink, R. A., Cooper, D. C. 1947. The endosperm in seed development. *Bot. Rev.* 13:423–541
23. Broyles, S. B., Wyatt, R. 1990. Paternity analysis in a natural population of *Asclepias exaltata:* multiple paternity, functional gender, and the "pollen donation hypothesis". *Evolution* 44:1454–68
24. Campbell, D. R. 1989. Measurement of selection in a hermaphroditic plant: variation in male and female pollination success. *Evolution* 43:318–34
25. Casper, B. B. 1988. Evidence of selective embryo abortion in *Cryptantha flava*. *Am. Nat.* 132:318–26
26. Casper, B. B., Sayigh, L. S., Lee, S. S. 1988. Demonstration of cryptic incompatibility in distylous *Amsinckia douglasiana*. *Evolution* 42:248–53
27. Cave, M. S., Brown, S. W. 1954. The detection and nature of dominant lethals in *Lilium*. II. Cytological abnormalities in ovules after pollen irradiation. *Am. J. Bot.* 41:469–83
28. Cave, M. S., Brown, S. W. 1957. The detection and nature of dominant lethals in *Lilium*. III. Rates of early embryogeny in normal and lethal ovules. *Am. J. Bot.* 44:1–8
29. Charlesworth, B. 1987. The heritability of fitness. In *Sexual Selection: Testing the Alternatives,* ed. J. W. Bradbury, M. Andersson, pp. 21–40. New York: Wiley
30. Charlesworth, D., Schemske, D. W., Sork, V. L. 1987. The evolution of plant reproductive characters; sexual versus natural selection. In *Evolution of Sex,* ed. S. C. Stearns, pp. 317–35. Basel: Birkhauser
31. Charnov, E. L. 1979. Simultaneous hermaphroditism and sexual selection. *Proc. Natl. Acad. Sci. USA* 76:2480–84
32. Cheliak, W. M., Skroppa, T., Pitel, J. A. 1987. Genetics of the polycross. 1. Experimental results from Norway spruce. *Theor. Appl. Genet.* 73:321–29
33. Cooper, D. C., Brink, R. A., Albrecht, H. R. 1937. Embryo mortality in relation to seed formation in alfalfa *(Medicago sativa)*. *Am. J. Bot.* 24:203–13
34. Cruzan, M. B. 1989. Pollen tube attrition in *Erythronium grandiflorum*. *Am. J. Bot.* 76:562–70
35. Cruzan, M. B. 1990. Pollen-pollen and pollen-style interactions during pollen tube growth in *Erythronium grandiflorum* (Liliaceae). *Am. J. Bot.* 77:116–22
36. Cruzan, M. B., Neal, P. R., Willson, M. F. 1988. Floral display in *Phyla incisa:* consequences for male and female reproductive success. *Evolution* 42:505–15
37. Currah, L. 1981. Pollen competition in onion (*Allium cepa* L.). *Euphytica* 30:687–96
38. Devlin, B., Ellstrand, N. C. 1990. Male and female fertility variation in wild radish, a hermaphrodite. *Am. Nat.* 136:87–107
39. Dumas, C., Knox, R. B., Gaude, T. 1984. Pollen-pistil recognition: New concepts from electron microscopy and cytochemistry. *Int. Rev. Cytol.* 90:239–72
40. Eames, A. 1961. *Morphology of Angiosperms*. New York: McGraw-Hill
41. Eenink, A. H. 1982. Compatibility and incompatibility in witloof-chickory (*Cichorum intybus* L.) 3. Gametic competition after mixed pollinations and double pollinations. *Euphytica* 31:773–86
42. Elgersma, A., Stephenson, A. G., den Nijs, A. P. M. 1989. Effects of genotype and temperature on pollen tube growth in perennial ryegrass (*Lolium perenne* L.). *Sexual Plant Reprod.* 2:225–30
43. Ellstrand, N. C. 1984. Multiple paternity within the fruits of the wild radish, *Raphanus sativus*. *Am. Nat.* 123:819–28
44. Ellstrand, N. C., Marshall, D. L. 1986. Patterns of multiple paternity in populations of *Raphanus sativus*. *Evolution* 40:837–42
45. Ennos, R. A., Clegg, M. T. 1982. Effect of population substructuring on estimates of outcrossing rate in plant populations. *Heredity* 48:283–92
46. Epperson, B. K., Clegg, M. T. 1987. First-pollination primacy and pollen selection in the morning glory, *Ipomoea purpurea*. *Heredity* 58:5–14
47. Fenster, C. B. 1991. Effect of male pollen donor and female seed parent on allocation of resources to developing seeds and fruit in *Chamaecrista fasciculata* (Leguminosae). *Am. J. Bot.* 78:13–23
48. Fenster, C. B., Sork, V. L. 1988. Effect of crossing distance and male parent on in vivo pollen tube growth in *Chamaecrista fasciculata*. *Am. J. Bot.* 75:1898–1903
49. Fisher, R. A. 1958. *The Genetical Theory of Natural Selection*. New York: Dover
50. Galen, C., Plowright, R. C. 1985. The effects of nectar level and flower development on pollen carry-over in inflorescences in fireweed *(Epilobium)*. *Can. J. Bot.* 63:488–91
51. Galen, C., Rotenberry, J. T. 1988. Variance in pollen carryover in animal-pollinated plants: implications for mate choice. *J. Theor. Biol.* 135:419–29

52. Galen, C., Shykoff, J. A., Plowright, R. C. 1986. Consequences of stigma receptivity schedules for sexual selection in flowering plants. *Am. Nat.* 127:462–76
53. Galil, J. 1990. Pollen germination in *Arisarum vulgare* Targ.-Tozz. *Flora* 184:51–61
54. Ganeshaiah, K. N., Uma Shaanker, R. 1988. Regulation of seed number and female incitation of mate competition by a pH-dependent proteinaceous inhibitor of pollen grain germination in *Leucaena leucocephala*. *Oecologia* 75:110–13
55. Ganeshaiah, K. N., Uma Shaanker, R. 1988. Seed abortion in wind-dispersed pods of *Dalbergia sissoo:* maternal regulation or sibling rivalry. *Oecologia* 75:135–39
56. Ganeshaiah, K. N., Uma Shaanker, R., Shivashanker, G. 1986. Stigmatic inhibition of pollen grain germination—its implication for frequency distribution of seed number in pods of *Leucaena leucocephala* (Lam) de Wit. *Oecologia* 70:568–72
57. Geber, M. A. 1985. The relationship of plant size to self-pollination in *Mertenzia ciliata*. *Ecology* 66:762–72
58. Haig, D., Westoby, M. 1988. Inclusive fitness, seed resources, and maternal care. See Ref. 13, pp. 60–79
59. Haig, D., Westoby, M. 1989. Parent-specific gene expression and the triploid endosperm. *Am. Nat.* 134:147–55
60. Handel, S. N. 1982. Dynamics of gene flow in an experimental population of *Cucumis melo* (Cucurbitaceae). *Am. J. Bot.* 69:1538–46
61. Harder, L. D., Thomson, J. D. 1989. Evolutionary options for maximizing pollen dispersal of animal-pollinated plants. *Am. Nat.* 133:323–44
62. Haring, V., Gray, J. E., McClure, B. A., Anderson, M. A., Clarke, A. E. 1990. Self-incompatibility: a self-recognition system in plants. *Science* 250:937–41
63. Hepher, A., Boulter, M. E. 1987. Pollen tube growth and fertilization efficiency in *Sapiglossis sinuata:* implications for the involvement of chemotropic factors. *Ann. Bot.* 60:595–601
64. Herrero, M., Arbeloa, A. 1989. Influence of the pistil on pollen tube kinetics in peach *(Prunus persica)*. *Am. J. Bot.* 76:1441–47
65. Herrero, M., Dickinson, H. G. 1979. Pollen-pistil incompatibility in *Petunia hybrida:* Changes in the pistil following compatible and incompatible intraspecific crosses. *J. Cell Sci.* 36:1–18
66. Heslop-Harrison, J. 1987. Pollen germination and pollen-tube growth. *Int. Rev. Cytol.* 107:1–78
67. Heslop-Harrison, J., Heslop-Harrison, Y. 1982. The specialized cuticles of the receptive surfaces of angiosperm stigmas. In *The Plant Cuticle,* ed. D. F. Cutler, K. L. Alvin, C. E. Price, pp. 99–119. New York: Academic
68. Heslop-Harrison, J., Heslop-Harrison, Y. 1985. Surfaces and secretions in the pollen-stigma interaction: A brief review. In *The Cell Surface in Plant Growth and Development,* ed. K. Roberts, A. W. B. Johnston, C. W. Lloyd, P. Shaw, H. W. Woolhouse, pp. 287–300. Supp. 2, *J. Cell Sci.,* Cambridge, UK: Company Biologists Ltd.
69. Heslop-Harrison, J., Heslop-Harrison, Y. 1986. Pollen-tube chemotropism: fact or delusion? In *Biology of Reproduction and Cell Motility in Plants and Animals,* ed. M. Cresti, R. Dallai, pp. 169–174. Siena, Italy: Univ. Siena Press
70. Heslop-Harrison, Y. 1977. The pollen-stigma interaction: pollen-tube penetration in Crocus. *Ann. Bot.* 41:913–22
71. Heslop-Harrison, Y., Heslop-Harrison, J., Reger, B. J. 1985. The pollen-stigma interaction in the grasses. 7. Pollen-tube guidance and the regulation of tube number in *Zea mays* L. *Acta Bot. Neerl.* 34:193–211
72. Heslop-Harrison, Y., Shivanna, K. R. 1977. The receptive surface of the Angiosperm stigma. *Ann. Bot.* 41:1233–58
73. Hill, J. P., Lord, E. M. 1986. Dynamics of pollen tube growth in the wild radish, *Raphanus raphanistrum* (Brassicaceae). I. Order of fertilization. *Evolution* 40:1328–33
74. Hodgkin, T., Lyon, G. D. 1983. Detection of pollen germination inhibitors in *Brassica oleraceae* tissue extracts. *Ann. Bot.* 52:781–89
75. Hodgkin, T., Lyon, G. D. 1984. Pollen germination inhibitors in extracts of *Brassica oleraceae* L. stigmas. *New Phytol.* 96:293–98
76. Horovitz, A., Meiri, L., Beiles, A. 1976. Effects of ovule positions in fabaceous flowers on seed set and outcrossing rates. *Bot. Gaz.* 137:250–54
77. Janzen, D. H. 1977. A note on optimal mate selection by plants. *Am. Nat.* 111:365–71
78. Johri, B. M., Ambegaokar, K. B. 1984. Embryology: then and now. In *Embryology of Angiosperms,* ed. B. M. Johri, pp. 1–52. New York: Springer-Verlag
79. Karron, J. D., Marshall, D. L. 1990. Fitness consequences of multiple patern-

ity in wild radish, *Raphanus sativus*. *Evolution* 44:260–68
80. Kelley, S. E. 1989. Experimental studies of the evolutionary significance of sexual reproduction. V. A field test of the sib-competition lottery hypothesis. *Evolution* 43:1054–65
81. Kevan, P. G., Eisikowitch, D., Rathwell, B. 1989. The role of nectar in the germination of pollen in *Asclepias syriaca* L. *Bot. Gaz.* 150:266–70
82. Klinkhamer, P. G. L., deJong, T. J., deBruyn, G. 1989. Plant size and pollinator visitation in *Cynoglossum officinale*. *Oikos* 54:201–4
83. Knox, R. B. 1984. Pollen-pistil interaction. *Encycl. Plant Physiol.* (New Ser.) 17:508–608
84. Kress, W. J. 1981. Sibling competition and evolution of pollen unit, ovule number, and pollen vector in angiosperms. *Syst. Bot.* 6:101–12
85. Lee, T. D. 1988. Patterns of fruit and seed production. See Ref. 13, pp. 177–202
86. Lee, T. D., Bazzaz, F. A. 1982. Regulation of fruit maturation pattern in an annual legume, *Cassia fasiculata*. *Ecology* 63:1374–88
87. Levin, D. A. 1975. Gametophytic selection in *Phlox*. In *Gamete Competition in Plants and Animals*, ed. D. L. Mulcahy, pp. 207–17. Amsterdam: North-Holland
88. Levin, D. A. 1981. Dispersal versus gene flow in plants. *Ann. Mo. Bot. Gard.* 68:233–53
89. Lewis, D., Verma, S. C., Zuberi, M. I. 1988. Gametophytic-sporophytic incompatibility in the Cruciferae—*Raphanus sativus*. *Heredity* 61:355–66
90. Lloyd, D. G., Yates, J. M. A. 1982. Intrasexual selection and the segregation of pollen and stigmas in hermaphrodite plants, exemplified by *Wahlenbergia albomarginata* (Campanulaceae). *Evolution* 36:903–13
91. Lord, E. M., Heslop-Harrison, Y. 1984. Pollen-stigma interaction in the Leguminosae: Stigma organization and the breeding system in *Vicia faba* L. *Ann. Bot.* 54:827–36
92. Lord, E. M., Kohorn, L. U. 1986. Gynoecial development, pollination, and the path of pollen tube growth in the tepary bean, *Phaseolus acutifolius*. *Am. J. Bot.* 73:70–78
93. Lu, J., Mayer, A., Pickersgill, B. 1990. Stigma morphology and pollination in *Arachis* L. (Leguminosae). *Ann. Bot.* 66:73–82
94. Lyons, E. L., Waser, N. M., Price, M. V., Antonovics, J., Motten, A. F. 1989. Sources of variation in plant reproductive success and implications for concepts of sexual selection. *Am. Nat.* 134:409–33
95. Maheshwari, P. 1950. *An Introduction to the Embryology of Angiosperms*. New York: McGraw-Hill
96. Malti, Shivanna, K. R. 1985. The role of the pistil in screening compatible pollen. *Oecologia* 70:684–86
97. Marshall, D. L. 1988. Post pollination effects on seed paternity: mechanisms other than microgametophyte competition operate in wild radish. *Evolution* 42:1256–66
98. Marshall, D. L. 1990. Non-random mating in a wild radish, *Raphanus sativus*. *Plant Species Biol.* 5:143–56
99. Marshall, D. L., Ellstrand, N. C. 1985. Proximal causes of multiple paternity in wild radish, *Raphanus sativus*. *Am. Nat.* 126:596–605
100. Marshall, D. L., Ellstrand, N. C. 1986. Sexual selection in *Raphanus sativus:* experimental data on non-random fertilization, maternal choice, and consequences of multiple paternity. *Am. Nat.* 127:446–61
101. Marshall, D. L., Ellstrand, N. C. 1988. Effective mate choice in wild radish: evidence for selective seed abortion and its mechanism. *Am. Nat.* 131:736–59
102. Marshall, D. L., Ellstrand, N. C. 1989. Regulation of mate number in fruits of wild radish. *Am. Nat.* 133:751–65
103. Marshall, D. L., Oliveras, D. M. 1990. Is regulation of mating within branches possible in wild radish (*Raphanus sativus*. L.)? *Func. Ecol.* 4:619–27
104. Marshall, D. L., Whittaker, K. L. 1989. Effects of pollen donor identity on offspring quality in wild radish, *Raphanus sativus*. *Am. J. Bot.* 76:1081–88
105. Mascarenhas, J. P. 1975. The biochemistry of angiosperm pollen development. *Bot. Rev.* 41:259–314
106. Mascarenhas, J. P., Machlis, L. 1962. The pollen-tube chemotropic factor from *Antirrhinum majus:* bioassay, extraction, and partial purification. *Am. J. Bot.* 49:482–89
107. Mazer, S. J. 1987. Maternal investment and male reproductive success in angiosperms: parent-offspring conflict or sexual selection? *Biol. J. Linn. Soc.* 30:115–33
108. Mazer, S. J. 1987. Parental effects on seed development and seed yield in *Raphanus raphanistrum:* implications for natural and sexual selection. *Evolution* 41:355–71
109. Mazer, S. J., Snow, A. A., Stanton, M. L. 1986. Fertilization dynamics and parental effects upon fruit development in

Raphanus raphanistrum: consequences for seed size variation. *Am. J. Bot.* 73:500–11

110. McCall, C., Mitchell-Olds, T., Waller, D. 1989. Fitness consequences of outcrossing in *Impatiens capensis:* tests of the frequency-dependent and sib-competition models. *Evolution* 43:1075–84
111. Meagher, T. R. 1986. Analysis of paternity within a natural population of *Chamelirium luteum.* 1. Identification of most-likely male parents. *Am. Nat.* 128:199–215
112. Mogensen, H. L. 1975. Ovule abortion in *Quercus* (Fagaceae). *Am. J. Bot.* 62:160–65
113. Moran, G. F., Griffin, A. F. 1985. Nonrandom contribution of pollen in polycrosses of *Pinus radiata* D. Don. *Silvae Genet.* 34:117–21
114. Morse, D. H., Fritz, R. S. 1985. Variation in the pollinaria, anthers, and allar fissures of common milkweed (*Asclepias syriaca* L.). *Am. J. Bot.* 72:1032–38
115. Mulcahy, G. B., Mulcahy, D. L. 1987. Induced pollen tube directionality. *Am. J. Bot.* 74:1458–59
116. Müller-Starck, G., Ziehe, M. 1984. Reproductive systems in conifer seed orchards. 3. Female and male fitnesses of individual clones realized in seeds of *Pinus sylvestris* L. *Theor. Appl. Genet.* 69:173–77
117. Murdy, W. H., Carter, M. E. B. 1987. Regulation of timing of pollen germination by the pistil in *Talinum mengesii* (Portulacaceae). *Am. J. Bot.* 74:1888–92
118. Nakamura, R. R. 1988. Seed abortion and seed size variation within fruits of *Phaseolus vulgaris:* pollen donor and resource limitation effects. *Am. J. Bot.* 75:1003–10
119. Nakamura, R. R., Stanton, M. L. 1989. Embryo growth and seed size in *Raphanus sativus* maternal and paternal effect in vivo and in vitro. *Evolution* 43:1435–43
120. Nason, J., Ellstrand, N. C. 1989. Pollen tube growth as a function of donor and recipient relatedness. In *Plant Reproduction from Floral Induction to Pollination,* ed. E. Lord, G. Bernier, p. 193. Rockville, Md: Am. Soc. Plant Physiol.
121. Nettencourt, D. de. 1977. *Incompatibility in Angiosperms.* Berlin: Springer-Verlag
122. Okamota, G., Shibuya, I., Furuichi, M., Shimamura, K. 1989. Inhibition of pollen tube growth by diffusate and extract of grape pistils. *J. Japan. Soc. Hortic. Sci.* 58:515–21
123. Ottaviano, E., Mulcahy, D. L. 1990. Genetics of angiosperm pollen. *Adv. Genet.* 26:1–64
124. Ottaviano, E., Sari-Gorla, M., Arenari, I. 1983. Male gametophytic competitive ability in maize: selection and implications with regard to the breeding system. In *Pollen: Biology and Implications for Plant Breeding,* ed. D. Mulcahy, E. Ottaviano, pp. 367–73. New York: Elsevier Sci.
125. Ottaviano, E., Sari-Gorla, M., Pe, E. 1982. Male gametophytic selection in maize. *Theor. Appl. Genet.* 63:249–54
126. Palmer, M., Travis, J., Antonovics, J. 1988. Seasonal pollen flow and progeny diversity in *Amianthium muscaetoxicum:* ecological potential for multiple mating in a self-incompatible, hermaphroditic perennial. *Oecologia* 77:19–24
127. Palmer, M., Travis, J., Antonovics, J. 1989. Temporal mechanisms influencing gender expression and pollen flow within a self-incompatible perennial, *Amianthium muscaetoxicum* (Liliaceae). *Oecologia* 78:231–36
128. Pimienta, E., Polito, V. S. 1982. Ovule abortion in 'nonpareil' almond (*Prunus dulcis* [Mill.] D. A. Webb). *Am. J. Bot.* 69:913–20
129. Pittman, K. E., Levin, D. A. 1989. Effects of parental identities and environment on crossing success in *Phlox drummondii. Am. J. Bot.* 76:409–18
130. Price, M. V., Waser, N. M. 1982. Population structure, frequency-dependent selection, and the maintenance of sexual reproduction. *Evolution* 36:35–43
131. Quesada, M., Schlichting, C. D., Winsor, J. A., Stephenson, A. G. 1991. Effects of pollen genotype on pollen performance in *Cucurbita pepo. Sexual Plant Reprod.* In press
132. Queller, D. C. 1983. Kin selection and conflict in seed maturation. *J. Theor. Biol.* 100:153–72
133. Queller, D. C. 1983. Sexual selection in a hermaphroditic plant. *Nature* 305:706–07
134. Queller, D. C. 1987. Sexual selection in flowering plants. In *Sexual Selection: Testing the Alternatives,* ed. J. W. Bradbury, M. G. Andersson, pp. 165–81. New York: Wiley
135. Reiss, H.-D., McConchie, C. A. 1988. Studies of *Najas* pollen tubes. Fine structure and the dependence of chlorotetracycline fluorescence on external free ions. *Protoplasma* 142:25–35
136. Rice, W. R. 1988. Heritable variation in fitness as a prerequisite for adaptive

female choice: the effects of mutation-selection balance. *Evolution* 42:817–20

137. Richards, A. J. 1986. *Plant Breeding Systems*. London: Allen & Unwin
138. Rocha, O. J., Stephenson, A. G. 1991. Effects of non-random seed abortion on progeny performance in *Phaseolus coccineus* L. *Evolution*. In press
139. Sampson, D. R. 1967. Frequency and distribution of self-incompatibility alleles in *Raphanus raphanistrum*. *Genetics* 56:241–51
140. Sanders, L. C., Lord, E. M. 1989. Directed movement of latex particles in the gynoecia of three species of flowering plants. *Science* 243:1606–8
141. Satina, S., Rappaport, J., Blakeslee, A. F. 1950. Ovular tumors connected with incompatible crosses in *Datura*. *Am. J. Bot*. 37:576–86
142. Sato, S., Yamada, N. 1988. Scanning electron microscope study on pollination of *Primula sieboldii* E. Morren. *Cytologia* 53:607–15
143. Schaal, B. A. 1980. Measurement of gene flow in *Lupinus texensis*. *Nature* 284:450–51
144. Schemske, D. W., Lande, R. 1985. The evolution of self-fertilization and inbreeding depression in plants. II. Empirical observations. *Evolution* 39:41–52
145. Schemske, D. W., Paulter, L. 1984. The effects of pollen composition on fitness components in a neotropical herb. *Oecologia* 62:31–36
146. Schlichting, C. D., Stephenson, A. G., Small, L. E. 1990. Pollen loads and progeny vigor in *Cucurbita pepo:* the next generation. *Evolution* 44:1358–72
147. Schmid-Hempel, P., Speiser, B. 1988. Effects of inflorescence size on pollination in *Epilobium angustifolium*. *Oikos* 53:98–104
148. Schmitt, J., Ehrhardt, D. W. 1987. A test of the sib-competition hypothesis for outcrossing advantage in *Impatiens capensis*. *Evolution* 41:579–90
149. Schoen, D. J., Clegg, M. T. 1985. The influence of flower color on outcrossing rate and male reproductive success in *Ipomoea purpurea*. *Evolution* 39:1242–49
150. Schoen, D. J., Stewart, S. C. 1986. Variation in male reproductive investment and male reproductive success in white spruce. *Evolution* 40:1109–20
151. Seavey, S. R., Bawa, K. S. 1986. Late-acting self-incompatibility in angiosperms. *Bot. Rev*. 52:195–219
152. Shivanna, K. R., Heslop-Harrison, J., Heslop-Harrison, Y. 1983. Heterostyly in *Primula*. 3. Pollen water economy: a factor in the intramorph-incompatibility response. *Protoplasma* 117:175–184
153. Snow, A. A. 1986. Pollination dynamics in *Epilobium canum* Onagraceace: consequences for gametophytic selection. *Am. J. Bot*. 73:139–51
154. Snow, A. A. 1990. Effects of pollen load size and number of donors on sporophyte fitness in wild radish *(Raphanus raphanistrum)*. *Am. Nat*. 136:742–58
155. Snow, A. A., Mazer, S. J. 1988. Gametophytic selection in *Raphanus raphanistrum:* a test for heritable variation in pollen competitive ability. *Evolution* 42:1065–75
156. Stanton, M. L., Preston, R. E. 1988. Ecological correlates of petal size variation in wild radish, *Raphanus sativus* (Brassicaceae). *Am. J. Bot*. 75:528–39
157. Stanton, M. L., Snow, A. A., Handel, S. N., Bereczky, J. 1989. The impact of a flower-color polymorphism on mating patterns in experimental populations of wild radish (*Raphanus raphanistrum* L.). *Evolution* 43:335–46
158. Stephenson, A. G. 1981. Flower and fruit abortion: proximate causes and ultimate functions. *Annu. Rev. Ecol. Syst*. 12:253-79
159. Stephenson, A. G., Bertin, R. I. 1983. Male competition, female choice, and sexual selection in plants. In *Pollination Biology,* ed. L. Real, pp. 109–49. Orlando: Academic
160. Stephenson, A. G., Devlin, B., Horton, J. B. 1988. The effects of seed number and prior fruit dominance on the pattern of fruit production in *Cucurbita pepo* (zucchini squash). *Ann Bot*. 62:653–61
161. Stephenson, A. G., Johnson, R. S., Winsor, J. A. 1988. Effects of competition on the growth of *Lotus corniculatus* L. seedlings produced by random and natural patterns of fruit abortion. *Am. Midl. Nat*. 120:102–7
162. Stephenson, A. G., Winsor, J. A. 1986. *Lotus corniculatus* regulates offspring quality through selective fruit abortion. *Evolution* 40:453–58
163. Thomson, J. D. 1986. Pollen transport and deposition by bumble bees in *Erythronium:* influences of floral nectar and bee grooming. *J. Ecol*. 74:329–41
164. Thomson, J. D., Plowright, R. C. 1980. Pollen carryover, nectar rewards, and pollinator behavior with special reference to *Diervilla lonicera*. *Oecologia* 46:68–74
165. Thomson, J. D., Price, M. V., Waser, N. M., Stratton, D. A. 1986. Comparative studies of pollen and fluorescent dye transport by bumble bees visiting

Erythronium americanum. *Oecologia* 55:251–57
166. Thomson, J. D., Stratton, D. A. 1985. Floral morphology and cross-pollination in *Erythronium grandiflorum* (Liliaceae). *Am. J. Bot.* 72:433–37
167. Tukey, H. B. 1933. Embryo abortions in the early ripening varieties of *Prunus avium*. *Bot. Gaz.* 94:433–68
168. Uma Shaanker, R., Ganeshaiah, K. N., Bawa, K. S. 1988. Parent-offspring conflict, sibling rivalry, and brood size patterns in plants. *Annu. Rev. Ecol. Syst.* 19:177–205
169. Vander Kloet, S. P., Tosh, D. 1984. Effects of pollen donors on seed production, seed weight, germination and seedling vigor in *Vaccinium carymbosum* L. *Am. Midl. Nat.* 112:392–96
170. Waser, N. M. 1983. The adaptive nature of floral traits: ideas and evidence. In *Pollination Biology*, ed. L. Real, pp. 242–85. Orlando: Academic
171. Waser, N. M. 1991. Population structure, optimal outbreeding, and assortative mating in angiosperms. In *The Natural History of Inbreeding and Outbreeding: Theoretical and Empirical Perspectives*, ed. N. W. Thornhill, pp. Chicago: Univ. Chicago Press. In press
172. Waser, N. M., Price, M. V. 1984. Experimental studies of pollen carryover: effects of floral variability in *Ipomopsis aggregata*. *Oecologia* 62:262–68
173. Waser, N. M., Price, M. V. 1991. Outcrossing distance effects in *Delphinium nelsonii:* pollen loads, pollen tubes, and seed set. *Ecology* 72:171–79
174. Waser, N. M., Price, M. V., Montalvo, A. M., Gray, R. N. 1987. Female mate choice in a perennial herbaceous wildflower, *Delphinium nelsonii*. *Evol. Trends Plants* 1:29–33
175. Welk, Sr. M., Millington, W. F., Rosen, W. G. 1965. Chemotropic activity and the pathway of the pollen tube in lily. *Am. J. Bot.* 52:774–81
176. Westoby, M., Rice, B. 1982. Evolution of the seed plants and inclusive fitness of plant tissues. *Evolution* 36:713–24
177. Wet, E. de, Robbertse, P. J., Coetzee, J. 1990. Ultrastructure of the stigma and style of *Mangifera indica* L. S.-Afr. Tydskr. Plantk. 56:206–13
178. Wiens, D., Calvin, C. L., Wilson, C. A., Davern, C. I., Frank, D., Seavey, S. R. 1987. Reproductive success, spontaneous embryo abortion, and genetic load in flowering plants. *Oecologia* 71:501–9
179. Willemse, M. T. M., Franssen-Verheijen. 1988. Ovular development and pollen tube growth in the ovary of *Gasteria verrucosa* (Mill.) H. Duval as condition for fertilization. In *Sexual Reproduction in Higher Plants*, ed. M. Cresti, P. Gori, E. Pacini, pp. 357–62. New York: Springer-Verlag
180. Willson, M. F. 1990. Sexual selection in plants and animals. *Trends Ecol. Evol.* 5:210–14
181. Willson, M. F., Burley, N. 1983. *Mate Choice in Plants: Tactics, Mechanisms and Consequences*. Princeton: Princeton Univ. Press
182. Winsor, J. A., Davis, L. E., Stephenson, A. G. 1987. The relationship between pollen load and fruit maturation and the effect of pollen load on offspring vigor in *Cucurbita pepo*. *Am. Nat.* 129:643–56
183. Wyatt, R. 1982. Inflorescence architecture: how flower number, arrangement, and phenology affect pollination and fruit-set. *Am. J. Bot.* 69:585–94
184. Young, H. J., Stanton, M. L. 1990. Influence of environmental quality on pollen competitive ability in wild radish. *Science* 248:1631–33
185. Zimmerman, M. 1988. Nectar production, flowering phenology, and strategies for pollination. In *Plant Reproductive Strategies*, ed. J. Lovett Doust, L. Lovett Doust, pp. 157–78. New York: Oxford Univ. Press
186. Zuberi, M. I., Lewis, D. 1988. Gametophytic-sporophytic incompatibility in the Cruciferae—*Brassica campestris*. *Heredity* 61:367–77

Annu. Rev. Ecol. Syst. 1991. 22:065–93

CANALIZATION: GENETIC AND DEVELOPMENTAL ASPECTS

Willem Scharloo

Department of Plant Ecology and Evolutionary Biology, University of Utrecht, Utrecht, The Netherlands

KEY WORDS: genetic assimilation, artificial selection, mutant expression, phenotypic reactions, genetic variation

---

INTRODUCTION

Canalization

The concept of canalization was used by Waddington (66) in the context of developmental biology. He emphasized two important points in the development of higher organisms. First, the end products of development, i.e. the adult tissues, are of sharply distinct types without intergradation. Second, the normal course of development is in his view a preferred path. Deviations of this path due to disturbances in the internal or external environment are corrected by regulatory processes. This occurs not only in the development of distinct types of tissue, but also on the organismic level in the realization of morphological patterns, in size and shape of organs and in matters of growth and determination of size of whole organisms. He depicted these phenomena in what he called the epigenetic landscape. Development starts in the egg. From there numerous developmental pathways branch out, leading to a great variety of distinct end results. These pathways are represented as a system of branching valleys in a descending slope. The developmental process is visualized as balls rolling through the valleys to their end point. The steeper the valley and the larger the ridges separating the valleys, the stronger the tendency of the ball, when it is pushed from its course along the valley bottom by internal or external disturbances, to go back to its original course. This

0066-4162/91/1120-0065$02.00

picture suggested the concept of canalization of developmental pathways. Waddington envisaged that the epigenetic landscape is generated by the interaction of a large number of gene-controlled processes. This view was derived from his embryological work on mutants of 38 different gene loci which caused wing abnormalities in *Drosophila* (67). The interactions of their normal alleles are necessary for the development of a normal wing.

In developmental biology the concept of canalization has had only a limited appeal. This is perhaps due to the fact that regulatory processes have always stood in the center of embryological research and to the feeling that the concept of canalization could not add much to existing theory.

In evolutionary biology, the concept became important after Waddington's experiments (69, 71) with *Drosophila* on genetic assimilation of environmentally induced phenotypes. These experiments drew much attention because they showed how seemingly Lamarckian processes suggesting the inheritance of acquired characters could be explained according to genuinely neo-Darwinistic principles.

Canalization and the Inheritance of Acquired Characters

Waddington's experiments on genetic assimilation (69, 71) were generated by the concepts expounded in his paper, "Canalization of development and the inheritance of acquired characters" (68). With this title Waddington emphasized the relation between these two phenomena: "The first step in the argument is one which will scarcely be denied but is perhaps often overlooked. The capacity to respond to an external stimulus by some developmental reaction. . . . must itself be under genetic control" (68, p. 563).

This train of thought begins with the assumption of genetic variability for the sensitivity of characters to environmental factors: "The occurrence of an adaptive response to an environmental stimulus depends on the selection of a suitable genetically controlled reactivity in the organism. If it is an advantage, as it usually seems to be for developmental mechanisms, that the response should obtain an optimal value more or less independently of the intensity of the environmental stimulus received by a particular animal, the reactivity will become canalized, again under the influence of natural selection. Once the developmental path has been canalized, it is to be expected that many different agents, including a number of mutations available in the germ plasm of the species, will be able to switch development into it; and the same considerations which render the canalization advantageous will favour the supersession of the environmental stimulus by a genetic one" (68, p. 563).

Genetic variation in environmental response had been revealed by Goldschmidt (26, 27) and by Landauer (29) in phenocopy experiments, i.e. experiments in which normal genotypes were exposed during their development to extreme values of normal environmental factors or to environmental factors—e.g. chemicals—that are not a part of their normal environment. Such treatment too often produces morphological abnormalities similar to abnormalities caused by known mutants (phenocopies). The kind of abnormalities, their expression level, and their frequency were revealed to be different in different stocks.

CANALIZATION AND GENETIC ASSIMILATION

The First Experiments

Wings of wild-type *Drosophila* possess a second cross-vein forming a connection between two longitudinal veins. Treatment of a population with heat shock (40°C) for 4 hr, 20–23 hr after puparium formation produces flies that have an incomplete cross-vein or none at all. This character, which is a phenocopy of several mutants found by *Drosophila* geneticists, was induced with a frequency of 40%. Waddington (69) started two selection lines. In the first line he selected as parents for the next generation flies that showed after the heat treatment the *cross-veinless* phenotype. In the second line he selected flies that had not reacted to the heat treatment with loss of the second cross-vein. The frequency of reaction increased in the line in which in every generation heat shock–induced *cross-veinless* phenotypes were selected and decreased in the line in which flies were selected that showed the normal phenotype after the heat shock. After 12 generations the difference in the induction of *cross-veinless* phenotypes between the two lines had become some 55%. In each generation, in addition to the pupae submitted to heat shock, a large number of flies were grown without heat shock. In the line selected for increased sensitivity in this control group, flies appeared without cross-veins. In these flies the erstwhile phenotypic reaction to the heat shock appeared now without environmental stimulus, i.e. appeared to have become genetically determined. Waddington (69) called this phenomenon of genetic fixation of a phenotypic reaction to an environmental factor *genetic assimilation*. He (69, 73, 77) gave an explanation in neo-Darwinian terms. The selection response was shown to be based on genetic variation present in the base population at the start of the experiment. Long inbred lines which are supposed to approach genetic homogeneity did not show any change in the frequency of the heat shock–induced *cross-veinless* phenotype under the same selection regime (2). Moreover, Bateman (2) repeated the first Waddington *cross-veinless* assimilation experiment with a different base population and obtained similar results. However, when she screened her base population for occurrence of *cross-veinless* phenotypes without applying heat treatment, she found that a low percentage was present spontaneously.

Waddington (71, 72) performed similar experiments with a far more spectacular phenotypic reaction. Gloor (24) had obtained phenocopies of the *bithorax* mutant of *Drosophila* by submitting eggs in an early stage of their development to a treatment with ether vapor. This is a typical threshold reaction seemingly determined more by the structure of the developmental system than by a specific chemical action of the ether: heat shock applied at the same stage of development caused similar phenotypes (31). Extreme phenotypes obtained by these treatments showed a transformation of the third haltere-carrying thorax segment into a copy of a full-blown second thorax segment with a large mesonotum and a pair of wings, a phenotypic change of almost macro-evolutionary scope.

Twenty generations of selection caused a large increase in the frequency of the *bithorax* phenotype as a reaction to the ether treatment. Then *bithorax* phenotypes started to appear from control eggs not submitted to the environmental treatment. This happened in two selection experiments started from two different base populations.

In addition to *cross-veinless* phenocopies Bateman (2) induced by a heat treatment—2–4 hr at 40°C, 18 hr after puparium formation—three other aberrant phenotypes: absence of the anterior cross-vein, an extra cross-vein between longitudinal veins 3 and 4, and an extra cross-vein between longitudinal veins 2 and 3. Moreover, she induced phenocopies of the well-known *dumpy* mutant (1). By artificial selection for higher reactivity for each type of abberation in separate selection lines, she succeeded in increasing the frequency of phenocopies and obtained assimilation in each case.

Canalization and Assimilation: A Necessary Linkage?

The question arose whether the canalization concept is necessary for the explanation of the assimilation experiments. Bateman (1), a PhD-student of Waddington, explained the results of a repeat of Waddington's *cross-veinless* experiment in terms of a simple threshold model. She assumed in the population a normal distribution for the tendency to have a broken cross-vein. The percentage of the *cross-veinless* phenotype in the population will then be represented by the part of the distribution above the threshold. An increase of the frequency of this phenotype can occur as a consequence of a shift of the mean of the distribution, by an increase of its variance, or by a shift of the threshold. She depicts a model in which the position of the threshold is shifted to a lower value by the heat-shock. As a consequence part of the population-distribution now transgresses the new threshold, and *cross-veinless* phenotypes appear in the population. Selection of these phenotypes causes an accumulation of genes promoting this phenotype, thereby causing a shift to higher values of the population distribution for breakage of the cross-vein. After further progress, part of the distribution exceeds the threshold for normal temperature, and the *cross-veinless* character appears without heat-shock. The same model was presented by Falconer (11). A probit analysis of the selection response of her two *cross-veinless* selection lines showed indeed that the selection response could be explained as a shift of the normal distribution for the tendency to cross-vein breaks without a change of its variance.

Curt Stern (61) suggested a similar explanation in a paper, "Selection for subthreshold differences and the origin of pseudoexogenous adaptations." He recognized that the phenomenon of genetic assimilation was deduced by Waddington (68) from the picture of his epigenetic landscape and that the experiments (69, 71) showing that genetic assimilation could occur were designed on this basis. But, as Stern stated, this does not mean that Waddington's basic assumptions were valid.

Suppose a population possesses genes that promote a certain phenotype in two environments, but their selection can only occur in the second environment, because a threshold prevents their expression in the first environment. Then selection for these genes causing their accumulation in the population will ultimately lead to their expression in that first environment. Stern presented a simple single locus model involving the recessive *Drosophila* mutant *cubitus interruptus (ci)*.

In a reply, Waddington (74) conceded that genetic assimilation can be explained by such a threshold model, but he regarded it as a "told to the children version." He maintained that the model could not explain why selection would go much further than is necessary in the environment that induces the new character, i.e. fixing it also in the original environment. Therefore, the threshold model would not be sufficient; canalization would have to be brought in. Bateman (2) already dealt with that argument by showing that such was not the case in her cross-veinless assimilation experiments, although she could imagine that when assimilation occurred in nature a phase of canalization by stabilizing selection might follow.

In 1961 Waddington (77) still maintained his viewpoint as is clearly revealed in a subheading "Genetic assimilation as a consequence of canalization" of a paragraph in which he attacked Stern (61) again. However, his arguments are related to a hypothetical evolutionary history of the callosities on the ventral surface of the ostrich. The sufficiency of Bateman's and Stern's model for the results of the assimilation experiments is not touched upon.

Genetic Variation in Phenotypic Reactions: The Anal Papillae

In his reaction to Stern's paper, Waddington (74) had already stated that Stern's model did not accommodate differences in reactivity between individuals. As we have seen, these were not required for the explanation of the experiments published before, but they are involved in Waddington's experiments on the genetic assimilation of changes in size of the anal papillae of *Drosophila* larvae.

In the title of the 1959 *Nature* paper, "Canalization of development and genetic assimilation of acquired characters" (76), Waddington linked these two concepts again, although they were not integrated in the text. The first part of the paper dealt with an experiment on what Waddington named *canalizing selection,* i.e. artificial selection trying to change the sensitivity for temperature of eye-facet number in the *Drosophila* mutant *Bar*. The second part is on genetic assimilation of the purported increase in size of the anal papillae in larvae of *Drosophila melanogaster* by exposure to salt in the food medium.

Waddington gave two reasons for this experiment: (*a*) The previous experiments on genetic assimilation were all done with artificial selection on characters without clear adaptive significance, and (*b*) the characters used earlier were all threshold characters. It would be interesting if assimilation could be demonstrated with a quantitative, continuously varying character.

The anal papillae were described by Gloor & Chen (25) as regions of modified, strongly enlarged epidermal cells at either side of the anus. They suggested that the anal papillae were involved in osmoregulation. At pupation the papillae fold inwards, and the remnants of the papillae can be seen in the pupae. Waddington measured the size of (what he thought were) the remnants of the anal papillae in pupae. When he grew larvae on food with different amounts of salt (between 2% and 7%) he found that his measurements increased when the salt concentration increased. He interpreted this as an adaptive reaction: Larger papillae would perform better in osmoregulation. He did an experiment in which three different populations could adapt by *natural* selection to increasing concentrations of salt in the food medium. The increase of the salt concentration was gradual so that 20–30% of the eggs laid on the selection medium grew up to produce flies. After 21 generations of adaptation, the survival and the size of the anal papillae of the adapted populations and of their controls held on normal medium were determined on a range of salt concentrations. The results showed that survival was higher on salted medium, in particular at higher salt concentrations, in all three populations than in the corresponding controls, that the change of the purported length measurement of the anal papillae was larger in all three adapted populations than in the controls, and that at the lowest salt concentration (2%), the measurement in the adapted populations was still higher than in the control populations. Waddington's interpretation was that the outcome of the adaptation process was an increase of the strength of the phenotypic adaptive reaction of the length measurement, and fixation of part of the phenotypic response at the lowest salt concentration, i.e. genetic assimilation. However, this last point is perhaps not valid. If we extrapolate the curves of the phenotypic reaction to 0% salt addition, it seems possible that there the curves of the adapted population would intersect the curves of their controls. This would mean that the adaptation was based only on genetic change that increased the strength of the phenotypic reaction of the anal papillae, an explanation not involving the concept of genetic assimilation.

A more serious objection is that of te Velde and Scharloo (53, 62–65) who discovered that Waddington measured not the remnants of the anal papillae in pupae but the remnants of the inactive epidermal zone *between* the anal papillae. A comparison, at the one hand, of the careful drawings of Gloor (25) of the photographs of larvae with papillae stained with silver-nitrate and of the scanning pictures made by te Velde et al (63) with, at the other hand, the drawings in Waddington (76, reprinted in 78) makes this very clear.

The anal papillae, measured on *Drosophila* larvae, increased in size with *decreasing* salt concentration. This is in agreement with results with other Dipteran species (e.g. *Aedes* and *Culex,* 79). The size of the inactive zone between the papillae which Waddington mistakenly measured as papillae-size is negatively correlated with the size of the papillae (64). Therefore Waddington's data are at best indirect measurements of the size of the papillae.

The phenotypic response of the anal papillae to decreased salt concentrations is twofold; one response is in size as a consequence of an increase in cell number, and the other is a rapid response of the ultrastructure of the papillae cells (3, 63). The papillae cells are not covered by the normally rather thick epidermal cuticula. The outward cell membrane is folded in large so-called apical lamellae which frequently are connected with mitochondria. This type of cells is well known, e.g. in many aquatic animals where they serve for the intake of ions. These lamellae are far better developed and in contact with more mitochondria when the salt concentrations in diluted food medium are low and an active intake of ions has to occur. Genetic variance occurs in the structure of the papillae which is related to the ultrastructure of the papillae cells. Normally the papillae are retracted at puparium formation. In populations grown on a low salt concentration (2%) an aberrant type was found in low frequency. In this type the papillae were not retracted during puparium formation. While in the normal retracted (R) type the papillae were present in the pupae as narrow ridges at either side of the interpapillar zone, in the stretched (S) type the papillae are still quite large organs. In populations adapting to salted media the S-type increased. Moreover, by artificial selection for this S-type on a low salt medium it was possible to increase S-type frequency to almost 100%, i.e. to obtain its fixation.

Comparison of the two types indeed showed that the S-type was better adapted to high salt concentrations and the R-type to diluted media. The size of the papillae decreased in both types with increasing salt concentration to the same extent. However, in the R-cells the apical lamellae were well developed and reacted strongly on changing salt concentrations. In S-cells the lamellae were practically absent on normal medium and increased only slightly on diluted medium.

On normal medium the two types performed equally well. They maintained the osmotic value of their hemolymph equally well with moderate changes of salt concentration. When the salt concentration increased to higher values, the internal osmotic value could be better maintained by the S-type while on diluted medium the R-type did better. This was expressed in survival, in development time, in feeding rate, and in competition (53, 62).

CANALIZATION AND QUANTITATIVE GENETICS

Introduction

In evolutionary biology the concept of canalization became known by the results of the assimilation experiments. The importance of these experiments for an explanation of the evolution of pseudo-exogenous adaptations was widely accepted. The concept of canalization was, as it were, hitchhiking on the assimilation concept to recognition in evolutionary genetics. However, we have seen that several investigators (2, 61) showed that the concept of canalization was not necessary to explain the results of the assimilation experiments. A simpler explanation, in terms of expression of subthreshold genes, is sufficient.

How could it be shown that canalization really exists, that the organization of development could affect the expression of genetic variation in quantitative genetic characters and thereby the course of selection response?

Variation at Either Side of the Wild-Type

Waddington (70) attempted this in a paper "On a case of quantitative inheritance on either side of the wild-type" in which he reported on crosses with *Drosophila* stocks which showed either an interruption of the second cross-vein or extra venation attached to this vein. He used two stocks derived from his assimilation experiments which differed in the size of cross-vein interruption and two stocks with a different addition of extra vein material to the second cross-vein. With the normal phenotype included, he had five different levels of cross-vein expression. When he crossed the two lines with different levels of cross-vein interruption, the inheritance of the size of the interruption was additive. A similar result was obtained in crosses between the two lines with extra vein material. But when he made crosses between stocks "either side of the wild-type" the results were widely different from additivity. He concluded that "the wild phenotype could conceal within it a much greater range of dosages of vein producing genes than can any other phenotype" (77, p. 279). This implies that "the canalization of the normal vein pattern is such that it is highly resistant to the disturbing effects of changes in the dosage of genes tending either to make more or to make less vein. In individuals in which the buffering capacity of the normal developmental course is exceeded, the phenotype does become altered, and reflects not too inaccurately the actual dosage of genes contained in them" (77, p. 279).

But there were complications: The quantity of wing-vein material used in the posterior cross-vein region could be arranged in different ways. There could be a complete cross-vein, or the same amount of vein material could be used for an interrupted cross-vein with a piece of extra vein attached. The character "amount of cross-vein" is not causally homogeneous; it is clearly the result of more than one participating developmental pathway, but each of the constitutive pathways of the character is, in Waddington's view, itself canalized. Here there is clearly no simple linear relation between effects of the genes involved and their expression in the amount of vein material; considerable interaction must occur between these genes.

Selection on Mutant Expression

Dun & Fraser (7, 8, 17–19), Rendel (40), Sondhi & Maynard Smith (56, 60, 34) and Scharloo (49, 51, 54) performed artificial selection on morphological characters which in wild-type individuals were constant or almost so. Instead of creating, as Waddington did, phenotypic variability by extreme environmental factors such as heat-shock or chemical agents, they introduced mutants producing variable phenotypes.

VIBRISSAE NUMBER IN MICE Dun & Fraser (7, 8) selected on the number of secondary vibrissae in mice. This set of vibrissae is arranged in three paired groups on the left and right side of the facial part of the head, one unpaired group under the chin, and a group on each forelimb. The total number of 19 is very stable; in a survey of 3000 normal mice, approximately one abnormality per 500 groups was observed. The sex-linked gene *Tabby* reduces vibrissae number to an average of 15 in *Ta*/+ females, and to 8 in *Ta*/*Ta* females and in the *Ta*/*Y* males. However, in contrast with the wild-type mice, in the *Tabby* genotypes considerable variation occurs. Artificial selection for vibrissae number could therefore now be applied on both *Ta*/+ females and *Ta*/*Y* males. The base population was formed by crossing three stocks with the mutant *Tabby* and two inbred lines. The selected populations—one selected for high vibrissae number and one selected for a low number—consisted throughout of some 30 females and 10 males. The mating system was such that in every generation there was segregation of +/*Y* and *Ta*/*Y* males and of +/+ and *Ta*/+ females. They differ only in the *Tabby* gene and its surrounding X-chromosome material. Such a scheme makes it possible to observe in every generation the effect of the genes accumulated by artificial selection on the three levels of expression of the character, i.e. wild-type (i.e. males and females) with 19 vibrissae, *Ta*/+ females with 15 and *Ta*/*Y* males with 8 bristles.

The selection proved to be very successful when monitored in the *Tabby* genotypes. After seven generations the divergence between the two lines was at the *Ta*/+ level approximately 5 vibrissae and increased to about 8 vibrissae at the end of the experiment in generation 19. In the *Ta*/+ animals the change in the low line was twice the change in the high line; expressed in the *Ta*/*Y* males the response is approximately equal in the low and high selection lines while their divergence is smaller.

Also on the wild-type level there is divergence. The low line showed a small response in the first generations that accelerated in the second half of the experiment. In addition to a shift of the mean to 17 vibrissae, variability increased strongly: Instead of the almost invariable number of 19, the last generation showed a range of 4 vibrissae—15–19. The change in wild-type mice of the high line is small: The first sign was that animals with extra vibrissae started to appear more frequently; this happened in particular in the last generations.

Dun & Fraser concluded that there is genetic variation for vibrissae number behind the uniform genotype of wild-type mice which is revealed only in the mutant phenotype. This makes possible artificial selection which has an immediate response in the mutant types. Accumulation of the genetic change caused by selection on the mutant types is sufficient to change the invariable wild-type. The mutant types are far more sensitive for genetic change than are normal types, i.e. the development of the normal types is canalized. In their first publication Dun & Fraser explained the absence of phenotypic variability in wild-type mice, despite the presence of genetic variability, by assuming a sigmoid relationship between gene action and phenotypic effects. They assumed that (*a*) there is a basic genetic system, the vibrissae number system, forming a vibrissae substance varying in the population according to a normal distribution; (*b*) a separately determined genetic system governs the relationship between change in vibrissae substance and phenotypic change and (*c*) this relationship is a sigmoid function and the *Ta*-gene affects the mean and the slope of this function. Thus, each *Tabby* genotype has its sigmoid function, each with its own characteristic slope which would reflect the strength of its canalization, and its own mean. In a later paper (8) Dun & Fraser suggest that their results could also be explained according to a model suggested by Rendel, explaining the results of selection on the expression of the mutant scute in Drosophila (40). That model assumes (*a*) a polygenic system causing variation of a vibrissae forming substance, and (*b*) a genetic system determining one sigmoid relationship between the amount of vibrissae substance and vibrissae number with a steep inflection zone around the normal vibrissae number representing the canalization of that phenotype. The *Tabby* gene would affect not canalization but only the amount of vibrissae substance. The selection would change only the amount of vibrissae substance.

In 1962, Fraser & Kindred (18) accepted Rendel's model because in the last generations of their selection experiment the *Ta*/+ females in the high line start to overlap with the frequency distribution of the normal phenotype. The *Ta*/+ females and the wild-type animals can be distinguished by the effect of *Tabby* on coat color. The variation of the *Ta*/+ females went down strongly because the frequency distribution shrinks at its lower end and animals seem to accumulate in the wild-type class of 19 vibrissae. Progress in this genotype then begins to halt while progress in the *Ta*/*Y* males continues. This shows that the strength of canalization is not controlled by the *Tabby* gene but is determined by the phenotypic value of the character vibrissae number.

SCUTELLAR BRISTLES IN DROSOPHILA Rendel (40) performed a similar experiment with the sex-linked mutant *scute* of *Drosophila melanogaster*. *Scute* is located almost at the distal end of the X-chromosome. Homozygous and hemizygous it causes a decrease of bristle number in several bristle groups and in particular of the scutellar bristles. Wild-type flies have 4 bristles on the scutellum in a fixed pattern of 2 anterior bristles and 2 posterior bristles. This is virtually an invariable pattern; it is diagnostic for the genus *Drosophila*. In wild populations, either in mass culture in the laboratory or in nature, only occasionally are flies observed with 3 or 5 bristles. The *scute* mutant lowers the mean in females to two and in males to one bristle. In addition, in flies carrying the mutant the number of scutellar bristles is variable; in males it varies from 0 to 3 and in females from 0 to 4.

Rendel practiced selection for a higher number of scutellar bristles on males. In his selection line, females were heterozygous for the *scute* mutant. So, in each generation there was segregation of *scute* and wild-type genotypes of both sexes, in which Rendel could monitor the phenotypic effect of genetic change caused by the selection on the *scute* males. The experiment consisted of 10 lines. In each line the 5 *scute* males with the highest number of scutellar bristles were mated with 5 randomly selected heterozygous females with 4 bristles.

The mean number of scutellar bristles in *scute* males is approximately 1 in the base population; this increases to 2.8 after 22 generations of selection. In *scute* females the original mean is a bit higher than 2 and increases to 3.3. At the start the sex-difference is around 1 bristle; it is halved at the end at generation 22. In *scute* males the variation in bristle number is rather constant. In the 0 bristle class and in the 1 bristle class there are almost an equal number of flies, while only in later generations of selection do flies with 4 bristles start to appear in an appreciable number. In *scute* females the variation in bristle number declines because the frequency distribution shrinks at the lower values and the flies accumulate in the class with the normal wild-type number of 4 bristles. Notwithstanding the strong progress in the *scute* animals, the mean of 4 bristles of the wild-type males and the $+/sc$ heterozygote females scarcely changes. Only in the last 5 generations do flies with more than 4 bristles appear with a certain regularity; in the males there are even some with 6 bristles.

This pattern is similar to that found for the vibrissae in the *Tabby* mice. Selection on a mutant character toward the normal type is very successful until the expression of the mutant comes near to normal expression: Progress then stops and the variability declines. The genetic change causing large phenotypic change in the mutant phenotype scarcely changes the phenotypes of the nonmutant genotypes segregating in the selection line. For scutellar bristle number also, it was concluded that the normal phenotype is highly canalized.

Rendel interpreted his experiments according to a model depicted in Dun & Fraser (8) which can be summarized in the following points (*a*) Bristle formation is dependent on a gene product or morphogenetic substance (later called "make") which is normally distributed in populations, (*b*) the relation between the amount of morphogenetic substance and bristle number is not linear but is sigmoid, with a region around the normal, wild-type number where change of morphogenetic substance does not cause a phenotypic effect. Beyond this region, both at lower and at higher bristle numbers, change is easy and there is always phenotypic variability.

Rendel designed a method to measure the strength of the canalization of the different bristle classes based on the theory of the normal distribution and the assumption that the morphogenetic substance has a normal distribution in populations. When the amount of morphogenetic substance in an individual transgresses a threshold, 1 bristle is formed, and when it transgresses a second threshold, 2 bristles are formed, and so on. When a change in the amount of morphogenetic substance does not lead to a change in bristle number, clearly in that region the distance between the thresholds separating bristle classes must be large. The problem of quantifying the strength of canalization of a bristle class is measuring the width of a bristle class, i.e. the distance between two thresholds that form the boundaries of that class. When the morphogenetic substance has a normal distribution in the population and the frequencies of flies with different bristle numbers are known, the distance between the thresholds and the mean of the underlying normal distribution can be expressed in its standard deviation as probits. This makes it possible to compare the relative width of the different bristle classes within a population. Comparisons between populations or between different generations of a selection experiment can only be made when the variance of the morphogenetic substance is not different. In many publications this seems to be taken for granted. Rendel could calculate that to move from 3 bristles to 5 it takes eight times the genetic change that it does to move from 1 bristle to 3 bristles.

OCELLI WITH BRISTLES IN DROSOPHILA The sex-linked mutant *ocelli-less* in *Drosophila subobscura* affects the ocelli and 3 pairs of adjacent bristles on the top of the head. (34, 59, 60). In the base population the homozygous *ocelli-less* mutant removed all bristles and ocelli in some flies, while the other flies could show a diversity of combinations. In some flies the ocelli were slightly displaced from their normal position, and the size of bristles and ocelli could be affected. Between the presence of bristles and ocelli, a correlation of $r = 0.46$ was found. This was used as a justification to express the degree of expression of the *ocelli-less* mutant by counting the presence of an ocellus or a bristle as a unit. A fly deprived of any ocelli or bristles got a 0 score, a fly with 3 ocelli and 6 bristles scored 9.

Sondhi applied artificial selection for a lower and for a higher number. There were two high selection lines and two low selection lines. The lines of each pair of lines had a different mating system; in one line inbreeding was prevented, in the other line matings were between siblings. In the base population founded by outcrossing of some laboratory stocks, the mean score in females was 1.76 and for males 1.44. Each line consisted of 5 pairs of flies. In the low lines the selection response was small; less than 1 bristle. In the high lines there was an immediate response which was more rapid in the inbred than in the outbred population. After 13 generations, in both lines progress stopped and so was the selection. In the upward selection in the last generations some flies were seen with extra ocelli and the frequency of extra bristles had increased.

Sondhi (59, 60) designed a model to explain "the bounded distributions of phenotypes in terms of genes which in their primary effects on an ocelli-bristle-forming substance are additive throughout their range." He assumes that this morphogenetic substance is normally distributed in the population. When this distribution does not transgress a threshold T_1, all flies will lack bristles and ocelli; when it straddles the threshold, part of the flies will miss all bristles and ocelli, but the part with amounts of substance higher than the threshold will possess ocelli and bristles in various combinations. When the distribution extends beyond a threshold T_2, flies with all 6 bristles and 3 ocelli will appear; when a third threshold T_3 is passed, extra ocelli and/or bristles will appear. However, it is not clear how Sondhi explains the constancy of the wild-type flies. He depicts for the *ocelli-less* stocks—the foundation population and the downward and upward selected lines—frequency distributions of the ocelli-bristle forming substance with equal variances which only differ in their means. For the wild-type population he gives a distribution with very low variance, thereby not extending beyond the thresholds T_2 and T_3. This implies that the *ocelli-less* mutant affects both mean *and* variance of the distribution, although Sondhi only remarks that "the amount of precursor in flies carrying the $+^{oc}$ gene is unlikely to be absolutely constant." He then states that "canalizing selection during the past evolution of the species would ensure that the same adult phenotype would develop, provided that the amount of precursor is within this range," i.e. the range between T_2 and T_3. In this model canalization seems to act independently of the components of the model.

This is in contrast with the models designed by Dun & Fraser (8, 18) and Rendel (40) where canalization is related to the degree of phenotypic expression of the character and is presented as the slope of the function relating phenotypic change to genetic change in that region or in the probit-width of the wild-type class, i.e. the distance between the thresholds which set the limits of phenotypic wild-type class; 4 scutellar bristles in *Drosophila* and 19 vibrissae in the mouse.

WING-VEIN INTERRUPTION IN DROSOPHILA The selection experiments described above all dealt with characters consisting of discrete elements which were added to obtain a total score; vibrissae in mice, bristles or bristles and ocelli in *Drosophila*. In the mouse it was a character composed of elements of similar structure located in groups on different parts of the body; in *Drosophila* with the *scute* mutant it was similar structures in a fixed pattern on the scutellum; and with the *ocelli-less* mutant with dissimilar structures but with a partly common underlying developmental system.

Scharloo (49, 50, 54) selected on the relative length of the fourth wing-vein in the presence of the mutant ci^D *(cubitus interruptus dominant)* in *Drosophila melanogaster*. The mutant ci^D is a recessive lethal gene on the tiny fourth chromosome (< 0.2% of the total genetic map) and has as dominant morphological effects terminal interruptions of the fourth and of the fifth longitudinal wing-vein.

This character was chosen because the location of ci^D on the fourth chromosome made it possible to introduce the mutant into the genome of populations with the help of marked balancer chromosomes without disturbing the rest of the genome. A character is thereby created in the population that did not exist before. Therefore, it could not have been shaped in a long history of natural selection, and this has consequences for the structure of its genetic variation (55).

The mutant was introduced into three wild-type populations of different origin. The character relative length of the fourth longitudinal wing-vein was expressed as the percentage ratio of the length of this vein to the length of the third longitudinal vein, both measured distally of the first cross-vein. From each base population a low line was started in which flies with the shortest veins were selected, and another line was begun in which the flies with the longest veins were used as parents for the next generations. Each selection line consisted of 3 bottle cultures. From each bottle the wing-veins of 20 flies from each sex were measured, and from each sample the 4 with the most extreme measurements were selected. A rotational mating system was used to minimize inbreeding.

In all lines there was an immediate response. Realized heritabilities in the three pairs of lines were between 0.3 and 0.4. In the low lines there was a regular shift of the frequency distributions to lower values. Variability decreased when the fourth vein became shorter and the end of the vein approached its attachment to the second cross-vein.

In the high lines the response pattern was unexpected. In all three lines selected for a longer fourth vein, bimodal frequency distributions appeared. After passing this phase there was a concentrated unimodal distribution and some slow progress until a plateau was maintained. The appearance of bimodal frequency distributions was described earlier by Clayton et al (5) as a consequence of segregation of genes with large effect. But there were several arguments that this could not have occurred here: (*a*) In all three high lines the bimodal distributions occurred in the same range of phenotypic values between 70 and 80. (*b*) When selection was reversed (back-selection) from the unimodal distribution around 80, the bimodal distribution reappeared in the same range as in the upward selection and became unimodal again after shifting below the 70 value. (*c*) In this phase of bimodality, a high frequency of asymmetric flies appeared with a long fourth vein in one wing and a short vein in the other wing. (*d*) When the length of the fourth vein was shifted by temperature (in the lines involved it becomes longer when the larvae and pupae are reared at lower temperature) through the same range where change by selection generated bimodal frequency distributions, the bimodal distributions appeared again. This occurred when the distribution shifted through this region from below, i.e. when the back selection line with a unimodal distribution below 70 was submitted to low temperature. It occurred too when the high line with the unimodal distribution above 80 was grown at higher temperatures. When bimodality reappeared there was again a high frequency of asymmetry. (*e*) When the mean vein-length is plotted against temperature there is a thresholdlike relation when the frequency distributions go through their bimodal phase.

These points together prove that properties of the developmental system determine the pattern of response to disturbances. The disturbances can be caused by genetic differences as in selection, by such environmental differences as temperature, or by developmental error—the lack of precision in development which causes fluctuating asymmetry. This is the kind of situation predicted by Waddington in canalized systems; buffering against disturbances generated by genetic variability, by environmental factors, or by developmental noise.

To explain these phenomena, the following model was made: (*a*) In the wing-anlagen of larvae there is a vein-forming substance. The level of vein-forming substance varies between individual larvae according to a normal distribution. It is subject to change by genetic and environmental factors, and by developmental error. (*b*) Vein formation depends on the competence of the cells along the track of the fourth vein to react to the vein-forming substance with the formation of vein-material. (*c*) There is a gradient of competence in the wing-anlagen. This gradient decreases from wing-base to wing-tip. When the level of vein-forming substance increases, vein length increases from wing-base to wing-tip. The slope of the gradient explains that the interruption of the fourth wing-vein in ci^D is always terminal. (*d*) The gradient is not linear but shows a thresholdlike change in the region 70–80. This explains the bimodality. When wild-type is approached, i.e. when the fourth vein becomes nearly complete, the gradient is supposed to become steeper.

The increasing steepness of the gradient when the fourth vein approaches completion is based on the following observations: The variability of the frequency distributions decreased when the mean vein length increased between the zone of bimodality and completeness of the fourth vein. Low temperature had little effect though it was very effective in lengthening the fourth vein when ci^D stocks had lower mean values i.e. lower than the region of bimodality. Plateaus in the selection lines were established at similar values just higher than 80, and there was little progress thereafter even with long continued further selection, i.e. in one line 30 and in the other 2 lines 10 generations.

The lack of response to selection in the direction of wild-type (i.e. complete) fourth vein was not due to exhaustion of genetic variability. The three possible crosses among the three high lines were made to create new genetic variability. From the F_3, selection was started for a shorter and for a longer vein, respectively. While the short vein lines made good progress showing the presence of genetic variability in the F_3 base populations, the progress in the high lines was very small indeed, and only in one line were a few flies obtained with a complete fourth vein. (54).

The important features of these experiments are: The developmental system of wing-venation influences the effect of factors that change this continuously variable character; genetic factors, environmental factors, and accidents of development (developmental error or developmental noise) are affected to the same extent; when the mutant phenotype is approaching wild-type, change becomes progressively more difficult.

CONCLUSIONS In the classic experiments on artificial selection of quantitative characters in *Drosophila* and mice (4, 5, 9, 10, 32, 33, 48, 51), the scale of measurement was more or less linear at least on a logarithmic scale. Factors that changed the character had similar effects on different parts of the scale, and if selection was not proceeding smoothly, this was a consequence of genetic processes (delayed recombination, involvement of factors with large effects) and not of limitations set by physiological or developmental processes.

In all the experiments in which quantitative characters, affected by mutant-expression, were artificially selected to normal type, progress of selection became more difficult in the neighborhood of wild-type. This was seen as the consequence of canalization of the wild-type phenotype and as a confirmation of Waddington's theory. Moreover, also at other levels of expression there were regions in which change is more difficult or where it is facilitated. Therefore, these experiments showed clearly the influence of developmental systems on the expression of genetic variability and thereby on selection response.

In the experiments with the mutants *scute* and *ocelli-less* in *Drosophila,* and the *Tabby* mutant in mice, it was possible to break through the wild-type barrier and to proceed at the other side of wild-type. This did not go far but did suggest that there was one continuous function, relating factors that changed a character and their phenotypic effects. This function would encompass a range beginning far below the wild phenotype where change was easy; then change would become progressively more difficult when passing wild-type, followed again by a region of facilitated change beyond the bounds of normality. This function represented canalization; change of canalization would occur by change of the developmental processes underlying this function. This hypothesis is depicted in the model of the *scute* experiments first given by Dun & Fraser (8) and adopted by Fraser & Kindred (18) for the mice vibrissae. It was later accepted by Waddington and is now featured in several reputable text books. The function would have got its shape by the action of natural selection of the stabilizing type. However, Fraser & Kindred (8) suggested as an alternative possibility "regularities in the basic pattern of development of the tissue concerned," foreshadowing the discussion on the role of developmental constraints in evolution.

The original concept of canalization implied that buffering would involve all factors pushing development out of course. Only in ci^{D} is there unequivocal evidence that the effects of genetic differences and nongenetic factors, i.e. environmental differences and the developmental indeterminacies generating asymmetry, are similarly affected. The other side of wild-type of the fourth vein interruption is, of course, extra venation. This occurred in two of the three lines selected for a long fourth vein. The extra-venation appeared before the fourth vein was complete and was in one line widespread and caused by a *plexus* allele. Although the variation at the other side of wild-type, i.e. extra venation, was affected by the same genetic differences acting at the other side, i.e. on the length of the interrupted fourth vein, they are not acting according to a single continuous curve, as envisaged in Rendel's model.

Complications

Discussion concerning the significance and general validity of models with a single canalization function for scutellar bristle number and other characters arose from theoretical considerations and further analysis and experiments.

Alan Robertson (47) suggested that the probit transformation could not be used in a biologically meaningful way because of the nature of some characters. This would mean that some of the conclusions drawn about genotype-phenotype relationship were no more than statistical artefacts. In the models a morphogenetic substance shows continuous variation in its concentration between individuals. It determines the phenotype on a measurable scale; as the concentration transgresses a threshold, the score is 1; when it transgresses the next threshold the score becomes 2; and so on. Robertson states that such a model is only meaningful when the scores are really cumulative or sequential.

SCUTELLAR BRISTLES Scutellar bristle number is not such a cumulative character. There are four separate sites which are not filled up in any sequence; the total score is the sum of four independent events. Robertson suggested a model that starts with probabilities for the presence of a bristle on each separate site. The frequency distribution of individuals with particular bristle numbers can then be calculated. This model implies that we are not dealing with one frequency distribution of morphogenetic substance underlying the total number of scutellar bristles, but with four separate frequency distributions. Each site would have its own frequency distribution of morphogenetic substance and its own threshold which has to be transgressed for the appearance of one bristle on the particular site. Determining bristle class width in probits for total scutellar bristle number is then a meaningless procedure. Class width would have no significance because according to this "per site" model it does not indicate the amount of change of an underlying variable necessary for change of bristle number. Crucial here is the independence of bristle formation at the separate sites and the possibility of calculating frequency distributions for total bristle number from the probabilities of bristle formation for the particular sites.

Rendel (41) compared calculated distributions with distributions observed in four *scute* stocks. In one stock with the single site probability model, flies with 5 bristles were predicted and were not found.

Canalization of total bristle number would imply negative correlations between bristles on different sites. In a stock canalized around 2 bristles by a special mode of stabilizing selection (45), the correlation between the presence of anterior and posterior bristles was −0.6. Latter (30) found negative correlations in a wild-type stock selected for high bristle number. However Latter & Scowcroft (30a) found that at the normal 4-bristle level the genetic correlation could be zero or positive. While these observations show the occurrence of mutual dependence of the bristle sites in the presence of scute, it does not exclude the possibility of a partly independent determination.

Alan Robertson (47) suggested that scutellar bristle number could better be considered as two pairs of bristles, an anterior and a posterior pair, each at least partly under specific control. He mentioned that extra bristles tend to appear first at the anterior sites, and bristles tend to be lost first at the posterior sites.

Scowcroft et al (57) indeed showed that a single developmental scale does not provide an adequate description of the phenomenon of canalization of scutellar bristles. The effect of the addition of extra *scute* loci was earlier reported by Fraser & Green (16) and by Rendel et al (44). They had found that additional *scute* loci always caused an increase in bristle number. Rendel et al observed that this effect became smaller measured on the underlying probit scale when approaching wild-type. They interpreted this as the "switching off" of the *scute* gene as the normal level of morphogenetic substance comes closer.

Scowcroft et al studied the effect of varying the number of *scute* loci from 1 to 3 in males and from 2 to 4 in females. They constructed special chromosomes with 3 different *scute* alleles, one mutant allele with extreme expression, one mutant approaching wild-type, and an allele that produces a normal number of scutellars. Scutellar bristle number always increased with increasing dosage of the *scute* locus, although the increase became smaller when coming closer to wild-type. A separate analysis was made of the change at anterior and posterior bristle sites. The responses to increasing dosage are strikingly different when measured at the anterior or at the posterior bristle sites. Starting from a low bristle number in the extreme mutant the gain is predominantly at the posterior sites, starting from wild-type level and increasing the dosage of the normal allele the number of anterior bristles increases without change of the posteriors. Moreover, the posterior bristles are more strongly canalized than the anteriors. Finally, the posterior bristles play a different role than anteriors in defining the 3 to 4 and the 4 to 5 thresholds which would determine the canalization at 4 bristles. In fact, the lower border of the 4-class is determined by posterior bristle number while the upper border is determined by the number of anterior bristles.

This is confirmed in experiments of Scowcroft (56) who applied selection for anterior or for posterior bristle number and for total scutellar bristle number. He used two populations, one straddling the 3/4 threshold and the other the 4/5 threshold. The changes of total bristle number round the 3/4 threshold involve addition or loss at the posterior site while changes around the 4/5 threshold are changes of anterior bristles. The independent control of posterior and anterior bristles is also revealed by temperature experiments; while the anterior bristles decline in number at higher temperatures, the posterior bristles increased (23, 37).

Fraser & his associates (12–15) have produced further evidence showing that the relation between genetic and environmental factors in changing scutellar bristle number and phenotype cannot be represented by one morphogenetic substance acting according to one single function controlling bristle number at either side of the wild-type. First, Fraser et al point out that it is possible simultaneously to add to and to subtract from the basic pattern of four bristles. Such flies are very rare in wild populations but are more frequently present in lines selected for scutellar bristle number. This situation is comparable to Waddington's and Scharloo's flies with simultaneous vein-interruption and extra-venation (70, 52). It could mean that the character is not causally homogeneous.

Secondly, Fraser et al studied the effect of genetic variation accumulated in lines selected for higher scutellar bristle number in the presence of *scute* on non-*scute* individuals and vice versa. Fraser & Green (16) used wild-type lines selected for higher numbers of scutellars. They substituted a *scute* mutant for the wild-type *scute* allele in their lines. There was no correlation between the number of extra bristles in the wild-type lines and the number of bristles after the introduction of the *scute* mutant. Moreover, they constructed Y-chromosomes with the extreme tip of an X-chromosome which includes the wild-type *scute* gene attached to it. When these chromosomes were segregating in their lines, flies were obtained with different numbers and different combinations of alleles of the *scute* locus. Positive correlations for bristle number between genotypes were found only when both genotypes had either more than four bristles or less than four bristles. Notwithstanding their identical genotypes (besides the situation on the *scute* locus) there was no correlation when one of the lines possessed more than four bristles per fly and the other had a mean below four. Their conclusion was that there were two different sets of genes changing scutellar bristle number, one set acting in *scute* flies, the other in wild-type flies.

This is in contrast with Rendel's results on which his model was based. There he reported an, albeit small, increase of scutellar bristles in the wild-type sibs of the *scute* flies on which selection was practiced. Fraser (13) repeated Rendel's selection for higher number of scutellars in *scute* flies. He had four replicate selection lines. The results are roughly in agreement; the selection on scutellar bristles in *scute* flies caused higher bristle numbers in non-*scute* sibs. However, Fraser (14, 15) found in his wild-type lines selected for higher bristle number a recessive gene *extraverticals* (*x-vert,* probably an allele of *polychaetoid*) which contributed about 3 bristles to the selection response of 5 bristles. Backcrossing this gene into the base population decreased its effect: it is dependent on the genetic background. This gene and its modifiers are not expressed in the presence of scute; on the other hand the modifiers of scute are not expressed in the homozygous presence of *x-vert.*

Sheldon & Milton (58) selected wild-type flies for high scutellar bristle number and obtained two lines with about 11 and 13 scutellar bristles after 140 generations. Backcrossing of a *scute* mutant into both lines produced means of 8.1 and 6.8 respectively for the sc/sc^{+} wild-type females and 3.2 and 3.1 respectively for the *sc/sc* females, compared to means of 4 and 1.2 for the same genotypes in the control population. The genes causing an increase of bristles in the non-*scute* selection lines do not have much effect in the presence of *scute*. The wild-type sibs of *scute* flies selected for a higher bristle number obtained means of only 4.07 and 4.3, although the mean of the *scute* sibs was not much different from the flies with *scute* in the background of the selection lines. Haskell (28) earlier reported that modifiers of *scute* expression in abdominal bristles did not affect the same character in wild-type flies.

Scharloo (1988) obtained canalization of 8 scutellar bristles in a wild-type stock by stabilizing selection. He found that it was not based on canalization of total bristle number but on an increased precision per bristle site.

VIBRISSAE IN MICE Alan Robertson's (47) objections against the use of probits for genephysiological interpretations of canalization apply not only to the scutellar bristle pattern but also to the ocellar apparatus and to vibrissae number in mice. In fact, Pennycuik & Rendel (38) found for vibrissae number a good fit between expectations, on the basis of Robertson's theory and observation. The frequency distributions of total vibrissae number in mice agreed with the prediction made on the basis of the vibrissae numbers on the five separate sites. The vibrissae number is clearly controlled independently on each site.

An indication that in this character different processes can be involved in changes in vibrissae number is the different embryological basis for an increase in vibrissae number in the high selection line of Dun & Fraser and for a decrease of vibrissae number in a back selection line. In the first there was a change of the number of hair follicles, in the second there was not a decrease of the number of follicles, but a suppression of the formation of hairs (Jacobsen quoted in Fraser—14).

This was confirmed in selection experiments by Kindred (28a). She selected for a higher number of vibrissae in wild-type (i.e. non-*Tabby* mice) from the experiment in which mice with the *Tabby* mutant were selected for higher vibrissae number (7, 8, 18). In that experiment after 34 generations of selection the mean vibrissae number of the wild-type sibs of the *Tabby* mice was scarcely higher than normal (19.1 instead of 19). Selection for higher number from these wild-type sibs had an immediate response without any effect on the *Tabby* sibs which still segregated in the line. Selection caused changes of genetic variability acting on a different aspect of the development of the vibrissae number than when selection was performed on the mutant character with values below normal. This does not fit the earlier model (18) which has one continuous canalization function translating genetic differences in bristle number over the whole range of variation of the character at either side of wild-type.

WING-VEIN INTERRUPTIONS In the early experiments with selection to normal phenotypes of morphological mutants in mice and *Drosophila,* changing expression became more difficult when wild-type was near. This was found too in the selection for a longer fourth longitudinal wing vein of the mutant ci^D. However, selection toward wild-type and temperature experiments with a different allele, *cubitus interruptus dominant of Gloor* (ci^{D-G}), revealed a strikingly different pattern. The temperature experiments showed an almost perfect linear relation between temperature and the relative length of the fourth vein. The frequency distributions were unimodal throughout over a large part of the possible range. At the lower temperatures (20°, 17.5°, 15°) wild-type was approached easily, and overlap with wild-type was obtained as a continuation of the linear shift of the frequency distributions (50, 51). While with the ci^D wild-type could never, or only in a few flies, be obtained after 40 to 60 generations of selection, on the contrary, with ci^{D-G}, overlap with wild-type occurred after only 4 or 5 generations (55).

Scharloo (51) selected the same character in the mutant *Hairless. Hairless* (*H*) is a homozygous lethal, and has as a heterozygote dominant morphological effects: It removes bristles on the top of the head and causes interruptions of the fourth and the fifth longitudinal wing-vein. These interruptions are only present in part of the flies. Selection for shorter length of the fourth vein was done in five selection lines. In all lines after a few generations all flies showed the fourth vein interruption, and the veins became shorter rapidly. The variability of the length of the vein was quite large in the beginning, but it became smaller when the vein became shorter and its end was approaching the the second cross-vein. Then selection progress, i.e. change of the mean vein-length, became smaller.

The length of the fourth vein in *Hairless* is sensitive to the rearing temperature. The fourth vein becomes longer at lower temperatures. When the vein-length was shifted through the same range as was transgressed by the selection, a similar variation pattern was observed: large variability when the vein approaches completion and smaller variability in regions nearer to the attachment of the fourth vein with the second cross-vein. Change per degree Celsius is greater when the vein is nearing completion than in regions nearer to the second cross-vein (50, 51). Introduction of chromosomes of unselected stocks gives a similar picture; genetic change is easy in the neighborhood of wild-type and becomes more difficult in regions of more extreme expression of the mutant (51). Fluctuating asymmetry was approximately proportional to total variance.

Also in this case all factors which change this character—genetic factors, environmental factors, and developmental error—acted according to the same scale. However, the scale for the same character differs: there is a difference not only between mutants of different genes but also between alleles of the same gene locus.

CONCLUSIONS

The following conclusions can be drawn from the outcomes of the early assimilation experiments:

1. If artificial selection favors environmentally induced aberrant phenotypes, an increase of the frequency of the phenotypic reaction is obtained.
2. Selection for the induced phenotypes causes accumulation of genes which first, in conjunction with the environmental factor, increase the frequency of the phenotypic reaction and later realize the same phenotype on their own without environmental interference.
3. While specific environmental treatments induce often a variety of morphological aberrations, artificial selection for a specific type increases specifically that type of reaction and ultimately its assimilation.
4. Because of a lack of response to such a selection in genetically homogeneous inbred lines; it must be concluded that the response may be based on genetic variation present in the base population before the selection started.
5. Waddington (68, 69, 73, 74, 77) suggested that genetic assimilation has to be explained in terms of canalization and genetic variation of reactivity to environmental factors. Canalization would be a consequence on systems of interacting genes; genetic variation in reactivity implies gene-environment interaction. Bateman (2) and Stern (61) showed that a simple threshold model with additive effects of the genes involved and without gene-environment interaction could explain the experiments on genetic assimilation.

Change of a phenotypic reaction of a measurable character was shown in Waddington's experiments (75) on what he thought to be the size of the anal papillae in *Drosophila* larvae. Because he measured not the anal papillae in larvae but the remnants of the inactive epidermal region between the papillae, he made at best an inverse measurement of an organ with adaptive reactions. The real reaction of the anal papillae is a decrease of their size with increasing salt concentration. Te Velde et al (64, 65) did not show genetic variation in this reaction which is based on changes in cell number. A rapid phenotypic reaction in the ultrastructure of the papillae cells (63) showed genetic variation in populations which had clear adaptive significance and could be fixed by artificial selection.

Introduction of morphological mutants revealed genetic variation for characters that were invariable in normal phenotypes. Artificial selection on such characters thereby became possible. The patterns of change and variation of the selected characters revealed the effect of the organization of developmental systems on the expression of genetic variability and thereby on selection response.

In the early experiments it was found for all characters that, when the character was approaching the normal, phenotypic change became progressively more difficult. In all experiments the modifier genes for mutant expression accumulated by selection toward wild-type could also change the character at the other side of wild-type, where variability increased again. This picture was interpreted as an expression of the canalization of the normal phenotype.

This suggested models in which the expression of the character was determined by two genetically independent components: a morphogenetic substance supposed to have a normal frequency distribution in the population, and one mapping or canalization function, in the *scute* and *Tabby* models ranging from extreme mutant expression, beyond wild-type, into a region with supernumerary structures. The function maps the effect of changes in morphogenetic substance into the character expression. The change in morphogenetic substance could be caused by genes or by environmental factors. The mutants were supposed to change the mean of the normal distributions of morphogenetic substance and not the mapping function. Artificial selection changing mutant expression would act via polygenic variability which is partly responsible for the normal distribution of morphogenetic substance. When environmental factors would only act via the morphogenetic substance, the mapping function would be a gene-environmental factor/phenotype mapping function (GEMP, 51). Then, the mapping function would represent real canalization, and the developmental system would govern the effects of genetic and environmental differences in the same way. Such a situation was revealed for the *cubitus interruptus* dominant mutant; genetic differences, environmental factors, and within-fly variance all acted according to the same mapping function (49, 51).

However, the concept of one mapping function characteristic for a specific morphological character could not be maintained in the light of further analysis. In the first place, it was shown that such pattern characters as the scutellar bristle pattern and ocellar pattern in *Drosophila* and the vibrissae in mice could not be considered unitary characters. Their components, the separate bristles, vibrissae, and ocelli, were shown to have a partially independent determination which has important consequences for the interpretation of the experiments in which they are involved.

It has now been shown that there is not one homogeneous developmental process governing the expression of a character at either side of wild-type. In scutellar bristle number and for venation characters, effects at either side of wild-type, i.e. loss and addition, could occur simultaneously. This shows that the characters involved are causally not homogeneous. In this respect it is important that in the scutellar bristles modifiers effective below the normal score of 4 bristles are often not effective above the four level, and modifiers effective above the four level are suppressed below this level. A similar situation exists in vibrissae number in mice. Further, in the experiments on the interrupted fourth vein in *Drosophila,* the mapping function for the same character was strikingly different in different mutants. This shows that characters do not have specific mapping functions. Different mutants changing the same character affect and sensitize different steps in its developmental pathway, each with its own mapping function.

Does this mean that we have to abandon the canalization concept? Of course not. It is a fact that wild-type is relatively constant compared to mutants and to phenocopies. We cannot escape the conclusion that development in wild-type individuals is geared to produce constant phenotypes notwithstanding the presence of genetic variability and the omnipresence of environmental differences. This is the case not only in the constant morphological patterns which were the predominant object of research on canalization. It also occurs as we saw in this review, in adaptable physiological processes for the regulation of osmotic value of the hemolymph of Drosophila larvae. Forbes Robertson (47a) revealed it for developmental time and growth of *Drosophila* larvae. He showed that abnormal food conditions revealed large amounts of genetic variability for growth. They were not expressed under normal conditions, and the genetic variants involved showed interaction which generated normal growth and development.

However, we have to recognize that our understanding of this fascinating subject is still limited. Further research has to analyze the developmental processes underlying the characters involved. It is of course a sobering thought that we do not know how the mutants involved in the experiments described earlier realize their morphological effects nor what underlies the changes made by artificial selection.

There were promising starts some 30 years ago, for instance, the work of Forbes Robertson (47b, 47c) on the role of cell size and cell number in the determination of body size in *Drosophila*. Another example is the work of Spickett (60a) on the changes in the development of sternopleural chaetae in which both pattern formation and the timing of differentiation of bristles were related. That timing cannot be neglected is indicated in the work of Poodry (39) who showed that there is a clear difference in the time of differentiation of anterior and posterior scutellar bristles.

Comparative analysis of patterns in different species as done by Garcia Bellido (21), studies on the genetic analysis with mutants changing complex patterns (6), and changing patterns by artificial selection (51, 52) can be important tools. We may hope that not too long from now we will be able to build a synthetic view from the building stones of quantitative genetics and developmental biology. Quantitative genetics will have to unravel their overall characters in components amenable for developmental and physiological analysis. Developmental biology will have to make the link between the theories of pattern formation (20, 36, 46) and the underlying molecular structures and processes (22).

This will ultimately give the basis for understanding the relation between constraints generated by inherent, structural properties of developmental processes and reversible developmental constraints as canalization built by natural selection (35).

ACKNOWLEDGMENTS

This review was written during a stay in the Department of Ecology and Evolutionary Biology of the University of Arizona. I would like to thank Dr. Bill Heed and Dr. Conrad Istock for their hospitality and help.

Literature Cited

1. Bateman, K. G. 1959. The genetic assimilation of the dumpy phenocopy. *J. Genet.* 56:341–51
2. Bateman, K. G. 1959. The genetic assimilation of four venation phenocopies. *J. Genet.* 56:443–47
3. Chen, P. S., Brugger, C. 1973. An electron microscope study of the anal organs of *Drosophila melanogaster*. *Experientia* 29:233–35
4. Clayton, G. A., Robertson, A. 1957. An experimental check on quantitative genetical theory. II. The long-term effects of selection. *J. Genet.* 55:152–70
5. Clayton, G. A., Morris, J. A., Robertson, A. 1957. An experimental check on quantitative genetical theory. I. Short-term responses to selection. *J. Genet.* 55:131–51
6. Diaz-Benjumea, F. J., Garcia-Bellido, A. 1990. Genetic analysis of a wing vein pattern of Drosophila. *Roux's Arch. Dev. Biol.* 198:336–54
7. Dunn, R. B., Fraser, A. S. 1958. Selection for an invariant character—'vibrissae number'—in the house mouse. *Nature* 181:1018–19
8. Dunn, R. B., Fraser, A. S. 1959. Selection for an invariant character, vibrissae number in the house mouse. *Aust. J. Biol. Sci.* 12:506–23
9. Falconer, D. S. 1953. Selection for large and small size in mice. *J. Genet.* 51:470–501
10. Falconer, D. S. 1955. Patterns of response in selection experiments with mice. *Cold Spring Harbor Symp. Quant. Biol.* 20:178–96
11. Falconer, D. S. 1960. *Introduction to Quantitative Genetics*. Edinburgh: Oliver and Boyd. 335 pp.
12. Fraser, A. S. 1963. Variation of scutel-

lar bristles in Drosophila I. Genetic leakage. *Genetics* 48:497–514
13. Fraser, A. S. 1966. Variation of scutellar bristles in Drosophila. XII. Selection in scute lines. *Aust. J. Biol. Sci.* 19:147–54
14. Fraser, A. S. 1968. Variation of scutellar bristles. XV. Systems of modifiers. *Aust. J. Biol. Sci.* 57:919–34
15. Fraser, A. S. 1970. Variation of scutellar bristles in Drosophila XVI. Major and minor genes. *Genetics* 65:305–09
16. Fraser, A. S., Green, M. M. 1964. Variation of scutellar bristles in Drosophila. III. Sex-dimorphism. *Genetics* 50:351–62
17. Fraser, A. S., Kindred, B. M. 1960. Selection for an invariant character, vibrissae number in the house mouse. II. Limits to variability. *Aust. J. Biol. Sci.* 13:48–58
18. Fraser, A. S., Kindred, B. M. 1962. Selection for an invariant character, vibrissae number in the house mouse. III. Correlated responses. *Aust. J. Biol. Sc.* 15:188–206
19. Fraser, A. S., Nay, T., Kindred, B. 1959. Variation of vibrissae number in the mouse. *Aust. J. Biol. Sci.* 12:331–33
20. Garcia-Bellido, A. 1981. From the gene to the pattern. In *Cellular Controls in Differentiation,* ed. C. W. Lloyd, D. A. Rees, pp. 281–304. London: Academic
21. Garcia-Bellido, A. 1983. Comparative anatomy of cuticular patterns in the genus Drosophila. In *Development and Evolution. Sixth Symposium of the Br. Soc. Dev. Biol.*, ed. B. C. Goodwin, N. Holder, C. C. Wylie, pp. 227–55. Cambridge: Cambridge Univ. Press.
22. Ghysen, A., Dambly-Chaudiere, C. 1988. From DNA to form: The achaete-scute complex. *Genes Dev.* 2:495–501
23. Gibson, J. 1970. Effects of temperature on selection for scutellar bristle number. *Heredity* 25:591–607
24. Gloor, H. J. 1947. Phaenokopie Versuche mit Aether an Drosophila. *Rev. Suisse Zool.* 54:637–712
25. Gloor, H. J., Chen, P. S. 1950. Ueber ein Analorgan bei Drosophila-larven. *Rev. Suisse Zool.* 57:571–76
26. Goldschmidt, R. B. 1938. *Physiological Genetics*. New York: McGraw Hill. 375 pp.
27. Goldschmidt, R. B. 1955. *Theoretical Genetics*. Berkeley: Univ. Calif. Press
28. Haskell, G. M. L. 1943. The polygenes affecting the manifestation of scute in *Drosophila melanogaster*. *J. Genet.* 45:269–76
28a. Kindred, B. 1967. Selection for an invariant character, vibrissae in the house mouse. V. Selection on non-Tabby segregants from Tabby selection lines. *Genetics* 55:365–73
29. Landauer, W. 1957. Phenocopies and genotype, with special reference to sporadically-occurring developmental variants. *Am. Nat.* 91:79–89
30. Latter, B. D. H. 1970. Selection for a threshold character in Drosophila III. Genetic control of variability in plateaued populations. *Genet. Res.* 15: 285–300
31. Maas, A. H. 1948. Ueber die Ausloesbarkeit von Temperatur-Modifikationen waehrend der Embryonalentwicklung von Drosophila melanogaster. *W. Roux Arch.* 143:515–44
32. Mather, K 1941. Variation and selection of polygenic characters. *J. Genet.* 41: 159–93
33. Mather, K., Harrison, B. J. 1949: The manifold effects of selection. *Heredity* 3:1–52, 131–62
34. Maynard Smith, J., Burian, R., Kauffman, S., Alberch, J., Campbell, B., Goodwin, B., Lande, R., Raup, D., Wolpert, L. 1985. Developmental constraints and evolution. *Q. Rev. Biol.* 60:265–87
35. Maynard Smith, J., Sondhi, K. C. 1960. The genetics of a pattern. *Genetics* 45:1039–50
36. Meinhardt, H. 1982. *Models of Biological Pattern Formation*. London: Academic. 230 pp.
37. Pennycuik, P., Fraser, A. S. 1964. Variation of scutellar bristles in Drosophila. II. Effects of temperature. *Aust. J. Biol. Sci.* 17:764–70
38. Pennycuik, P. R., Rendel, J. M. 1977. Selection for constancy of score and pattern of secondary vibrissae in Ta/Ta-Ta/Y and Ta/I mice. *Aust. J. Biol. Sci.* 30:303–17
39. Poodry, C. A. 1975. A temporary pattern in the development of sensory bristles in Drosophila. *W. Roux Arch.* 178:203–13
40. Rendel, J. M. 1959. Canalization of the scute phenotype of Drosophila. *Evolution* 13:425–39
41. Rendel, J. M. 1965. Bristle pattern in scute stocks of Drosophila melanogaster. *Am. Nat.* 99:25–32
42. Deleted in proof
43. Deleted in proof
44. Rendel, J. M., Sheldon B. L., Finlay, D. E. 1965. Canalization of development of scutellar bristles in Drosophila by control of the scute locus. *Genetics* 52:1137–51
45. Rendel, J. M., Sheldon, B. L., Finlay, D. E. 1966. Selection for canalization of

the scute phenotype. II. *Am. Nat.* 100:13–31
46. Richelle, J., Ghysen, A. 1979. Determination of sensory bristles and pattern formation in Drosophila. I. A model. *Dev. Biol.* 70:418–37
47. Robertson, A. 1965. Variation in scutellar bristle number—an alternative hypothesis. *Am. Nat.* 99:19–24
47a. Robertson, F. W. 1964. The ecological genetics of growth in Drosophila 7. The role of canalization in the stability of growth relations. *Genet. Res.* 5:107–26
47b. Robertson, F. W. 1959. Studies in quantitative inheritance. XII. Cell size and cell number in relation to genetic and environmental variation of body size in *Drosophila melanogaster. Genetics* 44:869–96
47c. Robertson, F. W. 1959. Studies in quantitative inheritance. XIII. Interrelationships between genetic behaviour and development in the cellular constitution of the Drosophila wing. *Genetics* 44:1113–30
48. Robertson, F. W., Reeve, E. C. R. 1952. Studies in quantitative inheritance. The effects of selection on wing and thorax length in *Drosophila melanogaster. J. Genet.* 50:414–48
49. Scharloo, W. 1962. The influence of selection and temperature on a mutant character (ci^D) in *Drosophila melanogaster. Arch. Neerl. Zool.* 14:431–512
50. Scharloo, W. 1964. Mutant expression and canalization. *Nature* 203:1095–96
51. Scharloo, W. 1987. Constraints in selection response. In *Genetic Constraints on Adaptive Evolution.* ed. V. Loeschke, pp. 125–49. Berlin: Springer
52. Scharloo, W. 1988. Selection on morphological patterns. In *Population Genetics and Evolution,* ed. G. de Jong, pp. 230–50. Berlin: Springer
53. Scharloo, W. 1989. Developmental and physiological aspects of reaction norms. *Bioscience* 39:465–71
54. Scharloo, W. 1990. The effect of developmental constraints on selection response. In *Organisational Constraints on the Dynamics of Evolution,* ed. G. Vida, J. M. Maynard Smith, pp. 197–210. Manchester: Manchester Univ. Press
55. Scharloo, W., Hoogmoed, M. S., Kuile, A. ter 1976. Stabilizing and disruptive selection on a mutant character in Drosophila. I. The phenotypic variance and its components. *Genetics* 60:373–88
56. Scowcroft, W. R. 1973. Scutellar bristle components and canalisation in *Drosophila melanogaster. Heredity* 30:289–301
57. Scowcroft, W. R., Green, M. M., Latter, B. D. H. 1968. Dosage at the scute locus, and canalisation of anterior and posterior scutellar bristles in *Drosophila melanogaster. Genetics* 60:373–88
58. Sheldon, B. L., Milton, M. K. 1972. Studies on the scutellar bristles of *Drosophila melanogaster.* II. Long-term selection for high bristle number in the Oregon RC strain and correlated responses in abdominal chaetae. *Genetics* 71:567–95
59. Sondhi, K. C. 1960. Selection for a character with a bounded distribution of phenotypes in *Drosophila subobscura. J. Genet.* 57:193–221
60. Sondhi, K. C. 1961. Developmental barriers in a selection experiment. *Nature* 189:249–50
60a. Spickett, S. G. 1963. Genetic and developmental studies of a quantitative character. *Nature* 189:870–73
61. Stern, C. 1958. Selection for subthreshold differences and the origin of pseudoexogenous adaptations. *Am. Nat.* 2:313–16
62. te Velde, J. 1985. *The significance of the anal papillae in salt adaptation of* Drosophila melanogaster. Thesis. Univ. Utecht, The Netherlands. 199 pp.
63. te Velde, J. H. Gordens, H., Scharloo, W. 1987. The genetic fixation of phenotypic response of an ultrastructural character in the anal papillae of *Drosophila melanogaster. Heredity* 60:47–53
64. te Velde, J. H., Molthoff, C. F. M. 1988. The function of anal papillae in salt adaptation of *Drosophila melanogaster* larvae. *J. Evol. Biol.* 1:139–53
65. te Velde, J. H., Scharloo, W. 1988. Natural and artificial selection on a deviant character of the anal papillae in *Drosophila melanogaster* and their significance for salt adaptation. *J. Evol. Biol.* 1:155–64
66. Waddington, C. H. 1940. *Organisers and Genes.* Cambridge: Cambridge Univ. Press. 223 pp.
67. Waddington, C. H. 1941. The genetic control of wing development. *J. Genet.* 41:75–139
68. Waddington, C. H. 1942. The canalization of development and the inheritance of acquired characters. *Nature* 150: 563
69. Waddington, C. H. 1953. The genetic assimilation of an acquired character. *Evolution* 7:118–26
70. Waddington, C. H. 1955. On a case of quantitative inheritance of either side of

the wild-type. *Z. ind. Abstammungs-u. Vererb Lehre* 87:208–28

71. Waddington, C. H. 1956. Genetic assimilation of the bithorax phenotype. *Evolution* 10:1–13
72. Waddington, C. H. 1956. The genetic basis of the assimilated bithorax stock. *J. Genet.* 55:241–45
73. Waddington, C. H. 1957. *The Strategy of the Genes*. London: Allen & Unwin. 262 pp.
74. Waddington, C. H. 1958. Comment on Professor Stern's letter. *Am. Nat.* 92: 375–376
75. Waddington, C. H. 1959. Canalization of development and genetic assimilation of acquired characters. *Nature* 183: 1654–55
76. Waddington, C. H. 1959. Evolutionary adaptation. In *Evolution after Darwin. II. The Evolution of Life,* ed. Sol Tax Pp. 381–402. Chicago: Chicago Univ. Press
77. Waddington, C. H. 1961. Genetic assimilation. *Adv. Genet.* 10:257–93
78. Waddington, C. H. 1975. *The evolution of an evolutionist*. Ithaca, NY: Cornell Univ. Press. 328 pp.
79. Wigglesworth, V. B. 1933. The adaptation of mosquito larvae to salt water. *J. Exp. Biol.* 10:27–37

Annu. Rev. Ecol. Syst. 1991. 22:95–114

HERBICIDE RESISTANCE IN WEEDY PLANTS: Physiology and Population Biology

Suzanne I. Warwick

Biosystematics Research Centre, Agriculture Canada, Research Branch, Ottawa, Canada, KlA OC6

KEY WORDS: ecological fitness, genetics, population models, resistant crops, selection pressure

---

INTRODUCTION—HISTORY AND DISTRIBUTION OF HERBICIDE RESISTANCE

Predictions of the likely occurrence of herbicide resistance in plants (34) were followed by the first discovery of a triazine-resistant biotype of *Senecio vulgaris* in the United States in 1968 (66). Since then, and particularly in recent years, the number of reported cases of herbicide resistant weed species has rapidly increased, as has the variety of herbicides to which resistance has evolved. There are over 100 weed species with biotypes known to be resistant to herbicides (Table 1, see also 5, 27, 38, 44); these have been the subject of several symposia, reviews, etc (e.g. 26, 38, 39, 43). Given the enormous literature in this field, citations in this review are limited to key papers, reviews, and current papers.

Herbicide resistance may be defined as the condition whereby a plant withstands the normal field dose of a herbicide, as a result of selection and genetic response to repeated exposure to herbicides with a similar mode of action. Susceptible plants are normally killed by recommended field doses. For the remainder of this review susceptible and resistant biotypes, plants, populations, etc, of a given species will be designated as S- and R-, respectively. The weed population/herbicide model provides an excellent example

0066-4162/91/1120-0095$02.00

Table 1 Occurrence and distribution of triazine-resistant weedy biotypes

Taxa	Location[+] (Reference)
Abutilon theophrasti Medic.	EUR (44), USA (3)
Alopecurus myosuroides Huds.	CZC (54), ISR (89)
Amaranthus arenicola S. Wats	USA (5)
A. bouchonii Thell.	HUN (72), SWI (27)
A. blitoides L.	SWI (27)
A. graezicans L.	SWI (27)
A. hydridus L.	FRA (27), SWI (27), USA (5)
A. lividus L.	SWI (27)
A. powellii S. Wats	CAN (5, 74), USA (5)
A. retroflexus L.	AUS (27), CAN (5, 74), FRA (27), HUN (27, 70), POL (46), SWI (27)
Ambrosia artemisiifolia L.	CAN (5, 74), USA (44)
Arenaria patula L.	EUR (44)
Atriplex patula L.	GER (27)
Bidens tripartita L.	AUS (27)
Brassica rapa L.(= *B. campestris*)	CAN (5), EUR (44)
Brachypodium distachyon (L.) Beauv.	ISR (29)
Bromus tectorum L.	EUR (44), USA (5)
Capsella bursa-pastoris L.	POL (46)
Chamomilla suaveolens	UK (12)
Chenopodium album L.	AUS (27), BEL (11), CAN (5, 74), CZC (54), FRA (27), GER (27), HUN (71), NZD (44), POL (46), SPN (18), SWI (27), USA (5), UK (12)
C. ficifolium Sm.	GER (27)
C. missouriense Allen	USA (5)
C. polyspermum L.	AUS (27), FRA (27), SWI (27)
C. strictum Roth.	CAN (5, 74)
Conyza(Erigeron) bonariensis (L.) Cronq	SPN (18)
C. canadensis (L.) Cronq.	EUR (44), HUN (45), SWI (27), UK (12)
Digitaria sanguinalis (L.) Scop.	EUR (44)
Echinochloa crus-galli (L.) Beauv.	CAN (5, 74), EUR (44), FRA (27), USA (5)
Epilobium ciliatum Raf.	UK (12)
Kochia scoparia L. Schrad.	USA (5, 68)
Lolium rigidum Gaud.	AST (60), ISR (89)
Matricaria matricariodes (Less.) Port.	EUR (44)
Panicum capillare L.	CAN (5, 74), USA (5)
Phalaris paradoxa L.	ISR (89)
Poa annua L.	BEL (10). EUR (44), FRA (27), POL (46), USA (5)
Polygonum convolvulus L.	AUS (27)
P. lapathifolium L.	CZC (54), FRA (27)
P. persicaria L.	FRA (27)
P. monspeliensis (L.) Desf.	ISR (44)
Senecio vulgaris L.	CAN (5, 74), FRA (27), USA (5), UK (27)
Setaria faberi Hermann	FRA (27), USA (5, 65)
S. glauca (L.) Beauv.	SPN (18), USA (44)

(continued)

Table 1 (*Continued*)

Taxa	Location^{+} (Reference)
S. pumila (Poir.) Rets.	FRA (27)
S. viridis (L.) Beauv.	FRA (27), USA (5)
Sicyos angulatus L.	USA (44)
Sinapis arvensis L.	CAN (74)
Solanum nigrum L.	BEL (11), FRA (27), GER (27), UK (12)
Sonchus asper (L.) Hill	FRA (27)
Stellaria media (L.) Vill.	EUR (44), GER (27)

$^{+}$ AUS: Austria; AST: Australia; CAN: Canada; CZC: Czechoslovakia; EGY: Egypt; EUR: Europe; FRA: France; GER: Germany; HUN: Hungary; ISR: Israel; NZD: New Zealand; POL: Poland; SPN: Spain; UK: United Kingdom; USA: United States

for studying adaptation and evolution because herbicide selection is one of the rare selection pressures that is easy to study within a specific environment. This paper reviews the extent of herbicide resistance in plants, mechanisms of their resistance, and the direct and indirect physiological effects associated with resistance. In addition, aspects of their population biology are evaluated, including factors affecting the origins, establishment, and spread of herbicide resistance in natural populations, the status of population models simulating the evolution of herbicide resistance, and the ecological implications of the release of herbicide resistant crops. It concludes with a summary of future research directions.

Triazines

Over the past 20 years, numerous reports of triazine-resistant weeds in widely separated locations have made triazine resistance the best known and most studied class of herbicide resistance. Fifty-five weed species, including 40 dicots and 15 grasses, representing 35 genera, are known to have triazine R-biotypes (Table 1, see also 5, 27, 38, 44). One or more resistant species have arisen in 31 states in the United States, four provinces of Canada, 18 countries in Europe, and in Israel, Japan, Australia, and New Zealand.

Nontriazine Herbicides

Resistance to nontriazine herbicides has been more recently detected; it has been restricted in distribution but is now becoming more widespread (5, 27, 38, 44). To date, resistance has been documented to an additional 15 herbicide families, most frequently to bipyridyliums and sulfonylureas. Bipyridyliums (e.g. paraquat) resistant biotypes (reviewed in 22, 27, 39) have been reported in ten weedy species: *Arctotheca calendula* (L.) Levyns, *Hordeum glaucum* Steud., and *H. leporinum* Link in Australia (60); *Conyza (Erigeron) bonariensis* in Egypt (27) and Hungary (39); *C. canadensis* in Hungary (59) and Japan (39); *C. philadelphicus* (L.) Cronq., *C. sumatrensis* (Retz.) Walker, and *Youngia japonica* (L.) DC. in Japan (39); *Poa annua* in the United Kingdom (12); and *Epilobium ciliatum* in Belgium (39) and the United Kingdom (12). Resistance to the substituted ureas is present in the United Kingdom, West Germany, and Hungary (38, 42). Since 1987, biotypes of at least six weedy species resistant to the sulfonylurea herbicides have arisen in the United States, Canada, Australia, and Costa Rica (44, 49). In North America these include: *Lactuca serriola* L. in winter wheat fields in Idaho (49); *Kochia scoparia,* first identified from Kansas wheat fields and now found in other locations in the United States (49, 66); and *Salsola iberica* Sennen & Pau in the United States (49); and *Stellaria media* from western Canada (33). Dinitroanilines (e.g. trifluralin) resistance, first reported in *Eleusine indica* (L.) Gaertner from cotton fields in South Carolina in 1984, now occurs widely in the southeastern United States (56, 79), and more recently was confirmed in *Setaria viridis* from both cereal and oilseed crops in western Canada (55). Diclofop-methyl resistant weed species, including *Lolium rigidum* and *Avena fatua* L. (60), are problems in cereal production in Australia (60) and have also been found in Oregon, South Africa, and the United Kingdom (38). Resistance to phenoxy herbicides (e.g. mecoprop) is rarely reported, e.g. *Tripleurospermum inodorum* L. (27) and *Stellaria media* (48) in the United Kingdom, and *Ranunculus acris* L. and *Carduus nutans* L. in New Zealand (9).

Multiple herbicide resistance, i.e. the evolution of populations resistant to more than one class of chemically unrelated herbicides with different modes of action, has been documented relatively recently. Examples include: *Alopecurus myosuroides* Huds., resistant to the phenylurea chlortoluron and diclofop-methyl, in several sites in the United Kingdom (42); and *Lolium rigidum,* resistant to diclofop-methyl with cross-resistance to sulfonylurea and dinitroaniline, in crops and pastures throughout the cropping zones of southern Australia (60). A new triple-resistant form of *Chenopodium album* in Hungary has evolved in response to postemergence herbicide applications of pyridate and pyrazon to existing triazine R-populations (71).

MECHANISMS OF RESISTANCE AND PHYSIOLOGICAL EFFECTS

Herbicide resistance in weeds is most often due to an alteration of the site of action of the herbicide at the cellular level. In contrast, herbicide tolerance, i.e. to low herbicide doses, is usually the result either of differences in herbicide uptake and translocation at the plant level or of differences in plant metabolism and herbicide detoxification.

Triazines

Triazine herbicides are strong inhibitors of photosynthesis; and resistance in most weedy species results from a loss of herbicide binding ability. This is due to an alteration of the binding site on the thylakoid membrane of the chloroplast, i.e. at a 32-kd protein (Q_B = D_1 protein) of photosystem II (4). Site alteration results from a point mutation in the chloroplast *psbA* gene that codes for this protein, resulting in a single base change: serine to glycine at amino acid position 264 (6, 7, 35). As a result of this mutation photosynthetic electron transport is markedly (1000-fold) slower in resistant as compared to wild-type chloroplasts.

Triazine resistance mechanisms involving herbicide detoxification through enhanced metabolism have been reported in a few cases, e.g. *Abutilon theophrasti* from the United States (32) and *Echinochloa crus-galli* from France (44). In *Brachypodium distachyon* from Israel (29), resistance is apparently achieved both by enhanced metabolism and by alteration of the site of action at the plastid level.

A variety of physiological, biochemical, and anatomical conditions are associated with the evolution of triazine resistance (reviewed in 37, 64). Numerous data support the possibility of reduced efficiency of the photosynthetic light reactions in triazine R-biotypes, providing evidence for lower photosynthetic rates (36, 77), reduced levels of CO_2 fixation, and/or lower rates of photosystem II (PSII) electron transport (1, 36, 37, 53, 57, 64, 77). Several triazine R-biotypes are reported to have a shade-adapted chloroplast ultrastructure (47, 78, 80), including an association of a greater portion of the chlorophyll with the light-harvesting chlorophyll a/b proteins, a lower chlorophyll a/b ratio, and an increased volume of grana lamellae and greater lamellae stacking. Consistent with these ultrastructural changes, differential temporal organization of photosynthetic function has been demonstrated in R- and S-biotypes of *Brassica napus* (17). The R-biotype was better adapted to the low illuminance regimes of the early and late portions of the light-period in the diurnal, while the S-biotype had maximum photosynthetic function at peak illuminance periods. Several studies have concluded that the thylakoid membranes of triazine R-biotypes are richer in unsaturated fatty acids (45, 47, 58, 76), the presence of which has been implicated in enhanced resistance to low-temperature stress. These indirect physiological effects may give the R-biotype an adaptive advantage in adverse environments (17, 47, 58).

As is discussed later, triazine R-biotypes are generally less productive than the S-biotypes. The mechanism by which the productivity of R-biotypes is reduced may be a combination of the direct effects of altered photosystem II functioning and the indirect or compensatory physiological effects described above (37). For example, the synthesis of increased PSII reaction centers (47) and modification in thylakoid membrane lipid composition may represent a diversion of carbon resources from other areas of growth and may partially explain reduced productivity of R-biotypes.

Nontriazine Herbicides

Fewer data are available on the mechanisms and physiological effects of resistance to nontriazine herbicides, including multiple herbicide resistance, but appreciable information has been accumulated as is noted in the following. Dinitroaniline herbicides inhibit the formation of microtubules and thereby block mitosis in S-plants. Resistance in *Eleusine indica* is conferred by an altered form of tubulin that results in microtubule insensitivity to the dinitroanilines (79). Mechanisms of bipyridylium resistance include either the rapid sequestration of the herbicide resulting in reduced herbicide levels at the site of action in the chloroplast, and/or the rapid enzymatic detoxification of superoxide and other toxic forms of oxygen due to elevated levels of superoxide dismutase (22). Sulfonylurea herbicides inhibit acetolactate synthase, the first enzyme specific to the branched-chain amino acid biosynthetic pathway. Resistance is due to an altered site of action, accomplished by production of a form of acetolactate synthase that is insensitive to inhibition (33, 67). In a way that is similar to triazine resistance, single amino acid substitutions in acetolactate synthase confer resistance (90). Phenoxy herbicide resistance in *Stellaria media* appears to be associated with some form of compartmentalization or herbicide detoxification at the site of action (48). In contrast, rather than an altered site of action, the most likely mechanism of multiple herbicide resistance is a common enhanced metabolism resulting in detoxification or degradation of several herbicide classes (42, 60). Studies suggest an elevated activity of the plant's mono-oxygenase system (60).

POPULATION BIOLOGY

Evolution of Herbicide Resistance in Natural Populations

Several factors govern the evolution of herbicide resistance in plants (30, 31, 52, 62), including: intensity of selection imposed by herbicides and effects of such agronomic practices as crop rotation and cultivation; generation time of plants and soil seed reserve and seed carryover; genetic factors, such as levels of genetic variation to herbicide response, initial mutation frequencies, mode of inheritance of resistance, gene flow or genetic exchange with S-populations; relative ecological fitness of R- and S-biotypes, and their response to density dependent and independent factors. Although progress has been made, much more basic research is needed on the relative contribution of each of these components. Little is known concerning levels of genetic variation of herbicide resistance in natural plant populations, particularly the occurrence and spread of R-plants before any selection due to herbicides, field mutation rates, and the establishment and spread of resistance within populations.

SELECTION PRESSURE OF HERBICIDES AND OTHER AGRONOMIC PRACTICES

Characteristics of herbicide use that would increase selection pressure and the probability of the evolution of herbicide resistance, include herbicides with: a single target and specific mode of action, increased activity and effectiveness in killing a wide range of weed species, long soil residual and season-long control of germinating weeds, and those which are applied frequently, over several growing seasons of the weed population without rotating, alternating, or combining with other types of herbicides (44). Indeed, the majority of herbicides for which there are documented cases of resistance fill these requirements. Nonpersistent herbicides will usually exert a low selection pressure on annual species, since both R- and S- seedlings can establish after spraying in the same year, slowing the build-up of resistance. They will have more effect on perennials, characterized by infrequent seedling establishment (9). Preemergence herbicides would favor the R-biotypes, since S-seedlings are killed and the competition with R-biotype prevented (9).

Few field data are available on the selective effects of other agronomic practices, such as crop rotation, herbicide mixtures, post- versus pre- emergence herbicide applications, inter-row cultivation, crop use as silage and extent of manure return to the soil, on R-populations. Differences in such practices are often considered predominant forces influencing the regional distribution and spread of resistant weeds, as described for example in Ontario, Canada (74). Studies of triazine resistant *Amaranthus retroflexus* in Hungary (70) showed no significant differences between the proportion of R-plants in 7 to 20 yr maize monocultures (c. 81%), compared with a maximum of only 25% for crops grown in rotation over a 7-yr period. Studies of the behavior of seedling and adult plants in a mixed simazine R- and S-population of *Senecio vulgaris* in a fruit farm in the United Kingdom (63, 85) revealed the influence of alternative weed management practices on the phenology and life history of plants. Seasonal variation in selection pressure was associated with herbicide degradation in the soil.

INHERITANCE, MUTATION RATES, AND FREQUENCY OF RESISTANT GENES

The frequency of mutations conferring herbicide resistance depends on several genetic factors (20), including: gene number (mono- versus polygenic traits, ploidy level), inheritance mechanism, gene type, and nature of modification to the gene.

Information on the mode of inheritance of herbicide resistance is available for triazine and paraquat resistance but is lacking for other cases of single and multiple herbicide resistance. Triazine resistance, encoded in the chloroplast genome, is usually maternally inherited (reviewed in 73), although nuclear DNA involvement may also have to be considered for *Poa annua* and *Solanum nigrum,* where inheritance is mainly maternal with some paternal transfer (13, 73). In contrast, genetic studies of *Abutilon theophrasti* indicated that triazine resistance was controlled by a single, partially dominant nuclear gene (3). Inheritance of paraquat resistance (22, 39) is known to be controlled by a single dominant nuclear gene in *Conyza bonariensis* and *Erigeron philadelphicus,* and a semi-dominant nuclear gene in *Hordeum glaucum*. In contrast, inheritance of multiple herbicide resistance may occur as a linked or pleiotropic trait or as a secondarily evolved trait (44).

With the exception of studies of triazine-resistant *Chenopodium album* in France, little is known about mutation rates and frequency of resistant genes in field populations of weedy species. Initially it was assumed that naturally occurring genetic variation or allelic variation for herbicide resistance is likely present in populations of the weed before extensive herbicide selection. The extended use of the same herbicide would favor R- plants and result in the selective build-up of these genotypes over time. The proposed frequency levels of such resistant alleles were similar to other naturally occurring nuclear-inherited mutations, ranging from 10^{-5} for dominant alleles to 10^{-11} for recessively controlled traits, and slightly lower for plastid inherited traits (30). However, the frequency of triazine resistant weeds led Duesing & Arntzen (20) to predict (i) the likely presence of a mutator gene in triazine R-populations that increased the mutation rate of chloroplast DNA mutations, and (ii) that an enrichment for triazine resistance would carry enrichment for resistance to other photosystem II–inhibiting herbicides, as in the newly discovered co-resistant biotype of *Chenopodium album* in Hungary (71).

Studies by French researchers (7, 14, 15, 23) have provided the first evidence that the appearance of triazine resistance in *Chenopodium album* is not a purely random event. The evolution of R-populations appears to be a two-step process (Sp-I-R pathway). First is the required presence of "resistant-precursor" genotypes (Sp genotype) in natural populations. These have a higher than normal random mutation rate (i.e. 10^{-4} to 3×10^{-3}) for the chloroplast *psbA* gene and have the same isozyme phenotype as R-plants from the same region. The Sp genotype produces an intermediate (I)-biotype which in the presence of herbicides produces progeny that show typical resistance (R-plants). Intermediacy was also maternally inherited and was due to the same chloroplast *psbA* gene mutation as typical R-plants. The mechanism for this genetic change of I to R is not yet clear, but preliminary studies would suggest an extrachloroplastic mechanism or nuclear control. Further studies are required to determine whether this mutational mechanism is true for other triazine-resistant species. Similar I-biotypes of *Chenopodium polyspermum* and *Amaranthus bouchonii* were described in Hungary (72). Their occurrence also suggested nonrandom mutational events, since one enzyme phenotype represented more than 50% of the I-plants.

Data on field mutation rates for resistance to nontriazine herbicides are not available. Random mutation may account for resistance in species that show only one or a few R-populations. A dinitroaniline resistant I-biotype of *Eleusine indica* (50-fold) was also found under continuous cotton production in South Carolina with the highly R-biotype (100–10,000 fold). Unlike the triazine-resistant I-biotypes, the I-biotype of *E. indica* would appear to have a unique mechanism of resistance from the R-biotype (79).

Very little is known about genetic variability for herbicide response in field populations of weedy species. Some studies suggest that selection may actually favor genes conferring herbicide tolerance in natural populations. A study (61) of Californian populations of *Avena barbata* Brott., *A. fatua,* and *Clarkia williamsonii* Lewis & Lewis, not previously exposed to the herbicides barban (for *Avena* ssp.) and bromoxynil (for *Clarkia*), indicated significant amounts of inter- and intra-population variability for herbicide reaction. The amount of genetic variance for herbicide reaction in the *Avena* ssp. was also higher than expected on the basis of random mutation alone. Studies of intra-specific variation in susceptibility to simazine in 46 populations of *Senecio vulgaris* from fruit farms in the United Kingdom (63) indicated significant genetically based population differences in susceptibility. There was a positive linear relationship between percent survival of a population and the number of consecutive years of simazine application. Studies of *Avena fatua* populations in Canada (75) also demonstrated significant differences in variability for triallate reaction within and among exposed and unexposed populations. Within population variation was higher in unexposed than in exposed populations. Populations with recurrent exposure to triallate were more tolerant than unexposed populations.

POPULATION STRUCTURE, GENETIC VARIATION, AND GENE FLOW Population response to selection depends on population structure. The more heterogeneous the population, the more likely that it can adapt to new agricultural stresses. Populations of agricultural weeds, in general, tend to exhibit lower levels of genetic variation compared with other plant species (81). In addition, populations of the same weedy species from ruderal habitats are likely to be more variable than those from cultivated fields (e.g. 2, 81).

The genetic structure of triazine R-populations varies among resistant weed species studied (13, 83, 84). Loss of isozyme polymorphism was very marked in the triazine R-populations of *Chenopodium album* in Canada (83, 84), France (3, 24), and Hungary (71). These were always monomorphic, suggesting founder or bottleneck effects. In all three countries, surveys revealed regional differences, i.e. the same isozyme phenotype was found in R-populations in the same area with different phenotypes in different regions, indicating multiple origins of resistance. In contrast, of the R-populations of *Poa annua* from roadsides in France (13), *Senecio vulgaris* from fruit plantations in the United Kingdom (83), and *Amaranthus retroflexus* (83) and *Brassica rapa* (S. I. Warwick, paper in preparation) from cultivated fields in Canada, none were completely homogeneous, although all were significantly less polymorphic than the S-populations. Reduced variability of the R-populations compared to S-populations was also evident in growth characteristics (84, 88).

Greater genetic variation is expected in R-populations occurring in more heterogeneous habitats, as for example in *Poa annua* from roadsides, and in habitats where herbicide selection pressure may be more intermittent, e.g. in *Senecio vulgaris* in fruit plantations. The mating system and differences in the degree of spatial and temporal isolation of R- and S- populations will also contribute to species differences in genetic structure (13, 83). Crop density appears to impede gene flow or exchange of pollen between plants in cultivated field populations and any S-plants outside the crop. Studies indicate that pollen flow and hybridization between individuals of *C. album* growing in a maize field is one tenth that observed in a pure stand of the weed (14). Greater outcrossing is expected in plants growing in more open roadside populations, e.g. *Poa annua,* or in an obligate outcrosser such as *Brassica rapa.*

To date no studies of isozyme variation have been conducted on species resistant to nontriazine herbicides. However, dinitroaniline R- and S-populations of *Eleusine indica* generally exhibited similar ranges of variability in growth characteristics (56).

ECOLOGICAL FITNESS Fitness measures describe the potential evolutionary success of a genotype. It may be defined as the reproductive success or the proportion of genes an individual leaves in the gene pool of a population, with the most fit leaving the greatest number of offspring. Fitness differences between R- and S-biotypes are usually inferred from measures of relative plant productivity and/or competitiveness. The earliest fitness studies of herbicide R- and S-biotypes were conducted with only single populations, these often originating from widely separated areas. Later studies conducted more critical tests, including estimates of both inter- and intra-population variability of each biotype and effects of geographical origin (56, 84, 88). Recent studies include comparisons of biotypes of similar genetic background, either from the same field population or as isogenic lines. Fitness studies have usually restricted measurements to vegetative and reproductive growth, while recent population models (see below) emphasize evaluation for a number of life-history stages: dormancy, germination, establishment, survival, growth, pollination, and seed production.

Triazines Triazine R-plants are generally less fit than S-plants (reviews 37, 64). One would expect the frequency of R-biotypes to decrease in the absence of herbicide treatments. Reduced productivity and/or competitiveness of the R-biotypes are reported for several species, including: *Amaranthus hybridus* (1), *A. powellii* (64, 87, 88), *A. retroflexus* (87, 88), *Brassica rapa* L. (50), *Chenopodium album* (11, 71, 82, 84), *Poa annua* (10), *Senecio vulgaris* (37, 64), and *Solanum nigrum* (11). The correlation of reduced plant vigor with triazine resistance has also been observed in engineered triazine-resistant crops, for example 20–30% yield reductions have been observed in resistant *Brassica* cultivars (8) and in resistant *Setaria italica* (L.) Beauv. (16).

The use of nuclear-isogenic lines allows one to compare the effects of the chloroplast mutation separately from other genetic differences. Similar yield reductions in the R-biotype were evident in triazine R- and S-cultivars of *Brassica napus* when examined in either a similar (28) or common nuclear-genetic background (8). Isogenic studies of triazine resistant *Senecio vulgaris* conducted on F_1 hybrids and backcrossed plants (37, 53) also supported the reduced productivity of R-biotypes; the actual percent reduction in yield was dependent on the nuclear genome. Relative fitness studies of isogenic lines of intermediate (I) and R-biotypes of *C. album* (14) indicated that both I and R–plants flowered later and produced fewer seed than did the S-biotype. However, I plants at the seedling stage grew more vigorously than either S- or R-plants.

Not all triazine R-biotypes demonstrated reduced vigor and competitiveness. In *Chenopodium strictum,* a late-flowering, slower-growing species (possibly not limited by photosynthesis), R- and S-biotypes produced similar amounts of biomass in both competitive and noncompetitive conditions (82). One study of *C. album* reported similar rates of photosynthetic activity for the two biotypes and that the R-biotype was a better competitor than the S-biotype (41). Net photosynthesis was also reported to be similar in R- and S-biotypes of *C. album, Poa annua, Polygonum lapthifolium, Solanum nigrum* and *Stellaria media* (77). The R-biotype of *Phalaris paradoxa* was either equal or superior to the S-biotype in photosynthetic potential and growth under non-competitive conditions (69). An increase in the light saturated electron transport rate of the latter R-biotype appeared to compensate for the reduction of the quantum yield of electron transport, which is characteristic of triazine R-biotypes, and may explain the similarity between biotypes in photosynthetic potential and growth (69).

There is increasing evidence that the relative productivity of S- and R-biotypes of a single species may depend upon environmental conditions, including temperature and light quality. (See also physiological section above). Results of field trials testing differential growth of R- and S-biotypes to light vary. For example, under noncompetitive conditions at 100%, 40%, and 10% light levels, the growth rate of *Amaranthus hybridus* at 10% light did not differ between biotypes, while at the two higher light intensities, biomass of the R-biotype was 40% less when compared to the S-biotype (1). In contrast, the R-biotypes of *Brassica* spp. were always less productive than the S-biotypes regardless of light quality conditions (36).

Reports also vary with respect to differential biotype response to temperature. An increased heat and pH sensitivity of the photosynthetic apparatus was evident in several triazine R-biotypes (including isogenic nuclear lines of resistant *S. nigrum*), suggesting a greater instability of the oxygen evolving system in R-plants (19). Several studies have suggested less of a competitive disadvantage of triazine R-biotypes at lower temperatures. Thus, triazine R-cultivars of *Setaria italica* indicated no biotype differences at 17°C compared to 27°C, at which the usual differential between biotypes was observed (16). Similarly, for *Polygonum lapthifolium,* the 20–50% growth differential of R- and S-biotypes in total biomass observed under warm greenhouse conditions was reduced to 4% under cool conditions (25). The latter result was consistent with enhanced low temperature germination in R-biotypes of *P. lapthifolium* and *A. retroflexus* observed in the same study. In contrast, studies of *Amaranthus hybridus* indicated that the S-biotypes were more vigorous at lower temperatures than were R-biotypes (80). Other studies have indicated no correlation between differential biotype productivity and temperature, for example, in R-biotypes of *Chenopodium album* from different geographical origins (80). Similarly, photosynthetic differences between triazine R- and S-biotypes of *Brassica* spp. (36) and the relative growth of R- and S-biotypes of *Solanum nigrum* (40) (the latter evaluated in a common nuclear genetic background by parental lines and reciprocal F_1 crosses) were consistent across the range of temperatures tested. Additional studies are clearly needed on a wide range of species of contrasting growth form and geographical origin to test R- and S-biotype response to light, temperature, and other environmental conditions.

Nontriazine herbicides Compared to that on triazine-resistant species, few data are available on the relative fitness of plants resistant to nontriazine herbicides. Definitive fitness and competition experiments are required to produce reliable population models. Preliminary results suggested that the differences in fitness between sulfonyl-urea R- and S-biotypes of *Kochia scoparia* were less than that observed for triazine-resistant weeds (67). Growth of paraquat S-biotypes of *Erigeron canadensis* and *Hordeum glaucum* was more vigorous than the R-biotype under noncompetitive conditions in the absence of paraquat (39). Reduced fitness of diclofop-methyl R-biotypes of *Lolium rigidum* was also reported, where the relative fitness of the R-biotype was 0.81 when grown in pure stands and 0.65 when grown in mixed stands (60). Fitness studies of several populations of dinitroaniline R- and S-biotypes of *Eleusine indica* under noncompetitive conditions (56) have indicated no significant differences between biotypes in most growth and development characteristics, with the exception of significantly greater inflorescence weight in the S-biotype. Additional competition studies of the above species (reviewed in 37) also indicated a lower reproductive weight in the R-biotype, and that the R-biotype was less competitive than S-plants, responding to competition by reduced reproductive output.

ESTABLISHMENT AND SPREAD OF RESISTANT PLANTS There are few field studies which attempt to estimate the chance of establishment and spread of herbicide-resistant mutants within plant populations (14, 63, 85). In most instances with triazines, resistance appeared after 7 or more yr (5, 27), with dinitroanilines after 10 yr (79), with paraquat after 5 or more yr (22), while sulfonylurea resistance has appeared after only 3 to 5 yr (49). A field study simulating establishment from a single founding R-plant of *C. album* grown with maize and sprayed once a year with atrazine was followed over a 4-yr period, and effects of density-dependent regulation and cultivation monitored (14). The observed rate of population buildup could not account for the heavy and rapid infestation rates characteristic of R-populations, suggesting that numerous R-plants must have been present before the first triazine treatment. The lack of triazine resistance in certain areas is also equally puzzling and may reflect the absence of resistant precursor genotypes in these populations (14) and/or differences in agronomic practices (74).

The presence of a seed bank would contribute to the delay in build-up of herbicide resistance by keeping S-plants in a population (9, 30). Little is known about seed bank dynamics in resistant field populations. Reports of differences in dormancy between R- and S- biotypes vary depending on species and growth conditions. Some studies have shown that the R-biotype has greater dormancy and/or later germination than S-biotypes (10, 11, 50, 86). Others suggest no correlation in dormancy with resistance (11, 86) or report variable results depending on conditions (25). Recent studies (85) have described variation patterns in the seed bank of R- and S-biotypes of *Senecio vulgaris* under three management practices: simazine treated, rototilling to a depth of 15 cm, and an untreated control. Two seed banks were identified, each having a different effect on biotype maintenance. Achenes of the R-biotype had greater longevity than the S-biotype in the lower seedbank (> 2 cm depth), while in the surface bank (0–2 cm) biotypes showed different longevity according to management practice, i.e. substantial net gain to the achene seed bank by the R-biotype with simazine treatment and by the S-biotype under rototilling. Cultivation practices that place achenes at greater depths will result in a depletion of the S-biotype at faster rates than the R-biotype. In the surface seed bank, the fate of the adult plants and their relative seed yield determined the relative success of the two biotypes.

Interspecific competition with the crop and associated weed species may be important in determining the greater spread of some and not other R-biotypes. For example, within the two weedy genera *Chenopodium* and *Amaranthus* in Ontario, triazine R-populations of *C. album* and *A. powellii* have continued to increase in frequency and abundance, while *C. strictum* and *A. retroflexus* have not spread from the original resistant sites (74). Ecological studies have shown that both S- and R-biotypes of *C. album* are more competitive than *C. strictum* (82) and also that *A. powellii* is more competitive than *A. retroflexus* (87). Additional interspecific studies are required where two or more R-taxa co-exist and follow-up studies are required on their relative spread.

Population Models for Herbicide Resistance

In 1978, Gressel & Segel (30) developed the first simulation model to predict the rate at which populations could be expected to develop herbicide resistance. Their model integrated the following factors: (i) the selection pressure of the herbicide (based on the rate used, its effectiveness with particular weeds, and its persistence); (ii) the germination dynamics of the weeds (over the season and from the soil seedbank); (iii) the initial frequency of R-plants deriving from natural mutations in the S-populations; (iv) the fitness of the evolved R-biotypes in competition with the wild type under field conditions; and (v) the number of generations (seasons) the herbicide was used. At the time, in the absence of field ecological data, the biotype fitness differential was incorrectly considered to be of minor importance. It was concluded that only at high selection pressure should resistance appear within 10 years, and that resistance would develop slowly in response to nonpersistent herbicides.

More complex and comprehensive simulation models have been developed recently (30), including an expansion of this early model (31) and two new models (52, 62). These resistance models include the influences of additional gene flow and fitness factors and suggest alternative management options for dealing with herbicide resistance.

Contrary to predictions from their original model, no populations of triazine-resistant weeds have appeared in maize where rotations of crops and herbicide mixtures were used. Gressel & Segel revised their model (31) to include the fitness differences of R- and S-biotypes and their effect when herbicide mixtures and rotations are used. The model describes how these factors would reduce the resistant individuals to extremely low frequencies during rotation, since reduced competitive fitness of the R-biotype would only be expressed during rotational cycles. It also suggests that rotations or mixtures are not likely to delay the rate of the appearance of resistance to herbicides where there are no fitness differences between biotypes.

The recent model by Putwain & Mortimer (62) integrates the effects of both density-dependent and independent forces which regulate the population sizes of biotypes in mixtures. It includes the relative ecological fitnesses of both mature plants (seed yield) and seeds. The model equation describes changes over generations for R- and S-biotypes of the same species, with an optional extension to include the competitive influence of a second biotype or species. It applies to mixtures of competing R- and S-biotypes where resistance alleles do not segregate as in triazine resistance. The simulation model has been found to adequately describe rates of change in populations of annual weeds under a range of cropping practices and to qualitatively predict relative abundance (62).

The model by Maxwell et al (52) incorporates plant population demographics with the Hardy-Weinberg concept for gene segregation. It is designed to simulate the evolution, spread, and subsequent dynamics of resistance in the presence and absence of a herbicide. The model identifies and incorporates two key biological factors: gene flow processes of seed and pollen immigration and processes that influence ecological fitness of R-biotypes relative to the S-biotype and to the crop. The simulation model organizes the life-history of a weed population into seed, seedling, mature plant, and pollen stages, and defines the fitness and gene flow processes that govern the transitions between each stage. The model can be used to evaluate the potential influences of inheritance mechanisms, mating system, and size and distance from a susceptible source.

More field data are needed to test the validity and predictive value of these proposed models. Future models will have to consider evidence for nonrandom mutation frequencies of resistant genes in populations (14). They will also have to incorporate the potential effects of negative cross-resistance factors, i.e. those factors contributing to the greater sensitivity of R-plants as compared to S-plants (31). For example, certain alternative herbicides or herbicide mixtures, pests, and control practices appear to have a more severe effect on the R-biotype (reviewed in 31).

Ecological Implications of Herbicide Resistant Crops

The agronomic, environmental, and economic implications of herbicide-resistant crops have been extensively considered (e.g. 15, 21, 51). Two potential problems with the use of resistant crops exist. The first deals with the potential transfer of herbicide resistance via pollen from a resistant crop to formerly susceptible weedy relatives, as well as the potential establishment of weeds as escapes from a resistant crop. The potential for exchange with wild and weedy relatives is not only possible but likely (15, 21), particularly if the trait confers an advantage to the wild species. Recent studies of pollen exchange between neighboring populations have shown that interpopulation gene flow for many plant species can proceed over greater distances (1000 m) and at higher rates than previously suggested. Knowledge of breeding systems and the potential for hybridization between crop and adjacent weed populations is important, and much more field data is required. For example, potential crossing is high between triazine resistant *Brassica* spp. and other mustard weeds, such as *Sinapis arvensis* (a species with known triazine R-biotype). Studies of spontaneous hybridization between triazine resistant *Setaria italica* and *S. viridis* were estimated at 0.2% which, given the high seed yield, suggests a high probability that resistant hybrid seed could be released (15). The development of containment strategies, such as extra modifications for increased self-fertilization and decreased pollen longevity in engineered crop plants, may be necessary to safeguard against such escape (21). The second potential problem is the selection of resistant weeds with an increased risk of developing cross-resistance to more than one herbicide, leading to the need to apply new herbicides to the weedy populations (e.g. 71).

CONCLUSIONS AND FUTURE RESEARCH DIRECTIONS

There is, in general, a need for more population biology/population genetic studies in agricultural weeds (81). The increasing and widespread occurrence of herbicide-resistant weeds has reached the level where more extensive basic research is needed to develop effective management strategies. After much competent research, a sound framework of knowledge now exists on the topic of triazine resistance. A clear need exists for similar studies on nontriazine herbicides and multiple herbicide resistance, including research on the mechanisms and physiological effects of resistance. Field genetic studies are required that evaluate the magnitude of genetic variability for herbicide response, the capacity of weed populations to respond to herbicide selection pressure, and the effects of the plant's mating system. Knowledge of the modes of inheritance of nontriazine resistance are lacking, as are processes for the origins and evolution of mutations with resistance to herbicides in natural populations. More field ecological data are needed for all cases of herbicide resistance, including the effects of other agronomic practices on the establishment and spread of resistant weed populations. To understand why some resistant species spread while others do not, fitness data are required for the entire life cycle of R- and S-biotypes, including measures of relative fitness with the crop and associated weed species. It is not clear why most, but not all, triazine R-biotypes have reduced fitness compared to the S-biotype, and whether this is true for different environmental conditions or for nontriazine herbicides. The basic information described above will be useful in evaluating the ecological and environmental consequences of releasing herbicide resistant crops.

ACKNOWLEDGMENTS

I wish to thank L. Black for her help in searching the literature and preparation of the final copy; and for their helpful suggestion on revising the manuscript: Mr. C. Crompton and Dr. E. Small, Biosystematics Research Centre, and Dr. S. Weaver, Agriculture Canada, Harrow Research Station, Ontario.

Literature Cited

1. Ahrens, W. H., Stoller, E. W. 1983. Competition, growth rate, and CO_2 fixation in triazine-susceptible and -resistant smooth pigweed *(Amaranthus hybridus). Weed Sci.* 31:438–44
2. Al Mouemar, A., Gasquez, J. 1983. Environmental conditions and isozyme polymorphism in *Chenopodium album* L., *Weed Res.* 23:141–49
3. Andersen, R. N. 1987. Noncytoplasmic inheritance of atrazine tolerance in velvetleaf *(Abutilon theophrasti). Weed Sci.* 36:496–98
4. Arntzen, C. J., Pfister, K., Steinback, K. E. 1982. The mechanism of chloroplast triazine resistance: alterations in the site of herbicide action. See Ref. 43, pp. 185–214
5. Bandeen, J. D., Stephenson, G. R., Cowett, E. R. 1982. Discovery and distribution of herbicide-resistant weeds in North America. See Ref. 43, pp. 9–30
6. Barros, M. D. C., Dyer, T. A. 1988. Atrazine resistance in the grass *Poa annua* is due to a single base change in the chloroplast gene for the D_1 protein of photosystem II. *Theor. Appl. Genet.* 75:610–16
7. Bettini, P., McNally, S., Sevignac, M., Darmency, H., Gasquez J., et al. 1987. Atrazine resistance in *Chenopodium album:* low and high levels of resistance to the herbicide are related to the same chloroplast psbA gene mutation. *Plant Physiol.* 84:1442–46
8. Beversdorf, W. D., Hume, D. J., Donnelly-Vanderloo, M. J. 1988. Agronomic performance of triazine-resistant and susceptible reciprocal spring canola hybrids. *Crop Sci.* 28:932–34
9. Bourdot, G. W., Harrington, K. C., Popay, A. I. 1989. The appearance of phenoxy-herbicide resistance in New Zealand pasture weeds. *Proc. Br. Crop. Prot. Conf., Weeds,* pp. 309–16. Surrey, England
10. Bulcke, R., De Praeter, H., Van Himme, M., Stryckers, J. 1984. Resistance of annual meadow-grass, *Poa annua* L., to 2-chloro-1,3,5-triazines. *Meded. Rijksfac. Landbouwwet. Gent* 46:1041–50
11. Bulcke, R., De Vleeschauwer, J., Vercruysse, J., Stryckers, J. 1985. Comparison between triazine-resistant and -susceptible biotypes of *Chenopodium album* L. and *Solanum nigrum* L. *Meded. Rijksfac. Landbouwwet. Gent* 47:211–20
12. Clay, D. V. 1989. New developments in triazine and paraquat resistance and coresistance in weed species in England. *Proc. Br. Crop Prot. Conf. Weeds,* pp. 317–24. Surrey, England
13. Darmency, H., Gasquez, J. 1983. Interpreting the evolution of a triazine resistant population of *Poa annua* L., *New Phytol.* 95:299–304
14. Darmency, H., Gasquez, J. 1990. Appearance and spread of triazine resistance in common lambsquarters *(Chenopodium album). Weed Technol.* 4:173–77
15. Darmency, H., Gasquez, J. 1990. Fate of herbicide resistance genes in weeds. See Ref. 26, pp. 353–63
16. Darmency, H., Pernes, J. 1989. Agronomic performance of a triazine resistant foxtail millet (*Setaria italica* (L.) Beauv.). *Weed Res.* 29:147–50
17. Dekker, J., Westfall, B. 1987. A temporal phase mutation of chlorophyll fluorescence in triazine-resistant *Brassica napus. Z. Naturforsch.* 42c:775–78
18. De Prado, R., Dominguez, C., Tena, M. 1989. Characterization of triazine-resistant biotypes of common lambs quarters *(Chenopodium album),* hairy

fleabane *(Conyza bonariensis),* and yellow foxtail *(Setaria glauca)* found in Spain. *Weed Sci.* 37:1–4

19. Ducruet, J. M., Ort, D. R. 1988. Enhanced susceptibility of photosynthesis to high leaf temperature in triazine-resistant *Solanum nigrum* L., Evidence for photosystem II D-1 protein site of action. *Plant Sci.* 56:39–48
20. Duesing, J. 1983. Genetic analysis of herbicide resistance. *Proc. North Cent. Weed Control Conf.*, pp. 143–47. Columbus, Ohio
21. Ellstrand, N. C., Hoffman, C. A. 1990. Hybridization as an avenue of escape for engineered genes. *Bioscience* 40:438–42
22. Fuerst, E. P., Vaughn, K. C. 1990. Mechanisms of paraquat resistance. *Weed Technol.* 4:150–56
23. Gasquez, J., AlMouemar, A., Darmency, H. 1985. Triazine herbicide resistance in *Chenopodium album* L.: Occurrence and characteristics of an intermediate biotype. *Pesticide Sci.* 16:392–96
24. Gasquez, J., Compoint, J. P. 1981. Isoenzymatic variation in populations of *Chenopodium album* L. resistant and susceptible to triazines. *Agro-ecosystems* 7:1–10
25. Gasquez, J., Darmency, H., Compoint, J. P. 1981. Comparaison de la germination et de la croissance de biotypes sensibles et résistants aux triazines chez quatre espèces de mauvaises herbes. *Weed Res.* 21:219–25
26. Green, M. B., LeBaron, H. M., Moberg, W. K., eds. 1990. *Managing Resistance to Agrochemicals—from Fundamental Research to Practical Stategies.* Washington: Am. Chem. Soc. Symp. Ser. 421, ASC Books. 496 pp.
27. Gressel, J., Ammon, H. U., Fogelfors, H., Gasquez, J., Kay, Q. O. N., et al. 1982. Discovery and distribution of herbicide-resistant weeds outside North America. See Ref. 43, pp. 31–55
28. Gressel, J., Ben-Sinai, G. 1985. Low intraspecific competitive fitness in a triazine-resistant, nearly nuclear-isogenic line of *Brassica napus. Plant Sci.* 38:29–32
29. Gressel, J., Regev, Y., Malkin, S., Kleifeld, Y. 1983. Characterization of an s-triazine-resistant biotype of *Brachypodium distachyon. Weed Sci.* 31:450–56
30. Gressel, J., Segel, L. A. 1978. The paucity of plants evolving genetic resistance to herbicides: possible reasons and implications. *J. Theor. Biol.* 75:349–71
31. Gressel, J., Segel, L. A. 1990. Modelling the effectiveness of herbicide rotations and mixtures as strategies to delay or preclude resistance. *Weed Technol.* 4:186–98
32. Gronwald, J. W., Andersen, R. N., Yee, C. 1989. Atrazine resistance in velvetleaf *(Abutilon theophrasti)* due to enhanced atrazine detoxification. *Pesticide Biochem. Physiol.* 34:149–63
33. Hall, L. M., Devine, M. D. 1990. Cross-resistance of a chlorsulfuron-resistant biotype of *Stellaria media* to a triazolopyrimidine herbicide. *Plant. Physiol.* 93:962–66
34. Harper, J. L. 1956. The evolution of weeds in relation to resistance to herbicides. *Proc. Br. Weed Control Conf.*, pp. 179–88. Surrey, England
35. Hirschberg, J., Bleeker, A., Kyle, D. J., McIntosh, L. Arntzen, C. J. 1984. The molecular basis of triazine-herbicide resistance in higher-plant chloroplasts. *Z. Naturforsch.* 39c:412–20
36. Hobbs, S. L. A. 1987. Comparison of photosynthesis in normal and triazine-resistant *Brassica. Can. J. Plant Sci.* 67:457–66
37. Holt, J. S. 1990. Fitness and ecological adaptability of herbicide-resistant biotypes. See Ref. 26, pp. 419–29
38. Holt, J. S., LeBaron, H. M. 1990. Significance and distribution of herbicide resistance. *Weed Technol.* 4:141–49
39. Itoh, K., Matsunaka, S. 1990. Parapatric differentiation of paraquat resistant biotypes in some Compositae species. In *Biological Approaches and Evolutionary Trends in Plants,* ed. S. Kawano, pp. 33–49. London: Academic.
40. Jacobs, B. F., Duesing, J. H., Antonovics, J., Patterson, D. T. 1988. Growth performance of triazine-resistant and -susceptible biotypes of *Solanum nigrum* over a range of temperatures. *Can. J. Bot.* 66:847–50
41. Jansen, M. A. K., Hobé, J. H., Wesselius, J. C., Van Rensen, J. J. S. 1986. Comparison of photosynthetic activity and growth performance in triazine-resistant and susceptible biotypes of *Chenopodium album. Physiol. Veg.* 24:475–84
42. Kemp, M. S., Moss, S. R., Thomas, T. H. 1990. Herbicide resistance in *Alopecurus myosuroides.* See Ref. 26, pp. 376–93
43. LeBaron, H. M., Gressel, J., eds. 1982. *Herbicide Resistance in Plants.* New York: Wiley. 401 pp.
44. LeBaron, H. M., McFarland, J. 1990. Herbicide resistance in weeds and crops: an overview and prognosis. See Ref. 26, pp. 336–52

45. Lehoczki, E., Pölös, E., Laskay, G., Farkas, T. 1985. Chemical compositions and physical states of chloroplast lipids related to atrazine resistance in *Conyza canadensis*. *Plant. Sci.* 42:19–24
46. Lipecki, J. 1984. Development of resistance to chlorotriazines in weeds. *Postepy Nauk Roln.* 31:47–52
47. Lemoine, Y., Dubacq, J-P., Zabulon, G., Ducruet, J-M. 1986. Organization of the photosynthetic apparatus from triazine-resistant and -susceptible biotypes of several plant species. *Can. J. Bot.* 64:2999–3007
48. Lutman, P. J. W., Heath, C. R. 1990. Variations in the resistance of *Stellaria media* to mecoprop due to biotype, application method and 1-aminobenzotriazole. *Weed Res.* 30:129–37
49. Mallory-Smith, C. A., Thill, D. C., Dial, M. J. 1990. Identification of sulfonylurea herbicide-resistant prickly lettuce *(Lactuca serriola)*. *Weed Technol.* 4:163–68
50. Mapplebeck, L. R., Souza Machado, V., Grodzinski, B. 1982. Seed germination and seedling growth characteristics of atrazine-susceptible and resistant biotypes of *Brassica campestris*. *Can. J. Plant Sci.* 62:733–39
51. Marshall, G. 1987. Implications of herbicide-tolerant cultivars and herbicide-resistant weeds for weed control management. *Proc. Br. Crop Prot. Conf. Weeds*, pp. 489–98. Surrey, England
52. Maxwell, B. D. Roush, M. L., Radosevich, S. R. 1990. Predicting the evolution and dynamics of herbicide resistance in weed populations. *Weed Technol.* 4:2–13
53. McCloskey, W. B., Holt, J. S. 1990. Triazine resistance in *Senecio vulgaris* parental and nearly isonuclear back crossed biotypes is correlated with reduced productivity. *Plant Physiol.* 92:954–62
54. Mikulka, J. 1987. Development of resistant weed biotypes following long-term application of herbicides. *Int. Z. Landwirtsch.* 1:59–62
55. Morrison, I. N., Todd, B. G., Nawolsky, K. M. 1989. Confirmation of trifluralin-resistant green foxtail (*Setaria viridis*) in Manitoba. *Weed Technol.* 3:544–51
56. Murphy, T. R., Gossett, B. J., Toler, J. E. 1986. Growth and development of dinitroaniline-susceptible and -resistant goosegrass *(Eleusine indica)* biotypes under noncompetitive conditions. *Weed Sci.* 34:704–10
57. Ort, D. R., Ahrens, W. H., Martin, B., Stoller, E. W. 1983. Comparison of photosynthetic performance in triazine resistant and susceptible biotypes of *Amaranthus hybridus*. *Plant Physiol.* 72:925–30
58. Pillai, P., St. John, J. B. 1981. Lipid composition of chloroplast membranes from weed biotypes differentially sensitive to triazine herbicides. *Plant Physiol.* 68:585–87
59. Pölös, E., Mikulás, J., Szigeti, Z., Laskay, G., Lehoczki, E. 1987. Cross resistance to paraquat and atrazine in *Conyza canadensis*. *Proc. Br. Crop Prot. Conf. Weeds*, pp. 909–16. Surrey, England
60. Powles, S. B., Howat, P. D. 1990. Herbicide-resistant weeds in Australia. *Weed Technol.* 4:178–85
61. Price, S. C., Hill, J. E., Allard, R. W. 1983. Genetic variability for herbicide reaction in plant populations. *Weed Sci.* 31:652–57
62. Putwain, P. D., Mortimer, A. M. 1989. The resistance of weeds to herbicides: rational approaches for containment of a growing problem. *Proc. Br. Crop Prot. Conf. Weeds*, pp. 285–94. Surrey, England
63. Putwain, P. D., Scott, K. R., Holliday, R. J. 1982. The nature of resistance to triazine herbicides: case histories of phenology and population studies. See Ref. 43, pp. 99–115
64. Radosevich, S. R., Holt, J. S. 1982. Physiological responses and fitness of susceptible and resistant weed biotypes to triazine herbicides. See Ref. 43, pp. 163–83
65. Ritter, R. L., Kaufman, L. M., Monaco, T. J., Novitzky, W. P., Moreland, D. E. 1989. Characterization of triazine-resistant giant foxtail *(Setaria faberi)* and its control in no-tillage corn *(Zea mays)*. *Weed Sci.* 37:591–95
66. Ryan, G. F. 1970. Resistance of common grounsel to simazine and atrazine. *Weed Sci.* 18:614–16
67. Saari, L. L., Cotterman, J. C., Primiani, M. M. 1990. Mechanism of sulfonylurea herbicide resistance in the broadleaf weed, *Kochia scoparia*. *Plant Physiol.* 93:55–61
68. Salhoff, C. R., Martin, A. R. 1985. *Kochia scoparia* growth response to triazine herbicides. *Weed Sci.* 34:40–42
69. Schönfeld, M., Yaacoby, T., Michael, O., Rubin, B. 1987. Triazine resistance without reduced vigor in *Phalaris paradoxa*. *Plant Physiol.* 83:329–33
70. Solymosi, P., Kostyal, S. 1985. Mapping of atrazine resistance for *Amaran-*

thus retroflexus L. in Hungary. *Weed Res.* 25:411–14

71. Solymosi, P., Lehoczki, E. 1989. Characterization of a triple (atrazine-pyrazon-pyridate) resistant biotype of common lambsquarters (*Chenopodium album* L.). *J. Plant Physiol.* 134:685–90
72. Solymosi, P., Kostyal, Z., Lehoczki, E. 1986. Characterization of intermediate biotypes in atrazine-susceptible populations of *Chenopodium polyspermum* L. and *Amaranthus bouchonii* Thell. in Hungary. *Plant Sci.* 47:173–79
73. Souza Machado, V. 1982. Inheritance and breeding potential of triazine tolerance and resistance in plants. See Ref. 43, pp. 257–73.
74. Stephenson, G. R., Dykstra, M. D., McLaren, R. D., Hamill, A. S. 1990. Agronomic practices influencing triazine-resistant weed distribution Ontario. *Weed Technol.* 4:199–207
75. Thai, K. M., Jana, S., Naylor, J. M. 1985. Variability for response to herbicides in wild oat *(Avena fatua)* populations. *Weed Sci.* 33:829–35
76. Tremolieres, A., Darmency, H., Gasquez, J., Dron, M., Connan, A. 1988. Variation of transhexadecenoic acid content in two triazine resistant mutants of *Chenopodium album* and their susceptible progenitor. *Plant Physiol.* 86:967–70
77. Van Oorschot, J. L. P., Van Leeuwen, P. H. 1984. Comparison of the photosynthetic capacity between intact leaves of triazine-resistant and -susceptible biotypes of six weed species. *Z. Naturforsch.* 39c:440–42
78. Vaughn, K. C., Duke, S. O. 1984. Ultrastructural alterations to chloroplasts in triazine-resistant weed biotypes. *Physiol. Plant.* 62:510–20
79. Vaughn, K. C., Vaughan, M. A. 1990. Structural and biochemical characterization of dinitroaniline-resistant *Eleusine*. See Ref. 26, pp. 364–75
80. Vencill, W. K., Foy, C. L., Orcutt, D. M. 1987. Effects of temperature on triazine-resistant weed biotypes. *Environ. Exp. Bot.* 27:473–80
81. Warwick, S. I. 1990. Genetic variation in weeds—with particular reference to Canadian agricultural weeds. In *Biological Approaches and Evolutionary Trends in Plants,* ed. S. Kawano, pp. 3–18. London: Academic.
82. Warwick, S. I., Black, L. D. 1981. The relative competitiveness of atrazine susceptible and resistant populations of *Chenopodium album* and *C. strictum*. *Can. J. Bot.* 59:689–93
83. Warwick, S. I., Black, L. D. 1986. Electrophoretic variation in triazine-resistant and susceptible populations of *Amaranthus retroflexus* L. *New Phytol.* 104:661–70
84. Warwick, S. I., Marriage, P. B. 1982. Geographical variation in populations of *Chenopodium album* resistant and susceptible to atrazine. I. Between-and within-population variation in growth and response to atrazine. *Can. J. Bot.* 60:483–93
85. Watson, D., Mortimer, A. M., Putwain, P. D. 1987. The seed bank dynamics of triazine resistant and susceptible biotypes of *Senecio vulgaris*—implications for control strategies. *Proc. Br. Crop. Prot. Conf. Weeds,* pp. 917–24. Surrey, England
86. Weaver, S. E., Thomas, A. G. 1986. Germination responses to temperature of atrazine-resistant and -susceptible biotypes of two pigweed *(Amaranthus)* species. *Weed Sci.* 34:865–70
87. Weaver, S. E., Warwick, S. I. 1982. Competitive relationships between atrazine resistant and susceptible populations of *Amaranthus retroflexus* and *A. powellii* from southern Ontario. *New Phytol.* 92:131–39
88. Weaver, S. E., Warwick, S. I., Thompson, B. K. 1982. Comparative growth and atrazine response of resistant and susceptible populations of *Amaranthus* from southern Ontario, Canada. *J. Appl. Ecol.* 19:611–20
89. Yaacoby, T., Schönfeld, M., Rubin, B. 1986. Characteristics of atrazine-resistant biotypes of three grass weeds. *Weed Sci.* 34:181–84
90. Yadav, N., McDevitt, R. E., Benard, S., Falco, S. C. 1986. Single amino acid substitutions in the enzyme acetolactate synthase confer resistance to the herbicide sulfometuron methyl. *Proc. Natl. Acad. Sci. USA* 83:4418–22

45. Lehoczki, E., Pölös, E., Laskay, G., Farkas, T. 1985. Chemical compositions and physical states of chloroplast lipids related to atrazine resistance in *Conyza canadensis*. *Plant. Sci.* 42:19–24
46. Lipecki, J. 1984. Development of resistance to chlorotriazines in weeds. *Postepy Nauk Roln.* 31:47–52
47. Lemoine, Y., Dubacq, J-P., Zabulon, G., Ducruet, J-M. 1986. Organization of the photosynthetic apparatus from triazine-resistant and -susceptible biotypes of several plant species. *Can. J. Bot.* 64:2999–3007
48. Lutman, P. J. W., Heath, C. R. 1990. Variations in the resistance of *Stellaria media* to mecoprop due to biotype, application method and 1-aminobenzotriazole. *Weed Res.* 30:129–37
49. Mallory-Smith, C. A., Thill, D. C., Dial, M. J. 1990. Identification of sulfonylurea herbicide-resistant prickly lettuce *(Lactuca serriola)*. *Weed Technol.* 4:163–68
50. Mapplebeck, L. R., Souza Machado, V., Grodzinski, B. 1982. Seed germination and seedling growth characteristics of atrazine-susceptible and resistant biotypes of *Brassica campestris*. *Can. J. Plant Sci.* 62:733–39
51. Marshall, G. 1987. Implications of herbicide-tolerant cultivars and herbicide-resistant weeds for weed control management. *Proc. Br. Crop Prot. Conf. Weeds,* pp. 489–98. Surrey, England
52. Maxwell, B. D. Roush, M. L., Radosevich, S. R. 1990. Predicting the evolution and dynamics of herbicide resistance in weed populations. *Weed Technol.* 4:2–13
53. McCloskey, W. B., Holt, J. S. 1990. Triazine resistance in *Senecio vulgaris* parental and nearly isonuclear back crossed biotypes is correlated with reduced productivity. *Plant Physiol.* 92:954–62
54. Mikulka, J. 1987. Development of resistant weed biotypes following long-term application of herbicides. *Int. Z. Landwirtsch.* 1:59–62
55. Morrison, I. N., Todd, B. G., Nawolsky, K. M. 1989. Confirmation of trifluralin-resistant green foxtail (*Setaria viridis*) in Manitoba. *Weed Technol.* 3:544–51
56. Murphy, T. R., Gossett, B. J., Toler, J. E. 1986. Growth and development of dinitroaniline-susceptible and -resistant goosegrass *(Eleusine indica)* biotypes under noncompetitive conditions. *Weed Sci.* 34:704–10
57. Ort, D. R., Ahrens, W. H., Martin, B., Stoller, E. W. 1983. Comparison of photosynthetic performance in triazine resistant and susceptible biotypes of *Amaranthus hybridus*. *Plant Physiol.* 72:925–30
58. Pillai, P., St. John, J. B. 1981. Lipid composition of chloroplast membranes from weed biotypes differentially sensitive to triazine herbicides. *Plant Physiol.* 68:585–87
59. Pölös, E., Mikulás, J., Szigeti, Z., Laskay, G., Lehoczki, E. 1987. Cross resistance to paraquat and atrazine in *Conyza canadensis*. *Proc. Br. Crop Prot. Conf. Weeds,* pp. 909–16. Surrey, England
60. Powles, S. B., Howat, P. D. 1990. Herbicide-resistant weeds in Australia. *Weed Technol.* 4:178–85
61. Price, S. C., Hill, J. E., Allard, R. W. 1983. Genetic variability for herbicide reaction in plant populations. *Weed Sci.* 31:652–57
62. Putwain, P. D., Mortimer, A. M. 1989. The resistance of weeds to herbicides: rational approaches for containment of a growing problem. *Proc. Br. Crop Prot. Conf. Weeds*, pp. 285–94. Surrey, England
63. Putwain, P. D., Scott, K. R., Holliday, R. J. 1982. The nature of resistance to triazine herbicides: case histories of phenology and population studies. See Ref. 43, pp. 99–115
64. Radosevich, S. R., Holt, J. S. 1982. Physiological responses and fitness of susceptible and resistant weed biotypes to triazine herbicides. See Ref. 43, pp. 163–83
65. Ritter, R. L., Kaufman, L. M., Monaco, T. J., Novitzky, W. P., Moreland, D. E. 1989. Characterization of triazine-resistant giant foxtail *(Setaria faberi)* and its control in no-tillage corn *(Zea mays)*. *Weed Sci.* 37:591–95
66. Ryan, G. F. 1970. Resistance of common grounsel to simazine and atrazine. *Weed Sci.* 18:614–16
67. Saari, L. L., Cotterman, J. C., Primiani, M. M. 1990. Mechanism of sulfonylurea herbicide resistance in the broadleaf weed, *Kochia scoparia*. *Plant Physiol.* 93:55–61
68. Salhoff, C. R., Martin, A. R. 1985. *Kochia scoparia* growth response to triazine herbicides. *Weed Sci.* 34:40–42
69. Schönfeld, M., Yaacoby, T., Michael, O., Rubin, B. 1987. Triazine resistance without reduced vigor in *Phalaris paradoxa*. *Plant Physiol.* 83:329–33
70. Solymosi, P., Kostyal, S. 1985. Mapping of atrazine resistance for *Amaran-*

thus retroflexus L. in Hungary. *Weed Res.* 25:411–14

71. Solymosi, P., Lehoczki, E. 1989. Characterization of a triple (atrazine-pyrazon-pyridate) resistant biotype of common lambsquarters (*Chenopodium album* L.). *J. Plant Physiol.* 134:685–90
72. Solymosi, P., Kostyal, Z., Lehoczki, E. 1986. Characterization of intermediate biotypes in atrazine-susceptible populations of *Chenopodium polyspermum* L. and *Amaranthus bouchonii* Thell. in Hungary. *Plant Sci.* 47:173–79
73. Souza Machado, V. 1982. Inheritance and breeding potential of triazine tolerance and resistance in plants. See Ref. 43, pp. 257–73.
74. Stephenson, G. R., Dykstra, M. D., McLaren, R. D., Hamill, A. S. 1990. Agronomic practices influencing triazine-resistant weed distribution Ontario. *Weed Technol.* 4:199–207
75. Thai, K. M., Jana, S., Naylor, J. M. 1985. Variability for response to herbicides in wild oat *(Avena fatua)* populations. *Weed Sci.* 33:829–35
76. Tremolieres, A., Darmency, H., Gasquez, J., Dron, M., Connan, A. 1988. Variation of transhexadecenoic acid content in two triazine resistant mutants of *Chenopodium album* and their susceptible progenitor. *Plant Physiol.* 86:967–70
77. Van Oorschot, J. L. P., Van Leeuwen, P. H. 1984. Comparison of the photosynthetic capacity between intact leaves of triazine-resistant and -susceptible biotypes of six weed species. *Z. Naturforsch.* 39c:440–42
78. Vaughn, K. C., Duke, S. O. 1984. Ultrastructural alterations to chloroplasts in triazine-resistant weed biotypes. *Physiol. Plant.* 62:510–20
79. Vaughn, K. C., Vaughan, M. A. 1990. Structural and biochemical characterization of dinitroaniline-resistant *Eleusine*. See Ref. 26, pp. 364–75
80. Vencill, W. K., Foy, C. L., Orcutt, D. M. 1987. Effects of temperature on triazine-resistant weed biotypes. *Environ. Exp. Bot.* 27:473–80
81. Warwick, S. I. 1990. Genetic variation in weeds—with particular reference to Canadian agricultural weeds. In *Biological Approaches and Evolutionary Trends in Plants,* ed. S. Kawano, pp. 3–18. London: Academic.
82. Warwick, S. I., Black, L. D. 1981. The relative competitiveness of atrazine susceptible and resistant populations of *Chenopodium album* and *C. strictum. Can. J. Bot.* 59:689–93
83. Warwick, S. I., Black, L. D. 1986. Electrophoretic variation in triazine-resistant and susceptible populations of *Amaranthus retroflexus* L. *New Phytol.* 104:661–70
84. Warwick, S. I., Marriage, P. B. 1982. Geographical variation in populations of *Chenopodium album* resistant and susceptible to atrazine. I. Between-and within-population variation in growth and response to atrazine. *Can. J. Bot.* 60:483–93
85. Watson, D., Mortimer, A. M., Putwain, P. D. 1987. The seed bank dynamics of triazine resistant and susceptible biotypes of *Senecio vulgaris*—implications for control strategies. *Proc. Br. Crop. Prot. Conf. Weeds,* pp. 917–24. Surrey, England
86. Weaver, S. E., Thomas, A. G. 1986. Germination responses to temperature of atrazine-resistant and -susceptible biotypes of two pigweed *(Amaranthus)* species. *Weed Sci.* 34:865–70
87. Weaver, S. E., Warwick, S. I. 1982. Competitive relationships between atrazine resistant and susceptible populations of *Amaranthus retroflexus* and *A. powellii* from southern Ontario. *New Phytol.* 92:131–39
88. Weaver, S. E., Warwick, S. I., Thompson, B. K. 1982. Comparative growth and atrazine response of resistant and susceptible populations of *Amaranthus* from southern Ontario, Canada. *J. Appl. Ecol.* 19:611–20
89. Yaacoby, T., Schönfeld, M., Rubin, B. 1986. Characteristics of atrazine-resistant biotypes of three grass weeds. *Weed Sci.* 34:181–84
90. Yadav, N., McDevitt, R. E., Benard, S., Falco, S. C. 1986. Single amino acid substitutions in the enzyme acetolactate synthase confer resistance to the herbicide sulfometuron methyl. *Proc. Natl. Acad. Sci. USA* 83:4418–22

ecological questions" (36, p. 631). MacMahon et al (98) attempted to distinguish "functional group" from "guild" by defining the former to be species performing the same function, but by 1984 the terms were often used synonymously (67). A classification at that time of functional groups of aquatic insects incorporated ways of using a resource (e.g. shredders, scrapers, piercers) but did not refer to "guilds" (108). There is surely still confusion. For example, Menge et al (107) depicted functional groups rather than guilds on the grounds that guilds are simply groups of species using the same resource, whereas functional groups focus on the method of foraging! In fact, the original "guild" paper focussed on the method of using resources, while the original "functional groups" paper focussed on resources themselves.

Because most studies of guilds considered food as the resource, the burgeoning literature on trophic web structure used entities very much like guilds. The resemblance was inevitable because many tabulated trophic webs lumped groups of species using the same food into "trophic species" or the like (120), while much research on guilds de-emphasized "in a similar way" and emphasized the same food. Noting this resemblance, Yodzis (164) compared concepts from food web studies to the guild. A clique is a set of species in which every species pair shares some resource, and a dominant clique is a clique that is not a subset of another clique. Yodzis suggested dominant cliques might be viewed as "trophic guilds" but are not guilds in the original sense because how resources are used is not considered; thus a dominant clique or trophic guild might contain several guilds defined on the basis of method of feeding. Burns (20) appropriated "trophic guild" for a different group—an aggregation of species with similar trophic resources; this use corresponded to the traditional concept of "trophic level," which he saw as too imprecise.

Conflation of trophic structure concepts with the guild concept has led to profound confusion about the status and importance of the latter. For example, Heatwole & Levins (68) detected an emergent community property (133) in the apportionment of mangrove island insect communities into crude trophic categories—herbivores, predators, scavengers, etc. In fact, the apparent regularity of this apportionment is probably an artifact of the statistic used to characterize differences in apportionment (142). Others (1, 31, 69, 112) resuscitated this argument in the context of the role of guilds, referring to the original papers as being about guilds, though neither of them mentioned guilds or addressed in more than cursory fashion how resources are gathered.

The Motivation Behind the Concept

Root (130) suggested three advantages of using guilds in the study of ecological communities:

1. Guilds focus attention on all sympatric competing species, regardless of their taxonomic relationship.

2. "Guild" eliminates the dual usage of the term "niche" as meaning both the functional role of a species in a community and the set of conditions that permits a species to exist in a particular biotope. This duality had led to controversy, which Root hoped to resolve by limiting the term "niche" to the latter purpose and recognizing that groups of species having very similar ecological roles within a community are members of the same guild, not occupants of the same niche.

3. Guilds are useful in comparative study of communities. Since it is usually impossible to study all species living in an ecosystem at once, guilds enable us to concentrate on specific groups with specific functional relationships. This is preferable to studying taxonomic groups, within which different species may perform unrelated roles.

Investigators often cite the first and third advantages when studying guilds. The almost universal need to limit research on communities to manageable units makes the use of guilds so attractive. More recently, as the term has become popular, a fourth goal is often articulated for study of guilds—they might represent the "basic building blocks" (67) of communities, and the partitioning of communities into guilds might reveal a structure not attributable simply to species numbers and identities (67, 148). The view that species in communities fill fixed, basic functional roles is one of the most venerable in ecology, stretching back at least 200 years (104). This view, in turn, has spawned attempts to depict different communities in terms of these roles and thus to demonstrate their underlying similarity. "The view that there exist fixed ecological roles is supported by the observation that ecological communities often bear striking resemblances to each other even though their constituent species have very different evolutionary origins" (104, p. 256–57). Until the last decade, discussions of this sort were mostly cast in terms of niches, which were construed as individual species' functional roles. However, the recognition that different species may have very similar functional roles, and Root's proposition that groups of such species, namely guilds, can be compared among communities, led to a shift from niches to guilds as the potential fundamental units of communities. If guilds really are fundamental units, the nature of guilds forming a community might be an emergent property, the elusive grail of community ecologists (133). This hope has generated the prediction that "guilds will become the standard currency of ecologists in their efforts to understand community relationships of many kinds" (148, p. 90).

THE IMPORTANCE OF CLEARLY DEFINING AND DELINEATING GUILDS

"Guild" has come to be used in many different senses, as we will document in later sections. Often, it seems to be used colloquially to mean all species using some resource, with rather little attention paid to how the resource is used. This usage probably arose because no other term has taken this meaning, yet much current ecological research is aimed at such groups of species. Colloquial usage of scientific terms that were originally narrowly defined is hardly unique to *guild,* and the fact that its first ecological usage was metaphoric almost ensured that it would acquire a variety of meanings. So long as all readers bear in mind a distinction between the narrow and colloquial definitions, the word can perhaps usefully fill the lexical void just noted and still function in precise scientific discourse. However, the inadvertent confounding of different meanings of guild can lead to confusion and misunderstanding.

Root (130) viewed members of a guild as molded by adaptation to the same resources and by competition. This is itself a statement of belief in how ecological communities are structured. Use of the same resources need not imply competition for them. For example, Duggins (50) studied three congeneric sea urchins occupying virtually the same habitat and eating similar food. Experiments showed that addition of the largest species, *Strongylocentrotus franciscanus,* did not decrease the populations of the other species and, if anything, resulted in an increase in gonad size, a good measure of fitness. Apparently, interspecific facilitation of two sorts overshadowed possible consumptive competition (*sensu* 136). First, *S. franciscanus* is especially adept at snagging drift algae, which the other two species then share. Second, presence of *S. franciscanus* results in decreased abundance of the predatory starfish *Pycnopodia helicanthoides,* which consumes or drives out the other two species. Ironically, this example seems to conform closely to the primary dictionary definition of *guild:* "a confraternity, brotherhood, or association formed for the mutual aid and protection of its members, or for the prosecution of some common purpose" (117, p. 1225). Thus, if guilds are to be used to test hypotheses on the relative importance of various processes in structuring communities, they should be defined independently of the mechanisms by which the members may interact, lest the entire enterprise drift toward circularity (cf 99, 100).

It is only natural, during a decade in which the ecological and evolutionary role of competition has been intensely argued, that the *guild,* originally conceived as molded by interspecific competition, would generate considerable interest and debate. The view of guilds as coevolved entities, as "arenas of intense competition" (122), is one reason the clear delineation of guilds is critical. An alternative view of ecological communities as groups of sympatric species shaped primarily by response to autecological pressures (e.g. 60, 157) implies that guilds are groups of independent species sharing a resource but not likely to be subject to much coevolution. This view necessarily detracts from the importance of identifying guilds. If this view is correct, or correct for some species assemblages, then perhaps the term "functional group" for species that perform a certain functional role in the ecosystem is more appropriate, and functional groups, not guilds, deserve more attention. Perhaps, in the original spirit in which the term was coined, guilds as coevolved entities are the exception rather than the rule in ecological communities. Only studies of groups of species using similar resources in similar ways, whatever they would be called, will resolve the question of how prevalent competitively driven coevolution is in nature.

Many studies of guilds seek determinants of community organization in particular competitive interactions. The search for such determinants has been a cornerstone of recent community ecology. Morphological patterns and ecological data implying niche partitioning within guilds frequently result from such searching (e.g., 9, 11 12, 47, 62). The interpretation of such data and the results drawn from them rest heavily on guild assignments. Lack of stated, unambiguous criteria for these assignments can potentially lead to ambiguous results and controversial interpretations. For example, Diamond (47) studied several guilds of New Guinea birds and concluded that a series of competitively determined assembly rules governs the distribution of all species on islands. While he listed some reasons for grouping certain species into guilds, he did not refer to supporting ecological studies, nor did he discuss the possibility that other species share these resources or why they were excluded from particular guilds. These guilds form a small part of the New Guinea avifauna, but their choice was not rationalized. When Gilpin & Diamond briefly discussed membership in one guild (fruit pigeons), they based it on "extensive dietary data published by Crome (1975) and other authors" (58, p. 335). Crome studied seven species of frugivorous pigeons in Queensland, made no claims concerning guild membership, and suggested that "a characteristic of tropical rainforests is the importance of fruit as food for birds, so much that large groups such as parrots, pigeons, manakins, and hornbills are adapted to a partly or wholly frugivorous diet" (32, p. 155). Indeed, in a later study of a wider (though by no means exhaustive) assemblage of birds in the same region, Crome (33) divided bird species on the basis of their height in the canopy, foraging site, and foraging behavior, and found six other avian species with which the fruit pigeons of his previous study should be grouped. Thus, the exclusive treatment of fruit pigeons as a guild bears further discussion. A large part of the heated debate on assembly rules (30, 48, 58) rested precisely on how to delimit the guilds within which patterns would be sought. Had a thorough rationale for guild membership been available at the outset to readers, much of this "unpleasant" and "bitter" debate (64) might have been avoided.

The study of single guilds within a whole fauna (e.g. 100) is not uncommon. While investigators usually at least partly justify guild designations, they often give the reader no clue about other sympatric species. Although frequently appealing intuitively, and quite possibly biologically sound, such guilds cannot be critically evaluated. It is true that studies of complete faunas are often technically impossible, or impractical, while guilds are small, manageable units, more amenable to ecological study. A brief outline of the sympatric species and their trophic or other relationship to the guild studied may, however, give other investigators a better grasp of the system studied and potential alternative designations. Root (130) himself listed the other bird species that occasionally take insects, or those that are insectivorous but differ in foraging behavior, and explained his guild designation. This enables readers to evaluate this study critically. Case et al (22) discussed only four of the guilds of West Indies birds but reported the species that are not included in these guilds and mentioned their habits. Cody's study of South African forest birds (28) is similarly detailed. Thus, while not designating guilds for the other species, they afforded the reader the chance to assess their conclusions.

WHICH SPECIES ARE GUILD ASSOCIATES

Guilds in the Original Sense

The guild concept was specifically meant to relate to species using the same class of resources, but the possibility of different partitions based on different resource classifications was explicit: "For instance, the Plain Titmouse *(Parus inornatus),* while belonging to the foliage gleaning guild with respect to its foraging habits, is also a member of the hole nesting guild by virtue of its nest-site requirements" (130, p. 335). However, the emphasis in animal community ecology has increasingly been on food resources as generating interspecific competition: "In any study of evolutionary ecology, food relations appear as one of the most important aspects of the system of animate nature" (77, p. 147). Accordingly, in most guild designations the shared resource is food. Exceptions are not infrequent: e.g. habitat guilds (61, 86), nesting guilds (89, 102, 159), reproductive guilds (5). Some guild designations are not based on resource use and clearly violate the definition of guilds: body mass guilds and mobility guilds (63) are striking examples.

Many studies define a guild as the group of species (usually within a taxon) that inhabit a certain microhabitat. For example, Sedgwick & Knopf (139) defined as a guild the group of birds directly dependent on the grass-herb-shrub layer of vegetation for foraging, nesting, or both. The way they are dependent ("in a similar way") played no role in this guild designation. Similarly, Cruz studied effects of microhabitat change on the avifauna of a neotropical mahogany plantation by dividing it into guilds using different heights of vegetation. He concluded that "removal or reduction in the understory vegetation probably will cause pronounced decreases in the members of the understory guild" (34, p. 286). The resource here is in fact the microhabitat.

Some authors, in delineating guilds, attempt to stay close to the original criteria of both the same resource and a similar way. For example, Bush (21) assigned all intestinal helminth parasites of birds to four guilds based on how they garner resources (absorbing across their body surface or engulfing material), where they gather resources (absorbing organ in the mucosa or lumen of the gut), and what the resources are (gut tissue or gut contents). The guilds largely parallel taxonomy at either the phylum or class level because feeding biology differs greatly between the groups. However, cestodes include some species in a mucosal absorber guild and others associated with acanthocephalans in a lumenal absorber guild. Osborne et al (119) partitioned bird species at one site into 18 guilds based on their primary food, foraging substrate, and foraging behavior. Humphrey et al (75) described a guild of nine surface-gleaning, primarily beetle-eating bats. They provided extensive dietary data, while detailed field research by Bonaccorso (7) documented both behavior and diet of these species plus sympatric bats.

The Guild as a Taxonomic Group

One advantage of the guild concept stressed by Root (130) is that it focusses attention on all sympatric species involved in a potentially competitive interaction, regardless of their taxonomic relationships. While most previous studies of interspecific competition had considered only sympatric congeners, on the assumption that closely related species tend to compete most strongly, Root cited several studies demonstrating intergeneric competition, and he applied this term to groups of bird species in different families.

Despite its formal definition, guild designations still tend to include closely related species. Schoener (137) terms a group of closely related species using the same resource a "taxon-guild." Taxonomically limited guild designations stem, in part, from insufficient biological data. For example, Hanski & Koskela (66) defined six guilds of dung-inhabiting beetles; they adopted a conservative within-genus approach in the absence of detailed information on the feeding and breeding biology of most species. Likewise, Walter & Ikonen (156) in their study of nematophagous arthropods rationalized the use of genus level groupings. They argued that feeding behavior (as well as other functional parameters) is rarely predictably similar beyond the generic level, and they advocated intensive behavioral studies. Lambert & Reid divided the herpetofauna of Colorado into guilds and reasoned that, "as a set of rather specialized consumer species, the guilds inevitably reflect taxonomy" (91, p. 145). Partly, of course, the tendency to delineate closely related guilds springs from the knowledge and taste of researchers, who are often particularly expert in restricted taxa.

MacNally & Doolan suggested an alternative guild definition: "a set of closely related species (usually coordinal or confamilial) that are both sympatric and synchronously active, and that forage on similar items in similar ways" (100, p. 34). They argued that by limiting guilds to closely related species we also limit other differences, because closely related species often overlap in foraging methods, habitat preference, etc. Thus, MacNally (99) formalized the guild as a taxonomic construct but did not circumvent the ambiguity inherent in judging degree of similarity in foraging method and food. This definition is limited to feeding guilds.

Guilds of Distantly Related Species Sharing Resources

The expectation that competition will be strongest among congeners has an illustrious pedigree: "As the species of the same genus usually have, though by no means invariably, much similarity in habits and constitution, and always in structure, the struggle will generally be more severe between them, if they come into competition with each other, than between the species of distinct genera" (37, p. 87). Nevertheless, as Darwin noted, this pattern is by no means invariable, and examples abound of distantly related taxa engaging in consumptive competition. An oft-cited example is granivorous rodents and ants of American southwestern deserts (15, 19, 40). In this instance suggestive dietary data were buttressed by controlled experimental removal. Similarly, introduction and removal experiments showed that insects compete with frogs for pond periphyton (113), while lizards and spiders compete for insect prey (138), as do ducks and fishes (54, 70). Non-experimental evidence implied competition for nectar among all bees and finches (135) and between wasps and a parrot (4, 10, 111), for insect prey between lizards and birds (163), and for zooplankton between flamingoes and fishes (76).

Oddly, though an entire symposium was devoted to competition between distantly related taxa (125) and most recent papers (e.g. 71) repeat a litany of well-known examples, almost all point to the phenomenon as an oddity worth noting, and there has yet to be a thorough review of the phenomenon. Without such a review, it is difficult to surmise to what extent consumptive competition is predominantly between closely related as opposed to distantly related species.

Rarely have researchers formally identified distantly related species as guild associates. Jaksic (79) argued that difficulty in studying interactions among very many species leads researchers to study "community ecology" within taxonomic groups that they know well—lizard "communities" (121), bird "communities" (143), small mammal "communities" (83), etc—and for the same reason many ecologists describe guilds within arbitrarily chosen taxonomic boundaries. Jaksic suggested a distinction between "true" (resource-based) community guilds, which unite species exploiting an investigator-based resource in a similar manner, and taxonomically based assemblage guilds. He proceeded to study guild structure of predators in various regions, grouping falconiforms, owls, snakes, and carnivores as potential guild associates, through a study of the actual prey composition of the different species (80, 81); he recognized 50% overlap as the minimum value for guild association.

Some guilds thus defined encompass species from different orders and even different classes. For example, the insectivore guild of Spain comprises one falconiform, two strigiforms, and four carnivores (81). However, the authors conceded that this guild may be divided into three groups: avian predators and two carnivore groups. The distinction between "groups" and "guilds" stems from the use of the arbitrary 50% overlap threshold and from the fact that prey were not weighted according to their mass. Thus, a red fox with 90% of its prey items insects is liable to appear as an insectivore, despite having over 80% of its mammalian prey items rabbits, with a mean weight of 900 g, so that mammals constitute the bulk of its diet. Similarly, the insectivore guild of Chile appears to comprise one falconiform and one owl (81), with insects constituting 64.8% and 76.8% of their diets, respectively (80). Closer scrutiny, however, reveals that diurnal birds and lizards which are over 20% of the falcon's prey, by count of items, comprise, unsurprisingly, only 1% by count of the owl's diet. Conversely, over 15% of the owl's prey items were identified mammals; this is three times as much as in the falcon's diet. The weight of these vertebrate prey must be much greater than that of the insects taken, so partitioning vertebrate prey must be more significant than this guild assignment implies. Without information on prey weights, this analysis must be questioned.

Even though many instances are known in which taxonomically distant species share a resource and probably (or surely) compete, Jaksic's is one of the few attempts to classify such species as guild associates. It is quite possible that differences in foraging behavior, possibly in relation to morphological differences between distantly related species (such as birds and mammals, or even different families in the same order) reduce the ecological overlap between them; this notion is discussed in later sections. If this were so, the relative rarity of published multitaxon guilds would not simply result from lack of knowledge or difficulties in studying wide ranges of species, but would reflect nature.

QUANTITATIVE METHODS OF PARTITIONING A COMMUNITY INTO GUILDS

Most studies of guilds do not use quantitative methods to divide a group of species into guilds, relying instead on taxonomy plus intuition. Root viewed the guild as having fuzzy borders; the precision of specifying guild boundaries would be inherently subjective and would rest on the sensitivity and judgment of the investigator: "As with the genus in taxonomy, the limits that circumscribe the membership of any guild must be somewhat arbitrary" (130, p. 335). It is often difficult to determine which species were considered for membership in a particular guild and which omitted, and why. For example, MacNally & Doolan (100) used MacNally's definition (99) to produce a guild of nine cicadas, though the precise criteria for limiting the guild are obscure. The study focussed on structure *within* this guild, and one result exemplifies the ambiguity. A "tall forest subguild" of three species was detected and subsequently became a "tall forest guild" (p. 43). The fact that results rest heavily on particular guild assignments plus an unease with qualitative procedures has fostered attempts to produce objective algorithms that automatically delineate guilds.

Quantitative methods used for guild assignment include nearest neighbor statistics (78, 162), cluster analysis (33, 93), principal components analysis (72, 141, 150), canonical correlation (55), and Monte Carlo techniques (84). All of these approaches, though explicit, do not unambiguously determine guilds because the investigator sets arbitrary levels for clustering. Various Monte Carlo methods (e.g. 146) can allow tests of hypotheses such as whether potential guild associates have diets more similar than would be expected given specified randomizations of the data (e.g. 82), but the level of nonrandomness required to qualify for membership in the same guild is still arbitrary. Further, so long as the basic data for these analyses consist of relative amounts of some resources used by each species, the classification depends on which resources are selected for analysis (148) and neglects the "similar way" aspect of guild assignment.

Adams (1) suggested psychophysical unfolding as a means of delineating and, incidentally, defining guilds. For a set of candidate guild associates offered the same set of resource states, preference data were examined to see if they could be "resolved" to fit a single axis. If so, all species were construed as forming a guild. If not, at least some of the species were interpreted as using the resources in a different way. The degree of resolution to a single axis was tested by Monte Carlo methods, so there remained the arbitrary decision of how much deviation from perfect resolution is required to disqualify a group as a single guild. This approach is attractive in that it formalizes the definition of "similar way," and it accords with a long tradition of ordering groups of species along single axes (e.g. bill length, body weight, size of seed eaten) and seeking patterns in that order that might be interpreted as reflecting competition (12, 47). But there is no consensus on whether the abstract definition of "similar way" corresponds to the commonsensical interpretation.

It seems unlikely that any of these quantitative methods of apportioning species to guilds will become widely accepted. Intuition usually seems to play an important role. For example, Emlen (53) studied six avian foraging guilds. These included two guilds of terrestrial grain gleaners, one comprising doves and the other comprising galliforms. These two guilds appear to differ only in their members' systematics. Emlen apparently separated them on the basis of differences in structure and body size that he judged important, but he did not discuss why. Similarly, for MacMahon's research in North American deserts, "the guilds chosen here were selected on the basis of subjective familiarity with desert mammals" (97, p. 144). Probably the investigator's understanding of the biology of the species in question will continue to guide guild assignment, but unless more explicit discussion of the criteria is routinely provided, conclusions will be suspect and arguments will abound.

AMBIGUITIES IN PARTITIONING A COMMUNITY INTO GUILDS

Root's (130) definition of a guild as a group of species using the same class of resources in a similar way leaves two basic points wide open for interpretation: What constitutes the same class of resources, and just how similar should a similar way be.

Same Class of Resources

The definition of the same class of resources has probably been the less equivocal. The standard practice has been to define general classes of resources—insects on leaves (e.g. 130), fruit (e.g. 144), seeds (e.g. 14)—and then to study groups of species exploiting these resources. The species are usually a subset chosen for a certain foraging behavior, or simply winnowed on the basis of taxonomy.

But the same class of resources may be defined more or less broadly. "For example, seeds are a resource for a variety of desert consumers. Is the resource class all of the seeds on an area? Seeds of only some of the plant species? Seeds of particular sizes? Seeds in particular microsites? Clearly any of these might be the appropriate resource class, depending on the investigator's frame of reference" (98, p. 302). Correspondingly, some workers based guild assignments on study of actual food items taken by potential guild associates (80, 81). However, over a wide range of body sizes, there is a general correlation between sizes of consumers and their resources (e.g. 59). Also, species that may decrease overlap by different foraging methods, and so qualify as members of different guilds, may still overlap substantially in prey items taken. For example, the largest members of different predatory guilds may use many of the same prey. Conversely, species that feed in similar ways on similar resources may exhibit low or no overlap in use of specific prey items because they differ greatly in body size and, therefore, prey-handling capabilities. A large raptor is likely to take rabbits and thus overlap more in its resource use with medium-sized mammalian carnivores than with a small raptor that takes small rodents. Should we then consider the large raptor and the carnivore, but not the small raptor, members of one guild? How narrowly can one define the same class of resources? At an extreme, are rabbits a resource class? We think not. One could conceivably construe a single prey species as a resource class if a predator were so highly adapted to it that it is nearly restricted to this species. Among vertebrate predators, this condition is, at best, extremely rare. The likelihood of assigning more than one such species to the same guild must be negligible. Because different species' resource distributions tend to comprise more than one prey species, and different predators' ranges of prey species tend to overlap only partially, the class of resources is best viewed as the sum of the prey spectra of the different predators.

A more difficult exercise is defining a guild on the basis of shared resources in cases of omnivory or either opportunistically or regularly changing diet. Can organisms be viewed as adapted generalists and thus placed in an omnivore or generalist guild? For species whose diet changes with resource availability, does guild membership change accordingly? Species whose life history stages use different resources raise a similar question. So can migratory species. For example, DuBowy (49) assigned seven dabbling duck species on their breeding grounds in North Dakota to one guild on the grounds that they forage in close proximity and take similar food items, although four foraged primarily by dabbling and three by "head-under," so they might have been construed as forming two guilds (see next section). On their wintering grounds in California, six of these species are again found together (the seventh is replaced by a single similar species), but they are generally more specialized in food, habitat, and behavior, and these traits often differ from those of the summer. For example, *Anas strepera* primarily dabbles in the summer but forages head-under in the winter. Primarily a dabbler in the summer, *Anas clypeata* in the winter has two distinct foraging modes and food types: alternately dabbling in the water column on small swimming invertebrates or "tipping-up" on the bottom for seeds and chironomid larvae. Diet, habitat, and behavioral overlaps between pairs of species are significantly less in winter than in summer. Should the guild partition be the same in the two seasons? There is no consensus on guild assignments in such matters. Because, as noted in the section on precursor and parallel concepts, guild designation without consideration of how resources are gathered is very similar to determining a food web, it is not surprising that most of these situations are also seen as problematic in depicting food webs (120).

In a Similar Way

Root (130) himself addressed the question of how similar is "similar." He excluded the western flycatcher *(Empidonax difficilis)* from the foliage gleaning guild on the basis of differences in its foraging behavior: This bird hunts by searching a large area from a "sentinel" position on an exposed perch. Most prey are taken in long, sweeping aerial attacks on flying insects or arthropods that alight momentarily on the foliage, so the western flycatcher concentrates on active insects (130). Thus, by its foraging method this bird, an important avian insectivore in the oak woodland, is likely to capture a greater proportion of Hymenoptera and Diptera (130). In fact, the blue-gray gnatcatcher, the member of the foliage gleaning guild that was the focus of Root's (130) study, also takes hymenopterans and dipterans and was observed also hawking flying insects. Its more typical foraging method—hopping from perch to perch and closely scrutinizing leaves—allows for a higher proportion of stationary or slow moving prey. The different, though partially overlapping, foraging methods of these two birds that coexist in the oak woodland, both exploiting insectivorous prey found on leaves, reduce (but do not eliminate) the overlap between prey species taken, and so merit their separation into two guilds.

Many studies of avian species follow Root (130) in assigning guild membership through foraging method. Case et al (22) discussed four major foraging guilds: frugivores, foliage gleaning insectivores, flycatching insectivores (including hawking, sallying, and gleaning forms), and nectarivores. Similarly, Cody (28) perceived three major foraging guilds: foliage insectivores, sallying flycatchers, and slow-searching omnivores. However, assessment of similarity in foraging behavior remains quite subjective. Landres & MacMahon (93) divided oak woodland avifauna into foraging guilds that are relatively more specialized, and include foliage gleaning, wood gleaning, wood probing, air sallying, and ground sallying. Other studies deemphasize foraging behavior and use coarser categories of feeding guilds; they seem to incorporate the nature of the resources but not the way resources are gathered. For example, Martin (102) divided island avifaunas into three guilds: granivore, omnivore, and insectivore. Meserve (109, 110) studied three feeding guilds of the Chilean semi-arid small mammal community: herbivore, insectivore, and granivore. Strand (145) described two guilds of California reef fishes: herbivores and predators. It appears that the investigator's notion of similarity, based probably on the particular system studied, strongly affects guild designations.

Different investigators have divided the same or ecologically closely related taxa into guilds in different ways that reflect a different grasp of this term, and a different impression of the importance of the "similar way" in guild definition. Consider predator guilds as an example. At one extreme

Jaksic (79) considered grouping taxonomically related species into guilds, a clear abuse of Root's (130) original definition of the term. Following MacMahon et al (98), Jaksic believed that the criterion for "similar manner" should be based on the effect of resource use on the resource itself: ". . . it does not matter whether an organism removes a tree leaf for nesting material, for food, or as a substrate to grow fungi which in turn are eaten; the leaf is gone and the leaf users belong to a common guild" (98, p. 301). Consequently Jaksic et al (80) and Jaksic & Delibes (81) described predatory guilds whose members include mammalian carnivores, falconiforms, owls, and snakes. A "similar way" does not play much of a role in these guild definitions.

A narrower definition of predatory guilds, to include only extant and extinct mammalian predators, was used by Van Valkenburgh (152, p. 407): "As defined here, the guild of large land predators includes the nonaquatic, nonvolant mammal species within a community that take prey and potentially compete for food." Van Valkenburgh (152) limited guild membership to species above jackal size (7 kg) because she expected heightened competition among large predators, but also because these are better represented in the fossil record.

Dayan et al (46) based guild designations of recent carnivorous mammals on limb morphology, which reflects locomotor function and affects foraging behavior. Underlying these guild designations was the assumption that similarities in locomotor behavior imply greater ecological overlap and thus the potential for increased competition. The three guilds are cursorial carnivores of open areas (canids), plantigrade, relatively slow species (mustelids and viverrids), and species that stalk their prey (felids). These carnivores differ also in their killing behavior. Mustelids and viverrids kill by using an upper canine to drive apart two cervical vertebrae, a specialized method allowing them to take prey that are quite large relative to themselves. Felids kill similarly, further assisted by their ability to seize prey with their forepaws. Canids are more omnivorous than the other carnivores. They use a series of slashing bites or shred prey in their mouth. Consequently, canids while hunting alone take prey that are relatively small. Dayan et al hypothesized that these differences in foraging behavior reduce overlap between prey species composition of the three guilds; members of different guilds are less likely to encounter, pursue, or take overlapping prey items than are members of the same guild. However, much as with Root's birds, some overlap may well occur. Thus "similar way" plays a key role in these guild designations. Unfortunately, basic data on food habits of the different species are unavailable. A coevolutionary morphological response within each guild is evidenced (42–45), and this pattern does not obtain when all species are placed in one guild, but the case for inter-guild resource partitioning remains indirect and not empirically tested.

To test the significance of differences in foraging methods one must study a resource-based grouping of species that show different foraging behaviors and whose resource use pattern is at least reasonably well understood. The granivores of North American deserts constitute exactly such an assemblage and exemplify problems in partitioning a biota into guilds.

AN EXAMPLE—GRANIVORES OF THE AMERICAN SOUTHWEST

Rodents, ants, and birds all eat seeds in the same areas (19). Many of the birds eat seeds opportunistically and/or seasonally, but most habitats support at least a quail and a dove that are resident granivores. Their impact on rodents and ants has scarcely been studied. The rodents consist of (*a*) heteromyids (kangaroo rats, kangaroo mice, and pocket mice) that are highly specialized physiologically and morphologically for eating seeds, and are primarily or exclusively granivores, and (*b*) cricetids that are opportunistic omnivores, lacking such specializations (14, 19). The percentage of seeds in cricetid diets is unknown. One kangaroo rat, *Dipodomys microps,* has a mixed diet of leaves and seeds that varies spatially and temporally (35).

Brown (11, 12, 14) and Davidson et al (40) lumped all heteromyids plus some cricetids in one guild. Bowers & Brown (9), aiming for a finer partition based on the trophic classifications of Morton (114), placed *D. microps* in a separate herbivore guild, the other heteromyids in a granivore guild, and the cricetids in an omnivore guild. Kangaroo rats and kangaroo mice jump bipedally, while pocket mice and the cricetids are quadrupedal (12). Kangaroo rats forage mostly in open areas between scattered perennial plants and hop quickly between widely spaced seed clumps, while pocket mice forage mostly on more evenly distributed seeds under shrubs (14, 19). "Rodent species feeding under shrub canopies and those feeding in open habitats may share only the subset of seed resources that cross habitat boundaries. This subset may be particularly insignificant for buried seeds that are less likely to be moved by physical forces" (19, p. 216). Because of these behavioral differences, we viewed bipedal and quadrupedal heteromyids as separate guilds and suggested that *D. microps* is a guild associate of the other kangaroo rats because it is at least partly granivorous and its foraging behavior and morphology resemble that of its congeners (41). We separated cricetids based on their different physiology, morphology, and behavior; the cricetids, lacking the large external cheek pouches of the heteromyids, appear to forage differently from heteromyids (126). Hallett (65) assigned the cricetids and pocket mice to one guild and kangaroo rats to another. Brown & Heske (18) also called the kangaroo rats a separate guild, on the grounds that they are ecologically and taxonomically similar. These classificatory dilemmas embody how subjective guild assignments can be. Even if cricetid diets were studied more thoroughly, the decision on how much dietary overlap qualifies species for guild association would be arbitrary.

The ants have also been variously apportioned into guilds. Some researchers (24, 106) perceived a guild of seed-foragers and an omnivore guild containing only *Novomessor cockerelli*. Others (23, 40) included the latter species in the seed-eating guild. Davidson et al (40), based on Davidson (38, 39), nevertheless distinguished within this guild between species that forage in columns and those that forage individually. The former tend to feed on energetically rich clumps of seeds, while the latter feed primarily on dispersed seed. However, unlike the rodents, the ants seem not to include species that forage particularly close to shrub cover (39). Ant species also differ in how they grasp and carry seeds (24). In sum, as with the rodents, one might argue, based on diet and foraging behavior, for at least three guilds (omnivore, single forager, column forager) among these species.

These rodents and ants are the most frequently cited example of competition between distantly related organisms: "the competitive interaction among the rodents may be weaker than that between rodents and ants" (51, p. 343). Though Brown & Davidson (15, Figure 1) and Davidson et al (40, Table 1) emphasized the broad overlap in sizes and species of seeds eaten by the two groups, in fact the differences are enormous. For seed size, a Smirnov 2-sample test between ants and rodents yields $T_2 = 0.387$, $P << 0.01$. For seed species, a variety of null models would be possible, but it is difficult to imagine any of them not finding rodents and ants remarkably different. For example, of 20 seed species recorded by Davidson et al (40), 13 are eaten only by ants or only by rodents. If one omits the 8 seed species eaten by only one granivore species, 5 of the remaining 12 seed species are eaten only by ants or only by rodents. A controlled experiment (15, 17) was equivocal (16, 57, 74): Removal of rodents resulted in a significant increase in ants only for the genus *Pheidole,* while, over the long term, rodent presence may facilitate ants by favoring growth of small seeded annual plants, the main resource for the ants. Even Brown & Davidson "no longer place so much confidence" (16, p. 1424) in this particular experimental implication that ant removal led to an increase in rodent density. However, experimentally demonstrating competition in the field even when it exists is extraordinarily difficult. The ambiguous results may stem from a reduction in competition effected by different foraging methods.

These ants and rodents may greatly lessen competition even though sizes and species of seeds overlap. As noted above, heteromyids are seed specialists while cricetids are omnivores, so overlap between these groups is reduced. The pocket mice specialize in seeds near shrub cover, while no ants do. Ants forage only on the surface, while the rodents, especially the kangaroo rats, can dig for buried seeds, which may remain in dense clumps long after surface seeds are gone (19, 127). In sum, though all these ants and rodents have been lumped together with birds in one guild (118, 137), they probably do not form one guild *sensu* Root (130). Possibly they form six. The differences in foraging behavior may suffice to reduce competition between species in different groups substantially.

PLANT GUILDS

Guilds defined for plants usually have not been tied to resources as obviously as animal guilds have, perhaps because of the difficulty in reconciling plant diversity with notions that resource partitioning structures communities. Field work in plant communities indicates that at most three or four resources are limiting in any community (149). "In terms of many of the theories of resource competition, it seems hard to imagine how hundreds of plant species may coexist when limited by a few resources, all of which are required for plant growth" (149, p. 8). Also, perhaps because researchers are animals, differences in ways that plants use resources do not seem as obvious as they do for animals. Some classifications seem little more than versions of life forms and other traditional schemes (cf 88) of classifying plants by vertical stratification. For example, Wilson (161) pointed to the difficulty in defining guilds *sensu* Root (130) in plant communities and suggested that some plant communities may be structured into guilds by stratification, especially of forests into canopy, small-tree stratum, herbs, etc. Resources, such as light and space, that might be partitioned by such stratification are tacit. Some designated tropical tree guilds (73)—shade-tolerant shrubs, understory trees—fall in this category. Others were defined on the basis of habitat—species restricted to slopes or swamps. Yet others may comprise species with similar, specialized conditions for regeneration. There is no single resource classification here; rather, the goal seems akin to the division of the entire community into "basic building blocks." Platt (124) assigned plants to a "fugitive species guild" on the grounds that they exploit a common resource (open space with increased soil moisture) in a similar manner, but he did not define what that manner was or suggest what might have constituted different manners. Fowler (56) doubted that a grassland plant community is divisible into well-defined guilds, on the grounds that all component species interact weakly and quite equally with one another.

Pollination guilds very much in the spirit of Root's definition are clearly possible for plants pollinated by animals. Because distantly related plants can compete for the same pollinators (90), pollination guilds can be taxonomically diverse. For example, Murray et al (115) discussed two tropical pollination guilds spanning several families. Here the resource was pollinating birds, shared by the guild associates: One guild is pollinated by long-billed hummingbirds and the other by short-billed hummingbirds. Wheelwright (158) conceived of plant guilds more narrowly, defining the Lauraceae as a guild that share avian seed dispersers and insect pollinators: "a group of species providing similar fruit resources and attracting the same seed dispersers. They also share similar floral morphologies and flower visitors" (158, p. 466). The original guild definition might have apportioned these species into guilds according to two separate classifications, and in each of these Lauraceae might have associated with heterofamilial species.

Fowler (56) designated warm- and cool-season grasses as two distinct "temporal guilds" and pointed out that, in her system, the cool-season species are all C_3 plants while the warm-season ones are all C_4 plants. The different photosynthetic pathways seem not so much different ways of getting a resource (light) as different ways of using it once it is captured. In much the same way, cow and horse digestive systems process grass differently, though the animals may be gathering the same resource in the same way. Digestion is not foraging. The parasitic habit of some flowering plants may lead to guilds more in the spirit of animal guilds whose associates gather similar foods in a similar way. No doubt detailed consideration of how parasitic plants find and use their hosts would lead to the demarcation of more than one such guild, but it seems clear that even hemiparasites are using at least some different resources than those of free-living plants and are garnering their resources very differently from free-living species.

Possibly because competition-structured guilds are so difficult to detect for plants, Atsatt & O'Dowd (3) used the term in a different sense entirely, one quite close to the dictionary definition of "guild." They identified "plant defence guilds" as groups of plant species that aid one another in defense against herbivores. For example, some plant species may have nectaries that attract predatory or parasitic insects to an entire plant assemblage. Or groups of plant species may contain alternate herbivorous insects that all function as hosts for a parasitoid that could not complete its life cycle on just one of them. This usage of "guild" has not prevailed.

GUILDS AS FUNDAMENTAL COMMUNITY BUILDING BLOCKS

The niche is usually viewed as a property of populations or species, but Elton (52) occasionally discussed niches as traits of groups of species not unlike guilds (67). As noted in the section on The Motivation Behind the Concept, Root hoped to reserve "niche" for the habitat requirements of a species and establish "guild" as the functional role of a group of species. This hope is unfulfilled, but "guild" and "niche" have come to have different connotations, "niche" referring to roles of species and "guild" to groups of species with similar roles. However, Terborgh & Robinson refer to "guild niches" (148, p. 89) as the sum of the niches of guild associates. Both guilds and niches in the "guild niche" sense have been prime candidates for fundamental elements of communities. Pianka suggested the possibility of a "periodic table of niches" (123, p. 264) completely analogous to the periodic table of elements, with entries that are groups of species corresponding to broadly delineated guilds (flycatchers, insectivorous bats, etc). Elton (52) anticipated the notion of community convergence of ecological roles and function, arguing that the divisions of different communities into roles is quite similar even when species compositions differ (67). Terborgh & Robinson (148) construed the many efforts at demonstrating morphological convergence as essentially in the same vein, with morphology reflecting ecological function. Various studies compared the structure of parallel guilds in different regions, and even in different periods. Van Valkenburgh (153, 154), for example, compared locomotor and trophic diversity within guilds of extant and extinct large predatory mammals, while Terborgh & Robinson (148) compared the partition into guilds of several tropical forest bird assemblages. Apparent constancy in proportions of species with different roles (e.g. 29, 68), quite similar to "guild signatures" (85), has been perceived as possibly indicating fundamental structural constraints on the relationship of basic community building blocks.

May was quite sanguine about this idea, seeing "the convergence of the structure of entire communities of plants and animals in spatially separate but climatically similar regions" (103, p. 151), of which the six identical guilds of insectivorous birds perceived by Cody (26, 27) in similar habitat of California, Chile, and South Africa were a prime example. Terborgh & Robinson (148), though optimistic, recognized two key problems that might lead researchers to see convergence that would suggest fundamental community units—first, it is not clear what aspects of communities to compare; and second, degree of similarity is largely in the eye of the beholder, and few descriptions entail statistical tests. Orians & Paine (118) were much more skeptical. They found little evidence of community convergence in the way terrestrial or benthic marine communities are structured into guilds; they see most anecdotal evidence of coevolutionary convergence as more parsimoniously explained by more or less similar physiological responses of individual species to similar environments.

Because this search for convergent community properties has focussed so heavily on ecological roles as integral to community constraints, and thus to how communities are likely to converge, guilds inevitably became one currency of this search. One would naturally start by focussing on which groups of species use the same resource (e.g. nectar) in different regions with different biotas. If one either omits the "similar way" part of the definition of "guild," or interprets it quite loosely, groups of species that use a resource more or less in the same way can often be construed as convergent guilds (e.g. sunbirds, honeycreepers, honey-eaters, and hummingbirds). However, observing similar sets of guilds defined in this way does not confirm a strong convergence in fundamental community properties. Evolutionary constraints may well restrict similarity of foraging methods between distantly related species to the extent that whether or not they are viewed as convergent guilds is a judgment call. Sunbirds usually perch and hummingbirds usually hover while feeding. Whether such differences in how they use a resource would lessen resource sharing could only be determined if they co-occurred. Preliminarily, all one can say is that birds have evolved to use nectar in different biogeographic regions. This is not to say that objective tests of convergence are impossible. For example, Niemi (116) found similar direction of morphological evolutionary change between congeneric shrub- and forest-dwelling peatland birds in both North America and Europe. But the division of species into more or less subjectively defined guilds in different regions does not seem to have aided the search for convergent emergent properties; this search remains frustrating (133).

GUILDS IN ENVIRONMENTAL ASSESSMENT

The popularity of the guild concept inevitably led to its use in environmental assessment and management (67). The key notion was that, if a guild consists of species using the same resources in similar fashion, "actions that affect environmental resources will similarly affect the members of the guilds using those resources. Once the impact on any one species in a guild is determined, the impact on every other species in the guild is known" (140, p. 187). The guild concept thus became inextricably linked with the idea of indicator species (94), except that indicators were transformed to "guild-indicators" (155) rather than indicators of entire communities. The assumed solid status of the guild concept in academic ecology justified the relatively easy but controversial management practice of using censuses of single species' populations as indicators of population trends of other species and of habitat quality. Severinghaus also suggested that one might predict the effect of an environmental action on a guild in one region from its effect on a similar guild in another region.

Traditional methods of classifying birds into guilds led to different partitions and, no matter what the partition, guild associates differ sufficiently in ways other than resource use that they need not respond identically to habitat change (92, 155). Verner (155) went so far as to suggest that, if guilds are to be used in management, they should be transmogrified to reflect Severinghaus's supposition. He defined a "management guild" as exactly those species that respond similarly to changes in their environment. Yet guilds are normally assigned, as noted in the section on Guilds in the Original Sense, on habitat and feeding grounds, in the absence of empirical research on how they respond to environmental changes (147). Further, if guilds as traditionally defined really do include species that most strongly compete with one another, one might expect that a common response among all guild associates to physical environmental change overlays compensatory, opposite-direction population changes among specific pairs of guild associates. It is therefore difficult to believe that a guild member can unambiguously represent a guild, even if all species have been studied for their individual responses to habitat change.

Probably because of its innate economic appeal (6), the use of guilds for management is remarkably persistent, in spite of the above objections and several critical studies of specific communities (e.g. 6, 101, 147). The guild concept seems to have taken on a life of its own (67). Roberts (129, p. 473) argued that, "Guilds can be used to select evaluation species to extrapolate information to nonstudied species." Karr (87) recommended comparing proportional representation among guilds of birds and fishes ("guild signature" [85], "guild spectrum" [31]) to indicate the health of biological systems, while warning against simplistic use of aggregate statistics such as various diversity indices to indicate environmental quality. However, it is difficult to imagine that guild signatures, guild indicators, and other guild-based shortcuts would not be used in exactly this way. Already, "response guilds" and the claim that individual guilds can indicate community-wide effects of disturbance are appearing in newspapers (e.g. 2). It would be sad if environmental managers, assuming ecologists know how to assign species to guilds easily and understand how guild associates behave with respect to one another and the surrounding milieu, based decisions about the fates of populations and communities on what is, in fact, an incompletely understood concept that is used inconsistently.

CONCLUDING REMARKS

Hawkins & MacMahon suggest that behavior—the way resources are used—is important only for descriptive cataloging of the components of a community: "If one uses guilds only as a classification scheme for community components, however, then a similar manner of resource use is an important criterion. Root's original intention seems to be related to classification" (67, p. 443). On the other hand, they feel that foraging method is irrelevant if one is seeking to understand how species interact in nature: "To state the question with a bit of hyperbole, does it matter that a particular insect species is captured by a silken spider web as opposed to a bird's beak? The ecosystem and community consequences are similar—one less insect of that species—and manner is irrelevant from the specific perspective" (67, p. 443).

We cannot accept this view. It is important to bear in mind that all species using a resource are potential competitors if that resource is limiting. But, as we noted in the desert granivore example, even if a resource is limiting—a point that is usually not demonstrated—to a great extent different ways of using a resource can reduce, though not necessarily eliminate, the effect that different species using that resource have on one another. It is quite possible that the "similar way" part of the original definition is just as important as the "same class," but it is certainly much more difficult to study and has thus received far less attention. Because related species often use resources in similar fashion, partitions into guilds, *sensu* Root, probably will often reflect taxonomy. But taxocenes taken as guilds without extensive information on which resources they use and how they use them should be viewed as, at best, preliminary constructs. Similarly, distantly related species are not automatically guild associates simply because they use similar resources.

That a search of titles and abstracts revealed 432 references to "guild" through 1986, most of them since 1980 (67), means that this word answered a felt need among ecologists. Had "guild" not come along, an analogous word would have. A major theme of ecology was resource partitioning and potential competition between species, and no prior term quite expressed the concept of the group of species that comprise the potential competitors or partitioners of some resource. We show in the section on Precursor and Parallel Concepts how "functional group" has evolved almost simultaneously to carry the same connotations, but the metaphor of the "guild" must have seemed more elegant than this term, because "guild" became far more popular. It is interesting and perhaps instructive that the metaphor was not quite apt from the outset, emphasizing competitive aspects of individuals doing the same thing, rather than the mutual aid and protection functions that typified craft guilds. Possibly the reason "guild" was the word that became popular, and not other more or less contemporaneous and synonymous terms, is the economic basis of its referent. A key attraction of the notion of communities regulated by competition among their populations may be its suggestion that unfettered capitalism is "natural" and has self-regulatory traits analogous to those of a supposed "balance of nature" (8, 95). If this is so it may not be surprising that the strong connotation of mutualism in the word "guild" was deemphasized. Others (8, 128) have argued that a similar deemphasis of ecological mutualism relative to competitive interactions and predation reflects a social background and worldview as much as the nature of nature.

The problem with "guild" now is that it has too many connotations. Root's original definition was very precise, but others adapted the term to their own needs so that different authors may connote very different things. Even researchers who know the original paper, supporting meticulous research and strict definition, lapse into idiosyncratic, casual use of the term. For the same reasons that the term caught on in the first place, it is still needed, but it is well on its way to becoming a panchreston, a term of such flexible meaning that its use is as likely to generate confusion as to communicate information. However, if "guild" is used in the narrow sense, the notion of the group of all species using a resource is an anonym. The need for a word for such a group is so great that this meaning has almost co-opted "guild;" now that "guild" is used in a very broad sense in textbooks as the basis for community ecology (51), it will be difficult for the narrow meaning to recapture it and save it for scientific use. If the term is to be used fruitfully in research, two conditions must be met:

1. A clear statement is always required of the criteria and considerations that have led to a particular guild assignment, in order to enable other investigators to consider the validity of this designation. To say "We divided the birds into four guilds based on microhabitat use," or "We divided the birds into six guilds based on the habitat and dietary information in Smith and Wesson (1913)," is insufficient. The role of foraging method should be emphasized because of its potential importance in effecting differences in resource use.

2. Sympatric related biota should always be listed and an explanation provided for why they were excluded from a designated guild, in the many cases where the grounds for this exclusion are not self-evident and consist of intuition or simply taxonomy.

ACKNOWLEDGMENTS

We thank Jim Brown, Fran James, Judy Rhymer, Sharon Strauss, Mary Tebo, and Joe Travis for helpful comments on this manuscript.

Literature Cited

1. Adams, J. 1985. The definition and interpretation of guild structure in ecological communities. *J. Anim. Ecol.* 54:43–59
2. Anonymous. 1990. Bird songs offer clues to habitats. *New York Times*, Sept. 4, 1990, p. B7
3. Atsatt, P. R., O'Dowd, D. J. 1976. Plant defence guilds. *Science* 193:24–29
4. Beggs, J. 1988. *Energetics of kaka in a South Island beech forest*. M.Sc. Thesis, Univ. Auckland, N.Z.
5. Berkman, H. E., Rabenin, C. F. 1987. Effects of siltation on stream fish communities. *Environ. Biol. Fishes* 18:285–294
6. Block, W. M., Brennan, L. A., Gutierrez, R. J. 1986. The use of guilds and guild-indicator species for assessing habitat suitability. In *Wildlife 2000. Modeling Habitat Relationships of Terrestrial Vertebrates*, ed. J. Verner, M. L. Morrison, C. J. Ralph, pp. 109–13. Madison, Wisc: Univ. Wisc. Press
7. Bonaccorso, F. J. 1979. Foraging and reproductive ecology in a Panamanian bat community. *Bull. Fla. State Mus.* 24:359–408
8. Boucher, D. H. 1985. The idea of mutualism, past and future. In *The Biology of Mutualism*, ed. D. H. Boucher, pp. 1–28. London: Croom Helm

9. Bowers, M. A., Brown, J. H. 1982. Body size and coexistence in desert rodents: chance or community structure? *Ecology* 63:391–400
10. Boyd, S. A. 1987. *Patterns of use of beech honeydew by birds and insects.* M.Sc. thesis. Univ. Auckland, N.Z.
11. Brown, J. H. 1973. Species diversity of seed-eating desert rodents in sand dune habitats. *Ecology* 54:775–87
12. Brown, J. H. 1975. Geographical ecology of desert rodents. In *Ecology and Evolution of Communities,* ed. M. L. Cody, J. M. Diamond, pp. 315–41. Cambridge, Mass: Harvard Univ. Press
13. Brown, J. H. 1981. Two decades of homage to Santa Rosalia: Toward a general theory of diversity. *Am. Zool.* 21:877–88
14. Brown, J. H. 1987. Variation in desert rodent guilds: patterns, processes, and scales. In *Organizations of Communities: Past and Present,* ed. J. H. R. Gee, P. S. Giller, pp. 185–203. Oxford: Blackwell Sci.
15. Brown, J. H., Davidson, D. W. 1977. Competition between seed-eating rodents and ants in desert ecosystems. *Science* 196:880–82
16. Brown, J. H., Davidson, D. W. 1986. Reply to Galindo. *Ecology* 67:1423–25
17. Brown, J. H., Davidson, D. W., Reichman, O. J. 1979. An experimental study of competition between seed-eating desert rodents and ants. *Am. Zool.* 19:1129–43
18. Brown, J. H., Heske, E. J. 1990. Control of a desert-grassland transition by a keystone rodent guild. *Science* 250: 1705–7
19. Brown, J. H., Reichman, O. J., Davidson, D. W. 1979. Granivory in desert ecosystems. *Annu. Rev. Ecol. Syst.* 10:201–27
20. Burns, T. P. 1989. Lindeman's contradiction and the trophic structure of ecosystems. *Ecology* 70:1355–62
21. Bush, A. O. 1990. Helminth communities in avian hosts: determinants of pattern. In *Parasite Communities: Patterns and Processes,* G. W. Esch, A. O. Bush, J. M. Aho, pp. 197–232. London: Chapman & Hall
22. Case, T. J., Faaborg, J., Sidell, R. 1983. The role of body size in the assembly of West Indian bird communities. *Evolution* 37:1062–74
23. Chew, R. 1977. Some ecological characteristics of the ants of a desert-shrub community in southeastern Arizona. *Am. Midl. Natl.* 98:33–49
24. Chew, R. M., De Vita, J. 1980. Foraging characteristics of a desert ant assemblage: functional morphology and species separation. *J. Arid Environ.* 3:75–83
25. Clements, F. E. 1905. *Research methods in ecology.* Univ. Publ. Lincoln, Nebraska
26. Cody, M. L. 1974. *Competition and the Structure of Bird Communities.* Princeton, NJ: Princeton Univ. Press
27. Cody, M. L. 1975. Towards a theory of continental species diversities. In *Ecology and Evolution of Communities,* ed. M. L. Cody, J. M. Diamond, pp. 214–57. Cambridge, Mass: Harvard Univ. Press
28. Cody, M. L. 1983. Bird diversity and density in South African forests. *Oecologia* 59:201–15
29. Cohen, J. 1977. Ratio of prey to predators in community food webs. *Nature* 270:165–67
30. Connor, E. F., Simberloff, D. 1984. Neutral models of species' cooccurrence patterns. In *Ecological Communities, Conceptual Issues and the Evidence,* ed. D. R. Strong, Jr., D. Simberloff, L. G. Abele, and A. B. Thistle pp. 316–31, 341–43. Princeton, NJ: Princeton Univ. Press
31. Cornell, H. V., Kahn, D. M. 1989. Guild structure in the British arboreal arthropods: Is it stable and predictable? *J. Anim. Ecol.* 58:1003–20
32. Crome, F. H. J. 1975. The ecology of fruit pigeons in tropical Northern Queensland. *Aust. Wildl. Res.* 2:155–85
33. Crome, F. H. J. 1978. Foraging ecology of an assemblage of birds in lowland rainforest in northern Queensland. *Austr. J. Ecol.* 3:195–212
34. Cruz, A. 1987. Avian community organization in a mahogany plantation on a neotropical island. *Carib. J. Sci.* 23(2):286–96
35. Csuti, B. A. 1979. Patterns of adaptation and variation in the Great Basin kangaroo rat *(Dipodomys microps). Univ. Calif. Publ. Zool.* 111:1–69
36. Cummins, K. W. 1974. Structure and function of stream ecosystems. *BioScience* 24:631–641
37. Darwin, C. 1872. *The Origin of Species.* (Reprinted 1962) New York: Collier
38. Davidson, D. W. 1977. Species diversity and community organization in desert seed-eating ants *Ecology* 58:711–24
39. Davidson, D. W. 1977. Foraging ecology and community organization in desert seed-eating ants *Ecology* 58:725–37
40. Davidson, D. W., Brown, J. H., Inouye, R. S. 1980. Competition and the structure of granivore communities. *BioScience* 30:233–38

41. Dayan, T., Simberloff, D. 1991. Morphological relationships among coexisting heteromyids: An incisive dental character. *Am. Nat.* submitted
42. Dayan, T., Simberloff, D., Tchernov, E., and Yom-Tov, Y. 1989. Inter- and intraspecific character displacement in mustelids. *Ecology* 70(5):1526–39
43. Dayan, T., Simberloff, D., Tchernov, E., Yom-Tov, Y. 1990. Feline canines: community-wide character displacement in the small cats of Israel. *Am. Nat.* 136(1):39–60
44. Dayan, T., Simberloff, D., Tchernov, E., Yom-Tov, Y. 1991. Canine carnassials: Character displacement among the wolves, jackals, and foxes of Israel. *Biol. J. Linn. Soc.* In press
45. Dayan, T., Tchernov, E., Yom-Tov, Y., Simberloff, D. 1989. Ecological character displacement in Saharo-Arabian *Vulpes:* Outfoxing Bergmann's Rule. *Oikos* 55:263–72
46. Dayan, T., Simberloff, D., Tchernov, E., Yom-Tov, Y. 1991. Tooth size: function and coevolution in carnivore guilds. In *Function and Evolution of the Teeth,* ed. P. Smith. Jerusalem: Freund. In press
47. Diamond, J. M. 1975. Assembly of species communities. In *Ecology and Evolution of Communities,* ed. M. L. Cody, J. M. Diamond, pp. 342–444. Cambridge, Mass: Harvard Univ. Press
48. Diamond, J. M., Gilpin, M. E. 1982. Examination of the "null" model of Connor and Simberloff for species co-occurrences on islands. *Oecologia* 52:64–74
49. DuBowy, P. J. 1988. Waterfowl communities and seasonal environments: Temporal variability in interspecific competition. *Ecology* 69:1439–53
50. Duggins, D. O. 1981. Interspecific facilitation in a guild of benthic marine herbivores. *Oecologia* 48:157–63
51. Ehrlich, P. R., Roughgarden, J. 1987. *The Science of Ecology.* New York: Macmillan
52. Elton, C. S. 1927. *Animal Ecology.* London: Sidgwick & Jackson
53. Emlen, J. T. 1986. Land-bird diversities in matched habitats on six Hawaiian islands: a test of resource regulation theory. *Am. Nat.* 127(2):125–39
54. Eriksson, M. O. G. 1979. Competition between freshwater fish and goldeneyes *Bucephala clagula* (L.) for common prey. *Oecologia* 41:99–107
55. Folse, L. T. Jr. 1981. Ecological relationships of grassland birds to habitat and food supply in east Africa. In *The Use of Multivariate Statistics in Studies of Wildlife Habitat. U.S.D.A. Forest Service, Gen. Techn. Report RM-87,* ed. D. E. Capen, pp. 160–66. Ft. Collins, Colo: Rocky Mountain For. Range Exp. Stat.
56. Fowler, N. 1981. Competition and coexistence in a North Carolina grassland. *J. Ecol.* 69:843–54
57. Galindo, C. 1986. Do desert rodent populations increase when ants are removed? *Ecology* 67:1422–23
58. Gilpin, M. E., Diamond, J. M. 1984. Are species co-occurrences on islands non-random, and are null hypotheses useful in ecology? In *Ecological Communities: Conceptual Issues and the Evidence.* ed. D. R. Strong, D. Simberloff, L. G. Abele, A. B. Thistle, pp. 296–315. Princeton, NJ: Princeton Univ. Press
59. Gittleman, J. L. 1985. Carnivore body size: ecological and taxonomic correlates. *Oecologia* 67:540–54
60. Gleason, H. A. 1926. The individualistic concept of the plant association. *Torrey Bot. Club Bull.* 53:7–26
61. Gorman, O. T. 1988. The dynamics of habitat use in a guild of Ozark minnows. *Ecol. Monogr.* 58:1–18
62. Grant, P., Schluter, D. 1984. Interspecific competition inferred from patterns of guild structure. In *Ecological Communities, Conceptual Issues and the Evidence,* ed. D. R. Strong, Jr., D. Simberloff, L. G. Abele, A. B. Thistle, pp. 201–33. Princeton, NJ: Princeton Univ. Press
63. Guillet, A., Crowe, T. M. 1985. Patterns of distribution, species richness, endemism and guild composition of water-birds in Africa. *Afr. J. Ecol.* 23:89–120
64. Hairston, N. G. Sr. 1987. *Community Ecology and Salamander Guilds.* Cambridge: Cambridge Univ. Press
65. Hallett, J. G. 1982. Habitat selection and the community matrix of a desert small-mammal fauna. *Ecology* 63:1400–10
66. Hanski, I., Koskela, H. 1979. Resource partitioning in six guilds of dung-inhabiting beetles (Coleoptera). *Annal. Entomol. Fennici* 45:1–12
67. Hawkins, C. P., MacMahon, J. A. 1989. Guilds: the multiple meanings of a concept. *Annu. Rev. Entomol.* 34:423–51
68. Heatwole, H., Levins, R. 1972. Trophic structure stability and faunal change during recolonization. *Ecology* 53:531–34
69. Hendrix, S. A., Brown, V. K., Dingle, H. 1988. Arthropod guild structure during early old field succession in a New

and Old World site. *J. Anim. Ecol.* 57:1053–65

70. Hill, D., Wright, R., Street, M. 1987. Survival of Mallard ducklings *Anas platyrhynchos* and competition with fish for invertebrates on a flooded gravel quarry in England. *Ibis* 129:159–67
71. Hochberg, M. E., Lawton, J. H. 1990. Competition between kingdoms. *Trends Ecol. Evol.* 5:367–71
72. Holmes, R. T., Bonney, R. E. Jr., Pacala, S. W. 1979. Guild structure of the Hubbard Brook bird community: a multivariate approach. *Ecology* 60:512–20
73. Hubbell, S. P., Foster, R. B. 1986. Biology, chance, and history and the structure of tropical rain forest tree communities. In *Community Ecology,* ed. J. Diamond, T. J. Case, pp. 314–29. New York: Harper & Row
74. Hulley, P. E., Walter, G. H., Craig, A. J. F. K. 1988. Interspecific competition and community structure, II. The recognition concept of species. Rivista di Biologia—*Biol. For.* 81:263–85
75. Humphrey, S. R., Bonaccorso, F. J., Zinn, T. L. 1983. Guild structure of surface-gleaning bats in Panama. *Ecology* 64(2):284–94
76. Hurlbert, S. H., Loayza, W., Moreno, T. 1986. Fish-flamingo-plankton interactions in the Peruvian Andes. *Limnol. Oceanogr.* 31:457–68
77. Hutchinson, G. E. 1959. Homage to Santa Rosalia, or why are there so many kinds of animals? *Am. Nat.* 93:145–59
78. Inger, R. F., Colwell, R. K. 1977. Organization of contiguous communities of amphibians and reptiles in Thailand. *Ecol. Monogr.* 47:229–53
79. Jaksic, M. F. 1981. Abuse and misuse of the term "guild" in ecological studies. *Oikos* 37:397–400
80. Jaksic, F. M., Greene, H. W., Yanez, J. L. 1981. The guild structure of a community of predatory vertebrates in central Chile. *Oecologia* 49:21–28
81. Jaksic, F. M., Delibes, M. 1987. A comparative analysis of food-niche relationships and trophic guild structure in two assemblages of vertebrate predators differing in species richness: causes, correlations, and consequences. *Oecologia* 71:461–72
82. Jaksic, F. M., Medel, R. G. 1990. Objective recognition of guilds: testing for statistically significant species clusters. *Oecologia* 82:87–92
83. Jaksic, F. M., Yanez, J. L., Fuentes, E. R. 1981. Assessing a small mammal community in central Chile. *J. Mammal.* 62:391–96
84. Joern, A., Lawlor, L. R. 1981. Guild structure in grasshopper assemblages based on food and microhabitat resources. *Oikos* 37:93–104
85. Karr, J. R. 1980. Geographical variation in the avifaunas of tropical forest undergrowth. *Auk* 97:283–98
86. Karr, J. R. 1982. Avian extinction on Barro Colorado Island, Panama: a reassessment. *Am. Nat.* 119(2):220–39
87. Karr, J. R. 1987. Biological monitoring and environmental assessment: a conceptual framework. *Environ. Manage.* 11:249–56
88. Kershaw, K. A. 1964. *Quantitative and Dynamic Ecology*. New York: Am. Elsevier
89. Knopf, F. L., Olson, T. E. 1984. Naturalization of Russian-olive: implications to Rocky Mountain wildlife. *Wildl. Soc. Bull.* 12:289–98
90. Kodric-Brown, A., Brown, J. H. 1979. Competition between distantly related taxa in the coevolution of plants and pollinators. *Am. Zool.* 19:1115–27
91. Lambert, S., Reid, W. H. 1981. Biogeography of the Colorado herpetofauna. *Am. Midl. Nat.* 106:145–56
92. Landres, P. B. 1983. Use of the guild concept in environmental impact assessment. *Enviro. Manage.* 7:393–98
93. Landres, P. B., MacMahon, J. A. 1980. Guilds and community organization: analysis of an oak woodland avifauna in Sonora, Mexico. *Auk* 97:351–65
94. Landres, P. B., Verner, J., Thomas, J. W. 1988. Ecological uses of vertebrate indicator species: a critique. *Conserv. Biol.* 2:316–28
95. Leigh, E. G. Jr. 1971. The energy ethic. *Science* 172:664–66
96. MacArthur, R. H., Levins, R. 1967. The limiting similarity, convergence, and divergence of coexisting species. *Am. Nat.* 101:377–85
97. MacMahon, J. A. 1976. Species and guild similarity of North American desert mammal faunas: A functional analysis of communities. In *Evolution of Desert Biota,* ed. D. W. Goodall, pp. 133–48. Austin: Univ. Texas Press
98. MacMahon, J. A., Schimpf, D. J., Andersen, D. C., Smith, K. G., Bayn, R. L. 1981. An organism based approach to some community and ecosystem concepts. *J. Theor. Biol.* 88:287–307
99. MacNally, R. C. 1983. On assessing the significance of interspecific competition to guild structure. *Ecology* 64:1646–52
100. MacNally, R. C., Doolan, J. M. 1986. An empirical approach to guild structure: habitat relationships in nine species of eastern-Australian cicadas. *Oikos* 47:33–46

101. Mannan, R. W., Morrison, M. L., Meslow, E. C. 1984. The use of guilds in forest bird management. *Wildl. Soc. Bull.* 12:426–30
102. Martin, T. E. 1981. Limitation in small habitat islands: chance or competition? *Auk* 98:715–34
103. May, R. M. 1976. Patterns in multispecies communities. In *Theoretical Ecology, Principles and Applications,* ed. R. M. May, pp. 142–62. Philadelphia: Saunders
104. May, R. M., Seger, J. 1986. Ideas in ecology. *Am. Sci.* 74:256–67
105. McIntosh, R. P. 1985. *The Background of Ecology*. Cambridge: Cambridge Univ. Press
106. Mehlhop, P., Scott, N. J. 1983. Temporal patterns of seed use and availability in a guild of desert ants. *Ecol. Entomol.* 8:69–85
107. Menge, B. A., Lubchenco, J., Ashkenas, L. R., Ramsey, F. 1986. Experimental separation of effects of consumers on sessile prey in the low zone of a rocky shore in the Bay of Panama: direct and indirect consequences of food web complexity. *J. Exp. Mar. Biol. Ecol.* 100:225–69
108. Merritt, R. W., Cummins, K. W., Burton, T. M. 1984. The role of aquatic insects in the processing and cycling of nutrients. In *The Ecology of Aquatic Insects,* ed. V. H. Resh, D. M. Rosenberg, pp. 134–63. New York: Praeger
109. Meserve, P. L. 1981. Trophic relationships among small mammals in a Chilean semi-arid thorn scrub community. *J. Mammal.* 62:304–14
110. Meserve, P. L. 1981. Resource partitioning in a Chilean semi-arid small mammal community. *J. Anim. Ecol.* 50:745–57
111. Moller, H., Tilley, J. A. V. 1989. Beech honeydew: Seasonal variation and use by wasps, honeybees, and other insects. *N.Z. J. Ecol.* 16:289–302
112. Moran, V. C., Southwood, T. R. E. 1982. The guild composition of arthropod communities in trees. *J. Anim. Ecol.* 51:289–306
113. Morin, P. J., Lawler, S. P., Johnson, E. A. 1988. Competition between aquatic insects and vertebrates: Interaction strength and higher order interactions. *Ecology* 69:1401–9
114. Morton, S. R. 1979. Diversity of desert-dwelling mammals: a comparison of Australia and North America. *J. Mammal.* 60:253–64
115. Murray, K. G., Feinsinger, P., Busby, W. H., Linhart, Y. B., Beach, J. H., Kinsman, S. 1987. Evaluation of character displacement among plants in two tropical pollination guilds. *Ecology* 68 (5):1283–93
116. Niemi, G. J. 1985. Patterns of morphological evolution In bird genera of New World and Old World peatlands. *Ecology* 66:1215–28
117. *Oxford English Dictionary* (Compact Edition). 1971. Oxford: Oxford University Press
118. Orians, G. H., Paine, R. T. 1983. Convergent evolution at the community level. In *Coevolution,* ed. D. J. Futuyma, M. Slatkin, pp. 431–58. Sunderland, Mass: Sinauer
119. Osborne, D. R., Beissinger, S. R., Bourne, G. R., 1983. Water as an enhancing factor in bird community structure. *Carib. J. Sci.* 19(1–2):35–38
120. Paine, R. T. 1988. Food webs: Road maps of interactions or grist for theoretical development? *Ecology* 69:1648–54
121. Pianka, E. R. 1973. The structure of lizard communities. *Annu. Rev. Ecol. Syst.* 4:53–74
122. Pianka, E. R. 1980. Guild structure in desert lizards. *Oikos* 35:194–201
123. Pianka, E. R. 1988. *Evolutionary Ecology*. New York: Harper & Row 4th ed.
124. Platt, W. J. 1975. The colonization and formation of equilibrium plant species associations on badger disturbances in a tall-grass prairie. *Ecol. Monogr.* 45: 285–305
125. Reichman, O. J. 1979. Introduction to the symposium: Competition between distantly related taxa. *Am. Zool.* 19: 1027
126. Reichman, O. J. 1981. Factors influencing foraging in desert rodents. In *Foraging Behavior: Ecological, Ethological, and Psychological Approaches,* ed. A. C. Kamil, T. D. Sargent, pp. 195–213. New York: Garland STPM
127. Reichman, O. J., Oberstein, D. 1977. Selection of seed distribution types by *Dipodomys merriami* and *Perognathus amplus*. *Ecology* 58:636–43
128. Risch, S., Boucher, D. 1976. What ecologists look for. *Bull. Ecol. Soc. Amer.* 57(3):8–9
129. Roberts, T. H. 1987. Construction of guilds for habitat assessment. *Environ. Manage.* 11:473–77
130. Root, R. B. 1967. The niche exploitation pattern of the blue-gray gnatcatcher. *Ecol. Monogr.* 37:317–50
131. Salt, G. W. 1953. An ecologic analysis of three California avifaunas. *Condor* 55:258–73
132. Salt, G. W. 1957. An analysis of avifaunas in the Teton Mountains and Jackson Hole, Wyoming. *Condor* 59:373–93

133. Salt, G. W. 1979. A comment on the use of the term "emergent properties." *Amer. Natur.* 113:145–61
134. Schimper, A. F. W. 1903. *Plant geography upon a physiological basis.* Oxford, Eng: Clarendon
135. Schluter, D. 1986. Character displacement between distantly related taxa? Finches and bees in the Galapagos. *Am. Nat.* 127:95–102
136. Schoener, T. W. 1983. Field experiments on interspecific competition. *Am. Nat.* 122:240–85
137. Schoener, T. W. 1986. Overview: kinds of ecological communities—ecology becomes pluralistic. In *Community Ecology,* ed. J. Diamond T. J. Case New York: Harper & Row
138. Schoener, T. W., Spiller, D. A. 1987. Effect of lizards on spider populations: Manipulative reconstruction of a natural experiment. *Science* 236:949–52
139. Sedgwick, J. A., Knopf, F. L. 1987. Breeding bird response to cattle grazing of a cottonwood bottomland. *J. Wildlife Manage.* 51(1):230–37
140. Severinghaus, W. D. 1981. Guild theory development as a mechanism for assessing environmental impact. *Environ. Manage.* 5:187–90
141. Short, H. L., Burnham, K. P. 1982. Techniques for structuring wildlife guilds to evaluate impacts on wildlife communities. *U.S.D.I. Fish and Wildlife Service, Special Science Report—Wildlife* 244
142. Simberloff, D. 1976. Trophic structure determination and equilibrium in an arthropod community. *Ecology* 57:395–98
143. Smith, K. G., MacMahon, J. A. 1981. Bird communities along a montane sere: structure and energetics. *Auk* 98:8–28
144. Smythe, N. 1986. Competition and resource partitioning in the guild of Neotropical terrestrial frugivorous mammals. *Annu. Rev. Ecol. Syst.* 17:169–88
145. Strand, S. 1988. Following behavior: interspecific foraging associations among Gulf of California reef fishes. *Copeia* 1988(2):351–57
146. Strauss, R. E. 1982. Statistical significance of species clusters in association analysis. *Ecology* 63:634–39
147. Szaro, R. C. 1986. Guild management: An evaluation of avian guilds as a predictive tool. *Environ. Manage.* 10:681–88
148. Terborgh, J., Robinson, S. 1986. Guilds and their utility in ecology. In *Community Ecology: Pattern and Process,* ed. J. Kikkawa, D. J. Anderson, pp. 65–90. Palo Alto: Blackwell
149. Tilman, D. 1982. *Resource Competition and Community Structure.* Princeton, NJ: Princeton Univ. Press
150. Toda, M. J. 1984. Guild structure and its comparisons between two local drosophilid communities. *Physiol. Ecol. Jpn* 21:131–72
151. Turpaeva, E. P. 1957. Food interrelationships of dominant species in marine benthic biocoenoses. Akad. nauk SSSR. *Trans. Inst. Okeanologii* v. 20 f. 171. (Am. Inst. Biol. Sci.: Washington DC Transl., 1959.) pp 137–148
152. Van Valkenburgh, B. 1985. Locomotor diversity within past and present guilds of large predatory mammals. *Paleobiology* 11:406–28
153. Van Valkenburgh, B. 1988. Trophic diversity in past and present guilds of large predatory mammals. *Paleobiology* 14: 155–73
154. Van Valkenburgh, B. 1989. Carnivore dental adaptations and diet: a study of trophic diversity within guilds. In *Carnivore Behavior, Ecology, and Evolution,* ed. J. L. Gittleman, pp. 410–36. Ithaca: Cornell Univ. Press
155. Verner, J. 1984. The guild concept applied to management of bird populations. *Environ. Manage.* 8:1–14
156. Walter, D. E., Ikonen, E. K. 1989. Species, guilds, and functional groups: taxonomy and behavior in nematophagous arthropods. *J. Nematol.* 21(3):315–27
157. Walter, G. 1988. Competitive exclusion, coexistence, and community ecology. *Acta Biotheoret.* 37:281–313
158. Wheelwright, N. T. 1985. Competition for dispersers, and the time of flowering and fruiting in a guild of tropical trees. *Oikos* 44(3):465–77
159. Whittam, T. S., Siegel-Causey, D. 1981. Species interactions and community structure in Alaskan seabird colonies. *Ecology* 62:1515–26
160. Wilson, J. B. 1989. Relations between native and exotic plant guilds in the Upper Clutha, New Zealand. *J. Ecol.* 77:223–35
161. Wilson, J. B. 1989. A null model of guild proportionality applied to stratification of a New Zealand temperate rain forest. *Oecologia* 80(2):263–67
162. Winemiller, K. O., Pianka, E. R. 1990. Organization in natural assemblages of desert lizards and tropical fishes. *Ecol. Monogr.* 60:27–55
163. Wright, S. J. 1981. Extinction-mediated competition: The *Anolis* lizards and insectivorous birds of the West Indies. *Am. Nat.* 117:181–92
164. Yodzis, P. 1982. The compartmentation of real and assembled ecosystems. *Am. Nat.* 120:551–70

Annu. Rev. Ecol. Syst. 1991. 22:145–75

LEMUR ECOLOGY

Alison F. Richard[1]

[1]Department of Anthropology and Peabody Museum of Natural History, Yale University, New Haven, Connecticut 06520

Robert E. Dewar[2]

[2]Department of Anthropology, University of Connecticut, Storrs, Connecticut 06269

KEY WORDS: Madagascar, seasonality, extinctions, history, biogeography

---

INTRODUCTION

If ever there were a place where nonhuman primates can be said to be ecologically dominant, Madagascar is that place. An entire suborder evolved on the island, including a minimum of 15 genera and 44 species that ranged in size from mice to gorillas. They lived amidst a flora as diverse as any tropical flora in the world (38) and, before the arrival of people, they shared it with few other mammals. There is evidence of only one family of insectivores (the Tenrecidae), an endemic subfamily of rodents (the Nesomyinae), a handful of small, endemic viverrids, a few bats, hippos, and a suid (1, 2, 6, 60, 65, 68, 109, 112). The bird fauna of Madagascar is depauperate too, not only compared to continental bird communities of similar latitude but also in comparison with island bird populations such as that of Borneo (80). Whole guilds, notably woodpeckers, nectar-feeders, and fruit-eaters, are absent or sparsely represented (17, 80; S. Zack, personal communication).

The reason for the remarkable predominance of primates on Madagascar is uncertain. It has been estimated that the island reached its current, isolated position 120 million years ago (128), making a sweepstake crossing of the Mozambique Channel the only route available to most colonizers. The lemur ancestors may simply have been luckier rafters than most other mammals, or the history of landlinks between Madagascar and the mainland may have been

0066-4162/91/1120-0145$02.00

more recent and, perhaps, more complex than now supposed (summarized in 94, 164). It is still unclear whether the lemurs are a monophyletic or polyphyletic group (94). Whatever their origins, the diversification of the lemurs was evidently favored by the wide range of environments and vegetation types on the 590,000 km^2 island, and also by its emptiness: the Holocene record shows that the only other large animals certainly present were ratites and tortoises. There is no pre-Holocene fossil record.

Two thousand years ago or less, people settled Madagascar (16, 24–27). The extinction of at least 25 species of mammals, ratites, and tortoises followed rapidly, all of them larger than their closest surviving relatives (if any) on the island. Among these were 8 genera and at least 15 species of lemurs (3). The causes of the extinctions probably varied regionally, but human activities are implicated, including hunting, the destruction and fragmentation of habitat by fire, cutting, and the introduction of domesticated bovids, as well as the effects of late Holocene climate shifts (16, 24, 26).

The importance of past events, as distinct from evolutionary processes, in shaping modern communities has received increasing attention in recent years (9, 28, 31, 188). Known and inferred historical events certainly complicate the interpretation of modern lemur ecology. Moreover, while comparative data are sorely needed to understand, for example, the ecological consequences of the recent and massive extinctions on Madagascar, there are no good models for the unique combination of diversity and depauperateness that characterized intact communities on the island.

We have no solution to this conundrum, but historical considerations have certainly influenced the scope of this review quite strongly. We consider the extinct lemurs as part of the ecological present, because their extinction is so recent (109, 151, 164, 183). Accordingly, our account of the lemurs as an adaptive array includes the extinct forms whenever morphological remains and context can reasonably be used to infer behavioral ecology. The recent history of local extinctions and invasions cannot yet be reconstructed, and so we can only try to inform our discussion of community structure and function with the general knowledge that even the least evidently disrupted communities are unlikely to be intact or at equilibrium.

Issues around which this review is organized are, first, the historical and evolutionary intepretation of lemur biogeography and, second, the ecological distinctiveness of this primate suborder: in what ways is lemur ecology appropriately viewed as the product of a long process of parallel and convergent evolution with haplorhine primates (i.e. tarsiers, monkeys, and apes), and in what ways is it the product of unique historical events? Both issues urgently need study. With the rapidly accelerating destruction of habitat in Madagascar much of the pertinent evidence may soon be lost for ever (47, 53, 101, 139, 167).

SYSTEMATICS AND TAXONOMY

The way in which lemurs are classified systematically has a crucial bearing on the interpretation of their ecology and biogeography (18, 42, 49, 50, 61, 148, 158, 164, 166, 168, 171). We use Tattersall's (164) 1982 classification (Table 1), with some modifications. Specifically, among living forms, two new species of lemurs have been discovered (95, 147), six species of *Lepilemur* are now usually recognized (166), the genus *Lemur* has been divided (50, 148); there have been revisions in the subspecies of *Propithecus verreauxi* (167); two subspecies are recognized in *Eulemur macaco* (76), and four in *Phaner furcifer* (51), both previously regarded as monotypic. One new extinct form, *Babakotia radofilai,* has been discovered (45). The taxonomy of other groups remains unsettled, and there may be unrecognized diversity in the nocturnal forms (51). Our higher level taxonomy follows that of Schwartz & Tattersall (145), except that *Mesopropithecus* is assigned to the Paleopropithecinae (42).

At present, then, there are 29 living species and 15 others that have become extinct during the Holocene (3, 24). All but two of the 29 species are found only in Madagascar. *Eulemur mongoz* is also found on the Comoro Islands of Ndzuwani and Mwali, and *E. fulvus mayottensis* on Maore, where their presence is most likely attributable to relatively recent human activity (165). The extinct species are not a random sample, being in every case at least as large-bodied as the largest surviving species (Figure 1), and disproportionately drawn from the family Indridae. All the extant lemurs are almost or completely sexually monomorphic with respect to body size and canine tooth morphology (8, 40, 62, 73) and, in surprising contrast to other primates, the large-bodied extinct species may also have been monomorphic (L. R. Godfrey, S Lyon, M. Sutherland, in preparation). Body weights used in this review, therefore, are combined means for males and females.

THE ENVIRONMENTAL CONTEXT

Madagascar's climates and environments are dominated by a mountain chain that rises steeply from the east coast to peaks over 2000 m, and runs almost the length of the island. West of this escarpment lies a hilly plateau which rolls gently to the western and southern coasts. The island's north-south mountainous spine intercepts the trade winds from the Indian Ocean and creates a steep rainfall gradient from the northeast to the southwest of the island. The rainfall is seasonally distributed, with particularly long dry seasons in the west, south, and extreme north. These patterns are reflected in the island's vegetation zones, which today range from humid evergreen forests at

Table 1 The existing and recently extinct (identified with †) lemurs.[a]

Family Lemuridae
- *Lemur catta* Ring-tailed lemur.
- *Eulemur mongoz* Mongoose lemur.
- *E. macaco* Black lemur.
  - *E. m. macaco*
  - *E. m. flavifrons*
- *E. fulvus* Brown lemur
  - *E. f. fulvus*
  - *E. f. albifrons*
  - *E. f. rufus*
  - *E. f. collaris*
  - *E. f. mayottensis*
  - *E. f. sanfordi*
  - *E. f. albocollaris*
- *E. coronatus* Crowned lemur.
- *E. rubriventer* Redbellied lemur.
- *Varecia variegata* Ruffed lemur.
  - *V. v. variegata*
  - *V. v. rubra*
- *Hapalemur griseus* Gentle lemur.
  - *H. g. griseus*
  - *H. g. alaotrensis*
  - *H. g. occidentalis*
- *H. simus* Broad-nosed gentle lemur.
- *H. aureus* Golden gentle lemur.
- †*Pachylemur insignis*
- †*P. jullyi*

Family Megaladapidae

Subfamily Lepilemurinae
- *Lepilemur mustelinus* Weasel lemur.
- *L. ruficaudatus* Red-tailed sportive lemur.
- *L. dorsalis* Grey-backed sportive lemur.
- *L. leucopus* White-footed sportive lemur.
- *L. edwardsi* Milne-Edward's sportive lemur.
- *L. septentrionalis* Northern sportive lemur.

Subfamily Megaladapinae
- †*Megaladapis edwardsi*
- †*M. grandidieri*
- †*M. madagascariensis*

Family Indridae[b]

Subfamily Indrinae[b]
- *Indri indri* Indri.
- *Avahi laniger* Wooly lemur
  - *A. l. laniger*
  - *A. l. occidentalis*
- *Propithecus diadema* Diademed sifaka
  - *P. d. diadema*
  - *P. d. candidus*
  - *P. d. edwardsi*
  - *P. d. perrieri*
- *P. verreauxi* Verreaux's sifaka
  - *P. v. verreauxi*
  - *P. v. coquereli*
  - *P. v. deckeni*
  - *P. v. coronatus*
- *P. tattersalli* Tattersall's sifaka

Subfamily Archaeolemurinae
- †*Archaeolemur majori*
- †*A. edwardsi*
- †*Hadropithecus stenognathus*

Subfamily Palaeopropithecinae
- †*Palaeopropithecus ingens*
- †*P. maximus*
- †*Mesopropithecus pithecoides*
- †*M. globiceps*
- †*Archaeoindris fontoynontii*[b]

Incertae sedis: †*Babakotia radofilai*

Family Daubentoniidae
- *Daubentonia madagascariensis* Aye-aye.
- †*D. robusta*

Family Cheirogaleidae
- *Cheirogaleus major* Greater dwarf lemur
- *C. medius* Fat-tailed dwarf lemur
- *Microcebus murinus* Gray mouse lemur
- *M. rufus* Brown mouse lemur
- *Mirza coquereli* Coquerel's dwarf lemur
- *Allocebus trichotis* Hairy-eared dwarf lemur
- *Phaner furcifer* Fork-marked lemur
  - *P.f. furcifer*
  - *P.f. pallescens*
  - *P.f. electromontis*
  - *P.f. parienti*

[a] Sources: (42, 45, 51, 76, 95, 147, 148, 164, 166, 167.)
[b] Spelling follows Jenkins (61.)

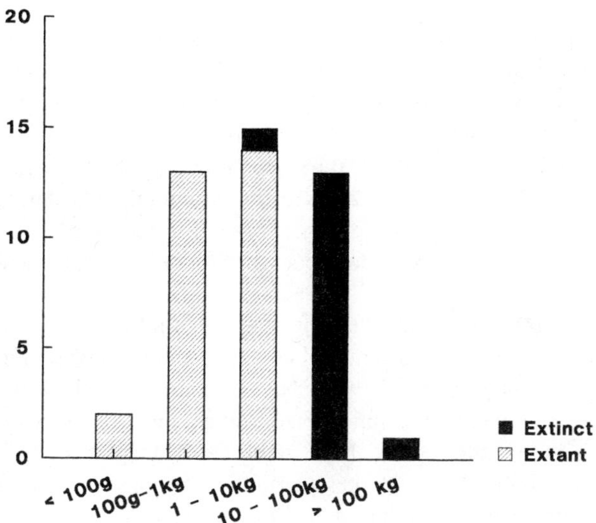

Figure 1 Adult body weights of extinct and extant lemurs. Data from Table 2 and (43).

sea level in the east, to dry spiny forests receiving only 300 mm of rainfall per year in the south and southwest (48). A primary division of the flora into humid eastern and drier western domains was made by Humbert (57) and has been followed by many (90, 109, 164, 182). The central region is today largely treeless; two thousand years ago, more continuous and clinally changing communities may have united east and west with less stark contrast.

Palynological analysis of sediments from many areas on the island has made it possible to reconstruct changes in vegetation zones, brought about by global climate shifts in the late Pleistocene and Holocene and by human activities in the last 2000 years (12–16). Changes continue today (47). The following inferences are of particular significance for lemur biogeography: (*a*) the central plateau was a mosaic of open country, woodland, and forest during the mid to late Holocene (ca. 5000–2000 years B.P.), but during the late Pleistocene (> 10,000 years B.P.) much of it had been heath and grassland (12, 13); (*b*) in the past 1500 years, the vegetation of the center changed to its modern treeless condition (12, 13, 85); (*c*) naturally occurring fires were common in some parts of the island during the late Pleistocene and throughout the Holocene (14); (*d*) in the arid southwest at least, there were significant vegetational changes probably associated with mid-to-late Holocene rainfall changes (15, 84). Thus, Madagascar is now known to have had a history of significant environmental change even before the arrival of humans, though this history is still poorly known for some regions.

LEMUR BIOGEOGRAPHY

In 1972, Martin (90) reviewed lemur biogeography and rebutted the then-prevalent view that sympatric speciation was the primary mechanism driving diversification (109). Like others, he identified a major biogeographical division between the east and the west. Low temperatures on the central plateau were proposed as a physical barrier for many species, resulting in the formation of coastal populations as "wreaths" around the plateau, with seven major rivers constituting barriers between different segments of these wreaths. Speciation was posited to occur when these barriers coincided with a bioclimatic boundary.

Tattersall (164) pointed out that the status of the central highlands as a barrier to lemur migration was untestable with available paleontological data, and that the rivers were unlikely to have functioned as barriers for long periods. He also noted that this model (like others more recent—49) assumes that biogeographic zones have been stable in Madagascar. He proposed instead a model that invoked the intermittent shrinking of Madagascar's forests to fragmented refugia during the Pleistocene; paleoecological evidence (see above) supports the idea of long-term climatic change.

The 1980s have yielded much new information on the distribution of extant lemurs, but their biogeography remains difficult to characterize: Forests continue to undergo fragmentation by human activities, and many species now have disjunct ranges; we can only imperfectly infer past distributions of extant species, and the ranges of the extinct species are poorly known; many forests remain to be thoroughly surveyed, although this is changing fast (96, 101). A further problem is presented by the lack of consensus on systematics; species level classifications are unstable for closely related forms with parapatric distributions, such as *Eulemur fulvus,* and *Lepilemur* sp., and some might be better regarded as "superspecies" (52).

The distribution of living species by region (Figure 2) shown in Table 2 is based on well-attested collection records, and on the recovery of identified specimens from Holocene archaeological and paleontological sites. Distributions are not presented for the extinct forms, since the subfossil sites are very unevenly distributed. To help visualize the relationships between regions, Figure 3 presents a nonmetric multidimensional scaling of the presence/absence of species within regions. Several patterns emerge: (*a*) There is a strong contrast between the west and south on one hand, and the eastern forests on the other (90, 109); (*b*) the arid north and the more humid Sambirano regions lie in an intermediate position between the East and the West in Figure 3, and their lemur communities are more similar to their respective neighbors than to communities in regions of similar climate (see also 113); (*c*) with respect to the center, the species found there today and those known from paleontological sites clearly align this region with the east (164). Finally, (*d*) the sharpest ecological divide on the island, between the dry south and the humid southeast, is also the clearest in terms of lemur species distributions (90).

These patterns are not clear cut. For instance, two subspecies of *Eulemur fulvus (E. f. fulvus* and *E. f. rufus*) have disjunct ranges in the east and the west. This strongly implies a recent link between these regions, perhaps along water courses, for this one species at least. Such exceptions notwithstanding, we suggest that the observed patterns support two general propositions. First, it appears that the east was the Pleistocene refugium for lemur species of the center. Second, except for the divide between the south and southeast, lemur communities of the coastal regions are all more similar to their neighbors than they are to more distant communities. This pattern is consistent with Tattersall's proposed pattern of post-Pleistocene range expansion.

The modern biogeography of lemurs must be viewed, thus, as a complex product of very recent range fragmentation, Holocene range expansion, and

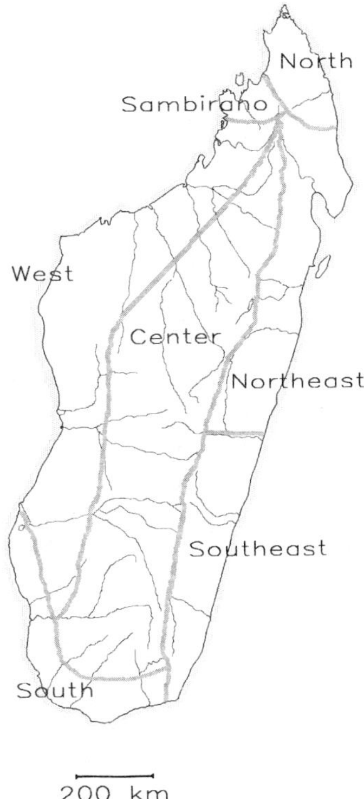

Figure 2 Ecogeographic regions of Madagascar (adapted from (57, 90, 101, 164).

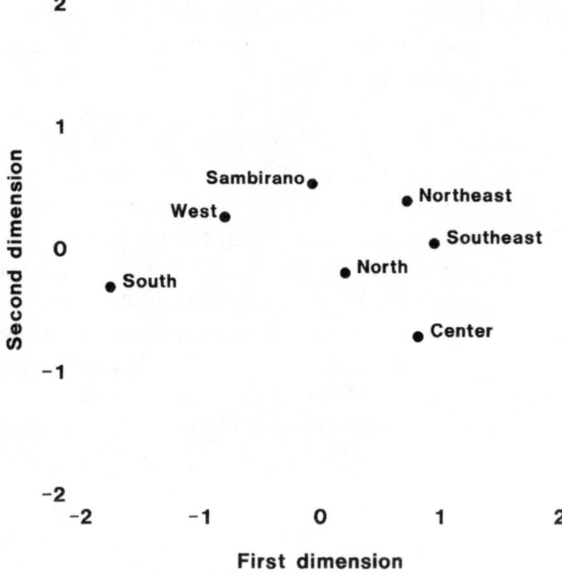

Figure 3 Nonmetric multidimensional scaling of the presence/absence of lemur species across the ecogeographic regions. Data from Table 2; similarity matrix based on Jaccard's coefficient; scaling method is Guttman's.

earlier Pleistocene range contraction. Moreover, the chronology of range shifts among the extant and extinct species is poorly known. While climatic and associated vegetation changes are likely to have played a major role in lemur range shifts, the extinction of the larger forms may also have triggered range shifts in some extant species.

Albrecht et al (3) have shown that there is clear ecogeographic patterning in size differences of sister taxa of both extant and extinct lemurs (see also 46): in general, lemurs from the south are smaller than their sister taxa in the west, north and Sambirano, and sister taxa from the east are larger than all of these. Sister taxa from the Center are usually the largest of all. It is unclear whether these differences are best explained genetically or as plastic phenotypic responses.

ECOLOGICAL DIVERSITY

Activity Cycle

The only strepsirrhines (i.e. lemurs, lorises, and galagos) active by day are found in Madagascar, and the lemurs have conventionally been viewed as almost equally divided between nocturnal and diurnal forms. Efforts to

Table 2 Distribution[a] and body weight[b] of extant species of lemurs. X = present in a region; (X) = present in a restricted area of a region; O = once present, but not known to be present today

Species	Body wt.	Center	Southeast	Northeast	North	Sambirano	West	South
Lemur catta	2700						(X)	X
Eulemur mongoz	1700						X	
E. macaco	2500					X	(X)	
E. fulvus	2400	(X)	X	X	X	X	X	
E. coronatus	1700				X			
E. rubriventer	2000		X	X				
Varecia variegata	3500		X	X				
Hapalemur griseus	930		X	X	(X)	X	(X)	
H. simus	~2400	O	(X)		O			
H. aureus	~1600		(X)					
Lepilemur sp.[c]	550–800	O	X	X	X	X	X	X
Indri indri	>6000	O	(X)	X				
Avahi laniger	1000	(X)	X	X	(X)	(X)	X	
Propithecus diadema	5800	O	X	X	X			
P. verreauxi	3600						X	X
P. tattersalli	3100				(X)			
Daubentonia madagascariensis	2700		(X)	X	(X)	(X)	(X)	
Cheirogaleus major	510	(X)	X	X	X	X		
C. medius	280					(X)	X	X
Microcebus murinus	60–100[d]				(X)		X	X
M. rufus	45–80[c]	(X)	X	X	X	X	(X)	
Mirza coquereli	310					(X)	X	
Allocebus trichotis	~130[e]			(X)				
Phaner furcifer	~400			(X)	(X)	(X)	X	(X)

[a] Sources: 3, 44, 101, 164.
[b] Mean adult body weight in grams. Data from 41, 73, 164.
[c] Species level taxonomy and distribution remain very uncertain.
[d] Body weights of captive and wild populations appear to differ significantly (164.)
[e] No reported weights; estimated from cranial length and head and body length.

develop supporting evidence for this behavioral division using morphological distinctions, notably eye socket size (e.g. 183), do not yield clear results (75). As a result, the activity rhythm of the subfossil lemurs is unknown.

Recent field studies suggest a more complex picture. First, some populations of most species in the Lemuridae are now known to be active by day and by night, at least seasonally (5, 23, 32, 54, 98, 103, 160–164, 169, 189). Second, there is a poor correspondence between activity pattern and degree of development of the reflective *tapetum lucidum* in the retina (107, 108, 164).

A cathemeral activity pattern, which indicates a significant level of activity during daylight hours and darkness (169), is rare in birds and mammals. This is probably because nighttime and daytime conditions exert different selective pressures, and most animals are specialized for one or other set of conditions and presumably are unable to switch between them (192).

One interpretation of the fluid activity cycle and the mismatch between visual specialization and activity cycle among lemurs is that they indicate a transitional stage from nocturnality, the primitive condition, to diurnality, newly made possible by the emptying of diurnal niches by the extinct lemurs (164). It is difficult to fit the degenerative modification of the tapetum in *Eulemur* into the nocturnal-to-diurnal directionality of this scenario, however, and Tattersall's preferred interpretation is that a cathemeral rhythm is a stable pattern that may even be ancestral for the Lemuridae. A third possibility is that formerly predominantly nocturnal and diurnal forms have both expanded their dietary repertoire, since the extinction of the megafauna, to include newly abundant bulk foods that must be ingested and processed by relatively small-bodied and unspecialized lemurs throughout the 24-hour cycle (32).

Positional Behavior

The diversity of locomotor and postural adaptations present among the extant and subfossil lemurs is arguably without parallel among all other members of the Order Primates (71, 159, 184, 185). The living forms are predominantly arboreal although a few, notably *Lemur catta* and *Daubentonia madagascariensis,* spend appreciable amounts of time foraging on the ground (11, 106, 153, 186; E. Sterling, personal communication). The Lemuridae, Daubentoniidae, and Cheirogaleidae exhibit a range of quadrupedalism, from slow climbing to branch walking and branch-scurrying in the smallest species (106). Among the Megaladapidae, *Lepilemur* spp. are slow climbers and leapers, and for the much larger *Megaladapis,* koala-like slow climbing on vertical supports has been proposed (70, 71, 182). The indrids qualify for an adaptive radiation all by themselves (71), with locomotor modes that include vertical clinging and leaping in the extant species (100), sloth-like hanging or antipronogrady in *Palaeopropithecus* (83, 87), and perhaps ground sloth-like quadrupedalism in the closely related *Archaeoindris* (71, 79, 181), potto-like, slow arboreal climbing and hanging in *Mesopropithecus* (42, 180), and predominantly terrestrial but noncursorial quadrupedalism in *Archaeolemur* and *Hadropithecus* (42, 71, 184).

Diet

The extant lemurs eat much the same range of items as haplorhine primates, with young leaves, leaf buds, fruit, flowers, and insects making up the bulk of the diet of all but a handful of species (55, 123). Nectar-feeding appears to be more common among lemurs than among other primates. Many primates eat flowers, but few exploit only the nectar (135). Among lemurs, in contrast, reports of nectarivory are mounting: *Eulemur mongoz, E. rubriventer, E. fulvus, Varecia variegata, Daubentonia madagascariensis, Microcebus murinus, Phaner furcifer, Cheirogaleus medius,* and *C. major* have now all been observed licking nectar or eating nectaries (22, 105, 114, 118, 154, 156; E. Sterling, P. Wright, personal communication). In some species, nectar-feeding is seasonally the single most important feeding activity: *E. mongoz* spent 84% of feeding time on nectar at the height of the dry season at Ampijoroa, and 74% of feeding records for *V. variegata* in one month were of nectar-feeding (150). Sussman & Raven (157) proposed that nectarivorous lemurs are important pollinators, and postulated that nonvolant mammalian pollinators have arisen primarily in areas like Madagascar, where nectarivorous bats are rare or absent. To the scarcity of bats can be added a scarcity of birds: Madagascar has only four clearly nectarivorous bird species (S. Zack, personal communication).

It has been suggested that lemur diets are unspecialized compared to other primates and show more overlap, with most species being either folivorous or mixed folivore-frugivores (33). Certainly, some species exhibit remarkable seasonal and regional dietary variation. For example, in eastern forest at Ranomafana *Eulemur fulvus rufus* was almost completely frugivorous (D. Overdorff, personal communication), but spent 77% of feeding time on leaves during the dry season in a western gallery forest (153). (We note that this species lacks specializations for mechanically or chemically processing leaves—146). Such different results may partly reflect seasonal and methodological differences, but a comparative study of *Propithecus verreauxi* in contrasting habitats revealed wide seasonal and regional differences in diet (133, 134); it therefore seems unlikely that the labile quality of extant lemur diets is only a methodological artifact.

Body size is broadly predictive of diet in many groups of primates, with insectivory characterizing the smallest species and increasing dependence on leaves appearing with increasing body size (36, 74). These trends are not clearly evident among the extant lemurs. The largest of the living lemurs, *Propithecus diadema* and *Indri,* exhibit dental and physiological specializations for folivory, but leaves comprise only about 53% and 35%, of their respective diets (121, 122, 191), percentages far below those of folivorous monkeys (135). One of the smaller lemurs, *Lepilemur mustelinus,* exhibits a similar (51%) reliance on leaves (21) and *Avahi laniger* feeds seasonally, at least, almost exclusively on leaves (35). The most frugivorous of all lemurs yet studied, *Varecia variegata,* weighs three times as much as these small folivores (150). Finally, the Cheirogaleidae, the smallest lemurs, are by no means all strongly insectivorous (20, 91). Dietary diversity among extant lemurs appears to be patterned taxonomically rather than by body size. The Indridae and Megaladapidae eat more leaves than do other lemurs, and the Cheirogaleidae and Daubentoniidae eat a mix of gums, fruit, insects and insect secretions (20, 22, 114, 118). The Lemuridae exhibit the widest range of diets, interspecifically as well as intraspecifically, from almost pure frugivory in *Varecia* to bamboo-feeding in *Hapalemur* (193), with a mixed frugivory and folivory in between. All *Hapalemur* spp. feed heavily on bamboo, and *H. aureus* daily consumes bamboo containing about 12 times the lethal dose of cyanide for humans. How it avoids cyanide poisoning is unknown (41).

These recent observations in the wild refute the idea that all lemurs are dietary generalists, and indications of dietary specialization have long been seen in the massive, procumbent, and continuously growing incisors of *Daubentonia madagascariensis,* its spindly third digit, and structural specializations of the brain case and face (e.g. 17). Accumulating observations show that the range of items eaten by aye-ayes is actually quite wide, including insects in adult and larval stages, kernel meats, nectar, sap, and gums (58, 115–117, 127, 190; E. Sterling, personal communication), and that the specialization may not be for a particular item but rather for a wide range of plant and animal foods that are abundant but difficult to extract (39, 58; E. Sterling, personal communication).

Inferences about the dietary habits of the extinct lemurs have been based on dental specializations and considerations of size. While their body weights have yet to be securely established, these animals ranged from the sloth-sized *Mesopropithecus* (8–10 kg) to the gorilla-sized *Archaeoindris* (200–240 kg) (43, 72). If size is a good indicator of diet, most of the subfossil lemurs were folivores or grazer/browsers; but, as noted above, folivory occurs in small as well as large-bodied extant lemurs. The extant relatives of most of the subfossil forms are predominantly folivorous, however, and dental morphology has been used as an independent argument for folivory in the extinct Megaladapidae and Indridae except *Hadropithecus* and *Archaeolemur*. Seed-eating has been proposed for *Hadropithecus* (69), and *Archaeolemur* is inferred to have been a semi-terrestrial frugivore (159a). *Pachylemur* spp. have been likened to robust, *Varecia*-like frugivores (42). *Daubentonia robusta* likely had a diet just as odd as *D. madagascariensis* (86).

Reproductive Ecology and Life History Patterns

Lemur populations have only one birth season each year, lasting one to two months. The timing of the birth season varies among species, but its brevity does not: Although sample sizes are still small, births seem to be as tightly clustered in the eastern rain forest (149, 191) as they are in the highly seasonal forests of the southwest (140, 143, 155). Females are seasonally mono- or polyestrous (125, 178). Males of some species experience seasonal cycles in testicular size and function (125, 129), and male sexual behavior is highly seasonal, but the endocrinology of male cycles is unknown. The onset of mating is broadly controlled by day length; lemurs in captivity in the northern hemisphere continue to reproduce seasonally, but with a six month shift in timing (125). The effects of photoperiodic change are not the same in all species. For example, *Microcebus murinus* become sexually active during long photoperiods whereas *L. catta* does so during short photoperiods. Artificially changing the rate of daylength in captivity alters the timing of reproduction: by doubling the rate, mouse lemurs can be brought twice into reproductive condition in one year (92, 119, 120, 125, 164). The relationship between daylength and reproductive activity in the wild is not fixed, and presumably ecological and social factors interact with photoperiodicity to determine onset of breeding. Little is known about the mechanisms involved. Social stimuli from male *L. catta* can cause photoinhibited females to start reproductive cycling in captivity (176, 177). Females in semi–free-ranging groups of *Lemur catta* all come into estrus within the same few weeks, but they show statistically significant day-to-day estrus asynchrony (110).

Lemurs have short gestation lengths compared to lorises or haplorhine primates of similar size, with the shortest gestation lengths of all primates found in the litter-producing species (*Microcebus, Mirza, Cheirogaleus,* and *Varecia*) (194). Using the ratio of neonatal litter weight to gestation length as an index of maternal investment, Young et al (194) found that strepsirrhines invest in the prenatal development of their offspring at a lower rate than haplorhines. This can be attributed to the lower basal metabolic rate and lower rate at which energy is available to strepsirrhines, and when prenatal maternal investment rate is expressed as a function of BMR, the distinction between strepsirrhines and haplorhines disappears. However, lemur females invest energy in their offspring at a higher rate than lorises.

Male and female life tables may also be distinctive in some lemurs. The number of adult males reportedly is equal to or greater than the number of adult females in many lemur populations (62–64, 67, 97, 102, 113, 134, 140). At Beza Mahafaly in southwestern Madagascar, the adult sex ratio was biased in favor of *Propithecus verreauxi* males throughout a five-year study (140), and the number of adult male *Lemur catta* equaled or was greater than the number of adult females over a two-year period in the same forest

(155). In neither case did the sex ratios depart significantly from 0.5, but the direction and magnitude of their departure from ratios reported for haplorhine populations make these ratios unusual: In most haplorhines, like most mammals, the adult sex ratio favors females (19, 30, 140). Just why adult sex ratios are so distinctive in some lemur populations is unclear, but causal factors may include birth cohorts intermittently skewed in favor of males, and intermittent high mortality among reproductively active females (140).

Some aspects of lemur life histories vary considerably between species. For example, *Propithecus verreauxi* females at Beza Mahafaly give birth for the first time at five years. 44% of females gave birth each year, and two years was the most common interval between the births of surviving offspring. Mean infant survival during the first year of life was 0.61 (140). In contrast, *Lemur catta* females in the same forest give birth for the first time at the age of three years, and 80% or more of females gave birth in two consecutive years. Infant survival was 0.48 (155). In both species females were philopatric and males dispersed, but *P. verreauxi* males typically transferred to neighboring groups whereas tagged *L. catta* males were sighted in forest 2–3 km from the reserve (R. W. Sussman, in preparation; A. F. Richard, in preparation). The adaptive significance of these differences is not understood.

Socioecology

Lemurs live in small social groups (123, 164). Groups containing more than 15 individuals are rare, and the largest group ever counted in the wild numbered only 24 animals (63). This stands in marked contrast to many haplorhine species, in which groups of more than 30 animals are common. Feeding competition has been proposed as a major constraint on group size in primates, and predator pressure as a factor favoring large groups (174). The determinants of group size among lemurs are unclear. Recent studies provide weak support, at best, for the idea that group size is strongly constrained by feeding patch size (33, 34, 105, 150). If the risk of predation is indeed low for lemurs, as widely assumed, this would reduce the advantage of large group size, but this assumption has recently been challenged (142).

An alternative hypothesis invokes "intrinsic" social and demographic factors rather than "extrinsic" ecological ones (174) to explain the small size of lemur groups; this hypothesis implies that social relationships among lemurs are more narrowly bounded than among haplorhines. Vick & Pereira (179) propose that lemur females only tolerate other females of the same matriline; when group size increases and the level of genetic relatedness among its female members drops, females target unrelated or distantly related females for expulsion from the group. This hypothesis receives support from research on semi–free-ranging groups of *Lemur catta* and *Eulemur fulvus rufus* but has yet to be evaluated in the wild.

The range of lemur group sizes is low, but their social and spatial organization is quite variable. Most of the nocturnal species have dispersed or nongregarious social systems (20); animals communicate by loud calls or scent-marks, and rarely meet. *Indri indri* and *Eulemur rubriventer* are consistently found in adult pairs with young (105, 121). In two of three populations studied, *E. mongoz* was always seen in pairs with young, but in the third population, groups contained several adult males and females (156, 160, 162, 170). Several species, including *Propithecus verreauxi* and *Varecia variegata,* are sometimes found in pairs but more commonly aggregate in larger groups (63, 134, 187). On Nosy Mangabe, *V. variegata* live in communities comprising networks of females and loosely affiliated adult males. Community members share a home range that encompasses several core areas used preferentially by groups of 2–5 males and females. Females defended the communal home range against females from neighboring communities; males took no part in these encounters (150).

Little is known about lemur mating systems, because mating is rarely observed. For example, *Indri indri* is generally assumed to be monogamous, but neither the duration nor the exclusiveness of the pair-bond is known. In fact, there are no published reports of mating in this species. Serial matings with several males have been reported in *Lemur catta, Propithecus verreauxi, Varecia variegata* and *Daubentonia madagascariensis* (10, 63, 77, 132, 137, 143, 150; E. Sterling, personal communication). Social group boundaries are highly permeable in *L. catta* and *P. verreauxi,* and mating does not conform to the conventional view of "promiscuity" within the social group: Both female *L. catta* and *P. verreauxi* sometimes mate with visiting males from other groups (137, 143). On the other hand, stable polygyny appears to characterize many social groups of *P. verreauxi*.

With one exception, female feeding priority characterizes all group-living lemurs studied thus far in the wild (66, 124, 136, 150). Females are consistently able to displace males from feeding sites, and in species in which mating has been observed, females exercise strong mate choice. The exception is *Eulemur fulvus rufus,* in which gender has no systematic effect on agonistic relations in captivity or in the wild (111). Female feeding priority is rare among mammals, and its presence among lemurs is unexplained (56, 66, 126, 136). It may help females cope with an unusually high energetic cost of reproduction (66, 138). An allometric analysis showed that female feeding priority is most pronounced in strepsirrhines with low basal metabolic rates and particularly high prenatal maternal investment rates (194). However, this model fails to predict the absence of female feeding priority in *E.*

f. rufus, possibly because postnatal investment rates are more important than prenatal rates.

COMMUNITY ECOLOGY

Community Structure and Resource Partitioning

The species richness of extant primate communities in Madagascar is among the greatest anywhere, and the lemur assemblage at Ampasambazimba, a Holocene subfossil site on the plateau, is richer still (Figure 4). The body size range in extant communities is narrow, comparable to Neotropical primate communities. When the subfossils are included, however, the size range expands dramatically to equal or surpass that of Old World primate communities. Moreover, in Madagascar all primates come from a single diversified superfamily, whereas in the continental Old World three superfamilies comprise the small, medium-sized, and large forms, respectively. Total primate biomass in the eastern rain forest of Madagascar today has been estimated at 400 kg/km^2; this is within the range for primate biomass in the Neotropics but is lower than estimates for tropical rainforests in Africa and Asia (33). Studies in South America, Africa, and Southeast Asia have consistently demonstrated resource partitioning between sympatric primate species (135). Two modes of partitioning are particularly evident: year-round dietary differentiation, and extensive or total dietary overlap most of the year with specialization on different foods during periods of food scarcity. Other dimensions of partitioning include vertical and horizontal stratification within the forest, foraging technique, and activity pattern (172). Differences in body size or food-processing specializations underlie some, but not all, the observed differences.

The recency and scale of primate extinctions in Madagascar raise the possibility that resources are not stably partitioned and that there may be, rather, an ecology of chaos. In our view, the evidence currently available cannot yet rule out the latter possibility, although dietary differentiation has been shown between some species (34, 41, 55, 153), often in conjunction with temporal or spatial separation of activity (32, 34, 153, 164, 169).

Ecological separation through food chemistry, plant parts eaten, and microhabitat utilization has been studied in seven sympatric lemur species living in eastern rain forest at Analamazaotra (*Avahi laniger, Cheirogaleus major, Hapalemur griseus, Indri, Eulemur fulvus, Lepilemur mustelinus* and *Microcebus murinus*) (33, 34). Broad food categories and tree size were insufficient to separate species, but phytochemical analyses of over 400 plant parts in the diets of these species revealed differences with respect to tannins, fiber, alkaloids and protein concentration. Discriminant analysis segregated

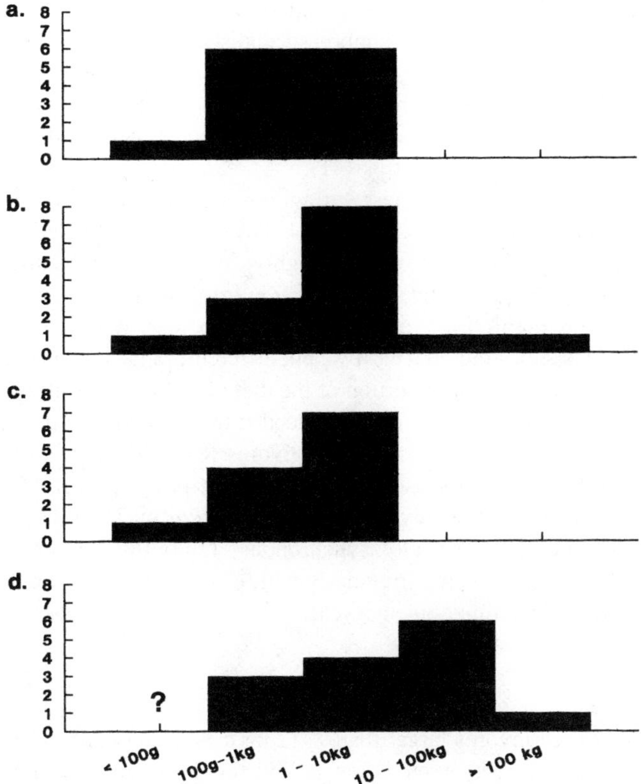

Figure 4 Body weight distributions for sympatric primate communities: (*a*) Coca Cashu, Peru (9 genera, 11 spp.); (*b*) Makoukou, Gabon (11 genera, 15 spp.); (*c*) Ranomafana, Madagascar (9 genera, 12 spp.); (*d*) from the Holocene palaeontological site of Ampasambazimba, central Madagascar (14 genera, no. spp. unknown). (Sources: 24, 37, 41, 43, 73, 164, 172; P. Wright, personal communication).

three groups of two species based on food chemistry, and species paired with one another on phytochemical criteria exhibited other forms of separation, notably differentiation of microhabitat and activity rhythms. The interaction of these factors is complex and variable. For example, *Avahi laniger* and *Lepilemur mustelinus,* both nocturnal folivores of similar size and digestive morphology (106), could be distinguished with respect to feeding height and tree crown diameter at Analamazaotra but were indistinguishable by these criteria in western forest at Ampijoroa. However, phytochemical analysis showed that at Ampijoroa *A. laniger* ate significantly fewer leaves with alkaloids than did *L. mustelinus.*

Clear evidence of separation was found during 153 hours of observation of the three sympatric *Hapalemur* bamboo specialists, *H. simus, H. griseus,* and *H. aureus*. They consistently selected different plant parts or species: *H. griseus* ate the leaf bases and blades of *Cephalostachyum perrieri; H. simus* fed preferentially on the mature culms of *C. cf. viguieri;* and *H. aureus* ate its leaf bases, pith, and growing tips (41).

Seasonal separation of otherwise strongly overlapping diets has been found in *Eulemur fulvus rufus* and *E. rubiventer* during a 13-month study at Ranomafana (105). Fruit production was highly seasonal, with a three-month period during which just a few species were in fruit at a time. *E. fulvus* and *E. rubriventer* were highly frugivorous throughout the year. When fruit was least abundant each species fed on a narrow, but different, range of fruits. Curiously, fruit made up a higher proportion of the diet of both species at this time of least abundance. Plant species that produced fruit out of season may play the role of keystone resources for these frugivores (81, 105, 173). Year-round frugivory despite marked seasonal cycles in fruit production has also been found during a two-year study of *Varecia variegata* on Nosy Mangabe (4, 150). Cross-cutting the broadly synchronous, rainfall-driven phenological cycle, a few species flowered irregularly within and across years so that some fruit was available throughout the year.

Predation

The absence of carnivores larger than 7–12 kg in Madagascar has been taken to indicate that the risk of predation is low for the large-bodied Malagasy primates (144). This ignores the inherent vulnerability of young (142), as well as a variety of direct and indirect evidence of predation. The harrier hawk *(Polyboroides radiatus)* has been seen taking an infant *Lemur catta* (131). Bones of immature lemurs have been found scattered around nests of *Cryptoprocta ferox* (an endemic, puma-like viverrid) in the south (2), and partly consumed bodies of adult *L. catta* and *P. verreauxi* at Beza Mahafaly showed signs of predation by mammalian predators (140, 142). Lemurs respond with vocalizations and agitation to the presence of snakes, carnivores, and raptors, and the alarm calls of some species differentiate between aerial and terrestrial predators (54, 63, 82, 121, 134, 142). Sauther (142) has argued that the recent emphasis on habitat destruction as the primary threat to the continuing survival of the extant lemurs has detracted from the study of factors regulating lemur populations in the wild and may, indeed, have obscured the importance of predation. She suggests that avian predators are a threat to young animals, especially infants, while mammalian predators pose a threat to infants and adults.

THE DISTINCTIVENESS OF LEMUR ECOLOGY

The ecological distinctiveness of lemurs derives from characteristics of the animals themselves, the uniquely constituted communities to which they belong, and the recent and major changes to which some or all these communities have been subjected.

Lemurs have few characteristics not found in other primates. What is distinctive is the widespread co-occurrence of these traits in lemurs. Compared to most other primates, lemurs have low basal metabolic rates and markedly seasonal reproduction. Compared to the lorisiform strepsirrhines, lemurs exhibit high maternal investment rates and high postnatal growth rates. Distinctions between males and females found in other primates are differently configured in lemurs: even the largest species were not significantly sexually dimorphic, female feeding priority is widespread, and male and female survivorship curves differ from those of most other primates. The absence of large social groups in even the most gregarious diurnal species is striking.

One interpretation of these characteristics is that they are primitive mammalian traits. This is implausible: strict breeding seasonality occurs in many primate populations that experience marked seasonal fluctuations in food availability (90, 93, 94); from the distribution of hypometabolism among mammals, it appears to be not a primitive trait (99) but rather a dietary adaptation (88) or an energy-conserving adaptation to a range of nutritional stresses (78, 130). In the case of female feeding priority, it is unclear that the trait is homologous across species: the ascendancy of female lemurs over males is widespread but not universal, and its form varies considerably among species (111).

It has been repeatedly suggested that many of the traits distinguishing lemurs are adaptively linked to the marked seasonality of their environment, now or in the past (56, 66, 90, 93, 138, 194). Taking rainfall as a rough indicator of seasonality, there is no support for the idea of island-wide seasonality: the south, west, and center experience prolonged annual dry seasons (Figure 5), but rainfall in the eastern forests is patterned much like that in other tropical forests supporting diverse primate communities (Figure 6). Mean annual temperatures range from 23–26°C at sea level to about 20°C on the plateau. Annual ranges in mean temperature are relatively small, but the diurnal range seasonally reaches 15°C in the south and inland, and 7–8°C in the east (48).

There are several possible explanations for the island-wide breeding seasonality of Madagascar's primates, even with the island's range of climatic regimes: (*a*) Lemurs evolved in the more seasonal environments of the west,

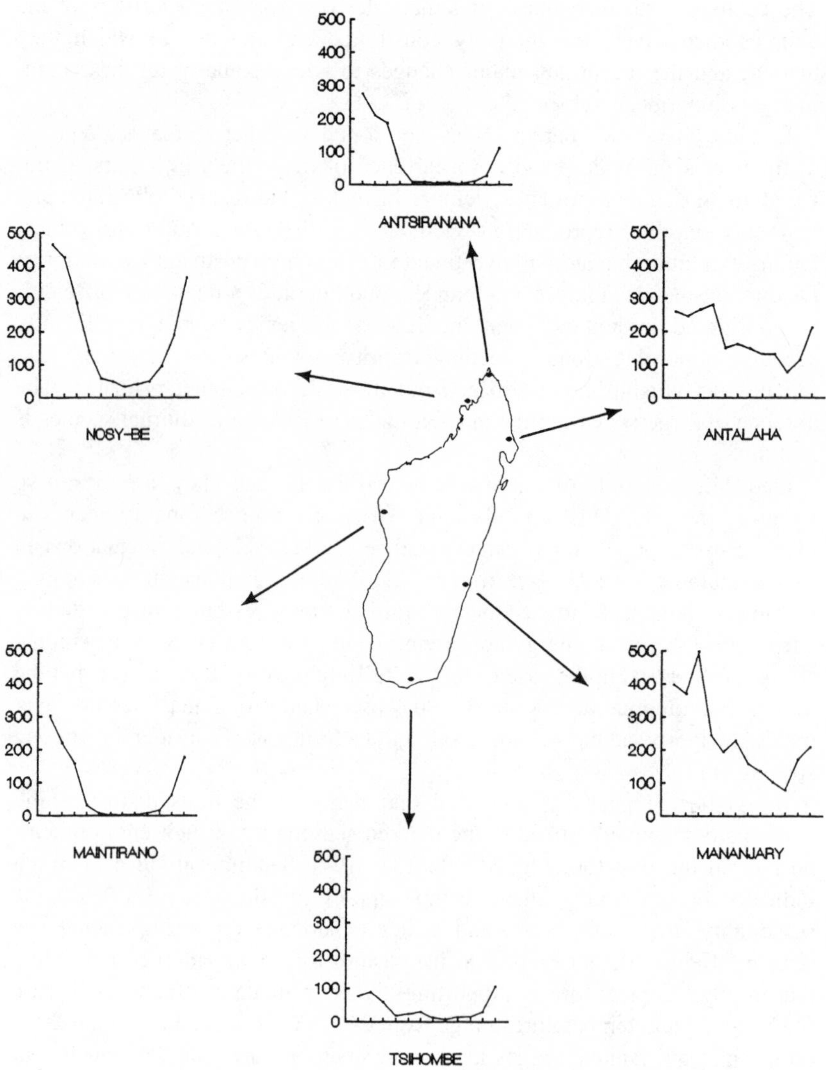

Figure 5 Distribution of rainfall, by month, in each of the six coastal ecogeographic regions of Madagascar. (Data from 48).

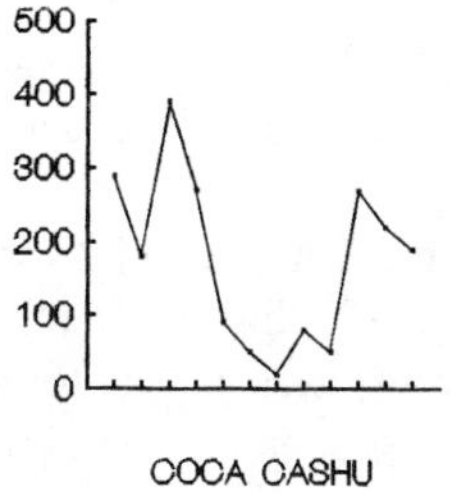

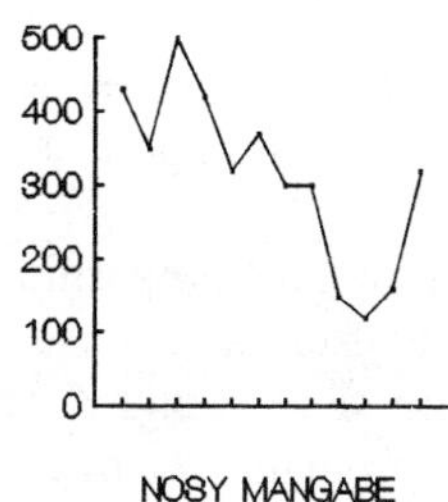

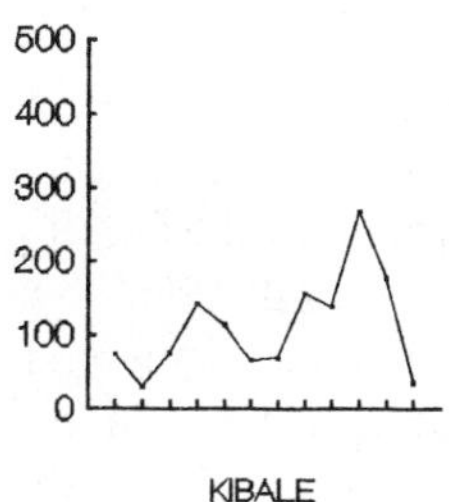

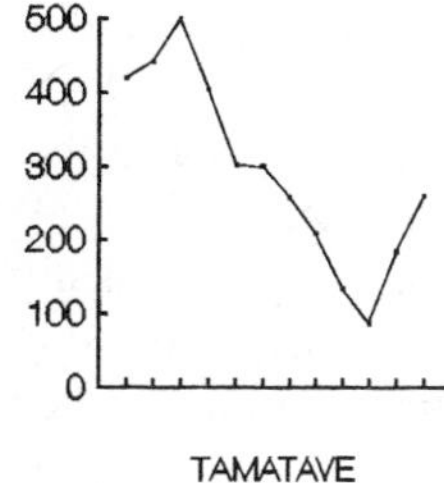

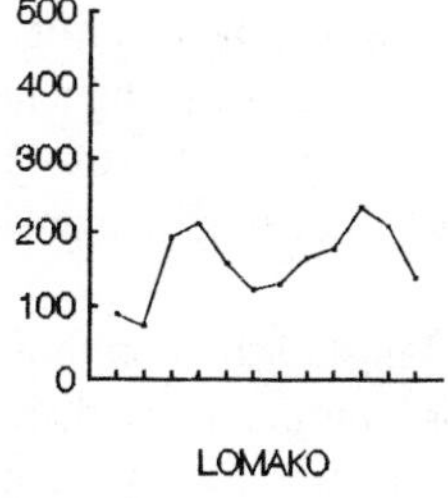

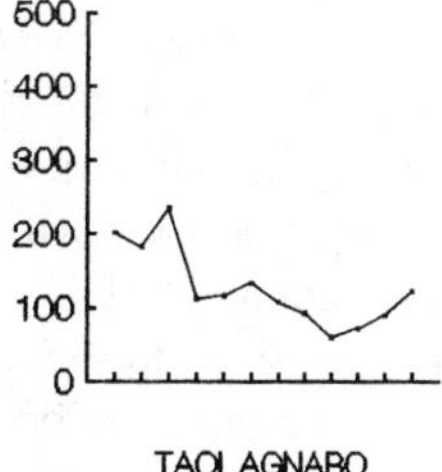

Figure 6 Distribution of rainfall, by month, for (left side) three tropical forests with diverse primate communities: Coca Cashu (Peru), Kibale (Uganda) and Lomako (Zaire); (right side) three locations along the east coast of Madagascar. (Data from 48, 89, 152, 172).

south, and center, and later expanded into the eastern forests; (*b*) Madagascar was more seasonal when lemurs evolved; (*c*) seasonal cycles of food production exist that are not evident from the climatic data; (*d*) seasonal food shortage is not an important ecological determinant of breeding seasonality among lemurs.

The first of these propositions cannot be evaluated. The second is implausible. Late Pleistocene climates were likely harsher (12), but the speciosity of the modern flora strongly indicates that rain forest was never completely eliminated from the island, and that the lemurs were never without climatically equable refugia. Moreover, global climates during most of the Cenozoic were warmer and less seasonal. The third possibility has yet to be explored. Terborgh & van Schaik (175) predicted an unusual phenological cycle in eastern Madagascar, with more or less continuous leafing and sharply seasonal fruiting. The premise on which they based this prediction, the apparent prominence of folivory among lemurs, is probably more a historical consequence of the nonrandom selection of species for intensive study than a real trend. However, a period when fruit is much less abundant has been reported in two eastern forests (4, 105), and in one of these forests leaf loss and production were found to be continual in most species during a ten-month study of the phenology of 248 trees from 32 genera (E. Carlson, in preparation). More data are needed to document these cycles fully, and to establish their implications for breeding seasonality and diet among lemurs. The case for breeding seasonality being ecologically determined would be greatly strengthened, for example, if it could be shown that survival is lower among infants born early or late in the birth season (e.g. 29).

Alternative adaptive explanations for breeding seasonality among lemurs have received little attention. Martin (94) suggested that it might represent a predator satiation strategy, and Pollock (123, 126) emphasized the direct and indirect evidence of imperfect temperature homeostasis among lemurs, including torpor and aestivation in the Cheirogaleidae (20, 141), huddling and sunning behavior in the Indridae and Lemuridae (63, 134), and the regular use of sleeping holes or nests by *Daubentonia, Lepilemur* and the Cheirogaleidae; during the birth season, *Varecia variegata* construct nests for their newborn infants (149). The possibility that seasonal thermoregulatory stress, rather than or in addition to nutritional stress, underlies breeding seasonality deserves exploration.

Ancient omissions in the array of animals making up the communities to which lemurs belong may be more significant for lemur ecology than the list of those present, abundant and potentially in competition with them, for it seems that the omissions permitted lemurs to play ecological roles taken by other animals elsewhere (17, 157). In particular, the many species of birds, bats, and rodents that serve variously as pollinators and seed dispersal agents in other tropical communities are poorly represented on Madagascar. In conjunction with mounting evidence of frugivory and nectarivory by lemurs, this suggests that lemurs play unusually important roles in pollination and seed dispersal. Coevolutionary relationships between plants and their pollinators and seed dispersal agents are a subject of practical as well as theoretical importance: Conservation will not work if forests are protected but lack the fauna needed to ensure their persistence (7, 59). The exploration of these mutualistic interactions in Madagascar is a high priority.

If extant lemurs play unusual ecological roles, it is likely that the extinct lemurs did as well. It is possible that some of Madagascar's plant species lost important dispersal agents with the extinction of the megafauna in the last two thousand years (J. Ratsirarson, in preparation). Deciding how to manage these species today is but one facet of a larger issue confronting all ecological research in Madagascar: What are the continuing effects of events of the recent past on the structure and function of remaining communities? These events include the addition of new species as well as the elimination of species already present: When people arrived, they brought with them exotic plants, and also exotic animals, notably mice, rats, cattle, pigs, and goats.

In order to consider seriously the ecological implications of the extinctions, we need better information about the geographical ranges of the extinct species. At present, important questions remain about which communities lost species and which, if any, did not. The subfossil sites are relatively few in number and clustered in the north, center, and southwest of the island. This reflects a collecting bias, in that subfossil sites have been rarely found in the east and less survey work has been done there. Still, it is uncertain whether any of the extinct lemur species occurred in the eastern forests. The presence of extinct lemur species at Ampasambazimba, and the general affinity to the east of extant lemur communities of the eastern highlands, suggest that giant lemurs were present in the east. But this is a surmise, and more paleontological research is needed to resolve the issue.

The addition by people of new species further complicates ecological study and interpretation. In particular, while the presence of bovids may have little immediate bearing on lemur ecology, their introduction undoubtedly had a major impact on the vegetation on which lemurs depend for their survival (24, 26), an impact continued and possibly magnified today. Human activities continue to affect ecological communities in Madagascar and are a focus of concern and research. For practical as well as theoretical reasons, it is essential that a knowledge of past events inform this research: it would serve as little purpose to study the "livestock problem" without regard to the extinction of the endemic grazer-browser community as it would to save forests without regard to the extinction of lemurs that helped ensure their persistence.

CONCLUSION

Heightened interest in the ecology of Madagascar's diverse communities has come not a moment too soon, for these communities are increasingly threatened by the need to sustain the rapidly growing population of one of the poorest nations in the world. Biological and ecological research should be a high priority: our knowledge of lemur systematics and their past and present biogeography is incomplete; we are far from understanding the seasonal cycles of growth and reproduction that characterize Madagascar's complement of plants and animals, or the coevolutionary relationships between them. The unique configuration of Madagascar's natural communities makes them a valuable source of insights into community structure and function and into the ecology of their component parts. In some cases, this research is doubly important because it will document communities in the process of disappearing. On a more positive note, there are mounting efforts to conserve and manage remaining communities; the likelihood that these efforts will succeed can surely be enhanced by a fuller understanding of the island's past and present ecology than has yet been achieved.

ACKNOWLEDGMENTS

We thank Robin Absher, Laurie Godfrey, Michelle Sauther, Marion Schwartz, and Steve Zack for comments on an earlier draft. There is much exciting new work being done in Madagascar at this time, and we are most grateful to the following for giving us access to unpublished manuscripts and data: Betsy Carlson, Laurie Godfrey, Martin Nicoll, Deborah Overdoff, Joelisoa Ratsirarson, Eleanor Sterling, Bob Sussman, and Pat Wright. Recriminations over the classification used in this review should be addressed to us, but we thank Ian Tattersall and Laurie Godfrey for their advice. The preparation of this paper was supported in part by NSF BSR-8501079, BSR-8708350 and INT-8410362.

Literature Cited

1. Albignac, R. 1972. The carnivora of Madagascar. In *Biogeography and Ecology in Madagascar,* ed. R. Battistini, G. Richard-Vindard, pp. 667–82 The Hague: Junk
2. Albignac, R. 1973. Mammifères carnivores. *Faune de Madagascar* 36:1–209
3. Albrecht, G. H., Jenkins, P. D., Godfrey, L. R. 1990. Ecogeographic size variation among the living and subfossil prosimians of Madagascar. *Am. J. Primatol.* 22:1–50
4. Andrianisa, J. 1989. Observations phenologiques dans la reserve speciale de Nosy-Mangabe, Maroantsetra. *Memoire de Fin d'Etudes*. Univ. d'Antananarivo Madagascar
5. Andriatsarafara, R. 1988. Note sur les rhythmes d'activité et sur le régime alimentaire de *Lemur mongoz* Linnaeus, 1766 à Ampijoroa. In *L'Equilibre des Ecosystèmes forestiers à Madagascar: Actes d'un Séminaire International,* ed. L. Rakotovao, V. Barre, J. Sayer, pp. 103–6. Gland, Switzerland: IUCN
6. Battistini, R., Richard-Vindard, G., eds. 1972. *Biogeography and Ecology of Madagascar*. The Hague: Junk

7. Bawa, K. S. 1990. Plant-pollinator interactions in tropical rain forests. *Annu. Rev. Ecol. Syst.* 21:399–422
8. Beauchamp, G. 1989. Canine tooth size variability in primates. *Folia Primatol.* 52:167–77
9. Brooks, D. R. 1985. Historical ecology: a new approach to studying the evolution of ecological associations. *Ann. Missouri Bot. Gard.* 72:660–80
10. Budnitz, N., Dainis, K. 1975. *Lemur catta:* ecology and behavior. In *Lemur Biology,* ed. I. Tattersall, R. Sussman, pp. 219–36. New York: Plenum
11. Buettner-Janusch, J. 1963. In *Evolutionary and Genetic Biology of Primates,* ed. J. Buettner-Janusch pp. 1–64. New York: Academic
12. Burney, D. A. 1987. Pre-settlement vegetation changes at Lake Tritrivakely, Madagascar. *Paleoecol. Africa* 18:357–81
13. Burney, D. A. 1987. Late Holocene vegetational change in central Madagascar. *Quat. Res.* 28:130–43
14. Burney, D. A. 1987. Late Quaternary stratigraphic charcoal records from Madagascar. *Quat. Res.* 28:274–80
15. Burney, D. A., MacPhee, R. D. M. 1988. Mysterious island. *Nat. Hist.* 97:46–55
16. Burney, D. A., MacPhee, R. D. M., Rafamantanantasoa, J. G., Rakotondrazafy, T., Kling, G. 1991. The role of natural factors and human activities in the environmental changes and faunal extinctions of late Holocene Madagascar. In *Early Man and Island Environments,* ed. P. Sondaar, M. Soares. In press
17. Cartmill, M. 1974. *Daubentonia, Dactylopsila,* woodpeckers and klinorhynchy. In *Prosimian Biology,* ed. R. D. Martin, G. A. Doyle, A. C. Walker, pp. 655–70. London: Duckworth
18. Cartmill, M. 1975. Strepsirhine basicranial structures and the affinities of the Cheirogaleidae. In *Phylogeny of the Primates: A Multidisciplinary Approach,* ed. W. P. Luckett, F. S. Szalay, pp. 313–54. New York: Plenum
19. Caughley, G. 1977. *Analysis of Vertebrate Populations.* New York: Wiley
20. Charles-Dominique, P., Cooper, H. M., Hladik, A., Hladik, C. M., Pages, E., et al, eds. 1980. *Nocturnal Malagasy Primates: Ecology, Physiology and Behavior.* New York: Academic
21. Charles-Dominique, P., Hladic, C. M. 1971. Le Lépilémur du sud de Madagascar: écologie, alimentation, et vie sociale. *Terre Vie* 25:3–66
22. Charles-Dominique, P., Petter, J. J. 1980. Ecology and social life of *Phaner furcifer*. See Ref. 20, pp. 75–96
23. Conley, J. M. 1975. Notes on the activity pattern of *Lemur fulvus rufus. J. Mammal* 56:712–15
24. Dewar, R. E. 1984. Recent extinctions in Madagascar: the loss of the subfossil fauna. In *Quaternary Extinctions: A Prehistoric Revolution,* ed. P. S. Martin, R. G. Klein, pp. 574–93. Tucson: Univ. Arizona Press
25. Dewar, R. E. 1991. The archeology of the early settlement of Madagascar. In *The Indian Ocean in Prehistory,* ed. J. Reade. London: Routledge. In press
26. Dewar, R. E., Rakotoarisoa, J.-A. 1990. The human transformation of Madagascar. Pres. Int. Cong. Syst. Evol Biol.; Beltsville, Md.
27. Dewar, R. E., Rakotovololona, H. F. S. 1991. Hunting camps in northern Madagascar in the twelfth century. In *Early Man in Island Environments,* ed. P. Sondaar, M. Sanges. In press
28. Duellman, W. E., Pianka, E. R. 1990. Biogeography of nocturnal insectivores: historical events and ecological filters. *Annu. Rev. Ecol. Syst.* 21:57–68
29. Dunbar, R. I. M. 1980. Demographic and life history variables of a population of gelada baboons *(Theropitheus gelada). J. Anim. Ecol.* 49:485–506
30. Dunbar, R. I. M. 1987. Demography and reproduction. In *Primate Societies,* ed. B. B. Smuts, D. L. Cheney, R. M. Seyfarth, R. W. Wrangham, T. T. Struhsaker, pp. 240–49. Chicago: Univ. Chicago Press
31. Endler, J. A. 1982. Problems in distinguishing historical from ecological factors in biogeography. *Am. Zool.* 22: 441–52
32. Engvist, A., Richard, A. F. 1991. Diet as a possible determinant of cathemeral activity patterns in primates. *Folia Primatol.* In press
33. Ganzhorn, J. U. 1988. Food partitioning among Malagasy primates. *Oecologia* 75:436–50
34. Ganzhorn, J. U. 1989. Niche separation of seven lemur species in the eastern rainforest of Madagascar. *Oecologia* 79:279–86
35. Ganzhorn, J., Abraham, J. P., Razanahoera-Rakotomalala, M. 1985. Some aspects of the natural history and food selection of *Avahi laniger. Primates* 26:452–63
36. Gaulin, S. J. C. 1979. A Jarman/Bell model of primate feeding niches. *Hum. Ecol.* 7:1–19
37. Gautier-Hion, A. 1978. Food niches and coexistence in sympatric primates

in Gabon. In *Recent Advances in Primatology,* vol. 1, *Behaviour,* ed. D. J. Chivers, J. Herbert, pp. 269–86. New York: Academic

38. Gentry, A. H. 1988. Changes in plant community diversity and floristic composition on environmental and geographic gradients *Ann. Miss. Bot. Gard.* 75(1):1–34
39. Gibson, K. R. 1986. Cognition, brain size and the extraction of embedded food resources. In *Primate Ontogeny, Cognition and Social Behavior,* ed. J. G. Else, P. C. Lee, pp. 93–103, Cambridge: Cambridge Univ. Press
40. Gingerich, P. D., Ryan, A. S. 1979. Dental and cranial variation in living Indriidae. *Primates* 20:141–59
41. Glander, K. E., Wright, P. C., Seigler, D. S., Randrianasolo, V., Randrianasolo, B. 1989. Consumption of cyanogenic bamboo by a newly discovered species of bamboo lemur. *Am. J. Primatol.* 18:1–7
42. Godfrey, L. R. 1988. Adaptive diversification of Malagasy strepsirrhines. *J. Hum. Evol.* 17:93–134
43. Godfrey, L. R. 1991. *How big were the 'giant' subfossil lemurs of Madagascar?* Pres. Ann. Meet., Am. Assoc. Phys. Anth., 60th, Milwaukee
44. Godfrey, L. R., Vuillaume-Randriamanantena, M. 1986. *Hapalemur simus:* endangered lemur once widespread. *Primate Conserv.* 7:92–96
45. Godfrey, L. R., Simons, E. L., Chatrath, P. J., Rakotosamimanana, B., 1990. A new fossil lemur (*Babakotia,* Primates) from Northern Madagascar. *C. R. Acad. Sci. Paris,* t. 310, ser. II, pp. 81–7
46. Godfrey, L. R., Sutherland, M. R., Peto, A. J., Boy, D. S. 1990. Size, space and adaptation in some subfossil lemurs from Madagascar. *Am. J. Phys. Anthropol.* 81:45–66
47. Green, G. M., Sussman, R. W. 1990. Deforestation history of the eastern rain forests of Madagascar from satellite images. *Science* 248:212–15
48. Griffiths, J. J., Ranaivoson, R. 1972. Madagascar. In *Climates of Africa,* ed. J. F. Griffiths, pp. 461–98. New York: Elsevier
49. Groves, C. P. 1989. *A Theory of Human and Primate Evolution.* Oxford: Clarendon
50. Groves, C. P., Eaglen, R. H. 1988. Systematics of the Lemuridae (Primates, Strepsirhini). *J. Hum. Evol.* 17:513–38
51. Groves, C. P., Tattersall, I. 1991. Geographical variation in the fork-marked lemur, *Phaner furcifer* (Primates, Cheirogaleidae). *Folia Primatol.* 56:39–49
52. Haffer, J. 1986. Superspecies and species limits in vertebrates. *Z. Zool. Syst. Evolut.-forsch* 24:169–90
53. Harcourt, C., Thornback, J. 1990. *Lemurs of Madagascar and the Comoros. The IUCN Red Data Book.* Gland, Switzerland and Cambridge, UK: IUCN
54. Harrington, J. E. 1978. Diurnal behavior of *Lemur mongoz* at Ampijoroa Madagascar. *Folia Primatol.* 29:291–302
55. Hladik, C. M. 1979. Diet and ecology of prosimians. In *The Study of Prosimian Behavior,* ed. G. A. Doyle, R. D. Martin, pp. 307–339. New York: Academic
56. Hrdy, S. B. 1981. *The Woman that Never Evolved.* Cambridge: Harvard Univ. Press
57. Humbert, H. 1955. Les térritoires phytogéographiques de Madagascar, leur cartographie. *Ann. Biol.* (3 ser.) 31:195–204
58. Iwano, T., Iwakawa, C. 1988. Feeding behavior of the aye-aye *(Daubentonia madagascariensis). Folia Primatol.* 50: 136–42
59. Janzen, D. H. 1987. Insect diversity of a Costa Rican dry forest: why keep it, and how? *Biol. J. Linn. Soc.* 30:343–56
60. Jenkins, M. D. 1987. *Madagascar: an Environmental Profile.* Cambridge: IUCN Monitoring Cent.
61. Jenkins, P. D. 1987. *Catalogue of the Primates in the British Museum (Natural History) and Elsewhere in the British Isles. IV. Suborder Strepsirrhini, including the Subfossil Madagascan Lemurs and Family Tarsiidae.* London: Br. Mus. (Nat. Hist.)
62. Jenkins, P. D., Albrecht, G. H. 1991. Sexual dimorphism and sex ratios in Madagascan prosimians. *Am. J. Primatol.* In press
63. Jolly, A. 1966. *Lemur Behavior.* Chicago: Univ. Chicago Press
64. Jolly, A. 1972. Troop continuity and troop spacing in *Propithecus verreauxi* and *Lemur catta* at Berenty (Madagascar). *Folia Primatol.* 17:335–62
65. Jolly, A. 1980. *A World Like Our Own: Man and Nature in Madagascar.* New Haven: Yale Univ. Press
66. Jolly, A. 1984. The puzzle of female feeding priority. In *Female Primates: Studies by Women Primatologists,* ed. M. Small, pp. 197–215. New York: Liss
67. Jolly, A., Gustafson, H., Oliver, W. L. R., O'Connor, S. M. 1982. Population and troop ranges of *Lemur catta* and *Lemur fulvus* at Berenty, Madagascar:

1980 census. *Folia Primatol.* 39:115–23

68. Jolly, A., Oberlé, P., Albignac, R. 1984. *Key Environments: Madagascar.* Oxford: Pergamon
69. Jolly, C. J. 1970. *Hadropithecus,* a lemuroid small-object feeder. *Man* (n.s.) 5:525–29
70. Jungers, W. L. 1977. Hindlimb and pelvic adaptations to vertical climbing and clinging in *Megaladapis,* a giant subfossil prosimian from Madagascar. *Yrbk. Phys. Anthropol.* 20:508–24
71. Jungers, W. L. 1980. Adaptive diversity in subfossil Malagasy prosimians. *Z. fur Morphol. Anthropol.* 71:177–86
72. Jungers, W. L. 1990. Problems and methods in reconstructing body size in fossil primates. In *Body Size in Mammalian Paleobiology,* ed. J. Damuth, B. MacFadden, pp. 103–18. Cambridge: Cambridge Univ. Press
73. Kappeler, P. M. 1990. The evolution of sexual size dimorphism in prosimian primates. *Am. J. Primatol.* 21:201–214
74. Kay, R. F. 1984. On the use of anatomical features to infer foraging behavior in extinct primates. In *Adaptations for Foraging in Nonhuman Primates,* ed. P. S. Rodman, J. G. H. Cant, pp. 21–53. New York: Columbia Univ. Press
75. Kay, R. F., Cartmill, M. 1977. Cranial morphology and adaptations of *Palaechthon nacimienti* and other Paramomyidae (Plesiadapoidea, Primates) with a description of a new genus and species. *J. Hum. Evol.* 6:19–53
76. Koenders, L., Rumpler, Y., Ratsirarson, J., Peyrieras, A. 1985. *Lemur macaco flavifrons* (Gray, 1867): A rediscovered subspecies of primates. *Folia Primatol.* 44:210–15
77. Koyama, N. 1988. Mating behavior of ring-tailed lemurs *(Lemur catta)* at Berenty, Madagascar. *Primates* 29:163–74
78. Kurland, J. A., Pearson, J. D. 1986. Ecological significance of hypometabolism in nonhuman primates: allometry, adaptation, and deviant diets. *Am. J. Phys. Anthropol.* 71:445–57
79. Lamberton, C. 1946. Contribution à la connaissance de la faune subfossile de Madagascar. XX. Membre posterieur des Néopropithèques et des Mésopropithèques. *Bull. Acad. Malg.* 27:24–28
80. Langrand, O. 1990. *Guide to the Birds of Madagascar.* New Haven: Yale Univ. Press
81. Leighton, M., Leighton, D. R. 1981. Vertebrate responses to fruiting seasonality within a Bornean rain forest. In *Tropical Rain Forest: Ecology and Management,* ed. T. C. Whitmore, A. C. Chadwick, pp. 181–96. Oxford: Blackwell Sci.
82. Macedonia, J. M. 1990. Vocal communication and antipredator behavior in the ringtailed lemur *(Lemur catta),* with a comparison to the ruffed lemur *(Varecia variegata).* PhD thesis. Duke Univ. Durham, N. Carolina
83. MacPhee, R. D. E. 1985. Primates. *1986 Yrbk. Sci. Tech.,* pp. 362–64. New York: McGraw-Hill
84. MacPhee, R. D. E. 1986. Environment, extinction, and Holocene vertebrate localities in southern Madagascar. *Natl. Geo. Res.* 2:441–55
85. MacPhee, R. D. E., Burney, D. A., Wells, N. A. 1985. Early Holocene chronology and environment of Ampasambazimba, a Malagasy subfossil site. *Int. J. Primatol.* 6:463–89
86. MacPhee, R. D. E., Raholimavo, E. M. 1988. Modified subfossil aye-aye incisors from southwestern Madagascar: Species allocation and palaeoecological significance. *Folia Primat.* 51:126–42
87. MacPhee, R. D. E., Simons, E. L., Wells, N. A., Vuillaume-Randriamanantena, M. 1984. Team finds giant lemur skeleton. *Geotimes* 29:10–11
88. McNab, B. K. 1986. The influence of food habits on the energetics of eutherian mammals. *Ecol. Monogr.* 56:1–19
89. Malenky, R. K. 1990. Ecological factors affecting food choice and social organization in *Pan paniscus.* PhD thesis, SUNY Stony Brook
90. Martin, R. D. 1972. Adaptive radiation and behavior of the Malagasy lemurs. *Philos. Trans. Roy. Soc. Lond.* (series B) 264:295–352
91. Martin, R. D. 1972. A preliminary field-study of the Lesser Mouse Lemur (*Microcebus murinus* J. F. Miller, 1777). *Z. Comp. Ethol.,* Suppl. 9:43–89
92. Martin, R. D. 1973. A review of the behavior and ecology of the lesser Mouse Lemur (*Microcebus murinus* J. F. Miller, 1777). In *Comparative Ecology and Behavior of Primates,* ed. R. P. Michael, J. H. Crook, pp. 1–68. London: Academic
93. Martin, R. D. 1975. The bearing of reproductive behavior on the ontogeny of strepsirhine phylogeny. In *Phylogeny of the Primates,* ed. W. P. Luckett, F. S. Szalay, pp. 265–97. New York: Plenum
94. Martin, R. D. 1990. *Primate Origins and Evolution.* Princeton: Princeton Univ. Press
95. Meier, B., Albignac, R., Peyrieras, A., Rumpler, Y., Wright, P. 1987. A new

species of *Hapalemur* (Primates) from South East Madagascar. *Folia Primatol.* 48:211–15

96. Meier, B., Albignac, R. 1991. Rediscovery of *Allocebus trichotis* Gunther 1875 (Primates) in Northeast Madagascar. *Folia Primatol.* 56:57–63
97. Mertl-Millhollen, A. S., Gustafson, H. L., Budnitz, N., Dainis, K., Jolly, A. 1979. Population and territory stability of the *Lemur catta* at Berenty, Madagascar. *Folia Primatol.* 31:106–22
98. Meyers, D. M. 1988. Behavioral ecology of *Lemur fulvus rufus* in rain forest in Madagascar. *Am. J. Phys. Anthropol.* 75:250
99. Muller, E. F. 1985. Basal metabolic rates in primates: the possible role of phylogenetic and ecological factors. *Comp. Biochem. Physiol. A. Comp. Physiol.* 81:707–11
100. Napier, J. R., Walker, A. C. 1967. Vertical clinging and leaping—a newly recognized category of locomotor behavior in primates. *Folia Primatol.* 6:204–19
101. Nicoll, M. E., Langrand, O. 1989. *Madagascar: Revue de la Conservation et des Aires Protégées*. Gland, Switzerland: WWF
102. O'Connor, S. M. 1987. The effect of human impact on vegetation and the consequences to primates in two riverine forests, southern Madagascar. PhD thesis. Univ. Cambridge, England
103. Overdorff, D. 1988. Preliminary report on the activity cycle and diet of the Red-Bellied Lemur *(Lemur rubriventer)* in Madagascar. *Am. J. Primatol.* 16:143–53
104. Overdorff, D. 1990. Flower predation and nectarivory in *Lemur fulvus rufus* and *Lemur rubriventer*. *Am. J. Phys. Anthropol.* 81:276
105. Overdorff, D. 1991. Seasonal patterns of frugivory in *Lemur fulvus rufus* and *Lemur rubriventer*. *Am. J. Phys. Anthropol.* In press
106. Oxnard, C. E., Crompton, R. H., Lieberman, S. S. 1990. *Animal Lifestyles and Anatomies: The Case of the Prosimian Primates*. Seattle: Univ. Wash. Press
107. Pariente, G. F. 1975. Observation ophthalmologique de zones fovéales vraies chez *Lemur catta* et *Hapaleur griseus,* primates de Madagascar. *Mammalia* 39:487–97
108. Pariente, G. F. 1976. Les differents aspects de la limite du tapétum lucidum chez les prosimiens. *Vision Res.* 16: 387–91
109. Paulian, R. 1961. La zoogéographie de Madagascar et des îles voisines. *Faune de Madagascar* 13:1–485
110. Pereira, M. E. 1990. Asynchrony within estrous synchrony among ringtailed lemurs (Primates: Lemuridae). *Physiol. Behav.*:1–6
111. Pereira, M. E., Kaufman, R., Kappeler, P. M., Overdorff, D. J. 1990. Female dominance does not characterize all of the Lemuridae. *Folia Primatol.* 55:96–103
112. Petter, F. 1972. The rodents of Madagascar: the seven genera of Malagasy rodents. In *Biogeography and Ecology in Madagascar,* ed. R. Battistini, G. Richard-Vindard, pp. 661–65. The Hague: Junk
113. Petter, J. -J. 1962. *Recherches sur l'Ecologie et l'Ethnologie des Lémuriens Malgaches*. Mem. Mus. Nat. d'Hist. Nat. (Paris) n. s., Sér. A, t. 27, fasc. 1, 146 pp.
114. Petter, J. -J. 1975. Observations on the behavior and ecology of *Phaner furcifer*. In *Lemur Biology,* ed. I. Tattersall, R. W. Sussman, pp. 209–18. New York: Plenum
115. Petter, J. -J. 1977. The aye-aye. In *Primate Conservation,* ed. H. S. H. Prince Rainier III and G. H. Bourne, pp. 38–59. New York: Academic
116. Petter, J. -J., Petter-Rousseaux, A. 1967. The aye-aye of Madagascar. In *Social Communication Among Primates,* ed. S. A. Altmann, pp. 195–205. Chicago: Univ. Chicago Press
117. Petter, J. -J., Peyrieras, A. 1970. Nouvelle contribution à l'étude d'un lémurien malgache, le aye-aye (*Daubentonia madagascariensis* E. Geoffroy). *Mammalia* 34:167–93
118. Petter, J. -J. Schilling, A., Pariente, G. 1971. Observations éco-éthologiques sur deux lémuriens malgaches nocturnes: *Phaner furcifer* et *Microcebus coquereli*. *Terre Vie* 25:287–327
119. Petter-Rousseaux, A. 1964. Reproductive physiology and behavior of the Lemuroidea. In *Evolution and Genetic Biology of the Primates,* ed. J. Buettner-Janusch, pp. 91–132. New York: Academic
120. Petter-Rousseaux, A. 1980. Seasonal activity rhythms, reproduction and body weight variations in five sympatric nocturnal prosimians, in simulated light and climatic conditions. In *Nocturnal Malagasy Primates,* ed. P. Charles-Dominique et al., pp. 137–52. New York: Academic
121. Pollock, J. I. 1975. Field observations on *Indri indri:* a preliminary report. In

Lemur Biology, ed. I. Tattersall, R. W. Sussman, pp. 287–311. New York: Plenum

122. Pollock, J. I. 1977. The ecology and socioecology of feeding in *Indri indri*. In *Primate Ecology,* ed. T. H. Clutton-Brock, pp. 38–69. New York: Academic
123. Pollock, J. I. 1979. Spatial distribution and ranging behavior in lemurs. In *The Study of Prosimian Behavior,* ed. G. A. Doyle, R. D. Martin, pp. 359–409. New York: Academic
124. Pollock, J. I. 1979. Female dominance in *Indri indri*. *Folia Primatol.* 31:143–64
125. Pollock, J. I. 1986. The management of prosimians in captivity for conservation and research. In *Primates: the Road to Self-Sustaining Populations,* ed. K. Benirschke, pp. 269–87. New York: Springer-Verlag
126. Pollock, J. I. 1989. Intersexual relationships amongst prosimians. *Hum. Evol.* 4:133–43
127. Pollock, J. I., Constable, I. D., Mittermeier, R. A., Ratsirarson, J., Simons, H. 1985. A note on the diet and feeding behavior of the aye-aye *Daubentonia madagascariensis*. *Int. J. Primatol.* 6: 435–47
128. Rabinowitz, P. D., Coffin, M. F., Falvey, D. 1983. The separation of Madagascar and Africa. *Science* 220:67–69
129. Rasamimanana, P., Brun, B., Meyer, J. M., Roos, M. Rumpler, Y. 1990. Seasonal variation in the seminiferous epithelium cycle in Mayotte's brown lemur, *(Eulemur fulvus mayottensis)*. *Folia Primatol.* 55:193–99
130. Rasmussen, D. T., Izard, M. K. 1988. Scaling of growth and life history traits relative to body size, brain size, and metabolic rate in lorises and galagos (Lorisidae, Primates). *Am. J. Phys. Anthropol.* 75:357–67
131. Ratsirarson, J. 1985. Contribution à l'étude comparative de l'Ecoéthologie de Lemur catta dans deux habitats differents de la reserve Speciale de Beza Mahafaly. *Memoire de Fin d'Etudes*. Univ. de Madagascar
132. Richard, A. F. 1974. Patterns of mating in *Propithecus verreauxi*. In *Prosimian Biology,* ed. R. D. Martin, G. A. Doyle, A. C. Walker, pp. 49–74. London: Duckworth
133. Richard, A. 1977. The feeding behaviour of *Propithecus verreauxi*. In *Primate Ecology,* ed. T. H. Clutton-Brock, pp. 72–96. London: Academic
134. Richard, A. F. 1978. *Behavioral Variation: Case Study of a Malagasy Lemur*. Lewisburg, Pa: Bucknell Univ. Press
135. Richard, A. F. 1985. *Primates in Nature*. New York: W. H. Freeman
136. Richard, A. F. 1987. Malagasy prosimians: female dominance. In *Primate Societies,* ed. B. B. Smuts, D. L. Cheney, R. M. Seyfarth, R. W. Wrangham and T. T. Struhsaker, pp. 25–33. Chicago: Univ. Chicago Press
137. Richard, A. F. Sexual dimorphism and sexual selection: evidence and implications from a behavioral study. *J. Hum. Evol.* In press
138. Richard, A. F., Nicoll, M. E. 1987. Female social dominance and basal metabolism in a Malagasy primate. *Propithecus verreauxi*. *Am. J. Primatol.* 12:309–14
139. Richard, A. F., Sussman, R. W. 1987. Framework for primate conservation in Madagascar. In *Primate Conservation in the Tropical Rain Forest,* ed. C. W. Marsh, R. A. Mittermeier, pp. 329–41. New York: Liss
140. Richard, A. F., Rakotomanga, P., Schwartz, M. 1991. Demography of *Propithecus verreauxi* at Beza Mahafaly: sex ratio, survival and fertility, 1984–1989. *Am. J. Phys. Anthropol.* In press
141. Russell, R. J. 1975. Body temperatures and behavior of captive Cheirogaleids. In *Lemur Biology,* ed. I. Tattersall, R. Sussman, pp. 193–208. New York: Plenum
142. Sauther, M. L. 1989. Antipredator behavior in troops of free-ranging *Lemur catta* at Beza Mahafaly Special Reserve, Madagascar. *Int. J. Primatol.* 10:595–606
143. Sauther, M. L. 1991. Reproductive behavior of free-ranging *Lemur catta* at Beza Mahafaly Special Reserve, Madagascar. *Am. J. Phys. Anthropol.*
144. Deleted in proof.
145. Schwartz, J. H., Tattersall, I. 1985. Evolutionary relationships of living lemurs and lorises (Mammmalia, Primates) and their potential affinities with European Eocene Adapidae. *Anthropol. Pap., Am. Mus. Nat. Hist.* 60(1):1–100
146. Sheine, W. S. 1979. The effect of variations in molar morphology on masticatory effectiveness and digestion of cellulose in prosimian primates. PhD thesis. Duke Univ., Durham, N. Carolina
147. Simons, E. L. 1988. A new species of *Propithecus* (Primates) from northeast Madagascar. *Folia Primatol.* 50:143–51
148. Simons, E. L., Rumpler, Y. 1988. *Eulemur:* new generic name for species of *Lemur* other than *Lemur catta*. *C. R.*

Acad. Sci. Paris 307, Sér. 111:547–51
149. Simons-Morland, H. S. 1990. Reproductive activity in ruffed lemurs *(Varecia variegata)* in a northeast Madagascar rain forest: births, infant survival, and parental care. *Am. J. Primatol.* 20:253–65
150. Simons-Morland, H. S. 1991. *Social organization and ecology of black and white ruffed lemurs* (Varecia variegata variegata) *in lowland rain forest, Nosy Mangabe Island, Madagascar*. PhD thesis, Yale Univ., New Haven, Conn.
151. Standing, H. 1908. On recently discovered subfossil primates from Madagascar. *Trans. Zool. Soc. Lond.* 18:69–162
152. Struhsaker, T. T. 1975. *The Red Colobus Monkey*. Chicago: Univ. Chicago Press
153. Sussman, R. W. 1974. Ecological distinctions in sympatric species of lemurs. In *Prosimian Biology*, ed. R. D. Martin, G. A. Doyle, A. C. Walker, pp. 75–108. London: Duckworth
154. Sussman, R. W. 1978. Nectar-feeding by prosimians and its evolutionary and ecological implications. In *Recent Advances in Primatology*, vol. 3, *Evolution*, ed. D. J. Chivers, K. A. Joysey, pp. 119–24. New York: Academic
155. Sussman, R. W. 1991. Demography and social organization of free-ranging *Lemur catta* in the Beza Mahafaly Reserve, Madagascar. *Am. J. Phys. Anthropol.* 84:43–58
156. Sussman, R. W., Tattersall, I. 1976. Cycles of activity, group composition, and diet of *Lemur mongoz mongoz* Linnaeus, 1766 in Madagascar. *Folia Primatol.* 26:270–83
157. Sussman, R. W., Raven, P. H. 1978. Pollination by lemurs and marsupials: an archaic coevolutionary system. *Science* 200:731–36
158. Szalay, F. S., Delson, E. 1979. *Evolutionary History of the Primates*. New York: Academic
159. Tattersall, I. 1973. Subfossil lemuroids and the "adaptive radiation" of the Malagasy lemurs. *Trans. N. Y. Acad. Sci.* 35:314–24
159a. Tattersall, I. 1973. Cranial anatomy of the Archaeolemurinae (Lemuroidea, Primates). *Anthropol. Pap. Am. Mus. Nat. Hist.* 52:1–110
160. Tattersall, I. 1976. Group structure and activity rhythm in *Lemur mongoz* (Primates, Lemuriformes) on Anjouan and Mohéli islands, Comoro Archipelago. *Anthropol. Papers Am. Mus. Nat. Hist.* 53:367–80
161. Tattersall, I. 1977. Ecology and behavior of *Lemur fulvus mayottensis (Primates, lemuriformes)*. *Anthropol. Pap. Am. Mus. Nat. Hist.* 54:421–82
162. Tattersall, I. 1978. Behavioral variation in *Lemur mongoz* (= *L. m. mongoz*). In *Recent Advances in Primatology*, ed. D. J. Chivers, K. A. Joysey, pp. 127–32. London: Academic
163. Tattersall, I. 1979. Patterns of activity in the Mayotte lemur *(Lemur fulvus mayottensis)*. *J. Mammal.* 60:314–23
164. Tattersall, I. 1982. *The Primates of Madagascar*. New York: Columbia Univ. Press
165. Tattersall, I. 1983. Studies on the lemurs of the Comoro Archipelago. *Nat. Geo. Soc. Res. Rep.* 15:641–54
166. Tattersall, I. 1986. Systematics of the Malagasy Strepsirhine primates. In *Comparative Primate Biology*, Vol. 1: *Systematics, Evolution and Anatomy* ed. D. R. Swindler, J. Erwin, pp. 43–72. New York: Liss
167. Tattersall, I. 1986. Notes on the distribution and taxonomic status of some subspecies of *Propithecus* in Madagascar. *Folia Primatol.* 46:51–63
168. Tattersall, I. 1988. A note on nomenclature in Lemuridae. *Phys. Anthropol. Newslett.* 7(1):14
169. Tattersall, I. 1988. Cathemeral activity in primates: a definition. *Folia Primatol.* 49:200–202
170. Tattersall, I., Schwartz, J. A. 1991. Phylogeny and nomenclature in the "*Lemur*-Group" of Malagasy Strepsirhine primates. *Anthropol. Papers Am. Mus. Nat. Hist.* In press
171. Tattersall, I., Sussman, R. W. 1975. Observations on the ecology and behavior of the mongoose lemur *Lemur mongoz mongoz* Linnaeus (Primates, Lemuriformes) at Ampijoroa, Madagascar. *Anthropol. Papers Am. Mus. Nat. Hist.* 52:193–216
172. Terborgh, J. 1983. *Five New World Primates: A Study in Comparative Ecology*. Princeton: Princeton Univ. Press
173. Terborgh, J. 1986. Keystone plant resources in the tropical forest. In *Conservation Biology: the Science of Scarcity and Diversity*, ed. M. Soulé, pp. 330–44. Sunderland: Sinauer
174. Terborgh, J., Janson, C. H. 1986. The socioecology of primate groups. *Annu. Rev. Ecol. Syst.* 17:111–35
175. Terborgh, J., van Schaik, C. 1987. Convergence vs. nonconvergence in primate communities. In *Organization of Com-*

munities Past and Present, ed. J. H. R. Gee, P. S. Giller, pp. 205–26. Oxford: Blackwell Sci.

176. Van Horn, R. N. 1975. Primate breeding season: photoperiodic regulation in captive *Lemur catta*. *Folia Primatol.* 24:203–20
177. Van Horn, R. N. 1980. Seasonal reproductive patterns in primates. *Prog. Reprod. Biol.* 5:181–221
178. Van Horn, R. N., Eaton, G. G. 1979. Reproductive physiology and behavior in prosimians. In *The Study of Prosimian Behavior*, ed. G. A. Doyle and R. D. Martin, pp. 79–122. New York: Academic
179. Vick, L. G., Pereira, M. E. 1989. Episodic targeting aggression and the histories of *Lemur* social groups. *Behav. Ecol. Sociobiol.* 25:3–12

179a. Van Schaik, C. P., Van Hooff, J. A. R. A. M. 1983. On the ultimate causes of primate social systems. *Behaviour* 85:91–117

180. Vuillaume-Randriamanantena, M. 1982. *Contribution à l'étude des os longs des lémuriens subfossiles malgaches*. Thèse de Doctorat de 3e Cycle. Univ. de Madagascar
181. Vuillaume-Randriamanantena, M. 1988. The taxonomic attributions of giant sub-fossil bones from Ampasambazimba: *Archaeoindris* and *Lemuridotherium*. *J. Hum. Evol.* 17:379–91
182. Walker, A. C. 1967. *Locomotor adaptations in recent and fossil Madagascan lemurs*. PhD thesis. Univ. London
183. Walker, A. C. 1967. Patterns of extinction among the subfossil Madagascan lemuroids. In *Pleistocene Extinctions*, ed. P. S. Martin, H. E. Wright, Jr., pp. 425–32. New Haven: Yale Univ. Press
184. Walker, A. 1974. Locomotor adaptations in the past and present prosimian primates. In *Primate Locomotion*, ed. F. A. Jenkins, Jr., pp. 349–82. New York: Academic
185. Walker, A. C. 1979. Prosimian locomotor behavior. In *The Study of Prosimian Behavior*, ed. G. A. Doyle, R. D. Martin, pp. 543–65. New York: Academic
186. Ward, S. C., Sussman, R. W. 1979. Correlates between locomotor anatomy and behavior in two sympatric species of *Lemur*. *Am. J. Phys. Anthropol.* 50: 575–90
187. White, F. J. 1989. Diet, ranging behavior and social organization of the black and white ruffled lemur, *Varecia variegata variegata*, in southeastern Madagascar. *Am. J. Phys. Anthropol.* 78:323
188. Wiens, J. A. 1977. On competition and variable environments. *Am. Sci.* 65: 590–97
189. Wilson, J. M., Stewart, P. D., Ramangason, G. -S., Denning, A. M., Hutchings, M. S. 1989. Ecology and conservation of the crowned lemur, *Lemur coronatus*, at Ankarana, N. Madagascar. With notes on Sanford's Lemur, other sympatrics and subfossil lemurs. *Folia Primatol.* 52:1–26
190. Winn, R. 1989. The aye-ayes, *Daubentonia madagascariensis*, at the Paris Zoological Garden: maintenance and preliminary behavioural observations. *Folia Primatol.* 52:109–23
191. Wright, P. C. 1987. Diet and ranging patterns of *Propithecus diadema edwardsi* in Madagascar. *Am. J. Phys. Anthropol.* 72:218
192. Wright, P. C. 1989. The nocturnal primate niche in the New World, *J. Hum. Evol.* 18:635–58
193. Wright, P. C. 1989. Comparative ecology of three sympatric bamboo lemurs in Madagascar. *Am. J. Phys. Anthropol.* 78:327
194. Young, A. L., Richard, A. F., Aiello, L. C. 1990. Female dominance and maternal investment in strepsirhine primates. *Am. Nat.* 135:473–88

Annu. Rev. Ecol. Syst. 1991. 22:177–91

THE CAUSES OF TREELINE

George C. Stevens

Department of Biology, University of New Mexico, Albuquerque, New Mexico 87131-1091

John F. Fox

Institute of Arctic Biology, and Department of Biology and Wildlife, University of Alaska Fairbanks, Fairbanks, Alaska 99775-0180

KEY WORDS: timberline, limits, range, root, architecture

INTRODUCTION

The word "treeline" generally evokes images of alpine or arctic ecotones, but treelines appear under a much wider variety of circumstances. Wet treelines occur along the margins of bogs or swamps (42, 43, 64, 74, 99, 100, 101). Dry treelines mark the transitions between forest and grassland habitats (18, 89, 103, 104). Cold treelines include the familiar arctic and alpine (31, 65, 71), but also maritime (31) and mountain meadow (25, 58) ecotones. While all treeline studies focus on the replacement of trees by nontrees along an environmental gradient, rarely is this shared theme explicitly developed to make comparisons between different treeline types.

Our goal is to develop general principles of treeline ecology based on the ecological correlates of differences in plant growth form (35, 90). To do this, we use a review of the current explanations for treelines to highlight those features of the tree growth form that appear to limit tree distribution. Our suggestions for future research arise from the gaps in knowledge that appear once the literature is reorganized in this fashion.

Explanations for Treeline

Many of the current explanations for treelines arise from the correlation between tree distribution and climate (3, 6, 28, 39, 46, 62, 63, 66, 97, 105,

00664162/91/1120-0177$2.00

109). The migration of treeline in response to global climate change (1, 13, 29, 32, 33, 49, 50, 59, 60, 61, 75, 82, 85, 86, 91, 94) suggests that some aspect of climate is important to treeline and that trees must differ from nontrees in their ability either to avoid or to tolerate unfavorable climate.

STATURE-RELATED HYPOTHESES At cold treelines the damaging effects of blowing ice (52, 68), strong winds (20, 22, 36, 111), and the drying of plant parts above snowpack (the "winter drought" hypothesis—4, 37, 52, 87, 97, 107, 110, 111) all relate to the inability of trees to avoid harsh climatic conditions. At cold treeline, smaller plants are able to exploit the narrow zone of favorable conditions near the ground while upright trees cannot (27, 38, 107).

A stature-related disadvantage of trees may also account for grassland boundaries. Where fires are an important feature of local ecology, the ratio of woody to nonwoody plants is correlated with fire frequency (5, 98). Both fire avoidance and snowpack exploitation favor those plants with predominantly lateral or subterranean growth.

Even at treelines where fires are uncommon, the upright growth of trees may be indirectly responsible for treeline. Since the wind loading of trees favors those individuals with deep roots, trees have deeper and more massive root systems than do grasses or smaller plants. The need to anchor the tree's crown in soggy ground near wet treelines conflicts with the need to place roots where nutrients are accessible. As a result smaller plants are more efficient at harvesting what limited nutrients are available at wet treelines (93).

At dry treelines, the deep roots of trees may put them at a competitive disadvantage, compared to grasses, when rains are infrequent and fail to fully saturate the soil (90, 102–104). The shallow and diffuse root systems of grasses are probably better at harvesting water under these circumstances (53, but also see 84, 95).

GROWTH-RELATED HYPOTHESES To grow to a large size, trees accumulate tissues over many seasons. The accumulating modular growth of trees (90) carries with it disadvantages not experienced by smaller plants or those that more frequently shed their aboveground parts. As trees grow, the proportion of their tissues that are photosynthetic declines (14, 55, 57). The accumulation of nonphotosynthetic tissue through time may limit the survival of trees beyond treeline, especially if carbon fixation tends to be a limiting factor near treeline. There are two current explanations for cold treelines that take carbon limitation as their central theme.

The "carbon balance" hypothesis (10, 22, 41) uses the decline in photosynthetic rate with temperature to explain why treeline occurs. Under this hypothesis, treeline is situated where the annual carbon fixation of trees does not balance the carbon lost to respiration. Given the accumulation of nonphotosynthetic tissues just mentioned, trees near treeline are probably at a disadvantage relative to plants that more readily shed nonphotosynthetic parts. With the passage of time it becomes harder and harder to balance the energy budget in trees as the nonphotosynthetic but respiring tissue accumulates through life.

The "seasonal compression" hypothesis (97, 108, 110) also emphasizes the carbon limitation that puts trees at a competitive disadvantage. As cold treeline is approached, the growing season shortens. Eventually there comes a point beyond which trees are unable to fix enough carbon during the growth season to support them through the entire year. The "seasonal compression" hypothesis proposes that treeline marks that spot.

TESTING TREELINE HYPOTHESES Demonstrating that any one of these hypotheses for treeline is the correct one is probably not possible given our inability to isolate and separately test their predictions. As just one example of the kinds of problems experimentalists face, consider that while individual plants growing under controlled circumstances may respond in ways compatible with a particular hypothesis (21), at treeline the position of the individual relative to the rest of the vegetation may strongly influence its success. Neighboring trees may modify air flow patterns (111), snow accumulation (8, 40, 73, 107), and seedling survival (44, 108) and as a result enhance tree success. On the other hand, neighbors may reduce survivorship by increasing stem breakage due to snow pressure (19, 52, 78, 110), increasing the frequency of attacks by snow mold (45, 72, 107), or reducing the growth of trees through the slow melting of accumulated snow (73, 79, 107). At cold arctic treelines, clusters of trees can modify site microclimate and local nutrient cycling, leading to a buildup of mosses, a rise in the permafrost table, and formation of hummocks (115, 116) or ice lenses (100). Whether neighboring trees produce cohesion or promote scattering of individuals probably depends on the particular treeline.

What can be said with certainty is that the structure of the vegetation near treeline complicates tests of treeline hypotheses. Most researchers monitor the damage to plants near treeline to provide clues for what climatic factors actually cause treeline. If clusters of individuals experience conditions far different than those of isolated plants, transplant experiments may indicate why isolated individuals fail to survive (34, 108), but not why treeline fails to advance at its margin.

Monitoring the damage to established plants near treeline can also be misleading for other reasons. To see how, imagine that treeline is produced by limited recruitment of seedlings. The top-heavy age structure of many treeline populations (25, 26, 33, 34, 61, 76, 79, 82) suggests this idea, as does evidence that trees near treeline have relatively low fecundity (2, 11, 47, 79, 106) and produce seeds infrequently (24, 45, 62, 81, 106) and of low viability (106). If treeline is the result of limited recruitment, then any attempt to use adult growth or survivorship to determine the cause of treeline is likely to miss the true climate-related cause.

A similar insight comes from considering the life stages of patches of treeline trees (16) instead of individuals. Patches may form because once established, trees facilitate colonization by others. These patches may use up local resources, and then die back. Their offspring may form new patches with the rate of colonization depending on the availability of safe germination sites. If favorable germination sites are limiting beyond treeline (11, 62, 65, 105), and the rate of formation of new clusters falls below the rate at which old clusters die back, then the local population will eventually go extinct.

Under this scenario, individual plants of all ages and life stages would appear to be healthy and vigorous even at the limit of their distribution (16). While the failure of individuals may set treeline, the success of individuals does not necessarily mean that a population can be maintained.

A major difficulty with using demographic explanations for treeline is that very precise estimates of population parameters are needed. This is because a low rate of recruitment must be distinguished from no recruitment. Trees might persist at the present treeline, for example, after climate worsens, as a result of their longevity (24, 34, 65, 76). Conversely, preemptive competition by tundra vegetation or grasses might prevent tree seedlings from establishing (23, 77) after an improvement in climate (91). These cases of apparent vegetation inertia in the short term (18) do not explain what limits tree distribution, because on a long timescale the present treeline would eventually move (48).

The role of history is illustrated most clearly in the studies of the treelines associated with bald mountains in the eastern United States (9, 69). Correlations between present day elevational limits of trees and the altitudes of contemporary balds suggest that past climate change is responsible for some of today's treeless peaks. During warm interglacial periods of the past, trees that today occupy high mountain areas were driven locally extinct on intermediate-sized mountains by advancing lowland tree species. In the present cool period, the lowland tree species have retreated, creating bald spots on intermediate-sized peaks. On larger mountains the upward movement of lowland trees did not drive more highland species to extinction because there was room at the top. Further support for this historical explanation comes from noting that the maintenance of these balds appears to be due to lack of seed sources for colonization (9, 69), not climatic intolerance or seedling mortality.

Site history can also complicate treeline studies in less obvious ways.

Difficulties in establishing trees near treelines have been associated with a lack of sufficient mycorrhizal inoculum of the suitable species in the soil (70, 108). Since established trees may act as a source for mycorrhizal infections, present treelines may be caused by the history of tree occupation. A similar argument can be constructed by noting the soil enrichment beneath trees on grassland borders (51, 112). As before, these historical explanations for treelines cannot be easily tested or interpreted as a consequence of tree structure, but then they are not the outcome of contemporary ecological processes either.

Given the strong evidence that fecundity declines near treeline, coupled with the observation that many of the individuals found there are propagules from areas more central to the plant's distribution (2), many of the traits expressed by trees near treeline may not have evolved in the treeline habitat and may even be maladaptive there (108).

For example, the krummholz growth form may (38) or may not be adaptive, even though it is only expressed near treeline. The stunting of trees may reflect the very functional damage response of trees growing under less marginal circumstances, but at treeline these responses may not result in higher relative fitness as compared to nonstunted forms.

One might imagine that the degree to which treeline populations have adapted to local conditions could be resolved through genetic studies of their distinctiveness compared to other populations of their species. In the past, the different growth forms of the same tree species were considered genetically distinct (17; for a contrasting view, see 61), but both common garden experiments (2, 15, 114) and recent electrophoretic tests of this view have yielded ambiguous results (30, 96). Further ambiguity is added by noting that reproduction near treeline is patchy in both time and space (mentioned earlier) with the result that a random sample of the genetic variation in the population is difficult to obtain. This makes any estimate of the distinctiveness of the treeline population suspect on purely statistical grounds.

Synthesis of Treeline Studies

Two aspects of tree size are hidden within the wide variety of current explanations for treelines: upright stature and the accumulating growth habit of trees. That trees continue to increase in size throughout their lives means both that they grow out of favorable microclimates near the ground and that they become less and less efficient at fixing carbon owing to their accumulation of nonphotosynthetic material. Large size is obviously advantageous when competing for light in a forest, but near treeline, large size may be maladaptive.

Testing treeline hypotheses requires that the stability of the treeline be verified before any experimental tests are carried out. If the treeline is in the process of migrating either forward or back, then there is no sense in asking how treeline is being maintained. Searching for adaptive explanations for treeline may also be nonproductive since treeline populations are not necessarily adapted to treeline conditions.

NEW SIZE-RELATED HYPOTHESES FOR TREELINE Consider three functional differences between trees and smaller plants in regard to their size. The first is the diminishing returns to increasing size as the plant extends the radius of its root "territory" away from its single main trunk. Colonial plants are able to spread their aboveground parts across a large area and are able to do so without greatly increasing the distance between root tips and leaves (7). Annual plants can accomplish the same spread through time, with the local dispersal of offspring. The diminishing returns associated with root transport in trees are likely to be magnified under harsh conditions.

A second difference is that trees average their resource acquisition over a mosaic of small patches, some rich and some poor in resource. In contrast, smaller plants or semi-independent clones may exploit small patches rich in resources while avoiding poor ones. At treelines, the spatial scale of the variability in resource patches relative to tree-root-crown area, and the ratio of resource yield in rich versus poor patches, may favor smaller growth forms.

Finally, having a single stem connecting the root and shoot systems makes trees vulnerable to damage of this vital link. To grow large, trees sacrifice lateral branches through apical dominance to produce a large crown with little self-shading. Other perennial growth forms may reduce or avoid self-shading by being clonal or herbaceous. Annuals and other relatively short-lived plants avoid this problem altogether by retreating each year to the seed. The key variables here are the return time of events destructive to tree stems or crowns, and the time required for trees to mature and reproduce.

CENTRAL PLACE FORAGING Trees extending roots to collect water and nutrients should experience increasing resistance to flow and increasing costs (to enlarge vessels nearer the stem), with increasing radius of extension. This follows from hydraulic laws and the simplified but qualitatively correct analogy between flow in roots and flow in pipes and electrical circuits (56, 92). As root length (and net flow) increase, either vessel diameter must increase nearer the stem to accommodate an increasing flow, or flow (and nutrient uptake or transport) will be restricted. This trade-off exists because flow in tubes is proportional to the fourth power of the tube diameter (56). Thus, for each unit of water or nutrient flux added, an increasing cost must be paid by increasing the diameters of roots.

The most limiting resistance to water uptake is along the pathway from soil to root. Uptake across root cells is especially sensitive to low soil temperature, soil drought, and to lack of oxygen in waterlogged soils (56, 92). These three conditions are encountered, respectively, at cold, dry, and wet treelines, where not only trees but all plants experience decreased water and nutrient uptake per unit of root length. Treelines on serpentine soils may also reflect these inherent difficulties in extracting nutrients from toxic or nutrient impoverished soils (18, 104, 105).

This coincidence between the limits to uptake in roots and the variety of treelines is hard to dismiss. Because the decrease in yield per unit investment in root will weigh more heavily on larger plants (like trees), these conditions should favor smaller plants like those found beyond treeline. Since large plants require root crowns of larger radius than smaller plants, large plants may be unable to exploit successfully the area beyond treeline as a result of the increasing costs of long-distance root transport. This problem is roughly similar to that of central place foraging by animals (80).

A graphical analysis of the diminishing returns associated with root transport illustrates the measurable consequences of this constraint (Figure 1). "Size" can represent either the plant's net primary productivity or the area it mines for resources, since these are assumed to be proportional. "Cost" represents investment in construction and maintenance of the root system, and benefit represents the return on this investment. Cost and benefit may have units of carbon, water, or nutrients depending on whatever resource is limiting. By using a linear cost curve, we assume a fixed root:shoot ratio and fixed ratio of investment in roots per unit size. If instead we assumed that per-unit costs increase with increasing size, the cost curve would be bowed upward. This would not qualitatively alter our predictions. Benefit curves are bowed downward due to the diminishing effectiveness of the root system at its periphery. In a harsher (treeline) environment, the benefit curve is lower for the reasons stated above. As a result, the limiting plant size (at which cost equals benefit) is smaller.

The central-place-foraging hypothesis provides several ad hoc explanations for treeline phenomena. The smaller stature of trees near both cold and wet treelines may be due to the decreasing return on investment in roots (93) even in the absence of mechanical damage due to ice and winter drought. Growth beyond a small size is not possible since transport is the limiting factor. This might also be reflected in the tendency for trees near treeline to "walk," by forming adventitious roots (33, 72, 78, 83). Finding that seedlings survive, but that there is very little reproduction at adulthood near treeline may also reflect the diminishing return on investment in roots (79), as might the report that the maximum size of supra-treeline trees decreases with distance from cold treeline (82).

One approach to testing the central-place-foraging hypothesis would be to

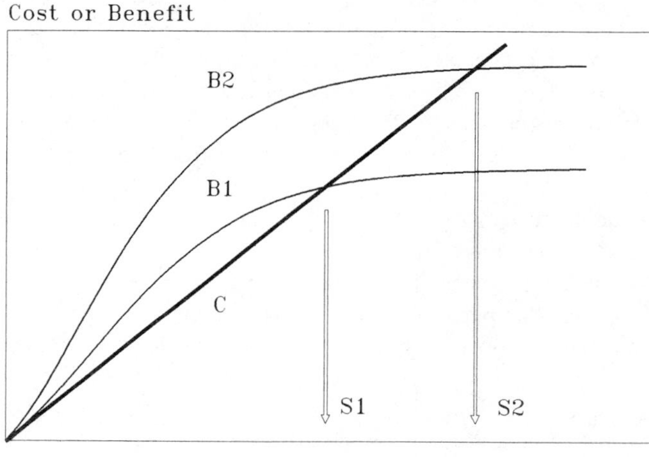

Figure 1 Cost (C) and benefit (B) of increasing plant size in central (B2) and marginal (B1) environments. The return (flow of water and nutrients) per unit investment in root is assumed to be lower in marginal treeline environments where soil is cold, dry, or wet (low in oxygen). The limiting plant size (S1, S2) occurs where cost equals benefit.

ask if treeline populations have evolved toward higher rates of reproduction at a smaller size relative to more central populations. Another would be to look for unusual root traits at treeline that partially compensate for the transport constraints assumed by the central-place-foraging hypothesis. If such tendencies were found it would counter the argument (given previously) for the nonadaptive nature of tree responses near treeline. Experimental tests of the central-place-foraging hypothesis through watering, fertilization, and soil warming could also be attempted and would be the most convincing of all. Unfortunately, neither manipulative experiments or correlative observations would separate the central-place-foraging from the resource-averaging hypothesis.

RESOURCE AVERAGING Suppose that the area of ground included within a tree's branching or rooting area consists of a mosaic of patches of differing concentration or rate of supply of resources. That is, the environment is fine-grained (67) for a large tree, but coarse-grained for a smaller plant. The tree will average the different resource levels, while a small plant will experience the resource level in one patch. Since increases in plant performance (e.g. relative growth rate) exhibit diminishing returns to increases in resource concentration or supply rate (35, 88), the potential exists for the patchiness of the subsurface soil environment to influence optimal plant size in the habitat.

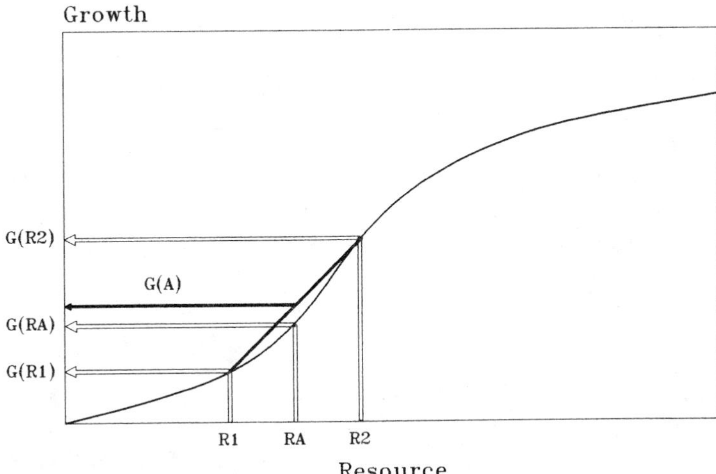

Figure 2 Growth (or other fitness-related trait) as a function of patch resource concentration. Large plants (trees) average over patches with resource levels R1 and R2 obtaining growth at an average resource level, RA. The offspring of small plants (or clones) experience levels R1 and R2 separately. Their average growth rate is G(A).

In this case, one can compare the performance of large and small plants, experiencing the habitat as fine- or coarse-grained, respectively, by a well-known graphical method (Figure 2; 88). The average large plant obtains approximately the average resource supply, RA, and performs accordingly with, say, growth G(RA) (Figure 2). Each small plant experiences, say, either a low (R1) or high (R2) resource level and grows accordingly. The average performance (growth) of small plants depends on the variance in resource availability between patches.

There is a benefit to being small in a resource poor environment (Figure 2). Spatial heterogeneity may enhance this benefit. At cold treelines, this may express itself as a great sensitivity to the location of snow fields or large stones that may influence local soil moisture supply. Patches of forest might also reflect increasing patchiness in resource availability along a transect.

To separate central place foraging from resource averaging, experiments are needed since both have similar predictions as to the constraints on plant size. However, some indication of the role each plays might be found by studying the spatial distribution of the plants that replace trees beyond treeline. In the central-place-foraging hypothesis, the replacement plants are expected to be a solid carpet of highly fecund individuals, while under resource-averaging conditions the success of replacement plants should be patchy (reflecting the underlying heterogeneity of soil resources). While both types of replacements occur near treeline (e.g. forest to grassland versus krummholz to alpine cushions) this correlative approach depends on the assumption that both trees and their replacements are limited by the same heterogeneous resources.

A more direct approach for separating central-place foraging from resource averaging would be to manipulate the variance in soil resources around treeline species. While this may be needed, at this stage in our understanding it would be better to lump the two hypotheses and focus on the role of plant size in limiting tree distribution. If the effects of plant size can be eliminated, both the central-place-foraging and the resource-averaging hypotheses are eliminated.

APICAL DOMINANCE If plant growth is likened to a financial investment strategy (12), trees take the long view of the market, while annual plants are today's programmed traders. Along the spectrum of plant investment strategies, intermediate-sized, multistemmed woody plants fall somewhere between these two extremes, minimizing risk by discounting long-term returns through early profit-taking (35, 90, 103). In this analogy, the branching patterns and maximum size attained by woody plants reflects their investment strategy. If initial lateral growth is inhibited through apical dominance so that future self-shading and structural costs are minimized (113), early reproduction through rapid domination of space is compromised.

Focusing on the disruption of long-term investment strategies of trees as a cause of treeline suggests that studying the growth response of treeline plants to pruning may shed light on the underlying causes of treeline. If treeline populations have adapted to treeline conditions, then individual plants from treeline should tolerate the frequent loss of their growing tips better than nontreeline populations (54). The tendency for treeline species to be found in early successional stages at lower elevations (79, 109) gives weak but consistent support for this prediction. Interactions between the tolerance of pruning and water availability, temperature, or length of growing season should also be explored.

If treeline populations are adapted to local conditions, then the apical dominance hypothesis predicts a breakdown in the relationship between age and size with increasing proximity to treeline (as is observed, 11, 25, 79, 82). This would result from the increased frequency with which individual plants divide their crowns through the formation of adventitious roots (mentioned earlier) or through repeated shedding of a few of their multiple stems. However, this difference in growth strategy would have to be genetic and not simply environmentally induced, for the apical dominance hypothesis to be supported.

CONCLUSION

Emphasizing the structural features of treeline leads to a new emphasis on the multiple levels of analysis necessary to understand treeline. The physiological tolerances of individuals alone may not set distributional limits if the structure of the vegetation strongly influences microclimate. Conversely, studying just the dynamics of the treeline populations may not lead to ultimate causes if treeline is seen to drift, or if history and not contemporary processes sets distributional boundaries.

The size of trees—both the tendency for trees to grow up rather than out and their accumulating modular growth habit—appears to be a relevant factor in the determination of all treelines. The effects of programmed growth to large size may at times be only an indirect cause of treeline, but it is nevertheless a theme that runs through all treeline studies. The need to develop this theme in more detail leads to the following questions:

1. What is the relationship between plant size and fecundity near treeline? Do smaller plants appear to invest more in reproduction for their size than larger plants? If they do, a root-based limit to tree size and distribution is likely.

2. What is the spatial distribution of limited nutrients or water near treeline? Does the patchiness of important resources increase near treeline to the point where a resource averaging strategy is no longer successful? If it does, a root-based limit to tree size and distribution is likely.

3. What is the relationship between pruning tolerance and treeline? Does the apical dominance required by growth to large size ultimately set a limit to treeline by indirectly affecting pruning tolerances? If it does, then treeline can be best understood as an indirect consequence of selection favoring the competitive ability of trees in closed-canopy forests, not at treeline.

Literature Cited

1. Ågren, J., Isaksson, L., Zackrisson, O. 1983. Natural age and size structure of *Pinus sylvestris* and *Picea abies* on a mire in the inland part of northern Sweden. *Holarctic Ecol.* 6:228–37
2. Arno, S. F. 1984. *Timberline: Mountain and Arctic Forest Frontiers.* Seattle, Wash: The Mountaineers
3. Arris, L. L., Eagleson, P. S. 1989. Evidence of a physiological basis for the boreal-deciduous forest ecotone in North America. *Vegetatio* 82:55–58
4. Baig, M. N., Tranquillini, W. 1980. The effects of wind and temperature on cuticular transpiration of *Picea abies* and *Pinus cembra* and their significance in desiccation damage at the alpine treeline. *Oecologia* 47:252–56
5. Barnes, D. L. 1979. Cattle ranching in the semi-arid savannas of east and southern Africa. In *Management of Semi-Arid Ecosystems,* ed. B. H. Walker, pp. 9–51. New York: Elsevier
6. Becwar, M. R., Rajashekar, C., Hansen Bristow, K. J., Burke, M. J. 1981. Deep undercooling of tissue water and winter hardiness limitations in timberline flora. *Plant Physiol.* 68:111–14
7. Bell, A. D., Tomlinson, P. B. 1980. Adaptive architecture in rhizomatous plants. *Bot. J. Linn. Soc.* 80:125–60
8. Billings, W. D. 1969. Vegetational pat-

tern near alpine timberline as affected by fire-snowdrift interactions. *Vegetatio* 19:192–207
9. Billings, W. D., Mark, A. F. 1957. Factors involved in the persistence of montane treeless balds. *Ecology* 38:140–42
10. Billings, W. D., Mooney, H. A. 1968. The ecology of arctic and alpine plants. *Biol. Rev.* 43:481–529
11. Black, R. A., Bliss, L. C. 1980. Reproductive ecology of *Picea mariana* (Mill.) BSP., at tree line near Inuvik, Northwest Territories, Canada. *Ecol. Monogr.* 50:331–54
12. Bloom, A. J., Chapin, F. S. III, Mooney, H. A. 1985. Resource limitation in plants—an economic analogy. *Annu. Rev. Ecol. Syst.* 16:363–92
13. Brink, V. C. 1959. A directional change in the subalpine forest-heath ecotone in Garibaldi Park, British Columbia. *Ecology* 40:10–16
14. Büsgen, M., Münch, E. 1929. Trans. T. Thomson, The structure and life of forest trees. New York: Wiley
15. Callaham, R. Z., Liddicoet, A. R. 1961. Altitudinal variation at 20 years in Ponderosa and Jeffrey pines. *J. For.* 59:814–20
16. Carter, R. N., Prince, S. D. 1988. Distribution limits from a demographic viewpoint. In *Plant Population Ecology: the 28th Symposium of the British Ecological Society, Sussex, 1987,* ed. A. J. Davy, M. J. Hutchings, A. R. Watkinson, pp. 165–84, Oxford, Eng: Blackwell Sci.
17. Clausen, J. 1965. Population studies of alpine and subalpine races of conifers and willows in the California High Sierra Nevada. *Evolution* 19:56–68
18. Collinson, A. S. 1977. *Introduction to World Vegetation.* London: Unwin
19. Daly, C. 1984, Snow distribution patterns in the alpine krummholz zone. *Progr. Phys. Geogr.* 8:157–75
20. Daly, C., Shankman, D. 1985. Seedling establishment by conifers above tree limit on Niwot Ridge, Front Range, Colorado, U. S. A. *Arct. Alp. Res.* 17:389–400
21. Daubenmire, R. F. 1943. Soil temperature versus drought as a factor determining lower altitudinal limits of trees in the Rocky Mountains. *Bot. Gaz.* 105:1–13
22. Daubenmire, R. 1954. Alpine timberlines in the Americas and their interpretation. *Butler Univ. Bot. Stud.* 11:119–36
23. Dunwiddie, P. W. 1977. Recent tree invasion of subalpine meadows in the Wind River mountains, Wyoming. *Arct. Alp. Res.* 9:393–99
24. Elliott, D. L. 1979. The current regenerative capacity of the northern Canadian trees, Keewatin, N.W.T., Canada: some preliminary observations. *Arct. Alp. Res.* 11:243–51
25. Franklin, J. F., Moir, W., Douglas, G., Wiberg, C. 1971. Invasion of subalpine meadows by trees in the Cascade Range, Washington and Oregon. *Arct. Alp. Res.* 3:215–24
26. Franklin, J. F., Dyrness, C. T. 1988. *Natural Vegetation of Oregon and Washington.* Corvallis, Ore: Oregon State Univ. Press
27. Frey, W. 1983. The influence of snow on growth and survival of planted trees. *Arct. Alp. Res.* 15:241–51
28. George, M. F., Burke, M. J., Pellet, H. M., Johnson, A. G. 1974. Low temperature exotherms and woody plant distribution. *Hort. Science* 9:519–22
29. Gorchakovsky, P. L., Shiyatov, S. G. 1978. The upper forest limit in the mountains of the boreal zone of the USSR. *Arct. Alp. Res.* 10:349–63
30. Grant, M. C., Mitton, J. B. 1977. Genetic differentiation among growth forms of Englemann spruce and Subalpine fir at tree line. *Arct. Alp. Res.* 9:259–63
31. Griggs, R. F. 1934. The edge of the forest in Alaska and the reasons for its position. *Ecology* 15:80–96
32. Griggs, R. F. 1937. Timberlines as indicators of climatic trends. *Science* 85:251–55
33. Griggs, R. F. 1942. Indications as to climatic changes from the timberline of Mount Washington. *Science* 95:515–19
34. Griggs, R. F. 1946. The timberlines of northern America and their interpretation. *Ecology* 27:275–289
35. Grime, J. P. 1979. *Plant Strategies and Vegetation Processes.* New York, NY: Wiley
36. Hadley, J. L., Smith, W. K. 1983. Influence of wind exposure on needle desiccation and mortality for timberline conifers in Wyoming, U. S. A. *Arct. Alp. Res.* 15:127–35
37. Hadley, J. L., Smith, W. K. 1986. Wind effects on needles of timberline conifers: seasonal influence on mortality. *Ecology* 67:12–19
38. Hadley, J. L., Smith, W. K. 1987. Influence of krummholz mat microclimate on needle physiology and survival. *Oecologia* 73:82–90
39. Halliday, W. E. D., Brown, A. W. A. 1943. The distribution of some important forest trees in Canada. *Ecology* 24:353–73

40. Hare, F. K., Ritchie, J. C. 1972. The boreal bioclimates. *Geogr. Rev.* 62:333–65
41. Häsler, R. 1982. Net photosynthesis and transpiration of *Pinus montana* on east and north facing slopes at alpine timberline. *Oecologia* 54:14–22
42. Heilman, P. E. 1968. Relationship of availability of phosphorus and cations to forest succession and bog formation in interior Alaska. *Ecology* 49:331–36
43. Heinselman, M. L. 1970. Landscape evolution, peatland types, and the environment in the Lake Agassiz Peatlands Natural Area, Minnesota. *Ecol. Monogr.* 40:235–61
44. Höllerman, P. W. 1978. Geoecological aspects of the upper timberline in Tenerife, Canary Islands. *Arct. Alp. Res.* 10:365–82
45. Holtmeier, F. -K. 1973. Geoecological aspects of timberlines in northern and central Europe. *Arct. Alp. Res.* 5:A45–A54
46. Hopkins, D. M. 1959. Some characteristics of the climate in forest and tundra regions of Alaska. *Arctic* 12:214–20
47. Hustich, I. 1953. The boreal limits of conifers. *Arctic* 6:149–62
48. Hustich, I. 1983. Tree-line and tree growth studies during 50 years: some subjective observations. In *Tree-Line Ecology, Proceedings of the Northern Québec Tree-Line Conference*, ed. P. Morisset, S. Payette, pp, 181–88, Québec, Canada: Cent. d'études nordiques, Univ. Laval
49. Jungerius, P. D. 1969. Soil evidence of postglacial tree line fluctuations in the cypress hills area, Alberta, Canada. *Arct. Alp. Res.* 1:235–46
50. Kearney, M. S., Luckman, B. H. 1983. Holocene timberline fluctuations in Jasper National Park, Alberta. *Science* 221:261–63
51. Kellman, M. 1979. Soil enrichment by neotropical savanna trees. *J. Ecol.* 67:565–77
52. Klikoff, L. G. 1965. Microenvironmental influence on vegetational pattern near timberline in the central Sierra Nevada. *Ecol. Monogr.* 35:187–221
53. Knoop, W. T., Walker, B. H. 1985. Interactions of woody and herbaceous vegetation in a southern African savanna. *J. Ecol.* 73:235–253
54. Koop, H. 1987. Vegetative reproduction of trees in some European natural forests. *Vegetatio* 72:103–110
55. Kozlowski, T. T. 1971. *Growth and Development of Trees,* Vol. 1. *Seed Germination, Ontogeny and Shoot-Growth*. New York: Academic
56. Kramer, P. J. 1982. *Water Relations of Plants*. New York: Academic
57. Kramer, P. J., Kozlowski, T. T. 1979. *Physiology of Woody Plants*. New York: Academic
58. Kuramoto, R. T., Bliss, L. C. 1970. Ecology of subalpine meadows in the Olympic Mountains, Washington. *Ecol. Monogr.* 40:317–47
59. LaMarche, V. C. Jr. 1973. Holocene climatic variations inferred from treeline fluctuations in the White Mountains, California, *Quat. Res.* 3:632–60
60. LaMarche, V. C. Jr., Mooney, H. A. 1967. Altithermal timberline advance in western United States. *Nature* 213:980–82
61. LaMarche, V. C. Jr., Mooney, H. A. 1972. Recent climatic change and development of the Bristlecone Pine (*P. longaeva* Bailey) krummholz zone, Mt. Washington, Nevada. *Arct. Alp. Res.* 4:61–72
62. Larsen, J. A. 1965. The vegetation of the Ennadai Lake area, N. W. T.: studies in subarctic and arctic bioclimatology. *Ecol. Monogr.* 35:37–59
63. Larsen, J. A. 1971. Vegetational relationships with air mass frequencies: boreal forest and tundra. *Arctic* 24:177–94
64. Larsen, J. A. 1982. *Ecology of the Northern Lowland Bogs and Conifer Bogs*. New York: Academic
65. Larsen, J. A. 1989. *The Northern Forest Border in Canada and Alaska: Biotic Communities and Ecological Relationships*. New York: Springer-Verlag
66. Lauer, W. 1978. Timberline studies in central Mexico. *Arct. Alp. Res.* 10:383–96
67. Levins, R. 1968. *Evolution in Changing Environments: Some Theoretical Explorations*. Princeton, NJ: Princeton Univ. Press
68. Marchand, P. J., Chabot, B. F. 1978. Winter water relations of tree-line plant species on Mt. Washington, New Hampshire. *Arct. Alp. Res.* 10:105–16
69. Mark, A. F. 1958. The ecology of the southern Appalachian grass balds. *Ecol. Monogr.* 28:293–336
70. Marks, G. C., Kozlowski, T. T. 1973. *Ectomycorrhizae: Their Ecology and Physiology*. New York: Academic
71. Marr, J. W. 1948. Ecology of the forest-tundra ecotone on the east coast of Hudson Bay. *Ecol. Monogr.* 18:117–44
72. Marr, J. W. 1977. The development and movement of tree islands near the upper limit of tree growth in the southern Rocky Mountains. *Ecology* 58:1159–64
73. Minnich, R. A. 1984. Snow drifting

and timberline dynamics on Mount San Gorgonio, California, U.S.A. *Arct. Alp. Res.* 16:395–412

74. Moore, P. D., Bellamy, D. J. 1974. *Peatlands*. London: ELEK Sci.
75. Morisset, P., Payette, S. 1983. *Tree-Line Ecology, Proceedings of the Northern Québec Tree Line Conference*. Québec, Canada: Centre d'études nordiques, Univ. Laval
76. Nichols, H. 1976. Historical aspects of the northern Canadian treeline. *Arctic* 29:38–47
77. Nobel, I. R. 1980. Interactions between tussock grass (*Poa* spp.) and *Eucalyptus pauciflora* seedlings near treeline in south-eastern Australia. *Oecologia* 45:350–53
78. Norton, D. A., Schönenberger, W. 1984. The growth forms and ecology of *Nothofagus solandri* at the alpine timberline, Craigieburn Range, New Zealand. *Arct. Alp. Res.* 16:361–70
79. Ohsawa, M. 1984. Differentiation of vegetation zones and species strategies in the subalpine region of Mt. Fuji. *Vegetatio* 57:15–52
80. Orians, G. H., Pearson, N. E. 1979. On the theory of central place foraging. In *Analysis of Ecological Systems,* ed. D. J. Horn, R. D. Mitchell, G. R. Stairs, pp. 155–77. Columbus, Ohio: Ohio State Univ. Press
81. Payette, S., Deshaye, J., Gilbert, H. 1982. Tree seed populations at the treeline in Rivière aux Feuilles area, northern Quebec, Canada. *Arct. Alp. Res.* 14:215–21
82. Payette, S., Filion, L. 1985. White Spruce expansion at the tree line and recent climatic change. *Can. J. For. Res.* 15:241–51
83. Plesník, P. 1973. Some problems of the timberline in the Rocky Mountains compared with Central Europe. *Arct. Alp. Res.* 5:A77–A84
84. Pressland, A. J. 1973. Rainfall partitioning by an arid woodland (*Acacia aneura* F. Muell.) in south-western Queensland. *Aust. J. Bot.* 21:235–45
85. Ritchie, J. C. 1976. The late-Quaternary vegetational history of the western interior of Canada. *Can. J. Bot.* 54:1793–1818
86. Ritchie, J. C., Hare, F. K. 1971. Late-Quaternary vegetation and climate near the arctic tree line of northwestern North America. *Quat. Res.* 1:331–42
87. Sakai, A. 1970. Mechanism of desiccation damage of conifers wintering in soil-frozen areas. *Ecology* 51:657–64
88. Salisbury, F. B., Ross, C. W. 1978. *Plant Physiology,* Belmont, Calif: Wadsworth 2nd ed.
89. Sarmiento, G. 1984. *The Ecology of Neotropical Savannas*. (Transl. O. Solbrig) Cambridge, Mass: Harvard Univ. Press
90. Schulze, E. -D. 1982. Plant life forms and their carbon, water and nutrient relations. In *Encyclopedia of Plant Physiology,* Volume 12B, *Physiological Plant Ecology II, Water Relations and Carbon Assimilation,* ed. O. L. Lange, P. S. Nobel, C. B. Osmond, H. Ziegler, pp. 616–76. Berlin: Springer-Verlag
91. Scott, P. A., Hansell, R. I. C., Fayle, D. C. F. 1987. Establishment of White Spruce populations and responses to climatic change at the treeline, Churchill, Manitoba, Canada. *Arct. Alp. Res.* 19:45–51
92. Slatyer, R. O. 1967. *Plant-Water Relationships*. New York: Academic
93. Small, E. 1972. Water relations of plants in raised *Sphagnum* peat bogs. *Ecology* 53:726–28
94. Sorenson, C. J., Knox, J. C., Larsen, J. A., Bryson, R. A. 1971. Paleosols and the forest border in Keewatin, N. W. T. *Quat. Res.* 1:468–73
95. Soriano, A., Sala, O. 1983. Ecological strategies in a Patagonian arid steppe. *Vegetatio* 56:9–15
96. Tigerstedt, P. M. A. 1973. Studies on isozyme variation in marginal and central populations of *Picea abies*. *Hereditas* 75:47–60
97. Tranquillini, W. 1979. *Physiological Ecology of the Alpine Timberline: Tree Existence at High Altitudes with Special Reference to the European Alps*. New York: Springer-Verlag
98. Trollope, W. S. W. 1982. Ecological effects of fire in South African savannas. In *Ecology of Tropical Savannas,* ed. B. J. Huntley, B. H. Walker, pp. 292–306. Berlin: Springer-Verlag
99. Van Cleve, K., Chapin, F. S. III, Flanagan, P. W., Viereck, L. A., Dyrness, C. T. 1986. *Forest Ecosystems in the Alaskan Taiga*. New York: Springer-Verlag
100. Viereck, L. A. 1970. Forest succession and soil development adjacent to the Chena River in interior Alaska. *Arct. Alp. Res.* 2:1–26
101. Viereck, L. A., Dyrness, C. T., Van Cleve, K., Foote, M. J. 1983. Vegetation, soils, and forest productivity in selected forest types in interior Alaska. *Can. J. For. Res.* 13:703–20
102. Walker, B. H., Ludwig, D., Holling, C. S., Peterman, R. M. 1981. Stability of

semi-arid savanna grazing systems. *J. Ecol.* 69:473–98

103. Walter, H. 1971. *Ecology of Tropical and Subtropical Vegetation,* Transl. D. Mueller-Dombois, Edinburgh: Oliver & Boyd
104. Walter, H. 1985. *Vegetation of the Earth, and Ecological Systems of the Geo-biosphere*. New York: Springer-Verlag. 2nd ed.
105. Walter, H., Breckle, S. 1986. *Temperate and Polar Zonobiomes of Northern Eurasia*. Berlin: Springer Verlag
106. Wardle, J. 1970. The ecology of *Nothofagus solandri:* 3. Regeneration. *N. Z. J. Bot.* 8:571–608
107. Wardle, P. 1968. Englemann spruce (*Picea engelmannii* Engel.) at its upper limits on the Front Range, Colorado. *Ecology* 49:483–95
108. Wardle, P. 1971. An explanation for alpine timberline. *N. Z. J. Bot.* 9:371–402
109. Wardle, P. 1973. New Zealand timberlines. *Arct. Alp. Res.* 5:A127–A135
110. Wardle, P. 1977. Japanese timberlines and some geographic comparisons. *Arct. Alp. Res.* 9:249–58
111. Warren Wilson, J. 1959. Notes on wind and its effects in arctic-alpine vegetation. *J. Ecol.* 47:415–27
112. White, D. P. 1941. Prairie soil as a medium for tree growth. *Ecology* 22:398–407
113. Whitney, G. G. 1976. The bifurcation ratio as an indicator of adaptive strategy in woody plant species. *Bull. Torrey Bot. Club* 103:67–72
114. Woodward, F. I. 1986. Ecophysiological studies on the shrub *Vaccinium myrtillus* L. taken from a wide altitudinal range. *Oecologia* 70:580–86
115. Zoltai, S. C. 1985. Structure of subarctic forests on hummocky permafrost terrain in northwestern Canada. *Can. J. For. Res.* 5:1–9
116. Zoltai, S. C., Pettapiece, W. W. 1974. Tree distribution on perennially frozen earth hummocks. *Arct. Alp. Res.* 6:403–11

Annu. Rev. Ecol. Syst. 1991. 22:193–228

PHYSIOLOGICAL DIFFERENTIATION OF VERTEBRATE POPULATIONS

Theodore Garland, Jr. and Stephen C. Adolph

Department of Zoology, University of Wisconsin, Madison, Wisconsin 53706

KEY WORDS: evolution, comparative methods, quantitative genetics, thermoregulation, metabolism, locomotion, adaptation

---

INTRODUCTION

> . . . The foundation of most evolutionary theory rests upon inferences drawn from geographic variation or upon the verification of predictions made about it.
>
> Gould & Johnston (120, p. 457)

> The study of geographic variation occupies a central position in evolutionary biology . . . because geographic variation is ordinarily the smallest amount of evolution that can be detected in nature and because evolutionary theory, in its strongest form, applies only to small evolutionary change.
>
> Arnold (7, p. 510)

The Importance of Population Differentiation

Although studies of geographic variation are relatively common, the vast majority have dealt with morphometric or meristic traits (13, 120, 297, 327), allozymes (150, 233, 327), or most recently, mitochondrial DNA (9). The relative rarity of studies of geographic variation in physiological traits (see 42, 245, 312 for reviews of early studies) is somewhat surprising, considering that a major focus within vertebrate "physiological ecology," "comparative physiology," "environmental physiology," and "biophysical ecology" is the study of adaptation (14, 15, 44, 75, 102, 112, 156, 181, 244, 246, 298, 299). Of course, populations may show differentiation on other than large geographic scales, such as "microgeographic" (150, 188, 213, 233, 290),

0066-4162/91/1120-0193$02.00

temporal (164, 165, 213, 253, 278, Armitage, this volume) or altitudinal; the last has received considerable attention from vertebrate physiologists.

Why are population differences in physiology not studied more often? We suggest four reasons: (i) Physiological measurements require living organisms. Collecting from multiple populations is formidable enough without the additional difficulty of making physiological measurements. "Expeditionary physiology" is possible, but requisite equipment often is considerably less portable than that needed for sampling tissues or for obtaining study skins or skeletons. Moreover, physiological measurements can be quite time-consuming and require maintaining animals in a healthy state, sometimes for weeks or months, to achieve a common state of acclimation. (ii) Physiologists and evolutionary biologists often share little in the way of training, interests, perspectives, techniques, and study organisms. Population geneticists routinely study *Drosophila* because they are small and so can be bred in large numbers. Physiological ecologists rarely study *Drosophila* because they are small and hence intractable for most physiological measurements (but see 63, 137, 158, 166). Nonetheless, some studies of population differentiation in physiological traits have been done. Many involve invertebrates, including *Drosophila* (42, 63, 137, 158, 223, 233, 245, 312 refs. therein). Another cohort of researchers has come from a background in population genetics, having moved toward physiology in an attempt to discover the adaptive significance of population differences in allozymes (104, 137, 191, 223, 224, 243, refs. therein). (iii) To increase the likelihood of finding differences and exemplary cases of adaptation, comparative physiologists have tended to focus on species expected to display extremes in physiological function (102, 255). Recognizing problems introduced by rampant acclimatization, physiologists may have been inhibited from searching for relatively small population differences. Consequently, the focus on proximate mechanisms of coping with environmental change (i.e. acclimatization) drew more attention than the possibility of genetic differentiation among populations. (iv) "Typological thinking" still exists among many physiologists (102). (The coexistence of typological thinking and ardent adaptationism is somewhat surprising!)

Scope of This Review and Some Definitions

We review population differences in physiological traits of vertebrates. As few physiological studies of natural variation allow direct inferences about genetics, we have not restricted ourselves to these. We avoid the issue of how to define a population and instead rely on each author's judgment. We have attempted to consider examples involving all physiological processes in all vertebrate classes except Chondrichthyes and Agnatha. We have excluded human examples, and our review reflects our particular physiological and taxonomic interests.

Physiology has to do with how organisms work. But where to draw lines between physiology on the one hand and behavior, morphology, or biochemistry on the other is not always clear (112, 245). For example, morphological (e.g. heart mass), biochemical (e.g. blood hemoglobin content, in vitro enzyme activity), and whole-animal performance traits (e.g. maximal sprint running speed) are routinely studied by "physiologists." We have therefore chosen to include certain morphological, biochemical, or performance measures wherever they seem closely tied to physiological function. Our holistic view is consistent with recent recommendations to "adopt and promulgate the definition of physiology as 'Integrative Biology' " (2). Studies relating growth or developmental rate to multilocus heterozygosity are reviewed elsewhere (104, 213, 223; refs. therein).

Many *life history traits* (e.g. fecundity, reproductive effort, growth rate, body size) are closely tied to physiology. However, we include only a few examples here, because variation in life history traits is reviewed extensively elsewhere (e.g. 27, 33, 82, 107, 118, 186, 200, 218, 226, 235, 253–255, 261, 289, 290; Godfray, this volume). The empirical validity of several well-known *biogeographic rules* pertaining to body size, proportions, and coloration (e.g. Allen's, Bergmann's, Gloger's) is highly questionable (39, 117, 139, 157, 173, 174, 252, 312, 327), and thorough "common garden" (60a) studies are rare (but see 77, 196, 252). Island populations also show characteristic patterns of gigantism or dwarfism (24, 25, 83, 226), but physiological correlates have not been studied. Variation in body size results partly from variation in growth rate, the genetic and physiological bases of which have been well studied in some domestic animals (99, 113, 256).

The term *adaptation* is used in two ways by physiologists (321). First, adaptation may have the traditional evolutionary meaning of genetic changes *in a population* occurring in response to either natural or artificial selection. Second, adaptation may refer to any change that occurs *within an individual* in response to changes in the environment, *and* that helps the organism function "better" under those changed conditions (e.g. 245). To quote Dejours (75, p. 14): "Adaptation consists in a change minimizing the physiological strain which results from a stressful environment." In the physiological literature, many authors are imprecise in their use of the word adaptation (321), and many are unclear as to whether they think observed population differences are in part genetically based or represent entirely acclimatization.

Nongenetic changes in physiology occurring during the lifetime of an individual generally are termed acclimation or acclimatization. *Acclimation* refers to changes that occur in response to change in any component of the environment in the laboratory. *Acclimatization* refers to the same, but as it occurs in nature. Most cases of acclimation or acclimatization are adaptive in the physiological sense (e.g. 55, 208), although physiologists have tended to be too liberal in interpreting specific cases of physiological adaptation as evolutionarily adaptive (43, 102, 321). Nevertheless, a general capacity for acclimation or acclimatization is certainly adaptive in the evolutionary sense, and such capacity may itself be genetically based and subject to adaptive evolution (114, 115, 300).

Changes as a result of acclimation or acclimatization are generally reversible and are examples of *phenotypic plasticity* (131, 267, 300). Nongenetic adaptation buys time for individuals and for populations and "restores flexibility to physiological responses much as learning restores flexibility to behavior" (285, p. 401). As such, physiological adaptation is but one component of the overall "graded response" system that organisms may use when faced with changing environmental conditions, and nongenetic adaptive modifications may "prepare the way" for "subsequent evolutionary advance" (114, p. 100). Thus, phenotypic plasticity in response to environmental changes may reduce selection pressures and so "discourage the building up of geographic physiological races" (131, p. 53).

NATURE VERSUS NURTURE

Population differences in physiological traits may be due to genetic and/or environmental effects or their interaction. From a physiological perspective, any comparison showing (or failing to show) population differences may be of interest. Once population differences in physiology are discovered, they may be related to behavioral and ecological variation (a common goal of physiological ecologists) or to underlying mechanistic bases at lower levels of biological organization (a common goal of comparative physiologists and biochemists). From an evolutionary perspective, however, comparisons that allow conclusions regarding genetic differentiation are of prime importance for most phenotypic traits (300).

Special Problems Posed by Physiology

Physiological traits are highly susceptible to a wide range of environmental factors (e.g. nutritional status, stress, acclimatization, early environmental effects, maternal effects); indeed, much of the physiological literature addresses these effects. For some traits, the magnitude of environmental effects can be quite large relative to the magnitude of genetic differences among populations; unfortunately, experimental data are rare (131, 132, 153, 155). When population differences in physiology are found in field-fresh animals, environmental effects are likely to be present (cf 300).

Seasonal variation in physiological traits has been documented innumerable times. Hence, populations might show differences simply because they are at different points on an annual cycle. Seasonal rhythms may be endogenous and/or driven by environmental variables such as photoperiod (19, 40, 55, 78, 100, 138, 186, 194, 200, 218, 226, 242, 294, 326). The variety of traits that vary seasonally is surprising; examples include life history traits (253), gut morphology (271), and allozymes of blood proteins (214) and various enzymes (156). Physiological traits can vary on a daily basis as well; daily cycles in metabolic rates, body temperature, heat tolerance, blood pressure, and physical performance ability are well known (e.g. 43, 170, 240, 242). Problems due to annual and daily cycles are not always recognized by researchers.

Potential complications arising from acclimatization and other environmental effects often are dealt with by acclimating animals in the laboratory to standard conditions before measuring physiological traits. But not all environmental effects on physiological, morphological, or behavioral traits are reversible, particularly if they occur during early or critical periods of development (10–12, 75, 131, 136, 173, 174, 181, 199, 201, 209, 222, 235, 252, 268, 275, 300, refs. therein: but see 256). For example, diet can affect gut morphology and function in a variety of vertebrates (e.g. 12, 124, 181, 215, 271, refs. therein). Diet can even affect patterns of hibernation in small mammals (116). Nonreversible changes are troublesome for studies of population differences, because they are not eliminated by acclimation.

"Common Garden" Experiments—No Design is Ideal

Although proximate environmental effects can confound population comparisons, few studies have used a "common garden" design (e.g. 60a, 251) to study physiological differences of vertebrate populations. The usual approach of acclimation to common laboratory conditions is not guaranteed to erase previous environmental influences. The next most common approach is to rear animals from birth (or near birth) in the laboratory, by collecting gravid females, eggs, or newborns from the wild. However, differences due to maternal (or paternal—11) effects may occur at any time during development, and in some cases prior to conception. Maternal effects may even persist for more than one generation (257); a well-known example involves litter size in *Mus musculus* (98, 99). Some parental effects might be partly controlled statistically by using parental traits (e.g. age, body mass, parity) as covariates in comparisons (e.g. 109, 110, 304).

As another way to eliminate possible maternal and other environmental effects, populations might be raised for multiple generations in captivity. Unfortunately, laboratory selection pressures, however unintentional, undoubtedly will differ from those in nature (10, 131) and may lead to changes in genetic composition. Consequently, comparisons among laboratory populations may not reliably reflect the genetic differences present in nature (e.g. 219). Incorporating time in captivity as a covariate may help to account for such effects (149, 219).

Thus, no "common garden" experimental design is ideal. In many cases, studying lab-reared offspring of field-collected adults is the best compromise. Comparisons of species may also be confounded by environmental effects. Biologists generally seem less concerned about this problem, however, and most interspecific comparative studies do not involve adequate common garden controls. This includes the vast majority of the classic studies cited in comparative physiology textbooks.

The disadvantage of genetic changes occurring over multiple generations in captivity (including the domestication process) can be turned into an advantage, as genetic drift and/or artificial selection can produce divergent (or convergent: 65) lines (populations) useful for a variety of purposes (e.g. 7, 11, 12, 21, 54, 63, 78, 81, 90, 96, 99, 113, 123, 166, 197, 198, 200, 211, 215, 239, 247, 256, 257, 259, 261, 266, 324).

DOES PHYSIOLOGY VARY AMONG VERTEBRATE POPULATIONS?

Mammalia

Many studies of population differences have involved *Peromyscus* or *Mus,* two speciose rodent genera that are easy to maintain in captivity and have short generation times. A thorough review of breeds of domestic mammals or of the quantitative genetics literature in general is beyond the scope of this review. Breed differences in physiology exist in domestic dogs (59, 108, refs. in 245, 324) and in cattle and sheep (refs. in 202, 203, 245, 322).

GROWTH, REPRODUCTION, AND LIFE HISTORY TRAITS Both migratory and nonmigratory (e.g. in Ngorongora Crater, Tanzania) populations of ungulates exist, and population differences in social structure may covary with habitat in some mammals. Reviews of population differences in reproductive characteristics of free-living mammals are available elsewhere (33, refs. in 40, 61, 62, 184, 218, 226). A variety of reproductive traits show geographic variation within species of *Peromyscus* (218). Millar (218, p. 196) concludes that "Little modification in basic reproductive and developmental patterns is evident within species that occupy diverse environments; differences in the timing of reproductive events constitute the major adaptation to different environments at the intraspecific level." Desjardins et al (78) have used artificial selection to alter the reproductive photoresponsiveness of deer mice, and Lynch et al (200) report differences in photoresponsiveness between two captive breeding stocks of the Djungarian hamster (see also 218). Comparisons of domestic mammals with their wild counterparts often show that the former are less seasonal in their reproduction (refs. in 40), and domestic *Rattus* and *Mus* are essentially aseasonal. Millar & Threadgill (219) conclude that, in general, no major differences in reproductive and developmental patterns exist between wild, captive, and domesticated stocks of *Peromyscus*. This is not to say that *no* differences exist. The relationships of various reproductive traits to maternal mass differ among populations of captive-bred *Peromyscus* (96), and differences in offspring size and the trade-off between number and size of offspring "may represent nothing more than laboratory artifacts" (219, p. 1719). In a common garden experiment, cotton rat populations differed in their response to food restriction; Kansas dams, with larger litters, were more severely affected than were Tennessee dams (209).

Montane-mesic and lowland-xeric populations of *Marmota flaviventris* held under constant conditions in the lab for >25 months differ in the circannual rhythms of food consumption and body mass (317); in the field, population differences in growth rates and activity times occur (216; Armitage, this volume; refs. therein).

Latitudinal clines in pupping data exist in a number of pinniped species (107, 295), but their genetic basis, if any, is unknown. For some species, variation in the date of birth is apparently cued by photoperiod acting through the mechanism of delayed implantation (294). Similar phenomena occur in other Carnivora such as weasels (186).

THERMAL BIOLOGY, THERMOREGULATION, AND RESTING METABOLISM Acclimation and acclimatization of mammalian thermoregulatory variables in response to temperature and/or photoperiod have been studied extensively (55, 71, 132, 138, 245, 322). For example, changes in response to cold exposure may include fur growth, increased resting metabolic rate, increased body temperature, shifts in critical temperatures, increased sensitivity to norepinephrine-induced nonshivering thermogenesis, increased amounts of brown fat, changes in masses of internal organs, increases in Na^+,-K^+-ATPase activity, and increased food consumption. Interestingly, changes induced by acclimation may differ substantially from those seen under natural acclimatization. Some seasonal cycles persist when animals are held in the lab under constant conditions (12, 71, 132, 138, 201, 317). However, we know of no longitudinal studies (following individuals) of natural seasonal variation in thermoregulatory parameters.

The lability of mammalian thermoregulatory traits indicates that common garden approaches are extremely important for population comparisons. Moreover, several studies demonstrate that environmental factors acting early in life may have large and lasting influences on thermoregulatory traits of mammals (11, 12, 136, 199, 201).

Much of the earlier work on geographic differences in thermoregulation of rodents is in Russian, and some provides evidence that geographic population differences may in part be genetically based (132). Unfortunately, many of the earlier studies on population differences are difficult to interpret, because (i) combinations of wild caught and laboratory born (sometimes to females captured while gravid) animals are studied, (ii) metabolic rate is expressed per gram (metabolic rate generally does not scale as [body mass]$^{1.00}$), and (iii) statistical analyses are lacking or inadequate. Some of the relevant studies completed since Hart's review (132) are discussed below, and many suffer from the same problems. "Thermal conductance," used to describe the relationship between heat production and ambient temperature below the thermal neutral zone, is misleading but still in common use.

Body temperature appears to show little among-population variation, and whether basal metabolic rate (BMR) shows climatic adaptation within species of small mammals is still controversial (39, 132, 312). MacMillen & Garland (204), analyzing a multispecies sample of 31 *Peromyscus* populations taken from the literature, found a significant partial regression (after accounting for body mass effects) of basal metabolic rate (BMR) on environmental temperature: populations from hotter habitats tended to have lower BMRs. Neither latitude, precipitation nor "desert index" added significantly to the predictive equation.

Thompson (296) shows significant subspecific differences in BMR and heat loss coefficient (a.k.a. "thermal conductance") in harvest mice *(Reithrodontomys megalotis)*. The subspecies differ also in propensity to enter shallow daily torpor, consistent with previous reports for this species and suggestions for *Peromyscus* subspecies. [Populations of pygmy possum *(Cercartetus nanus)* may also differ in torpor patterns (F. Geiser, personal communication).] Chaffee (54) artificially selected for tendency to hibernate in hamsters.

Two populations of golden spiny mice born and bred in the laboratory differ in nonshivering thermogenesis and possibly in thyroid status (32). Subsequently, Haim & Borut (128) reported differences in ability to maintain body temperature when exposed to 6°C and in resting O_2 consumption below 20°C, but not in O_2 consumption within the thermal neutral zone or in thermal conductance; these animals were acclimated to common conditions for at least one month. Two populations of bushy-tailed gerbil acclimated to common conditions for at least three weeks show differences in resting O_2 consumption and body temperature when measured at 6°C, but not within the thermal neutral zone (35°C); nor did thermal conductance differ (129). Desert and mesic striped mice differ in resting oxygen consumption, body temperature, and thermal conductance after three weeks of laboratory acclimation (130).

For field-fresh animals, Hulbert et al (167) showed differences in plasma thyroxine levels among coastal, desert, and "intermediate" populations of several species of California rodents. These differences were greatly reduced after 10 weeks in captivity. Minimal and summit metabolism, measured after 2–5 weeks in captivity, apparently did not differ among populations, although sample sizes were small. For all three physiological traits, desert populations tend to show the lowest values.

Scheck (263) studied laboratory-reared F_1 offspring of wild-caught adult cotton rats from northern Kansas and south-central Texas. Kansas animals have a lower BMR, lower critical temperatures (both upper and lower), and lower thermal conductance, but body temperatures and evaporative water loss (EWL) are similar. Scheck (263) interpreted these lower values in Kansas animals as adaptive in the evolutionary sense and possibly permissive for northward range expansion within the last century. However, Scheck (264) found no differences between the two populations in the development of thermoregulatory abilities in 1–18 day old F_2 animals. Finally, the populations differ in growth rates, and Kansas animals have lower metabolic rates at 10°C and increased pelage insulation (265). In one of the most thorough common garden studies, Derting & McClure (77) studied cotton rats from laboratory-bred populations representing seven distinct geographic locations. Unfortunately, the focus of this study was on correlations between BMR and indices of production; the significance of population differences is not clearly reported, nor is it clear whether population differences in BMR exist after accounting for differences in body mass.

Nevo (232) and colleagues have studied geographic variation among 12 populations of mole rats *(Spalax ehrenbergi)* from Israel. These populations comprise four chromosomal forms that are considered to "represent four sibling species at final stages of speciation" (149). Population and/or species differences in BMR, nonshivering thermogenesis in response to noradrenaline, thermoregulation, heart and respiration rates, and hematological characteristics (6, 232) have been demonstrated, although origin (wild-captured or lab-born) and length of captivity are not clearly stated in many of these papers. Many of the traits studied covary with environmental characteristics in a way that has been interpreted as adaptive and possibly related to speciation (6, 232).

Berry and colleagues have extensively studied population variation in *Mus musculus* (now termed *Mus domesticus*) (24, 172). Many physiological differences have been demonstrated, but not with common garden experiments. Berry et al (25, p. 73) suggest that many of the observed "large inter-island differences can be attributed primarily to 'instant sub-speciation' produced by each colonization depending on a probable small number of effective founders." Perhaps the most sophisticated study of geographic variation in thermoregulatory traits of small mammals is that of Lynch (196). *Mus musculus* were captured from populations in five states, representing a thermal cline along the eastern United States (Maine to Florida). Mice were bred in the laboratory, and their F_1 offspring were tested. Populations show significant differences in body temperature and amount of brown adipose tissue, but not in a clinal pattern expected from thermal conditions (see also 12, 172). This lack of evolutionarily adaptive clinal variation was related to previous studies of the quantitative genetics of these and other traits. Because the physiological traits show very low heritabilities, they were not predicted to show adaptive clinal variation (196). Nest building behavior and body mass, on the other hand, show relatively high heritabilities and, as predicted, show apparently adaptive clinal variation. Strain differences in resting or basal oxygen consumption have not been clearly demonstrated in *Mus musculus* (159, 198, 239), in part owing to inappropriate corrections for body size effects.

Ringtail cats *(Bassariscus astutus)* from a desert population show lower BMR and other thermoregulatory differences when compared to animals from a mesic habitat (58). Similar differences exist among breeds of cattle and sheep (322). Wild piglets show higher resistance to cold than do domestic piglets, apparently owing to both higher oxygen consumption (heat production) and extra pelage (106).

Variation in resting metabolic rate may have a variety of functional bases at lower levels of biological organization, but little information is available concerning population differences within species (but see 167 on thyroid function). Strain differences in body temperature, and a circadian dependence of these differences, have been demonstrated in *Mus musculus* (64). Mice selected for large and small size differ in a variety of ways, including the amount of brown adipose tissue, "strongly suggesting that much of the decreased efficiency of the small lines is due to heat production by brown fat" (198, p. 299). Strain differences in amount of brown adipose tissue and in glucose utilization (measured in vitro) have also been demonstrated for mice (159). In vitro preparations of calf muscles reveal breed differences in oxygen consumption, owing to the Na^+, K^+-ATPase-independent component of respiration (cf 71), and in rate of protein synthesis, but not in the Na^+, K^+-ATPase-dependent component of respiration (123).

In summary, the available data, although relatively abundant, are too fraught with difficulties to allow firm conclusions regarding the prevalence, let alone the adaptive significance, of genetically based population differences in basal metabolic rate. One pattern that does seem to recur is that desert populations of small mammals often have relatively low resting metabolic rates (e.g. 58, 204, cf 158). Some studies suggest that differences in organismal metabolic rate may be more prevalent outside as opposed to inside the thermal neutral zone.

EXERCISE PHYSIOLOGY AND ADAPTATION TO ALTITUDE Adaptation to altitude and hypoxia has been studied intensively by vertebrate physiologists, especially in mammals (56, 57, 133–135, 282, 321). Following considerable earlier work on *Peromyscus maniculatus* (reviews in 55, 132), M. A. Chappell, J. P. Hayes, & L. R. G. Snyder studied both BMR and maximal oxygen consumption ($\dot{V}O_2max$) of this species in detail (see below). Significant differences in both exercise- and cold-induced $\dot{V}O_2max$ exist among laboratory-bred populations of *Peromyscus maniculatus*, and cold acclimation (three months at 3°C) increases both measures (135). For animals measured within two days of capture, in situ, Hayes (133) found no differences in cold-induced $\dot{V}O_2max$ rates for a high- and a low-altitude *P. maniculatus* population. At times of the year when the thermal environment was similar (i.e. summer at high altitude and winter at low altitude), Hayes (134) found differences in BMR but not $\dot{V}O_2max$ between field-fresh high- and low-altitude *P. maniculatus*.

Chappell & Snyder (57) have shown that α-chain hemoglobin haplotypes affect blood oxygen affinity (P_{50}) and both exercise- and cold-induced $\dot{V}O_2max$ in *Peromyscus maniculatus*. Alternate α-globin haplotypes are strongly correlated with altitude, and the effects on $\dot{V}O_2max$ of the high- and low-altitude hemoglobin genotypes are in the direction expected if natural selection is maintaining the polymorphism. The foregoing work (56, 57, 133–135; see also 282) is one of the best-documented cases of adaptive variation at the protein level in a vertebrate (see also 228, 270 on feral *Mus*). Interestingly, hemoglobin loci may be affected by selection for body size in *Mus musculus* (113).

Geographic variation in swimming ability among the 12 populations of Israeli mole rats mentioned above (6, 232) appears to correlate "with the extent and level of flooding and free-standing water" (149, p. 29). Djawdan (80) reports no significant difference in treadmill running endurance for field-fresh kangaroo rats *(Dipodomys merriami)* from two sites in California. Laboratory rats raised under common conditions show strain differences in exercise $\dot{V}O_2max$/unit body mass (20).

Breed differences in muscle fiber composition exist in rabbits (234), dogs (280), and horses (280). A variety of characteristics, such as long legs, a large relative muscle mass, and a high percentage of muscle fibers with high myosin ATPase activity, distinguish greyhounds from other dog breeds (280, 324). Various morphological, physiological, and biochemical differences also correlate with athletic abilities among breeds of horses (280).

OSMOREGULATION AND WATER BALANCE Grubbs (125) reports a lack of difference in water turnover rates of free-living rodents from three adjacent sites in agricultural fields and undisturbed habitat. Ward & Armitage (318) report apparently adaptive differences in urine volume and concentration between montane-mesic and lowland-xeric populations of yellow-bellied marmots held in the laboratory. Water restriction did not reduce evaporative water loss (EWL) in these populations, although it does in some small mammals. Limited data suggest site differences in water influx for yellow-bellied marmots in the field (216; Armitage, this volume). Desert ringtail cats show lower EWL than does a mesic population (58). Population differences in urine concentrating ability, correlated with habitat aridity, occur in bats (18; refs. therein). Possible variation in kidney function in mole rats (6, 232) is currently under study (E. Nevo., in litt.). Differences in water turnover rates among breeds of domestic ungulates are reviewed elsewhere (202, 203, 322); feral island and domestic goats can also differ (87).

OTHER Kavaliers & Innes (182, 183; refs. therein), using a combination of laboratory-reared and wild-captured *Peromyscus maniculatus,* demonstrate population differences in opioid-mediated analgesia and locomotor activity, induced by physical restraint, exposure to scent of conspecifics, exposure to predators, or peripheral administration of opiate agonists and antagonists. According to these workers, the pattern of population differences can be interpreted in an adaptive context in relation to differential selection pressures on mainland and island populations of small mammals, thus encouraging further "pharmaco-ecological" studies. Relevant to these findings, strain differences in opioid activity and a variety of related behaviors are known in laboratory mice and rats.

Aves

Poultry strains often differ in such physiological traits as growth rate, metabolism, water balance, blood characteristics, and disease resistance (e.g. 90, 91, 99, 211).

LIFE HISTORY TRAITS AND ANNUAL RHYTHMS Hormonal cycles governing reproduction might be expected to vary with latitude, because the relationship between the optimal time for breeding and proximate cues (e.g. photoperiod) is likely to vary. In a common annual photoperiod regimen (48°N), redpolls from populations at 65° and 48°N latitude differ in (i) the timing of their circadian activity rhythm with respect to the daily solar cycle, (ii) the timing and extent of night activity, (iii) the date of postnuptial molt, and (iv) the pattern of seasonal change in body weight (242). Domesticated or semi-domesticated stocks of several bird species "have lost, to varying degrees, the mechanisms of their feral progenors [sic] for the use of environmental information in the control of reproductive function . . ." (326, p. 66).

White-crowned sparrow *(Zonotrichia leucophrys)* populations differ in migration distance, timing of reproduction, and number of broods per year. Correspondingly, populations have different patterns of seasonal change in gonad size and plasma levels of luteinizing hormone and steroid hormones (326).

THERMAL BIOLOGY, THERMOREGULATION, AND RESTING METABOLISM As in mammals, acclimation, acclimatization, and seasonality of avian thermoregulatory variables have been studied intensively; seasonal change in plumage insulation is especially common (55, 138, 208). Interspecific geographic variation in avian SMR has been documented (319), but intraspecific data on wild species are limited (208). Strain differences occur in chickens (99, 211).

In North America, northern populations of the introduced house sparrow have lower lethal temperatures, reduced thermal conductance, and higher resting and existence metabolic rates than do southern birds (29, 30, 163, 185). On the other hand, gross energy intake, metabolized energy, body temperature, and EWL do not vary geographically (29, 163). No common garden studies have been attempted. Some other species show similar patterns. Montane rufous-collared sparrows have broader thermal neutral zones than do those from the lowlands; however, birds from these populations do not differ in BMR or in the threshold O_2 level at which metabolic rate is reduced (52). House finches from colder environments can remain homeothermic at low air temperatures longer than do their conspecifics from warmer climates (74). However, house finches from California and Colorado do not differ in standard metabolic rate, measured at several seasons and temperatures (72). A desert subspecies of horned larks shows lower O_2 consumption and lower body temperatures, when measured at 45°C, than does a mesic, inland subspecies (301).

American goldfinches from Michigan and Texas increase their fat content during the winter, whereas individuals from southern California do not (73). Starlings likewise vary geographically in winter lipid levels and in several other body composition traits related to thermoregulation (31). In Gambel's white-crowned sparrow, however, the rate and degree of premigratory fattening do not vary with latitude, although northern populations begin to fatten at an earlier date (187).

EXERCISE PHYSIOLOGY Both migratory and nonmigratory populations exist within many bird species, including, formerly, the whooping crane. Long-distance migrants of the white-crowned sparrow show greater increases in hematocrit at the time of spring migration than do birds from a population that migrates only short distances (326). Montane and lowland populations of several bird species differ in heart size, lung size, and various hematological variables, usually in the expected (physiologically adaptive) direction (51, 86).

OSMOREGULATION AND WATER BALANCE Domesticated mallards are more tolerant to salt and have a developmentally more flexible salt gland than do wild mallards (268). Similarly, mallards found along the Greenland coast have larger salt glands than do those found inland in Europe (Stresemann 1934, cited in 268). In several species of phalacrocoracids, marine and freshwater populations differ in the size of nasal gland depressions (272). In three species of cormorants, marine birds have smaller nasal glands, whereas in imperial shags, marine birds have larger depressions; double-crested cormorant populations did not differ. Salt gland size and function can acclimate (268).

Salt marsh populations of the savannah sparrow have enhanced osmoregulatory abilities, including higher plasma and ureteral osmolalities, urine sodium concentrations, and tolerance of saline drinking water, compared with upland populations (17, 119). Anatomical modifications of the kidney may underlie these functional differences (119, cf 18). Desert horned larks have lower EWL than do inland valley birds, but they do not differ in rate of water consumption or ability to survive without water (301). Chickens from different strains differ in water intake and urine production (90, 91).

EGGS Intraspecific variation in the physiology of bird eggs and embryos is frequently studied in the laboratory. Montane coot embryos have lower oxygen consumption rates than do lowland embryos (50). However, red-winged blackbird embryos from different elevations have similar O_2 consumption and incubation periods (51a).

Eggs of red-winged blackbirds, robins, cliff swallows, barn swallows, domestic fowl, and the native chicken of India have lower gas conductances at high elevations (48, 237, 249, 283, 316). This would counteract the greater diffusivity of water vapor at high elevations. However, conductance increases with elevation in coot eggs (50), which would compensate for reduced oxygen availability but compound the problem of water loss. In Andean chickens, eggshell conductance first decreases with altitude, then increases (192), suggesting that the conflicting demands of water loss and O_2 requirements are resolved differently at different elevations. Pigeons from a dry habitat lay eggs with reduced gas conductance (5). Finally, black-billed magpie eggs show no consistent altitudinal variation in conductance (287).

Several workers have transferred birds from high to low elevations, or vice versa, to determine whether females can alter eggshell characteristics. In one instance, hens increased egg conductances by 30%, roughly the magnitude of among-population differences seen in wild birds (248). However, no consistent changes occurred in several other studies of quail, finches, and domestic fowl (49, 193). Hence, it is an open question whether altitudinal variation in eggshell gas conductances represents maternal adjustments or genetic differentiation among populations.

OTHER In marsh wrens, a population difference in the ability to learn songs is correlated with a difference in the sizes of brain regions involved in song learning (45).

Reptilia

GROWTH AND REPRODUCTION Reptiles often show population differences in life history traits, including field growth rates, but the genetic basis of population differences is rarely studied (4, 82). Lab-reared hatchlings show population differences in growth rate in the lizard *Sceloporus undulatus* (103). Similarly, reciprocal field transplant experiments suggest genetic differences between populations of *S. undulatus* (P. H. Niewiarowski, W. M. Roosenburg, personal communication). Sinervo & Adolph (1, 273, 274; Sinervo & Adolph, in preparation) found population differences in the thermal sensitivity of growth rate in lab-reared *S. occidentalis,* but not in *S. graciosus*.

Reptilian reproductive cycles vary among populations in the wild, often in relation to latitude (e.g. 82, 194, 217, 225; refs. therein); however, we know of no common garden studies. Previous reports of population differences in degree of placentation in the lizard *Sceloporus aeneus* (127) actually involve two separate sibling species (127a).

THERMAL BIOLOGY AND THERMOREGULATION Reptile populations frequently differ in body temperatures (T_b) during field activity; in most cases, this probably reflects environmental differences (8, 240). However, preferred T_b may differ between populations of the lizard *Lacerta vivipara* (310), and between some populations of *Sceloporus occidentalis* (273) but not others (320).

Critical thermal limits are frequently studied in reptiles; these are generally measured as the upper (CTMax) and lower (CTMin) T_b at which an animal loses its righting response. Traditionally, thermal physiology was thought to be evolutionarily conservative (i.e. relatively invariant within species and even genera; 69, 146). This view is supported by studies of lizards which found no significant population differences in critical thermal limits or other thermal physiological traits (69, 141, 146, 147, 307, 311). Similarly, eastern and western Canadian populations of the turtle *Chrysemys picta* (at the same latitude) do not differ in a variety of traits related to freeze-tolerance, including survival times, temporal changes in glucose concentration, supercooling points, and changes in lactate levels (in hatchlings, dug from overwintering sites, often in a frozen state; K. B. Storey, personal communication). On the other hand, CTMax and/or CTMin does vary geographically in several species of lizards (141, 143–145, 303, 325) and turtles (169).

Sex is determined by egg incubation temperature in some reptiles (41). The threshold temperature for a 1 : 1 sex ratio varies geographically in the turtles

Chrysemys picta and *Graptemys pseudogeographica* (41); surprisingly, southern populations have slightly lower threshold temperatures.

RESTING METABOLISM Both acclimation and acclimatization of resting metabolic rate have been demonstrated in various reptiles (22, 42, 303). In the tropical skink *Mabuya striata,* high-elevation lizards can acclimate their resting metabolic rate to temperature, whereas low-elevation lizards cannot (238).

In *Sceloporus occidentalis* and *S. malachiticus,* lizards from high elevations have higher resting rates of O_2 consumption ($\dot{V}O_2$); lab-reared *S. malachiticus* show the same pattern, suggesting a genetic basis (16). Similarly, Tsuji (302, 303) found latitudinal differences in the pattern of seasonal change in standard metabolic rate (SMR) and the direction of seasonal acclimation of SMR in *S. occidentalis*. The dependence of resting $\dot{V}O_2$ on partial pressure of O_2 differs between high- and low-altitude *S. occidentalis* (281). On the other hand, *S. occidentalis* from different elevations do not differ in minimum $\dot{V}O_2$; differences in 24-hr O_2 consumption were attributed to behavior rather than physiology (148, 177). Two New Jersey populations of *S. undulatus* do not differ in SMR (V. Pierce, H. B. John-Alder, personal communication). Whiptail lizards *(Cnemidophorus hyperythrus)* show interhabitat differences in field metabolic rates (measured by doubly-labeled water; 180). Painted turtles *(Chrysemys picta)* from a Canadian population accumulate less lactate when submerged in anoxic water than do those from Alabama (305).

EXERCISE PHYSIOLOGY Sprint speeds differ among lizard populations in at least eight species, mainly iguanids (69, 112, 146, 164, 165, 274–277, 279, 309; D. B. Miles, personal communication; J. Herron & B. S. Wilson, personal communication). Usually, populations at lower altitudes or latitudes are faster. Differences sometimes depend on age class or reproductive condition (see also 112). In *Sceloporus occidentalis,* sprint speeds of gravid females vary among populations, whereas those of nongravid females do not (277); in lab-reared hatchlings, sprint performance on various substrates varies among populations, paralleling field habitat use (276). By experimentally manipulating egg size in *S. occidentalis,* Sinervo & Huey (274, 275) showed that among-population differences in hatchling sprint speed, but not stamina, are largely due to differences in egg size; variation in stamina presumably reflects additional physiological differences (cf 111). Thermal sensitivity of sprint performance (e.g. optimal temperature) typically does not vary (or varies only slightly) among lizard populations (69, 146, 307, 309, 311).

Two New Jersey populations of *Sceloporus undulatus* show no differences in treadmill endurance or in $\dot{V}O_2max$ (V. Pierce, H. B. John-Alder, personal communication). Neither duration of maximal activity nor amount of lactate formed differs between two populations of *S. occidentalis* from different elevations (23). Population differences in locomotor performance occur in garter snakes (36, 37). Hydric conditions during incubation can affect locomotor performance of turtles (222).

Hemoglobin concentrations, erythrocyte counts, and/or hematocrits are higher in high-elevation populations in some lizard species but not in others (154, 281, 320). Similarly, erythrocyte count varies latitudinally in the turtle *Sternotherus odoratus* (168).

OSMOREGULATION AND WATER BALANCE Hertz & colleagues have studied intraspecific variation in evaporative water loss (EWL) rates and dehydration resistance in several species of *Anolis* lizards. Variation with altitude occurs in some species but not in others (142, 144). Similarly, Hillman et al (153) found among-population differences in EWL rate in two *Anolis* species, but not in *A. cristatellus* (152). Desert, montane, and coastal populations of the lizard *Uta stansburiana* have similar EWL rates (210, refs therein). EWL rates can acclimate rapidly, and the magnitude of acclimation effects can be several times as great as the difference between populations (153).

Estuarine populations of the water snake *Nerodia fasciata* have (i) lower rates of water and sodium influx, (ii) a lower rate of water efflux, and (iii) less skin permeability to water and sodium, compared to freshwater animals (88); however, these populations may comprise separate species (93). Hatchling snapping turtles *(Chelydra serpentina)* from saline water grow faster in saline water, but more slowly in fresh water, than do yearlings from a freshwater population (89).

OTHER In captivity, two *Sceloporus occidentalis* populations differ in blood osmolality, protein content, and hematocrit, and in plasma corticosterone levels in response to handling (K. Dunlap, personal communication). Garter snake populations sympatric with the toxic newt *Taricha granulosa* are more resistant to tetrodotoxin (37); resistance was measured as degree of locomotor impairment. The potential for multiple paternity may also differ among garter snake populations (269, refs. therein).

Amphibia

GROWTH, REPRODUCTION, AND LIFE HISTORY TRAITS Larval growth and developmental rates are strongly temperature-sensitive and density-dependent (27, 300). Hence, variation observed in the field may reflect proximate environmental effects (27). Because amphibian eggs are easy to collect and rear, a number of common garden or transplant experiments have been conducted (300). Berven & colleagues have documented genetic differences in the temperature dependence of larval differentiation rate and size at metamorphosis in two frog species *(Rana sylvatica, R. clamitans)* and measured heritabilities in some populations (26–28). In a seminatural common garden setting, populations of the salamander *Ambystoma maculatum* differ in larval survival, rate of metamorphosis, and activity (105). Temperature sensitivity of early development varies geographically in a number of other amphibians (3, 38, 94, 313); populations from colder environments are almost invariably cold-adapted. On the other hand, population differentiation is absent in some species (140, 227, 313, 314). Some previously described differences among populations of "*Rana pipiens*" (227) actually represent interspecific differences (151).

THERMAL BIOLOGY AND THERMOREGULATION As in reptiles, most cases of body temperature variation in nature probably represent environmental variation. However, montane *Bufo boreas* select higher T_b than do lowland toads (46).

Thermal tolerance (e.g. critical thermal maximum—CTMax) acclimates in response to both temperature and photoperiod, and varies diurnally, seasonally, and geographically in amphibians (34, 35, 42, 84, 245). Geographic variation in CTMax, and/or its acclimation rate, is common in frogs (34, 35, 160, 220, 221) and salamanders (70, 162), particularly in wide-ranging species (34, 35). However, larval and neotenic adult *Ambystoma tigrinum* from desert and mountain ponds have similar CTMax (76); likewise, two populations of the frog *Acris crepitans* do not differ in CTMax (122).

Several temperate frog species are amazingly freeze-tolerant. Although one might expect freeze tolerance to vary in wide-ranging species, K. B. Storey (personal communication) has found no differences among wood frogs *(Rana sylvatica)* from northern Ontario, southern Ontario, the middle United States, and even South Carolina: populations build up similar levels of glucose as a cryoprotectant, store similar amounts of glycogen in their livers, and possess the same nucleating proteins in their blood. Likewise, *R. sylvatica* varies latitudinally in heat tolerance but not in cold tolerance (207).

A few studies have compared thermal physiology at the tissue or cell level in amphibians. Thermal properties of cells and tissues from the frog *Rana ridibunda* do not differ between populations living in hot springs versus cool water (306). However, temperature tolerance of tissues from the frog *Limnodynastes tasmaniensis* differs with latitude (284).

RESTING METABOLISM SMR is higher in montane than in lowland toads *(Bufo boreas)*, under several thermal and acclimation conditions; however,

active metabolic rates are similar (47). Oxygen consumption rate, and its response to temperature, varies with latitude and altitude in the frog *Hyla regilla* (176).

EXERCISE PHYSIOLOGY Populations of the poison-arrow frog *Dendrobates pumilio* differ in aerobic capacity for locomotion and in degree of toxicity, but not in anaerobic capacity (288). Florida and New Jersey populations of the frog *Hyla crucifer* do not differ in sprint swimming abilities or in muscle contractile properties (except for twitch tension at low temperatures) (179). Populations of *Hyla regilla* differ in erythrocyte count and hemoglobin concentration per cell (176), and two populations of *Rana pipiens* differ in levels of liver and muscle glycogen, but not blood glucose (101). Hellbender populations differ in a variety of hematological characteristics (178).

OSMOREGULATION AND WATER BALANCE In the laboratory, Puerto Rican *Eleutherodactylus coqui* from low-elevation rehydrate faster than do highland frogs (308). Similarly, in the frog *Hyla versicolor,* desiccation tolerance differs between mesic and xeric habitats (250). Two Illinois populations of the frog *Acris crepitans* differ slightly in desiccation rates (122), as do several populations of *Hyla regilla* (175) and two populations of *Bufo arenarum* (53).

OTHER Vocalization characteristics (call duration, pulse rate) vary among populations of *Rana pipiens* (85). Changes in serum proteins at metamorphosis differ between two populations of the salamander *Dicamptodon ensatus* (236). Connecticut *Rana sylvatica* show population differences in acid tolerance of embryos and tadpoles; breeding studies suggest a genetic basis for the latter but not the former (241).

Osteichthyes

GROWTH, REPRODUCTION, AND LIFE HISTORY TRAITS A number of fish species show polymorphism, with relatively discrete morphs differing genetically and in such life history variables as growth and developmental rates (97, 206, 290, 293). Geographic variation in life history traits, including size-specific fecundity, occurs in many species (e.g. 235, 253–255, 290), and biogeographic "rules" may exist for fishes (252). Genetically based differences in growth rate occur among strains of various species of fishes, some of which have been produced by artificial selection (e.g. 81, 247). Differences in developmental and growth rates occur among hatchery strains of trout (refs. in 81) and may correlate positively with multilocus heterozygosity (104). We do not consider the general problem of stock discreteness and differentiation in marine fishes (see 290 for some refs.).

Reznick et al (253, 254) perturbed a natural population of guppies to test the effects of predation on life history evolution. Correlated population differences in mating preferences and coloration exist in guppies (161). The Atlantic silverside shows genetic variation for temperature-dependent sex determination (65).

The marine killifish *Fundulus heteroclitus* can show "clinal" geographic variation (Canada to Florida) in frequencies of alleles at several loci (79, 243, 244). Allelic isozymes of lactate dehydrogenase-B (LDH-B) are functionally nonequivalent, and relative catalytic efficiencies are reversed at different temperatures. Oxygen consumption of embryos, developmental rate, and hatching time all differ among allelic isozyme LDH-B genotypes, which differ also in erythrocytic ATP concentrations and blood oxygen affinity (see below). Developmental rate also correlates with allelic variation at the malate dehydrogenase-A and glucose phosphate isomerase-B loci. Population differences in the amount of LDH-B^b appear to be due both to acclimatization and genetic adaptation (68).

THERMAL BIOLOGY AND THERMOREGULATION Population differences in optimum temperatures for development (fastest development) exist in fishes (refs. in 10). Thermal tolerance (e.g. critical thermal maximum) acclimates in response to both temperature and photoperiod, and shows daily, seasonal, and geographic variation in fishes (42, 131, 212, refs. therein). A thorough common garden experiment demonstrated significant differences in both the limits and the range of thermal and oxygen tolerance between two pupfish populations, and showed intermediate values for hybrids (155). Apparently adaptive changes in allele frequencies have occurred rapidly in mosquitofish inhabiting areas receiving thermal effluent from a nuclear power plant (195, 258, 278; see also 188, 224).

Koehn (190) presented data suggesting thermally adaptive geographic variation in esterase alleles in the freshwater *Catostomus clarkii;* subsequent studies suggesting temperature-dependent selection in relation to population differences in allozyme frequencies have been reviewed elsewhere (79, 191, 243, 244, 290). The thermal stability of tissues from *Cobitus taenia* and *Carassius auratus* does not differ between populations living in hot springs versus cool water (306). Temperature acclimation of fish may result in changes in ATP concentrations in erythrocytes (244) and in amounts of various enzymes; changes in allozyme or isozyme expression are less common (68, 156).

RESTING METABOLISM Oxygen consumption of (relatively) inactive fishes is subject to acclimation and acclimatization, although some studies have shown a lack of acclimation (42, 245). Seasonal variation is apparently more common in adult than in juvenile fishes (19). Studies of population differences are uncommon, although differences among hatchery strains have been documented (refs. in 81). Cave-dwelling populations of *Astyanax fasciatus* do not differ from surface-dwelling populations in average daily metabolic rate (323).

EXERCISE PHYSIOLOGY Seasonal changes in blood parameters occur in some fishes but not in others (42, 121). Seasonal variation in swimming performance has been reported (262), and activities of various enzymes important for swimming performance may change before or during migration (126). Acclimation to low temperature of swimming performance has been demonstrated in fishes (41, 260). Nelson (229) reported acclimation to pH of critical swimming speed in perch; however, more than two years of aquarium housing alone had no effect.

Population differences in swimming performance between coastal and interior coho salmon are apparently genetically based, related to differences in body form, and possibly related to differences in the energetic demands of migration and/or predation (292). Similarly, Couture & Guderley (67) compared field-fresh fish from populations in two rivers differing in flow and temperature and hence in the difficulty of migration. As predicted, the population of anadromous cisco with greater swimming requirements during migration show higher aerobic capacity in their swimming musculature, whereas the whitefish population do not, possibly because whitefish are larger and so are less affected by flow differences. Field-captured adult anadromous and resident freshwater sticklebacks also differ in swimming performance (293). Lab-reared anadromous and nonanadromous *Oncorhynchus nerka* differ in critical swimming speeds (291), and some hatchery-reared strains of trout exhibit poor swimming performance (95).

Perch from a naturally acidic lake are less (negatively) affected by soft, acid water than are fish from a circumneutral lake, although the former do not have higher absolute critical swimming speeds (229). Hematocrit, hemoglobin, and mean corpuscular hemoglobin content also differ between perch populations from acidic and alkaline waters, but red blood cell numbers and volume, and buffering capacity of white muscle, do not (230, 231). Seasonal variation also occurs in buffering capacity and in some hematological variables.

In the killifish, erythrocytic ATP concentration correlates with LDH-B genotype, and this may be causally related to differences in critical swimming speed among LDH-B genotypes (79, 243, 244). Other possible allozymic correlates of population differences in swimming performance of fish have been reported by investigators (refs. in 292).

Population differences in hemoglobin concentrations and in red cell Hb-phosphate ratio exist in some air-breathing fishes and may correlate with habitat oxygen level (121). Wells (321) cautions against excessive adaptationism in interpreting patterns of variation in fish hemoglobins.

OSMOREGULATION AND pH TOLERANCE Measures of osmotic tolerance undergo acclimation and acclimatization in fishes (42). Racial differences in osmotic tolerance such as the ability to develop in saline waters occur in sticklebacks and correlate with differences in scalation (refs. in 10, 131, 245, 312). Common garden experiments indicate population and family differences in salinity tolerance in juvenile chinook salmon; the former but not the latter are largely a function of body size differences (289).

Strains of brook trout differ in pH tolerance (92, 259, 286), and yellow perch exhibit "ecotypes" with respect to low pH and oxygen concentration (205, 229–231).

OTHER Genetic differences in gas retention by swimbladders exist between two populations of lake trout (171). Cave-dwelling populations of *Astyanax fasciatus* differ from populations living in surface waters in the structure of the lateral line system and of the eye (323). Genetically based population differences in disease resistance occur in salmonids (290).

Two populations of pumpkinseeds, living in lakes with and without snails, differ in handling time (a measure of physiological performance) when feeding on snails, and in the motor patterns of one muscle, but not of several others (315). McLeese & Stevens (215) report differences between two strains of rainbow trout in relative caecal mass and in trypsin activity; moreover, one strain apparently produced a new trypsin isozyme in response to cold acclimation, whereas the other did not. E. D. Stevens (in litt.) also reports population differences, including "clear latitudinal gradients," "in the number of pyloric caeca and their function."

CONCLUSIONS AND SUGGESTIONS FOR FUTURE RESEARCH

Based on the examples reviewed here, physiological traits do indeed vary among populations. Unfortunately, few studies have been designed to indicate whether observed differences are to any extent genetic in origin, but those that have often suggest genetic differentiation. On the other hand, the extensive literature on physiological acclimation and acclimatization cautions that any population differences observed in the absence of proper common garden controls may not be genetically based.

To overcome this constraint, we need more common garden studies of physiological differentiation (e.g. 1, 27, 77, 155, 183, 196, 209, 218, 251, 252, 273, 274, 289). In addition, physiologists need to study more than two or a few populations (cf 241). Studies of several different types of traits (e.g. morphometric, biochemical, hormonal, organismal) would allow us to determine whether "physiology" is to any extent fundamentally different in terms of microevolution (cf 110, 131, 196, 212). Questions asked routinely of morphometric traits should be asked of physiological traits. How common are physiological clines (63, 158, 196, 295)? Do physiological traits show character displacement? How much of the among-population variation of animals in nature is genetically versus environmentally based? What is the range and scope of acclimation/acclimatization in physiological traits, as compared with the magnitude of genetically based differentiation seen among populations? We encourage more studies that use common garden designs and test a priori hypotheses based on quantitative genetic or other information (e.g. 158, 196). On a more topical note, the study of population differentiation in physiological traits will be important for understanding the ecological and evolutionary implications of global warming (e.g. 21, 166, 188, 278) and of other changes wrought by human activities, such as acid rain and lake acidification (cf 205, 229–231, 241, 259, 286) and other types of pollution.

Another virtually unexplored area is the year-to-year repeatability of population differences in physiological traits (253), although some studies are now in progress with lizards (164, 165) and with snakes (S. J. Arnold & A. F. Bennett, personal communication). Multiyear studies of the physiology of single populations (other than humans) are more common but still quite rare (cf 66, 118; Armitage, this volume). Thus, long-term studies of populations should be encouraged (82, 165, 254, 278).

Regardless of their origin (genetic or environmental), population differences in physiological (or in any) traits cannot be assumed to be "adaptive." Demonstration of adaptive significance requires further work, often including experimental manipulations and comparative studies (e.g. 27, 43, 44, 56, 60, 68, 102, 112, 137, 156, 191, 196, 224, 232, 233, 243, 251, 254, 273, 276, 278, 279, 282, 285, 300, 315, 321). Notwithstanding, studies reporting an apparent lack of adaptation among populations are in the minority (cf 251). On the other hand, the predilection of comparative physiologists for choosing extreme species for comparison (15, 44, 75, 102, 156, 181, 298) may have led to a bias in our data base and hence in our view regarding the commonness of physiological adaptation to the environment (43, 60, 102, 251, 321). A close matching at the population level is perhaps much less likely, in part due to the homogenizing effects of gene flow. Moreover, patterns of (co)variation among populations may differ fundamentally from those among species (e.g. 82, 112, 255). As well, we may expect to find multiple, rather than single optimal, solutions to adaptive problems (that is, specific answers to general physiological questions; 14, 15), especially as chance differences in the initial genetic constitution of populations may predispose them to different responses to uniform selection (63).

With more and more practitioners employing quantitative genetic (e.g. 21, 36, 63, 78, 109, 110, 166, 196, 261, 273, 304), population/biochemical genetic (104, 191, 243), and rigorous comparative (111a, 112) analyses, it has been rumored that a new field of "evolutionary physiology" is at hand (cf 44, 43, 102). Soon, it would seem, Dobzhansky's 1949 statement (267, p. xv) that "Among the major subdivisions of modern biology only physiology and biochemistry still remain largely unaffected by evolutionary ideas, doubtless to mutual detriment," or Prosser's (246, p. 4) statement that "evolutionists pay little attention to physiology, and most physiologists have only a superficial knowledge of evolution," will no longer ring true. We believe that an important thrust of this new evolutionary physiology will be studies of population differentiation. To quote Prosser (245, pp. 254–255; and cf 208), there exists "an immediate need for the description of phenotypic and genotypic variation of physiological characters within . . . species . . . according to history and distribution of natural populations. . . ."

ACKNOWLEDGMENTS

We thank H. V. Carey and A. R. Ives for reviewing the manuscript, and B. W. Grant, J. P. Hailman, M. Kreitman, P. Licht, and the students in Zoology 962 (fall 1990) for helpful discussions. J. R. Baylis, F. W. H. Beamish, D. G. Blackburn, R. Burton, C. Carey, J. J. Childress, W. R. Dawson, K. Dunlap, W. A. Dunson, F. Geiser, H. C. J. Godfray, H. Guderley, J. P. Hayes, P. W. Hochachka, R. B. Huey, H. B. John-Alder, L. B. Keith, J. F. Kitchell, R. K. Koehn, J. J. Magnuson, G. Mayer, M. C. Moore, D. L. G. Noakes, D. M. Powers, D. N. Reznick, M. Rochelle, E. D. Stevens, K. B. Storey, E. B. Taylor, G. R. Ultsch, and a number of other colleagues kindly answered our requests for information and relevant references; we apologize for citation omissions due to space limitations! Supported in part by National Science Foundation Grant BSR-9006083 to T. Garland and by the Department of Energy through contract DE-FG02-88ER60633 to W. P. Porter.

Literature Cited

1. Adolph, S. C., Sinervo, B. 1990. Thermal sensitivity of growth rate in hatchling *Sceloporus* lizards. *Am. Zool.* 30:55A (Abstr.)
2. Am. Physiological Society, Long Range Planning Committee. 1990. A "White Paper" on the future of physiology and the role of the American Physiological Society in it. *Physiologist* 33:161–80
3. Anderson, J. D. 1972. Embryonic temperature tolerance and rate of development in some salamanders of the genus *Ambystoma*. *Herpetologica* 28:126–30
4. Andrews, R. M. 1982. Patterns of growth in reptiles. In *Biology of the Reptilia,* Vol. 13, *Physiology D, Physiological Ecology,* ed. C. Gans, F. H. Pough, pp. 273–320. New York: Academic. 345 pp.
5. Arad, Z., Gavrieli-Levin, I., Marder, J. 1988. Adaptation of the pigeon egg to incubation in dry hot environments. *Physiol. Zool.* 61:293–300
6. Arieli, R., Heth, G., Nevo, E., Hoch, D. 1986. Hematocrit and hemoglobin concentration in four chromosomal species and some isolated populations of

actively speciating subterranean mole rats in Israel. *Experientia* 42:441–43

7. Arnold, S. J. 1981. Behavioral variation in natural populations. II. The inheritance of a feeding response in crosses between geographic races of the garter snake, *Thamnophis elegans*. *Evolution* 35:510–15
8. Avery, R. A. 1982. Field studies of body temperatures and thermoregulation. In *Biology of the Reptilia,* Vol. 12. *Physiology* C, ed. C. Gans, F. H. Pough, pp. 93–166. New York: Academic. 536 pp.
9. Avise, J. C., Arnold, J., Ball, R. M., Bermingham, E., Lamb, T., et al. 1987. Intraspecific phylogeography: the mitochondrial DNA bridge between population genetics and systematics. *Annu. Rev. Ecol. Syst.* 18:489–522
10. Barlow, G. W. 1961. Causes and significance of morphological variation in fishes. *Syst. Zool.* 10:105–17
11. Barnett, S. A., Dickson, R. G. 1987. Hybrids show parental influence in the adaptation of wild house mice to cold. *Genet. Res., Camb.* 50:199–204
12. Barnett, S. A., Dickson, R. G. 1989. Wild mice in the cold: some findings on adaptation. *Biol. Rev.* 64:317–40
13. Barrowclough, G. F., Johnson, N. K., Zink, R. M. 1985. On the nature of genic variation in birds. In *Current Ornithology*. Vol. 2, ed. R. F. Johnston, pp. 135–54. New York: Plenum. 364 pp.
14. Bartholomew, G. A. 1986. The role of natural history in contemporary biology. *Bioscience* 36:324–29
15. Bartholomew, G. A. 1987. Interspecific comparison as a tool for ecological physiologists. See Ref. 102, pp. 11–37
16. Bartlett, P. N. 1970. *Seasonal and elevational comparisons of oxygen consumption rates in the lizard* Sceloporus occidentalis. PhD thesis. Univ. Calif., Riverside. 105 pp.
17. Basham, M. P., Mewaldt, L. R. 1987. Salt water tolerance and the distribution of South San Francisco Bay song sparrows. *Condor* 89:697–709
18. Bassett, J. E. 1982. Habitat aridity and intraspecific differences in the urine concentrating ability of insectivorous bats. *Comp. Biochem. Physiol.* 72A:703–8
19. Beamish, F. W. H. 1990. Metabolism and photoperiod in juvenile lake charr and walleye. *Env. Biol. Fish.* 29:201–07
20. Bedford, T. G., Tipton, C. M., Wilson, N. C., Oppliger, R. A., Gisolfi, C. V. 1979. Maximum oxygen consumption of rats and its changes with various experimental procedures. *J. Appl. Physiol.: Respirat. Environ. Exercise Physiol.* 47:1278–83
21. Bennett, A. F., Dao, K. M., Lenski, R. E. 1990. Rapid evolution in response to high-temperature selection. *Nature* 346: 79–81
22. Bennett, A. F., Dawson, W. R. 1976. Metabolism. In *Biology of the Reptilia,* Vol. 5, ed. C. Gans, W. R. Dawson, pp. 127–223. New York: Academic
23. Bennett, A. F., Ruben, J. A. 1975. High altitude adaptation and anaerobiosis in sceloporine lizards. *Comp. Biochem. Physiol.* 50A:105–8
24. Berry, R. J. 1981. Town mouse, country mouse: adaptation and adaptability in *Mus domesticus (M. musculus domesticus). Mammal Rev.* 11:91–136
25. Berry, R. J., Jakobson, M. E., Peters, J. 1978. The house mice of the Faroe Islands: a study in microdifferentiation. *J. Zool., Lond.* 185:73–92
26. Berven, K. A. 1987. The heritable basis of variation in larval developmental patterns within populations of the wood frog *(Rana sylvatica). Evolution* 41:1088–97
27. Berven, K. A., Gill, D. E. 1983. Interpreting geographic variation in life-history traits. *Am. Zool.* 23:85–97
28. Berven, K. A., Gill, D. E., Smith-Gill, S. J. 1979. Countergradient selection in the green frog, *Rana clamitans. Evolution* 33:609–23
29. Blem, C. R. 1973. Geographic variation in the bioenergetics of the house sparrow. *Ornithol. Monogr.* 14:96–121
30. Blem, C. R. 1974. Geographic variation of thermal conductance in the house sparrow *Passer domesticus. Comp. Biochem. Physiol.* 47A:101–8
31. Blem, C. R. 1981. Geographic variation in mid-winter body composition of starlings. *Condor* 83:370–76
32. Borut, A., Haim, A., Castel, M. 1978. Non-shivering thermogenesis and implication of the thyroid in cold labile and cold resistant populations of the golden spiny mouse *(Acomys russatus). Effectors of Thermogenesis, Experientia Supplement* Vol. 32, ed. L. Givardiel, J. Seydoux, pp. 219–27. Basel: Birkhauser. 345 pp.
33. Boyce, M. S. 1988. *Evolution of Life Histories of Mammals.* New Haven, Conn: Yale Univ. Press. 373 pp.
34. Brattstrom, B. H. 1968. Thermal acclimation in anuran amphibians as a function of latitude and altitude. *Comp. Biochem. Physiol.* 24:93–111
35. Brattstrom, B. H. 1970. Thermal acclimation in Australian amphibians. *Comp. Biochem. Physiol.* 35:69–103

36. Brodie, E. D. III. 1989. Behavioral modification as a means of reducing the cost of reproduction. *Am. Nat.* 134:225–38
37. Brodie, E. D. III, Brodie, E. D. Jr. 1990. Tetrodotoxin resistance in garter snakes: an evolutionary response of predators to dangerous prey. *Evolution* 44:651–59
38. Brown, H. A. 1975. Embryonic temperature adaptations of the Pacific treefrog, *Hyla regilla*. *Comp. Biochem. Physiol.* 51A:863–73
39. Brown, J. H., Lee, A. K. 1969. Bergmann's rule and climatic adaptation in woodrats *(Neotoma)*. *Evolution* 23:329–38
40. Bubenik, G. A., Brown, R. D., Schams, D. 1990. The effect of latitude on the seasonal pattern of reproductive hormones in the male white-tailed deer. *Comp. Biochem. Physiol.* 97A:253–57
41. Bull, J. J. 1983. *Evolution of Sex Determining Mechanisms*. Menlo Park, Calif.: Benjamin-Cummings. 316 pp.
42. Bullock, T. H. 1955. Compensation for temperature in the metabolism and activity of poikilotherms. *Biol. Rev.* 30:311–42
43. Burggren, W. W., Bemis, W. E. 1990. Studying physiological evolution: paradigms and pitfalls. In *Evolutionary Innovations*, ed. M. H. Nitecki, pp. 191–238. Chicago: Univ. Chicago Press. 304 pp.
44. Calow, P., ed. 1987. *Evolutionary Physiological Ecology*. Cambridge: Cambridge Univ. Press. 239 pp.
45. Canady, R. A., Kroodsma, D. E., Nottebohm, F. 1984. Population differences in complexity of a learned skill are correlated with brain space involved. *Proc. Natl. Acad. Sci. USA* 81:6232–34
46. Carey, C. 1978. Factors affecting body temperatures of toads. *Oecologia* 35:197–219
47. Carey, C. 1979. Effect of constant and fluctuating temperatures on resting and active oxygen consumption of toads, *Bufo boreas*. *Oecologia* 39:201–12
48. Carey, C., Garber, S. D., Thompson, E. L., James, F. C. 1983. Avian reproduction over an altitudinal gradient. II. Physical characteristics and water loss of eggs. *Physiol. Zool.* 56:340–52
49. Carey, C., Hoyt, D. F., Bucher, T. L., Larson, D. L. 1984. Eggshell conductances of avian eggs at different altitudes. In *Respiration and Metabolism of Embryonic Vertebrates*, ed. R. S. Seymour, pp. 259–70. Dordrecht: Junk. 445 pp.
50. Carey, C., Leon-Velarde, F., Dunin-Borkowski, O., Bucher, T. L., de la Torre, G., et al. 1989. Variation in eggshell characteristic and gas exchange of montane and lowland coot eggs. *J. Comp. Physiol.* B 159:389–400
51. Carey, C., Morton, M. L. 1976. Aspects of circulatory physiology of montane and lowland birds. *Comp. Biochem. Physiol.* 54A:61–74
51a. Carey, C., Thompson, E. L., Vleck, C. M., James, F. C. 1982. Avian reproduction over an altitudinal gradient: incubation period, hatchling mass, and embryonic consumption. *Auk* 99:710–18
52. Castro, G., Carey, C., Whittembury, J., Monge, C. 1985. Comparative responses of sea level and montane rufous-collared sparrows, *Zonotrichia capensis*, to hypoxia and cold. *Comp. Biochem. Physiol.* 82A:847–50
53. Cei, J. M. 1959. Ecological and physiological observations on polymorphic populations of the toad *Bufo arenarum* Hensel, from Argentine. *Evolution* 13: 532–36
54. Chaffee, R. R. J. 1966. On experimental selection for super-hibernating and non-hibernating lines of Syrian hamsters. *J. Theoret. Biol.* 12:151–54
55. Chaffee, R. R. J., Roberts, J. C. 1971. Temperature acclimation in birds and mammals. *Annu. Rev. Physiol.* 33:155–202
56. Chappell, M. A., Hayes, J. P., Snyder, L. R. G. 1988. Hemoglobin polymorphisms in deer mice *(Peromyscus maniculatus)*: physiology of beta-globin variants and alpha-globin recombinants. *Evolution* 42:681–88
57. Chappell, M. A., Snyder, L. R. G. 1984. Biochemical and physiological correlates of deer mouse α-chain hemoglobin polymorphisms. *Proc. Natl. Acad. Sci. USA* 81:5484–88
58. Chevalier, C. D. 1991. *Aspects of thermoregulation and energetics in the Procyonidae (Mammalia: Carnivora)*. PhD thesis. Univ. Calif., Irvine. 202 pp.
59. Christensen, K., Arnbjerg, J., Andresen, E. 1985. Polymorphism of serum albumin in dog breeds and its relation to weight and leg length. *Hereditas* 102:219–23
60. Clarke, A. 1980. A reappraisal of the concept of metabolic cold adaptation in polar marine invertebrates. *Biol. J. Linn. Soc.* 14:77–92
60a. Clausen, J., Keck, D. D., Hiesey, W. M. 1948. Experimental studies on the nature of species. III. Environmental responses of climatic races of Achillea. *Carnegie Inst. Wash. Publ. 581*. Baltimore: Lord Baltimore Press. 129 pp.

61. Cockburn, A. 1988. *Social Behaviour in Fluctuating Populations*. London: Croom Helm. 239 pp.
62. Cockburn, A., Lee, A. K. 1985. *Evolutionary Ecology of Marsupials*. Cambridge: Cambridge Univ. Press. 274 pp.
63. Cohan, F. M., Hoffmann, A. A. 1989. Uniform selection as a diversifying force in evolution: evidence from *Drosophila*. *Am. Nat.* 134:613–37
64. Connolly, M. S., Lynch, C. B. 1983. Classical genetic analysis of circadian body temperature rhythms in mice. *Behav. Genetics* 13:491–500
65. Conover, D. O., Van Voorhees, D. A. 1990. Evolution of a balanced sex ratio by frequency-dependent selection in a fish. *Science* 250:1556–58
66. Cooke, F. 1987. Lesser snow goose: a long-term population study, in *Avian Genetics*, ed. F. Cooke, P. A. Buckley, pp. 407–32. London: Academic. 488 pp.
67. Couture, P., Guderley, H. 1990. Metabolic organization in swimming muscle of anadromous coregonines from James and Hudson bays. *Can. J. Zool.* 68:1552–58
68. Crawford, D. L., Powers, D. A. 1989. Molecular basis of evolutionary adaptation at the lactate dehydrogenase-B locus in the fish *Fundulus heteroclitus*. *Proc. Natl. Acad. Sci. USA* 86:9365–69
69. Crowley, S. R. 1985. Thermal sensitivity of sprint-running in the lizard *Sceloporus undulatus:* support for a conservative view of thermal physiology. *Oecologia* 66:219–25
70. Cupp, P. V., Brodie, E. D. 1972. Intraspecific variation in the critical thermal maximum of the plethodontid salamander, *Eurycea quadridigitatus*. *Am. Zool.* 12:689 (Abstr.)
71. Davis, J. P., Hillyard, S. D. 1983. Seasonal variation in Na^+-K^+-ATPase in tissues of field acclimatized woodrats, *Neotoma lepida*. *Comp. Biochem. Physiol.* 76A:115–21
72. Dawson, W. R., Buttemer, W. A., Carey, C. 1985. A reexamination of the metabolic response of house finches to temperature. *The Condor* 87:424–27
73. Dawson, W. R., Marsh, R. L. 1986. Winter fattening in the American Goldfinch and the possible role of temperature in its regulation. *Physiol. Zool.* 59:357–68
74. Dawson, W. R., Marsh, R. L., Buttemer, W. A., Carey, C. 1983. Seasonal and geographic variation of cold resistance in house finches *Carpodacus mexicanus*. *Physiol. Zool.* 56:353–69
75. Dejours, P. 1987. What is a stressful environment? An introduction. In *Adaptive Physiology to Stressful Environments*, ed. S. Samueloff, M. K. Yousef, pp. 113–20. Boca Raton, Fla: CRC. Inc. 212 pp.
76. Delson, J., Whitford, W. G. 1973. Critical thermal maxima in several life history stages in desert and montane populations of *Ambystoma tigrinum*. *Herpetologica* 29:352–55
77. Derting, T. L., McClure, P. A. 1989. Intraspecific variation in metabolic rate and its relationship with productivity in the cotton rat, *Sigmodon hispidus*. *J. Mammal.* 70:520–31
78. Desjardins, C., Bronson, F. H., Blank, J. L. 1986. Genetic selection for reproductive photoresponsiveness in deer mice. *Nature* 322:172–73
79. DiMichele, L., Powers, D. A., DiMichele, J. A. 1986. Developmental and physiological consequences of genetic variation at enzyme synthesizing loci in *Fundulus heteroclitus*. *Am. Zool.* 26:201–08
80. Djawdan, M. 1989. *Locomotor performance of bipedal and quadrupedal rodents*. PhD thesis. Univ. Calif., Irvine. 95 pp.
81. Doyle, R. A., Talbot, A. J. 1986. Artificial selection on growth and correlated selection on competitive behaviour in fish. *Can. J. Fish. Aquat. Sci.* 43:1059–64
82. Dunham, A. E., Miles, D. B., Reznick, D. N. 1988. Life history patterns in squamate reptiles. In *Biology of the Reptilia*, Vol. 16, *Ecology B, Defense and Life Histories*, ed. C. Gans, R. B. Huey, pp. 441–522. New York: Alan R. Liss. 659 pp.
83. Dunham, A. E., Tinkle, D. W., Gibbons, J. W. 1978. Body size in island lizards: a cautionary tale. *Ecology* 59:1230–38
84. Dunlap, D. G. 1969. Evidence for a daily rhythm of heat resistance in the cricket frog, *Acris crepitans*. *Copeia* 1969:852–54
85. Dunlap, D. G., Platz, J. E. 1981. Geographic variation of proteins and call in *Rana pipiens* from the northcentral United States. *Copeia* 1981:876–79
86. Dunson, W. A. 1965. Adaptation of heart and lung weight to high altitude in the robin. *Condor* 67:215–19
87. Dunson, W. A. 1974. Some aspects of salt and water balance of feral goats from arid islands. *Am. J. Physiol.* 226: 662–69
88. Dunson, W. A. 1980. The relation of sodium and water balance to survival in sea water of estuarine and freshwater

races of the snakes, *Nerodia fasciata, N. sipedon* and *N. valida. Copeia* 1980: 268–80

89. Dunson, W. A. 1986. Estuarine populations of the snapping turtle *(Chelydra)* as a model for the evolution of marine adaptations in reptiles. *Copeia* 1986:741–56
90. Dunson, W. A., Buss, E. G. 1968. Abnormal water balance in a mutant strain of chickens. *Science* 161:167–69
91. Dunson, W. A., Buss, E. G., Sawyer, W. H., Sokol, H. W. 1972. Hereditary polydipsia and polyuria in chickens. *Am. J. Physiol.* 222:1167–76
92. Dunson, W. A., Martin, R. R. 1973. Survival of brook trout in a bog-derived acidity gradient. *Ecology* 54:1370–76
93. Dunson, W. A., Mazzotti, F. J. 1989. Salinity as a limiting factor in the distribution of reptiles in Florida Bay: A theory for the estuarine origin of marine snakes and turtles. *Bull. Mar. Sci.* 44:229–44
94. DuShane, D. P., Hutchinson, C. 1944. Differences in size and developmental rate between eastern and midwestern embryos of *Ambystoma maculatum. Ecology* 25:414–23
95. Duthie, G. G. 1987. Observations of poor swimming performance among hatchery-reared rainbow trout, *Salmo gairdneri. Environ. Biol. Fish.* 18:309–11
96. Earle, M., Lavigne, D. M. 1990. Intraspecific variation in body size, metabolic rate, and reproduction of deer mice *(Peromyscus maniculatus). Can. J. Zool.* 68:381–88
97. Echelle, A. A., Kornfield, I., eds. 1984. *Evolution of Fish Species Flocks.* Orono, Maine: Univ. Maine Press
98. Falconer, D. S. 1965. Maternal effects and selection response. In *Genetics Today, Proc. XI Int. Congress Genet.*, The Hague, The Netherlands, September 1963, pp. 763–74. Oxford: Pergamon. 1089 pp.
99. Falconer, D. S. 1989. *Introduction to Quantitative Genetics.* London: Longman. 3rd. ed.
100. Farner, D. S. 1985. Annual rhythms. *Annu. Rev. Physiol.* 47:65–82
101. Farrar, E. S., Frye, B. E. 1979. Factors affecting normal carbohydrate levels in *Rana pipiens. Genet. Comp. Endocr.* 39:358–71
102. Feder, M. E., Bennett, A. F., Burggren, W. W., and Huey, R. B., eds. 1987. *New Directions in Ecological Physiology.* New York: Cambridge Univ. Press. 364 pp.
103. Ferguson, G. W., Brockman, T. 1980. Geographic differences of growth rate of *Sceloporus* lizards (Sauria: Iguanidae). *Copeia* 1980:259–64
104. Ferguson, M. M., Danzmann, R. G., Allendorf, F. W. 1985. Developmental divergence among hatchery strains of rainbow trout *(Salmo gairdneri).* I. Pure strains. *Can. J. Genet. Cytol.* 27:289–97
105. Figiel, C. R. Jr., Semlitsch, R. D. 1990. Population variation in survival and metamorphosis of larval salamanders *(Ambystoma maculatum)* in the presence and absence of fish predation. *Copeia* 1990:818–26
106. Foley, C. W., Seerley, R. W., Hansen, W. J., Curtis, S. E. 1971. Thermoregulatory responses to cold environment by neonatal wild and domestic pigs. *J. Animal Science* 32:926–29
107. Food and Agriculture Organization of the United Nations. 1979. *Mammals in the Seas.* Vol. II. *Pinniped Species Summaries and Report on Sirenians.* Rome: F.A.O. and United Nations Environment Program. 151 pp.
108. Fuller, J. L. 1951. Genetic variability in some physiological constants of dogs. *Am. J. Physiol.* 166:20–24
109. Garland, T. Jr., Bennett, A. F. 1990. Quantitative genetics of maximal oxygen consumption in a garter snake. *Am. J. Physiol.* 259:R986–92
110. Garland, T. Jr., Bennett, A. F., Daniels, C. B. 1990. Heritability of locomotor performance and its correlates in a natural population. *Experientia* 46:530–33
111. Garland, T. Jr., Else, P. L. 1987. Seasonal, sexual, and individual variation in endurance and activity metabolism in lizards. *Am. J. Physiol.* 252:R439–R49

111a. Garland, T. Jr., Huey, R. B., Bennett, A. F. 1991. Phylogeny and coadaptation of thermal physiology in lizards: a reanalysis. *Evolution.* In press

112. Garland, T. Jr., Losos, J. B. 1992. Ecological morphology of locomotor performance in reptiles. In *Ecological Morphology: Integrative Organismal Biology,* ed. P. C. Wainwright, S. M. Reilly, pp. xx. Chicago: Univ. Chicago Press. xx pp. In press
113. Garnett, I., Falconer, D. S. 1975. Protein variation in strains of mice differing in body size. *Genet. Res., Camb.* 25:45–57
114. Gause, G. F. 1942. The relation of adaptability to adaptation. *Q. Rev. Biol.* 17:99–114
115. Gause, G. F. 1947. Problems of evolution. *Trans. Conn. Acad. Arts Sci.* 37:17–68
116. Geiser, F. 1990. Influence of polyun-

saturated and saturated dietary lipids on adipose tissue, brain and mitochondrial membrane fatty acid composition of a mammalian hibernator. *Biochim. Biophys. Acta* 1046:159–66

117. Geist, V. 1987. Bergmann's rule is invalid. *Can. J. Zool.* 65:1035–38
118. Gibbons, J. W., ed. 1990. *Life History and Ecology of the Slider Turtle*. Washington, DC: Smithsonian Inst. 368 pp.
119. Goldstein, D. L., Williams, J. B., Braun, E. J. 1990. Osmoregulation in the field by salt-marsh savannah sparrows *Passerculus sandwichensis beldingi*. *Physiol. Zool.* 63:669–82
120. Gould, S. J., Johnston, R. F. 1972. Geographic variation. *Annu. Rev. Ecol. Syst.* 3:457–98
121. Graham, J. B. 1985. Seasonal and environmental effects on the blood hemoglobin concentrations of some Panamanian air-breathing fishes. *Env. Biol. Fish.* 12:291–301
122. Gray, R. H. 1977. Lack of physiological differentiation in three color morphs of the cricket frog *(Acris crepitans)* in Illinois. *Trans., Ill. State Acad. Sci.* 70:73–79
123. Gregg, V. A., Milligan, L. P. 1982. In vitro energy costs of Na^+, K^+ATPase activity and protein synthesis in muscle from calves differing in age and breed. *Br. J. Nutr.* 48:65–71
124. Gross, J. E., Wang, Z., Wunder, B. A. 1985. Effects of food quality and energy needs: changes in gut morphology and capacity of *Microtus ochrogaster*. *J. Mammal.* 66:661–67
125. Grubbs, D. E. 1980. Tritiated water turnover in free-living desert rodents. *Comp. Biochem. Physiol.* 66A:89–98
126. Guderley, H., Blier, P., Richard, L. 1986. Metabolic changes during the reproductive migration of two sympatric coregonines, *Coregonus artedii* and *Coregonus clupeaformis*. *Can. J. Fish. Aquat. Sci.* 43:1859–65
127. Guillette, L. J., Jr. 1982. The evolution of viviparity and placentation in the high elevation, Mexican lizard *Sceloporus aeneus*. *Herpetologica* 38:94–103

127a. Guillette, L. J. Jr. 1991. The evolution of viviparity in amniote vertebrates: new insights, new questions. *J. Zool.* (London) 223:521–26

128. Haim, A., Borut, A. 1981. Heat production and dissipation in golden spiny mice, *Acomys russatus* from two extreme habitats. *J. Comp. Physiol.* 142: 445–50
129. Haim, A., Borut, A. 1986. Reduced heat production in the bushy-tailed gerbil *Sekeetamys calurus* (Rodentia) as an adaptation to arid environments. *Mammalia* 50:27–33
130. Haim, A., Fairall, N. 1986. Geographical variations in heat production and dissipation within two populations of *Rhabdomys pumilio* (Muridae). *Comp. Biochem. Physiol.* 84A:111–12
131. Hart, J. S. 1952. Geographic variations of some physiological and morphological characters in certain freshwater fish. *Univ. Toronto Biol. Ser. No. 60; Publ. Ont. Fish. Res. Lab.* 72:1–79
132. Hart, J. S. 1971. Rodents. In *Comparative Physiology of Thermoregulation,* Vol. II, *Mammals,* ed. G. C. Whittow, pp. 1–149. New York: Academic Press. 410 pp.
133. Hayes, J. P. 1989. Field and maximal rates of deer mice *(Peromyscus maniculatus)* at low and high altitudes. *Physiol. Zool.* 62:732–44
134. Hayes, J. P. 1989. Altitudinal and seasonal effects on aerobic metabolism of deer mice. *J. Comp. Physiol.* B 159: 453–59
135. Hayes, J. P., Chappell, M. A. 1986. Effects of cold acclimation on maximum oxygen consumption during cold exposure and treadmill exercise in deer mice, *Peromyscus maniculatus*. *Physiol. Zool.* 59:473–81
136. Heath, M. E. 1983. The effects of rearing-temperature on body composition in young pigs. *Comp. Biochem. Physiol.* 76A:363–66
137. Hedrick, P. W. 1986. Genetic polymorphism in heterogenous environments: A decade later. *Annu. Rev. Ecol. Syst.* 17:535–66
138. Heller, H. C., Musacchia, X. J., Wang, L. C. H., eds. 1986. *Living in the Cold: Physiological and Biochemical Adaptations*. New York: Elsevier. 587 pp.
139. Hengeveld, R. 1990. *Dynamic Biogeography*. Cambridge: Cambridge Univ. Press. 249 pp.
140. Herreid, C. F. II, Kinney, S. 1967. Temperature and development of the wood frog, *Rana sylvatica,* in Alaska. *Ecology* 48:579–90
141. Hertz, P. E. 1979. Sensitivity to high temperatures in three West Indian grass anoles (Sauria, Iguanidae), with a review of heat sensitivity in the genus *Anolis*. *Comp. Biochem. Physiol.* 63A: 217–22
142. Hertz, P. E. 1980. Responses to dehydration in *Anolis* lizards sampled along altitudinal transects. *Copeia* 1980: 440–46
143. Hertz, P. E. 1981. Adaptation to altitude in two West Indian anoles (Reptilia: Iguanidae): field thermal biology and

physiological ecology. *J. Zool., Lond.* 195:25–37

144. Hertz, P. E., Arce-Hernandez, A., Ramirez-Vazquez, J., Tirado-Rivera, W., Vazquez-Vives, L. 1979. Geographical variation of heat sensitivity and water loss rates in the tropical lizard, *Anolis gundlachi*. *Comp. Biochem. Physiol.* 62A:947–53
145. Hertz, P. E., Huey, R. B. 1981. Compensation for altitudinal changes in the thermal environment by some *Anolis* lizards on Hispaniola. *Ecology* 62:515–21
146. Hertz, P. E., Huey, R. B., Nevo, E. 1983. Homage to Santa Anita: thermal sensitivity of sprint speed in agamid lizards. *Evolution* 37:1075–84
147. Hertz, P. E., Nevo, E. 1981. Thermal biology of four Israeli agamid lizards in early summer. *Isr. J. Zool.* 30:190–210
148. Heusner, A. A., Jameson, E. W. Jr. 1981. Seasonal changes in oxygen consumption and body composition of *Sceloporus occidentalis*. *Comp. Biochem. Physiol.* 69A:363–72
149. Hickman, G. C., Nevo, E., Heth, G. 1983. Geographic variation in the swimming ability of *Spalax ehrenbergi* (Rodentia: Spalacidae) in Israel. *J. Biogeography* 10:29–36
150. Highton, R. 1977. Comparison of microgeographic variation in morphological and electrophoretic traits. *Evol. Biol.* 10:397–436
151. Hillis, D. M. 1988. Systematics of the *Rana pipiens* complex: puzzle and paradigm. *Annu. Rev. Ecol. Syst.* 19:39–63
152. Hillman, S. S., Gorman, G. C. 1977. Water loss, desiccation tolerance, and survival under desiccating conditions in 11 species of Caribbean *Anolis*. *Oecologia* 29:105–16
153. Hillman, S. S., Gorman, G. C., Thomas, R. 1979. Water loss in *Anolis* lizards: evidence for acclimation and intraspecific differences along a habitat gradient. *Comp. Biochem. Physiol.* 62A:491–94
154. Hillyard, S. D. 1980. Respiratory and cardiovascular adaptations of amphibians and reptiles to altitude. In *Environmental Physiology: Aging, Heat and Altitude,* ed. S. M. Horvath, M. K. Yousef, pp. 363–77. New York: Elsevier/North-Holland. 468 pp.
155. Hirshfield, M. F., Feldmeth, C. R., Soltz, D. L. 1980. Genetic differences in physiological tolerances of Amargosa pupfish *(Cyprinodon nevadensis)* populations. *Science* 207:999–1001
156. Hochachka, P. W., Somero, G. N. 1984. *Biochemical Adaptation*. Princeton: Princeton Univ. Press.
157. Hock, R. J. 1965. An analysis of Gloger's rule. *Hvalradets Skrifter* 48:214–26
158. Hoffmann, A. A., Parsons, P. A. 1989. An integrated approach to environmental stress tolerance and life-history variation: desiccation tolerance in *Drosophila*. *Biol. J. Linn. Soc.* 37:117–36
159. Hoover-Plow, J., Nelson, B. 1985. Oxygen consumption in mice (I strain) after feeding. *J. Nutr.* 115:303–10
160. Hoppe, D. M. 1978. Thermal tolerance in tadpoles of the chorus frog *Pseudacris triseriata*. *Herpetologica* 34:318–21
161. Houde, A. E., Endler, J. A. 1990. Correlated evolution of female mating preferences and male color patterns in the guppy *Poecilia reticulata*. *Nature* 248:1405–08
162. Howard, J. H., Wallace, R. L., Stauffer, J. R. Jr. 1983. Critical thermal maxima in populations of *Ambystoma macrodactylum* from different elevations. *J. Herpetol.* 17:400–02
163. Hudson, J. W., Kimzey, S. L. 1966. Temperature regulation and metabolic rhythms in populations of the house sparrow, *Passer domesticus*. *Comp. Biochem. Physiol.* 17:203–17
164. Huey, R. B., Dunham, A. E. 1987. Repeatability of locomotor performance in natural populations of the lizard *Sceloporus merriami*. *Evolution* 41: 1116–20
165. Huey, R. B., Dunham, A. E., Overall, K. L., Newman, R. A. 1990. Variation in locomotor performance in demographically known populations of the lizard *Sceloporus merriami*. *Physiol. Zool.* 63:845–72
166. Huey, R. B., Partridge, L., Fowler, K. 1991. Thermal sensitivity of *Drosophila melanogaster* responds rapidly to laboratory natural selection. *Evolution*. 45: 751–56
167. Hulbert, A. J., Hinds, D. S., MacMillen, R. E. 1985. Minimal metabolism, summit metabolism and plasma thyroxine in rodents from different environments. *Comp. Biochem. Physiol.* 81A: 687–93
168. Hutchison, V. H., Szarski, H. 1965. Number of erythrocytes in some amphibians and reptiles. *Copeia* 1965:373–75
169. Hutchison, V. H., Vinegar, A., Kosh, R. J. 1966. Critical thermal maxima in turtles. *Herpetologica* 22:32–41
170. Iberall, A. S. 1984. An illustration of the experimental range of variation of blood pressure. *Am. J. Physiol.* 246: R516–R532

171. Ihssen, P. E., Tait, J. S. 1974. Genetic differences in retention of swimbladder gas between two populations of lake trout *(Salvelinus namaycush). J. Fish. Res. Bd. Can.* 311:1351–54
172. Jakobson, M. E. 1981. Physiological adaptability: the response of the house mouse to variations in the environment. *Symp. Zool. Soc. Lond.* 47:301–335
173. James, F. C. 1983. Environmental component of morphological differentiation in birds. *Science* 221:184–86
174. James, F. C., NeSmith, C. 1986. Nongenetic effects in geographic differences among nestling populations of red-winged blackbirds. In *Acta XIX Congressus Internationalis Ornithologici,* Vol. II, ed. H. Ouelett, pp. 1424–33. Ottawa: Univ. Ottawa Press. 2815 pp.
175. Jameson, D. L. 1966. Rate of weight loss of treefrogs at various temperatures and humidities. *Ecology* 47:605–13
176. Jameson, D. L., Taylor, W., Mountjoy, J. 1970. Metabolic and morphological adaptation to heterogeneous environments by the Pacific tree frog, *Hyla regilla. Evolution* 24:75–89
177. Jameson, E. W. Jr., Heusner, A. A., Arbogast, R. 1977. Oxygen consumption of *Sceloporus occidentalis* from three different elevations. *Comp. Biochem. Physiol.* 56A:73–79
178. Jerrett, D. P., Mays, C. E. 1973. Comparative hematology of the hellbender, *Cryptobranchus alleganiensis* in Missouri. *Copeia* 1973:331–37
179. John-Alder, H. B., Barnhart, M. C., Bennett, A. F. 1989. Thermal sensitivity of swimming performance and muscle contraction in northern and southern populations of tree frogs *(Hyla crucifer). J. Exp. Biol.* 142:357–72
180. Karasov, W. H., Anderson, R. A. 1984. Interhabitat differences in energy acquisition and expenditure in a lizard. *Ecology* 65:235–47
181. Karasov, W. H., Diamond, J. M. 1988. Interplay between physiology and ecology in digestion. *BioScience* 38:602–11
182. Kavaliers, M. 1990. Responsiveness of deer mice to a predator, the short-tailed weasel: population differences and neuromodulatory mechanisms. *Physiol. Zool.* 63:388–407
183. Kavaliers, M., Innes, D. 1989. Population differences in benzodiazepine sensitive male scent-induced analgesia in the deer mouse, *Peromyscus maniculatus. Pharm. Biochem. Behav.* 32:613–19
184. Keith, L. B. 1990. Dynamics of snowshoe hare populations. In *Current Mammalogy,* Vol. 2, ed. H. H. Genoways, pp. 119–95. New York: Plenum. 577 pp.
185. Kendeigh, S. C., Blem, C. R. 1974. Metabolic adaptation to local climate in birds. *Comp. Biochem. Physiol.* 48A: 175–87
186. King, C. 1989. *The Natural History of Weasels and Stoats.* Ithaca, NY: Comstock. 253 pp.
187. King, J. R., Mewaldt, L. R. 1981. Variation of body weight in Gambel's white-crowned sparrows in winter and spring: latitudinal and photoperiodic correlates. *Auk* 98:752–64
188. King, T. L., Zimmerman, E. G., Beitinger, T. L. 1985. Concordant variation in thermal tolerance and allozymes of the red shiner, *Notropis lutrensis,* inhabiting tailwater sections of the Brazos River, Texas. *Env. Biol. Fish.* 13:49–57
189. Kirkland, G. L. Jr., Layne, J. N., eds. 1989. *Advances in the Study of* Peromyscus *(Rodentia).* Lubbock: Texas Tech Univ. Press. 367 pp.
190. Koehn, R. K. 1969. Esterase heterogeneity: dynamics of a polymorphism. *Science* 163:943–44
191. Koehn, R. K. 1987. The importance of genetics to physiological ecology. See Ref. 102, pp. 170–88
192. Leon-Velarde, F., Whittembury, J., Carey, C., Monge, C. 1984. Permeability of eggshells of native chickens in the Peruvian Andes. In *Respiration and Metabolism of Embryonic Vertebrates,* ed. R. S. Seymour, pp. 245–57. Dordrecht: Junk
193. Leon-Velarde, F., Whittembury, J., Carey, C., Monge, C. 1984. Shell characteristics of eggs laid at 2,800 m by hens transported from sea level 24 hours after hatching. *J. Exp. Zool.* 230:137–39
194. Licht, P. E. 1984. Reptiles. In *Marshall's Physiology of Reproduction,* Vol. 1, ed. G. E. Lamming, pp. 206–82. Edinburgh: Churchill Livingstone. 842 pp.
195. Liu, E. H., Smith, M. H., Godt, M. W., Chesser, R. K., Lethco, A. K., Henzler, D. J. 1985. Enzyme levels in natural mosquitofish populations. *Physiol. Zool.* 58:242–52
196. Lynch, C. B. 1986. Genetic basis of cold adaptation in laboratory and wild mice, *Mus domesticus.* See Ref. 138, pp. 497–504
197. Lynch, C. B., Roberts, R. C. 1984. Aspects of temperature regulation in mice selected for large and small size. *Genet. Res., Camb.* 43:299–306
198. Lynch, C. B., Sulzbach, D. S. 1984. Quantitative genetic analysis of tempera-

ture regulation in *Mus musculus*. II. Diallel analysis of individual traits. *Evolution* 38:527–40

199. Lynch, G. R., Lynch, C. B., Dube, M., Allen, C. 1976. Early cold exposure: effects on behavioral and physiological thermoregulation in the house mouse, *Mus musculus*. *Physiol. Zool.* 49:191–99
200. Lynch, G. R., Lynch, C. B., Kliman, R. M. 1989. Genetic analyses of photoresponsiveness in the Djungarian hamster, *Phodopus sungorus*. *J. Comp. Physiol.* A 164:475–81
201. Macari, M., Ingram, D. L., Dauncey, M. J. 1983. Influence of thermal and nutritional acclimatization on body temperatures and metabolic rate. *Comp. Biochem. Physiol.* 74A:549–53
202. Macfarlane, W. V. 1978. The ecophysiology of water in desert organisms. In *Water: Planets, Plants and People*, ed. A. K. McIntyre, pp. 108–43. Canberra: Austr. Acad. Sci. 182 pp.
203. Macfarlane, W. V., Howard, B. 1972. Comparative water and energy economy of wild and domestic mammals. *Symp. Zool. Soc. Lond.* 31:261–96
204. MacMillen, R. E., Garland, T. Jr. 1989. Adaptive physiology. See Ref. 189, pp. 143–68
205. Magnuson, J. J., Paszkowski, C. A., Rahel, F. J., Tonn, W. M. 1989. Fish ecology in severe environments of small isolated lakes in northern Wisconsin. In *Freshwater Wetlands and Wildlife*, ed. R. R. Sharitz, J. W. Gibbons, pp. 487–515. CONF-8603101, DOE Symp. Ser. No. 61. USDOE Office of Sci. Tech. Info., Oak Ridge, Tenn.
206. Magnusson, K. P., Ferguson, M. M. 1987. Genetic analysis of four sympatric morphs of Arctic charr, *Salvelinus alpinus*, from Thingvallavatn, Iceland. *Env. Biol. Fish.* 20:67–73
207. Manis, M. L., Claussen, D. L. 1986. Environmental and genetic influences on the thermal physiology of *Rana sylvatica*. *J. Therm. Biol.* 11:31–36
208. Marsh, R. L., Dawson, W. R. 1989. Avian adjustments to cold. In *Advances in Comparative and Environmental Physiology*, Vol. 4, ed. L. C. H. Wang, pp. 205–53. Berlin: Springer-Verlag. 441 pp.
209. Mattingly, D. K., McClure, P. A. 1985. Energy allocation during lactation in cotton rats *(Sigmodon hispidus)* on a restricted diet. *Ecology* 66:928–37
210. Mautz, W. J. 1982. Patterns of evaporative water loss. In *Biology of the Reptilia*, Vol. 12, *Physiology C*, ed. C. Gans, F. H. Pough, pp. 443–81. New York: Academic
211. McCarthy, J. C., Siegel, P. B. 1983. A review of genetical and physiological effects of selection in meat-type poultry. *Anim. Breed. Abstr.* 51:87–94
212. McCauley, R. W. 1958. Thermal relations of geographic races of *Salvelinus*. *Can. J. Zool.* 36:655–62
213. McClenaghan, L. R. Jr., Smith, M. H., Smith, M. W. 1985. Biochemical genetics of mosquitofish. IV. Changes of allele frequencies through time and space. *Evolution* 39:451–60
214. McGovern, M., Tracy, C. R. 1985. Physiological plasticity in electromorphs of blood proteins in free-ranging *Microtus ochrogaster*. *Ecology* 66:396–403
215. McLeese, J. M., Stevens, E. D. 1986. Trypsin from two strains of rainbow trout, *Salmo gairdneri*, is influenced differently by assay and acclimation temperature. *Can. J. Fish. Aquat. Sci.* 43:1664–67
216. Melcher, J. C., Armitage, K. B., Porter, W. P. 1989. Energy allocation by yellow-bellied marmots. *Physiol. Zool.* 62:429–48
217. Mendonça, M. T. 1987. Photothermal effects on the ovarian cycle of the musk turtle, *Sternotherus odoratus*. *Herpetologica* 43:82–90
218. Millar, J. S. 1989. Reproduction and development. See Ref. 189, pp. 169–232
219. Millar, J. S., Threadgill, D. A. L. 1987. The effect of captivity on reproduction and development in *Peromyscus maniculatus*. *Can. J. Zool.* 65:1713–19
220. Miller, K., Packard, G. C. 1974. Critical thermal maximum: ecotypic variation between montane and piedmont chorus frogs (*Pseudacris triseriatus*, Hylidae). *Experientia* 30:355–56
221. Miller, K., Packard, G. C. 1977. An altitudinal cline in critical thermal maxima of chorus frogs *(Pseudacris triseriata)*. *Am. Natur.* 111:267–77
222. Miller, K., Packard, G. C., Packard, M. J. 1987. Hydric conditions during incubation influence locomotor performance of hatchling snapping turtles. *J. Exp. Biol.* 127:401–12
223. Mitton, J. B., Grant, M. C. 1984. Associations among protein heterozygosity, growth rate, and developmental homeostasis. *Annu. Rev. Ecol. Syst.* 15:479–99
224. Mitton, J. B., Koehn, R. K. 1985. Genetic organization and adaptive response of allozymes to ecological variables in *Fundulus heteroclitus*. *Genetics* 79:97–111

225. Moll, E. O. 1973. Latitudinal and intersubspecific variation in reproduction of the painted turtle, *Chrysemys picta*. *Herpetologica* 29:307–18
226. Montgomery, W. I. 1989. *Peromyscus* and *Apodemus:* patterns of similarity in ecological equivalents. See Ref. 189, pp. 293–366
227. Moore, J. A. 1949. Geographic variation of adaptive characters in *Rana pipiens* Schreber. *Evolution* 3:1–24
228. Myers, J. H. 1974. Genetic and social structure of feral house mouse populations on Grizzly Island, California. *Ecology* 55:747–59
229. Nelson, J. A. 1989. Critical swimming speeds of yellow perch *Perca flavescens:* comparison of populations from a naturally acidic lake and a circumneutral lake in acid and neutral water. *J. Exp. Biol.* 145:239–54
230. Nelson, J. A., Magnuson, J. J. 1987. Seasonal, reproductive, and nutritional influences on the white-muscle buffering capacity of yellow perch *(Perca flavescens). Fish Physiol. Biochem.* 3: 7–16
231. Nelson, J. A., Magnuson, J. J., Chulakasem, W. 1988. Blood oxygen capacity differences in yellow perch from northern Wisconsin lakes differing in pH. *Can. J. Fish Aquat. Sci.* 45:1699–704
232. Nevo, E. 1986. Evolutionary behavior genetics in active speciation and adaptation of fossorial mole rats. In *Variability and Behavioral Evolution,* ed. G. Montalenti, G. Tecce, pp. 39–109. Rome: Accademia Nazionale dei Lincei. 275 pp.
233. Nevo, E. 1988. Genetic diversity in nature. Patterns and theory. In *Evolutionary Biology,* Vol. 23, ed. M. K. Hecht, B. Wallace, pp. 217–46. New York: Plenum
234. Nimmo, M. A., Snow, D. H. 1983. Skeletal muscle fiber composition in New Zealand white rabbits, wild rabbits and wild rabbits bred in captivity: effect of heredity. *Comp. Biochem. Physiol.* 74A:955–59
235. Noakes, D. L. G., Skulason, S., Snorrason, S. S. 1989. Alternative life-history styles in salmonine fishes with emphasis on arctic charr, *Salvelinus alpinus*. In *Alternative Life-History Styles of Animals,* ed. M. N. Burton, pp. 329–46. Dordrecht: Kluwer Academic. 616 pp.
236. Nussbaum, R. A. 1976. Changes in serum proteins associated with metamorphosis in salamanders of the family Ambystomatidae. *Comp. Biochem. Physiol.* 53B:569–73
237. Packard, G. C., Sotherland, P. R., Packard, M. J. 1977. Adaptive reduction in permeability of avian eggshells to water vapour at high altitudes. *Nature* 266:255–56
238. Patterson, J. E. 1984. Thermal acclimation in two subspecies of the tropical lizard *Mabuya striata. Physiol. Zool.* 57:301–06
239. Pennycuik, P. R. 1979. Selection of laboratory mice for improved reproductive performance at high environmental temperature. *Aust. J. Biol. Sci.* 32:133–51
240. Peterson, C. R. 1987. Daily variation in the body-temperatures of free-ranging garter snakes. *Ecology* 68:160–69
241. Pierce, B. A., Harvey, J. M. 1987. Geographic variation in acid tolerance of Connecticut wood frogs. *Copeia* 1987: 94–103
242. Pohl, H., West, G. C. 1976. Latitudinal and population specific differences in time of daily and seasonal functions in redpolls *(Acanthis flammea). Oecologia* 25:211–27
243. Powers, D. A. 1987. A multidisciplinary approach to the study of genetic variation within species. See Ref. 102, pp. 102–30
244. Powers, D. A., Dalessio, P. M., Lee, E., DiMichele, L. 1986. The molecular ecology of *Fundulus heteroclitus* hemoglobin-oxygen affinity. *Am. Zool.* 26:235–48
245. Prosser, C. L. 1955. Physiological variation in animals. *Biol. Rev.* 30:229–62
246. Prosser, C. L. 1986. The challenge of adaptational physiology. *Physiologist* 29:2–4
247. Purdom, C. E. 1979. Genetics of growth and reproduction in teleosts. *Symp. Zool. Soc. Lond.* 44:207–17
248. Rahn, H., Ledoux, T., Paganelli, C. V., Smith, A. H. 1982. Changes in eggshell conductance after transfer of hens from an altitude of 3,800 to 1,200 m. *J. Appl. Physiol.* 53:1429–31
249. Rahn, H., Carey, C., Balmas, K., Bhatia, B., Paganelli, C. V. 1977. Reduction of pore area of the avian eggshell as an adaptation to altitude. *Proc. Natl. Acad. Sci. USA* 74:3095–98
250. Ralin, D. B. 1981. Ecophysiological adaptation in a diploid-tetraploid complex of treefrogs (Hylidae). *Comp. Biochem. Physiol.* 68A:175–79
251. Rapson, G. L., Wilson, J. B. 1988. Non-adaptation in *Agrostis capillaris* L. (Poaceae). *Funct. Ecol.* 2:479–90
252. Ray, C. 1960. The application of Bergmann's and Allen's rules to poikilotherms. *J. Morph.* 106:85–108

253. Reznick, D. N. 1989. Life-history evolution in guppies: 2. Repeatability of field observations and the effects of season on life histories. *Evolution* 43:1285–97

254. Reznick, D. N., Bryga, H., Endler, J. A. 1990. Experimentally induced life-history evolution in a natural population. *Nature* 346:357–59

255. Reznick, D. N., Miles, D. B. 1989. Review of life history patterns in poeciliid fishes. In *Ecology and Evolution of Livebearing Fishes (Poeciliidae),* ed. G. K. Meffe, F. F. Snelson, Jr., pp. 125–48. Englewood Cliffs, NJ: Prentice Hall. 453 pp.

256. Riska, B., Atchley, W. R., Rutledge, J. J. 1984. A genetic analysis of targeted growth in mice. *Genetics* 107:79–101

257. Riska, B., Rutledge, J. J., Atchley, W. R. 1985. Covariance between direct and maternal genetic effects in mice, with a model of persistent environmental influences. *Genet. Res., Camb.* 45:287–97

258. Robbins, L. W., Hartman, G. D., Smith, M. H. 1987. Dispersal, reproductive strategies, and the maintenance of genetic variability in the mosquitofish *(Gambusia affinis). Copeia* 1987:156–63

259. Robinson, G. D., Dunson, W. A., Wright, J. E., Mamolito, G. E. 1976. Differences in low pH tolerance among strains of brook trout *(Salvelinus fontinalis). J. Fish. Biol.* 8:5–17

260. Rome, L. C., Loughna, P. T., Goldspink, G. 1985. Temperature acclimation: improved sustained swimming performance in carp at low temperatures. *Science* 228:194–96

261. Rose, M. R., Service, P. M., Hutchinson, E. W. 1987. Three approaches to trade-offs in life-history evolution. In *Genetic Constraints on Adaptive Evolution,* ed. V. Loeschcke, pp. 91–105. Berlin: Springer-Verlag. 188 pp.

262. Sandstrom, O. 1983. Seasonal variations in the swimming performance of perch (*Perca fluviatilis* L.) measured with the rotary-flow technique. *Can. J. Zool.* 61:1475–80

263. Scheck, S. H. 1982. A comparison of thermoregulation and evaporative water loss in the hispid cotton rat, *Sigmodon hispidus texianus,* from northern Kansas and south-central Texas. *Ecology* 63: 361–69

264. Scheck, S. H. 1982. Development of thermoregulatory abilities in the neonatal hispid cotton rat, *Sigmodon hispidus texianus,* from northern Kansas and south-central Texas. *Physiol. Zool.* 55: 91–104

265. Scheck, S. H. 1990. Divergence in thermoregulatory function during growth of young from two lineages of hispid cotton rat. *Am. Zool.* 30:32A (Abstr.)

266. Schlager, G., Freeman, R., El Seoudy, A. A. 1983. Genetic study of norepinephrine in brains of mice selected for differences in blood pressure. *J. Heredity* 74:97–100

267. Schmalhausen, I. I. 1949. *Factors of Evolution.* (1986 reprint) Chicago: Univ. Chicago Press. 327 pp.

268. Schmidt-Nielsen, K., Kim, Y. T. 1964. The effect of salt intake on the size and function of the salt gland of ducks. *Auk* 81:160–72

269. Schwartz, J. M., McCracken, G. F., Burghardt, G. M. 1989. Multiple paternity in wild populations of the garter snake, *Thamnophis sirtalis. Behav. Ecol. Sociobiol.* 25:269–73

270. Selander, R. K., Yang, S. Y., Hunt, W. G. 1969. Polymorphism in esterases and hemoglobin in wild populations of the house mouse *(Mus musculus).* In *Studies in Genetics V,* ed. M. R. Wheeler, pp. 271–338. Austin: Univ. Texas Press. 338 pp.

271. Sibly, R. M., Monk, K. A., Johnson, I. K., Trout, R. C. 1990. Seasonal variation in gut morphology in wild rabbits *(Oryctolagus cuniculus). J. Zool.* (London) 221:605–19

272. Siegel-Causey, D. 1990. Phylogenetic patterns of size and shape of the nasal gland depression in Phalacrocoracidae. *Auk* 107:110–18

273. Sinervo, B. 1990. Evolution of thermal physiology and growth rate between populations of the western fence lizard *(Sceloporus occidentalis). Oecologia* 83:228–37

274. Sinervo, B. 1990. The evolution of maternal investment in lizards: an experimental and comparative analysis of egg size and its effect on offspring performance. *Evolution* 44:279–94

275. Sinervo, B., Huey, R. B. 1990. Allometric engineering: an experimental test of the causes of interpopulational differences in performance. *Science* 248: 1106–09

276. Sinervo, B., Losos, J. B. 1991. Walking the tight rope: a comparison of arboreal sprint performance among populations of *Sceloporus occidentalis* lizards. *Ecology.* In press

277. Sinervo, B., Hedges, R., Adolph, S. C. 1991. Decreased sprint speed as a cost of reproduction in the lizard *Sceloporus*

occidentalis: variation among populations. *J. Exp. Biol.* 155:323–36

278. Smith, M. H., Smith, M. W., Scott, S. L., Liu, E. H., Jones, J. C. 1983. Rapid evolution in a post-thermal environment. *Copeia* 1983:193–97
279. Snell, H. L., Jennings, R. D., Snell, H. M., Harcourt, S. 1988. Intrapopulation variation in predator-avoidance performance of Galápagos lava lizards: the interaction of sexual and natural selection. *Evol. Ecol.* 2:353–69
280. Snow, D. H., Harris, R. C. 1985. Thoroughbreds and greyhounds: Biochemical adaptations in creatures of nature and of man. In *Circulation, Respiration, and Metabolism,* ed. R. Gilles, pp. 227–39. Berlin: Springer-Verlag. 568 pp.
281. Snyder, G. K., Weathers, W. W. 1977. Activity and oxygen consumption during hypoxic exposure in high altitude and lowland sceloporine lizards. *J. Comp. Physiol.* 117:291–301
282. Snyder, L. R. G., Hayes, J. P., Chappell, M. A. 1988. Alpha-chain hemoglobin polymorphisms are correlated with altitude in the deer mouse, *Peromyscus maniculatus. Evolution* 42:689–97
283. Sotherland, P. R., Packard, G. C., Taigen, T. L., Boardman, T. J. 1980. An altitudinal cline in conductance of cliff swallow *(Petrochelidon pyrrhonota)* eggs to water vapor. *Auk* 97:177–85
284. Stephenson, E. M. 1968. Temperature tolerance of cultured amphibian cells in relation to latitudinal distribution of donors. *Aust. J. Biol. Sci.* 21:741–57
285. Stini, W. A. 1979. *Physiological and Morphological Adaptation and Evolution*. The Hague: Mouton. 525 pp.
286. Swarts, F. A., Dunson, W. A., Wright, J. E. 1978. Genetic and environmental factors involved in increased resistance of brook trout to sulfuric acid solutions and mine acid polluted waters. *Trans. Am. Fisheries Soc.* 107:651–77
287. Taigen, T. L., Packard, G. C., Sotherland, P. R., Boardman, T. J., Packard, M. J. 1980. Water-vapor conductance of black-billed mapie *(Pica pica)* eggs collected along an altitudinal gradient. *Physiol. Zool.* 53:163–69
288. Taigen, T. L., Pough, F. H. 1985. Metabolic correlates of anuran behavior. *Am. Zool.* 25:987–97
289. Taylor, E. B. 1990. Variability in agonistic behaviour and salinity tolerance between and within two populations of juvenile chinook salmon, *Oncorhynchus tshawytscha,* with contrasting life histories. *Can. J. Fish. Aquat. Sci.* 47:2172–80
290. Taylor, E. B. 1991. A review of local adaptation in Salmonidae, with particular reference to Pacific and Atlantic salmon. *Aquaculture*. In press
291. Taylor, E. B., Foote, C. J. 1991. Critical swimming velocities of juvenile sockeye salmon and kokanee, the anadromous and non-anadromous forms of *Oncorhynchus nerka* (Walbaum). *J. Fish Biology* 38:407–19
292. Taylor, E. B., McPhail, J. D. 1985. Variation in burst and prolonged swimming performance among British Columbia populations of Coho salmon, *Oncorhynchus kisutch. Can. J. Fish. Aquat. Sci.* 42:2029–33
293. Taylor, E. B., McPhail, J. D. 1986. Prolonged and burst swimming in anadromous and freshwater threespine stickleback, *Gasterosteus aculeatus. Can. J. Zool.* 64:416–20
294. Temte, J. L. 1991. Precise birth timing in captive harbor seals *(Phoca vitulina)* and California sea lions *(Zalophus californianus). Mar. Mammal Sci.* 7:145–56
295. Temte, J. L., Bigg, M. A., Wiig, O. 1991. Clines revisited: the timing of pupping in the harbor seal *(Phoca vitulina). J. Zool. (London)*. In press
296. Thompson, S. D. 1985. Subspecific differences in metabolism, thermoregulation, and torpor in the western harvest mouse *Reithrodontomys megalotis. Physiol. Zool.* 58:430–44
297. Thorpe, R. S. 1989. Geographic variation: multivariate analysis of six character systems in snakes in relation to character number. *Copeia* 1989:63–70
298. Townsend, C. R., Calow, P. 1981. *Physiological Ecology: An Evolutionary Approach to Resource Use*. Sunderland, Mass: Sinauer. 393 pp.
299. Tracy, C. R., Turner, J. S. 1982. What is physiological ecology? *Bull. Ecol. Soc. Am.* 63:340–47
300. Travis, J. 1992. Evaluating the adaptive role of morphological plasticity. In *Ecological Morphology: Integrative Organismal Biology,* ed. P. C. Wainwright, S. M. Reilly, pp. 00000. Chicago: Univ. Chicago Press. In press
301. Trost, C. H. 1972. Adaptations of horned larks *(Eremophila alpestris)* to hot environments. *Auk* 89:506–27
302. Tsuji, J. S. 1988. Seasonal profiles of standard metabolic rate of lizards *(Sceloporus occidentalis)* in relation to latitude. *Physiol. Zool.* 61:230–40
303. Tsuji, J. S. 1988. Thermal acclimation of metabolism in *Sceloporus* lizards

from different latitudes. *Physiol. Zool.* 61:241–53

304. Tsuji, J. S., Huey, R. B., van Berkum, F. H., Garland, T. Jr., Shaw, R. G. 1989. Locomotor performance of hatchling fence lizards *(Sceloporus occidentalis):* quantitative genetics and morphometric correlates. *Funct. Ecol.* 3:240–52
305. Ultsch, G. R., Hanley, R. W., Bauman, T. R. 1985. Responses to anoxia during simulated hibernation in northern and southern painted turtles. *Ecology* 66: 388–95
306. Ushakov, V. 1964. Thermostability of cells and proteins of poikilotherms and its significance in speciation. *Physiol. Rev.* 44:518–60
307. van Berkum, F. H. 1988. Latitudinal patterns of the thermal sensitivity of sprint speed in lizards. *Am. Nat.* 132:327–43
308. van Berkum, F. H., Pough, F. H., Stewart, M. M., Brussard, P. F. 1982. Altitudinal and interspecific differences in the rehydration abilities of Puerto Rican frogs *(Eleutherodactylus). Physiol. Zool.* 55:130–36
309. Van Damme, R., Bauwens, D., Castilla, A. M., Verheyen, R. F. 1989. Altitudinal variation in the thermal biology and running performance in the lizard *Podarcis tiliguerta. Oecologia* 80:516–24
310. Van Damme, R., Bauwens, D., Verheyen, R. 1986. Selected body temperatures in the lizard *Lacerta vivipara:* variation within and between populations. *J. Therm. Biol.* 11:219–22
311. Van Damme, R., Bauwens, D., Verheyen, R. 1990. Evolutionary rigidity of thermal physiology: the case of the cool temperate lizard *Lacerta vivipara. Oikos* 57:61–67
312. Vernberg, F. J. 1962. Comparative physiology: latitudinal effects on physiological properties of animal populations. *Annu. Rev. Physiol.* 24:517–46
313. Volpe, E. P. 1953. Embryonic temperature adaptations and relationships in toads. *Physiol. Zool.* 26:344–54
314. Volpe, E. P. 1957. Embryonic temperature tolerance and rate of development in *Bufo valliceps. Physiol. Zool.* 30:164–76
315. Wainwright, P. C., Lauder, G. V., Osenberg, C. W., Mittelbach, G. G. 1991. The functional basis of intraspecific trophic diversification in sunfishes. In *Proc. 4th Int. Cong. Syst. Evol. Biol.*, ed. E. Dudley, Portland: Diascortes. In press
316. Wangensteen, O. D., Rahn, H., Burton, R. R., Smith, A. R. 1974. Respiratory gas exchange of high altitude adapted chick embryos. *Respir. Physiol.* 21:61–70
317. Ward, J. M. Jr., Armitage, K. B. 1981. Circannual rhythms of food consumption, body mass, and metabolism in yellow-bellied marmots. *Comp. Biochem. Physiol.* 69A:621–26
318. Ward, J. M. Jr., Armitage, K. B. 1981. Water budgets of montane-mesic and lowland-xeric populations of yellow-bellied marmots. *Comp. Biochem. Physiol.* 69A:627–30
319. Weathers, W. W. 1979. Climatic adaptation in avian standard metabolic rate. *Oecologia* 42:81–89
320. Weathers, W. W., White, F. N. 1972. Hematological observations on populations of the lizard, *Sceloporus occidentalis* from sea level and altitude. *Herpetologica* 28:172–75
321. Wells, R. M. G. 1990. Hemoglobin physiology in vertebrate animals: a cautionary approach to adaptationist thinking. In *Advances in Comparative and Environmental Physiology,* Vol. 6, *Vertebrate Gas Exchange,* ed. R. G. Boutilier, pp. 143–61. Berlin: Springer-Verlag. 411 pp.
322. Whittow, G. C. 1971. Ungulates. In *Comparative Physiology of Thermoregulation.* Vol. II. *Mammals,* ed. G. C. Whittow, pp. 191–281. New York: Academic. 410 pp.
323. Wilkens, H. 1988. Evolution and genetics of epigean and cave *Astyanax fasciatus* (Characidae, Pisces): support for the neutral mutation theory. *Evol. Biol.* 23:271–368
324. Willis, M. B. 1989. *Genetics of the Dog.* London: Witherby. 417 pp.
325. Wilson, M. A., Echternacht, A. C. 1986. Geographic variation in the critical thermal minimum of the green anole, *Anolis carolinensis* (Sauria: Iguanidae), along a latitudinal gradient. *Comp. Biochem. Physiol.* 87A:757–60
326. Wingfield, J. C., Farner, D. S. 1980. Control of seasonal reproduction in temperate-zone birds. *Prog. Reprod. Biol.* 5:62–101
327. Zink, R. M., Remsen, J. V., Jr. 1986. Evolutionary processes and patterns of geographic variation in birds. In *Current Ornithology,* Vol. 4, ed. R. F. Johnston, pp. 1–69. New York: Plenum. 324 pp.

Annu. Rev. Ecol. Syst. 1991. 22:229–56

NOVELTY IN EVOLUTION: RESTRUCTURING THE CONCEPT

Gerd B. Müller

Department of Anatomy, University of Vienna, A-1090 Wien, Austria

Günter P. Wagner

Zoological Institute, University of Vienna, A-1090 Wien, Austria

KEY WORDS: evolution, novelty, homology, developmental constraint

---

INTRODUCTION

> "In the frequent fits of anger to which the males especially are subject, the efforts of their inner feelings cause the fluids to flow more strongly towards that part of their head; in some there is hence deposited a secretion of horny matter, and in others of bony matter mixed with horny matter, which gives rise to solid protuberances: thus we have the origin of horns and antlers.

This statement by Jean Lamarck (41, p. 122) not only documents how the novelty problem arose immediately with the formulation of scientific theories about the evolution of life, it also exemplifies the early attempt to identify a mechanistic cause for the origin of new organs. It is of little importance that Lamarck failed to identify the mechanism correctly. Later, Darwin (20) also "felt much difficulty in understanding the origin of simple parts" (p. 194) which he thought could have "originated from quite secondary causes, independently of natural selection" (p. 196), and he had recourse to Lamarckian explanations to deal with the problem.

Consequently, the difficulty of how new characters could arise from a process of gradual variation and selection was at the center of the early critique of Darwin's theory (36, 62, 68, 83). At that time, novelty was treated both by its critics and by its advocates (e.g. 108), as a distinct problem of

0066-4162/91/1120-0229$02.00

organismic evolution. The rise of genetics again refuelled the debate and favored mutationist explanations for the origin of innovations (29, 89). Subsequently, however, with the broad acceptance of the neo-Darwinian synthesis, the issue of novelty became diffused in discussions of the origin of adaptations (31, 90) and in the concept of macroevolution (37, 93, and others). Novelties were seen increasingly as an aspect of the problem of speciation and of the origin of higher taxa and less as a problem of the primary causes responsible for the generation of new anatomical structures. Only Mayr (57, 58) identified novelty again as a distinct and neglected problem of evolutionary biology, but the prevalence of the adaptationist program, characteristic for the past decades of evolutionary research, largely prevented its further analysis.

Spurred by a recent trend toward organismic approaches in evolutionary biology, the issue of novelty has again come to the fore. Several recent publications and meetings were devoted to the problems of innovation (70, 107). The present understanding of novelty, however, is characterized by remarkable heterogeneity. The issue is linked on one hand to the character discussion in taxonomy and on the other hand to the Lamarkism-mutationism, microevolution-macroevolution, and gradualism-punctualism debates. These historical polarities in effect obscure the real problem. Therefore, the primary objective of this chapter is to liberate the novelty issue from its historical burden and to provide a new conceptual foundation for its analysis. After a brief review of traditional concepts and their deficiencies we proceed to analyze the empirical evidence for novelties at the character level. Based on this analysis we redefine the problem and investigate the possible generative mechanisms underlying the origin of new morphological structures. Particular emphasis is placed on the distinction between the generation of new characters and their fixation, which may eventually lead to the formation of novel body plans. In conclusion, we propose that an empirical approach to the problem of novelty has to focus on the organizational principles of developmental systems and their ability to generate new structures.

CONCEPTS OF MORPHOLOGICAL NOVELTY

We restrict our analysis here to the origin of new structures in morphological evolution. Even when limited to the morphological level, very different attitudes are taken toward the problem in current evolutionary biology. The prevailing one is a purely phenomenological treatment of novelty. This is embodied in the discussions of the rates of origination of novel characters (15, 93, 96), their significance as taxonomic characters (111), or their role as key triggers of diversification and adaptive radiation (46, 51). While important for each of these chapters of evolutionary theory, the phenomenological aspect of novelty is not the central problem and is not dealt with in our further discussion. We concentrate on the generative aspect of morphological innovations in the process of evolution. Although this aspect has figured less prominently in past discussions, it is possible to distinguish three conceptual approaches.

Functional Concepts

That a change of function may initiate the generation of new structures was already expressed by Darwin (20), and the concept was elaborated by Dohrn (23), Plate (73), Sewertzoff (91), and Mayr (57, 58). The basic idea is that environmental and behavioral changes induce the acquisition of new functions which in turn favor the selection of small variations that facilitate the exertion of the new function. This concept is based on a "duplication of function" and "duplication of structure" principle. As noted by Mayr (57), either the organ under question must initially be able to perform two distinct functions simultaneously, or two distinct organs must perform the same function over a transitional period. The classic example for the latter is the coexistence of gills and primitive lungs in the evolution of respiratory organs. Many such duplications of function are known and make a strong case for the change of function concept. Accordingly, Mayr (57, p. 351) defines novelty as "any newly acquired structure or property that permits the assumption of a new function."

Several problems arise both from a functional definition of novelties and from the mechanism proposed for their origination. Mainly when combined with the change of function principle, the definition harbors a danger of circularity. New structures arise from new functions, and new functions from new structures. Thus, it does not seem useful to restrict the definition only to those structures that permit a new function. Such a definition also excludes all those structures that might originate without association to a new function, e.g. exaptations (30).

More importantly, the change of function concept bypasses the generative problem. While the coexistence of old and new functions, as well as that of ancestral and new structures, represents an important principle of functional and morphological transition, it does not explain the first appearance of a new structure. Gills and lungs must coexist for a transitional period to permit the takeover of a new mode of respiration, but what mechanisms generated the lungs, or even the gills? The change of function principle is helpful only in so far as it indicates that a new structure must always arise in a different functional context than the one which eventually represents its adaptive advantage. But what we need to know is, what creates the heritable variation at the site where it is required? And what precisely are the mechanistic causes that are responsible for a specific morphological solution to a new functional and/or structural problem. And finally, if we accept that new functions are initiating factors for the generation of new structures, is this a necessary prerequisite or can new structures also arise without a change of function?

Genetic Concepts

Although natural selection may act on any kind of heritable phenotypic variation, irrespective of the cause of heritability (59), the majority of evolutionarily important phenotypic variation is ultimately linked to genetic variation and becomes finally established in a population by selection, drift, or genetic drive. Consequently, genetic concepts concerning the origin of morphological novelties have two aspects: first, the kind of genetic change that makes the phenotypic variation heritable, and, second, the population genetic mechanisms that lead to the fixation of these genetic variants.

THE KIND OF GENETIC CHANGE A recent critical review of the molecular concepts concerning the origin of morphological novelties was provided by John & Miklos (38) under the title "The Unsolved Problem." Below we briefly discuss their major conclusions. A number of specific molecular mechanisms have been proposed to explain the origin of novelties, including structural gene mutations, changes in genome size, chromosomal rearrangements, and regulatory mutations caused by diffusion of repeated sequences (9, 16, 24, 25, 54). However, the main problem is that no conclusive evidence is available to demonstrate a specific role of any of these molecular mechanisms in the origin of morphological novelties. There are at least two reasons for this situation, one biological, the other methodological.

The biological reason is that metazoan development is realized via the interaction between cells that communicate by utilizing their gene products. This self-referential structure of metazoan development (72) makes impossible a clear distinction between regulatory and structural genes (16, 74). Possibly the best example is found in the role of the extracellular matrix. Hyaluronic acid, laminin, and fibronectin, all products of structural genes or of secondary metabolism, play an important role in regulating the migration of neural crest cells and thus have a regulatory role in vertebrate development (34). Therefore, it is not sensible to expect genetic changes, responsible for the heritability of a novel morphological feature, to be of a particular molecular type. This means: Certain specific structural gene mutations are as plausible candidates for the genetic basis of a novelty as are changes in gene regulation networks or chromosomal rearrangements.

The methodological problem is critical for all problems of evolutionary genetics, namely, the question of how to distinguish between those genetic changes that are causative in the origin of novelties and those that merely coincide with the observed change. The short answer to this question given by

John & Miklos (38) is negative: "We won't know for certainty"—but this is true for all empirical sciences. On the positive side, their discussion clearly indicates that the way out is to study the role of gene products in development, i.e. to determine the biological role of observed genetic differences in the developmental mechanisms responsible for the realization of morphological differences.

POPULATION GENETIC PROBLEMS Once a heritable phenotypic change has been achieved by a mutation, it has to be integrated into the gene pool of the species. Two problems arise in this area: (*a*) If one assumes, as the neo-Darwinian orthodoxy does, that major changes are realized by the accumulation of many mutational steps with individually small effects, one is confronted with the problem of whether natural selection can deal with such a multitude of pleiotropically and functionally interrelated changes. (*b*) If one believes that new adaptations are initiated by a major genetic mutation (or threshold effect), then one has to deal with the question of how such drastic changes can be accommodated in a genetic background unprepared to compensate for unavoidable and possibly deleterious pleiotropic effects.

The general conclusion is that natural selection is easily able to produce phenotypic changes much faster than has been observed in the fossil record (16, 42). This has also been confirmed by recent studies on the evolution of functionally constrained phenotypes (i.e. the interaction of directional and stabilizing selection on two or more characters), although adaptation by natural selection does not appear as inevitable as in simpler models of selection (12, 101, 103). These studies are based on the assumption of additive genetic effects. In the case of strong epistatic effects, it is generally concluded that a combination of drift and selection (shifting balance) is sufficient to explain new adaptations (6, 17, 19, 45, 114).

Several concepts are available to explain the integration of discontinuous variation into the gene pool. One is the concomitant selection of modifier genes that can compensate for the deleterious pleiotropic effects of a discontinuous variation (44)—this selection works fine under certain conditions. The other concepts are less orthodox. According to West-Eberhard (109), the integration of a discontinuous variant has to pass through a stage in which a stable polymorphism exists with the original condition. This would allow coadaptive fine-tuning of a new structure while a working alternative is maintained. Erwin & Valentine (25) have suggested that horizontal gene transfer may increase the frequency of a new variant to a level where homozygous genotypes become available for selection in spite of a selective disadvantage of the heterozygous genotypes. Arthur (3) has suggested a magnitude effect of phenotypic change, where large changes are viable with a higher probability than changes with intermediate effects. Finally, molecular drive may be an alternative mechanism to natural selection to explain the first steps in the integration of a novelty into the gene pool (24).

In summary, the origination and fixation of a new genetic variant can be achieved via a multitude of mechanisms and does not appear to be an unresolved question with respect to the origin of morphological novelties.

Developmental Concepts

The currently most popular concept of how development relates to evolution is *heterochrony*—phylogenetic changes in the timing and rates of ontogenetic processes. Heterochrony has particularly been associated with the origination of structural novelty in a number of recent publications (2, 64, 75, 106). Earlier, De Beer (21) paid detailed attention to the ways in which changes of developmental timing can affect the appearance of embryonic structures and the introduction of novel characters, and recent studies demonstrate the pervasiveness of heterochronic alterations in the phylogeny of a large variety of taxa (61). We may safely assume that heterochrony is a fact in evolutionary biology, but not all heterochrony observed is necessarily causal in morphological evolution. Much of it could be a consequence of alterations that do not primarily affect the timing of developmental processes. Including these passive effects would rob the concept of heterochrony of its explanatory value (76). Therefore, ways must be found to distinguish between causal and secondary heterochrony. Also, the occurrence of heterochrony is rarely distinguished from the mechanistic processes through which changes of timing could generate new structures. This however is the central problem if a generative role is to be assigned to heterochrony.

In several instances heterochrony could be related to specific processes of development and to the appearance of novel morphological features. Raff et al (75, 76) and Wray & Raff (113) were able to relate the evolution of direct development in sea urchins, which involves the appearance of several novel larval features, to heterochronic events in early development. Changes in the timing of cell lineage segregation in blastomeres of the direct developing embryos lead to novel forms of nonfeeding larvae, in which some of the features of the primitive pluteus larva are eliminated and other features make a very early appearance. These and other derived features of direct developers such as changes in cleavage pattern and mitotic rates are dependent on the heterochronic changes in developmental mode and not on adaptations in the traditional sense.

The sea urchin example shows that heterochrony can lead to the production of novel features through alterations in the timing of very early ontogenetic processes. But heterochrony is not confined to early ontogeny, and empirical evidence suggests that heterochronic alterations of the processes of pattern formation and morphogenesis are also causal in the generation of novelty. For instance, truncations of skeletal patterning processes, at the level both of chondrogenesis and of osteogenesis, underlie the transformations of skeletal patterns in vertebrate limb evolution. Nonsegmentations of mesenchymal arrays and secondary fusions of chondrogenic condensations, occurring at advanced stages of the embryonic period, result in the generation of novel skeletal elements (65).

Although such examples document the possibility of novelties being introduced through heterochrony at all stages of ontogeny, and although the specific processes affected by heterochronic alterations can sometimes be identified, as yet few concepts suggest why paedomorphic or peramorphic changes to developmental processes should result in new structural characters. Two kinds of solutions were recently proposed.

In a study based on an evolutionary analysis of visual-neuronal control, functional morphology, and development of the feeding system in plethodontid salamanders, Wake & Roth (106) suggest that novelties are generated through ontogenetic repatterning. Ontogenetic repatterning refers to the establishment of new sets of morphogenetic processes through dissociation and recombination of compartmentalized subsets of the developmental system. Heterochrony is seen as the process initiating the dissociation and recombination events, thus being ultimately responsible for the foundation of new patterns of developmental interaction that give rise to new morphological arrangements of the phenotype.

Another approach is based on the system properties of development (64). According to this concept, heterochronic and nonheterochronic mechanisms of evolution have a quantitatively modifying effect on developmental parameters, but the magnitude of these modifications is limited by system-specific thresholds. Modifications that go beyond such thresholds can cause nonlinear effects, e.g. by interrupting developmental interactions or by initiating new ones. The kind of resulting morphological effect depends on the developmental reaction norms of the affected cell populations and tissues. Initially inconspicuous structures arising from such a process may first assume an embryonic function and become fixated in the developmental network. In a possibly much later step such "caenogenetic" structures can be moved heterochronically into the postembryonic period and can be further elaborated. The threshold origin and the embryonic preexistence of novel structures is thought to underlie their often rapid phenotypic appearance in a phylogenetic lineage. According to this hypothesis, the first rudiments of morphological novelties appear as neutral by-products of evolutionary alterations to developmental processes. The causality for their appearance is thus proposed to lie in the system properties of development, which can transform gradual and quantitative evolution into qualitative phenotypic effects.

An approach that differs greatly from the two previous ones was taken by

Buss (14) who considered the origin of novelties as resulting from conflicts between levels of selection. Each multicellular organism is composed of units capable of self-replication. The primary evolutionary function of developmental interactions is to solve this conflict between levels of selection. Major developmental innovations are thus expected at those points where a transition between levels of selection occurs. But the fact that new organs and new anatomical elements can originate in the phylogeny of well-established multicellular organisms (e.g. vertebrates) indicates that this cannot be the only mode for the origin of morphological novelties.

In summary, although a number of attempts were made to conceptualize the contributions of developmental systems to the origin of novelties, the developmental concepts are the least elaborated. They also have a common weakness, which is their formulation rather independently from population genetics.

APOMORPHIES VERSUS NOVELTIES

We intend here to set the stage for a reformulation of the problem of novelties. The point of departure will be the least theory-laden definition of a novelty available in the literature. The definition consists simply of the statement that all traits characteristic of a supraspecific taxon were a novelty at some point in the evolution of that group (18, 27).

To obtain an objective picture of the kinds of characters that have been identified as apomorphies of supraspecific taxa, we listed the morphological apomorphies of the higher taxa of mammals (Table 1). The table is based on a recent summary of mammalian characters, used to illustrate the cladistic approach (4). From the list of characters in Table 1, it becomes immediately clear that this set of apomorphies comprises a number of traits whose origin is quite unproblematic and easy to explain on the basis of known evolutionary mechanisms. For instance, a number of characters are negative traits, i.e. the absence of certain structures is characteristic for a clade. Negative characters are legitimate apomorphies in cladistic analyses (4). Among these are the reduction of the nucleus in the erythrocytes (Mammalia), the loss of teeth (Monotremata), the reduction of the coracoid bone (Theria), and the reduction of the marsupial bone in the Placentalia. Although there is no conclusive evidence concerning the causes of reductive evolution, little doubt exists that it can be explained by Darwinian mechanisms because the genetic basis of reduction is largely additively polygenic (112).

Another class of apomorphies that are quite unproblematic are shape characters. For instance, a bent cochlea is apomorphic for the class of Mammalia, but these characters are rare among those characteristic of higher taxa. The great majority of apomorphies is less easily classified with respect to the kind of processes underlying their origin. A tentative but by no means exhaustive classification would include (*a*) characters that result from differentiations of repeated elements, (*b*) new elements, (*c*) change of context, and (*d*) differentiations caused by the synorganization of plesiomorph traits. Others are hard to classify, such as the differentiation of trophoblast and embryoblast (Placentalia), or the appearance of prismatic enamel (Theria). To determine whether there are specific difficulties in explaining the remaining novelties, some examples are discussed in detail below.

Differentiation of Repeated Elements

A key innovation of mammals with profound functional and adaptive consequences is the differentiation of the teeth. The plesiomorph status is homodont conical teeth that all look basically the same. Tooth differentiation allows the use of a broader spectrum of prey and is considered as one factor responsible for the tremendous success of mammals (39). From a morphological point of view the origin of heterodont teeth is a differentiation of serially homologous elements. Other characters of that kind are the differentiation of the cervical vertebrae (Mammalia) and the origin of whiskers, which are apomorphic for the taxon Theria. Differentiation must be considered to be a major mode of morphological evolution (54, 79, 80).

The explanation of these characters appears similar to simple shape changes, especially when the result is of such obvious adaptive value as are heterodont teeth or whiskers. Adaptively sensible shape changes should be easy to explain given the extensive amount of heritable phenotypic variation available for almost every quantitative character (16, 63). However, the differentiation of homonomous (iteratively homologous) elements is not as easily explainable. Repeated anatomical elements are most probably due to the repeated expression of the same genetic instructions (80, 84, 99). There is no reason to expect that two hairs from the head or two erythrocytes from the blood are due to the activity of different sets of genes.

If the development of repeated elements is only controlled by identical sets of genes, their genetic variation will be highly, if not perfectly, correlated. However, it has been shown that correlation caused by early developmental events can be repatterned during later developmental stages (115). To what extent repeated elements are genetically correlated in species belonging to a taxon that is ancestral to a species with differentiated homonomous structures is an empirical issue. Of relevance would be measurements of genetic and phenotypic correlations of corresponding elements from the left and the right side of the body, of segmentally repeated but undifferentiated structures such as fish vertebrae, or any other class of repeated elements, such as scales and fin rays of fishes.

The available evidence is equivocal. In a study of genetic correlations

Table 1 Apomorphic characters of the major mammalian taxa classifies according to the evolutionary transformation underlying their origin.[a]

Type of change	Apomorph character	Plesiomorph character	Taxon
Loss of elements	Erythrocytes without nucleus	Erythrocytes with nucleus	Mammalia
	Teeth absent	Teeth present	Monotremata
	Coracoid absent	Coracoid present	Theria
	No oil droplets in retinal cones	Oil droplets in retinal cones	Placentalia
	Marsupial bones absent	Marsupial bones present	Placentalia
Change of shape	Lateral temporal skull opening	No lateral temporal skull opening	Mammalia
	Cochlea bent	Cochlea straight	Mammalia
	Cochlea coiled	Cochlea not coiled	Theria
	Penis simple	Penis forked	Placentalia
Differentiation of repeated elements	Teeth heterodont	Teeth homodont	Mammalia
	Qualitative differentiation of cervical and thoracic vertebrae	Only gradual difference between cervical and thoracic vertebrae	Mammalia
	Some hairs specialized as whiskers	No whiskers	Theria

New elements	Marsupial bones	—	Mammalia
	Hair	—	Mammalia
	Muscular diaphragm	—	Mammalia
	Lips and facial musculature	—	Mammalia
	Glans penis	—	Mammalia
	Thigh glands	—	Monotremata
	Pseudo vagina	—	Marsupialia
	Corpus callosum	—	Placentalia
Change of context	Centrum of first vertebra fused to second	Centrum of first vertebra a part of first vertebra	Mammalia
	Secondary jaw joint	Primary jaw joint	Mammalia
	Angular (tympanic) fused to temporale	Angular part of lower jaw	Mammalia
	Yolk sack attached to uterus (placenta)	Yolk sack not attached to maternal body	Theria
	Separate opening of gut and urogenital sinus	Common opening of gut and urogenital sinus	Theria
	Scapular origin of supracoracoid muscle	Coracoidal origin of supracoracoid muscle	Placentalia
Combination of plesiomorph elements	Secondary palate	Separate maxillary processes and palatines	Mammalia
	Nipples	Dispersed external orifices of milk glands	Theria

[a] The list of characters is based on Ax (4). Several characters could not be classified, especially the physiological ones, such as homeothermy, sucking movements, vivipary, and the type of tooth replacement. Further, the histological type of enamel, the sweat glands, the phalangeal formula, the yolk content of the egg, and the location of the embryoblast relative to the trophoblast were not taken into account.

between bilaterally represented nonmetric cranial traits of rhesus macaques, McGrath et al (60) found high correlations between left and right characters and no significant heritability of directed asymmetry in 11 out of 13 traits. Phenotypic correlations between osteometric traits from fore- and hindlimbs of *Myotis sodalis* are higher among corresponding (homologous) structures within and between limbs, while the overall correlations were rather low (5). Repeated elements were also measured in fossil specimens of the teleost *Knightia* (length of centra of four vertebrae and three neural spines). The average pooled nonparametric correlation of these characters is 0.921 (centra) and 0.903 (neural spines), while the overall average correlation was 0.876 (calculated from data in Olson & Miller; 71). However, there are also less convincing results, e.g. about antennal segments of alate *Pemphigus populitransversus* (82, 95). Hence, some evidence suggests that repeated elements are strongly correlated, but the data are far from conclusive. This question will have to be examined with especially designed experiments, comparing species that have undifferentiated repeated traits with species in which differentiations of these traits have occurred.

If we assume that repeated characters are most probably highly correlated genetically, it becomes more difficult to explain the origin of new characters by differentiation of repeated elements. Technically speaking, the problem is that differentiation of repeated elements is a multivariate process, for which the univariate measures of heritability are inadequate to predict the evolutionary potential. Even if the heritability of each trait were positive, differentiation would be difficult as long as the characters are highly correlated genetically. This was shown by Maynard Smith & Sondhi (56) who demonstrated that it is impossible to select for directional asymmetry in laboratory strains of *Drosophila melanogaster*. All that selection led to was an increase in the level of fluctuating asymmetry; no stable difference between left and right could be achieved. Whether this result is representative of repeated elements in general needs to be tested with other characters, such as snake vertebrae or teleost fin rays.

An interesting fact is that the directional asymmetry of the internal organs in mammals can be converted into fluctuating asymmetry by a single autosomal recessive mutation in mice (47). This shows that specific mechanisms are necessary to realize directional asymmetry in addition to the genetic information required for the development of the traits themselves. The developmental mechanisms are unknown, but see Brown & Wolpert (10) for a recent hypothesis. It is at least not self-evident that there is always ample genetic variation, allowing selection to differentiate repeated elements. To what extent genetic variation is available in natural populations for independent heritable variation of homonomous traits needs to be examined.

New Elements

Often apomorphic characters are anatomical structures that have no predecessors as repeated elements in the plesiomorphic state. Examples are the corpus callosum of the placental mammals, the so-called marsupial bone of the mammals (which became reduced in the Placentalia), and glands, such as sweat glands and sebacous glands (Mammalia). At lower taxonomic levels examples are equally frequent. They include, to name a few, the famous thumb of the giant panda or the horns and antlers found among artiodactyl placentals. Again, these characters are of obvious functional and adaptive significance, but the main problem is whether one can expect significant amounts of heritable phenotypic variation for these characters in the ancestral lineage. For two characters, namely, the corpus callosum and new bony elements, extensive developmental data are available and are discussed below.

The corpus callosum is a massive fiber tract that connects the two telencephalic hemispheres of placental mammals. It is autapomorphic for the taxon Placentalia. In subplacental mammals, the telencephalic hemispheres are connected only via the anterior commissure. This commissure is also present in placental mammals, but the majority of cortical areas are connected via callosal connections (87). Embryologically, the corpus callosum is not derived from the rudiment of the anterior commissure (crossing the medial plane via the lamina terminalis) but is a new structure that bridges the interhemispheric fissure (77). The first cellular elements that bridge the gap between the hemispheres are a specific population of glial cells, called glial sling (92). If these glial cells are experimentally destroyed, the majority of callosal fibers fail to reach the contralateral side, and they never compensate by entering the anterior commissure (49, 50). Acallosal states are also known as congeneric malformations in humans and mice (28, 92). The independent embryological origin, its dependency on a specific set of radial glial cells, and the lack of regulation of the anterior commissure in acallosal brains speak for the fact that the corpus callosum is a true novelty and not simply a part of the anterior commissure.

The development of the corpus callosum passes through a critical stage, a kind of epigenetic needle's eye, where a certain population of glial cells must be present after the septal regions of the telencephalic hemispheres become fused (92). The glial sling is not known from marsupials and acallosal strains of mice. It is not reasonable to assume great amounts of heritable variation for the presence or absence of fibers that cross the interhemispheric fissure in species ancestral to the placentalia. Of course, at some time in the phylogeny of the placentals such a population must have existed, but it is not evident that the presence of the glial sling is within the range of variation typical for subplacental mammals. Some special but unknown conditions must have been attained in the placental lineage that allowed the expression of these characters.

A similar needle's eye situation has to be realized in the ontogeny of new bony elements. These conditions are best known from the fibular crest that appears in the archosaur lineage and are discussed later in this paper.

Other Nontrivial Novelties

Much less is known about the genetics and development of other characters that originate from a change of context (such as the separation of the angulare from the dentals and its fusion with the temporale in mammals), or from the synorganization of elements already present in the plesiomorph state. Few structures listed by Ax (4) are combinations of plesiomorphic characters, but such characters can be found in all higher taxa. For instance, multicellular epidermal mucous glands are rare in teleosts. In fish, mucus is usually produced by singular mucous cells. The multicellular glands seen in ripe male blennies are composed of goblet cells (40), a cell type usually found dispersed within the epidermis of fish (110). It would be highly interesting to know more about the developmental conditions necessary to realize these traits.

Common Features of Nontrivial Novelties

Differentiation of repeated elements and new elements such as new bones or new fiber tracts are certainly innovations with profound adaptive value. Hence, there is every reason to think that the fixation of these characters in the population was due to natural selection. This, however, does not solve the problem completely. In all these cases the main problem is to explain why and how heritable phenotypic variation for that character became available in the first place. Independent genetic variation of repeated elements is not always present and the critical embryological features necessary for the development of the corpus callosum are absent in primarily acallosal mammals.

Common to nontrivial novelties is their origin in spite of strong developmental constraints against their realization in the ancestral taxon. Developmental constraint on natural variation is a prevailing feature in morphological evolution (1, 55), but shifts of developmental constraints are quite common (81). For instance, in each salamander species the majority of carpal variants are due to one or two fusions between neighboring elements. But which of the fusions prevails is more or less genus specific. For instance, the fusion between the distal carpal 4 and 3 is a common variant in *Bolitoglossa* species and is even a fixed trait in at least two *Bolitoglossa* species (105), but it is completely unknown in natural populations of *Plethodon cinereus* from Maine and Virginia, and is very rare in the highly polymorphic Nova Scotian population (35). The ultimate causes of these apparent shifts of constraints are unknown.

Nontrivial novelties appear to become realized in spite of developmental constraints in the ancestral lineage. If one is willing to accept this premise, one must conclude that an adequate explanation of the origin of anatomical novelties has to account for the fact that these constraints were overcome at some stage of the phylogeny of the group.

A REFORMULATION OF THE NOVELTY PROBLEM

In this section we discuss a definition of morphological novelty that meets two objectives: (*a*) the definition is not based on assumptions about the mechanistic bases of novelties, since we are not in the position to provide an empirically justified and general explanation as yet; (*b*) the definition has to be specific enough to highlight the important unsolved biological problems.

If we consider the table of apomorphies discussed in the last section, one realizes that some of the apomorphies can hardly qualify as novelties. For instance, negative characters that result from the loss of certain elements cannot be considered as novelties. The same is true of size and shape characters. On the other hand, it is quite obvious that new elements, like the corpus callosum, or new bones and cartilages, are proper novelties. But there are other phenotypic variations that are difficult to classify as novelties or nonnovelties. This is the case with variation in the number of repeated elements, such as bristle number of an insect, or the number of vertebrae and fin rays. If a species has two more pectoral fin rays than the parental species, the two additional rays are something new. But do we want to call these additions novelties? In a certain sense they are, but one may also consider this meristic change as a case of quantitative variation (i.e. more of the same). What is then the difference between the additional digit of the panda and an additional bristle of a drosophila? The following definition is an attempt to avoid this dilemma.

DEFINITION *A morphological novelty is a structure that is neither homologous to any structure in the ancestral species nor homonomous to any other structure of the same organism*. This definition is less restrictive than previous ones (57, 64). In accordance with our considerations above, it excludes simple quantitative variation or negative traits. In addition, it allows a distinction between meristic variation, e.g. additional bristles or fin rays, and novelties like the marsupial bone or the panda's thumb. Additional bristles are both homologous to the bristles already present in the source population and homonomous to all other bristles on the same fly. But there is nothing that can be meaningfully identified in reptiles with the marsupial bone or in subplacental mammals with the corpus callosum.

The situation is more subtle with regard to other kinds of apomorphies classified in Table 1, e.g. the differentiation of repeated elements. Molar teeth are both homologous to the conodont teeth of reptiles and homonomous to the other tooth types of mammals. However, in these cases the hierarchical nature of homology (80) must be taken into account. The molars are homologous to conodont teeth of reptiles, but nevertheless, reptiles do not possess teeth that can be identified as molars. Hence, a "molar tooth" in mammals is a new anatomical entity that originated from the differentiation of preexisting repeated elements and thus counts as proper novelty.

The same argument holds for new structures that are composed of elements already present in the ancestral lineage. For instance, the main body parts (tagmata) of insects (head—thorax—abdomen) consist of segments already present in the annelid-like or myriapod-like ancestors of insects (94). But tagmata are units that result from the synorganization of several segments and cannot be identified with any body part of an annelid or a myriapod (102).

More problematic is the last category of apomorphies, those that result from a change of context. One may argue that the fusion of the centrum of the first cervical vertebra with the second cervical vertebra leads to an anatomical element that is a new unit of the phenotype, comparable to the case of multisegmental body parts. On the other hand, the fusion of the angular (tympanic) with the temporale does not change the character of the latter, since the angular simply becomes integrated into the preexisting unit. Without further information, these cases must be accepted as gray areas in the range of application of the above definition, but the difficulties point to interesting biological problems.

Although the definition helps to clarify the terminological question of what one may want to call a proper novelty and what is just a modification of the given design, it also leads to conceptual costs because of the reference to homology. The biological basis of homology is still a matter of debate and unfortunately of little positive evidence (85, 99, 104). But it is not necessary to wait for a solution to the homology problem. It is sufficient to rely on the accepted methods to establish homology between body parts on the basis of structural and developmental similarity (78, 80). Note that the homology concept used in this definition is more restrictive than the one used in systematics. In systematics, any discernable structural difference may be homologized. In evolutionary biology it is more useful to restrict the homology concept to anatomical units (104). This excludes merely quantitative variation, changes of proportion, and topological relationships among body parts.

To identify the relevant research questions, it is useful to recall that the set of characters described as novelties is, according to the above definition, the same as those apomorphies that became realized in spite of apparent developmental constraints in the ancestral lineage. In the light of this concordance, the most obvious questions in relation to the study of morphological novelties are the following:

1. What is the generative potential of the developmental mechanisms in the members of the ancestral taxon? Only in rare cases does the ancestral species still exist, but the conservatism of developmental mechanisms justifies the comparative analysis of species that are members of the same supraspecific taxon as the supposed ancestral species. Hence, it is appropriate to examine crocodilian development to learn about the generative potential of the ancestral bird lineage, or to study salamander development of the genus *Plethodon* to understand the preconditions for the evolution of more derived plethodontid taxa.

2. What are the critical changes in generative mechanisms of development that allowed the realization of the derived feature, i.e. the novelty? This can be achieved by comparative experimental studies of derived and ancestral ontogenies (66).

3. Which genetic changes were the reason for the heritability of morphological novelties? This is essentially the same question as raised by John & Miklos (38), but with an important methodological difference. We propose that we first need to understand the biological context in which the genes play a role, before a sensible distinction can be attempted between causally relevant genetic changes and genetic changes that simply happened to occur at the same stage of phylogeny, but that were not causative in the transformation to be explained. The least understood context of genetic change, but obviously the most relevant, is that of its developmental expression.

GENERATIVE MODES FOR THE ORIGIN OF MORPHOLOGICAL NOVELTY

Given that the emphasis of the open questions lies on developmental biology, we propose that the study of the developmental modes associated with the appearance of new characters is the critical step for further elucidation of the novelty problem. We have already presented the arguments for why this approach now seems more relevant than a genome-centered one. Here we identify particular properties of developmental systems that could promote the origination of novelty. Our approach, however, resides in a strictly neo-Darwinian frame, assuming that morphological evolution proceeds through gene substitutions that primarily affect cell behavior in developmental processes, leading primarily to changes in relative proportions and positions of embryonic characters. If these classic processes can produce novelties in the anatomical structure of organisms, one is led to hypothesize that the causality for their appearance lies in very basic and general properties of developmental systems that are affected by gene substitutions. We briefly review the evidence in support of this assumption.

Hierarchical Organization

It is commonplace to understand organisms as a hierarchy of building blocks from molecules to organs. However, with few exceptions (3, 7, 69, 74), evolutionary concepts rarely take into account that development, as the process of deployment of this hierarchical order, is itself organized largely hierarchically. Underlying are geometric hierarchies of cell and tissue organization, but also, and most importantly, hierarchies of stepwise successions of qualitatively different kinds of processes. The products of each step form the starting point for the next, and modifications introduced at one level of the developmental hierarchy can be assumed to have profound effects at very distant levels. For instance, the studies of sea urchin development mentioned above (75, 76, 113) show that the novelties in the larvae of direct developers are a consequence of very early modifications in cell lineage segregation, an alteration much higher up in the hierarchy of developmental decisions than the level of anatomical effect.

A similar and equally well-documented example comes from detailed comparative and experimental studies of spiralian development in protostomes. In some spiralian lineages novel larval types appear, such as the veliger of molluscs or the setiger of annelids. The work of Freeman & Lundelius (26) indicates that the origination of the derived larval types is dependent on a change of mechanism in early blastomere specification, the first major event in spiralian embryogenesis, establishing the axis of bilateral symmetry. This process is determined by the specification of the "D quadrant," the blastomere responsible for the formation of large parts of the mesodermal and endodermal structures of the embryo. In primitive forms the D quadrant is specified by inductive interactions between certain macromeres and micromeres that result from several sets of cleavages. In the derived forms the D quadrant is specified through cytoplasmic inheritance from the vegetal pole causing unequal cleavage and resulting in one of the first four macromeres being larger than the other three. This macromere invariably becomes the D quadrant, a sheer effect of size, which could be mimicked experimentally (26). The cytoplasmic specification of the D quadrant occurs earlier in the developmental sequence than the inductive specification, and it has a series of consequences down the hierarchy. The larger macromere gives rise to larger micromeres, and these lead to a further acceleration of development, which in turn results in the appearance of larvae with adult features in some lineages, while others lose the larval stages completely and become direct developers. Thus, in effect, larval morphology is profoundly altered through the acquisition of a mechanism that modifies the sequence of cell specification. The acceleration of D quadrant specification through cytoplasmic inheritance seems to have played a causal role in the origination of novel larval forms during spiralian evolution.

It is obvious that heterochrony has been an initiating factor in both examples, but the specificity of its phenotypic consequences depends on the hierarchical arrangement of the processes that were affected. However, the generative qualities of hierarchical organization lie not only in its cascading and amplifying effects. The hierarchical succession of processes also contains the possibility of changing qualitatively the patterns and structures of previous levels of organization. Each switch-over from one mechanism to the next represents an opportunity for structural change, a principle that has been proposed to underlie many qualitative transformations in morphological evolution (64). In avian limb development, for example, the switching from chondrogenesis to osteogenesis generates the unique tarsometatarsal bone from the cartilaginous rudiments of one tarsal and three metatarsals. Thus, the basic mechanisms of ontogenetic patterning can remain conserved in the evolution of an organismal lineage while the final phenotypes can be substantially altered through the expansion of secondary and tertiary level processes.

Interactivity and Dissociability

Developmental systems are characterized not only through sequential hierarchies but also by the interactivity between parts of different hierarchies, a condition described by the terms "epigenetic cascades" (34, 97) and "ontogenetic networks" (86). It is generally thought that an increase in the number of interactive events in which a morphological character takes part leads to an increasing phylogenetic stability of this trait. This forms the basis of the concepts of "burden" (80) and of "epigenetic traps" (104). With regard to the origin of novelties it is crucial whether and how interactive networks can be dissociated and whether new sets of interaction can be causal in the generation of new structures. We restrict the discussion to the cellular level.

The best understood epigenetic cascades lie in the domain of epithelial-mesenchymal interactions that are involved in the differentiation and patterning of a great number of anatomical structures, such as the inductive cascades leading to the formation of vertebrate eyes, limbs, and epidermal appendages (88). The variety of epidermal structures, all developmentally initiated by a similar sequence of inductions, is a good example of how the progressive elaboration of a primitive mechanism of integumental differentiation has led to the generation of greatly different structures, such as hair, glands, or teeth. This indicates that it is not so much the establishment of new kinds of interactions that is generatively important for the formation of new structures but the change of context in which the conservative and long established interactive mechanisms take place.

Not many empirical examples are available for the kind of contextual change that could have provoked new routes of interaction. Nevertheless, some of the few cases of novelty that were analyzed from a developmental perspective are instructive. One is the origin of the turtle carapace. The carapace is a unique association of ribs and vertebrae with a specialized dermis. This arrangement also represents a profound deviation from the tetrapod Bauplan because the elements of the limb girdles lie inside the rib cage, as opposed to an outside position in all other tetrapods. Studies of the developmental events that generate this arrangement indicate that epithelial-mesenchymal interactions, which when primitive produced only integumental features, were expanded to affect deeper layers of the mesenchyme (13). Through this mechanism the prospective costal cells are oriented toward a more lateral pathway than in other tetrapods, causing the superficial position of the ribs. The primary contextual change in this process seems to have been the timing of the epithelial-mesenchymal interaction. It takes place much earlier than the primitive interactions that lead to purely dermal differentiation and thus affects a much smaller embryo. Burke (13) suggests that the precocious inductive activity in a smaller embryo would have a relatively deeper penetration into the mesenchyme, reaching the skeletogenic cells that migrate from the somites, reorienting their pathway and causing the ribs to form superficially to the limb girdles.

A second instructive case is the origin of external cheek pouches in pocket gophers and kangaroo rats, a novelty in the evolution of rodents (52). In contrast to the primitive internal cheek pouches known from other rodent taxa, the external pouches open outside of the mouth cavity and their inner surface is not covered with buccal epithelium but with fur. Both pouch types arise from an invagination of the buccal epithelium of the oral cavity, close to the corner of the mouth. A detailed comparison of these processes shows that the externalization of the derived pouch types is initiated developmentally by a slight anterior shift of the invagination, leading to the inclusion of the lip epithelium (11). As a consequence the pouch not only acquires an external opening, but the epithelium of the pouch rudiment grows into a dermal environment which has the capacity to induce hair follicle formation. Fur-lining of the pouch is thus a consequence of a change of context, i.e. a shift of invaginated epithelium into an area with inductive capacity. The shift itself is possibly a mere allometric consequence of evolutionary modifications in facial proportions.

Both examples illustrate that a change of context, initiated by temporal or spatial shifts, can lead to the formation of novel morphologies on the basis of preexisting interactive capacities. The historically established networks of developmental interactivity, in particular those of epithelial-mesenchymal inductions, thus not merely constrain morphological evolution, they also represent an important generative source for the origination of new structures.

Equilibria and Thresholds

Ontogenies can be understood as systems of temporary equilibria or steady states between developmental entities (7, 53, 69, 100). This is not the place to discuss the various formalisms that were developed in this regard, but we want to emphasize the principal importance of these properties for the origin of novelty. They explain why continuous variation of developmental parameters can result in discontinuous phenomena. Upon transgression of certain thresholds a developmental system can fall into a different steady state resulting in different phenotypic expressions. Thresholds must and do exist at all levels of development and have been demonstrated in a variety of experiments (e.g. 8, 22, 32, 98). Conceptually, the realization of discontinuous forms of morphological variation has been attributed to thresholds in development (48), and polygenic models of digital reduction have been based on threshold concepts (43). Only recently, however, has it been proposed that threshold effects may represent a generative factor in the origination of morphological novelties (64, 66).

Developmental thresholds can lie in molecular and physical parameters of pattern formation, in critical cell number or blastema size, in inductive or spatial relationships, in physiological or biomechanical factors, etc. A spatial threshold effect, for example, was proposed to have initiated the formation of external cheek pouches discussed above (11). Here, we shall focus on simple biomechanical changes that are associated with continuous developmental variation. It is well known that embryonic movement is required for the formation of sesamoids and of secondary cartilage (33). We can assume that evolutionary changes in the proportions of embryonic structures also modify pressures and tensions that arise from embryonic movements. As a consequence, when these biomechanical forces transgress a threshold intensity, we should expect the appearance of sesamoid cartilages in connective tissue structures or of secondary cartilage in the vicinity of dermal bone. These reactive structures provide an important source of skeletal novelty and can be elaborated during the further course of evolution. That this is indeed the case is supported by a large number of cases in which skeletal neomorphs are based on sesamoids or on secondary ossifications (Table 2).

An example studied in more detail is the fibular crest of theropod dinosaurs (67). The fibular crest is a neomorph on the tibia that appears first in theropod dinosaurs and is synapomorphic in birds. Developmentally, the formation of the osseous crest is based on a separate cartilaginous sesamoid that is later ossified and eventually becomes incorporated into the tibia. Paralysis experiments in bird embryos demonstrate the dependence of the sesamoid's formation on embryonic movement and the consecutive loss of the crest in paralyzed embryos. Müller & Streicher (67) propose a scenario in which the

Table 2 Examples of skeletal novelties in vertebrates that are based on reactive cartilage and bone formation.

Skeletal novelty	Taxon	Based on
Fibular crest	Theropods, birds	Sesamoid
Preglossale	Passerine birds	?
Panda's "thumb"	Panda bears	Sesamoid
Panda's "7th digit"	Panda bears	Sesamoid
Rüsselknochen	Boars	?
Calcar	Bats	?
Falciforme	Moles	Sesamoid
Third forearm bone	Golden mole	Ossified tendon
Naviculare	Horses	Sesamoid
Patella	Birds, mammals	Sesamoid

evolutionary reduction of the reptilian fibula leads to an increased mechanical instability during embryonic movement of the limbs, which could have initiated the formation of the sesamoid, on the basis of the reactive potential of connective tissue to form cartilage under pressure stresses. The origination of this novelty is thus based on a number of very specific conditions, such as skeletal proportions, biomechanical changes, and the reactive potential of connective tissues.

We are aware that the formal separation of the three generative modes is to some extent artificial. Most examples would fit into all three categories. However, we do believe that these are three fundamental properties of ontogenetic systems that must be taken into account when we think about evolutionary modifications of developmental parameters and their role in the origination of novelty. Common to all three modes is their potential for rapid morphological transitions, and the fact that their effects have an indirect and removed relation to the level of genome evolution.

FROM NOVELTY TO BAUPLAN

A discussion of evolutionary novelties would be incomplete without mentioning the most profound innovations that occurred in the history of life—the origin of the basic design principles underlying the major supraspecific taxa, i.e. the bauplans of phyla and classes. So far, we have been concerned only with the origin of new morphological characters but not with the origin of supraspecific taxa, even if this is often considered as the same problem (27). While the origin of new body plans and the origin of new characters are linked processes, they are not necessarily the same. This insight is mainly due to Riedl (80), and we discuss his concept below.

The notochord is an axial rod of cells representing the functional precursor of the vertebral column, both ontogenetically and phylogenetically. In mammals this structure has lost most of its adult function and persists only rudimentarily as the nucleus pulposus of the intervertebral discs. Nevertheless, the notochord is present in all ascidian larvae, in *Amphioxus,* and in the embryos of all vertebrates. The stability of this structure is best explained by its central role in embryogenesis, in organizing the differentiation of the central nervous system and of the axial mesoderm. Originally, however, the notochord was not as indispensable as it is for the derived members of the phylum. This is indicated by the complete lack of a notochord in two orders of the chordate class Thaliacea, which belongs to the subphylum of tunicates. The fact that the members of one order of Thaliacea, the Doliolida, do possess a notochord, indicates that it is most likely secondarily lost in the other orders.

Here the main point is that the notochord is a constant character of the acranian and vertebrate bauplan, but hardly was a bauplan character when it first arose. The essential characteristic of a bauplan is not the degree of similarity or dissimilarity to other forms of life, but the fact that each group of animals has its own characteristic patterns of constraints and opportunities. According to Riedl (80, p. 196), a bauplan (or morphotype) is defined by the "*pattern of freedom and fixations [constraints] formed by the collective of features of a phyletic group.*" From this definition it is obvious that the origin of a new character is not sufficient to change a bauplan. Only if the new character achieves an indispensable function, and becomes epigenetically integrated into the basic body design, does it become a bauplan character. The origin of new body plans requires the origin of morphological novelties, but it also requires the integration of this new character with the other parts of the organism. In this context it is irrelevant whether integration is due to functional necessities or due to epigenetic interdependencies. What counts is that some characters acquire an indispensable biological role that causes their conservation in spite of changing adaptive pressures.

CONCLUSIONS

Morphological novelty has the status of a distinct problem in evolutionary biology. Novelties are not synonymous with all taxonomically relevant apomorphies, and their emergence is not identical with the process of speciation or with the origin of novel body plans. Once new variants have occurred, their fixation by drift or selection is easily explained. But there are problems specific for the origin of novelties that are not the same as in the case of adaptive modifications of existing structures, namely the developmental realization of novelties depends on very specific epigenetic conditions. For these, no significant amounts of heritable variation have been demonstrated in taxa related to the ancestral groups. To the contrary, novelties apparently arise in spite of strong developmental constraints that generally canalize morphological evolution.

We conclude that the problem of novelty must be considered from a new perspective in order to be able to formulate adequate research questions. At the organismic level, morphological evolution can be described as a process of progressive origination, transformation, and loss of homologs. Therefore, we suggest a definition of novelty that is framed in the homology concept. The main properties of the definition are that it is independent from descriptive or mechanistic qualifiers, that it excludes merely quantitative or negative traits, and that it allows distinction between meristic variation and true novelties.

The new questions that arise from an organismic definition concentrate on the mechanistic basis of their generation. The genetic side of the generative problem does not seem to differ substantially from the classic mechanisms, and does not hold much promise for further advances with regard to the novelty problem. The majority of open questions, and the greatest potential for an increase in our understanding of novelty, lie in the realm of the developmental context in which genetic changes can trigger a change of structure. It is unlikely that explanations for the origin of morphological novelties can be successful without the inclusion of the generative properties of developmental systems.

A preliminary overview of the developmental modes associated with the origination of novelties point to a central role of heterochrony as the primary initiating factor. Heterochrony alone, however, can only modify processes that are already established. The specific morphological composition of novelties that arise as a consequence of heterochronic alterations of a developmental process will depend on the particular organization of the developmental network of which the process is a part. Hierarchical organization, interactive interdependency, and equilibrium conditions are basic properties of all developmental systems that will invariably be affected. Evolutionary modifications of any part of these systems that go beyond specific thresholds can automatically cause morphological effects that are only indirectly related to the causes of the primary modification. By-products of development will be "seen" by selection and can be further elaborated through neo-Darwinian processes. We need to learn through experimental and comparative studies what specific potentials exist in the developmental systems of an organismic lineage, to be able to identify the individual causes that lead to a particular novelty in evolution. In general, however, the available data strongly suggest that side effects of developmental organization represent the kernel of morphological novelty.

Literature Cited

1. Alberch, P. 1982. Developmental constraints in evolutionary processes. In *Evolution and Development*. ed. J. T. Bonner, pp. 313–32. Berlin: Springer-Verlag
2. Alberch, P. 1982. The generative and regulatory roles of development in evolution. In *Environmental Adaptation and Evolution*. ed. D. Mossakowski, G. Roth, pp. 19–36. Stuttgart: Gustav Fischer Verlag
3. Arthur, W. 1984. *Mechanisms of Morphological Evolution*. Chichester: Wiley
4. Ax, P. 1984. *Das phylogenetische System*. Stuttgart: Gustav Fischer Verlag
5. Bader, R. S., Hall, J. S. 1960. Osteometric variation and function in bats. *Evolution* 14:8–17
6. Barton, N. H., Charlesworth, B. 1984. Genetic revolutions, founder events, and speciation. *Ann. Rev. Ecol. Syst.* 15: 133–64
7. Bertalanffy, L. 1952. *Problems of Life*. New York: Harper & Brothers
8. Bretscher, A., Tschumi, P. 1951. Gestufte Reduktion von chemisch behandelten *Xenopus*-Beinen. *Rev. Suisse Zool.* 58:391–98
9. Britten, R. J., Davidson, E. H. 1971. Repetitive and non-repetitive DNA and a speculation on the origin of evolutionary novelty. *Q. Rev. Biol.* 46:111–33
10. Brown, N. A., Wolpert, L. 1990. The development of handedness in left/right asymmetry. *Development* 109:1–9
11. Brylski, P., Hall, B. K. 1988. Ontogeny of a macroevolutionary phenotype: The external cheek pouches of geomyoid rodents. *Evolution* 42:391–95
12. Bürger, R. 1986. Constraints for the evolution of functionally coupled characters: A nonlinear analysis of a phenotypic model. *Evolution* 40:182–93
13. Burke, A. C. 1989. Epithelial-mesenchymal interactions in the development of the chelonian Bauplan. In *Trends in Vertebrate Morphology*, ed. H. Splechtna, H. Hilgers, pp. 206–09. Stuttgart: Gustav Fischer Verlag
14. Buss, L. 1987. *The Evolution of Individuality*. Princeton: Princeton Univ. Press
15. Campbell, K. S. W., Day, M. F. 1987. *Rates of Evolution*. London: Allen & Unwin
16. Charlesworth, B., Lande, R., Slatkin, M. 1982. A neo-Darwinian commentary on macroevolution. *Evolution* 36:474–98
17. Charlesworth, B., Rouhani, S. 1988. The probability of peak shifts in a founder population II. An additive polygenic trait. *Evolution* 42:1129–45
18. Cracraft, J. 1990. The origin of evolutionary novelties: Pattern and process at different hierarchical levels. In *Evolutionary Innovations*. ed. M. H. Nitecki, pp. 304. Chicago: Univ. Chicago Press
19. Crow, J. F., Engels, W. R., Denniston, C. 1990. Phase three of Wrights shifting-balance theory. *Evolution* 44:233–47
20. Darwin, C. 1859. *On the Origin of Species by Means of Natural Selection, or Preservation of Favoured Races in the Struggle for Life*. London: Murray
21. De Beer, G. 1958. *Embryos and Ancestors*. Oxford: Oxford Univ. Press
22. Doherty, P., Fruns, M., Seaton, P., Dickson, G., Barton, C. H., et al. 1990. A threshold effect of the major isoforms of NCAM on neurite outgrowth. *Nature* 343:464–66
23. Dohrn, A. 1875. *Prinzip des Funktionswechsels*. Leipzig: Engelmann
24. Dover, G. 1986. Molecular drive in multigene families: How biological novelties arise, spread and are assimilated. *Trends Genet* 2:159–65
25. Erwin, D. H., Valentine, J. W. 1984. "Hopeful monsters," transposons, and metazoan radiation. *Proc. Natl. Acad. Sci. USA* 81:5482–83
26. Freeman, G., Lundelius, J. W. 1991. Evolutionary implications of the mode of D quadrant specification in coelomates with spiral cleavage. *J. Evol. Biol.* In press
27. Futuyma, D. J. 1986. *Evolutionary Biology*. Sunderland: Sinauer
28. Gazzaniga, M. S. 1970. *The Bisected Brain*. New York: Appleton-Century-Crofts
29. Goldschmidt, R. 1940. *The Material Basis of Evolution*. New Haven: Yale Univ. Press
30. Gould, S. J., Vrba, E. S. 1982. Exaptation—a missing term in the science of form. *Paleobiology* 8:4–15
31. Grant, V. 1963. *The Origin of Adaptations*. New York: Columbia Univ. Press
32. Grüneberg, H. 1952. Genetical studies on the skeleton of the mouse. *J. Genet.* 51:95–114
33. Hall, B. K. 1986. The role of movement and tissue interactions in the development and growth of bone and secondary cartilage in the clavicle of the embryonic

chick. *J. Embryol. Exp. Morph.* 93: 133–52
34. Hall, B. K., Hörstadius, S. 1988. *The Neural Crest.* Oxford: Oxford Univ. Press
35. Hanken, J., Dinsmore, C. E. 1986. Geographic variation in the limb skeleton of the red-backed salamander, *Plethodon cinereus. J. Herpetol.* 20:97–101
36. Hertwig, O. 1916. *Das Werden der Organismen.* Jena: Gustav Fischer Verlag
37. Huxley, J. 1942. *Evolution, the Modern Synthesis.* London: Allen & Unwin
38. John, B., Miklos, G. L. 1988. *The Eukaryote Genome in Development and Evolution.* London: Allen & Unwin
39. Kermack, D. M., Kermack, K. A. 1984. *The Evolution of Mammalian Characters.* Washington, DC: Kapitan Szabo
40. Kotrschal, K., Weise, H., Goldschmid, A. 1984. Mehrzellige Drüsen in der Epidermis der unpaaren Flossen bei den Blenniiden. *Z. Mikrosk. Anat. Forsch.* 98:184–92
41. Lamarck, J. B. 1809. (1984) (Trans. H. Elliot) *Zoological Philosophy.* Chicago: Univ. Chicago Press
42. Lande, R. 1976. Natural selection and random drift in phenotypic evolution. *Evolution* 30:314–34
43. Lande, R. 1978. Evolutionary mechanisms of limb loss in tetrapods. *Evolution* 32:73–92
44. Lande, R. 1983. The response to selection on major and minor mutations affecting a metrical trait. *Heredity* 50:47–65
45. Lande, R. 1985. Expected time for random genetic drift of a population between stable phenotypic states. *Proc. Natl. Acad. Sci. USA* 82:7641–45
46. Larson, A., Wake, D. B., Maxson, L. R., Highton, R. 1981. A molecular phylogenetic perspective on the origins of morphological novelties in the salamanders of the tribe plethodontini (Amphibia, Plethodontidae). *Evolution* 35:405–22
47. Layton, W. M. 1976. Random determination of a developmental process. *J. Hered.* 67:336–38
48. Lehmann, F. E. 1953. Konkurrenz- und Schwelleneffekte bei der Realisierung von Körper- und Organgestalten. *Rev. Suisse Zool.* 60:490–96
49. Lent, R. 1983. Cortico-cortical connections reorganize in hamsters after neonatal transaction of the callosal bridge. *Dev. Brain Res.* 11:137–42
50. Lent, R. 1984. Neuroanatomical effects of neonatal transaction of the corpus callosum in hamsters. *J. Comp. Neurol.* 223:548–55
51. Liem, K. F. 1974. Evolutionary strategies and morphological innovations: Chichlid pharyngeal jaws. *Syst. Zool.* 20:425–41
52. Long, C. A. 1976. Evolution of mammalian cheek pouches and a possibly discontinuous origin of a higher taxon (Geomyoidea). *Am. Nat.* 110: 1093–1111
53. Lotka, A. J. 1925. *Elements of Physical Biology.* Baltimore: Williams & Wilkins
54. Maynard Smith, J. 1983. The genetics of stasis and punctuation. *Annu. Rev. Genet.* 17:11–25
55. Maynard Smith, J., Burian, R., Kauffman, S., Alberch, P., Campbell, J., et al. 1985. Developmental constraints and evolution. *Q. Rev. Biol.* 60:265–87
56. Maynard Smith, J., Sondhi, K. C. 1960. The genetics of a pattern. *Genetics* 45:1039–50
57. Mayr, E. 1960. The emergence of evolutionary novelties. In *Evolution After Darwin.* ed. S. Tax, pp. 349–80. Chicago: Univ. Chicago Press
58. Mayr, E. 1976. *Evolution and the Diversity of Life.* Cambridge: Harvard Univ. Press
59. Mayr, E. 1982. *The Growth of Biological Thought.* Cambridge: Harvard Univ. Press
60. McGrath, J. W., Cheverud, J. M., Buikstra, J. E. 1984. Genetic correlations between sides and heritability of asymmetry for nonmetric traits in Rhesus Macaques on Cayo Santiago. *Am. J. Phys. Anthropol.* 64:401–11
61. McKinney, M. L. 1988. *Heterochrony in Evolution.* New York: Plenum
62. Mivart, S. G. J. 1871. *The Genesis of Species.* London
63. Mousseau, T. A., Roff, D. A. 1987. Natural selection and the heritability of fitness components. *Heredity* 59:181–97
64. Müller, G. B. 1990. Developmental mechanisms at the origin of morphological novelty: A side-effect hypothesis. In *Evolutionary Innovations,* ed. M. H. Nitecki, pp. 99–130. Chicago: Univ. Chicago Press
65. Müller, G. B. 1991. Evolutionary transformation of limb pattern: Heterochrony and secondary fusion. In *Developmental Patterning of the Vertebrate Limb,* ed. J. R. Hinchliffe, J. Hurle, D. Summerbell. London: Plenum. In press
66. Müller, G. B. 1991. Experimental strat-

egies in evolutionary embryology. *Am. Zool.* 31: In press

67. Müller, G. B., Streicher, J. 1989. Ontogeny of the syndesmosis tibiofibularis and the evolution of the bird hindlimb: A caenogenetic feature triggers phenotypic novelty. *Anat. Embryol.* 179:327–39
68. Nägeli, C. 1884. *Mechanisch-physiologische Theorie der Abstammungslehre*. Leipzig: Oldenbourg
69. Needham, J. 1936. *Order and Life*. Cambridge: M.I.T. Press
70. Nitecki, M. H. 1990. *Evolutionary Innovations*. Chicago: Univ. Chicago Press
71. Olson, E. C., Miller, R. L. 1958. *Morphological Integration*. Chicago: Univ. Chicago Press
72. Oster, G. F., Shubin, N., Murray, J. D., Alberch, P. 1988. Evolution and morphogenetic rules: The shape of the vertebrate limb in ontogeny and phylogeny. *Evolution* 42:862–84
73. Plate, L. 1913. *Selektionsprinzip und Probleme der Artbildung*. Leipzig: Engelmann
74. Raff, R. A., Kaufman, C. 1983. *Embryos, Genes, and Evolution*. New York: Macmillan
75. Raff, R. A., Parr, B. A., Parks, A. L., Wray, G. A. 1990. Heterochrony and other mechanisms of radical evolutionary change in early development. In *Evolutionary Innovations*. ed. M. H. Nitecki, pp. 71–98. Chicago: Univ. Chicago Press
76. Raff, R. A., Wray, G. A. 1989. Heterochrony: Developmental mechanisms and evolutionary results. *J. Evol. Biol.* 2:409–34
77. Rakic, P., Yakovlev, P. I. 1968. Development of the corpus callosum and cavum septi in man. *J. Comp. Neurol.* 132:45–72
78. Remane, A. 1971. *Die Grundlagen des natürlichen Systems der vergleichenden Anatomie und der Phylogenetik*. Königstein-Taunus: Koeltz
79. Rensch, B. 1959. *Evolution Above the Species Level*. New York: Columbia Univ. Press
80. Riedl, R. 1978. *Order in Living Organisms*. Chichester: Wiley
81. Rienesl, J., Wagner, G. P. 1991. Constancy and change of basipodal variation patterns: A comparative study of crested and marbled newts and their natural hybrids. *J. Evol. Biol.* In press
82. Riska, B. 1985. Group size factors and geographic variation of morphometric correlation. *Evolution* 39:792–803
83. Romanes, G. J. 1897. *Darwin, and After Darwin*. Chicago: Open Court
84. Roth, V. L. 1991. Homology and hierarchies: Problems solved and unsolved. *J. Evol. Biol.* 4:167–94
85. Roth, V. L. 1984. On homology. *Biol. J. Linnean Soc.* 22:13–29
86. Sander, K. 1983. The evolution of patterning mechanisms: Gleanings from insect embryogenesis and spermatogenesis. In *Development and Evolution*, ed. B. C. Goodwin, N. Holder, C. C. Wylie, pp. 137–59 Cambridge: Cambridge Univ. Press
87. Sarnat, H. B., Netsky, M. G. 1981. *Evolution of the Nervous System*. New York: Oxford Univ. Press
88. Sawyer, R. H., Fallon, J. F. 1983. *Epithelial-Mesenchymal Interactions in Development*. New York: Praeger
89. Schindewolf, O. H. 1950. *Grundfragen der Paläontologie*. Stuttgart: Schweizerbart
90. Schmalhausen, I. I. 1949. *Factors of Evolution*. Philadelphia: Blakiston
91. Severtzoff, A. N. 1931. *Morphologische Gesetzmäßigkeiten der Evolution*. Jena: Gustav Fischer
92. Silver, J., Lorenz, S. E., Wahlsten, D., Coughlin, J. 1982. Axonal guidance during development of the great cerebral commissures: Descriptive and experimental studies in vivo, on the role of preformed glial pathways. *J. Comp. Neurol.* 210:10–29
93. Simpson, G. G. 1953. *The Major Features of Evolution*. New York: Columbia Univ. Press
94. Snodgrass, R. E. 1935. *Principles of Insect Morphology*. New York: McGraw-Hill
95. Sokal, R. R. 1962. Variation and covariation of characters of alate *Pemphigus populitransversus* in eastern North America. *Evolution* 16:227–45
96. Stanley, S. M. 1979. *Macroevolution*. San Francisco: Freeman
97. Thorogood, P. V. 1983. Morphogenesis of cartilage. In *Cartilage*. Vol. 2, ed. B. K. Hall, pp. 223–54. New York: Academic
98. Tschumi, P. 1953. Ontogenetische Realisationsstufen der Extremitäten bei *Xenopus* und die Interpretation phylogenetischer Strahlenreduktionen bei Wirbletieren. *Rev. Suisse Zool.* 60:496–509
99. Van Valen, L. 1982. Homology and causes. *J. Morphol.* 173:305–12
100. Waddington, C. H. 1941. Evolution of developmental systems. *Nature* 147: 108–10

101. Wagner, G. P. 1984. Coevolution of functionally constrained characters: Prerequisites for adaptive versatility. *BioSystems* 17:51–5
102. Wagner, G. P. 1986. The systems approach: An interface between development and population genetic aspects of evolution. In *Patterns and Processes in the History of Life,* ed. D. M. Raup, D. Jablonski, pp. 149–65. Berlin: Springer-Verlag
103. Wagner, G. P. 1988. The influence of variation and of developmental constraints on the rate of multivariate phenotypic evolution. *J. Evol. Biol.* 1:45–66
104. Wagner, G. P. 1989. The origin of morphological characters and the biological basis of homology. *Evolution* 43:1157–71
105. Wake, D. B. 1966. Comparative osteology and evolution of the lungless salamanders, family Plethodontidae. *Mem. So. Calif. Acad. Sci.* 4:1–111
106. Wake, D. B., Roth, G. 1989. The linkage between ontogeny and phylogeny in the evolution of complex systems. In *Complex Organismal Functions: Integration and Evolution in Vertebrates.* ed. D. B. Wake, G. Roth, pp. 361–77. New York: Wiley
107. Wake, D. B., Roth, G. 1989. *Complex Organismal Functions: Integration and Evolution in Vertebrates.* New York: Wiley
108. Weismann, A. 1909. The selection theory. In *Darwin and Modern Science,* ed. A. C. Seward, pp. 18–65. Cambridge: Cambridge Univ. Press
109. West-Eberhard, M. J. 1986. Alternative adaptations, speciation, and phylogeny. *Proc. Natl. Acad. Sci. USA* 83:1388–92
110. Whitear, M. 1986. The skin of fishes including cyclostomes: Epidermis. In *Biology of the Integument.* ed. J. Bereiter-Hahn, A. G. Matoltsy, K. S. Richards, pp. 8–38. Heidelberg: Springer Verlag
111. Wiley, E. O. 1981. *Phylogenetics.* New York: Wiley
112. Wilkens, H. 1971. Genetic interpretation of regressive evolutionary processes: Studies on hybrid eyes of two *Astyanax* cave populations. *Evolution* 25:530–44
113. Wray, G. A., Raff, R. A. 1990. Novel origins of lineage founder cells in the direct-developing sea urchin *Heliocidaris erythrogramma. Dev. Biol.* 141:41–54
114. Wright, S. 1931. Evolution in Mendelian populations. *Genetics* 16:97–159
115. Zelditch, M. L. 1988. Ontogenetic variation in patterns of phenotypic integration in the laboratory rat. *Evolution* 42:28–41

Annu. Rev. Ecol. Syst. 1991. 22:257–79

EFFECTS OF NITROGEN LOADING ON WETLAND ECOSYSTEMS WITH PARTICULAR REFERENCE TO ATMOSPHERIC DEPOSITION

James T. Morris

Department of Biological Sciences, University of South Carolina, Columbia, South Carolina 29208

KEY WORDS: marsh, bog, species diversity, productivity, eutrophication

INTRODUCTION

Wetlands fulfill an important role in global biogeochemical cycles. For example, they transfer to the atmosphere globally significant quantities of CH_4 (60, 61) and reduced sulfur gases (127). Coastal marshes may function as net sinks for N_2O (46). Because of the anaerobic nature of their soils, wetlands act as sinks for organic carbon. It has been estimated that wetlands once sequestered a net of 57 to 83 $\times$ 10^6 metric tons of carbon per year worldwide, but recent widespread drainage of wetland soils has shifted the carbon balance (7). Although this rate of carbon uptake is small in comparison to other global carbon fluxes, such as the annual release of carbon from combustion of fossil fuel (5–6 $\times$ 10^9 metric tons yr^{-1}) (112) or the net uptake of CO_2-C by the ocean (1.6 $\times$ 10^9 metric tons yr^{-1}) (129), it is important when the net balance between large fluxes is considered, and it is certainly important over geologic time scales.

Locally, wetlands function as habitats for wildlife, as flood control systems, as stabilizers and sinks for sediments, as storage reservoirs for water, and as biological filters that maintain water quality. For example, riparian

0066-4162/91/1120-0257$02.00

forests exert a positive influence on the water quality of receiving streams by intercepting and removing nutrients from runoff (22, 99, 103, 149). As sediment traps, salt marshes like those on the Louisiana coast can accumulate annually an impressive 0.76 cm of sediment (41).

Wetlands also harbor a disproportionate (relative to habitat area) share of flora that is threatened by extinction. Of the 130 plant species from the conterminous United States that are formally listed as endangered or threatened (33), 14% occur principally in wetland habitats. On the national list of species identified as endangered, threatened, or potentially threatened, 1776 species are listed for the conterminous United States (48), and 17% of these occur principally in wetland habitats. From the national list of 6728 facultative and obligate wetland plant species (106), we can estimate conservatively that 4.5% of wetland plants are endangered or potentially threatened.

In the United States where it has been largely a consequence of agricultural development involving drainage (131), habitat loss is an important cause of species endangerment. Total wetland area including intertidal and palustrine areas in the conterminous United States totaled 437,609 km^2 during the mid-1950s and decreased to 400,567 km^2, or 5.1% of total land area, by the mid-1970s (53). The net loss of wetland habitats during these two decades was equivalent to 1852 km^2/yr.

A less obvious cause of species endangerment, and one which may change the function of wetland ecosystems, is nitrogen enrichment of the environment by atmospheric deposition and other causes. The rise in anthropogenic releases of nitrogen oxides and ammonia to the atmosphere increases the deposition of biologically available forms of nitrogen onto the landscape with potential effects on productivity (or other aspects of function) and community structure. Current rates of atmospheric nitrogen deposition in parts of Europe are great enough to alter the competitive relationships among plants and threaten wetland species adapted to infertile habitats. This paper reviews the effects of nitrogen eutrophication of wetland ecosystems with an emphasis on assessing the impacts of atmospheric nitrogen deposition.

ATMOSPHERIC NITROGEN INPUTS

Atmospheric nitrogen inputs occur as both wet (in precipitation) and dry (sorption of nitrogen gases by wet surfaces and particle deposition) deposition. Most studies of atmospheric nitrogen inputs to wetlands focus only on wet deposition or bulk deposition; the latter measure combines wet deposition with some component of dry deposition. The rate of bulk NO_3^- deposition correlates positively with the concentration of NO_2 in air (102), although bulk deposition varies among different ecosystems according to the type and density of foliage.

Where dry deposition has been carefully measured, it has been concluded that the relative importance of wet and dry deposition varies geographically, that dry deposition can exceed wet deposition (13), and that bulk precipitation samplers underestimate the combined dry plus wet deposition rate (45). The available wet surface area of vegetation, onto which nitrogen gases will diffuse, significantly affects the dry deposition rate (62). Dry deposition accounts for more than one half of the total nitrogen oxide deposition in North America, an estimate based on a model of the fate of nitrogen oxide emissions to the atmosphere (79).

A third, and rarely measured, mechanism of deposition that is locally important is the capture of fog or cloud droplets by vegetation. Cloud deposition of NO_3^- in an alpine habitat in New Hampshire was 10.15 g N m^{-2} yr^{-1}, compared to a bulk deposition rate of 2.34 g N m^{-2} yr^{-1} (83). Similarly, the fact that 45% more water was collected as throughflow passing through *Sphagnum* mats in an English bog compared to adjacent precipitation gauges suggests that fog capture provides a significant source of solutes (147).

Table 1 summarizes several studies that report rates of wet or bulk N deposition in North American wetlands. From the data presented it may be concluded that bulk deposition of NH_4^+, NO_3^-, and organic nitrogen varies geographically and that their relative importance varies. In general, however, inputs of NO_3^-, NH_4^+ and organic nitrogen are all of the same order of magnitude, and their combined rate of deposition varies from 0.55 to 1.21 g N m^{-2} yr^{-1}. Exceptions may occur in the midwestern and northeastern United States where bulk deposition of NO_3^- alone may be greater (151).

Rates of nitrogen deposition, and NH_4^+ deposition in particular, in areas of Western Europe are greater than in North America. Bulk deposition in some areas of England exceeds 4.0 g m^{-2} yr^{-1} (50, 101). The combination of NO_3^- and NH_4^+ deposition downwind of Manchester and Liverpool is

Table 1 Bulk deposition of nitrogen (g N m^{-2} yr^{-1}) in North American wetlands

Site	NH_4^+	NO_3^-	Org-N	Tot-N	Reference
Chesapeake Bay, riverine tidal emergent marsh	0.27	0.43	0.47	1.17	68
Massachusetts, salt marsh	0.14	0.23	0.39	0.76	134
Massachusetts, basin bog	0.25	0.50	NR		63
Minnesota, spruce bog	0.17	0.17	0.38	0.73	144
Minnesota, spruce bog	0.30	0.20	0.05	0.55	132
Iowa, prairie marsh	0.40	0.40	NR		37
Florida, Everglades	0.30	0.96	NR		51
Manitoba, emergent marsh	NR	NR	NR	0.66–1.21	69
Ontario, poor fen	NR	0.31	NR		8

NR = not reported

reported to be 3.2 g N m^{-2} yr^{-1} (77). Nitrogen deposition in fens near Utrecht, Netherlands was 2.1 g N m^{-2} yr^{-1} of inorganic nitrogen, 0.3 to 0.5 g N m^{-2} yr^{-1} of organic nitrogen in bulk precipitation, and 1.8 g N m^{-2} yr^{-1} of inorganic nitrogen in dry deposition (75). Wet deposition of nitrogen in the Netherlands averages 1.5 g N m^{-2} yr^{-1} and is as great as 2–6 g N m^{-2} yr^{-1} in areas of intensive stockbreeding; 75–90% of this is deposited as NH_4^+ (108). In Europe, 81% of total NH_3 emissions are from livestock wastes, with the greatest emission densities concentrated in the Netherlands and Belgium (24). Annual NH_3 emissions from animal excreta in the Netherlands are reported to be 230 kt yr^{-1} (138) or about 6 g m^{-2} yr^{-1} country-wide.

The chemistry and quantity of surface runoff from watersheds is probably of greater significance to most wetlands, because the nitrogen load of surface runoff can greatly exceed that deposited directly. However, atmospheric deposition accounts for a large fraction of the total nitrogen entering watersheds (107) and has apparently become a major source of NO_3^- to surface waters in North America, especially in the east and upper midwest (122). Although increases in total nitrogen concentration at stream monitoring stations are strongly associated with high levels of atmospheric NO_3^- deposition (123), the actual contribution of atmospheric deposition to the nitrogen load in surface water is unknown. Measurements of stream chemistry in Georgia indicated that 93% of the precipitation inputs of NH_4^+ and NO_3^- were retained by the watershed (23). A study of mass nutrient balances of a small watershed of the Rhode River estuary, Chesapeake Bay showed that total wet nitrogen deposition to 88 ha of tidal marshes and mudflats was 740 kg N (8.4 kg ha^{-1}) in 13 months compared to total N in runoff from 2,050 ha of watershed of 10,000 kg N (34). Thus, only about 7% (740 kg ÷ 10,740 kg) of the nitrogen entering the wetland was deposited directly. However, since N deposition onto the watershed (8.4 kg ha^{-1} × 2050 ha = 17220 kg) exceeded total runoff from the watershed to the wetland (10,000 kg), deposition could have contributed indirectly through runoff the majority of nitrogen entering the wetland. But the contributions of other N sources in runoff, such as fixation, fertilizer, and animal waste, were not given.

THE WETLAND NITROGEN CYCLE

The feature that sets wetlands apart from terrestrial ecosystems is the anaerobic nature of their waterlogged soils which alters the relative importance of various microbial transformations of inorganic and organic nitrogen compounds. Generally, the absence of O_2 retards the decomposition of organic matter (32, 40, 56, 130, 139). Complex aromatic ring structures are more resistant to microbial attack under anoxic conditions (130), leading to the formation and buildup of peat in wetland environments. Soil anoxia also favors the rapid conversion of NO_3^- to N_2O or N_2 by denitrifying bacteria and, consequently, results in quantitatively important losses of nitrogen from wetland ecosystems. Finally, the hydrology of wetlands favors the physical transport of soluble and particulate nitrogen compounds within or among ecosystems. Thus, NH_4^+ can be transported from anoxic sediments to the oxidized sediment surface or water column, where nitrification can occur, and the newly formed NO_3^- can return to the anoxic sediment layers where denitrification occurs. The wetland nitrogen cycle has been reviewed recently (17, 105, 117).

It is possible to generalize that the significance of atmospheric nitrogen deposition increases in wetlands as rainfall increases as a fraction of the total water budget (Table 2). Ombrotrophic bogs that receive exogenous nutrients exclusively from precipitation occupy one end of the spectrum, and the species in them are adapted to low levels of nitrogen. Bogs develop where precipitation exceeds evapotranspiration and where there is some impediment to drainage of the surplus water (86). These ecosystems are dominated by *Sphagnum* spp. and may be sparsely forested. *Sphagnum* produces a dense mound of peat that, when elevated above the surrounding land, receives neither runoff from uplands nor inputs from ground water. Peat forming bog

Table 2 Nitrogen cycle fluxes (g N m^{-2} yr^{-1}) from wetlands classified according to the importance of direct precipitation to their water and nitrogen budgets

	Precipitation increasing in importance ——>		
Nitrogen cycle component	Intertidal fresh and salt water marshes †	Fens ‡	Wet heathland and bogs §
Gross Nitrogen Inputs			
Precipitation	0.8– 1.1	4.2– 4.4	0.7–0.9
Fixation	0.3– 6.8	0.2– 1.3	0.1–0.3
Aqueous surface and/or subsurface	56.5–66.8	0.7– 2.1	0
Total Inputs	57.6–74.4	5.3–7.6	0.9–1.0
Internal Nitrogen Cycle			
Plant assimilation	15.5–22.5	9.0–27.4	3.8–8.2
Mineralization	18.0–19.5	7.9–24.4	2.6–7.4
Gross Nitrogen Exports			
Denitrification	0.4–14.3	0.1	0.1–0.2
NH_4^+ Volatilization	trace	ND	trace
Aqueous Surface and/or Subsurface	53.9–65.4	4.6– 8.0	0.2–0.3
Total Exports	54.6–79.7	4.9– 8.8	0.4

Sources: † 1, 2, 18, 92, 134, 137; ‡ 75, 141; § 63, 113, 132

ecosystems are widely distributed throughout the northern hemisphere, but they are most common in formerly glaciated regions.

The nitrogen input to ombrotrophic bogs totals about 1 g N m^{-1} yr^{-1} (Table 2), and atmospheric deposition accounts for most of this (132, 63). Total nitrogen outputs from the system are about 0.4 g N m^{-2} yr^{-1} and are accounted for by denitrification (0.1 to 0.2 g N m^{-1} yr^{-1}) and by export in runoff of NH_4^+ and dissolved organic nitrogen (DON). No export of particulate organic nitrogen (PON) was reported, and it appears that nitrogen accumulated in plant tissues is largely recycled within the bog. Atmospheric deposition also was the greatest source of exogenous nitrogen to a wet heathland (113), and it accounted for 95% of the NH_4^+ and NO_3^- entering the 1000 km^2 Shark River Slough, the principal fresh water drainage of Everglades National Park (51). In these examples, precipitation is the major water input.

Fens have nitrogen input and output rates that are intermediate between those of bogs and intertidal marshes (Table 2). Ground and surface inputs of water to these fens are also moderately important and on the same order of magnitude as precipitation (74). The fens in these examples are influenced by their close proximity to heavily fertilized pastures, by high rates of atmospheric nitrogen deposition, by annual mowing and harvest of aboveground vegetation, and in some areas by ground water discharge (141).

The intertidal wetlands represented in Table 2 are characteristic of systems adapted to large nitrogen inputs, and the examples include coastal salt marshes from Massachusetts (134), dominated by the grass *Spartina alterniflora,* and from England (1, 2), dominated by the grass *Puccinellia maritima,* as well as a freshwater intertidal marsh from Massachusetts (18) dominated by *Typha latifolia, Carex* spp., and *Calamagrostis canadensis.* In these ecosystems the greatest nitrogen inputs are from tidal water and, in some cases, ground water. These sources exceed atmospheric inputs by one or two orders of magnitude (Table 2). Gross nitrogen inputs in tidal water in the Massachusetts salt marsh are largely as NH_4^+ (5.4 g N m^{-2} yr^{-1}), DON (33.7 g N m^{-2} yr^{-1}), and PON (13.9 g N m^{-2} yr^{-1}) (134). Sedimentation of PON and diffusive inputs of NO_3^- are probably the dominant processes by which these ecosystems assimilate nitrogen from surface water, since the concentration gradients normally favor the diffusion of NH_4^+ and DON out of the sediment. In contrast to the other intertidal systems (2, 18) ground water inputs of nitrogen are important in the Massachusetts salt marsh and account for 6.0 and 5.6 g N m^{-2} yr^{-1} of NO_3^- and DON, respectively, of the gross inputs (134). There are additional inputs and outputs, such as deposition of bird feces and shellfish harvest, but these are insignificant in comparison to other rates (134).

Large nitrogen inputs in the intertidal ecosystems are balanced by equally large outputs (Table 2), but important transformations take place within the marshes. Denitrification accounts for 17.9% of the total N-loss from the Massachusetts salt marsh. The fact that denitrification is greater than the combined inputs of NO_3^- (134) implies that rates of nitrification are large. The greatest N-losses occur in tidal water, and in the Massachusetts salt marsh important gross exports occur of NH_4^+ (7.3 g N m^{-2} yr^{-1}), NO_3^- (2.5 g N m^{-2} yr^{-1}), and DON (38 g N m^{-2} yr^{-1}), which result in a net loss of all forms of dissolved nitrogen in tidal water. Only PON showed a net import into the marsh (134).

By definition the rate of nitrogen assimilation by plants is proportional to primary production, and primary production is generally proportional to the rate of mineralization (eg. 140), although high rates of productivity can be supported by high external nutrient inputs when conditions are unfavorable for high mineralization rates (141). Mineralization rates differ greatly between the wetland types represented in Table 2. Nitrogen assimilation by the plant communities varies from 3.8 to 8.2 g N m^{-2} yr^{-1} in the bog and heathland ecosystems to as great as 22.5 to 27.4 g N m^{-2} yr^{-1} in the intertidal and fen ecosystems, respectively. The nitrogen cycle in the bog and heathland ecosystems is largely closed. In contrast, the nitrogen cycle in salt marshes and fens is open, and there is a great exchange of nitrogen with adjacent systems.

In all these ecosystems, the rate of nitrogen mineralization almost balances plant assimilation in the manner of a closed cycle (Table 2). However, it is unlikely that the salt marsh could function as a closed system and maintain its productivity or community structure. Likewise, it is unlikely that the bog ecosystem could maintain its community structure if the nitrogen inputs were greatly increased. In general, as the input rate of nitrogen increases, concomitant increases occur in the output rate and magnitude of the internal cycle (Table 2). In ecosystems with closed nutrient cycles and small rates of internal cycling, like bogs, if nitrogen loadings increase significantly, then we can predict that productivity will increase, but (as will be discussed later) the increased productivity will be accompanied by changes in species composition to those adapted to an elevated nutrient regime.

EFFECTS OF NITROGEN LOADING

Effects on Primary Production

Numerous field experiments have documented that primary production in a diversity of wetland ecosystems is commonly limited by the availability of nitrogen (26, 28, 29, 30, 42, 54, 59, 85, 93, 95, 97, 116, 133, 136, 142). Rates of nitrogen application in these studies have ranged from 0.7 to 312 g N m^{-2} yr^{-1} and in most cases have been one to two orders of magnitude greater than rates of atmospheric deposition (Table 1). Fertilization experiments in North American salt marshes (26, 91, 97, 133) and European fens and wet grasslands (142) involving applications of either N or P demonstrated that primary production was stimulated by N and not P. In contrast, aboveground biomass failed to respond on wet heathlands in the central Netherlands that were fertilized for 3 yr at a rate of 20 g N m^{-2} yr^{-1}, while sites fertilized with 4 g P m^{-2} yr^{-1} showed a significant increase in biomass (4). Thus, wetlands are not universally limited by nitrogen.

The magnitude of the response of primary production to nitrogen fertilization is a function of the in situ availability of nitrogen, the availability of other nutrients, and the influence of edaphic variables on the physiology of nutrient uptake. For example, the greatest absolute and relative increase in standing biomass of *Spartina alterniflora* was obtained in studies where the control biomass was low (Figure 1). Nitrogen fertilization has little effect on biomass when control biomass is high, which implies that nitrogen availability in some marshes is already near a threshold where other factors become limiting. This figure also shows the relative response expected if control biomass were increased in every case to a limit of 2 kg m^{-2}. Most empirical data lie below this curve which indicates that the rate of nitrogen uptake by *Spartina* on a majority of sites is reduced as a consequence of oxygen limitation, high

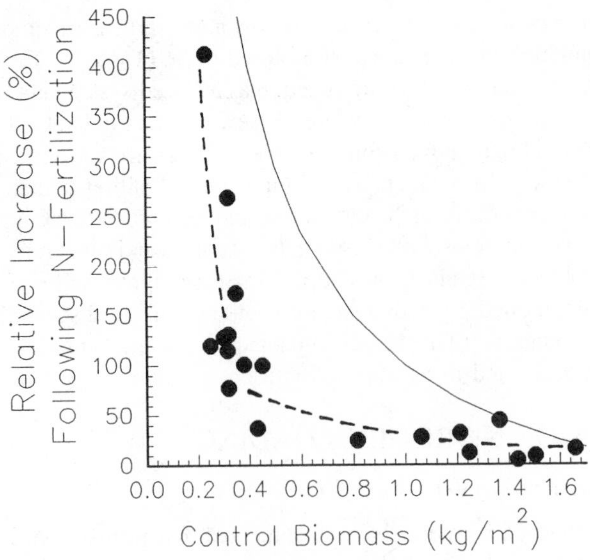

Figure 1 The relative increase (•) in standing, dry biomass of the salt marsh grass *Spartina alterniflora* that was achieved after 1 to 3 years of N-fertilization. Control biomass represents the maximum standing biomass on nonfertilized sites that was observed during the growing season. Sources: 26, 29, 30, 54, 59, 85, 91, 97, 133, 136. Also plotted is the relative response (——) that would be expected if control biomass from all sites were increased to a limit of 2 kg/m^2.

salinity, or sulfides (19, 20, 90, 92), that the rate of nitrogen fertilization was less than that necessary to achieve maximum production, or that other limitations are important beyond a threshold of nitrogen availability. For example, several studies have indicated that phosphorus became secondarily limiting after nitrogen applications reached a threshold (28, 116).

Several studies have investigated the effects of different nitrogen sources on primary productivity. For example, applications of NH_4^+ increased production of the grass *Puccinellia phryganodes* in a subarctic salt marsh by 175%, while equivalent applications of NO_3^- increased production by only 73% (28). Ammonium also stimulated higher productivity of *Spartina alterniflora* than equivalent applications of NO_3^- (85). Additions of NO_3^- were perhaps less effective than NH_4^+ because of denitrification of NO_3^- by bacteria in the anaerobic marsh sediments. This observation suggests that competition between plants and microbes for specific inorganic nitrogen compounds is important, and that plants compete more effectively for NH_4^+ than for NO_3^- in anaerobic soils.

The data discussed above pertain to growth of aboveground biomass only. Results of nitrogen fertilization studies where belowground biomass was measured have been variable; some studies showed a small decrease in living belowground biomass (135), while others showed small increases in belowground macroorganic matter (26, 54, 59) or no change (133). The evidence from controlled growth experiments (89, 126) suggests that the response of leaf growth to increased nitrogen supply is much greater than the response of roots.

It should be emphasized that the duration of a majority of fertilization studies is 3 yr or less, and it cannot be assumed that long-term nitrogen loadings will yield the same results. For example, one implication of a long-term increase in leaf growth is that the demand for mineral elements and water from the soil will increase. Higher rates of evapotranspiration occur on sites where aboveground biomass has been increased by nitrogen fertilization (65), and this phenomenon apparently accounted for the elevated soil salinities observed in a Georgia salt marsh (30). Moreover, increased evapotranspiration can influence the direction of wetland succession by altering the water balance of the soil. One modeling study indicated that nitrogen inputs greater than a threshold of 0.7 g N m^{-2} yr^{-1} can change the direction of succession from an open oligotrophic bog toward a mesotrophic bog dominated by trees (82). Thus, the long-term ecosystem and community responses to increased nitrogen loadings may differ significantly from results of short-term field experiments.

The Fate of Exogenous Nitrogen Supplies

Experiments in which mineral forms of ^{15}N were added to sediments in the absence of plants demonstrate that mineral nitrogen is rapidly used by the microbial community. Following a single addition of $^{15}NH_4^+$ equivalent to 10 g N/m^2 to sediments of a shallow saline lake, 20% of the added ^{15}N was converted to sediment organic nitrogen within 15 days, and this organic matter fraction remained constant for the remaining 337 days of the experiment (125). The amount of $^{15}NH_4^+$ in the sediment decreased exponentially to a nondetectable level by day 200. Diffusion of NH_4^+ into the water column and denitrification accounted for the loss of 80% of $^{15}NH_4^+$ from the sediment.

Twenty-seven days after the equivalent of 10 g N m^{-2} was added as either $^{15}NO_3^-$ or $^{15}NH_4^+$ to the floodwater in chambers containing swamp sediment, only 39.6% and 6.2% of the ^{15}N from NH_4^+ and NO_3^-, respectively, remained in the sediment and overlying water column (81). The remainder was apparently lost from the chambers by denitrification. The loss of 60% of the applied $^{15}NH_4^+$ indicated that NH_4^+ was rapidly converted to NO_3^- by nitrifying bacteria in the water column or surface sediment, and that the NO_3^- diffused into anaerobic zones of sediment where denitrification occurred. Nitrification was evidently the rate limiting step since the loss of ^{15}N by denitrification was more rapid when it was applied as NO_3^-.

Three weeks after making single additions of $^{15}NH_4^+$, equivalent to 1.5 g N m^{-2}, to the floodwater above sediment cores taken from swamps that had been receiving primary wastewater effluent for 2 and 50 years before the experiment, 0.5 to 2.3% of the added nitrogen was recovered in the flood water, largely as NO_3^-, and 13.6 to 17.8% in the sediment, largely as organic matter (39). The remaining fraction, ca. 80%, was apparently lost by denitrification, which indicates that conversion of NH_4^+ to NO_3^- and diffusion of NO_3^- to anaerobic sites of denitrification was rapid. Furthermore, the similarity of responses by the two sediment types suggests that the nitrification-denitrification potential of sediments is unchanged after 50 yr of high nitrogen loadings, provided that a continuous supply of suitable carbon substrates is also available to sustain continuous nitrification-denitrification reactions.

Short-term measurements of slurries of marl and peat sediments from the Florida Everglades, incubated under a N_2 atmosphere, demonstrated that 10–34% of NO_3^- added at levels of 10 and 100 μM was rapidly denitrified within 24 hr (57). Denitrification rates decreased following this initial burst of activity as the balance of the added NO_3^- was converted to NH_4^+. This experiment suggests that the process of dissimilatory nitrate reduction to NH_4^+ competes successfully with the denitrification process. However, if an aerobic-anoxic interface had existed, nitrification would have generated a continuous supply of NO_3^-, and denitrification would then have consumed a greater fraction of the NO_3^- over time.

The behavior of NH_4^+ applied to vegetated wetland sediments is quite different from the results described above and indicates that plants are able to compete successfully with microbes. Typically, there is a 75–94% recovery of ^{15}N, applied as NH_4^+, in plants and soil at the end of the first growing season following applications of 3 to 8 g N m^{-2} yr^{-1} to various plant-soil systems (27, 38, 41, 42). In a study of the fate of $^{15}NH_4^+$-N applied to marsh sediments supporting *Spartina alterniflora,* 28% of added ^{15}N was recovered in aboveground biomass and 65% in soil and belowground biomass after the first growing season (41). In soil and belowground biomass ^{15}N declined to 43% by the end of the third growing season and to 1.2% of original ^{15}N in aboveground biomass. The annual declines were postulated to have been due to the loss of nitrogen from the leaves, either by physical transport of aboveground plant material off the site or by decomposition of leaf material at the sediment surface followed by nitrification-denitrification reactions.

After a season of biweekly additions of $^{15}NH_4^+$, equivalent to 8.2 g N m^{-2} $season^{-1}$, to the flood water in sediment cores containing *Typha latifolia,* about 54% of original ^{15}N was recovered in the plants, including both above and belowground biomass, while 22% was contained in the soil (38). In cores that contained just sediments, only 35% of the added ^{15}N was recovered; most of this, 33% of the added ^{15}N, was in the sediment. The remaining 65% was thought to have been lost through nitrification-denitrification reactions.

The experiments discussed above indicate that plant biomass is the major sink for free NH_4^+, and that in the absence of plants the major fate is nitrification-denitrification. It should be emphasized that the nitrification-denitrification process can dominate only in environments, like wetlands, that have separate and distinct aerobic and anoxic zones of microbial activity between which solutes can freely diffuse.

Effects of Nitrogen Loading on Microbial Processes

Changes in deposition rate and the chemical form of nitrogen in deposition can influence microbial processes and details of the internal nitrogen cycle of wetlands. For instance, decomposition rate is sensitive to the nitrogen concentration of decomposing tissues and of the surrounding environment. Tissues with elevated nitrogen concentrations normally are observed to decompose at a faster rate than tissues containing low nitrogen concentrations (84, 96, 137). For example, litter from N-fertilized *Spartina alterniflora* decomposed 50% faster than control litter (84).

The temporal dynamics of nitrogen within decomposing litter is also sensitive to the nitrogen status of the original tissue. That is, litter of low original nitrogen content often acts as a net nitrogen sink during the first months of decomposition, whereas nitrogen-rich litter is likely to be a nutrient exporter rather than an accumulator during decomposition (96). There is some controversy about the mechanism of nitrogen immobilization (3, 14, 15), but its importance in conserving nitrogen within the wetland nitrogen cycle is recognized (21, 36, 94).

Microbial nitrogen transformations are also affected by the nitrogen status of the environment. It is well known that NH_4^+ inhibits the activity of nitrogen fixing bacteria (diazotrophs) (25). There is a repression of nitrogen fixation in salt marsh sediments enriched with either NH_4^+ or NO_3^- (43). It is thought that NH_4^+ represses synthesis of the nitrogenase enzyme. Alternatively, NH_4^+ may directly inhibit nitrogenase activity (150). Decreases in the proportion of diazotrophs among the heterotrophic bacteria after application of NH_4NO_3 have been observed, and these may result from a competitive suppression of diazotrophs by nondiazotrophs in the presence of combined nitrogen (76).

Acidification, which may be caused by deposition of NO_x or NH_4^+, can affect the nitrogen cycle. Decomposition rates are decreased by acidification (64, 78), but the degree of inhibition is dependent upon the buffering capacity of the litter (55). Similarly, nitrification was inhibited at pH 4–5 in cypress swamps (44), and at pH 5.4 to 5.7 in lakes (114). Acidification blocks the nitrogen cycle by inhibiting nitrification and leads to an accumulation of NH_4^+ (109, 114, 118, 119). In addition, the ratio of $N_2O:N_2$ produced by denitrifying bacteria is apparently pH sensitive with little N_2O production under anoxic conditions at pH 7 and almost 100% production at pH 5 (52).

Finally, NO_3^- and NH_4^+ have been shown to influence the relative and absolute production of end-products of dissimilatory nitrate reduction (12, 73, 100). High NO_3^- concentrations are thought to favor N_2O production and inhibit N_2 production, perhaps because of competition between the NO_3^- and N_2O terminal electron acceptors of the denitrification pathway (12, 31). Nitrate from sediment slurries incubated anaerobically with 250 μM NO_3^- (in the presence of acetylene) was reduced to approximately equal proportions of NH_4^+ and N_2O (72). As the nitrate concentration was increased up to 2 mM, the proportion of the nitrate denitrified to N_2O increased up to 83%. Higher ratios of $N_2O:N_2$ production and higher absolute amounts of N_2O are produced from eutrophic sediments than from unpolluted sediments of Narragansett Bay, RI (120, 121). And after salt marsh and brackish marsh soils were fertilized with 1.2–1.5 g NH_4^+-N m^{-2}, production of N_2O increased from 0.22 and 0.04 mg N_2O-N m^{-2} day^{-1}, respectively, to 1.5 and 2.9 mg N_2O-N m^{-2} day^{-1} (124). Others (11), however, failed to observe an inhibition of N_2O reduction in the presence of NO_3^-.

Quantitative information is scarce about the relationship between nitrogen eutrophication and rates of N_2O production in natural environments. Probably only a small fraction of depositional nitrogen inputs is likely to be evolved as N_2O. In one experiment, only 0.39% of fertilizer-N was recovered as N_2O from submerged soils amended with 34 g NO_3^--N m^{-2} and 12 g NH_4^+-N

m^{-2} in the laboratory (98). However, on a global scale, even a small change in the production of N_2O is potentially significant, considering the role of N_2O in the destruction of stratospheric O_3 (35, 58).

Effects on Diversity and Community Structure

In the introduction it was pointed out that wetlands harbor about 14% of the total number of plant species formally listed as endangered in the United States. While it is beyond the scope of this review to survey the physiological ecology of these wetland plants, several species on this list are widely recognized to be adapted to nitrogen-poor or infertile environments. These include the isoetids (16) and the insectivorous plants (70, 88, 146) like the endangered green pitcher plant, *Sarracenia oreophila*. In eastern Canadian wetlands, nationally rare species are found principally on infertile sites (88, 146).

These assertions are supported by research on floristic changes related to nitrogen deposition in central Europe. Based on a survey of the nitrogen requirements of 1805 plant species from West Germany, it was concluded that 50% can compete successfully only in habitats that are deficient in nitrogen supply (47). Furthermore, 75 to 80% of the endangered species are indicators of infertile habitats. It is also clear that the trend toward rare species occurring with greater frequency in nitrogen-poor habitats is a common phenomenon across many ecosystem types (47).

There is a history in Western Europe of changes in wetland community composition that are thought to result from deposition of atmospheric pollutants. *Sphagnum* species are largely absent from ombrotrophic peat bogs in areas of Britain where they were once common (49, 50, 77, 128). Ombrotrophic bogs downwind of Manchester and Liverpool have been extensively modified by atmospheric pollution for more than 200 yr, with the virtual elimination of the dominant peat-forming *Sphagnum* mosses from more than 60,000 ha of bog (77). This has led to a loss of water retention and widespread erosion. Atmospheric nitrogen deposition has been implicated in this process, although studies of this particular area should be interpreted cautiously because of its long history of exposure to multiple pollutants. The combination of NO_3^- and NH_4^+ deposition, about 3.2 g N m^{-2} yr^{-1}, is more than double the deposition rates in healthy *Sphagnum* communities of the Berwyn Mountains in North Wales and contributes to a supraoptimal nitrogen supply (77). In the Netherlands there has been a great decline during the past three decades in communities dominated by isoetids in soft water areas and their conversion to later successional stages dominated by grasslands or by *Juncus bulbosus* (rush) and *Sphagnum* spp. (108, 109, 110, 118).

Excess nitrogen has a direct toxic effect on some species. Decreased growth of *Sphagnum cuspidatum* accompanied an increase in tissue-N concentration to a level of 2.5% of dry weight when the plants were transplanted to a site of high N deposition in northern Britain (102). Similarly, among the elements N, S, Pb, Fe, and P examined in five *Sphagnum* species transplanted from a relatively clean-air site to a polluted site, the largest absolute increases were in nitrogen which ranged between 17.7 mg per gram of tissue in *Sphagnum recurvum* and 5.3 mg/g in *Sphagnum capillifolium* above control levels of about 10 mg g^{-1} (50). Nutrient supply at the polluted site, where N deposition is 4.3 g N m^{-2} yr^{-1}, is apparently supraoptimal for growth of ombrotrophic *Sphagnum* species, while another site with a nitrogen deposition rate of 2.0 g N m^{-2} yr^{-1} still supports *Sphagnum* (50).

Interspecies competitive relationships change with the nitrogen status of the environment. In weakly buffered ecosystems, a high deposition of NH_4^+ leads to acidification and nitrogen enrichment of soil. Consequently, plant species characteristic of poorly buffered environments disappear. Among the acid tolerant species there will be competition between slow growing and fast growing nitrophilous grasses or grass-like species. This process contributes to the observed change from heathlands into grasslands in areas of high nitrogen deposition. *Molinia caerulea* and/or *Deschampsia flexuosa* (grasses) expand at the expense of *Erica tetralix* or *Calluna vulgaris* (shrubs) and other heathland species (4, 5, 10, 110). In over 70 heathlands investigated, the shrub bogs dominated by *Erica tetralix* or *Calluna* had dissolved NH_4^+ levels in the soil water of 55 and 84 μM, while those dominated by the grasses *Deschampsia* and *Molinia* had average NH_4^+ concentrations of 248 and 429 μM (110). Changes in species composition or dominance also have been observed experimentally. For example, N fertilization increased the biomass and dominance of grasses at the expense of other species in fen and wet grassland communities (142). Biomass of some *Equisetum* spp. was reduced following fertilization. Species-specific changes in stem density in an English salt marsh fertilized with 61 g NO_3^--N m^{-2} yr^{-1} or 68 g NH_4^+-N m^{-2} yr^{-1} over a period of 3–4 years have also been observed (66).

Relationships among biomass, species diversity, and soil chemical characteristics were investigated in several fen and grassland communities from the Netherlands (143). In all the wetland types investigated, species number was greatest when the standing biomass of the site was intermediate and in the range of 400–500 g m^{-2}. Domination by a few species is associated with eutrophic conditions at the high end of the biomass scale as well as with conditions unfavorable for growth at the low end of the scale (143). Similarly, in wetlands of eastern Ontario and western Quebec, the greatest diversity of species (3–24 per 0.25 m^2) occurs at intermediate standing crops (60–500 g m^{-2}) and the lowest density of species (2–5 per 0.25 m^2) at standing crops greater than 1500 g m^{-2} (87, 146). In Great Britain species density in fens was greatest (about 12 per 0.25 m^2) at standing crops less than 1000 g m^{-2}

and lowest (3 per 0.25 m^2) when standing crop was 4000 g m^{-2} or greater (145). Exceptions to this trend are found where annual mowing and harvest of wetland vegetation minimize the accumulation of surface litter (141), and possibly where intense pressure from grazing animals favors domination by specific plant species (9, 67).

Several controlled studies have been carried out to identify the mechanisms of nitrogen control over community structure. This is a complex phenomenon because there are many direct and indirect effects of nitrogen eutrophication. For example, acidification resulting from the deposition of either NO_x, SO_4^{2-}, or NH_4^+ can decrease the availability of dissolved CO_2 in water which leads to the complete elimination of submerged plant species (109). Deposition of NH_4^+ and its subsequent nitrification or absorption by plants generates acidity. Biochemical conversions of SO_4^{2-} and NO_3^- generate alkalinity. These processes are mediated by bacteria, macrophytes, and algae (72, 104). Thus, atmospheric deposition can significantly affect the nitrogen and carbon budgets, the acidity, and consequently the community structure of wetland ecosystems.

Interactions among acidification, nitrogen supply and growth of seven common wetland plants from the Netherlands have been studied (119). All species utilized NH_4^+ and NO_3^- as a nitrogen source except *Sphagnum flexuosum,* which did not assimilate NO_3^-. When NH_4^+ and NO_3^- were offered simultaneously in equal amounts, NO_3^- uptake was the dominant form of nutrition (63–73%) in plants that are characteristic of soft waters (low Ca^{2+} and Mg^{2+}), while NH_4^+ strongly dominated the nutrition (85–90%) in species from acid waters. Differences in the site of uptake, either leaves or roots, among species were also found. High deposition of NH_4^+ and SO_4^{2-}, the most important sources of acidification in the Netherlands, is leading to an expansion of acid-tolerant nitrophilous plants (119).

The nutrition of *Sphagnum* is apparently species specific. Although *S. flexuosum* did not assimilate NO_3^- (119), the activity of nitrate reductase in *S. cuspidatum* (101) and in *S. fuscum* (148) shows that NO_3^- can be utilized by these species. *S. magellanicum* was shown to grow best when given the equivalent of 4.1 g NO_3^--N m^{-2} yr^{-1} plus 1.9 g NH_4^+-N m^{-2} yr^{-1} in simulated rain. When given 0.25 times that amount of NO_3^- and 1.5 times (and 4 times) as much NH_4^+, growth decreased (115). Growth of the dominant *Sphagnum* spp. (*S. angustifolium, S. fuscum,* and *S. magellanicum*) from oligotrophic sites in an Ontario fen that receives an atmospheric deposition of 0.31 g NO_3^--N m^{-2} yr^{-1} increased following the addition of 0.16 g NO_3^--N m^{-2} yr^{-1} applied in simulated acid rain, at least during the first year of the experiment (8). *Sphagnum* from minerotrophic sites did not respond to the simulated acid rain. One study showed that growth of *S. cuspidatum* was greatest in a medium containing 500 μM NH_4^+, and less at 1000 or 100 μM

NH_4^+ (111), while another showed that growth of this same species was best in N-free solutions, and that even small additions (10 μM) of either NH_4^+ or NO_3^- reduced growth (102). Some variations in results of nutritional studies are doubtless influenced by other variables like pH, prior treatment, ontogeny, or even genotype.

In a two-year greenhouse experiment designed to differentiate between acid and nitrogen effects, mixtures of different wetland plant species were exposed to simulated rain containing various combinations of SO_4^{2-}, NH_4^+, and NO_3^- (118). Marked changes were observed in systems receiving rain with 510 and 1585 μM NH_4^+. Plants typical of nutrient-poor soft waters, like the isoetids *Littorella uniflora* (shoreweed), *Luronium natans* (water plantain), and *Pilularia globulifera,* were adversely affected at this level of nitrogen input, while other species (*Juncus bulbosus, Sphagnum cuspidatum,* and the grass *Agrostis canina*) expanded. Acidification with little or no NH_4^+ addition had no clear effects, although biomass of *Sphagnum* was slightly higher. To preserve the remaining oligotrophic wetlands, it was recommended that acid inputs should not exceed 250 mol ha^{-1} yr^{-1}, and that nitrogen deposition should be limited to 1380 mol ha^{-1} yr^{-1} (1.94 g N m^{-2} yr^{-1}) or less if in the form of NH_4^+, because of the potential for this rate of NH_4^+ deposition to exceed the allowable acid input (118). This recommendation is consistent with a conclusion from studies of species distributions that the limit for many species is below 2.0 g N m^{-2} yr^{-1} and for oligotrophic bogs is probably about 1.0 g N m^{-2} yr^{-1} (80).

CONCLUSIONS

Bulk nitrogen deposition in North American wetlands ranges from 0.55 to 1.2 g N m^{-2} yr^{-1} and occurs in the form of NO_3^-, NH_4^+, and DON in roughly equal proportions (Table 1), although a recent study suggests that deposition may be greater in areas of the northeast and midwest (151). Dry deposition can exceed wet deposition and adds significantly to the total (13). Leaf-capture of nitrogen in fog droplets is a third form of deposition that is locally important (83).

Peat-forming *Sphagnum* spp. are largely absent from bogs in Western Europe where bulk deposition rates are about 4 g N m^{-2} yr^{-1} (49, 50, 77, 102, 128). Soft water communities once dominated by isoetids in the Netherlands have been converted to later successional stages dominated by *Juncus* spp. and *Sphagnum* spp. or to grasslands (108–110, 119). Heathlands dominated by shrubs have also converted to grasslands (110). Eutrophication of wetlands is associated with a decrease in plant species diversity (87, 88, 143, 146).

Nitrogen deposition can affect plant and microbial processes indirectly, by acidifying the environment or by altering the hydrologic cycle, or directly. An increase in nitrogen supply can alter the competitive relationships among plant species such that fast growing nitrophilous species are favored. Increased productivity associated with eutrophication is accompanied by increased rates of evapotranspiration (65) which can alter wetland hydrology and may influence the direction of wetland succession (82). Microbial rates of decomposition, nitrogen fixation, nitrification, and dissimilatory nitrate reduction are all affected. Acidification below pH 4–5.7 blocks the nitrogen cycle by inhibiting nitrification (44, 114), and the accumulation of NH_4^+ in the environment represses nitrogen fixation (25). The ratio of $N_2O:N_2$ produced by denitrification increases with decreasing pH below 7 (52), and both the absolute rate and proportion of N_2O produced increases with increasing eutrophication (120, 121).

Single additions to vegetated wetland soils of $^{15}NH_4^+$ at rates of about 10 g $N\ m^{-2}\ yr^{-1}$ indicate that 75–94% of the applied NH_4^+ is rapidly assimilated into organic matter within a single growing season (27, 38, 41, 42). The majority of the labelled nitrogen is lost from the system after 3 yr by the combined processes of physical transport of particulate and dissolved nitrogen, and denitrification. In the absence of plants, the major fate of inorganic nitrogen applied to wetland soils is loss to the atmosphere by denitrification (38, 39, 81, 125).

Primary production in wetlands is generally, but not universally, limited by nitrogen availability. Applications of nitrogen fertilizer in the field, which are typically one or two orders of magnitude greater than atmospheric inputs, have stimulated increases in standing biomass by as much as 413%. Edaphic variables such as oxygen availability may contribute to nitrogen deficiency by inhibiting nitrogen uptake (19, 20, 90, 92), and other nutrients, like phosphorus, may become secondarily limiting to primary production after nitrogen inputs reach a threshold (28, 116). Fertilization with nitrogen in the field has increased the dominance of grasses over other species in fens (142).

The sensitivity of wetland community structure to atmospheric nitrogen deposition increases as the water budget is increasingly dominated by direct precipitation. Ombrotrophic bogs characterize one extreme where exogenous nitrogen inputs totalling about 1 g $N\ m^{-2}\ yr^{-1}$ are largely from precipitation. The other extreme is characterized by intertidal wetlands with large surface water inputs and gross nitrogen inputs as high as 74 g $N\ m^{-2}\ yr^{-1}$ (Table 2). Primary production is proportional to the rate of internal nitrogen cycling, which is influenced by the quantity of mineralizable soil nitrogen and the exogenous supply of nitrogen to the ecosystem from the atmosphere or from surface and groundwater flow.

Fourteen percent (or 18) of the plant species from the conterminous United States that are formally listed as endangered, and an additional 284 species listed as potentially threatened, are found principally in wetland habitats. Some of the endangered plants, like the green pitcher plant, are known to be adapted to infertile habitats. There is growing evidence that plant species adapted to infertile habitats like ombrotrophic bogs cannot be maintained if exogenous nitrogen inputs exceed 2 g N m^{-2} yr^{-1}. Plant species that are threatened by high nitrogen deposition are not confined to wetland habitats, however, but are common across many ecosystem types (47).

ACKNOWLEDGMENTS

I thank Paul Bradley, Arne Jensen, Jos Verhoeven, and members of the EPA Peer Review Workshop for their thoughtful criticisms. This work was supported by the US Environmental Protection Agency.

Literature Cited

1. Abd. Aziz, S. A., Nedwell, D. B. 1986. The nitrogen cycle of an east coast, U.K., saltmarsh. I. Nitrogen assimilation during primary production; detrital mineralization. *Estuar. Coast. Shelf Sci.* 22:559–75
2. Abd. Aziz, S. A., Nedwell, D. B. 1986. The nitrogen cycle of an east coast, U.K. saltmarsh. II. Nitrogen fixation, nitrification, denitrification, tidal exchange. *Estuar. Coast. Shelf Sci.* 22:-689–704
3. Aber, J. D., Melillo, J. M. 1982. Nitrogen immobilization in decaying hardwood leaf litter as a function of initial nitrogen and lignin content. *Can. J. Bot.* 60:2263–69
4. Aerts, R., Berendse, F. 1988. The effect of increased nutrient availability on vegetation dynamics in wet heathlands. *Vegetatio* 76:63–69
5. Aerts, R., Berendse, F. 1989. An analysis of competition in heathland ecosystems. I. Competition for nutrients. In *Plant strategies and nutrient cycling in heathland ecosystems,* R. Aerts, pp. 147–61. Ph.D thesis. Univ. Utrecht, The Netherlands
6. Deleted in proof
7. Armentano, T. V., Menges, E. S. 1986. Patterns of change in the carbon balance of organic soil-wetlands of the temperate zone. *J. Ecol.* 74:755–74
8. Bayley, S. E., Vitt, D. H., Newbury, R. W., Beaty, K. G., Behr, R., Miller, C. 1987. Experimental acidification of a *Sphagnum*-dominated peatland: first year results. *Can. J. Fish. Aquat. Sci.* 44(Suppl. 1):194–205
9. Berendse, F. 1985. The effect of grazing on the outcome of competition between plant species with different nutrient requirements. *Oikos* 44:35–39
10. Berendse, F., Aerts, R. 1984. Competition between *Erica tetralix* L. and *Molinia caerulea* (L.) Moench as affected by the availability of nutrients. *Acta Oecol. Oecol. Plant.* 5:3–14
11. Betlach, M. R., Tiedje, J. M. 1981. Kinetic explanation for accumulation of nitrite, nitric oxide, and nitrous oxide during bacterial denitrification. *Appl. Environ. Microbiol.* 42:1074–84
12. Blackmer, A. M., Bremner, J. M. 1978. Inhibitory effect of nitrate on reduction of N_2O to N_2 by soil microorganisms. *Soil Biol. Biochem.* 10:187–91
13. Boring, L. R., Swank, W. T., Waide, J. B., Henderson, G. S. 1988. Sources, fates, and impacts of nitrogen inputs to terrestrial ecosystems: review and synthesis. *Biogeochemistry* 6:119–59
14. Bosatta, E., Berendse, F. 1984. Energy or nutrient regulation of decomposition: Implications for the mineralization-immobilization response to perturbations. *Soil Biol. Biochem.* 16:63–67
15. Bosatta, E., Staaf, H. 1982. The control of nitrogen turn-over in forest litter. *Oikos* 39:143–51
16. Boston, H. L. 1986. A discussion of the adaptations for carbon acquisition in relation to the growth strategy of aquatic isoetids. *Aquat. Bot.* 26:259–70
17. Bowden, W. B. 1987. The biogeochemistry of nitrogen in freshwater wetlands. *Biogeochemistry* 4:313–48
18. Bowden, W. B., Vörösmarty, C. J., Morris, J. T., Peterson, B. J., Hobbie, J. E. et al. 1991. Transport and processing of nitrogen in a tidal freshwater wetlands. *Water Resources Res.* In press

19. Bradley, P. M., Morris, J. T. 1990. Influence of oxygen and sulfide concentration on nitrogen uptake kinetics in *Spartina alterniflora*. *Ecology* 71:282–87
20. Bradley, P. M., Morris, J. T. 1991. The influence of salinity on the kinetics of NH_4^+ uptake in *Spartina alterniflora*. *Oecologia* 85:375–80
21. Brinson, M. M. 1977. Decomposition and nutrient exchange of litter in an alluvial swamp forest. *Ecology* 58:601–9
22. Brinson, M. M., Bradshaw, H. D., Kane, E. S. 1984. Nutrient assimilative capacity of an alluvial floodplain swamp. *J. Appl. Ecol.* 21:1041–57
23. Buell, G. R., Peters, N. E. 1988. Atmospheric deposition effects on the chemistry of a stream in northeastern Georgia. *Water Air Soil Pollut.* 39:275–91
24. Buijsman, E. 1987. Ammonia emission calculation—fiction and reality. In *Proc. Ammonia and Acidification EURASAP Symp.*, ed. W. A. N. Asman, S. M. A. Diederen, pp. 13–27. Netherlands: RIVM, TNO
25. Buresh, R. J., Casselman, M. E., Patrick, W. H. Jr. 1980. Nitrogen fixation in flooded soil systems, a review. *Adv. Agron.* 33:149–92
26. Buresh, R. J., DeLaune, R. D., Patrick, W. H. Jr. 1980. Nitrogen and phosphorus distribution and utilization by *Spartina alterniflora* in a Louisiana Gulf Coast marsh. *Estuaries* 3:111–21
27. Buresh, R. J., DeLaune, R. D., Patrick, W. H. Jr. 1981. Influence of *Spartina alterniflora* on nitrogen loss from marsh soil. *Soil Sci. Soc. Am. J.* 45:660–61
28. Cargill, S. M., Jefferies, R. L. 1984. Nutrient limitation of primary production in a sub-arctic salt marsh. *J. Appl. Ecol.* 21:657–68
29. Cavalieri, A. J., Hwang, A. H. C. 1981. Accumulation of proline and glycinebetaine in *Spartina alterniflora* Loisel. in response to NACl and nitrogen in the marsh. *Oecologia* 49:224–28
30. Chalmers, A. G. 1979. The effects of fertilization on nitrogen distribution in a *Spartina alterniflora* salt marsh. *Estuar. Coast. Shelf Sci.* 8:327–37
31. Cho, C. M., Sakdinan, L. 1978. Mass spectrometric investigation on denitrification. *Can. J. Soil Sci.* 58:443–57
32. Clark, M. D., Gilmour, J. T. 1983. The effect of temperature on decomposition at optimum and saturated soil water contents. *Soil Sci. Soc. Am. J.* 47:927–29
33. Code of Federal Regulations. 1987. Endangered and threatened wildlife and plants. Code Fed. Regul. 50:17.11, 17.12
34. Correll, D. L. 1981. Nutrient mass balances for the watershed, headwaters, intertidal zone, and basin of the Rhode River estuary. *Limnol. Oceanogr.* 26:-1142–49
35. Crutzen, P. J. 1970. The influence of nitrogen oxides on the atmospheric ozone content. *Q. J. R. Meteorol. Soc.* 96:320–25
36. Damman, A. W. H. 1988. Regulation of nitrogen removal and retention in Sphagnum bogs and other peatlands. *Oikos* 51:291–305
37. Davis, C. B., van der Valk, A. G., Baker, J. L. 1983. The role of four macrophyte species in the removal of nitrogen and phosphorus from nutrient-rich water in a prairie marsh, Iowa. *Madrono* 30:133–42
38. Dean, J. V., Biesboer, D. D. 1985. Loss and uptake of ^{15}N-ammonium in submerged soils of a cattail marsh. *Am. J. Bot.* 72:1197–1203
39. DeBusk, W. F., Reddy, K. R. 1987. Removal of floodwater nitrogen in a cypress swamp receiving primary wastewater effluent. *Hydrobiologia* 153:-79–86
40. DeLaune, R. D., Reddy, C. N., Patrick, W. H. Jr. 1981. Organic matter decomposition in soil as influenced by pH and redox conditions. *Soil Biol. Biochem.* 13:533–34
41. DeLaune, R. D., Smith, C. J., Patrick, W. H. Jr. 1983. Nitrogen losses from a Louisiana gulf coast salt marsh. *Estuar. Coast. Shelf Sci.* 17:133–41
42. DeLaune, R. D., Smith, C. J., Sarafyan, M. N. 1986. Nitrogen cycling in a freshwater marsh of *Panicum hemitomon* on the deltaic plain of the Mississippi River. *J. Ecol.* 74:249–56
43. Dicker, H. J., Smith, D. W. 1980. Physiological ecology of acetylene reduction (nitrogen fixation) in a Delaware salt marsh. *Microb. Ecol.* 6:161–71
44. Dierberg, F. E., Brezonik, P. L. 1982. Nitrifying population densities and inhibition of ammonium oxidation in natural and sewage-enriched cypress swamps. *Water Res.* 16:123–26
45. Dillon, P. J., Lusis, M., Reid, R., Yap, D. 1988. Ten-year trends in sulphate, nitrate and hydrogen deposition in central Ontario. *Atmos. Environ.* 22:901–5
46. Elkins, J. W., Wofsy, S. C., McElroy, M. B., Kolb, C. E., Kaplan, W. A. 1978. Aquatic sources and sinks for nitrous oxide. *Nature* 275:602–6
47. Ellenberg, H. 1988. Floristic changes due to nitrogen deposition in central Europe. In *Critical Loads for Sulphur and Nitrogen: Report from a Workshop;*

March; Skokloster, Sweden, ed. J. Nilsson, P. Grennfelt, pp. 375–83. Copenhagen, Denmark: Nordic Council of Ministers
48. Federal Register. 1985. Endangered and threatened wildlife and plants; review of plant taxa for listing as endangered or threatened species. *Fed. Reg.* 50:39526–83
49. Ferguson, P., Lee, J. A. 1980. Some effects of bisulphite and sulphate on the growth of *Sphagnum* species in the field. *Environ. Pollut. Ser.* A 21:59–71
50. Ferguson, P., Robinson, R. N., Press, M. C., Lee, J. A. 1984. Element concentrations in five *Sphagnum* species in relation to atmospheric pollution. *Bryologist* 13:107–14
51. Flora, M. D., Rosendahl, P. C. 1982. The impact of atmospheric deposition on the water quality of Everglades national park. In *Int. Symp. on Hydrometeorology,* pp. 55–61. Am. Water Res. Assoc.
52. Focht, D. D. 1974. The effect of temperature, pH, and aeration on the production of nitrous oxide and gaseous nitrogen—zero-order kinetic model. *Soil Sci.* 118:173–79
53. Frayer, W. E., Monahan, T. J., Dowden, D. C., Graybill, F. A. 1983. *Status and Trends of Wetlands and Deepwater Habitats in the Conterminous United States, 1950s to 1970s.* Ft. Collins, Colo: Colo. State Univ. Dep. For. Wood Sci.
54. Gallagher, J. L. 1975. Effect of an ammonium nitrate pulse on the growth and elemental composition of natural stands of *Spartina alterniflora* and *Juncus roemerianus. Am. J. Bot.* 62:644–48
55. Gallagher, J. L., Donovan, L. A., Grant, D. M., Decker, D. M. 1987. Interspecific differences in dead plant buffering capacity alter the impact of acid rain on decomposition rates in tidal marshes. *Water Air Soil Pollut.* 34:339–46
56. Godshalk, G. L., Wetzel, R. G. 1978. Decomposition of aquatic angiosperms. II. Particulate components. *Aquat. Bot.* 5:301–27
57. Gordon, A. S., Cooper, W. J., Scheidt, D. J. 1986. Denitrification in marl and peat sediments in the Florida Everglades. *Appl. Environ. Microbiol.* 52:-987–91
58. Hahn, J., Crutzen, P. J. 1982. The role of fixed nitrogen in atmospheric photochemistry. *Philos. Trans. R. Soc. Lond. B* 296:521–41
59. Haines, E. B. 1979. Growth dynamics of cordgrass, *Spartina alterniflora* Loisel., on control and sewage sludge fertilized plots in a Georgia salt marsh. *Estuaries* 2:50–53
60. Harriss, R. C., Sebacher, D. I., Day, F. P., Jr. 1982. Methane flux in the Great Dismal Swamp. *Nature* 297:673–74
61. Harriss, R. C., Gorham, E., Sebacher, D. I., Bartlett, K. B., Flebbe, P. A. 1985. Methane flux from northern peatlands. *Nature* 315:452–53
62. Heil, G. W., van Dam, D., Heijne, B. 1987. Catch of atmospheric deposition in relation to vegetation structures of heathland. In *Ammonia and Acidification,* ed. W. A. H. Asman, S. M. A. Diederen, pp. 107–23. Eur. Assoc. Sci. Air Pollution (EURASAP). Netherlands
63. Hemond, H. F. 1983. The nitrogen budget of Thoreau's Bog. *Ecology* 64: 99–109
64. Hendrickson, O. Q. 1985. Variation in the C:N ratio of substrate mineralized during forest humus decomposition. *Soil Biol. Biochem.* 17:435–40
65. Howes, B. L., Dacey, J. W. H., Goehringer, D. D. 1986. Factors controlling the growth form of *Spartina alterniflora:* Feedbacks between aboveground production, sediment oxidation, nitrogen and salinity. *J. Ecology* 74:-881–98
66. Jefferies, R. L., Perkins, N. 1977. The effects on the vegetation of the additions of inorganic nutrients to salt marsh soils at Stiffkey, Norfolk. *J. Ecol.* 65:867–82
67. Jensen, A. 1985. The effect of cattle and sheep grazing on salt-marsh vegetation at Skallingen, Denmark. *Vegetatio* 60:-37–48
68. Jordan, T. E., Correll, D. L., Whigham, D. F. 1983. Nutrient flux in the Rhode River: tidal exchange of nutrients by brackish marshes. *Estuar. Coast. Shelf Sci.* 17:651–67
69. Kadlec, J. A. 1986. Input-output nutrient budgets for small diked marshes. *Can. J. Fish. Aquat. Sci.* 43:2009–16
70. Keddy, P. A., Wisheu, I. C. 1989. Ecology, biogeography, and conservation of coastal plain plants: some general principles from the study of Nova Scotian wetlands. *Rhodora* 91:72–94
71. Deleted in proof
72. King, D., Nedwell, D. B. 1985. The influence of nitrate concentration upon the end-products of nitrate dissimilation by bacteria in anaerobic salt marsh sediment. *FEMS Microbiol. Ecol.* 31:23–28
73. Knowles, R. 1982. Denitrification. *Microbiol. Rev.* 46:43–70
74. Koerselman, W. 1989. Groundwater and surface water hydrology of a small groundwater-fed fen. *Wetlands Ecol. Manag.* 1:31–43

75. Koerselman, W., Bakker, S. A., Blom, M. 1990. Nitrogen, phosphorus and potassium mass balances for two small fens surrounded by heavily fertilized pastures. *J. Ecol.* 78:428–42
76. Kolb, W., Martin, P. 1988. Influence of nitrogen on the number of N_2-fixing and total bacteria in the rhizosphere. *Soil Biol. Biochem.* 20:221–25
77. Lee, J. A., Press, M. C., Woodin, S. J. 1986. Effects of NO_2 on aquatic ecosystems. In *Study for the Need for an NO_2 Long-Term Limit Value for the Protection of Terrestrial and Aquatic Ecosystems,* pp. 99–119. Luxembourg, Sweden: Commission Eur. Commun.
78. Leuven, R. S. E. W., Wolfs, W. J. 1988. Effects of water acidification on the decomposition of *Juncus bulbosus* L. *Aquat. Bot.* 31:57–81
79. Levy, H., II., Moxim, W. J. 1987. Fate of US and Canadian combustion nitrogen emissions. *Nature* 328:414–16
80. Liljelund, L.-E., Torstensson, P. 1988. Critical load of nitrogen with regard to effects on plant composition. In *Critical Loads for Sulphur and Nitrogen: Report from a Workshop; March; Skokloster, Sweden,* ed. J. Nilsson, P. Grennfelt, pp. 363–73. Copenhagen, Denmark: Nordic Council of Ministers.
81. Lindau, C. W., DeLaune, R. D., Jones, G. L. 1988. Fate of added nitrate and ammonium-nitrogen entering a Louisiana gulf coast swamp forest. *J. Water Pollut. Control Fed.* 60:386–90
82. Logofet, D. O., Alexandrov, G. A. 1984. Modelling of matter cycle in a mesotrophic bog ecosystem: II. Dynamic model and ecological succession. *Ecol. Modell.* 21:259–76
83. Lovett, G. M., Reiners, W. A., Olson, R. K. 1982. Cloud droplet deposition in subalpine balsam fir forests: hydrological and chemical inputs. *Science* 218:-1303–4
84. Marinucci, A. C., Hobbie, J. E., Helfrich, J. V. K. 1983. Effect of litter nitrogen on decomposition and microbial biomass in *Spartina alterniflora. Microb. Ecol.* 9:27–40
85. Mendelssohn, I. A. 1979. The influence of nitrogen level, form, and application method on the growth response of *Spartina alterniflora* in North Carolina. *Estuaries* 2:106–12
86. Mitsch, W. J., Gosselink, J. G. 1986. *Wetlands.* New York: Van Nostrand Reinhold
87. Moore, D. R. J., Keddy, P. A. 1989. The relationship between species richness and standing crop in wetlands: the importance of scale. *Vegetatio* 79:99–106
88. Moore, D. R. J., Keddy, P. A., Gaudet, C. L., Wisheu, I. C. 1989. Conservation of wetlands: do infertile wetlands deserve a higher priority? *Biol. Conserv.* 47:203–17
89. Morris, J. T. 1982. A model of growth responses by *Spartina alterniflora* to nitrogen limitation. *J. Ecol.* 70:25–42
90. Morris, J. T. 1984. Effects of oxygen and salinity on ammonium uptake by *Spartina alterniflora* Loisel. and *Spartina patens* (Aiton) Muhl. *J. Exp. Mar. Biol. Ecol.* 78:87–98
91. Morris, J. T. 1988. Pathways and controls of the carbon cycle in salt marshes. In *The Ecology and Management of Wetlands,* ed. D. D. Hook, et al, pp. 497–510. London: Croom Helm
92. Morris, J. T., Dacey, J. W. H. 1984. Effects of O_2 on ammonium uptake and root respiration by *Spartina alterniflora. Am. J. Bot.* 71:979–85
93. Morris, J. T., Houghton, R. A., Botkin, D. B. 1984. Theoretical limits of belowground production by *Spartina alterniflora:* an analysis through modelling. *Ecol. Modell.* 26:155–75
94. Morris, J. T., Lajtha, K. 1986. Decomposition and nutrient dynamics of litter from four species of freshwater emergent macrophytes. *Hydrobiologia* 131:215–23
95. Neely, R. K., Davis, C. B. 1985. Nitrogen and phosphorus fertilization of *Sparganium eurycarpum* Engelm. and *Typha glauca* Godr. stands. I. Emergent plant production. *Aquat. Bot.* 22:347–61
96. Neely, R. K., Davis, C. B. 1985. Nitrogen and phosphorus fertilization of *Sparganium eurycarpum* Engelm. and *Typha glauca* Godr. stands. II. Emergent plant decomposition. *Aquat. Bot.* 22:363–75
97. Patrick, W. H. Jr., Delaune, R. D. 1976. Nitrogen and phosphorus utilization by *Spartina alterniflora* in a salt marsh in Barataria Bay, Louisiana. *Estuar. Coast. Mar. Sci.* 4:59–64
98. Pedrazzini, F. R., Moore, P. A. 1983. N_2O emission and changings of redox potential and pH in submerged soil samples. *Z. Pflanzenernaehr. Bodenkd.* 146:-660–65
99. Peterjohn, W. T., Correll, D. L. 1984. Nutrient dynamics in an agricultural watershed: observations on the role of a riparian forest. *Ecology* 65:1466–75
100. Prakasam, T. B. S., Krup, M. 1982. Denitrification. *J. Water Pollut. Control Fed.* 54:623–31
101. Press, M. C., Lee, J. A. 1982. Nitrate

reductase activity of *Sphagnum* species in the South Pennines. *New Phytol.* 92:487–94

102. Press, M. C., Woodin, S. J., Lee, J. A. 1986. The potential importance of an increased atmospheric nitrogen supply to the growth of ombrotrophic Sphagnum species. *New Phytol.* 103:45–55
103. Qualls, R. G. 1984. The role of leaf litter nitrogen immobilization in the nitrogen budget of a swamp stream. *J. Environ. Qual.* 13:640–44
104. Raven, J. A. 1985. Regulation of pH and generation of osmolarity in vascular plants: a cost-benefit analysis in relation to efficiency of use of energy, nitrogen and water. *New Phytol.* 101:25–77
105. Reddy, K. R., Patrick, W. H. 1984. Nitrogen transformations and loss in flooded soils and sediments. *CRC Crit. Rev. Environ. Control* 13:273–309
106. Reed, P. G. Jr. 1988. National list of plant species that occur in wetlands: national summary. *US Fish Wildl. Serv. Biol. Rep.* 88(24)
107. Robertson, G. P., Rosswall, T. 1986. Nitrogen in West Africa: the regional cycle. *Ecol. Monogr.* 56:43–72
108. Roelofs, J. G. M. 1983. Impact of acidification and eutrophication on macrophyte communities in soft waters in the Netherlands: I. Field observations. *Aquat. Bot.* 17:139–55
109. Roelofs, J. G. M. 1986. The effect of airborne sulphur and nitrogen deposition on aquatic and terrestrial heathland vegetation. *Experientia* 42:372–77
110. Roelofs, J. G. M., Boxman, A. W., van Dijk, H. F. G. 1987. Effects of airborne ammonium on natural vegetation and forests. In *Ammonia and Acidification,* ed. W. A. H. Asman, S. M. A. Diederen, pp. 266–76. Eur. Assoc. Sci. Air Pollution. Netherlands
111. Roelofs, J. G. M., Schuurkes, J. A. A. R., Smits, A. J. M. 1984. Impact of acidification and eutrophication on macrophyte communities in soft waters. II. Experimental studies. *Aquat. Bot.* 18:-389–411
112. Rotty, R. M. 1983. Distribution of and changes in industrial carbon dioxide production. *J. Geophys. Res.* 88:1301–8
113. Rozé, F. 1988. Nitrogen cycle in Brittany heathland. *Acta Oecol. Oecol. Plant.* 9:371–79
114. Rudd, J. W. M., Kelly, C. A., Schindler, D. W., Turner, M. A. 1988. Disruption of the nitrogen cycle in acidified lakes. *Science* 240:1515–17
115. Rudolph, H., Voigt, J. U. 1986. Effects of NH_4^+-N and NO_3^--N on growth and metabolism of *Sphagnum magellanicum. Physiol. Plant* 66:339–43
116. Sanville, W. 1988. Response of an Alaskan wetland to nutrient enrichment. *Aquat. Bot.* 30:231–43
117. Savant, N. K., De Datta, S. K. 1982. Nitrogen transformations in wetland rice soils. *Adv. Agron.* 35:241–302
118. Schuurkes, J. A. A. R., Elbers, M. A., Gudden, J. J. F., Roelofs, J. G. M. 1987. Effects of simulated ammonium sulphate and sulphuric acid rain on acidification, water quality and flora of small-scale soft water systems. *Aquat. Bot.* 28:199–226
119. Schuurkes, J. A. A. R., Kok, C. J., den Hartog, C. 1986. Ammonium and nitrate uptake by aquatic plants from poorly buffered and acidified waters. *Aquat. Bot.* 24:131–46
120. Seitzinger, S. P., Nixon, S. W., Pilson, E. Q. 1984. Denitrification and nitrous oxide production in a coastal marine ecosystem. *Limnol. Oceanogr.* 29:73–83
121. Seitzinger, S. P., Pilson, M. E. Q., Nixon, S. W. 1983. Nitrous oxide production in nearshore marine sediments. *Science* 222:1244–46
122. Smith, R. A., Alexander, R. B., Wolman, M. G. 1987. Water-quality trends in the nations's rivers. *Science* 235:-1607–15
123. Smith, R. A., Alexander, R. B., Wolman, M. G. 1987. Analysis and interpretation of water quality trends in major U.S. rivers, 1974–1981. *US Geol. Surv. Water-Supply Pap. 2307*
124. Smith, C. J., DeLaune, R. D. 1983. Gaseous nitrogen losses from Gulf Coast marshes. *Northeast Gulf Sci.* 6: 1–8
125. Smith, C. J., DeLaune, R. D. 1985. Recovery of added ^{15}N-labelled ammonium-N from Louisiana Gulf Coast estuarine sediment. *Estuar. Coast. Shelf Sci.* 21:225–33
126. Steen, E. 1984. Root and shoot growth of *Atriplex litoralis* in relation to nitrogen supply. *Oikos* 42:74–81
127. Steudler, P. A., Peterson, B. J. 1984. Contribution of gaseous sulphur from salt marshes to the global sulphur cycle. *Nature* 311:455–57
128. Tallis, J. H. 1964. Studies on southern Pennine peats: III. The behaviour of *Sphagnum. J. Ecol.* 52:345–53
129. Tans, P. P., Fung, I. Y., Takahashi, T. 1990. Observational constraints on the global atmospheric CO_2 budget. *Science* 247:1431–38
130. Tate, R. L. III. 1979. Effect of flooding on microbial activities in organic soils:

carbon metabolism. *Soil Sci.* 128:267–73

131. Tiner, R. W. Jr. 1984. *Wetlands of the United States: Current Status and Recent Trends*. US Fish & Wildlife Serv. Washington: USGPO
132. Urban, N. R., Eisenreich, S. J. 1988. Nitrogen cycling in a forested Minnesota bog. *Can. J. Bot.* 66:435–49
133. Valiela, I., Teal, J. M. 1974. Nutrient limitation in salt marsh vegetation. In *Ecology of Halophytes,* ed. R. J. Reimold, W. H. Queen, pp. 547–63. New York: Academic
134. Valiela, I., Teal, J. M. 1979. The nitrogen budget of a salt marsh ecosystem. *Nature* 280:652–56
135. Valiela, I., Teal, J. M., Persson, N. Y. 1976. Production and dynamics of experimentally enriched salt marsh vegetation: belowground biomass. *Limnol. Oceanogr.* 21:245–52
136. Valiela, I., Teal, J. M., Sass, W. J. 1975. Production and dynamics of salt marsh vegetation and the effects of experimental treatment with sewage sludge: biomass, production and species composition. *J. Appl. Ecol.* 12:973–82
137. Valiela, I., Wilson, J., Buchsbaum, R., Rietsam, C., Bryant, D., et al. 1984. Importance of chemical composition of salt marsh litter on decay rates and feeding by detritivores. *Bull. Mar. Sci.* 35:261–69
138. Van der Molen, J., Bussink, D. W., Vertregt, N., Van Faassen, H. G., den Boer, D. J. 1989. Ammonia volatilization for arable and grassland soils. In *Nitrogen in Organic Wastes Applied to Soils,* ed. J. A. A. Hansen, K. Henriksen, pp. 185–201. New York: Academic
139. van der Valk, A. G., Attiwill, P. M. 1983. Above- and below-ground litter decomposition in an Australian salt marsh. *Aust. J. Ecol.* 8:441–47
140. Verhoeven, J. T. A., Arts, H. H. M. 1987. Nutrient dynamics in small mesotrophic fens surrounded by cultivated land: II. N and P accumulation in plant biomass in relation to the release of inorganic N and P in the peat soil. *Oecol.* 72:557–61
141. Verhoeven, J. T. A., Koerselman, W., Beltman, B. 1988. The vegetation of fens in relation to their hydrology and nutrient dynamics: a case study. In *Vegetation of Inland Waters,* ed. J. J. Symoens, pp. 249–82. Dordrecht, The Netherlands: Kluwer Academic
142. Vermeer, J. G. 1986. The effect of nutrients on shoot biomass and species composition of wetland and hayfield communities. *Acta Ecol. Acta Plant.* 7:31–41
143. Vermeer, J. G., Berendse, F. 1983. The relationship between nutrient availability, shoot biomass and species richness in grassland and wetland communities. *Vegetatio* 53:121–26
144. Verry, E. S., Timmons, D. R. 1982. Waterborne nutrient flow through an upland-peatland watershed in Minnesota. *Ecology* 63:1456–67
145. Wheeler, B. D., Giller, K. E. 1982. Species richness of herbaceous fen vegetation in Broadland, Norfolk in relation to the quantity of above-ground plant material. *J. Ecol.* 70:179–200
146. Wisheu, I. C., Keddy, P. A. 1989. The conservation and management of a threatened coastal plain plant community in eastern North America (Nova Scotia, Canada). *Biol. Conserv.* 48:229–38
147. Woodin, S. J., Lee, J. A. 1987. The fate of some components of acidic deposition in ombrotrophic mires. *Environ. Pollut.* 45:61–72
148. Woodin, S., Press, M. C., Lee, J. A. 1985. Nitrate reductase activity in *Sphagnum fuscum* in relation to wet deposition of nitrate from the atmosphere. *New Phytol.* 99:381–88
149. Yates, P., Sheridan, J. M. 1983. Estimating the effectiveness of vegetated floodplains/wetlands as nitrate-nitrite and orthophosphorus filters. *Agric. Ecosys. Environ.* 9:303–14
150. Yoch, D. C., Whiting, G. J. 1986. Evidence for NH_4^+ switch-off regulation of nitrogenase activity by bacteria in salt marsh sediments and roots of the grass *Spartina alterniflora. Appl. Environ. Microbiol.* 51:143–49
151. Zemba, S. G., Golomb, D., Fay, J. A. 1988. Wet sulfate and nitrate deposition patterns in eastern North America. *Atmos. Environ.* 22:2751–61

Annu. Rev. Ecol. Syst. 1991. 22:281–308

MOLECULAR CHANGES AT SPECIATION

Richard G. Harrison

Section of Ecology and Systematics, Cornell University, Ithaca, New York 14853

KEY WORDS: mitochondrial DNA, population structure, gene genealogy, founder effect, phylogeography

---

INTRODUCTION

The remarkably rapid growth of molecular genetics during the past two decades and concomitant advances in DNA technology have had an enormous impact in systematic and evolutionary biology (76, 89, 92, 110, 129). The ability to compare DNA sequences (either directly or indirectly) has resulted in a wealth of new, high resolution genetic markers appropriate for defining patterns of variation at all levels in the evolutionary hierarchy. Discoveries in molecular genetics have also fundamentally altered and expanded our understanding of genome structure and dynamics and of patterns and mechanisms of gene regulation. These discoveries, in turn, have fueled speculation about possible implications for evolutionary process (84, 127).

In this review I examine the consequences of speciation for patterns of molecular genetic variation within and among populations. I consider the following questions: (*i*) Does speciation leave a distinctive signature on patterns of molecular genetic variation? If so, can we use variation in DNA sequences (or allozymes) to gain insights into either the geography of speciation or the evolutionary forces that have been operating? (*ii*) Can we use estimates of genetic distance to make judgments about species status or to date speciation events?

Inferring process from pattern in evolutionary biology is notoriously difficult, and the use of molecular genetic variation to illuminate processes of

0066-4162/91/1120-0281$02.00

speciation is no exception. There is little doubt that molecular markers can be used to document patterns of variation at high resolution, but these patterns are sometimes ambiguous and often cannot provide definitive evidence for a particular population history.

Although clearly a major focus of evolutionary biology, the study of speciation has never emerged as a coherent discipline. It falls at the interface between population biology and systematic biology and does not fit comfortably into either domain. Systematic biologists define relationships among species (the units produced by past speciation events) and then infer process by examining the geographic distributions of sister species and the nature of the differences between them. An alternative approach is to consider speciation as a problem in population genetics, extrapolating from evolutionary dynamics within populations and establishing criteria for identifying the role of traditional evolutionary forces (mutation, recombination, drift, selection, gene flow) in promoting or inhibiting genetic isolation or cohesion. In order to reconstruct phylogeny, systematists document character state distributions of presumably homologous characters across nonanastomosing lineages. In contrast, population geneticists traditionally have relied on observations of frequencies of Mendelian markers (allelic variants) within and between populations of interbreeding individuals. Data from molecules, especially DNA sequences, clearly demonstrate that these approaches are complementary and that there is an obvious and direct connection between ancestor-descendant relationships within populations (genealogies) and phylogeny (5, 7). These data may ultimately allow a synthesis of population genetic and systematic (phylogenetic) approaches to the study of speciation.

MOLECULAR MARKERS IN POPULATION GENETICS AND SYSTEMATICS

Protein Electrophoresis

Protein electrophoresis, introduced to population biologists in the 1960s, provided the first easy access to an array of genetic markers that could be used to define amounts and patterns of variation within and between species. The great strength of protein gel electrophoresis is that, with relatively little investment of time and money, it is possible to characterize patterns of variation for many independent nuclear gene markers (109). Allozyme data are especially useful for documenting genetic differences among individuals or populations and for defining patterns of genetic exchange (or lack thereof). They have been used effectively to study population structure (both current and historical) (36, 39, 68, 69, 90, 130), to delineate species boundaries (65, 66, 105), and to characterize patterns of introgression (72). Allozymes have also been used to reconstruct phylogenies of closely related species, but considerably more controversy surrounds this application (24, 35, 137). First, there is clearly substantial "hidden variation" (41); bands with identical electrophoretic mobility cannot be assumed to represent identical alleles. Second, problems arise in defining characters and character states and in establishing reasonable criteria for ordering character states. It is impossible to determine genealogical relationships among alleles at single loci based on comparisons of electrophoretic mobility.

DNA Restriction Site and Sequence Data

Evolutionary and systematic biologists have increasingly come to rely on data from DNA sequences. At the level of conspecific populations or closely related species, most published comparisons of DNA sequences have been derived from analyses of restriction fragment patterns or restriction site maps (53). Although indirect, this approach provides estimates of DNA sequence similarity or difference averaged over many thousands of base pairs (e.g. entire organelle genomes). Following the introduction of the polymerase chain reaction (PCR), which permits selective amplification of specific sequences from large numbers of individuals (e.g. population samples), direct sequencing of DNA has become the method of choice for many (but certainly not all) applications (75). The great advantage of DNA data is that it can be used to extract detailed information on gene genealogies. If recombination rates are sufficiently low, a short length of DNA sequence (a gene) will have a shared evolutionary history, and a sample of alleles (from one or more populations) can be traced backwards through a series of coalescence events to a common ancestral gene (79; see Figure 1). Knowledge of the genealogy of sampled genes provides population geneticists with new opportunities for evaluating the importance of selection, drift, and gene flow in maintaining variation within populations and determining population structure (79, 134, 135).

Restriction site maps and DNA sequences also have obvious advantages for phylogenetic analysis (110, 137). Characters are virtually limitless and character states easy to define. Sequence comparisons provide exceedingly high resolution and therefore are especially attractive for tracing population histories and reconstructing phylogenies of closely related species.

Chloroplast DNA (cpDNA) and animal mitochondrial DNA (mtDNA) have received far more attention than most nuclear gene sequences, both because they are easy to isolate and purify and because of their relatively simple sequence organization (4, 40, 71, 107, 119). Organelle genomes are most often uniparentally inherited and nonrecombining. These characteristics make them ideal markers for phylogeny reconstruction because their sequences record the history of a lineage uncomplicated by recombination. Animal mtDNA exhibits a high rate of sequence divergence and therefore has become

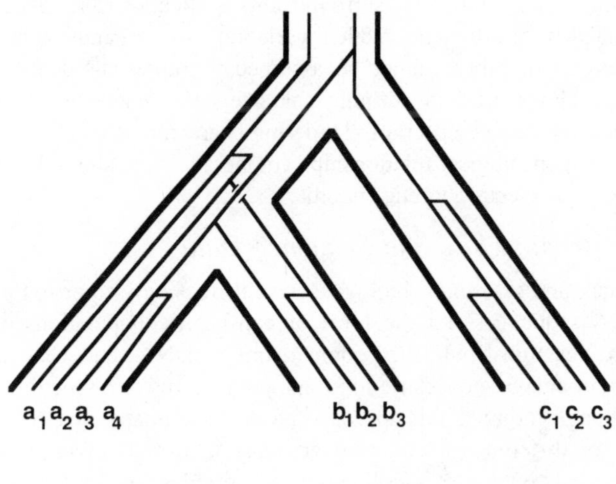

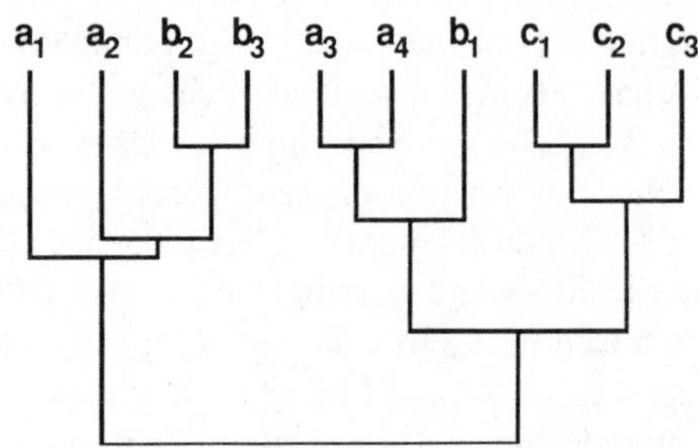

Figure 1 Genealogy of alleles in three populations (species), showing random sorting of ancestral polymorphisms. In the upper part of the figure the allele genealogy is superimposed on the pattern of dichotomous branching of populations leading to three extant populations (A, B, C). The lower figure shows the same allele genealogy but with alleles clustered by most recent common ancestor rather than by population. Some alleles from population A (a_3, a_4) are more closely related to alleles from population B and C than they are to other alleles in population A. Only population C appears to be monophyletic with respect to this genetic marker.

a favorite tool for studies of population structure, hybrid zones, and closely related species (4, 7, 71, 157). In contrast, the slow average rate of cpDNA evolution has meant that plant biologists have had greater difficulty identifying molecular markers appropriate for studies at the level of speciation (e.g. see 15, 119, 136).

It should be noted here that we can distinguish between tokogenetic (birth) relationships among individuals within populations and phylogenetic relationships of species or higher taxa. Some cladists (115) have argued that hierarchical phylogenetic methods are not appropriate for analysis of relationships within species, because of reticulation (interbreeding). However, these authors acknowledge that phylogenetic methods can be applied to clonally inherited molecules [mtDNA, cpDNA, short (nonrecombining) segments of nuclear DNA] (49, 115). Debates about the appropriate use of terminology or methodology should not obscure the essential point—that it is necessary to distinguish clearly between allele phylogenies and organismal phylogenies and to understand the correspondence between them.

Both differential introgression and random sorting of ancestral polymorphisms can lead to discordance between gene trees and species trees (71, 110, 114, 121, 139, 141, 143, 159). Hybridization between species (or gene exchange between populations) may result in incorporation (possibly fixation) of alleles from one species in the gene pool of the second species (58, 73). This pattern of introgression may be limited to one or a few markers—mtDNA in the case of populations of house mice (58, 67) or voles (145) in Scandinavia; use of these markers for phylogenetic analysis will obviously give a different view of population histories than would markers that show no evidence of introgression.

A more general (and perhaps more serious) problem is that, when there is allelic variation within species, an allele phylogeny will not necessarily have the same topology as the species phylogeny. If polymorphisms persist through speciation events, the probability that the gene tree and the species tree have the same topology may be quite small. This is considered in more detail later in the review.

Neutral Theory of Molecular Evolution

The neutral theory of molecular evolution (82) explains both polymorphisms within species and accumulation of amino acid or nucleotide substitutions over time in terms of mutation and random drift. It has provided an extremely important frame of reference for both molecular population geneticists and molecular systematists, and even its most severe critics agree that it is "the most widely held theory of molecular evolution" (64).

If alleles are strictly neutral, then variation within populations is a function of mutation rate (u) and effective population size (N); spatial patterns of variation are determined by these parameters and by the dispersal or migration rate (m). If population size and population structure are not constant over time, observed levels and patterns of variation will depend not only on current values of N and m, but also on long-term effective population size and past gene flow (110, 133). Therefore, differences between observed and "expected" values may provide important clues to population histories, including the nature of recent speciation events. Expected values are those calculated using neutral theory and observed (or assumed) values of u, N and m. For example, Nei & Graur (111) attribute the low levels of heterozygosity in many species (compared with expected values) to population bottlenecks during recent periods of glaciation. Similarly, the genetic distances among mtDNA haplotypes in red-winged blackbirds, American eels, and hardhead catfish are far lower than expected based on estimates of current population sizes (8). To explain the discrepancy, Avise et al (8) suggest either that mutation rates are dramatically reduced in these species (considered unlikely) or that long-term effective population sizes are significantly smaller than current population sizes.

Population bottlenecks and founder events are often invoked as critical elements in speciation processes (see below). The decline in genetic variability during a population bottleneck depends both on the size of the bottleneck and the rate of population growth (38, 113). If the bottleneck is brief, with population size increasing rapidly, the reduction in heterozygosity is small (although many or most rare alleles will be eliminated). Only if small population size persists for many generations will the reduction in heterozygosity be substantial (17). Population bottlenecks also lead to a rapid increase in genetic distance between populations, although this effect gradually disappears once the populations returns to the prebottleneck size (38, 110). Note that this result (based on a strictly neutral model) does not imply that founder events or population bottlenecks accelerate the rate of molecular evolution. The increase in genetic distance is a result of changes in gene frequencies and does not reflect a change in the rate of accumulation of new mutations (157). However, if most mutations are very slightly deleterious, rather than strictly neutral, the rate of evolution is inversely proportional to population size (82, 116, 117). In small populations, slightly deleterious mutations are effectively neutral and may drift to fixation; in large populations, these mutations are consistently eliminated by selection. Therefore, under the model of slightly deleterious alleles, population bottlenecks are expected to increase the rate of evolution (150).

Different genetic systems will respond differently to fluctuations in population size and to migration rates (150). Comparisons between organelle genes and nuclear genes may be especially instructive. Because organelle genes are generally homoplasmic (invariant within individuals) and inherited uniparentally, the effective number of organelle genes is one fourth that for nuclear genes when the sex ratio of breeding individuals (N_m/N_f) is 1 : 1 (28). As a consequence, in comparison with nuclear genes, the time to fixation or loss of neutral alleles is shorter for mitochondrial or chloroplast genes, the expected haplotype diversity at equilibrium is lower (assuming equal mutation rates), the extent of population subdivision is greater, and the loss of variability during a population bottleneck is more extreme (28, 150, 157). In both *Daphnia pulex* and *Drosophila mercatorum,* comparisons of mtDNA and allozyme data reveal greater population subdivision for mtDNA than for nuclear gene markers (45, 51). In the case of strict maternal inheritance (mtDNA in most animals and cpDNA in many plants) the fourfold difference is reduced or eliminated when $N_m/N_f<1$ (e.g. when males have harems). In contrast, male-biased dispersal will increase the difference between organelle and nuclear genes in their tendency to show population subdivision (28). Within the nuclear genome, X and Y chromosome markers will behave differently from autosomal markers, in general exhibiting greater sensitivity to founder events (150).

Even the most dedicated supporters of the neutral theory acknowledge that alleles at some loci are likely to be under strong directional and/or balancing selection. The frequencies of neutral alleles at one locus will be influenced by the impact of selection on linked loci. In particular, directional selection leading to the fixation of a new variant at one locus can result in a loss of neutral variation at linked loci (81, 96). The magnitude of this "hitchhiking effect" will depend upon the strength of selection (the rapidity with which the new mutation sweeps through the population) and the rate of recombination between the selected and neutral loci. For organelle genomes that lack recombination, hitchhiking effects may be of considerable significance (93). For example, recent selective sweeps could explain the lower than expected genetic distances among mtDNA haplotypes observed in the three wide-ranging vertebrate species studied by Avise et al (8). Even if relatively few variants are under direct selection at a given time, linkage and hitchhiking effects may extend their sphere of influence across significant portions of the genome (2).

Molecular Clocks

According to the neutral theory the rate of gene substitution is simply equal to the neutral mutation rate per locus (82). As long as mutation rates are constant over time, nucleotide substitutions will accumulate in clock-like fashion, and the number of substitutions (and other measures of genetic distance) will be proportional to time since lineage splitting. The reality and constancy of "molecular clocks" (160) are obviously of enormous interest to students of speciation, since they potentially provide a framework for dating speciation events. Substitution rates vary among genes (due in part to variation in selective constraints) and also among evolutionary lineages (explained by generation-time effects or by differences in the efficiency of DNA repair) (29, 89, 110).

Molecular clocks must be carefully calibrated, using divergence times based on the fossil record or on known vicariance events. A number of protein electrophoretic clocks have been proposed, each with a characteristic value of a constant k (where $t = kD$, t being time of divergence and D some measure of genetic distance). Unfortunately, values of k vary by a factor of 20, suggesting that protein electrophoretic clocks can at best give a very rough estimate of absolute times of divergence (6, 77). Calibrations of a mtDNA clock in a number of different vertebrate lineages suggest an initial average rate of divergence of 1% per million years per lineage (31, 131, 157). In primates, the relationship between time and amount of sequence divergence appears to be linear up to about 10–15% sequence divergence. In fact, there is significant heterogeneity in rates, among different coding regions, between synonomous and nonsynonomous substitutions, and between transitions and transversions (107). Thus, mtDNA sequences potentially harbor several different molecular clocks, each ticking at a characteristic rate. Unfortunately, it is not at all clear that clock rates are constant across lineages (107, 154). If we cannot extrapolate across lineages, it may be difficult (impossible) to obtain calibrations for organisms with poor fossil records.

Gillespie (61–64) has documented that rates of molecular evolution are often more variable than expectations based on a simple Poisson mutation process, leading him to suggest that patterns of both protein evolution and silent site evolution of DNA are not consistent with neutral theory. He proposes that molecular evolution is episodic (61) and is best explained by models invoking natural selection. Takahata (140) acknowledges that the molecular clock is "overdispersed" but suggests modifications of neutral theory to account for the observations. Regardless of which interpretation is correct, attempts to apply molecular clocks to comparisons of closely related species may be especially risky.

For much of what follows, it is convenient to start with the neutral theory as null hypothesis, as long as we do not fall into the trap of uncritically accepting it as truth (108).

THE IMPACT OF SPECIATION ON MOLECULAR GENETIC VARIATION

Evolutionary biologists have traditionally classified speciation events on the basis of the geographic context in which initial divergence occurs (34, 97, 99, 156). Allopatric, parapatric, and sympatric models of speciation describe situations in which divergence occurs in geographically isolated populations, in contiguous but nonoverlapping populations, or between subpopulations at a single locality. In fact, these situations represent points along a continuum defined by the amount of gene flow (m) between diverging populations (57). More recently, Templeton (146–149) stressed the need for a mechanistic taxonomy that can provide a population genetic framework for understanding speciation processes. He also pointed out that both ancestral population structure and type of split (geographic subdivision) influence the probability of particular modes of speciation and the extent of genetic differentiation among sister species (146, 147).

An appropriate context for understanding the impact of speciation on patterns of molecular genetic variation would seem to be an amalgam of the traditional and mechanistic taxonomies outlined above, augmented by an explicit phylogenetic approach. The important components of this framework are (*a*) ancestral population structure, (*b*) nature of the sampling event(s) giving rise to daughter species, (*c*) role of natural selection (consequences of "selective sweeps"), and (*d*) genetic architecture of speciation.

Population Structure

For any pair of sister species (or monophyletic group of species), the genetic structure of the common ancestral population(s) will be an important determinant of current patterns of variation. Genetic structure refers both to the total amount of variation and to how that variation is apportioned within and among populations. For DNA restriction site or sequence data, population structure can be characterized using an explicitly genealogical approach to calculate expected times to common ancestry for alleles from the same or different populations (79, 110). Taking advantage of the phylogenetic information that may be contained in DNA sequences, the genetic structure of a species can be clearly portrayed by superimposing a gene genealogy on a distribution map. Avise (5, Avise et al 7) has coined the term "intraspecific phylogeography" to describe this approach and has emphasized the utility of animal mtDNA for analyses of population structure.

The genetic structures of a wide variety of animal and plant species have been documented using as markers variants detected by protein electrophoresis (69, 90, 130). All possible structures have been found—ranging from high levels of polymorphism to virtual monomorphism, from homogeneity of allele frequencies across the entire range of widely distributed species to remarkable population subdivision on a very small spatial scale. Similarly, geographic surveys of mtDNA restriction site variation have revealed striking genetic structuring of populations in some animal species and absence of variation or homogeneity across vast distances in other taxa. For example, in American eels (10), red-winged blackbirds (14), monarch butterflies (40), crested newts (155) and sea urchins (120), there is relatively little sequence divergence among the observed mtDNA haplotypes and no evidence of geographic differentiation. In contrast, distinct phylogenetic assemblages (differing by 2–9% in mtDNA sequence) are geographically localized (often parapatrically distributed) in pocket gophers (9), field mice (11), grasshopper mice (125), desert tortoises (85), American oysters (124), and several species of sunfish (26). The observed "phylogenetic discontinuities" often correspond to current barriers to gene flow or to historical barriers inferred from regional geology and paleoclimatic reconstruction. Rarely do species show large amounts of sequence divergence among mtDNA haplotypes in the absence of geographic structure. Genetic structure revealed by mtDNA comparisons is not always reflected in patterns of allozyme variation; both oysters (32) and field mice (12) appear relatively homogeneous in allele frequencies for nuclear gene markers. It may be unwise to extrapolate from observations of mtDNA variation, given the difference between organelle and nuclear markers in extent of population subdivision, especially in animals with male-biased dispersal. Clearly genetic structure depends both on the biology of the organism and on the properties of the genetic marker.

Observations of current population structures can be used to infer ancestral population structure, but because "the demographies of populations have been remarkably dynamic and unsettled over space and recent evolutionary time" (5), there are obvious problems with this approach. Population structure certainly varies among congeneric species (2, 36, 39). However, data from extant populations at least provide insights into the range of population structures commonly encountered in particular groups of organisms. Using these data, we can begin to define how the amount of polymorphism within populations and the extent of population subdivision interact with mode of speciation to produce patterns of variation in descendant lineages (species).

How Is Variation Partitioned at Speciation?

In the process of speciation, a single lineage is split into two (or more) independent lineages, each of which potentially carries a different sample of the variation that existed in the ancestral species. For molecular markers that are "neutral" with respect to speciation, the geography of the partition and the amount of gene flow at the time of divergence interact with the ancestral population structure to determine the extent of genetic differentiation (genetic distance) and the phylogenetic relationships of the descendant lineages (114, 146; see Figure 2). The demographic histories of the diverging lineages (subsequent to the initial split) also influence amounts and patterns of variation.

It is difficult to construct a single, hierarchical classification scheme of sampling events. At one extreme, founders of a lineage leading to a new species may be drawn from a single local population. The founder population may be a relictual population isolated by a vicariance event or a group of dispersing individuals (a single female?) that successfully colonize previously unoccupied habitat. Sympatric speciation events may also involve founders drawn from one local source area; in this case gene flow between subpopulations persists during at least the initial stages of divergence. In all three models, the genetic variation present in the founder population will be a sample of the variation found in a single local population; the extent to which this sample reflects overall levels/patterns of variation in the ancestral species obviously depends on ancestral population structure. If there is significant heterogeneity among populations, the founder will represent only a small

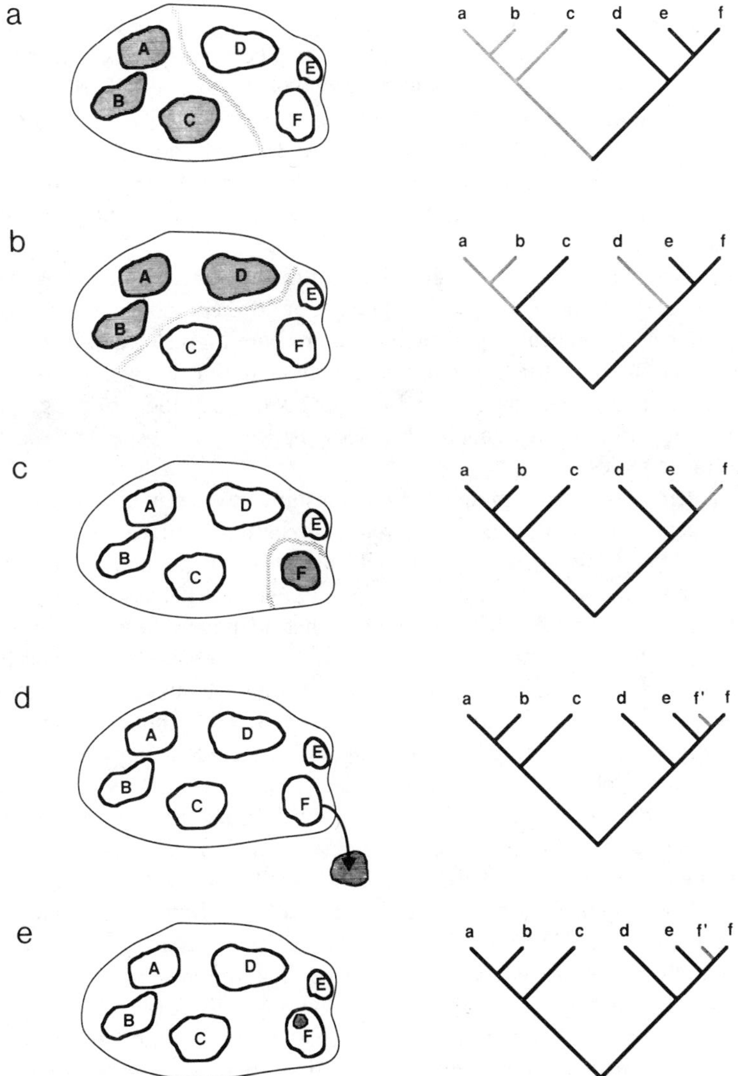

Figure 2 Five models of speciation and corresponding gene trees, showing distribution of alleles in the two daughter species. For simplicity, each population is represented as monomorphic, the gene tree in each case is (((a,b)(c))((d)(e,f))) and speciation is assumed to occur instantaneously. The five models are: (*a*) speciation by subdivision with the partition congruent with existing phylogenetic discontinuity; (*b*) speciation by subdivision with partition not congruent with existing phylogenetic discontinuity; (*c*) divergence of peripheral population; (*d*) colonization of new habitat by propagule(s) from single population; and (*e*) local sympatric speciation. In each allele phylogeny the dark lines represent lineages occurring in one daughter species, the light lines represent lineages found in the other daughter species.

fraction of the variation found in the ancestral species. Individuals giving rise to a new species need not be drawn from a single population but, at least in theory, could be a random sample of the entire ancestral species (i.e. including representatives from throughout the range of this species). Perhaps more realistic are models that invoke partitioning of a single species into two (or more) major geographic subdivisions (Figure 2). Such models may involve either allopatric or parapatric divergence. Each daughter lineage derives from many founders and contains much of the variation found within a broad geographic area.

Apart from considerations of the impact of founder events, there have been few general discussions of the consequences of different modes of speciation for patterns of molecular genetic variation. For a given combination of ancestral population structure and type of split (sampling event), it should be possible to define expectations for (*a*) amounts of variation, (*b*) geographic patterns of variation, (*c*) genetic distances (between sister species) and (*d*) allele phylogenies.

Templeton (146) has explored the relationship between mode of speciation and expected amount of genetic differentiation between the daughter species. For an adaptive divergence model, with the ancestral population split into large subpopulations, he predicts that initial genetic distances may be quite large. This follows from his assumptions that adaptive divergence is most likely when the ancestral population consists of many small demes and that the split occurs "along geographical lines." Templeton suggests that speciation via genetic transilience or founder effect (which should occur most frequently when founders are derived from large, panmictic populations) will result in relatively low initial genetic distances, comparable to those seen between demes of the ancestral species. However, he cautions that there is "no simple pattern between mode of speciation and genetic distance."

An alternative (and ultimately more powerful) approach is to examine expected gene genealogies when alleles are sampled from two recently diverged populations (species) (Figure 2). The simplest case is to assume that each population consists of N individuals, that there is no selection or recombination and no gene flow between the populations (which diverged t generations ago). Under these assumptions, the probability of obtaining certain types of evolutionary relationships (tree topologies) depends on the population size (N) and the time since divergence (t) (139, 143). When $t < 2N$, the probability that the allele phylogeny will accurately reflect the recent population history is low, but this probability increases with increasing t, the precise relationship depending on the number of alleles sampled from each of the extant populations (143). When the ancestral population is polymorphic and extinction of lineages is random, it may often be the case that haplotypes found in different species are more similar than are some haplotypes found within conspecific populations (i.e. interspecific coalescences will occur (going backwards in time) before some intraspecific coalescences). For example, analysis of *Adh* haplotypes in *Drosophila pseudoobscura* and *D. persimilis* indicates that the *D. persimilis* haplotype "falls within the cluster of *D. pseudoobscura* haplotypes" (128). If some type of balancing selection maintains variation within populations, polymorphisms may persist for millions of years (across several speciation events). This appears to be the case for polymorphisms at the major histocompatability complex loci in rodents and primates (59, 87, 100, 144) and at the self-incompatability locus in Solanaceae (80).

Clearly, both the nature of the divergence (speciation) event and the ancestral population structure will influence the probability of obtaining particular allele phylogenies. Neigel & Avise (114) have used computer simulations to explicitly model phylogenetic sorting of mtDNA lineages across speciation events. Founders were either chosen at random from the entire array of haplotypes present at the time of the split, at random from different halves of the array, or from opposite ends of the array. The latter two sampling regimes are equated with daughter species arising from geographically isolated populations, i.e. the ancestral species is assumed to have distinct genetic structure, such that geographically distant populations are characterized by haplotypes that are far apart on the mtDNA tree. Neigel & Avise (114) examined the "time to expected monophyly" (time until the allele phylogeny reflects the population history) and showed that it is dependent on both the geography of the original partition and the number of founder individuals for each of the daughter species. When the daughter species come from distinct geographic areas, there is a greater chance that at least one of the diverging lineages will appear monophyletic (lineage sorting will have occurred prior to speciation).

Population bottlenecks during speciation will decrease the time to expected monophyly, because N will be very small and even when time since divergence is short, it will likely be $> 4N$ ($4N$ generations is the expected coalescence time for neutral alleles in a population—110). Bottlenecks force most (or all) lineages to coalesce in a short period of time, the pattern of coalescence depending on the severity and duration of the bottleneck. Therefore, all intraspecific coalescences will likely occur prior to interspecific coalescences (again going back in time), and allele and species phylogenies will be equivalent (79).

Can we increase our confidence that observed allele phylogenies provide an accurate reflection of recent populations history? Pamilo & Nei (121) argue that the consistency probability between gene trees and species trees cannot be increased significantly by increasing the number of alleles sampled from each population (species). They conclude that it is necessary to obtain sequence data from several independent (unlinked) gene regions. Thus, the apparent advantages of DNA sequence data for reconstructing phylogenies of closely related species may only be fully realized when a number of unlinked genes are examined (42, 79, 121, 159; but see 141). Most published studies have involved comparisons of only one gene (usually one organelle genome).

Founder Effect Speciation

A long-standing debate in evolutionary biology concerns the genetic consequences of founder events and their implications for the origin of species (17, 37, 123). The debate centers not only on the possibility that founder effects are triggers for speciation in certain lineages, but on the nature of the genetic and demographic processes that can provide the appropriate trigger. Carson & Templeton (37) outline three distinct models of founder effect speciation: (*i*) Mayr's original model of "genetic revolution" (98), in which speciation occurs in peripheral populations, results in substantial loss of genetic variation and involves much of the genome; (*ii*) Carson's founder-flush models (available in several varieties—see 123) in which a population flush (and associated relaxation of selection) follows immediately on the heels of a bottleneck, resulting in minimal loss of genetic variation; and (*iii*) Templeton's genetic transilience model (147), in which founders derived from an outcrossed and highly variable ancestral population experience a brief bottleneck (an altered genetic environment), again with minimal loss of variation. Carson & Templeton (37) suggest that there is "little chance of speciation if a significant drop in levels of genetic variation occurs." In contrast, Barton & Charlesworth (17) reject the argument that founder events likely to lead to speciation will involve only minor loss of variation (models 2 and 3), asserting that "population bottlenecks small enough to cause peak shifts will inevitably cause a substantial and prolonged loss of variability at neutral loci." Therefore, inferences about founder effect speciation from patterns of molecular genetic variation depend critically on details of the model.

Featured prominently in the debate about founder effect speciation are the Hawaiian *Drosophilia,* which have been the inspiration for Carson's founder-flush models. The observation that recently derived Hawaiian species are no less variable at allozyme loci than many continental *Drosophila* suggests to Barton & Charlesworth (17) that founder events have not been important in speciation, whereas Carson & Templeton (37) see the relatively high heterozygosity as in conflict with Mayr's model but precisely the "pattern expected under the founder-flush and genetic transilience models." Barton (16) also views the relatively high mtDNA haplotype diversity (and sequence divergence among haplotypes) in *D. silvestris* and *D. heteroneura* (50) as evidence against founder effect speciation.

Neither the founder-flush nor the genetic transilience modes of speciation will leave an unambiguous signature on patterns of molecular genetic variation, although loss of mtDNA variation without accompanying decline in heterozygosity at nuclear gene loci may be indicative of a severe but very brief population bottleneck (118). Intense and prolonged population bottlenecks will have a major impact on variation in mtDNA and nuclear gene markers. In some taxa genetic variation in one species represents a "depauperate subset" of variation found in a more widely distributed sister species (91). In other cases, there is almost complete absence of variation in one member of a clade, whereas congeners have "normal" levels of heterozygosity (101). Such patterns suggest that population bottlenecks have been associated with speciation. In contrast, the remarkable persistence of MHC polymorphisms in primates (e.g. shared polymorphisms between humans and chimpanzees) argues against any significant bottlenecks in these lineages (142). A direct correspondence of population bottlenecks with speciation is virtually impossible to prove. A bottleneck subsequent to lineage splitting would erase any characteristic signature left by the speciation event.

Speciation by Subdivision

Vicariance events partition the range of a single species into two or more subdivisions and may ultimately lead to allopatric divergence of populations and to speciation. The consequences for patterns of genetic variation depend on the initial population structure and the nature of the partition. For species in which genetic variants are geographically localized—e.g. grasshopper mice (125), desert tortoises (85), sunfish species in southeastern United States (26)—speciation may involve further divergence between what are already distinct phylogenetic assemblages. Comparisons of mtDNA haplotypes in field mice indicate that *Peromyscus maniculatus* is paraphyletic with respect to *P. polionotus* and that the latter species is monophyletic (11). *P. polionotus* is restricted to the southeastern United States, whereas *P. maniculatus* is widespread throughout North America but does not overlap with *P. polionotus*. The observed pattern is, as one would expect if *P. polionotus* is a "peripheral isolate", recently diverged in allopatry (Figure 2). The *P. polionotus* clade harbors a diversity of haplotypes equivalent to that found within each of several genetically distinct geographical assemblages of *P. maniculatus,* suggesting a simple vicariance event without a significant bottleneck. In this case, geographic distribution and pattern of molecular genetic variation together provide strong support for a particular model of speciation.

It is not obvious that we should always expect the partition that occurs during speciation to be congruent with existing phylogenetic discontinuties (which may have been produced by previous vicariance events). The outcome of speciation, in terms of genetic distance or gene genealogies, is therefore not predictable, but neither daughter species may appear monophyletic (Figure 2). If allopatric speciation (subsequent to a vicariance event) occurred in species in which common haplotypes were widely distributed, the daughter species would be expected to share ancestral polymorphisms until lineage sorting resulted in monophyly with respect to genetic markers.

Sympatric Speciation

The consequences of sympatric speciation for patterns of molecular genetic variation depend on details of the speciation model. Virtually all scenarios for sympatric divergence start with a stable polymorphism affecting performance on different resources or in different habitats (52, 95). Speciation involves the evolution of an association between this polymorphism and alleles at loci affecting assortative mating (often mediated by habitat or resource selection). If sympatric speciation occurs at a single locality within the range of a widespread ancestral type, the daughter species, having shifted to a new habitat or resource, can then spread. The outcome, in terms of patterns of molecular genetic variation, may not be distinguishable from that produced by a founder event—i.e. one species will contain a subset of the variation found in the sister species, the precise pattern depending on the population structure of the ancestor. Alternatively, in cases of sister species sympatric over a broad area (e.g. host races of *Rhagoletis pomonella* (33)), current distributions may reflect multiple "speciation events" (host shifts). In *R. pomonella,* allelic diversity and heterozygosity are as high in the derived apple race as in the ancestral hawthorne race (56). This would seem to argue against a single local origin. However, it is possible that the scenario of the apple race arising in the Hudson Valley in the 1860s is correct (33), but that repeated episodes of introgression have introduced additional variation. Inferences about origins are further complicated by the the apparent importance of selection in determining allele frequencies at allozyme loci (56), producing latitudinal variation in allozyme frequencies in both host races.

Certain geographic patterns of variation are not consistent with sympatric origins. For example, based on mtDNA haplotype analysis, the normal and dwarf phenotypes of whitefish found sympatrically in northern lakes belong to distinct monophyletic assemblages that are primarily distributed allopatrically (27). This suggests that the sympatric pairs are the result of secondary contact of forms that differentiated in allopatry, rather than originating through divergence in situ.

Speciation by Hybridization

Molecular markers have proved especially informative in examining the hybrid origins of polyploid plants and unisexual animals. The markers have been used both to establish the identity of maternal and paternal parental species and to provide information on origins. Low levels of sequence variation within certain unisexual lizards—e.g. *Cnemidophorus tesselatus* and *C. neomexicanus* (48), *C. sexlineatus* (47)—suggest that these species are of recent origin. Lack of differentiation between the cpDNAs of polyploid and diploid representatives of the *Glycine tabacina* complex (54) suggests that not much time has elapsed since the hybridization events that gave rise to the polyploids. Indeed, recent origins appear to be the rule. Furthermore, in a number of different systems, allozymes, cpDNA, or mtDNA have provided evidence for multiple origins of unisexual (55, 78, 106) or polyploid (54, 136) lineages. Obligate parthenogenesis in *Daphnia pulex,* although not of hybrid origin, is also clearly polyphyletic (46). Finally, molecular markers have been used extensively to document species formation through introgression in plants (126).

Impact of Natural Selection

If speciation is accompanied or followed by episodes of directional selection, resulting in the fixation of new alleles (e.g. one of the daughter species occupies a "new" niche or habitat), hitchhiking will result in loss of variability for markers closely linked to the loci under selection (79, 81). The impact of natural selection will depend on the number of loci under selection and the relative magnitudes of selection coefficients and rates of recombination. If few loci are involved, only a small proportion of the genome will be affected. If many loci, distributed across all chromosomes, are involved, the loss of variation could be pervasive and the molecular genetic consequences could mimic those produced by a population bottleneck. If selective sweeps are very rapid, variation may be eliminated in the region surrounding the locus under selection (i.e. linkage disequilibrium will persist throughout the course of the selective sweep). If selection coefficients are small and divergence very gradual, the amount of neutral variation at linked loci may remain close to that expected at equilibrium.

Some genetic systems appear to lack recombination entirely (e.g. animal mtDNA) and will be particularly sensitive to periodic directional selection (93). In the Drosophila nuclear genome, recombination is reduced near the base and tip of the X-chromosome, and genes that map to these regions have reduced levels of variation within species (but not reduced levels of divergence between species) (2). This is the pattern expected if directional selection and hitchhiking are important in determining patterns of molecular genetic variation. At this juncture, we do not have sufficient data on the role of directional selection during speciation or on rates of recombination in organisms other than those favored by geneticists. It is clear, however, that the impact of directional selection can mimic that of a founder event (for a linked group of markers) and render suspect interpretations of recent population histories based on assumptions of neutrality. Analyses of patterns of variation for several unlinked molecular markers will be necessary if we are to have confidence in interpreting these patterns.

GENETIC ARCHITECTURE OF SPECIATION

The genetic architecture of speciation refers to the number, effect (large or small), and chromosomal distribution of genes that contribute to reproductive (genetic) isolation. Templeton (148) distinguished three alternative architectures: (*i*) many segregating units (each presumably of small effect) (type I), (*ii*) one or a few major segregating units, with many epistatic modifiers (type II), and (*iii*) complementary pairs of loci (type III). He argued (149) that only in cases of speciation by slow adaptive divergence are many genes likly to be involved and that more often speciation involves a type II architecture (especially in founder event speciation). These views are by no means universally accepted. Based on both theory and observations (from experimental crosses and hybrid zones), Barton & Charlesworth (16, 17) suggest that reproductive isolation usually depends on changes at many independent gene loci and that "there is no simple relationship between genetic architecture and the likely mode of speciation" (16). Using cline theory, estimates of the number of genes responsible for isolation are 150 for chromosomal races of the grasshopper *Podisma pedestris* (18, 19) and 50–300 for the hybridizing frogs *Bombina bombina*/*B. variegata* (138).

Patterns of differentiation (or introgression) for molecular markers depend on the underlying genetic architecture of speciation. If only a few major genes are involved, differentiation of neutral markers may only occur (or persist) for a small segment of the genome (regions closely linked to the few genes under selection) (19). If many genes, each of small effect, are involved, divergence will be more uniformly distributed across loci. In a hybrid zone, the flow of neutral alleles across the zone will be retarded to the extent that they are linked to loci that affect hybrid fitness and/or positive assortative mating (20, 21, 23, 72). Many hybrid zones show clear evidence of differential introgression, i.e. there is substantial variation in the extent to which molecular markers (allozymes, rDNA, mtDNA, Y-chromosome markers) penetrate hybrid zones (72). The tendency for mtDNA to introgress more extensively than nuclear gene markers is usually attributed to its segregating independently of all nuclear genes (22, 71). However, in no single hybrid zone have sufficient markers been examined to clearly define the underlying genetic architecture. With techniques borrowed from molecular genetics (identification of restriction fragment length polymorphisms either by traditional surveys using cloned probes or newer PCR based random amplification techniques), it will be possible not only to shed light on genetic architecture but ultimately to map genes responsible for components of the mate recognition system or for post-zygotic barriers (86, 122).

In a few cases, information on genetic architecture is available. For *Drosophila* there is convincing evidence for a major effect of the X chromosome on post-zygotic isolation and this "rule of speciation" may well extend to other groups of animals (44). The host races of *Rhagoletis pomonella* are consistently differentiated at six loci coding for soluble enzymes (out of a much larger number surveyed) (56). These six loci map to three discrete regions of the genome, suggesting that differentiation of the host races may depend on a limited number of genes. This is consistent with a model of recent sympatric origins.

INFERENCES FROM GENETIC DISTANCE

Genetic Distance as an Indicator of Species Status

The taxonomic status of allopatric populations is difficult to determine when we employ a species concept based on genetic exchange or field for recombination. In these situations a phylogenetic species concept (115) can be more consistently applied. However, if the evolution of genetic isolation were simply time-dependent and if there existed a reliable and universal molecular clock, genetic distance would be a good predictor of species status. In *Drosophila,* there is a significant correlation between genetic distance (based on allozyme data) and the extent of either pre- or post-zygotic isolation measured in the laboratory (43). Using this relationship and an estimate of the degree of isolation required to prevent fusion of sympatric populations, Coyne & Orr (43) calculated a threshold value of D beyond which populations of *Drosophila* appear to be distinct species. It is intriguing that this value is much lower for sympatric pairs than for allopatric pairs, suggesting that species status is achieved approximately twice as fast in sympatry as in allopatry.

In other groups of organisms, genetic distance is not a good predictor of degree of genetic (reproductive) isolation. For example, in sea urchins on either side of the Isthmus of Panama there is "no correlation between genetic divergence and strength of reproductive isolation" (88). In pocket gophers (genus *Thomomys*) speciation is "unrelated to the level of genetic differentiation between populations" (68).

Thorpe (152) plotted the distributions of genetic identities (based on allozyme data) for conspecific populations and congeneric species in a wide variety of taxa. He concluded that "if allopatric populations of dubious status have genetic identities below 0.85 it is improbable that they should be considered conspecific, while nominate species with I values above 0.85 should be considered doubtful if there is no other evidence of their specific status" (152). This suggestion has met with some approval, although support is certainly not unequivocal (13, 24, 102, 110). One obvious flaw is that birds were intentionally excluded from the analysis, because "their speciation processes seem to differ fundamentally from those of most other organisms" (152). This assessment was based on the observation that many avian sister species have values of I significantly greater than 0.85.

In fact many species pairs in a variety of groups of animals and plants show little genetic differentiation (1, 25, 60, 70, 74, 83, 102, 132, 153, 158). One of the most striking recent examples is an analysis of mtDNA variation in cichlid fishes from Lake Victoria (103). The "species flock" in Lake Victoria comprises some 200 species and yet appears to harbor little genetic variation (only about 2% of the surveyed nucleotide sites are variable in the 14 species examined). In contrast, subspecies of sunfish often differ by 6–10% in mtDNA sequence (26). In field crickets, closely related but clearly distinct species share mtDNA haplotypes, whereas geographically isolated populations of single species may have haplotypes that differ by 1–2% (R. G. Harrison, S. Bogdanowicz, unpublished data). Such observations suggest that extreme caution must be used in assessing species status on the basis of estimates of genetic differentiation. This is not unexpected, given the diverse genetic architectures and population histories that can lead to the evolution of barriers to gene exchange.

Can We Date Speciation Events?

One of the most appealing characteristics of electrophoretic, restriction site, or DNA sequence data is the possibility that they can provide reliable molecular clocks for estimating absolute divergence times between species. In fact, clock calibrations for allozyme data are remarkably variable (6), due in large part to the lack of independent estimates of the times of speciation events. Clocks based on animal mtDNA may be more reliable, although there is still considerable debate about whether rates are the same in all lineages (e.g. see 154).

However, for recent divergence events there are problems other than those associated with clock rates. Even if substitutions accumulate at a constant rate, lineage sorting due to sampling of alleles (haplotypes) at speciation and subsequent random lineage extinction may introduce significant errors, especially when little time has elapsed since speciation. Obviously, this is only a problem when the ancestral population (from which the daughter species are "sampled") is polymorphic. In such cases, it is clear that the divergence of gene sequences must precede population splitting and therefore current measures of sequence divergence can only overestimate the time since speciation (110, 143). Another way to think about the problem is to acknowledge that the amount of divergence at the time of speciation is not zero (151) and to apply a correction based on current levels of intraspecific variation (110, 112, 157). However, this correction assumes that levels of variation in the daughter species accurately reflect the amount of variation found in the ancestral population. Furthermore, the net number of nucleotide substitutions (*d*) has such a large variance when *d* is small that estimates of divergence times will be very unreliable (110, 143). This problem will be compounded if population bottlenecks have occurred. In addition, there may well be variation in the estimate of time since divergence based on comparisons of different molecular markers. This could reflect differences in the level of polymorphism at the time of speciation (151), or it could be a result of differential introgression or natural selection.

Finally, there is the possibility that speciation events themselves accelerate rates of divergence and therefore estimates of genetic differentiation will be proportional to the number of speciation events rather than to time. An early test of this hypothesis (3) compared rates of protein evolution in two groups of fishes of supposedly equal age, but with very different rates of speciation; it found no evidence that speciation accelerates protein divergence. This study has been criticized, primarily because the groups being compared do not in fact appear to satisfy the assumptions underlying the test (94). More recently, Mindell et al (104) presented evidence based on allozyme data in *Sceloporus* lizards and concluded that "punctuational change is at least a viable explanation" for their observations. This question obviously deserves more careful attention.

CONCLUSIONS

With recent advances in DNA technology, it is clear that we have arrived at a critical juncture in the application of molecular markers in systematic and evolutionary biology. DNA sequences can provide remarkably detailed views of patterns of variation within species and phylogenetic relationships among species. The challenge is to understand what these patterns reveal about evolutionary history and about evolutionary process.

In studies of speciation, evolutionary biologists must first clearly define how particular modes of speciation influence patterns of molecular genetic variation. Building on the work of others (5, 7, 79, 110, 114, 121, 139, 141, 143, 146, 148–150), I have attempted in this review to provide an introduction to major issues in molecular studies of speciation and an assessment of the problems encountered in resolving these issues. Debate over the nature and consequences of founder event speciation emphasizes how difficult it can be to infer process from pattern. It is evident that speciation events do not leave unique signatures and that it will be impossible to "prove" that a certain series of historical events has occurred. However, from patterns of molecular genetic variation, together with detailed knowledge of the biology of organisms, it will be possible to eliminate certain models or at least assign high probabilities to some scenarios and low probabilities to other scenarios. For example, it may be possible to show that recent bottlenecks or sympatric origins are unlikely explanations, based on both haplotype diversity and spatial distributions of haplotypes (e.g. see 142).

Especially important is the need to generate sequence data from several (many) unlinked genes, and to compare genealogies (phylogenies) based on these independent markers. Because a set of completely linked characters (e.g. an organelle genome) behaves as a single unit in random lineage extinction and introgression, an allele phylogeny based on a single gene may not reflect population history. However, unlinked markers will sort independently at divergence events and by using many such markers we can increase our chances of obtaining the "correct" species tree (121). Data from unlinked markers may also allow us to discriminate between reduction in variation due to selective sweeps and reduction in variation due to population bottlenecks. Such data should also allow evaluation of the importance of differential introgression as a confusion in constructing phylogenies and dating divergence events. It is essential both to build on the already substantial data base from cpDNA and animal mtDNA and to embark on a program of characterizing and comparing nuclear gene markers.

The ultimate frontier in studying the genetics of speciation is to be able to map (and eventually clone) genes responsible for prezygotic and postzygotic barriers to gene exchange. Producing detailed RFLP maps for species of interest is a first step in this process. Techniques for generating markers and for analyzing the results of crosses are now available (86, 122). The task is a formidable one, but provided with the necessary resources, evolutionary biologists should now be able to make significant progress in clarifying the genetics of the speciation process.

Acknowledgments

I am indebted to many colleagues who have challenged and inspired my thinking about molecular evolution and molecular systematics. Special thanks to Chip Aquadro and Jeff Doyle, with whom I have had many conversations about issues discussed in this review. I am also grateful for a long line of excellent graduate students—both past and present—who have contributed enormously to my education. My work in molecular evolution/molecular systematics has been generously funded by the Systematic Biology Program of the National Science Foundation.

Literature Cited

1. Apfelbaum, L., Reig, O. A. 1989. Allozyme genetic distances and evolutionary relationships in species of akodontine rodents (Cricetidae: Sigmodontinae). *Biol. J. Linn. Soc.* 38:257–80
2. Aquadro, C. F. 1991. Molecular population genetics of *Drosophila*. In *Molecular Approaches to Pure and Applied Entomology,* ed. J. Oakeshott, M. Whitten. Springer-Verlag
3. Avise, J. C. 1977. Is evolution gradual or rectangular? Evidence from living fishes. *Proc. Natl. Acad. Sci. USA* 74:5083–87
4. Avise, J. C. 1986. Mitochondrial DNA and the evolutionary genetics of higher animals. *Philos. Trans. R. Soc. Lond. B* 312:325–42
5. Avise, J. C. 1989. Gene trees and organismal histories: a phylogenetic approach to population biology. *Evolution* 43:1192–1208
6. Avise, J. C., Aquadro, C. F. 1982. A comparative summary of genetic distances in the vertebrates: patterns and correlations. *Evol. Biol.* 15:151–85
7. Avise, J. C., Arnold, J., Ball, R. M., Bermingham, E., Lamb, T., et al. 1987. Intraspecific phylogeography: The mitochondrial DNA bridge between population genetics and sytematics. *Annu. Rev. Ecol. Syst.* 18:489–522
8. Avise, J. C., Ball, R. M., Arnold, J. 1988. Current versus historical population sizes in vertebrate species with high gene flow: A comparison based on mitochondrial DNA lineages and inbreeding theory for neutral mutations. *Mol. Biol. Evol.* 5:331–44
9. Avise, J. C., Giblin-Davidson, C., Laerm, J., Patton, J. C., R. A. Lansman. 1979. Mitochondrial DNA clones and matriarchal phylogeny within and among geographic populations of the pocket gopher, *Geomys pinetis*. *Proc. Natl. Acad. Sci. USA* 76:6694–98
10. Avise, J. C., Helfman, G. S., Saunders, N. C., Hales, L. S. 1986. Mitochondrial DNA differentiation in North Atlantic eels: Population genetic consequences of an unusual life history pattern. *Proc. Natl. Acad. Sci. USA* 83:4350–54
11. Avise, J. C., Shapira, J. F., Daniel, S. W., Aquadro, C. F., Lansman, R. A. 1983. Mitochondrial DNA differentiation during the speciation process in *Peromyscus*. *Mol. Biol. Evol.* 1:38–56
12. Avise, J. C., Smith, M. H., Selander, R. K. 1979. Biochemical polymorphism and systematics in the genus *Peromyscus*. VII. Geographic differentiation in members of the *truei* and *maniculatus* species groups. *J. Mammal.* 60:177–92
13. Avise, J. C., Zink, R. M. 1988. Molecular genetic divergence between avian sibling species: King and clapper rails, long-billed and short-billed dowitchers, boat-tailed and great-tailed grackles, and tufted and black-crested titmice. *The Auk* 105:516–28
14. Ball, R. M. Jr., Freeman, S., James, F. C., Bermingham, E., Avise, J. C. 1988. Phylogeographic population structure of Red-winged Blackbirds assessed by mitochondrial DNA. *Proc. Natl. Acad. Sci. USA* 85:1558–62
15. Banks, J. A., Birky, C. W. Jr. 1985. Chloroplast DNA diversity is low in a wild plant, *Lupinus texensis*. *Proc. Natl. Acad. Sci. USA* 82:6950–54
16. Barton, N. H. 1989. Founder effect speciation. In *Speciation and its Consequences,* ed. D. Otte, J. A. Endler, pp. 229–56. Sunderland, Mass: Sinauer
17. Barton, N. H., Charlesworth, B. 1984. Genetic revolutions, founder effects, and speciation. *Annu. Rev. Ecol. Syst.* 15:133–64
18. Barton, N. H., Hewitt, G. M. 1981. The genetic basis of hybrid inviability in the grasshopper *Podisma pedestris*. *Heredity* 47:367–83
19. Barton, N. H., Hewitt, G. M. 1981. Hybrid zones and speciation. In *Evolution and Speciation,* ed. W. R. Atchley, D. S. Woodruff, pp. 109–45. Cambridge: Cambridge Univ. Press
20. Barton, N. H., Hewitt, G. M. 1985. Analysis of hybrid zones. *Annu. Rev. Ecol. Syst.* 16:113–48
21. Barton, N. H., Hewitt, G. M. 1989. Adaptation, speciation and hybrid zones. *Nature* 341:497–503
22. Barton, N. H., Jones, J. S. 1983. Mitochondrial DNA: new clues about evolution. *Nature* 306:317–18
23. Bazykin, A. D. 1969. Hypothetical mechanism of speciation. *Evolution* 23: 685–87
24. Berlocher, S. H. 1984. Insect molecular systematics. *Annu. Rev. Entomol* 29: 403–33
25. Berlocher, S. H., Bush, G. L. 1982. An electrophoretic analysis of *Rhagoletis* (Diptera: Tephritidae) phylogeny. *Syst. Zool.* 31:136–55
26. Bermingham, E., Avise, J. C. 1986. Molecular zoogeography of freshwater

fishes in the southeastern United States. *Genetics* 113:939–65
27. Bernatchez, L., Dodson, J. J. 1990. Allopatric origin of sympatric populations of lake whitefish *(Coregonus clupeaformis)* as revealed by mitochondrial-DNA restriction analysis. *Evolution* 44:1263–71
28. Birky, C. W. Jr., Maruyama, T., Fuerst, P. 1983. An approach to population and evolutionary genetic theory for genes in mitochondria and chloroplasts, and some results. *Genetics* 103:513–27
29. Britten, R. J. 1986. Rates of DNA sequence evolution differ between taxonomic groups. *Science* 231:1393–98
30. Brower, A. V. Z., Boyce, T. M. 1991. Mitochondrial DNA variation in monarch butterflies. *Evolution*. In press
31. Brown, W. M., Prager, E. M., Wang, A., Wilson, A. C. 1982. Mitochondrial DNA sequences of primates: Tempo and mode of evolution. *J. Mol. Evol.* 18:225–39
32. Buroker, N. E. 1983. Population genetics of the American oyster *Crassostrea virginica* along the Atlantic coast and the Gulf of Mexico. *Mar. Biol.* 75:99–112
33. Bush, G. L. 1969. Sympatric host race formation and speciation in frugivorous flies of the genus *Rhagoletis* (Diptera: Tephritidae). *Evolution* 23:237–51
34. Bush, G. L. 1975. Modes of animal speciation. *Annu. Rev. Ecol. Syst.* 6: 339–64
35. Buth, D. G. 1984. The application of electrophoretic data in systematic studies. *Annu. Rev. Ecol. Syst.* 15:501–22
36. Caccone, A., Sbordoni, V. 1987. Molecular evolutionary divergence among North American cave crickets. I. Allozyme variation. *Evolution* 41:1198–1214
37. Carson, H. L., Templeton, A. R. 1984. Genetic revolutions in relation to speciation phenomena: The founding of new populations. *Annu. Rev. Ecol. Syst.* 15:97–131
38. Chakraborty, R., Nei, M. 1977. Bottleneck effects on average heterozygosity and genetic distance with the stepwise mutation model. *Evolution* 31:347–56
39. Choudhary, M., Singh, R. 1987. A comprehensive study of genic variation in natural populations of *Drosophila melanogaster*. III. Variations in genetic structure and their causes between *Drosophila melanogaster* and its sibling species *Drosophilia simulans*. *Genetics* 117:697–710
40. Clegg, M. T., Leam, G. H., Golenberg, E. M. 1991. See Ref. 129, pp. 135–49
41. Coyne, J. A. 1982. Gel electrophoresis and cryptic protein variation. *Isozymes: Curr. Top. Biol. Med. Res.* 6:1–32
42. Coyne, J. A., Kreitman, M. 1986. Evolutionary genetics of two sibling species, *Drosophila simulans* and *D. sechellia*. *Evolution* 40:673–91
43. Coyne, J. A., Orr, H. A. 1989. Patterns of speciation in *Drosophila*. *Evolution* 43:362–81
44. Coyne, J. A., Orr, H. A. 1989. Two rules of speciation. See Ref. 16, pp. 180–207
45. Crease, T. J., Lynch, M., Spitze, K. 1990. Hierarchical analysis of population genetic variation in mitochondrial and nuclear genes of *Daphnia pulex*. *Mol. Biol. Evol.* 7:444–58
46. Crease, T. J., Stanton, D. J., Hebert, P. D. N. 1989. Polyphyletic origins of asexuality in *Daphnia pulex*. II. Mitochondrial-DNA variation. *Evolution* 43:1016–26
47. Densmore, L. D. III, Moritz, C. C., Wright, J. W., Brown, W. M. 1989. Mitochondrial DNA analyses and the origin and relative age of parthenogenetic lizards (genus *Cnemidophorus*). IV. Nine *sexlineatus*- group unisexuals. *Evolution* 43:969–83
48. Densmore, L. D. III, Wright, J. W., Brown, W. M. 1989. Mitochondrial DNA analyses and the origin and relative age of parthenogenetic lizards (genus *Cnemidophorus*). II. *C. neomexicanus* and the *C. tesselatus* complex. *Evolution* 43:943–57
49. De Queiroz, K., Donaghue, M. 1990. Phylogenetic systematics and species revisited. *Cladistics* 6:83–90
50. DeSalle, R., Giddings, L. V., Templeton, A. R. 1986. Mitochondrial DNA variability in natural populations of Hawaiian *Drosophila*. I. Methods and levels of variability in *D. silvestris* and *D. heteroneura* populations. *Heredity* 56:75–85
51. DeSalle, R., Templeton, A. R., Mori, I., Pletscher, S., Johnston, J. S. 1987. Temporal and spatial heterogeneity of mtDNA polymorphisms in natural populations of *Drosophila mercatorum*. *Genetics* 116:215–23
52. Diehl, S. R., Bush, G. L. 1989. The role of habitat preference in adaptation and speciation. See Ref. 16, pp. 345–65
53. Dowling, T. E., Moritz, C., Palmer, J. D. 1990. Nucleic acids II: Restriction site analysis. See Ref. 76, pp. 250–317
54. Doyle, J. J., Doyle, J. L., Brown, A. H. D., Grace, J. P. 1990. Multiple origins of polyploids in the *Glycine tabacina* complex inferred from chloroplast DNA

polymorphism. *Proc. Natl. Acad. Sci. USA* 714–17

55. Echelle, A. A., Dowling, T. E., Moritz, C. C., Brown, W. M. 1989. Mitochondrial-DNA diversity and the origin of the *Menidia clarkhubbsi* complex of unisexual fishes (Atherinidae). *Evolution* 43:984–93
56. Feder, J. L., Chilcote, C. A., Bush, G. L. 1990. The geographic pattern of genetic differentiation between host associated populations of *Rhagoletis pomonella* (Diptera: Tephritidae) in the eastern United States and Canada. *Evolution* 44:570–94
57. Felsenstein, J. 1981. Skepticism towards Santa Rosalia or why are there so few kinds of animals. *Evolution* 35:124–38
58. Ferris, S. D., Sage, R. D., Huang, C. M., Nielsen, J. T., Ritte, U., Wilson, A. C. 1983. Flow of mitochondrial DNA across a species boundary. *Proc. Natl. Acad. Sci. USA* 80:2290–94
59. Figueroa, F., Günther, E., Klein, J. 1988. MHC polymorphism pre-dating speciation. *Nature* 335:265–71
60. Gartside, D. F., Rogers, J. S., Dessauer, H. C. 1977. Speciation with little genic and morphological differentiation in the ribbon snakes *Thamnophis proximus* and *T. sauritus* (Colubridae). *Copeia* 1977:697–705
61. Gillespie, J. H. 1984. The molecular clock may be an episodic clock. *Proc. Natl. Acad. Sci. USA* 81:8009–13
62. Gillespie, J. H. 1986. Rates of molecular evolution. *Annu. Rev. Ecol. Syst.* 17:637–65
63. Gillespie, J. H. 1986. Variability of evolutionary rates of DNA. *Genetics* 113:1077–91
64. Gillespie, J. H. 1987. Molecular evolution and the neutral allele theory. *Oxford Surveys Evol. Biol.* 4:10–37
65. Grassle, J. P., Grassle, J. F. 1976. Sibling species in the marine pollution indicator *Capitella* (Polychaeta). *Science* 192:567–69
66. Guttman, S. I., Karlin, A. A. 1986. Hybridization of cryptic species of two-lined salamanders (*Eurycea bislineata* complex). *Copeia* 1986:96–109
67. Gyllensten, U., Wilson, A. C. 1987. Interspecific mitochondrial DNA transfer and the colonization of Scandinavia by mice. *Genet. Res.* 49:25–29
68. Hafner, M. S., Hafner, J. C., Patton, J. L., Smith, M. F. 1987. Macrogeographic patterns of genetic differentiation in the pocket gopher *(Thomomys umbrinus). Syst. Zool.* 36:18–34
69. Hamrick, J. L., Godt, M. J. 1990. Allozyme diversity in plant species. In *Plant Population Genetics, Breeding and Genetic Resources,* ed. A. H. D. Brown, M. T. Clegg, A. L. Kahler, B. S. Weir, pp. 43–63. Sunderland, Mass: Sinauer
70. Harrison, R. G. 1979. Speciation in North American field crickets: Evidence from electrophoretic comparisons. *Evolution* 33:1009–23
71. Harrison, R. G. 1989. Animal mitochondrial DNA as a genetic marker in population and evolutionary biology. *Trends Ecol. Evol.* 4:6–11
72. Harrison, R. G. 1990. Hybrid zones: windows on evolutionary process. *Oxford Surveys Evol. Biol.* 7:69–128
73. Harrison, R. G., Rand, D. M., Wheeler, W. C. 1987. Mitochondrial DNA variation in field crickets across a narrow hybrid zone. *Mol. Biol. Evol.* 4:144–58
74. Helenurm, K., Ganders, F. R. 1985. Adaptive radiation and genetic differentiation in Hawaiian *Bidens*. *Evolution* 39:753–65
75. Hillis, D. M., Larson, A., Davis, S. K., Zimmer, E. A. 1990. Nucleic acids III: Sequencing. See Ref. 76, pp. 318–370
76. Hillis, D. M., Moritz, C. 1990. *Molecular Systematics*. Sunderland, Mass: Sinauer
77. Hillis, D. M., Moritz, C. 1990. An overview of applications of molecular systematics. See Ref. 76, pp. 502–515
78. Honeycutt, R. L., Wilkinson, P. 1989. Electrophoretic variation in the parthenogenetic grasshopper *Warramaba virgo* and its sexual relatives. *Evolution* 43:1027–44
79. Hudson, R. R. 1990. Gene genealogies and the coalescent process. *Oxford Surveys Evol. Biol.* 7:1–44
80. Ioerger, T. R., Clark, A. G., Kao, T.-H. 1990. Polymorphism at the self-incompatability locus in Solanaceae predates speciation. *Proc. Natl. Acad. Sci. USA* 87:9732–35
81. Kaplan, N. L., Hudson, R. R., Langley, C. H. 1989. The "hitchhiking effect" revisited. *Genetics* 123:887–99
82. Kimura, M. 1983. *The Neutral Theory of Molecular Evolution*. Cambridge: Cambridge Univ. Press
83. Kirkpatrick, M., Selander, R. K. 1979. Genetics of speciation in lake whitefishes in the Allegash Basin. *Evolution* 33:478–85
84. Krieber, M., Rose, M. R. 1986. Molecular aspects of the species barrier. *Annu. Rev. Ecol. Syst.* 17:465–85
85. Lamb, T., Avise, J. C., Gibbons, J. W. 1989. Phylogeographic patterns in mitochondrial DNA of the desert tortoise

(*Xerobates agassizi*), and evolutionary relationships among the North American gopher tortoises. *Evolution* 43:76–87

86. Lander, E. S., Botstein, D. 1989. Mapping mendelian factors underlying quantitative traits using RFLP linkage maps. *Genetics* 121:185–99
87. Lawlor, D. A., Ward, F. E., Ennis, P. D., Jackson, A. P., Parham, P. 1988. HLA-A and B polymorphisms predate the divergence of humans and chimpanzees. *Nature* 335:268–71
88. Lessios, H. A., Cunningham, C. W. 1990. Gametic incompatibility between species of the sea urchin *Echinometra* on the two sides of the Isthmus of Panama. *Evolution* 44:933–41
89. Li, W.-H., Graur, D. 1991. *Fundamentals of Molecular Evolution*. Sunderland, Mass: Sinauer
90. Loveless, M. D., Hamrick, J. L. 1984. Ecological determinants of genetic structure in plant populations. *Annu. Rev. Ecol. Syst.* 15:65–96
91. Loveless, M. D., Hamrick, J. L. 1988. Genetic organization and evolutionary history in two North American species of *Cirsium*. *Evolution* 42:254–65
92. MacIntyre, R. J., ed. 1985. *Molecular Evolutionary Genetics*. New York: Plenum
93. Maruyama, T., Birky, C. W. Jr. 1991. Effects of periodic selection on gene diversity in organelle genomes and other systems without recombination. *Genetics* 127:449–51
94. Mayden, R. L. 1986. Speciose and depauperate phylads and tests of punctuated and gradual evolution: fact or artifact? *Syst. Zool.* 35:591–602
95. Maynard Smith, J. 1966. Sympatric speciation. *Am. Nat.* 100:637–50
96. Maynard Smith, J., Haigh, J. 1974. The hitchhiking effect of a favorable gene. *Genet. Res. Camb.* 23:23–35
97. Mayr, E. 1942., *Systematics and the Origin of Species*. New York: Columbia Univ. Press
98. Mayr, E. 1954. Change of genetic environment and evolution. In *Evolution as a Process,* ed. J. Huxley, A. C. Hardy, E. B. Ford, pp. 157–80. London: George Allen & Unwin
99. Mayr, E. 1963. *Animal Species and Evolution*. Cambridge: Belknap
100. McConnell, T. J., Talbot, W. S., McIndoe, R. A., Wakeland, E. K. 1988. The origin of MHC class II gene polymorphism within the genus *Mus*. *Nature* 332:651–54
101. Menken, S. B. J. 1987. Is the extremely low heterozygosity level in *Yponomeuta rorellus* caused by bottlenecks? *Evolution* 41:630–37
102. Menken, S. B. J., Ulenberg, S. A. 1989. Biochemical characters in agricultural entomology. In *Genetical and Biochemical Aspects of Invertebrate Crop Pests,* ed. G. E. Russell, pp. 129–184. Andover, Hampshire, UK: Intercept
103. Meyer, A., Kocher, T. D., Basasibwaki, P., Wilson, A. C. 1990. Monophyletic origin of Lake Victoria cichlid fishes suggested by mitochondrial DNA sequences. *Nature* 347:550–53
104. Mindell, D. P., Sites, J. W., Jr., Graur, D. 1989. Speciational evolution: a phylogenetic test with allozymes in *Sceloporus* (Reptilia). *Cladistics* 5:49–61
105. Miyamoto, M. M. 1983. Biochemical variation in the frog *Eleutherodactylus bransfordii:* geographic patterns and cryptic species. *Syst. Zool.* 32:43–51
106. Moritz, C., Donnellan, S., Adams, M., Baverstock, P. R. 1989. The origin and evolution of parthenogenesis in *Heteronotia binoei* (Gekkonidae): Extensive genotypic diversity among parthenogens. *Evolution* 43:994–1003
107. Moritz, C., Dowling, T. E., Brown, W. M. 1987. Evolution of animal mitochondrial DNA: Relevance for population biology and systematics. *Annu. Rev. Ecol. Syst.* 18:269–92
108. Moritz, C., Hillis, D. M. 1990. Molecular systematics: Context and controversies. See Ref. 76, pp. 1–12
109. Murphy, R. W., Sites, J. W. Jr., Buth, D. G., Haufler, C. H. 1990. Proteins I. Isozyme electrophoresis. See Ref. 76, pp. 45–126
110. Nei, M. 1987. *Molecular Evolutionary Genetics*. New York: Columbia Univ. Press
111. Nei, M., Graur, D. 1984. Extent of protein polymorphism and the neutral mutation theory. *Evol. Biol.* 3:418–26
112. Nei, M., Li, W.-H. 1979. Mathematical model for studying genetic variation in terms of restriction endonucleases. *Proc. Natl. Acad. Sci. USA* 76:5269–73
113. Nei, M., Maruyama, T., Chakraborty, R. 1975. The bottleneck effect and genetic variability in populations. *Evolution* 29:1–10
114. Neigel, J. E., Avise, J. C. 1986. Phylogenetic relationships of mitochondrial DNA under various demographic models of speciation. In *Evolutionary Processes and Theory,* ed. S. Karlin, E. Nevo, pp. 515–34. New York: Academic
115. Nixon, K. C., Wheeler, Q. D. 1990. An

amplification of the phylogenetic species concept. *Cladistics* 6:211–23
116. Ohta, T. 1973. Slightly deleterious mutant substitutions in evolution. *Nature* 246:96–98
117. Ohta, T. 1976. Role of very slightly deleterious mutations in molecular evolution and polymorphism. *Theor. Popul. Biol.* 10:254–75
118. Ovenden, J. R., White, R. W. G. 1990. Mitochondrial and allozyme genetics of incipient speciation in a landlocked population of *Galaxias truttaceus* (Pisces: Galaxiidae). *Genetics* 124:701–16
119. Palmer, J. D. 1987. Chloroplast DNA evolution and biosystematic uses of chloroplast DNA variation. *Am. Nat.* 130:S6–S29
120. Palumbi, S. R., Wilson, A. C. 1990. Mitochondrial DNA diversity in the sea urchins *Strongylocentrotus purpuratus* and *S. droebachiensis*. *Evolution* 44:403–15
121. Pamilo, P., Nei, M. 1988. Relationships between gene trees and species trees. *Mol. Biol. Evol.* 5:568–83
122. Paterson, A. H., Lander, E. S., Hewitt, J. D., Peterson, S., Lincoln, S. E., Tanksley, S. D. 1988. Resolution of quantitative traits into Mendelian factors by using a complete linkage map of restriction fragment length polymorphisms. *Nature* 335:721–26
123. Provine, W. B. 1989. Founder effects and genetic revolutions in microevolution and speciation: An historical perspective. In *Genetics, Speciation, and the Founder Principle,* ed. L. V. Giddings, K. Y. Kaneshiro, W. W. Anderson, pp. 43–76. New York: Oxford Univ. Press
124. Reeb, C. A., Avise, J. C. 1990. A genetic discontinuity in a continuously distributed species: mitochondrial DNA in the American oyster, *Crassostrea virginica*. *Genetics* 124:397–406
125. Riddle, B. R., Honeycutt, R. L. 1990. Historical biogeography in North American arid regions: an approach using mitochondrial DNA phylogeny in grasshopper mice (Genus *Onychomys*). *Evolution* 44:1–15
126. Rieseberg, L. H., Brunsfeld, S. J. 1991. Molecular evidence and plant introgression. In *Molecular Plant Systematics,* ed. P. S. Soltis, D. E. Soltis, J. J. Doyle. New York: Chapman & Hall
127. Rose, M. R., Doolittle, W. F. 1983. Molecular biological mechanisms of speciation. *Science* 220:157–162
128. Schaeffer, S. W., Aquadro, C. F., Anderson, W. W. 1987. Restriction-map variation in the alcohol dehydrogenase region of *Drosophila pseudoobscura*. *Mol. Biol. Evol.* 4:254–65
129. Selander, R. K., Clark, A. G., Whittam, T. S., eds. 1991. *Evolution at the Molecular Level.* Sunderland, Mass: Sinauer
130. Selander, R. K., Whittam, T. S. 1983. Protein polymorphism and the genetic structure of populations. In *Evolution of Genes and Proteins,* ed. M. Nei, R. K. Koehn, pp. 89–114. Sunderland, Mass: Sinauer
131. Shields, G. F., Wilson, A. C. 1987. Calibration of mitochondrial DNA evolution in geese. *J. Mol. Evol.* 24:212–17
132. Simon, C. M. 1979. Evolution of periodical cicadas: phylogenetic inferences based on allozyme data. *Syst. Zool.* 28:22–39
133. Slatkin, M. 1987. Gene flow and the geographic structure of natural populations. *Science* 236:787–92
134. Slatkin, M. 1989. Detecting small amounts of gene flow from phylogenies of alleles. *Genetics* 121:609–12
135. Slatkin, M., Maddison, W. P. 1989. A cladistic measure of gene flow inferred from the phylogenies of alleles. *Genetics* 123:603–13
136. Soltis, D. E., Soltis, P. S., Ness, B. D. 1989. Chloroplast DNA variation and multiple origins of autopolyploidy in *Heuchera micrantha* (Saxifragaceae). *Evolution* 43:650–56
137. Swofford, D. L., Olsen, G. J. 1990. Phylogeny reconstruction. See Ref. 76, pp. 411–501
138. Szymura, J. M., Barton, N. H. 1986. Genetic analysis of a hybrid zone between the fire-bellied toads, *Bombina bombina* and *Bombina variegata,* near Cracow in southern Poland. *Evolution* 40:1141–59
139. Tajima, F. 1983. Evolutionary relationship of DNA sequences in finite populations. *Genetics* 105:437–60
140. Takahata, N. 1987. On the overdispersed molecular clock. *Genetics* 116:169–79
141. Takahata, N. 1989. Gene genealogy in three related populations: Consistency probability between gene and population trees. *Genetics* 122:957–66
142. Takahata, N. 1990. A simple genealogical structure of strongly balanced allelic lines and trans-species evolution of polymorphism. *Proc. Natl. Acad. Sci. USA* 87:2419–23
143. Takahata, N., Nei, M. 1985. Gene genealogy and variance of interpopula-

tional nucleotide differences. *Genetics* 110:325–44

144. Takahata, N., Nei, M. 1990. Allelic genealogy under overdominant and frequency-dependent selection and polymorphism of major histocompatibility complex loci. *Genetics* 124:967–78
145. Tegelström, H. 1987. Transfer of mitochondrial DNA from the Northern Red-Backed Vole *(Clethrionomys rutilus)* to the Bank Vole *(C. glareolus)*. *J. Mol. Evol.* 24:218–27
146. Templeton, A. R. 1980. Modes of speciation and inferences based on genetic distances. *Evolution* 34:719–29
147. Templeton, A. R. 1980. The theory of speciation via the founder principle. *Genetics* 91:1011–38
148. Templeton, A. R. 1981. Mechanisms of speciation—a population genetic approach. *Annu. Rev. Ecol. Syst.* 12:23–48
149. Templeton, A. R. 1982. Genetic architectures of speciation. In *Mechanisms of Speciation,* ed. C. Barigozzi, pp. 105–121. New York; Liss
150. Templeton, A. R. 1987. Genetic systems and evolutionary rates. In *Rates of Evolution,* ed. S. W. Campbell, M. F. Day, pp. 208–17. London: Allen & Unwin
151. Templeton, A. R., DeSalle, R., Walbot, V. 1981. Speciation and inferences on rates of molecular evolution from genetic distances. *Heredity* 47:439–42
152. Thorpe, J. P. 1982. The molecular clock hypothesis: biochemical evolution, genetic differentiation and systematics. *Annu. Rev. Ecol. Syst.* 13:139–68
153. Turner, B. J. 1974. Genetic divergence of Death Valley pupfish species. *Evolution* 28:281–94
154. Vawter, L., Brown, W. M. 1986. Nuclear and mitochondrial DNA comparisons reveal extreme rate variation in the molecular clock. *Science* 234:194–96
155. Wallis, G. P., Arntzen, J. W. 1989. Mitochondrial-DNA variation in the crested newt superspecies: Limited cytoplasmic gene flow among species. *Evolution* 43:88–104
156. White, M. J. D. 1978. *Modes of Speciation*. San Francisco: Freeman
157. Wilson, A. C., Cann, R. L., Carr, S. M., George, M., Gyllenstein, U. B., et al. 1985. Mitochondrial DNA and two perspectives on evolutionary genetics. *Biol. J. Linn. Soc.* 26:375–400
158. Witter, M. S., Carr, G. D. 1988. Adaptive radiation and genetic differentiation in the Hawaiian silversword alliance (Compositae: Madiinae). *Evolution* 42: 1278–87
159. Wu, C.-I. 1991. Inferences of species phylogeny in relation to segregation of ancient polymorphisms. *Genetics* 127: 429–35
160. Zuckerkandl, E., Pauling, L. 1965. Evolutionary divergence and convergence in proteins. In *Evolving Genes and Proteins,* ed. V. Bryson, H. J. Vogel, pp. 97–166. New York: Academic

Annu. Rev. Ecol. Syst. 1991. 22:309–34

THE THEORY AND PRACTICE OF BRANCH AUTONOMY

D. G. Sprugel, T. M. Hinckley, and W. Schaap

College of Forest Resources, AR-10, University of Washington, Seattle, Washington 98195

KEY WORDS: carbon allocation, hydraulic architecture, source-sink relations, stress isolation, scaling

---

INTRODUCTION

The past 15 years have seen a surge of interest in modularity in plants; that is, in the implications of the fact that plants are composed of repetitive modules that may in some ways behave as in independent units (32). Much has been written about the importance of modularity in plant population biology (e.g. 33, 34, 113, 114), and also about the advantages and disadvantages of independence or interdependence among separate but connected modules (7, 31, 69). The interest in modularity is mainly at two scales. At the smallest scale, interest has focused on the "nutritional unit" (1) or "physiologically independent subunit" (111, 112), comprising a unit of foliage, the section of stem to which it is attached, and the subtending axillary bud. At the opposite end of the spectrum, research has focused on clonal herbs (3, 81, 82, 113), in which each module (ramet) contains all of the structural parts and physiological processes necessary for independent existence. However, this upsurge of interest in modularity has also led to renewed speculation about other, intermediately scaled, functional units that may also behave semiautonomously (112).

For woody plants, considerable interest has focused on the degree of autonomy of individual branches[1] on a tree or shrub (107, 112, 117).

[1]We define a branch as a unit attached directly to the main stem of the tree; it includes both leaves and woody tissues.

0066-4162/91/1120-0309$02.00

Although branches appear to be natural subunits into which a woody plant can be divided, there is no particularly obvious reason why branches should have any properties that set them apart from other possible subdivisions of a plant. Thus, one of the first questions one must ask about branch autonomy is: Is there something about the branch that makes it an especially appropriate unit for study? Or is the branch simply a convenient subunit in terms of physical scale—large enough to subsume most important physiological processes, but small enough to be a convenient experimental unit?

Obviously, a branch cannot be as completely autonomous as a ramet of a clonal plant; branches must depend on the root system for water and nutrients and on the stem for physical support. But since branches fix carbon, an established branch could easily be completely autonomous (not dependent on the tree) for carbohydrates, and independent of other branches as regards water and nutrient fluxes. Even this degree of autonomy has significant implications for woody plant structure and function.

Branch autonomy is also an interesting concept to investigate because it lies on the border between ecology and physiology. Many of the questions concerning branch autonomy are ecological in nature, relating to how a plant competes for limited resources, expends energy efficiently to maximize future gains, or protects itself against biotic or environmental stresses. The mechanisms of branch autonomy, however, are clearly in areas usually studied by physiologists: source-sink relations, hormonal control of growth and development, and hydraulic architecture. This linkage offers the possibility for productive cross-fertilization between the two fields.

Definitions

The working definition of branch autonomy depends upon the resource (carbon, water, or nutrients) being considered. Although we consider these individually, in reality they are structurally and functionally linked.

A branch is generally considered to be autonomous with regard to carbon if it fixes all the carbon it needs; that is, if it never imports carbon that was initially fixed in other branches (in energetically significantly quantities). A branch that imports carbon at some time during the year, even if it is an exporter at other times, is generally not considered to be autonomous with respect to carbon, at least during the period when import occurs.

Branches obviously cannot be autonomous with respect to water in the same sense as they are autonomous with respect to carbon, because water comes from the roots via the stem and is not recycled. A branch is generally considered to be autonomous with respect to water flux if it is insulated from factors that affect other branches (or, conversely, if other branches are insulated from it). That is, if one branch is heated by a sunfleck and begins to undergo significant moisture stress, do the other branches on the tree also become stressed? If so, the branches are closely coupled; if not, they are autonomous.

Autonomy with respect to nutrients has been little discussed, but a good working definition might be that a branch is autonomous to the extent that nutrient needs in one part of the branch are met by translocation from other locations within the branch. A branch that is rapidly growing and increasing its nutrient content, or that imports substantial quantities of nutrients from the stem and roots before and during bud-burst in the spring, would have to be considered nonautonomous. Autonomy with respect to nutrients is virtually unstudied, and we do not discuss it further.

The differences in these definitions reflect important contrasts in how these resources are transported throughout the plant. Carbon is fixed in the branch and then extensively recycled; thus to be autonomous with respect to carbon, a branch must fix all the carbon it requires. Water moves more or less linearly through the plant (with some storage), so a branch is autonomous if the flow through it is independent of flow through other branches. Minerals enter the plant through the root system and eventually leave by leaching or in litterfall, so their path through the plant is ultimately a one-way trip, but along the way they may be recycled extensively. Thus, the definition of autonomy with respect to minerals is a hybrid of water and carbon autonomy.

Applications of Branch Autonomy

SCALING The problem of scaling from measurements on individual leaves to estimates for larger units (trees, stands, or landscapes) has become a major concern of ecologists and ecophysiologists in recent years. Scaling directly from leaves to whole trees seems impractical because of the awesome diversity of leaf morphologies and light regimes in a tree and the complexity of the interactions among them. However, branches may form a useful intermediate step between leaves and whole trees, to allow researchers either to scale their studies of processes from the organ to the whole-canopy level more readily (23), or to reduce the seemingly complex crown of a tree to a much smaller number of biologically meaningful units for which experiments may be designed.

AIR POLLUTION AND GLOBAL CHANGE RESEARCH Experimental studies of the effects of air pollution on tree species have long been frustrated by a methodological dilemma. Most studies to date have been carried out on seedlings, but these have been criticized because seedling morphology and physiology are very different from that of mature trees (11, 21, 46), so it is difficult to know how to extrapolate results from one to the other. However, the most obvious alternative is simply impractical; the logistic and physical demands of fumigating mature trees (and getting a large enough sample size for results to be statistically persuasive) exceeds the capacity of even the largest research programs.

A number of investigators have suggested that branches represent repetitive and independent units of structure and function, suitable as experimental units for studies of the response of a tree to pollution (10, 92, 117). This is valid, however, only if the response of a branch to air pollution does not depend on the character of the atmosphere surrounding the rest of the tree. To quote Cregg (10), "If branches operate relatively autonomously and are essentially independent from each other and the tree to which they are attached for photosynthates, then expanding the responses from branch exposure studies to whole trees will be relatively straightforward. If, in contrast, compensatory movements of photosynthate occur in response to pollutant exposure, the degree of compensation for defense and repair must be estimated."

ECOLOGICAL REASONS FOR AUTONOMY

Hardwick (31) has presented an elegant and profound discussion of the consequences of autonomy vs close integration for different parts of the same plant. The major advantages of autonomy fit into two general areas:

Damage Control/Stress Isolation

Because higher animals are so thoroughly integrated, any stress or injury that affects one organ or body part usually has repercussions throughout the rest of the body. Because plants are less well integrated, this is not necessarily true. There are three major areas where some degree of branch autonomy might limit the damage that a localized stress or injury might do to the plant as a whole.

CARBON/ENERGY As branches get older, they usually become increasingly shaded by new branches developing above them. At some point the maintenance respiration costs of keeping the foliage and branch tissue alive exceed their energy capture rate, so older branches could become a net energy drain on the organism. If a tree were completely integrated, carbohydrate would be imported into the shaded branch to meet these respiratory demands. However, if branches were somewhat autonomous, the shaded branch might be isolated from carbohydrates in the stem, so that if it were unable to meet its own needs, it would starve instead of becoming a drain on the tree.

WATER Branch autonomy with respect to water could benefit plants in two ways. First, some isolation of individual branches might prevent a localized stress from decreasing the water potential of the plant as a whole (or at least decreasing it to some critical level). If trees were totally integrated, then if the leaves on one branch began to lose water at a high rate (e.g. due to a high radiation load), the water potential of the entire plant would decrease until stomatal closure reduced water loss from both the affected and unaffected organs. If branches were autonomous, the decreased water potential might be restricted to the affected branch, so that stomata there would close while the rest of the tree remained functional.

Another possible mechanism of damage control through branch autonomy deals with the danger of runaway cavitation. Cavitation occurs when liquid water is subjected to pressures less than its vapor pressure. Under these conditions gas bubbles form, which may block xylem vessels or tracheids. Cavitation in the xylem of transpiring plants historically was thought to occur only at very high tensions, typically greater than those observed except under unusually stressful conditions (or with freezing). However, recent work has shown that cavitation can occur at xylem water potentials greater than −2.0 MPa (104), which is well within the range encountered in moderately dry plants and soils. Runaway cavitation is the result of a negative feedback between cavitation events and the hydraulic conductivity of a vascular segment; increasing cavitation decreases hydraulic conductivity and thereby causes increased tensions and more cavitation events (unless stomatal closure occurs). In consideration of safety (avoidance of cavitation) vs efficiency (rapid transport of water), Zimmermann (119) developed the idea of segmentation in perennial higher plants. Because cavitation is irreversible for the short-term, and perhaps in some cases even for the long-term, an important design requirement is that such vapor blockage be isolated to relatively expendable tissues (e.g. leaves before twigs, twigs before branches, etc.). The main stem represents the greatest investment of the tree and should not under any circumstances be lost. Thus, the hydraulic isolation of individual branches could prevent long-term damage to the water-conducting system of the tree as a whole.

PATHOGENS Branch autonomy also has implications for the isolation of pathogens. In a heavily integrated animal, an infection in one organ can rapidly spread throughout the entire organism and thereby cause death. Similarly, an infection in a relatively isolated part of an animal necessitates a whole-organism response. The same is not necessarily true for trees. A few especially virulent introduced pathogens can spread very rapidly (e.g. *Ceratocystis ulmi,* which causes Dutch elm disease) or block vital conducting elements (e.g. *Cryphonectria parasitica,* which causes chestnut blight) and thereby kill the entire tree, but in most trees, the more common danger is from root rots or heart rot fungi, which can slowly weaken a tree from the inside until it breaks and falls. Bark and living sapwood provide an effective barrier that keeps heart rot fungi from getting into the dead heartwood at the center of the tree, but as branches die and break off, their dead stubs can provide an avenue into the heartwood through which fungi could invade. Mechanisms that isolate the branch from the rest of the tree, even after the branch dies, thus provide substantial safety benefits to the tree. Such mechanisms have been thoroughly reviewed by Zimmermann (119) and Shigo (85) and are not further discussed here.

Exploitation of Opportunities

A large and well-developed body of literature discusses foraging in animals, which can be loosely defined as the set of behaviors exhibited in searching for and acquiring food (e.g. 72, 83). No comparable literature exists for plants, mainly because plants do not forage in the classical sense of moving around to different possible prey locations. However, a small but growing group of researchers (3, 47, 79, 86–88) have pointed out that this distinction is essentially superficial; fundamentally, plants and animals must both expend energy in order to acquire more energy, and their fitness is increased if this energy is expended efficiently. For animals the energy is expended in searching or chasing prey, while in plants the energy is expended producing the leaves necessary to harvest light and the branch, stem, and root structures necessary to support them. In both cases, though, great benefits are to be gained by behavioral or structural patterns that ensure that search time or structural development will be carried out in places that offer the maximum possibility for additional energy capture.

Most of the work done on plant foraging has been carried out with clonal plants, apparently because of the mistaken belief that an individual plant (ramet) can harvest resources only from a single patch (3). However, for large woody plants this is manifestly untrue; the light distribution in a forest is notoriously patchy, so within the three-dimensional volume occupied by a single tree's canopy there will inevitably be areas of greater and lesser resource availability (40, 89, 115). A tree that forages efficiently (puts most of its new leaves in areas with high light levels) will clearly have an advantage over one that remains symmetric and puts out leaves equally in all accessible locations.

Branch (carbon) autonomy offers one mechanism for ensuring at least moderately efficient light foraging. If each branch retains all (or a constant fraction) of the energy it fixes, then the well-lit branches will grow bigger and shaded branches will stop producing new leaves and eventually die out. By contrast, if resources are redistributed among branches without regard to local light distribution, then a great deal of energy could be wasted in building leaves in shaded areas.

BRANCH AUTONOMY IN PRACTICE: PROCESSES, EVIDENCE, AND INTERPRETATION

Carbon

PROCESSES Carbon autonomy or integration in branches should be a fairly straightforward consequence of well-known physiological processes. Carbon is fixed through the photosynthetic process in green foliage (and to a much lesser extent in any tissue such as bark and cones containing chloroplasts) and then allocated to meet current respiratory needs, used for growth, stored as reserve material in the vacuole, or exported through the translocation process (27). Although there is still substantial debate above the exact mechanism by which assimilates are translocated (transported in the phloem), it is almost universally agreed that translocation is driven by gradients in assimilate concentration between sources (places where carbohydrate is produced or mobilized) and sinks (places where carbohydrate is used or stored) (14, 109). Active sinks are generally fed by the nearest source (17), but this pattern can be modified by variations in sink strength; if a source is located midway between a strong and a weak sink, its photosynthate will generally go toward the strong sink rather than the weak one (17, 109). In practice, this means that strong sinks can draw carbohydrate from sources substantial distances away, while weak sinks can pull in carbohydrate only from sources that are fairly close by. Kramer & Kozlowski (53) present a hierarchy of sink strengths at the height of the growing season as: fruits > young leaves and stem tips > mature leaves > cambia > roots > storage. Sink strength appears to be related to growth activity, which in turn may be largely a function of the type, ratio, and concentration of hormones present. For example, ^{14}C can be induced to move into a darkened *Vitis vinifera* leaf if benyladenine is applied (73) (although Turgeon (101) has questioned these results because of the very low quantities of label actually imported). Carbon autonomy of branches thus boils down to the question: "Are the carbohydrate sinks in the branch strong enough to draw carbon in from outside the branch, and if so, under what circumstances and to what degree?"

Studies of carbohydrate translocation fall into two broad categories: (i) nonintrusive studies (usually involving ^{14}C tracers) to see how and where photosynthate is translocated under normal conditions, and (ii) manipulative studies, in which sinks, sources, or pathways are artificially altered (e.g. by defoliating, debudding, application of hormones, girdling, etc) to see how translocation processes are changed. Neither of these approaches can provide answers to all of the questions about translocation. Nonintrusive studies provide good data about the "normal" patterns of photosynthate translocation; for example, a variety of studies (summarized in 19, 112) show that phyllotaxy exerts a strong influence over exactly which leaves feed which developing bud. However, these "normal" patterns can easily be changed by stress or other alterations in the normal source-sink relationships so the fact that a particular bud is normally fed by a particular leaf does not mean that if that leaf is injured, the bud will starve (17, 28, 109). On the other hand, manipulative studies can easily create conditions that exceed the bounds of anything that could occur under natural conditions, suggesting "possible" pathways that are never followed under such conditions (101). Moreover, manipulations may have unintended consequences (77); for example, removal of buds (to reduce local sinks) or girdling may disrupt normal patterns of hormonal production far more than they change carbohydrate distribution. Loosely, then, nonintrusive studies tell one what does happen under normal conditions, while manipulative studies may tell one what can happen when the system is stressed.

EVIDENCE *Direct measurements of the movement of ^{14}C-labelled carbon between branches* Sprugel & Hinckley (93) exposed the foliage of individual branches of *Abies amabilis* (Pacific silver fir) to $^{14}CO_2$, harvested the trees 3 or 30 days later, and then analyzed various tissues to see where the labelled photosynthate was located. Like many northern conifers and deciduous trees *A. amabilis* has a determinate growth habit and a single annual flush of growth; shoot extension is generally restricted to a relatively brief period (ca. 3 weeks) in the spring. Thereafter no new development of leaf tissues occurs, although cambial (diameter) growth may continue for several months.

During the course of the study, branches were labelled on several different occasions during the summer. In all cases carbon was exported to the stem and then downwards; ^{14}C was never found in any branch other than the labelled one. In one particular experiment two branches (one near the top of the tree and one near the bottom, but both below the labelled branch) were covered with black cloth before and after the labelling to eliminate photosynthesis and increase respiration (through heating by the sun), in the expectation that this might cause them to import carbon after their own reserves were exhausted. Although the shaded branches showed signs of significant stress after four days of shading and heating, they had not imported any labelled photosynthate. These studies indicate that even when stressed, Pacific silver fir branches are almost entirely autonomous (do not import carbon) during the summer.

A different result was obtained when branches on four Pacific silver fir trees were labelled before shoot growth started in the spring. In two trees that were harvested three days after exposure (still before bud break), labelled carbon was found in the trunk both above and below the labelled branch, rather than only below as after summer labelling. However, no ^{14}C was found in any branches other than the labelled ones. The other two trees were harvested 30 days later, after shoot expansion was well under way. At this time newly produced foliage everywhere on the tree contained labelled carbon. Concentrations of ^{14}C were highest in the new foliage on branches near the top of the tree, even though both the labelled branches and the initial deposition of ^{14}C in the stem were fairly far down on the trunk. Pacific silver fir branches are clearly not autonomous during the springtime shoot expansion period. Moreover, not all branches are equal; young, sunlit branches with a high ratio of current growth to old foliage are much less autonomous than older branches with a lower ratio.

Cregg (10) covered terminal shoots of *Pinus taeda* (loblolly pine) with shade cloth to reduce light by 30% or 60% (with an unshaded control), and then exposed lateral shoots on the same stem to $^{14}CO_2$ to see if the terminals would import carbon. In contrast to the single flush of growth seen in *A. amabilis,* extension growth in *P. taeda* involves multiple flushes throughout the growing season. When laterals were exposed in late June, at a time when needles of the first flush had finished expanding but those of the second flush had not yet begun, there was no import of labelled photosynthate into the terminals in any of the treatments. In a second experiment in late August, when second flush needle growth was still active, substantial ^{14}C was imported into the terminal under 60% shade and a lesser quantity into the terminal under 30% shade. The unshaded loblolly pine terminals appeared to be autonomous on both sampling dates, while artificially shaded terminals were autonomous even under stress when needles were not actively growing but could be induced to import carbon when needle elongation was underway.

Rangnekar et al (74) labelled individual branches of *Pinus resinosa* (red pine), which has a growth pattern similar to *A. amabilis,* during the spring shoot elongation period. Carbohydrate exported from the second whorl went up after it reached the stem, while carbohydrate exported from the third and fourth whorls went downward toward the roots. Small but measurable quantities of ^{14}C entered branches along the path of transport. Nevertheless, the authors concluded that "Each branch appears to be largely self-supporting," noting that at least some of the branches that imported small amounts of ^{14}C were in crown positions where they must also have been exporting large quantities of unlabelled carbohydrate.

Isebrands (41) and Dickson (18) labelled individual lateral branches of two-year-old *Populus* hybrids (which have an indeterminate growth pattern; i.e. they continue to elongate throughout the growing season) and found that most of the fixed ^{14}C was either retained in the treated branch or exported to the main stem. However, 2–10% of the recovered ^{14}C was consistently found in other lateral branches which, as in the Rangnekar et al study (74), must have been net exporters. No explanation was offered for this curious observation.

Pruning studies Pruning, the removal of dead, dying, or green branches in the lower portion of tree crowns, has long been a recognized silvicultural practice (and is, in effect, a very old application of the assumption of branch autonomy.) Although the primary reason for removing low branches is to improve tree quality for lumber (by eliminating knots), there has been a long-standing debate as to whether these low branches contribute carbohydrate to the rest of the tree, merely hold their own, or whether they actually parasitize carbohydrate produced by other branches. Hundreds (if not thousands) of pruning studies have been carried out for different species and intensities of pruning (summarized in 63, 105), and with a few exceptions (94, 96) growth has almost never been reported to increase after lower branches are removed (as it would if the low, shaded branches were draining carbohydrates from the stem). In many cases the lowest branches seem to be almost completely neutral; they do not put on annual growth rings (24, 35, 48, 75) and do not contribute carbohydrate to the rest of the tree (54, 58, 67, 105), but apparently they fix just enough carbon to meet their own needs. Rarely has removal of only the lower branches caused significant decreases in overall stem growth. The explanation for this prolonged period of borderline survival may lie in the fact that both deeply shaded leaves (4) and branches on which little new growth is occurring (91) have extremely low respiration rates that in part compensate for the low level of energy availability.

Studies of the role of stem- and root-based energy reserves in shoot elongation Kimura (49) analyzed carbohydrate storage in various organs and spring photosynthesis and respiration rates to determine the relative contribution of current photosynthesis and stored reserves to springtime shoot and needle elongation of *Abies veitchii*. He concluded that current photosynthesis (during the elongation period) was the primary source of energy for shoot elongation, but that reserves supplied one fourth to one third of the energy used. About 75% of these reserves were in the branches, and 25% (6–8% of the total energy cost of shoot elongation) came from the stem and roots. This would imply that, even during shoot expansion, *A. veitchii* branches as a group are largely autonomous under normal conditions. Loach & Little (59) carried out a similar study for *Abies balsamea* but looked only at current photosynthesis and storage in one-year-old foliage. They also concluded that current photosynthesis was the main energy source for elongation and that reserves within the branch could supply all, or nearly all, of the additional energy required. This suggests that balsam fir branches also are or could be largely autonomous. Kozlowski & Winget (52) approached this question more directly, defoliating all the branches on some eight-year-old red pine trees and girdling all the branches on others to prevent translocation of photosynthate from roots and stems. Again, they found that photosynthesis within the target branch during the shoot elongation period was the primary energy source for elongation, but girdling at the base of a branch reduced shoot growth by about 25% as compared with ungirdled controls, suggesting somewhat less autonomy than was predicted in the *Abies* species.

The results of the two photosynthesis/storage studies (49, 59) would appear to contradict the results of the girdling study (52), as well as our labelling study on *A. amabilis* (93). However, all of these must be interpreted with some care. The photosynthesis/storage analyses, which suggest that shoot elongation could be powered largely by current photosynthesis and within-branch reserves, lump all branches together. As our *Abies amabilis* work showed, this is deceptive since there is great branch-to-branch variation in the amount of carbohydrate imported. Thus while the Kimura and Loach & Little analyses suggest that the total current photosynthate and reserve mobilization storage in all the branches put together could almost supply the energy requirements for shoot elongation, they do not rule out the possibility that some branches import a substantial amount of the reserves used in shoot elongation, while others import little or none. On the other hand, the girdling/defoliation study, which suggests that a substantial proportion of the assimilate needed for elongation is imported, must also be interpreted carefully since these manipulations can interrupt hormonal interchange as well as carbohydrate flux.

Studies on the source of energy for fruit development Because of the economic importance of fruit and nut production, a great variety of studies have traced the sources of carbohydrate for developing fruit. Most of these (e.g. 16, 29, 30, 97) simply confirm that sinks are fed primarily by nearby sources, but do not provide information specifically on whether branches act wholly or partly as independent physiological units. However, a few have used branches as experimental units and thus provide information on branch autonomy. For example, Stephenson (95) removed 36 to 94% of the leaf area from small *Catalpa speciosa* (northern catalpa) branches (all initially containing 9 leaves), and found that fruit set and mean weight per fruit decreased, more or less in proportion to the leaf area removed. He inferred that the leaves on a given branch were the primary source of carbohydrate for ripening of fruit on that branch, and that relatively little of the loss of photosynthate due to defoliation was made up for by import from the main stem. Janzen (45) defoliated branches of *Gymnocladus dioicus* (Kentucky coffee-tree) after flowering and, again, found large increases in seed abortion and decreases in seed weight and the number of seeds per pod. However, even on completely defoliated branches, some seeds were set, suggesting that some carbohydrate import did occur (or possibly that there were substantial carbohydrate reserves in the branch before defoliation, although this seems unlikely in midsummer). Chalmers et al (8) girdled branches on *Prunus persica* (peach) at various distances from the trunk and found that fruits below the girdle gained less weight than fruits at comparable heights on ungirdled control branches. (Fruit located above the girdle gained more weight than fruit on control branches, presumably due to a superabundance of carbohydrate.) All of these studies seem to suggest that branches in these species are largely autonomous during fruit-bearing, and that even unusual stresses (defoliation, girdling) can not break down this autonomy to any significant degree.

Newell (66) carried out an especially informative study of fruit set in *Aesculus californica* (California buckeye), which is of particular interest because it fills its fruits during the leafless season, so fruit production is powered entirely by reserve materials. When she girdled some fruit-bearing branches where they met the trunk, mean fruit size was reduced by 50–65%, suggesting a large contribution from the main stem. However, she also found that ungirdled fruit-bearing shoots broke bud later and produced fewer new leaves in the following year than did nonfruiting shoots, and also had lower nitrogen concentrations (although there was no decrease in nonstructural carbohydrates; i.e. storage materials). This local carryover of the impact of fruit-bearing (which has also been observed in *Carya illinoiensis* (pecan) (62)) would seem to indicate that reserves within the fruit-bearing shoot are also a major source of energy for fruit production.

The work of Chan & Cain (9) raises significant questions about the interpretation of observations on the effect of fruiting on later vegetative growth. These authors examined the effect of fruiting on subsequent vegetative growth in "Spencer Seedless" apple trees, which normally produce seedless apples because the apetalous flowers do not attract pollinators, but which can be induced to produce seeded fruit if the flowers are fertilized with appropriate pollen. They found that 90–100% of shoots that produced seedless fruit flowered the following year, as compared to 5–26% of shoots on which seeded fruit was produced. Since the energy cost of producing seeds is trivial compared to the cost of producing an apple, they inferred that the suppression of flowering on branches that produced seeded fruit the previous year is probably due to a hormone produced by the developing seeds rather than by a lack of carbohydrate reserves. The fact that total nonstructural carbohydrates did not decline in fruiting *Aesculus* branches suggests that this effect might also have been important in Newell's study (66). The possibility exists that the patterns observed in many fruit development studies are only marginally related to carbohydrate levels, since seeds may produce hormones that inhibit growth activity on the branch and in the area surrounding the seeds.

Other studies Tuomi et al (99) found that leaves on catkin-bearing dwarf shoots of *Betula pendula* were 15–20% smaller than those on non-catkin-bearing shoots, but that leaf numbers were not reduced. They inferred that there was some competition for resources within the dwarf shoot, but also some supply of resources from the whole tree. They commented, "It seems obvious that birch is not a mere population of independent dwarf shoots, but that physiological processes underlying resource partitioning in individual dwarf shoots are coupled with the allocation processes of the whole tree." Later the same group found that simulated herbivory on *Betula pubescens* var. *tortuosa* branches stimulated increased phenol production and reduced nitrogen content in the leaves of the treated branch, but that other branches on the same tree were not affected (100). They noted that this particular example of branch autonomy seems rather peculiar, since phenols are usually assumed to be an antiherbivore defense, and in the presence of herbivore attack it makes little sense to defend one branch of a tree and leave the others undefended. They suggested that in this case phenol production and reduced foliar nitrogen concentration were simply local responses to reduced nitrogen availability in the attacked branches. Similar work on *Populus* has demonstrated that, when mature leaves are wounded, a signal is transmitted that induces production of wound proteins in phyllotaxically connected immature foliage (following the known paths of carbon transport) but not in mature foliage (15).

Chalmers et al (8) found that photosynthesis in peach leaves increased while fruits were expanding, and that the magnitude of the increase varied between parts of the tree. Areas where the leaf area:fruit mass ratio was high had the least increase in photosynthesis, while areas with low leaf-fruit ratios had much greater increases in photosynthesis. This suggests that photosynthetic compensation, like other physiological processes, may be quite localized.

Cook & Stoddart (13) removed all the current growth on half the branches of one group of *Artemisia tridentata* (big sagebrush) shrubs, and half the current growth on all branches of another group. In general, the branches in the first group that had all their current growth removed died, while the remaining branches grew faster than controls (doubtless due to an improved water supply; big sagebrush normally lives in desert habitats where water is very likely to be limiting). The authors commented that "The sagebrush plant separates rather easily into a number of self-supporting units, including branches and roots."

INTERPRETATION As was noted earlier, the question of whether branches are autonomous with respect to carbon translates to "Under what, if any, circumstances are the carbohydrate sinks in the branch strong enough to draw carbon in from outside the branch?" Some components of the answer are now clear:

1. *Old, shaded branches are on their own.* Maintenance respiration alone is apparently not a sufficiently strong sink to draw carbohydrates into a branch against the countervailing pull of the sinks in the main stem, since with the exception of a few questionable pruning studies (94, 96), there is no evidence that branches on the verge of being shaded out can ever import carbohydrate (63, 93, 105). Since photosynthate is routinely translocated into young, growing branches at certain times of the year (49, 52, 66, 93), there are probably no actual "one-way valves" preventing movement of photosynthate into old, shaded branches; rather, it is the balance of sink strengths that prevents this from happening. In mature leaves, however, there seem to be actual physical barriers to photosynthate import; the pathways through which photosynthate enters growing leaves are apparently destroyed as leaves mature (101).

2. *Under normal conditions, most branches are relatively autonomous during the growing season.* As long as photosynthate is being produced, the internal balance of sources and sinks seems to be such that most branch functions are self-supporting (10, 49, 59). Especially strong sinks such as growing shoots and ripening fruit might draw small quantities of photosynthate in from outside of the branch, but this appears to be the exception rather than the rule. However, when local sources of photosynthate are eliminated or much reduced (e.g. by defoliation or shading), or sinks are amplified (e.g. by application of hormones), there may be some import of photosynthate to growing tissues (10, 45, 73, 95). The amount imported is usually much smaller than that lost.

3. *Branches are least autonomous when carbohydrate reserves are involved.* Since carbohydrate reserves are stored in the stem in significant quantities and may thus be interchanged freely between branches, processes that draw significantly on reserves are the most important exceptions to branch autonomy. At times of the year when reserves are being mobilized to support shoot or fruit growth, these strong sinks are capable of drawing carbohydrate from sources outside the branch (49, 52, 66, 93).

The importance of reserve utilization in branch autonomy suggests some predictably consistent differences in the degree of branch autonomy between species and functional groups. For example in evergreen species that have indeterminate growth *(Tsuga, Thuja)* or which undergo multiple flushes (e.g. *Pinus taeda* and other southern pines), reserves in the stem and roots seem to play a relatively minor role in annual shoot growth (50). One would expect that under normal conditions branches on these species would be almost completely autonomous (although under stress they might be less so). In evergreen species which undergo only a single flush (e.g. *Pinus resinosa* and many other northern pines, *Abies* sp., *Picea* sp., etc), stem and to a lesser extent root reserves may play a substantial role in providing the energy for

shoot and needle elongation (50); one would expect branches of these species to be only moderately autonomous during the period of shoot and needle elongation. Deciduous trees with indeterminate growth patterns (e.g. *Populus, Alnus, Betula, Larix*) would be similar; somewhat autonomous in the spring (perhaps less so than evergreens, because deciduous trees tend to store more of their overwintering reserves in the root system and less in the branches—110) but largely autonomous during the summer. Deciduous trees with determinate growth patterns (e.g. *Quercus, Fagus, Acer saccharum),* would be the least autonomous of all, since at least the early stages of shoot expansion are almost entirely dependent on reserves stored in the roots and stem.

The ability of growing shoots (and in some cases fruits) to draw on reserves fixed elsewhere in the tree suggests that the factors controlling bud size and number are an important component of branch autonomy—because once buds are laid down, they apparently can draw in carbohydrates from well beyond the local area. It is well established that in species with determinate growth patterns, the size and distribution of buds formed in the late summer and early fall largely determine the amount of shoot elongation that will occur in the spring (2, 12, 24, 26, 43, 51, 55, 56, 68). (The main exception occurs when unusually bad growing conditions in the spring reduce shoot growth below the potential established by the buds; in most forested areas, however, such conditions are relatively rare—26, 55). Late summer drought can lead to reduced bud production over the whole tree (26, 55, 70), and summer fertilization can likewise increase bud production for the whole tree (60). However, relatively little is known about the mechanisms that control bud distribution *within* a tree; that is, what causes more buds to develop in one place than in another. What is known is relatively elementary: for example, Cannell et al (6) noted that the length of the current shoot tends to correlate with the number of buds produced on it, while Powell (71) reported that the number of buds is correlated with "vigour of growth."

Whatever the mechanism is, it must operate on a very local scale (33), since if one part of a branch is in the sun and another is in the shade, more and larger buds are produced on the sun half than on the shade half; thus, the sun half eventually gets bigger. It seems likely that high levels of photosynthesis in the sunlit area (or the transpiration that must accompany it) somehow lead to increased bud development, but the mechanism for this has never been demonstrated. Simple availability of photosynthate for bud production seems unlikely to be the answer, since as Wareing (108) pointed out, the actual construction of a bud requires only a very small amount of fixed carbon. Perhaps photosynthesis stimulates or depresses local production of some hormone [Franco (25) suggested that phytochromes might be involved], or transpiration brings some hormone or nutrient into the area—but again, no mechanism by which this happens has been demonstrated. This seems a fruitful area for future research into environmental controls on plant gorwth.

Our discussion to this point has focused on the traditional definition of carbon autonomy; that is, absence of carbon import. However, one might also imagine a stronger definition of autonomy with respect to carbon: for a branch to be truly autonomous (i.e. totally unaffected by the status of the rest of the tree), carbon uptake (photosynthesis) and export (translocation) would have to be completely independent of any factors affecting the tree as a whole. Carbon fixation would have to be unaffected by drought, nutrient stress, or any other factors whose influence is transmitted through the stem, and carbon export would be controlled only by the difference between carbon fixation and internally determined requirements for growth and maintenance respiration. This "strong form" of carbon autonomy has been referred to as "functional independence" of branches.

There appears to have been almost no work on functional independence with the exception of a study currently underway in our laboratory (80). In this study, open-top chambers were used to expose six whole trees (5–7 m tall) to either carbon-filtered ("clean") air or 200–300 ppb O_3. Four branch chambers were then installed within each open-top chamber so that the pollutant environment of individual branches could be manipulated independently of the rest of the tree. Comparisons were made between branches with the same treatment but different whole-tree treatments. Data from this study are still being analyzed, but preliminary results indicate that branches are largely but not entirely functionally independent where photosynthesis is concerned. Photosynthetic rates are determined primarily by the atmosphere surrounding the branch, but the atmosphere surrounding the rest of the tree and whole-tree vigor may also have some effect.

Water

PROCESSES Water movement from the main stem into a branch should first be considered in terms of the movement of water through the entire plant. Water movement from the soil through the plant and to the atmosphere must be treated as a series of interrelated, interdependent processes. For example, the rate of water absorption from the soil is affected by the rate of water loss from the leaves to the atmosphere, by the rate at which water can move from the soil to the root surface, and by the uptake properties of the root itself. Although the loss of water from the foliage can be treated comparatively simply as the product of the vapor density gradient from the foliage to the atmosphere and the conductance from the leaf to the atmosphere, the movement of water into the branch supporting the unit of foliage under consideration may not be treated so easily (38).

Historically, two models, based upon the catenary theory of water flow,

have been used to describe flow through the soil-plant-atmospheric continuum: unbranched (106) and branched (76, 102). In the first model, water was assumed to flow through a series of linked pathways; that is, from all of the roots to the stem to all of the branches to all of the leaves, and the rate of flow was considered to be a function of the water potential gradient across a path segment divided by the resistance to flow (equal to 1/hydraulic conductivity) across that segment. As a consequence the rate of water flow through all of the roots should equal the rate of flow through all the leaves. Unfortunately, and perhaps as a consequence of sampling difficulties, plants were often experimentally assumed to be composed of a root, a stem, a branch, and a leaf. Richter (76) recognized the development of this conceptual error and noted that, as originally suggested by Huber (30), plants are branched organisms and the equations proposed by van den Honert (106) inadequately addressed this situation. With the recognition of the elastic behavior (i.e. water storage capacity) of plant tissues (65), the observation of hydraulic constrictions between leaves and branches or branches and the stem (42, 57, 64, 118), and the introduction of the segmentation concept (119), more realistic models of water flow have been forthcoming (e.g. 102, see review in 116). Water autonomy of branches is reduced to the question: "When water is lost from a branch, to what extent is the water potential in other branches impacted?"

Although Richter (76), by using the conceptual framework of a network of branches, provided an interpretation of a variety of data sets demonstrating various logical and illogical patterns of within-crown water potential values, it was not until the work of Hellkvist et al (36) that the structural nature of the branch network was inferred. They noted marked gradients in water potential between tertiary and secondary and between secondary and primary branches. Detailed anatomical and physiological work followed in the laboratories of P. Larson and M. Zimmermann. This work laid the foundation for our current understanding of branches and for the recent statement of Tyree (102) that "branches may be treated as small, independent seedlings rooted in the main bole."

EVIDENCE

Anatomical evidence Recent work provides strong evidence of a constriction zone, located proximal to the abscission zone of the petiole, in leaves of *Populus deltoides* (42, 57) and in *Acer pensylvanicum* and *Populus grandidentata* (118). The diameter distribution of vascular elements in the constriction zone shifts downward to smaller diameter elements; this fact has important implications with regard to the safety versus efficiency of water movement in vascular tissue. For example, in both *Populus* species, the relative conductance through the constriction zone is about 25% of that through parts of the path on either side of the constriction zone. Perhaps one of the most dramatic examples of the role of this constriction zone in promoting vascular safety is in palms (e.g. *Rhapis excelsa*) (90). Although the presence of emboli, as a result of cavitation, is rather rare in palms, when they do occur they are confined to the relatively expendable leaf xylem. Tyree & Sperry (104) have unequivocally demonstrated that between-species comparisons of vascular element size do not relate to the relative likelihood of cavitation, as was historically believed; however, within a species, smaller diameter elements are less likely to cavitate than large diameter elements. Therefore, these constriction zones serve an important function in isolating leaf tissue from stem tissue.

Anatomical and hydraulic conductivity measures from a number of woody species including *Acer pensylvanicum, Populus grandidentata,* and *Abies balsamea* have demonstrated the presence of another constriction zone between branches and the stem (22, 78, 118). Detailed anatomical data and studies of patterns of pathogen infection by Shigo (85) support these observations. This suggests that branches are also hydraulically isolated from the stem. However, Salleo & LoGullo (78) have noted that nodal restrictions are proximal to the branch:stem junction in olive *(Olea oleaster),* distal in grape *(Vitis vinifera),* but lacking in carob *(Ceratonia siliqua).* How these restriction zones are formed and why there is such large variability in their distribution among species are not presently known.

Physiological evidence A number of authors have described patterns of within crown variation in water potential (e.g. 37, 84, 98). For example, Hinckley & Ritchie (37) observed that branches at the base of the crown had water potentials as much as 0.6 MPa more negative than those at the top of a 8-m-tall *Abies amabilis* tree, a pattern contrary to that expected based only on hydrostatic considerations. Differences between branches in stomatal aperture and hydraulic coupling between shaded and sunny sides of the crown were cited by these authors as possible explanations for their observations. None of these authors placed their values into a proper whole plant context as presented by Richter (76). Indeed, very crude measures of crown microclimate resulted in a spurious conclusion by Hinckley & Ritchie (37): that water potential on one side of the tree was solely coupled with that on the other side rather than to its own microclimate. Their data, however, did suggest that what was occurring in the sunlit portion of this 8-m-tall open grown tree was affecting the shaded portion. In contrast, Richter (76) hypothesized that cuvettes enclosing small branches provided an accurate estimate of physiological processes in neighboring unenclosed branches in spite of clear differences in cuvette microclimate. He demonstrated that this occurred because a large

change in the rate of water loss from a small branch would only have a minimal impact on the total flux of water through the branch to which it was attached.

Hellkvist et al (36) found significant gradients in water potential in the stems and branches of *Picea sitchensis* (Sitka spruce) indicating that hydraulic conductivity in the xylem has a role in determining differences in leaf water potential between a lateral and a main branch. Hydraulic conductivities in spruce were so low that a six-hour delay in the time for change in foliar water potential to be transmitted to the roots was observed. Recently, a number of authors (e.g. 78, 102–104, 118, 119) have shown that the hydraulic conductivity is relatively high in the main stem, intermediate in branches, and lowest at junctions between the stem and branches or between primary and secondary branches. Therefore, the presence of these constriction zones and the associated local drop in hydraulic conductivity strongly affect the gradients of water potential that exist within a plant, and, possibly, the rate at which the gradients are propagated from the twigs to branches and from the branches to the stem. Richter's (76) conclusions regarding the ability of cuvette data to represent values expected from unenclosed branch samples may be in error although his interpretation of the effect of the cuvette microenvironment on the rest of the branch is probably correct.

These observations of hydraulic isolation led Zimmermann (119) to hypothesize that trees are segmented in such a way that branches and individual leaves are shed preferentially to the main axis or apex of the tree. Thus as water stress develops during periods of high transpiration, vascular constrictions at branches or leaf petiole junctions restrict any lethal water stress to these lateral plant structures and reduce the likelihood of the main axis of the tree being endangered. As these leaves or branches are lost, less demand is placed on the root system to supply water to the shoot, and water stress in the more vital regions of the tree is minimized. In other words, as a safety feature to ensure continued survival of the main investment tissues of the plant, branch hydraulic autonomy is favored.

INTERPRETATION As was noted earlier, the question of whether branches are autonomous with respect to water translates to: "When water is lost from a branch, to what extent is the water potential in other branches impacted?" Some components of the answer are now clear:

1. *Branches in most woody plants are hydraulically isolated from the stem.* Extremely large gradients in water potential have been noted in branches, and because of known points of vascular constriction, there is a pattern to these gradients. For example, changes in a leaf's water potential during a day will exceed changes in water potential of the branch to which it is attached. Changes in branch water potential will exceed those in the stem. Because of

the hydraulic isolation of leaves and branches, a doubling of the rate of water loss from a single leaf or an individual branch will have a dramatic effect on the water potentials within that leaf or branch and a far smaller effect on the parts to which those organs are attached. However, the caution of Salleo & LoGullo (78) is critical: the presence and location of restriction zones are paramount in the treatment of branches as being hydraulically isolated and, therefore, autonomous. If no vascular constriction exists, then parts will not be hydraulically isolated. However, the considerations of Richter (76) then become important. If only a small portion of the system is being treated or affected, then the remainder of the system will be largely unaffected. Clearly hydraulic maps of the crown of many more different species of trees need to be determined. In addition, the role that aging and branch position have on hydraulic linkages needs to be elucidated.

2. *Under normal diurnal conditions, a branch may be regarded as hydraulically autonomous*. The strongest version of this statement occurs when the branch is sunlit, when this branch represents a relatively small proportion of the total branches present, and when a vascular constriction exists between the branch and the unit to which the branch is attached. Because the water potential at one location in the stem is the result of water moving to and from that point, changing individual branch water loss must affect the water potential in the stem. However, if the branch is small and the tree is large, this impact will be small. On the other hand, in open grown trees where 60% or more of the crown is in the sun, the water potential in the shaded portion of the tree appears to track the water potential of the sunlit portion, albeit at a much higher water potential. However, under stand conditions and under experimental conditions where only a limited number of branches would be affected, the hydraulic integration of the tree would be far less.

3. *Under seasonal or long-term conditions, a branch cannot be regarded as hydraulically autonomous*. As soil water potential or whole tree water potential declines, so must the water potential of a branch. Although an individual branch may not have either a large or even a detectable effect on this change, the change in the system will be significant to the behavior of the branch.

CONCLUSIONS

We can now provide partial answers to some of the ecological questions involving branch autonomy that were discussed in the introduction.

1. *Stress isolation: carbon*. Older branches seem to be sufficiently isolated that when their respiration exceeds their ability to fix carbon, they do not draw carbon out of the rest of the tree.

2. *Stress isolation: water*. Vascular constrictions at the base of branches normally keep the water system of a branch somewhat isolated from that of

the rest of the tree. This has two consequences: (*a*) when one branch becomes subject to water stress, the stress is not immediately transmitted to the rest of the tree; and (*b*) should stress in a single branch become so severe that cavitation occurs, the cavitation is restricted to that branch, so the continuity of water in the main stem is not compromised.

3. *Exploitation of opportunities*. In trees with determinate, single-flush growth patterns, branch autonomy does not appear to be the primary mechanism ensuring that energy expended in construction of new shoots is channelled into areas of greatest opportunity for light harvesting. The primary mechanism generating efficient foraging for light seems to be the tendency for numerous large buds to be produced in favorable areas; in the following year, these buds are able to draw on reserves throughout the tree as they expand and elongate. Thus, buds on small branches near the top of the tree are not handicapped by the lack of large, nearby photosynthetic surface; these buds simply draw on reserves in the trunk to a greater degree than do buds on larger branches farther down the tree.

In trees with indeterminate or multiple flushing growth patterns, branch autonomy may be significantly more important in determining where growth occurs and where it doesn't. It has been shown repeatedly (16, 18, 41) that the primary source of photosynthate for new leaves developing at the top of a shoot is the older leaves on the same shoot. In such a case, branch autonomy would likely play a significant role in the allocation of photosynthate.

4. *Special properties of branches*. The answer to the quesion of whether branches have special properties that make them especially appropriate subunits for study seems to be: for water, yes; for carbon, no. Vascular constrictions at the point where branches enter the stem do isolate each branch hydraulically from the rest of the tree, making the branch a "special" subunit. There do not appear to be any comparably specific restrictions on carbon flow into and out of the branch; however, the arrangement of carbon sources and sinks does tend to make branches relatively autonomous subunits despite the lack of a specific mechanism enforcing autonomy. This is, however, a function of branch size; young, small branches, which have strong sinks and relatively little photosynthetic surface, are likely to be significantly less autonomous than larger, older branches with a higher ratio of photosynthetic area to current growth.

5. *Utility of branches as subunits for scaling to whole trees*. Branches may well be useful subunits to facilitate scaling physiological measurements from the leaf level to the whole tree or stand, particularly for short-term studies. As noted above, branches are somewhat hydraulically isolated from the rest of the tree and are often large enough to internalize many important physiological processes. This is especially true for large branches, for evergreens, for trees with indeterminate growth patterns, and for determinate-growth trees after initial shoot elongation has ceased. Studies relying on branch autonomy

but carried out during the shoot expansion period on small branches of determinate-growth deciduous trees will not commend themselves to the careful experimenter.

Use of branches as experimental subunits is more questionable for studies where an artificial stress is imposed on one or two branches on a tree. As noted above, "normal" patterns of translocation can be rather easily altered by changes in source-sink relationships, especially where the stressed branch includes strong sinks such as growing shoots or ripening fruits. Special care is needed in interpreting studies where a long-term stress is imposed after buds are set in the fall but before bud burst and the period of initial shoot elongation in the spring. This timing could alter the balance between the potential growth of the shoots as determined by the size and number of buds set and the photosynthate available to support their expansion, with unpredictable consequences for carbohydrate translocation patterns.

It is important to note that critical whole-tree responses that might affect branch responses will not occur when branches alone are treated. For example, a treatment that caused increased water loss, if applied to the whole tree, would eventually cause a drop in the water potential of the stem, despite the fact that the stem is somewhat hydraulically isolated from the branches. However, that same treatment applied to a single branch might not cause a significant drop in stem water potential, which might affect the branch's response. This point is especially important for studies carried out over longer periods of time, when whole-tree responses such as alteration in root-shoot allocation patterns are unlikely to be seen if only single branches are treated.

Thus branches alone are useful but imperfect surrogates for studies on whole trees. Individual branches can be extremely useful experimental units, as long as the experiments are designed so that the known exceptions to branch autonomy do not complicate the interpretation of the results. For long-term studies or those involving stresses that might be expected to cause whole-tree responses, some type of modeling or other integration technique is necessary to integrate for results on individual branches.

Acknowledgments

The authors thank Renée Brooks, Bert Cregg, Ted DeJong, Phil Dougherty, Tim Fahey, David Ford, Jud Isebrands, Ted Kozlowski, Jim Lassoie, Jerry Leverenz, Beth Newell, John Owens, Peter Reich, Steve Ross, Bob Teskey, and Dick Waring for fruitful discussions and criticisms leading to the formation of this article. We are particular grateful to Mr. Dirk van der Wal who provided the initial ideas and data for many of the ideas and much of the research conducted by the authors to date. Research supported by NSF grants BSR-8415590 and BSR-8717450, EPA Exploratory Research Grant R814209-01, USDA Competitive Research Grant 86-FSTY-0-0220, and a grant from the National Council for Air and Stream Improvement.

Literature Cited

1. Adams, M. W. 1967. Basis of yield component compensation in crop plants with special reference to the field bean, *Phaseolus vulgaris*. *Crop. Sci.* 7:505–10
2. Allen, R. M., Scarborough, N. M. 1970. Morphology and length correlated in terminal flushes of longleaf pine saplings. *USDA For. Serv. Res. Paper SO-53*. 15 pp.
3. Bell, A. D. 1984. Dynamic morphology: a contribution to plant population ecology. See Ref. 20, pp. 48–65
4. Brooks, J. R. 1987. *Foliage respiration of* Abies amabilis. MS thesis. Univ. Wash., Seattle. 85 pp.
5. Cannell, M. G. R., Last, F. T., eds. 1976. *Tree Physiology and Yield Improvement*. London: Academic. 567 pp.
6. Cannell, M. G. R., Thompson, S., Lines, R. 1976. An analysis of inherent differences in shoot growth within some north temperate conifers. See Ref. 5, pp. 173–203
7. Caraco, T., Kelly, C. K. 1991. On the adaptive value of physiological integration in clonal plants. *Ecology* 72:81–93
8. Chalmers, D. J., Canterford, R. L., Jerie, P. H., Jones, T. R., Ugalde, T. D. 1975. Photosynthesis in relation to growth and distribution of fruit in peach trees. *Aust. J. Plant Physiol.* 2:635–45
9. Chan, B. G., Cain, J. C. 1967. The effect of seed formation of subsequent flowering in apple. *Proc. Am. Soc. Hortic. Sci.* 91:63–68
10. Cregg, B. M. 1990. *Net photosynthesis and carbon allocation of loblolly pine* (Pinus taeda *L.) branches in relation to three levels of shade*. PhD thesis. Univ. Georgia, Athens. 149 pp.
11. Cregg, B. M., Halpin, J. E., Dougherty, P. M., Teskey, R. O. 1989. Comparative physiology and morphology of seedling and mature forest trees. In *Air Pollution Effects on Vegetation: Proceedings of the 2nd US-USSR Symposium,* ed. R. D. Noble, J. L. Martin, K. F. Jensen. pp. 111–18. Washington, DC: USDA For. Serv./U.S. EPA
12. Clements, J. R. 1970. Shoot responses of young red pine to watering applied over two seasons. *Can. J. Bot.* 48:75–80
13. Cook, C. W., Stoddart, L. A. 1960. Physiological responses of big sagebrush to different types of herbage removal. *J. Range Manage.* 13:14–16
14. Cronshaw, J., Lucas, W. J., Giaquinta, R. T., eds. 1986. *Phloem Transport*. New York: Alan R. Liss, Inc. 650 pp.
15. Davis, J. M., Gordon, M. P., Smit, B. A. 1991. Assimilate movement dictates remote sites of wound-induced gene expression in populars. *Proc. Natl. Acad. Sci. USA*. 88:2393–96
16. Davis, J. T., Sparks, D. 1974. Assimilation and translocation patterns of carbon-14 in the shoot of fruiting pecan trees, *Carya illinoiensis*. *J. Am. Soc. Hortic. Sci.* 99:468–80
17. Denny, M. J. 1984. Translocation of nutrients and hormones. In *Advanced Plant Physiology,* ed. M. B. Wilkins. pp. 277–296. London: Pitman
18. Dickson, R. E. 1986. Carbon fixation and distribution in young *Populus* trees. In *Crown and Canopy Structure in Relation to Productivity,* ed. T. Fujimori, D. Whitehead. pp. 409–26. Ibaraki, Japan: For. and For. Products Res. Inst.
19. Dickson, R. E., Isebrands, J. G. 1991. Role of leaves in regulating structure-functional development in plant shoots. In *Integrated Responses of Plants to Stress,* ed. H. A. Mooney, W. E. Winner, C. J. Pell, pp. 3–34. New York: Academic
20. Dirzo, R., Sarukhan, J., eds. 1985. *Perspectives on Plant Population Ecology*. Sunderland, Mass: Sinauer. 478 pp.
21. Eamus, D., Jarvis, P. G. 1989. The direct effects of increases in the global atmospheric CO_2 concentration on natural and commercial temperate trees and forests. *Adv. Ecol. Res.* 19:1–49
22. Ewers, F. W., Cruiziat, P. 1991. Measuring water transport and storage. In *Techniques and Approaches in Forest Tree Ecophysiology,* ed. J. P. Lassoie, T. M. Hinckley, pp. 91–115. Boca Raton, Fla: CRC
23. Ford, E. D., Avery, A., Ford, R. 1990. Simulation of branch growth in the *Pinaceae:* Interactions of morphology, phenology, foliage productivity, and the requirement for structural support, on the export of carbon. *J. Theor. Biol.* 146:13–36
24. Forward, D. F., Nolan, N. J. 1961. Growth and morphogenesis in the Canadian forest species. IV. Radial growth in branches and main axis of *Pinus resinosa* Ait. under conditions of open growth, suppression, and release. *Can. J. Bot.* 39:385–409
25. Franco, M. 1986. The influence of neighbours on the growth of modular organisms with an example from trees. *Philos. Trans. R. Soc. Lond. B* 313:209–25
26. Garrett, P. W., Zahner, R. 1973. Fascicle density and needle growth responses of red pine to water supply over two seasons. *Ecology* 54:1328–34
27. Geiger, D. R. 1986. Processes affecting

carbon allocation and partitioning among sinks. See Ref. 14, pp. 375–88
28. Gifford, R. M., Evans, L. T. 1981. Photosynthesis, carbon partitioning, and yield. *Annu. Rev. Plant Physiol.* 32: 485–509
29. Hansen, P. 1967. ^{14}C studies in apple trees. I. The effect of the fruit on the translocation and distribution of photosynthates. *Phys. Plant.* 20:382–91
30. Hansen, P. 1969. ^{14}C studies in apple trees. IV. Photosynthate consumption in fruits in relation to the leaf-fruit ratio and to the leaf-fruit position. *Phys. Plant.* 22:186–98
31. Hardwick, R. C. 1986. Physiological consequences of modular growth in plants. *Philos. Trans. R. Soc. Lond. B* 313:161–73
32. Harper, J. L. 1977. *Population Biology of Plants*. New York: Academic. 892 pp.
33. Harper, J. L. 1985. Modules, branches, and capture of resources. See Ref. 44, pp. 1–33
34. Harper, J. L., Bell, A. D. 1979. The population dynamics of growth form in organisms with modular construction. In *Population Dynamics, 20th Symp. Br. Ecol. Soc.*, ed. R. M. Anderson, B. D. Turner, L. R. Taylor, pp. 29–52. Oxford: Blackwell
35. Hartig, R. 1872. Einfluss verschieden starker Ausastung und Entnadelung auf den Zuwachs der Weymouthskiefer und gemeinen Kiefer. *Z. Forst-u Jagdw.* 4:240–54
36. Hellkvist, J., Richards, G. P., Jarvis, P. G. 1974. Vertical gradients of water potential and tissue water relations in Sitka spruce trees measured with the pressure chamber. *J. Appl. Ecol.* 11: 637–68
37. Hinckley, T. M., Ritchie, G. A. 1970. Within-crown patterns of transpiration, water stress, and stomatal activity in *Abies amabilis*. *Forest Sci.* 16:490–93
38. Hinckley, T. M., Schulte, P. J., Richter, H. 1991. Water relations. In *Physiology of Trees*, ed. A. K. Ragavendra, pp. 137–62. New York: Wiley
39. Huber, B. 1928. Weitere quantitative Untersuchungen über das Wasserleitungssystem der Pflanzen. *Jb. Wiss. Bot.* 67:877–959
40. Hutchison, B. A., Matt, D. R. 1977. The distribution of solar radiation within a deciduous forest. *Ecol. Monogr.* 47:185–207
41. Isebrands, J. G. 1982. Toward a physiological basis of intensive culture of poplar. *Proc. 1982 TAPPI Res. & Dev. Div. Conf.*, pp. 81–90
42. Isebrands, J. G., Larson, P. R. 1977. Vascular anatomy of the nodal region in eastern cottonwood. *Am. J. Bot.* 64: 1066–77
43. Isik, K. 1990. Seasonal course of height and needle growth in *Pinus nigra* grown in summer-dry Central Anatolia. *For. Ecol. Manage.* 35:261–70
44. Jackson, J. B., Buss, L. W., Cook, R. E. eds. 1985. *Population Biology and the Evolution of Clonal Organisms*. New Haven, Conn: Yale Univ. Press. 530 pp.
45. Janzen, D. H. 1976. Effect of defoliation on fruit-bearing branches of the Kentucky coffee-tree, *Gymnocladus dioicus* (Leguminosae). *Am. Midl. Nat.* 95:474–78
46. Jarvis, P. G. 1989. Atmospheric carbon dioxide and forests. *Philos. Trans. R. Soc. Lond. B* 324:369–92
47. Kelly, C. K. 1990. Plant foraging: a marginal value model and coiling response in *Cuscuta subinclusa*. *Ecology* 71:1916–25
48. Kershaw, J. A. Jr., Maguire D. A., Hann, D. W. 1990. Longevity and duration of radial growth in Douglas-fir branches. *Can. J. For. Res.* 20:1690–95
49. Kimura, M. 1969. Ecological and physiological studies on the vegetation of Mt. Shimagare, VII. Analysis of production processes of young *Abies* stand based on the carbohydrate economy. *Bot. Mag. Tokyo* 82:6–19
50. Kozlowski, T. T., Keller, T. 1966. Food relations of woody plants. *Bot. Rev.* 32:293–382
51. Kozlowski, T. T., Torrie, J. H., Marshall, P. E. 1973. Predictability of shoot length from bud size in *Pinus resinosa*. *Can. J. For. Res.* 3:34–38
52. Kozlowski, T. T., Winget, C. H. 1964. The role of reserves in leaves, branches, stems, and roots of shoot growth in red pine. *Am. J. Bot.* 51:522–29
53. Kramer, P. J., Kozlowski, T. T. 1979. *Physiology of Woody Plants*. New York: Academic. 811 pp.
54. Labyak, L. F., Schumacher, F. X. 1954. The contribution of its branches to the mainstem growth of loblolly pine. *J. For.* 52:333–37
55. Lanner, R. M. 1971. Shoot growth patterns in loblolly pine. *For. Sci.* 17:486–87
56. Lanner, R. M. 1976. Patterns of shoot development in *Pinus* and their relationship to growth potential. See Ref. 5, p. 223–43
57. Larson, P. R., Isebrands, J. G. 1978. Functional significance of the nodal constricted zone in *Populus deltoides*. *Can. J. Bot.* 56:801–04
58. Lehtpere, R. 1957. The influence of

high pruning on the growth of Douglas-fir. *Forestry* 30:9–20
59. Loach, K., Little, C. H. A. 1973. Production, storage, and use of photosynthate during shoot elongation in balsam fir *(Abies balsamea). Can. J. Bot.* 51:1161–68
60. Luckwill, L. C. 1970. The control of growth and fruitfulness of apple trees. See Ref. 61, pp. 237–54
61. Luckwill, L. C., Cutting, C. V., eds. 1970. *Physiology of Tree Crops*. New York: Academic. 382 pp.
62. Malstrom, H. L., McMeans, J. L. 1982. Shoot length and previous fruiting affect subsequent growth and nut production of "Moneymaker" pecan. *HortScience* 17:970–72
63. Mar: Møller, C. 1960. The influence of pruning on the growth of conifers. *Forestry* 33:37–53
64. Meyer, F. J. 1928. Die Begriffe "stammeigene Bündel" und "Blattspurbündel" im Lichte unserer heutigen Kenntnisse vom Aufbau und der physiologischen Wirkungsweise der Leitbündel. *Jb. Wiss. Bot.* 69:237–63
65. Molz, F. J., Klepper, B. 1972. Radial propagation of water potential in stems. *Agron. J.* 64:469–73
66. Newell, E. A. 1991. Direct and delayed costs of reproduction in *Aesculus californica. J. Ecol.* In press
67. Onaka, F. 1950. The longitudinal distribution of radial increments in trees. *Kyoto Univ. For Bull.* 18:1–53
68. Owens, J. N. 1968. Initiation and development of leaves in Douglas-fir. *Can. J. Bot.* 46:271–78
69. Pitelka, L. F., Ashmun, J. W. 1985. Physiology and integration of ramets in clonal plants. See Ref. 44, pp. 399–436
70. Pollard, D. F. W., Logan, K. T. 1977. The effect of light intensity, photoperiod, soil moisture potential, and temperature on bud morphogenesis in *Picea* species. *Can. J. For Res.* 7:415–21
71. Powell, G. R. 1988. Shoot elongation, leaf demography, and bud formation in relation to branch position on *Larix laricina* saplings. *Trees* 2:150–64
72. Pyke, G. H. 1984. Optimal foraging theory: a critical review. *Annu. Rev. Ecol. Syst.* 15:523–75
73. Quinlan, J. D., Weaver, R. J. 1969. Influence of benzyladenine, leaf darkening, and ringing on movement of ^{14}C-labeled assimilates into expanded leaves of *Vitis vinifera* L. *Plant Physiol.* 44:1247–52
74. Rangnekar, P. V., Forward, D. F., Nolan, N. J. 1969. Foliar nutrition and wood growth in red pine: the distribution of radiocarbon photoassimilated by individual branches of young trees. *Can. J. Bot.* 47:1701–11
75. Reukema, D. L. 1959. Missing annual rings in young growth Douglas-fir. *Ecology* 40:480–82
76. Richter, H. 1973. Frictional potential losses and total water potential in plants: a re-evaluation. *J. Exp. Bot.* 24:983–94
77. Ross, S. D. 1972. *The seasonal and diurnal source-sink relationships for photoassimilated ^{14}C in the Douglas-fir branch*. PhD thesis. Univ. Wash., Seattle. 98 pp.
78. Salleo, S., LoGullo, M. A. 1989. Xylem cavitation in nodes and internodes of *Vitis vinifera* L. plants subjected to water stress: limits of restoration of water conduction in cavitated xylem conduits. In *Structural and Functional Responses to Environmental Stresses: Water Shortage,* ed. K. H. Kreeb, H. Richter, T. M. Hinckley. pp. 33–42. The Hague: SPB Academic
79. Salzman, A. 1985. Habitat selection in a clonal plant. *Science* 228:603–04
80. Schaap, W., Hinckley, T. M., Sprugel, D. G., Wang, D. 1990. Functional independence of Douglas-fir branches under ozone stress. *Bull. Ecol. Soc. Am.* 71(2 suppl.):315
81. Schmid, B. 1986. Spatial dynamics and integration within clones of grassland perennials with different growth forms. *Proc.* R. *Soc. Lond B* 228:173–86
82. Schmid, B., Bazzaz, F. A. 1987. Clonal integration and population structure in plants: effects of severing rhizome connections. *Ecology* 68:2016–22
83. Schoener, T. W. 1971. Theory of feeding strategies. *Annu. Rev. Ecol. Syst.* 2:369–404
84. Scholander, P. F., Hammel, H. T., Bradstreet, D., Hemmingsen, E. A. 1965. Sap pressure in vascular plants. *Science* 143:339–46
85 Shigo, A. L. 1985. How tree branches are attached to trunks. *Can. J. Bot.* 63:1391–1401
86. Slade, A. J., Hutchings, M. J. 1987. The effects of nutrient availability on foraging in the clonal herb *Glechoma hederacea. J. Ecol.* 75:95–112
87. Slade, A. J., Hutchings, M. J. 1987. The effects of light intensity on foraging in the clonal herb *Glechoma hederacea. J. Ecol.* 75:639–50
88. Slade, A. J., Hutchings, M. J. 1987. Clonal integration and plasticity in foraging behaviour in *Glechoma hederacea. J. Ecol.* 75:1023–36
89. Smith, W. K., Knapp, A. K., Reiners, W. A. 1989. Penumbral effects on sun-

light penetration in plant communities. *Ecology* 70:1603–09

90. Sperry, J. S. 1986. Relationship of xylem embolism to xylem pressure potential, stomatal closure, and shoot morphology in the palm *Rhapis excelsa*. *Plant Physiol*. 80:110–16
91. Sprugel, D. G. 1990. Components of woody-tissue respiration in young *Abies amabilis* (Dougl.) Forbes trees. *Trees* 4:88–98
92. Sprugel, D. G., Hinckley, T. M. 1988. The branch autonomy theory. See Ref. 117, pp. 1–19
93. Sprugel, D. G., Hinckley, T. M. 1990. Carbon autonomy in *Abies amabilis* branches—growth vs. survival. *Bull. Ecol. Soc. Am.* 71(2 suppl.):333–34
94. Stein, W. I. 1955. Fruit set, herbivory, fruit reduction, and the fruiting strategy of *Catalpa speciosa* (Bignoniaceae). *Ecology* 61:57–64
95. Stephenson, A. G. 1980. Fruit set, herbivory, fruit reduction, and the fruiting strategy of *Catalpa speciosa* (Bignoniaceae). *Ecology* 61:57–64
96. Stiell, W. M. 1969. Stem growth reaction in young red pine to the removal of single branch whorls. *Can. J. Bot.* 47:1251–55
97. Takeda, F., Ryuga, K., Crane, J. C. 1980. Translocation and distribution of ^{14}C-photosynthate in bearing and nonbearing pistachio branches. *J. Am. Soc. Hortic. Sci.* 105:642–44
98. Tobiessen, P., Rundel, P. W., Stecker, R. E. 1971. Water potential gradient in a tall *Sequoiadendron*. *Plant Physiol.* 48:303–04
99. Tuomi, J., Niemala, P., Manuila, R. 1982. Resource allocation on dwarf shoots of birch *(Betula pendula):* reproduction and leaf growth. *New Phytol.* 91:483–87
100. Tuomi, J., Niemala, P., Rousi, M., Siren, S., Vuorisalo, T. 1988. Induced accumulation of foliage phenols in mountain birch: branch response to defoliation? *Am. Nat.* 132:602–08
101. Turgeon, R. 1989. The sink-source transition in leaves. *Annu. Rev. Plant Physiol. Plant Mol. Biol.* 40:119–38
102. Tyree, M. T. 1988. A dynamic model for water flow in a single tree: evidence that models must account for hydraulic architecture. *Tree Physiol.* 4:195–217
103. Tyree, M. T., Sperry, J. A. 1988. Do woody plants operate near the point of catastrophic xylem dysfunction caused by dynamic water stress? Answers from a model. *Plant Physiol.* 88:574–80
104. Tyree, M. T., Sperry, J. A. 1989. Vulnerability of xylem to cavitation and embolism. *Annu. Rev. Plant Physiol. Plant Mol. Biol.* 40:19–38
105. Underwood, R. J. 1967. *A study of the effects of pruning on the longitudinal distribution of radial growth in Douglas-fir*. MF thesis. Univ. Wash., Seattle. 111 pp.
106. van den Honert, T. J. 1948. Water transport in plants as a catenary process. *Disc. Faraday Soc.* 3:146–53
107. van der Wal, D. W. 1985. *A proposed concept of branch autonomy and non-ring production in branches of Douglas-fir and grand fir*. MS thesis. Univ. Wash. Seattle. 96 pp.
108. Wareing, P. F. 1970. Growth and its co-ordination in trees. See Ref. 61, pp. 1–21
109. Wareing, P. F., Patrick, J. 1975. Source-sink relations and the partition of assimilates in the plant. In *Photosynthesis and Productivity in Different Environments,* ed. J. P. Cooper, pp. 481–99. Cambridge: Cambridge Univ. Press
110. Wargo, P. M. 1979. Starch storage and radial growth in woody roots of sugar maple. *Can. J. For. Res.* 9:49–56
111. Watson, M. A. 1986. Integrated physiological units in plants. *Trends Ecol. Evol.* 1:119–23
112. Watson, M. A., Casper, B. B. 1984. Morphogenetic constraints on patterns of carbon distribution in plants. *Annu. Rev. Ecol. Syst.* 15:233–58
113. White, J. 1979. The plant as a metapopulation. *Annu. Rev. Ecol. Syst.* 10:109–45
114. White, J. 1984. Plant metamerism. See Ref. 20, pp. 15–47
115. Whitehead, D., Grace, J. C., Godfrey, M. J. S. 1990. Architectural distribution of foliage in individual *Pinus radiata* D. Don crowns and the effects of clumping on radiation interception. *Tree Physiol.* 7:135–55
116. Whitehead, D., Hinckley, T. M. 1991. Models of water flux through forest stands: critical leaf and stand parameters. *Tree Physiol.* In press
117. Winner, W. E., Phelps, L. B. 1988. eds. *Responses of Trees to Air Pollution: the Role of Branch Studies. Proc. of a Workshop, Boulder, CO, Nov. 5–6, 1987*. Washington, DC: U.S. EPA. 248 pp.
118. Zimmermann, M. H. 1978. Hydraulic architecture of some diffuse-porous trees. *Can. J. Bot.* 56:2286–95
119. Zimmerman, M. H. 1983. *Xylem Structure and the Ascent of Sap*. Berlin: Springer-Verlag. 143 pp.

Annu. Rev. Ecol. Syst. 1991. 22:335–55

SPATIAL ANALYSIS OF GENETIC VARIATION IN PLANT POPULATIONS

John S. Heywood

Department of Biology, Southwest Missouri State University, Springfield, Missouri 65804

KEY WORDS: isolation by distance, spatial structure, genetic variation, plants

---

INTRODUCTION

The documentation of spatially restricted dispersal in a wide variety of plant species has led to the prediction that plant populations will often be genetically subdivided on a local scale, as the result either of spatial variation in selection (environmental heterogeneity) or of local genetic drift (isolation by distance) (4, 5, 32, 49). The potential for environmental heterogeneity to generate significant spatial structuring of genotypes in plant populations that occupy pronouncedly patchy habitats is well documented (e.g. 38, 88). In contrast, the effects of isolation by distance (IBD) on the internal structuring of plant populations are less well studied. This review critically examines the methodologies currently available for the analysis of spatial genetic variation within continuously distributed populations, with special emphasis on plant populations and on the detection of structure attributable to IBD. Except where explicitly stated otherwise, conclusions will refer to diploid nuclear genes.

The extensive literature on the detection of spatial associations between genetic and environmental variables has been reviewed several times (see 34) and is specifically omitted here. Therefore, in reviewing the empirical evidence for local structure, studies deliberately conducted with populations occupying manifestly heterogeneous habitats have been excluded.

0066-4162/91/1120-0335$02.00

Isolation by Distance

Isolation by distance (IBD) is defined here in the broadest sense as local deviations from the globally expected gene frequencies due to nondirected processes. Results from the many mathematical models of IBD have been summarized in several reviews (19, 82, 83). Of particular relevance here are the models of Wright (104, 106, 108) and Malecot (52) that concern continuously distributed populations. Environmental heterogeneity may enhance the process by generating local extinction/recolonization episodes (81, 102), but spatial variation in individual fitnesses imposed by the environment must be genotype-independent. Spatial genetic structuring that develops through IBD is necessarily also kinship structuring (52).

The IBD process can have profound evolutionary consequences for plant populations. The diversity of local gene and genotype frequencies generated over space and time may significantly enhance the potential for adaptive evolution as envisioned by Wright in his shifting-balance model (109, 110). Kinship structuring in conjunction with local pollen dispersal results in biparental inbreeding, which may in turn influence the evolution of mating systems (9, 43, 94). Because ecological interactions are primarily among neighbors in populations of immobile plants, kinship structuring also results in ecological interactions among relatives, and kin (group) selection could therefore contribute to the evolutionary dynamics of the population (54, 56).

DETECTING AND QUANTIFYING SPATIAL STRUCTURE

Indirect Measures of Structure

Considerable indirect evidence suggests the spatial structuring of genetic variation in plant populations. Outcrossing species often are deficient in allozyme heterozygosity relative to Hardy-Weinberg expectations, presumably due to biparental inbreeding (reviewed by Brown—5; there are numerous more recent examples). Breeding system studies on partial selfers likewise suggest biparental inbreeding in many cases (1, 7, 18, 40, 71, 72, 95, 99). For several natural populations, crosses between neighbors are less successful and/or result in offspring of lower quality than do crosses between more distant individuals (14, 27, 45, 46, 64, 65, 89, 97, 98). Reduced offspring quality from near-neighbor parents has been interpreted as biparental inbreeding depression, and reduced crossing success between near neighbors has been interpreted as spatial structuring of *S*-alleles. Polymorphisms maintained by selection-mutation balance develop less spatial structure through IBD than do neutral polymorphisms (23), so biparental inbreeding depression is actually a conservative indicator of IBD.

These results, while compelling, rely on assumptions about the determinants of offspring quality and mating success, or on accurate estimates of the breeding system. More importantly, they provide only qualitative measures of spatial structure. To assess the potential contributions of selection and drift to spatial structure, and to assess the magnitude of these processes, quantitative measures of both the magnitude and scale of spatial structure are required.

Geographically Subdivided Populations

If a geographical region occupied by a population is partitioned into subdivisions of smaller area, then spatial genetic structure may be reflected, in part, by variation in genotype frequencies among the subdivisions. The null hypothesis of no allele-frequency (or genotype-frequency) heterogeneity among subdivisions is usually tested under the assumption of sampling from a multinomial distribution, employing either Pearson's Chi-square test or a generalized likelihood ratio test (G-test) (e.g. 75). Proper application of the test requires unbiased estimation of the expected frequencies under the null hypothesis (12, 13). With the multinomial assumption, the standing population is considered to be a sample from an underlying theoretical distribution. If, instead, the goal is to make inferences about the standing population itself, then the sampling distribution is hypergeometric. The Chi-square test (or G-test) will still perform well in this case unless the samples are significant in size relative to the standing populations within subdivisions, in which case the error variance will be overestimated and power will be sacrificed. The same testing procedure can be used with multiple alleles and multiple loci, but the large number of genotypic (or gametic) categories will often require an impractically large sample size. In such cases, it is desirable to combine data from different alleles and different loci into a single summary statistic of spatial structure. Statistical tests then rely on obtaining an approximation to the distribution of the summary statistic.

For a single allele, frequency differences among subdivisions can be summarized as an allele-frequency variance among subdivisions, σ_p^2. Wright (104, 107) introduced the normalized parameter $F_{ST} = \sigma_p^2/p(1-p)$, which measures the degree of differentiation of subdivisions relative to complete fixation, subject to a constant global allele frequency, p. The numerator and denominator of F_{ST} are both variance components that can be estimated from an analysis of variance (12, 13, 111). Weir & Cockerham (101) provide estimators for these variance components that are unbiased if the subdivisions are independent samples from a theoretical distribution.

Under strict isolation by distance in an infinite population, F_{ST} will be the same for all alleles at all loci (108, 111), so that a single parameter is sufficient to describe spatial genetic structure. With this assumption, a generally best estimator of F_{ST} (again, based on independent samples) is given by

$$\hat{\theta} = \frac{\sum_i \sum_j \hat{h}_{ij}\, \hat{\theta}_{ij}}{\sum_i \sum_j \hat{h}_{ij}} = \frac{\sum_i \sum_j \hat{\sigma}^2_{p(ij)}}{\sum_i \sum_j \hat{h}_{ij}}, \qquad 1.$$

where the sum is over each allele j at each locus i, $\widehat{h_{ij}}$ is an unbiased estimator of $p_{ij}(1-p_{ij})$, $\widehat{\theta_{ij}}$ is an unbiased estimator of F_{ST} based on allele j at locus i, and $\widehat{\sigma}^2_{p(ij)}$ is an unbiased estimator of $\sigma^2_{p(ij)}$ (67, 101). Wright (111) proposed a general definition of F_{ST} that is identical in form to equation (1), with the estimators replaced by the population parameters. This provides a single generalized parameter of spatial structure even when individual alleles are characterized by unique F_{ST} values due to unique selection histories, and preserves the relation $1-F_{IT} = (1-F_{IS})\ (1-F_{ST})$between Wright's fixation indexes (111). The estimator $\widehat{\theta}$ is still appropriate for this generalized F_{ST}. Nei's G_{ST} (57, 60) is identical in form to the generalized F_{ST}, except that it is defined as a sample statistic rather than a parameter. Whether $\widehat{\theta}$ or a simpler estimator such as G_{ST} is most appropriate depends on the evolutionary processes involved and on whether interest lies in the standing population or in an underlying theoretical distribution from which the population is a sample (58, 85). It should be borne in mind that, although a generalized F_{ST} (or G_{ST}) value may be useful, it is representative of individual alleles and loci only under strict IBD.

The sampling distributions of $\widehat{\theta}$ and G_{ST} under IBD depend on the mating structure of the population and consequently cannot be specified a priori (100, 101), so statistical tests must in general be based on resampling techniques. Jackknifed estimates of the variance of $\widehat{\theta}$ (67, 101) can be used to generate approximate tests by assuming asymptotic normality of $\widehat{\theta}$. Under the assumption of IBD, Weir & Cockerham (101) recommend jackknifing across loci. When this assumption is not appropriate or the number of loci studied is few, it is necessary to jackknife across population subdivisions. Alternatively, the null sampling distribution can be generated by randomization of sampled genotypes among subdivisions. This distribution may then be used to test the null hypothesis that the sampled genotypes (as opposed to all genotypes in the population) are randomly distributed among subdivisions. Randomization tests are discussed more fully in the next section.

Spatial inferences based on σ_p^2 (or F_{ST}) are limited in two important ways. First, structure cannot be detected at any spatial scale smaller than the subdivision size (15). Second, because σ_p^2 is independent of the spatial arrangement of the subidivisions, no information is provided about the spatial scale of detected structure, except that it is at the scale of the subdivision or larger. These limitations are unimportant when interest is focused specifically on natural subdivisions, such as distinct habitat types in a patchy environment. In general, however, considerable information of interest may be lost. Much of this information can be regained by decreasing the size of the subdivisions and partitioning σ_p^2 into components attributable to different spatial scales. This is done with a nested analysis of variance with subdivisions nested into progressively larger spatial clusters (12, 13, 101, 105, 111). The generalized F-statistics and their estimators are easily extended to any degree of nesting (101, 111). Wright (105, 111) used this procedure to provide a detailed spatial analysis of the data of Epling & Dobzhansky (21) on flower color morphs in *Linanthus parryae*.

Wright's F_{ST} is an intraclass correlation coefficient (12, 13, 39) between alleles within subdivisions relative to the entire population. A hierarchical set of F_{ST} estimates obtained from a nested analysis of variance thus provides a set of intraclass correlation coefficients that describe genetic similarity as a function of distance (108). In this regard, the spatially nested ANOVA bears a close relationship to spatial autocorrelation analysis, described below.

Spatial Autocorrelation Analysis

Spatial autocorrelation analysis is a set of statistical procedures designed to detect and quantify spatial dependency in a variable based on sampled values from multiple mapped locations (10, 11, 91). In general, every possible pair of sample locations i and j ($i \neq j$) is assigned a relative weight w_{ij} that determines its contribution to the estimation of spatial dependency. The matrix of weights **W** specifies the type of spatial dependency to be investigated and represents the alternative hypothesis against which the null hypothesis of spatial independence is tested. For a nominal variable X, each possible pairing of two sample values (x_k, x_l) is characterized by a "join count" which is the sum of weights in **W** for all actual pairs of sample locations for which the sampled values were x_k and x_l. The observed join count is then compared to the expected join count under the null hypothesis of spatial independence. For ordinal and interval variables, an average index of similarity between measured values at paired locations (the autocorrelation coefficient) is calculated, with pairs of locations contributing to the calculation according to the weightings in **W.** The coefficient of spatial autocorrelation used in most biological applications in Moran's I (55), defined as

$$I = \frac{\sum_{i \neq j}^{n} \sum^{n} w_{ij} z_i z_j \,/\, \sum_{i \neq j}^{n} \sum^{n} w_{ij}}{\sum^{n} z_i^2/n} \qquad 2.$$

where $z_i = x_i - \bar{x}$, $\bar{x} = \Sigma x_i / n$, and n is the number of sample locations. It is important to bear in mind that I is a sample statistic, not a population parameter.

It is common practice to use binary weights, with each element of **W** being either 0 or 1, so that autocorrelation is assessed for a specific set of paired sample locations (weight 1), with all other possible pairs of locations not contributing to the calculation (weight 0) (11, 91). With binary weights, the matrix **W** can be represented by a map of the sample locations with line segments connecting paired samples (those given a weight of 1), thus generating a two-dimensional network. In the following discussion of autocorrelation techniques, binary weightings are assumed, and the entire binary weighting matrix is referred to as a "network."

For a random variable that is defined over continuous space and is mathematically continuous, autocorrelation will clearly be positive when the distance between paired samples becomes small relative to the scale of spatial variations in the variable, and it will increase to a maximum as the distance between paired samples is decreased to zero. This observation suggests that the ability of autocorrelation analysis to detect spatial structure is maximized when paired samples consist of nearest-neighbors within the sample. However, at any point in time, genotypes within a population are not defined over continuous space, but rather only at discrete points occupied by individuals. Under IBD, nearest-neighbors should still provide the highest autocorrelation because kinship is expected to diminish monotonically with distance (41, 52). It seems likely that this conclusion would also be valid for most types of selection, but exceptions could be encountered. For example, if two genotypes in a plant population preferentially utilize different soil horizons, competitive interactions might lead to an alternation of genotypes in space and thus to a negative autocorrelation among nearest-neighbors.

A frequently used algorithm for choosing pairs of neighboring sample locations is that of Gabriel (28). This algorithm has the advantage of generating significantly more sample pairs than would be obtained by including only strict nearest-neighbor pairs, thereby lowering the standard errors of the estimated autocorrelation statistics. It has the disadvantage of increasing the mean distance between paired locations relative to that for strict nearest-neighbors. The Gabriel method is thus preferable to strict nearest-neighbors only when the average distance between paired locations is small relative to the scale of spatial variations to be quantified, that is, whenever the sampling locations are sufficiently dense. In this case, the Gabriel and nearest-neighbor networks will provide similar estimates of autocorrelation, but the Gabriel network will provide a more precise estimate. In such a situation, Delaunay triangulation (69) would appear to provide an even better network, achieving a greater number of connections between neighboring sample locations.

Regardless of which of these algorithms is used to define a network of paired neighbors, I refer to the resulting value of Moran's I as I_{NN}.

It is often of interest to assess the relationship between the degree of autocorrelation and the distance between sampling locations. For this purpose, a series of networks are constucted, each including all pairs of sampling points separated by a distance that falls within a defined distance class. Although various distance measures could be used, applications in population genetics have generally employed euclidian distance. A graph of an autocorrelation coefficient against distance class is referred to as a correlogram (11).

Two approaches have been taken in applying spatial autocorrelation techniques to the analysis of spatial genetic variation among diploid individuals within continuous populations. Individual genotypes may be treated as nominal classes, with their spatial distributions described in terms of join count statistics (24, 25, 80). Alternatively, for each allele, individuals can be characterized by allele frequencies of 0, 0.5, or 1, according to whether they carry 0, 1, or 2 copies of the allele in their genotype, and the spatial distribution of alleles can be described in terms of an autocorrelation coefficient, usually Moran's I (8, 15, 78, 79). Calculated in this latter fashion, Moran's I is identical in form to Wright's coefficient of relationship, ρ (12, 103). Barbujani (2) made a similar observation for I based on population allele frequencies. Lacking any correction for sampling error, I is a biased estimator of ρ.

One disadvantage to using genotypic join counts is that information about the overall spatial pattern for an individual allele is distributed across a number of different join-count statistics representing all possible genotype pairs in which that allele is represented at least once. Thus, a test for a random spatial distribution of an allele would require the simultaneous testing of several join-counts, with a concommitant loss of statistical power. Therefore, if the spatial distribution of alleles provides a complete description of spatial structure, autocorrelation statistics based on individual allele frequencies are more powerful and will also be easier to interpret than join counts. This advantage is increased for polyploid populations. However, if the spatial distribution of genotypes is not completely determined by the spatial distribution of alleles, then the genotypic join-counts may provide useful additional information. This would be the case, for example, if there was spatial variation in dominance for fitness, or spatial variation in the mating system. Join counts are necessary when the autocorrelation analysis is based on discrete phenotypic classes that cannot be ascribed to unique genotypes because of dominance or epistasis (25).

Under the null hypothesis of spatial independence, the distributions of join counts and of Moran's I are both asymptotically normal, thus providing a simple test of this hypothesis when the number of sample locations is sufficiently large and the theoretical mean and variance of the statistic are known (11). To evaluate the mean and variance under the null hypothesis, it is necessary to assume either that the samples are drawn at random from an underlying distribution (sampling with replacement), or else that the sample is itself the population of interest (exhaustive sampling) (11). With the latter assumption, the mean and variance of each autocorrelation statistic under the null hypothesis may be obtained by repeated randomizations of the sampled values across the sample locations. Under the assumption of sampling with replacement, the mean and variance of Moran's I can only be evaluated if it is further assumed that the underlying distribution of the sampled variable is normal. The distribution of individual allele frequencies (which can assume only three different values in diploids) is decidedly nonnormal, so statistical tests based on Moran's I must utilize the theoretical mean and variance of I under the assumption of randomization. It should be kept in mind that conclusions based on randomization tests apply strictly only to the sample, and not to the population at large, although the distinction is not important for large samples. It is common practice to use the large-sample normal approximation to the distribution of I, together with the theoretical mean and variance of I under randomization, to convert I to a standard normal variate that is then compared to a standard normal distribution (e.g. 15, 25, 91). However, given the necessary randomization assumption, it would seem appropriate to perform a randomization test directly on a computer and thereby avoid invoking the large-sample normal approximation, especially for smaller sample sizes. Slatkin & Arter (84) provide an example of such a randomization test. The evaluation of the mean and variance of join counts does not require the randomization assumption (11), but, again, a direct randomization test is recommended to avoid errors due to deviations from the large-sample normal approximation, especially for smaller samples.

When there are no logical constraints on the possible patterns of deviation from complete spatial randomness among the samples, restricting the alternative hypothesis to a single network (such as the Gabriel network) can result in a serious loss of statistical power (63). In such cases it is more appropriate to perform a simultaneous test of all coefficients (or join counts) in a correlogram. Oden (63) discusses three different simultaneous testing procedures for Moran's I that consider all coefficients in a correlogram to be equally important. However, if the values of I at different distance classes are functionally related (beyond the statistical dependence that arises from there being one more coefficient than degrees of freedom), then the simultaneous tests have a much lower simultaneous type-I error rate than specified and thus are excessively conservative. Oden (63) specifically cautions against using the unweighted simultaneous test procedures when information about spatial causality is available. As discussed previously, genetic correlation is expected to be a decreasing function of distance under fairly general conditions, including IBD, in which case a test of autocorrelation in the smallest possible distance class (nearest neighbors in the population) is a de facto test of the entire correlogram.

For several ostensibly equivalent alleles, or for several ostensibly neutral loci, it is appropriate to combine results in order to obtain the most powerful simultaneous test possible for spatial structure resulting from IBD. To extend Moran's I, it is necessary to obtain a summary statistic by an appropriate average of I values across alleles and across loci. A test of significance is performed as before by generating the null distribution through randomization. Join counts are not readily combined across alleles or across loci into consensus structure statistics.

The Mantel Test

The method proposed by Mantel (53) has been applied extensively to samples from geographically distinct human populations in order to test for correspondence between a matrix of genetic distances and a matrix of some other type of distance measure based on geography, estimated migration parameters, linguistic similarity, etc (e.g. 87). In principle, the same procedure could be applied to samples of mapped individuals, rather than populations, by utilizing an appropriate measure of genetic distance between individuals. The statistic of correspondence between matrices is simply the regression coefficient of one distance measure on the other, or some simple transformation thereof, taken across all nondiagonal elements of the two distance matrices (53, 86). A test of significance is obtained by randomization of the rows and columns in one of the two matrices, so that no distributional assumptions are required (16, 86). An attractive feature of the Mantel test is that it uses all possible pairs of sample locations (all distance classes), thereby effectively increasing the sample size on which inferences are based. However, the Mantel test is sensitive only to linear relations between the two distance measures. This weakness can be compensated for by using a distance metric that is nonlinear with euclidian distance so as to reflect the expected spatial pattern, if such an a priori expectation exists. For example, under strict IBD an exponential function would be more appropriate (41, 52). A comparison of the relative powers of spatial autocorrelation analysis based on nearest-neighbors and the Mantel test would be useful for this case. In general, however, the Mantel test will suffer when genetic distance varies in a nonlinear fashion with the chosen distance metric. In the extreme case of a regular spatial pattern, such as a checkerboard, the Mantel statistic will not detect the spatial variation. In the present context, the key difference between spatial autocorrelation analysis and the Mantel test is that the former provides a test for the presence of spatial structure (based on I_{NN}), whereas the latter provides a test for a particular pattern of structure represented by the distance matrix.

Genetic distance measures are generally designed to deal with multiallele and multilocus data (e.g. 57), so the extension of the Mantel test is automatic. The cautionary comments made previously about combining data from different alleles and loci apply equally here.

The Evidence from Plant Populations

The results of experimental studies on spatial genetic structuring within continuous plant populations are summarized in Tables 1 and 2. Of the 32 species studied, 25 revealed significant structuring within populations. With the notable exception of *Linanthus parryae,* obligate outcrossers generally show weak differentiation among subpopulations, with F_{ST} estimates ($\tilde{F}_{ST}$) ranging from 0.004 to 0.08 (mean = 0.043, 14/18 significant) (Table 1). In contrast, partial or predominant selfers display remarkable local structure in some cases, with $\tilde{F}_{ST}$ values ranging from 0.026 to 0.78 (mean = 0.24, 4/6 significant). It is not possible to make the same comparison among all studies that have employed spatial autocorrelation analysis because of the use of join counts rather than Moran's I for many of the selfing species (Table 2). Nonetheless, the two selfers for which values of I have been reported (*Plantago major* and *P. coronopus*) show very pronounced local structure (I = 0.423 and 0.416; Table 2). This contrasts to values of I ranging from 0.054 to 0.17 (mean = 0.10) for five outcrossers (Table 2).

$\tilde{F}_{ST}$ values must be considered conservative indicators of structure, for the reasons discussed previously. In addition, both $\tilde{F}_{ST}$ and I_{NN} will underestimate structure if the study area is small relative to genetic patch size. This may be the case for *Liatris cylindracea,* in which genetic correlations were found between subpopulations separated by distances that approach the dimensions of the study area (74). Even in the absence of such biases, the relatively low reported values of $\tilde{F}_{ST}$ and I_{NN} for outcrossers should not be interpreted to mean that IBD is unimportant in these populations. As Wright (111) has shown, an F_{ST} value of only 0.05 implies that local demes will on occasion deviate considerably from the population mean allele frequency.

HOMOGENEITY TESTS

Kinship structuring affects all loci equally and independently. Consequently, the expected spatial distribution of allele frequencies under strict IBD is identical and independent for all alleles and all loci (50, 108, 111). In contrast, spatial patterns generated by selection may vary widely among alleles and among loci according to specific patterns and intensities of selection. Responses to common selective agents may cause nonhomologous

Table 1 Spatial genetic structure within continuous plant populations in the absence of apparent environmental heterogeneity, as measured by genetic divergence among subdivisions

Species	Life form[a]	Breeding system[b]	Community	Loci[c]	N_S[d]	$\bar{F}_{ST}$[e]	p[f]	Significant loci[g]	References
Linanthus parryae	A	O	Desert	1-P	1261	0.165	0.000	1	21, 105, 111
Liatris cylindracea	P	O	Dry prairie	15-E	66	0.0687	<.005	15	75
Desmodium nudiflorum	P	O	Temperate forest	2-E	3–5	0.012[i]	NS	0	77
Avena barbata	A	A	Grassland	2-P	30–60	0.29–0.78[h]	<.001	2	66
Cynosurus cristatus	P	O	Pasture	4-E	3	0.0114	NA	1	20
Plantago lanceolata	P	O	Pasture	8-E	8	0.0615[i]	NA	4/4[j]	3
Plantago lanceolata	P	O	Old field	4-E	11	0.07	<.05	3	S. Tonsor, unpublished
Fagus sylvatica	W	O	Second growth forest	4-E	3	0.004	NA	1	30
Impatiens capensis	A	M	Temperate forest	8-E	14	0.026[i]	NS	0	42
Delphinium nelsoni	P	O	Montane meadow	5-E	26	0.069	<.02	NA	96
Triticum dicoccoides	A	A	Grassland	13-E	10	0.408	NA	7/7[j]	29
Psychotria nervosa	W	O	Subtropical hammock	2-E	3	0.006	NS	0	15
Alseis blackiana	W	O	Wet tropical forest	25-E	4	0.034	<.01	NA	31
Brosimum alicastrum	W	O	Wet tropical forest	18-E	4	0.050	<.01	NA	31
Erythrina costaricensis	W	O	Wet tropical forest	30-E	4	0.029	<.01	NA	31
Hybanthus prunifolius	W	O	Wet tropical forest	28-E	4	0.065	<.01	NA	31

Table 1 (*Continued*)

Species	Life form[a]	Breeding system[b]	Community	Loci[c]	N_S[d]	$\bar{F}_{ST}$[e]	p[f]	Significant loci[g]	References
Platypodium elegans	W	O	Wet tropical forest	26-E	4	0.051	<.01	NA	31
Psychotria horizontalis	W	O	Wet tropical forest	11-E	4	0.030	<.01	NA	31
Rinorea sylvatica	W	O	Wet tropical forest	14-E	4	0.080	<.01	NA	31
Swartzia simplex	W	O	Wet tropical forest	27-E	4	0.037	<.01	NA	31
Tachigalia versicolor	W	O	Wet tropical forest	9-E	4	0.059	<.01	NA	31
Ipomopsis aggregata	P	O	Montane meadow	7-E	20–45	0.032	<.001	NA	8
Sanicula gregaria	P	A or M	Temperate forest	9-E	37	0.111	<.05	NA	C. F. Williams, unpublished
Osmorhiza claytonii	P	A or M	Temperate forest	6-E	19	0.058	NS	NA	C. F. Williams, unpublished
Cryptotaenia canadensis	P	A or M	Temperate forest	6-E	39	0.311	<.05	NA	C. F. Williams, unpublished

[a] Life forms: A, annual; P, perennial herb; W, woody perennial
[b] Breeding systems: O, predominantly or exclusively outcrossing; A, autogamous (predominantly selfing); M, mixed selfing and outcrossing; C, clonal (asexual)
[c] Number of polymorphic enzyme loci (E) or loci controlling phenotypic variants (P) used to calculate $\bar{F}_{ST}$
[d] Number of population subdivisions on which F_{ST} estimate is based
[e] Consensus estimate of F_{ST} based on all loci. A variety of estimators (including G_{ST}) and a variety of methods for combining across loci are represented. If more than one hierarchy of subdivisions was examined, differentiation between the smallest subdivisions relative to the next larger subdivisions is reported.
[f] Significance level for the consensus F_{ST} estimate. NS, not significant ($p > 0.05$); NA, information not available
[g] Number of individual loci with significant ($p < .05$) spatial variation in allele frequencies. NA, information not available
[h] Calculated from data in Figures 6–9 in Rai & Jain (66)
[i] Recalculated from published data
[j] Tests performed for highly polymorphic loci only

Table 2 Spatial genetic structure within continuous plant populations in the absence of apparent environmental heterogeneity, as assessed by spatial autocorrelation of individual genotypes

Species	Life form[a]	Breeding system[b]	Community	Loci[c]	I^d	p^e	Significant loci[f]	References
Ipomoea purpurea	A	M	Agricultural field	2-P	JC	NA	2	25
Psychotria nervosa	W	O	Subtropical hammock	2-E	0.075	NS	1	15
Plantago major	P	M	Disturbed fields	3-5-E	0.423[g]	NA	NA	94a
Plantago coronopus	P	M	Disturbed fields	6-E	0.416[g]	NA	NA	94a
Plantago lanceolata	A/P	O	Disturbed fields	6-E	0.093[g]	NA	NA	94a
Plantago lanceolata	P	O	Old field	5-E	0.17	<.05	3	S. Tonsor, unpublished
Pinus contorta	W	O	Monospecific forest	14-E	JC	NS	1-2	24
Impatiens capensis	A	M	Temperate forest	1-P	JC	<.05	1	80
Impatiens pallida	A	M	Temperate forest	2-P	JC	NA	2	80
Gleditsia triacanthos[h]	W	O,C	Second growth	19-E	0.084[i]	NA	7	78
				19-E	0.111[i]	NA	11	
Maclura pomifera	W	O,C	Second growth	4-E	0.054[i]	NA	2	79
Ipomopsis aggregata	P	O	Montane meadow	7-E	0.1	<.05	5	8
Chamaecrista fasciculata	A	M	Disturbed wet prairie	7-E	NA[j]	<.05	NA	26

[a] See footnote a in Table 1
[b] See footnote b in Table 1
[c] Number of polymorphic enzyme loci (E) or loci controlling phenotypic variants (P) used to calculate autocorrelation statistics
[d] Consensus value of Moran's I for all loci, based on the shortest distance class reported. If join counts were used, "JC" is entered in the table.
[e] Significance level of consensus I value, based on shortest distance class only. NS, not significant; NA, information not available
[f] See footnote g in Table 1.
[g] Transformed from F_r (0) values using the relation $I = 2F_r(0)/(1 + F_{IT})$. The resulting I values were then averaged across populations and subpopulations.
[h] Data from two populations
[i] Recalculated from published data
[j] A different measure of genetic similarity was used in this study.

alleles to covary in space, resulting in linkage disequilibrium across the populations (61). Thus, the null hypothesis of strict IBD can, in principle, be rejected in favor of an alternative that includes selection if statistics of spatial structure ($\hat{\theta}$, I_{NN}) are sufficiently heterogeneous among alleles or loci (50). However, when working with a sample from a finite population (or from a finite area within a much larger population), it is important to recognize that the standing population within the area of interest represents a single manifestation of a stochastic process, and not the expectation of that process (84, 101). For a single locus, the actual correlogram (or F_{ST} value) for a population will deviate from the expected correlogram due to stochastic variation. The statistical tests described below account for the sampling error that causes deviations between the sample correlogram and the actual correlogram, but they do not account for the stochastic error that causes deviations between the actual correlogram and the expected correlogram. The error variance is thus underestimated, and the type-I error rate for the test is greater than specified. This problem will affect the interpretation of tests based on any statistic of spatial structure, including Moran's I, join counts, and F_{ST} estimates. The magnitude of this stochastic variance can be estimated for particular population models (e.g. 59), but no general result is available.

However, if the stochastic process is spatially invariant, then the stochastic variance will be negligible if the population is large in area relative to the expected scale of spatial genetic structure, because the actual correlogram (or F_{ST} value) will be an average over many patches and hence will approach the expectation. Thus, the stochastic error can be minimized by spreading the sample over the greatest number of patches possible. If the sampling area is sufficiently large relative to the spatial scale of the underlying genetic structure, stochastic error will be inconsequential. However, in practice, obtaining a sample from a sufficiently large area may not be possible. In the following discussion of statistical tests, it is implicitly assumed that stochastic errors are minor in comparison to sampling errors. Type-I error rates will be inflated in porportion to the extent that this assumption is violated.

A second important consideration when interpreting the following statistical tests is the assumption of equilibrium. Inferences about current evolutionary processes can be drawn from the current population structure only if the population has achieved equilibrium with respect to current patterns of dispersal, recombination, and selection. If a population is not at equilibrium, then residual patterns of spatial genetic variation may persist from times when dispersal or selection was different. Unique historical events may impart transient patterns on many loci that would be misinterpreted under an assumption of equilibrium. For example, spatial variation at neutral loci that is generated by hitchhiking with a selected locus during species-range expansion may persist for a very long time (33, 35, 37). Thus, caution must always be exercised when attempting to infer connections between current patterns of genetic variation and current population dynamics.

Lewontin & Krakauer (50) proposed a test for the heterogeneity of F_{ST} estimates across loci based on the theoretical sampling variance of F_{ST} for independent samples from an infinite population at equilibrium. Stochastic variations in finite areas and non-equilibrium patterns have been shown to result in a significantly inflated type-I error rate for this test (e.g. 59, 62, 73), as would be expected from the previous discussion. Pairwise tests of homogeneity for two loci, i and j, can be developed from Moran's I by noting that I_i-I_j is asymptotically normal. To avoid assumptions about linkage disequilibrium, the variance of I_i-I_j can be estimated by jackknifing or bootstrapping across sample locations. Comparing entire correlograms for different loci presents a more difficult problem because of the strong dependency among I values for different distance classes. This problem may be partially overcome by collapsing the correlogram into a small number of summary statistics that are approximately independent. The theoretical monotonically decreasing form of a correlogram under IBD (41, 52) suggests that the major features of a correlogram might be expressed in terms of the *magnitude* of spatial structure (the maximum genetic correlation, I_{NN}) and the *scale* of the structure (the rate at which genetic correlation decays with distance). Estimates of scale parameters ("patch size") could be obtained from a weighted least-squares fit of the observed correlogram to an appropriate model (22). A bootstrapped or jackknifed estimate could then be obtained for the variance of patch size, although this would be quite computer-intensive since the correlogram and regression analysis would have to be regenerated for each resampling.

EXPERIMENTAL DESIGN

To minimize the effects of stochastic variation, sampling should be spread over as large an area as possible. Therefore, for a given sampling intensity, nearest-neighbor pairs scattered over a large area are more informative than concentrations of samples at relatively few locations. Either of these sampling schemes will provide a large number of nearest-neighbor pairs, but with clustered samples the structure revealed will characterize the relatively few areas of intensive sampling, and these local areas may show considerable stochastic deviation from the global expectation. The same consideration applies when spatial structuring is the deterministic consequence of environmental heterogeneity; samples concentrated at a few localities may be highly nonrepresentative of the overall spatial pattern. Ripley (69) discusses several stratified sampling designs that are compromises between spatially uniform and spatially random samples.

In selecting a population for study, several factors that influence the potential for IBD should be considered. First, both theoretical models (49, 52, 104) and computer simulations (92, 93) of IBD are scale-independent, indicating that effective dispersal is measured on a scale of $x\sqrt{D}$, where x is geographical distance and D the population density. A commonly used measure of effective dispersal is Wright's neighborhood size, $N_I = 4\pi D\sigma^2$, where σ^2 is the axial variance in dispersal locations from a point source (104, 108). Thus, for a given absolute dispersal schedule, the potential for IBD is greater in populations of lower density. In animal-pollinated species, this density effect may be partially negated by density-dependent pollinator behavior, with pollinators flying greater distances when neighboring plants are farther apart (47–49). However, seed dispersal generally is not density-dependent, so that low-density populations are still expected to experience greater IBD. Based on this criterion, subordinate members of a community are more likely to experience IBD than are dominants. Unusually diverse communities in which many species persist at relatively low densities (high alpha-diversity), such as temperate wet prairies (70) and moist tropical lowland forests (68), would seem to be particularly favorable in this regard. However, highly effective outcrossing mechanisms seem to counter this density effect in the moist tropical forest of Barro Colorado Island in Panama (31; Table 1).

The breeding system is another important factor in choosing a study population. Self-fertilization increases the potential for IBD both by decreasing average pollen dispersal distances and by increasing the sampling error when going from the local gamete pool to the local recruits (36, 51). However, care must be exercised when interpreting spatial structure observed in highly autogamous species, because the potential for genetic hitchhiking is greatly enhanced in such species (33, 35).

Asexual reproduction via agamospermy is similar to selfing in that there is no pollen dispersal phase. With vegetative reproduction, seed dispersal is also eliminated and the potential for IBD is very high (108). By chance, some genotypes will spread and others will be lost, resulting in a distinct patchwork of overlapping clones. However, because of the extremely restricted dispersal via vegetative reproduction, populations will have exceedingly long memories of initial colonization patterns and subsequent disturbance events. More importantly, in the absence of genetic recombination, different loci cannot establish independent equilibrium patterns, so the problem of genetic hitchhiking is extreme. Thus, all else being equal, populations of clonal species are the most likely to reveal spatial structure, but causation will be difficult or impossible to interpret purely on the basis of the observed spatial pattern.

Although this review has concentrated on nuclear genes, cytoplasmic DNA can also be analyzed for spatial structuring with the use of restriction fragment length polymorphisms (44). With sexual outcrossing, maternally-transmitted cytoplasmic DNA is dispersed over shorter distances, and paternally-transmitted cytoplasmic DNA is dispersed over greater distances, than is nuclear DNA. The most important feature of cytoplasmic DNA in the present context is the essential absence of genetic recombination, so that genetic hitchhiking will strongly influence spatial patterning of the entire cytoplasmic genome.

A final consideration is the history of the study population. Populations that are likely to have suffered recent disturbances, such as those on roadsides, pastures, and floodplains, may not be at genetic equilibrium with respect to contemporary population dynamics. Fortunately, however, computer simulations of isolation by distance (23, 90, 92, 93) indicate that relatively short periods of localized dispersal (10–20 generations) are sufficient for significant local structure to develop. All else being equal, species with the shortest generation time will have cycled through the greatest number of generations since the last major disturbance and will be most likely to show local spatial structure consistent with the current mating system.

CONCLUSIONS

Although most populations of outcrossing plants that have been studied reveal some microspatial genetic structuring, the magnitude of that structuring is relatively low. Recent evidence that gene dispersal via pollen can be considerably more extensive than previously thought (8, 17, 76) may provide a partial explanation for this. On the other hand, obligate outcrossing does not preclude substantial local genetic structuring due to IBD, as evidenced by *Linanthus parryae*. It would seem prudent to withhold judgement until additional outcrossing species with characteristics otherwise favorable for IBD have been studied using optimal sampling methods and statistical procedures.

In contrast to outcrossers, some autogamous species display remarkable levels of local genetic differentiation, consistent with theoretical expectations. All else being equal, low-density populations with reproductive systems characterized by partial selfing, or a mixture of sexual outcrossing and apomixis, are most likely to possess the combination of extensive sexual recombination and localized genetic drift envisioned by Sewall Wright in his shifting balance process.

Spatial autocorrelation analysis provides a more complete description of spatial genetic structure than does a nested ANOVA based on arbitrary subdivisions (*F*-statistics). The criticism of Slatkin & Arter (84) that statistical tests do not account for stochastic variation around the expectation in a finite population, while valid, applies to all statistics of spatial structure. For a given sampling intensity, autocorrelation statistics based on individual genotypes obtained with an optimal sampling scheme suffer less from stochastic variation than do statistics such as F_{ST} that are based on samples concentrated into population subdivisions. In large, continuous plant populations, there is essentially no stochastic variation if the study area is sufficiently large relative to the underlying structure.

Although much more work is needed, considerable progress has been made in the use of spatial statistics to infer evolutionary processes (22, 23, 50, 85, 89a, 90). If it is assumed that most allozyme variation is neutral or nearly so, consensus measures of spatial differentiation among sample sites can be used to obtain robust estimates of effective gene flow rates (85). Likewise, a consensus estimate of I_{NN} from multiple enzyme loci should provide a robust estimate of neighborhood size within a continuous population.

Literature Cited

1. Allard, R. W., Kahler, A. L., Clegg, M. T. 1977. Estimation of mating cycle components of selection in plants. In *Measuring Selection in Natural Populations,* ed. R. B. Christiansen, T. M. Fenchel, pp. 1–19. New York: Springer-Verlag
2. Barbujani, G. 1987. Autocorrelation of gene frequencies under isolation by distance. *Genetics* 117:777–82
3. Bos, M., Harmens, H., Vrieling, K. 1986. Gene flow in *Plantago*. I. Gene flow and neighborhood size in *P. lanceolata*. *Heredity* 56:43–54
4. Bradshaw, A. D. 1972. Some of the evolutionary consequences of being a plant. *Evol. Biol.* 5:25–43
5. Brown, A. H. D. 1979. Enzyme polymorphism in plant populations. *Theor. Popul. Biol.* 15:1–42
6. Brown, A. H. D., Clegg, M. T., Kahler, A. L., Weir, B. S., eds. 1990. *Plant Population Genetics, Breeding, and Genetic Resources*. Sunderland, Mass: Sinauer. 449 pp.
7. Brown, A. H. D., Matheson, A. C., Eldridge, K. G. 1975. Estimation of the mating system of *Eucalyptus obliqua* L'Herit. using allozyme polymorphisms. *Aust. J. Bot.* 23:931–49
8. Campbell, D. R., Dooley, J. L. 1991. The spatial scale of genetic differentiation in a hummingbird-pollinated plant: comparison with models of isolation by distance. *Am. Nat.* In press
9. Charlesworth, D., Charlesworth, B. 1987. Inbreeding depression and its evolutionary consequences. *Annu. Rev. Ecol. Syst.* 18:237–68
10. Cliff, A. D., Ord, J. K. 1973. *Spatial Autocorrelation*. London: Pion
11. Cliff, A. D., Ord, J. K. 1981. *Spatial Processes—Models and Applications*. London: Pion
12. Cockerham, C. C. 1969. Variance of gene frequencies. *Evolution* 23:72–84
13. Cockerham, C. C. 1973. Analysis of gene frequencies. *Genetics* 74:679–700
14. Coles, J. F., Fowler, D. P. 1976. Inbreeding in neighboring trees in two white spruce populations. *Silvae Genet.* 25:29–34
15. Dewey, S. E., Heywood, J. S. 1988. Spatial genetic structure in a population of *Psychotria nervosa*. I. Distribution of genotypes. *Evolution* 42:834–38
16. Dietz, E. J. 1983. Permutation tests for association between two distance matrices. *Syst. Zool.* 32:21–26
17. Ellstrand, N. C., Devlin, B., Marshall, D. L. 1989. Gene flow by pollen into small populations: data from experimental and natural stands of wild radish. *Proc. Natl. Acad. Sci. USA* 86:9044–47
18. Ellstrand, N. C., Torres, A. M., Levin, D. A. 1978. Density and the rate of apparent outcrossing in *Helianthus annuus* (Asteraceae). *Syst. Bot.* 3:403–7
19. Endler, J. A. 1977. *Geographic Variation, Speciation and Clines*. Princeton, NJ: Princeton Univ. Press
20. Ennos, R. A. 1985. The mating system and genetic structure in a perennial grass, *Cynosurus cristatus* L. *Heredity* 55:121–26
21. Epling, C., Dobzhansky, T. 1942. Genetics of natural populations: VI. Microgeographical races in *Linanthus parryae*. *Genetics* 27:317–32
22. Epperson, B. K. 1990. Spatial patterns of genetic variation within plant populations. See Ref. 6, pp. 229–53
23. Epperson, B. K. 1990. Spatial auto-

correlation of genotypes under directional selection. *Genetics* 124:757–71
24. Epperson, B. K., Allard, R. W. 1989. Spatial autocorrelation analysis of the distribution of genotypes within populations of lodgepole pine. *Genetics* 121:369–77
25. Epperson, B. K., Clegg, M. T. 1986. Spatial-autocorrelation analysis of flower color polymorphisms within substructured populations of morning glory *(Ipomoea purpurea)*. *Am. Nat.* 128:840–58
26. Fenster, C. B. 1988. *Gene flow and population differentiation in* Chamaecrista fasciculata *(Leguminosae)*. PhD thesis. Univ. Chicago, Ill.
27. Fenster, C. B. 1991. Gene flow in *Chamaecrista fasciculata*. II. Gene establishment. *Evolution* 45:410–22
28. Gabriel, K. R., Sokal, R. R. 1969. A new statistical approach to geographical variation analysis. *Syst. Zool.* 18:259–78
29. Golenberg, E. M. 1987. Estimation of gene flow and genetic neighborhood size by indirect methods in a selfing annual, *Triticum dicoccoides*. *Evolution* 41:1326–34
30. Gregorius, H.-R., Krauhausen, J., Muller-Stark, G. 1986. Spatial and temporal genetic differentiation among the seed in a stand of *Fagus sylvatica* L. *Heredity* 57:255–62
31. Hamrick, J. L., Loveless, M. D. 1989. The genetic structure of tropical tree populations: associations with reproductive biology. In *The Evolutionary Ecology of Plants,* ed. J. H. Bock, Y. B. Linhart, pp. 129–46. Boulder, Colo: Westview
32. Handel, S. N. 1983. Pollination ecology, plant population structure, and gene flow. In *Pollination Biology,* ed. L. Real, pp. 163–211. New York: Academic
33. Hedrick, P. W. 1980. Hitch-hiking: A comparison of linkage and partial selfing. *Genetics* 94:791–808
34. Hedrick, P. W. 1986. Genetic polymorphism in heterogeneous environments: A decade later. *Annu. Rev. Ecol. Syst.* 17:535–66
35. Hedrick, P. W., Holden, L. 1979. Hitch-hiking: An alternative to coadaptation for the barley and slender wild oat examples. *Heredity* 43:79–86
36. Heywood, J. S. 1986. The effect of plant size variation on genetic drift in populations of annuals. *Am. Nat.* 127:851–61
37. Heywood, J. S., Levin, D. A. 1985. Associations between allozyme frequencies and soil characteristics in *Gaillardia pulchella* (Compositae). *Evolution* 39:1076–86
38. Jain, S. K., Bradshaw, A. D. 1966. Evolutionary divergence among adjacent plant populations. 1. The evidence and its theoretical analysis. *Heredity* 20:407–41
39. Kendall, M., Stewart, A. 1979. *The Advanced Theory of Statistics,* Vol. 2. New York: Macmillan
40. Kesseli, R. V., Jain, S. K. 1985. Breeding system and population structure in *Limnanthes*. *Theor. Appl. Genet.* 71:292–99
41. Kimura, M., Weiss, G. H. 1964. The stepping stone model of population structure and the decrease of genetic correlation with distance. *Genetics* 49:561–76
42. Knight, S. E., Waller, D. M. 1987. Genetic consequences of outcrossing in the cleistogamous annual, *Impatiens capensis*. I. Population genetic structure. *Evolution* 41:969–78
43. Lande, R., Schemske, D. W. 1985. The evolution of self-fertilization and inbreeding depression in plants. I. Genetic models. *Evolution* 39:24–40
44. Learn, G. H. Jr., Schaal, B. A. 1987. Population subdivision for ribosomal DNA repeat variants in *Clematis fremontii*. *Evolution* 41:433–38
45. Levin, D. A. 1984. Inbreeding depression and proximity-dependent crossing success in *Phlox drummondii*. *Evolution* 38:116–27
46. Levin, D. A. 1989. Proximity-dependent cross-compatibility in *Phlox*. *Evolution* 43:1114–16
47. Levin, D. A., Kerster, H. W. 1969. Density-dependent gene dispersal in *Liatris*. *Am. Nat.* 103:61–74
48. Levin, D. A., Kerster, H. W. 1969. The dependence of bee-mediated pollen dispersal on plant density. *Evolution* 23:560–71
49. Levin, D. A., Kerster, H. W. 1974. Gene flow in seed plants. *Evol. Biol.* 7:139–220
50. Lewontin, R. C., Krakauer, J. 1973. Distribution of gene frequency as a test of the theory of the selective neutrality of polymorphisms. *Genetics* 74:175–95
51. Li, C. C. 1976. *First Course In Population Genetics*. Pacific Grove, Calif: Boxwood
52. Malecot, G. 1948. *The Mathematics of Heredity*. (Trans. D. M. Yermanis, 1969). San Francisco: W. H. Freeman (From French)
53. Mantel, N. A. 1967. The detection of disease clustering and a generalized

regression approach. *Cancer Res.* 27: 209–20
54. Maynard Smith, J. 1978. *The Evolution of Sex.* Cambridge: Cambridge Univ. Press
55. Moran, P. A. P. 1950. Notes on continuous stochastic phenomena. *Biometrika* 37:17–23
56. Nakamura, R. R. 1980. Plant kin selection. *Evol. Theor.* 5:113–17
57. Nei, M. 1973. Analysis of gene diversity in subdivided populations. *Proc. Natl. Acad. Sci. USA* 70:3321–23
58. Nei, M. 1986. Definition and estimation of fixation indices. *Evolution* 40:643–45
59. Nei, M., Chakravarti, A., Tateno, Y. 1977. Mean and variance of F_{ST} in a finite number of incompletely isolated populations. *Theor. Popul. Biol.* 11:291–306
60. Nei, M., Chesser, R. K. 1983. Estimation of fixation indices and gene diversities. *Ann. Hum. Genet.* 47:253–59
61. Nei, M., Li, W. H. 1973. Linkage disequilibrium in subdivided populations. *Genetics* 75:213–19
62. Nei, M., Maruyama, T. 1975. Lewontin-Krakauer test for neutral genes. *Genetics* 80:395
63. Oden, N. L. 1984. Assessing the significance of a spatial correlogram. *Geogr. Anal.* 16:1–16
64. Park, Y. S., Fowler, D. P. 1982. Effects of inbreeding and genetic variances in a natural population of Tamarack (*Larix laricina* (Du Roi) K. Koch) in eastern Canada. *Silv. Genet.* 31:21–26
65. Price, M. V., Waser, N. M. 1979. Pollen dispersal and optimal outcrossing in *Delphinium nelsoni. Nature* 277:294–97
66. Rai, K. N., Jain, S. K. 1982. Population biology of *Avena.* IX. Gene flow and neighborhood size in relation to microgeographic variation in *Avena barbata. Oecologia (Berl.)* 53:399–405
67. Reynolds, J., Weir, B. S., Cockerham, C. C. 1983. Estimation of the coancestry coefficient: basis for a short-term genetic distance. *Genetics* 105:767–79
68. Richards, P. W. 1952. *The Tropical Rain Forest.* Cambridge: Cambridge Univ. Press
69. Ripley, B. D. 1981. *Spatial Statistics.* New York: Wiley
70. Risser, P. G., Birney, E. C., Blocker, H. D., May, S. W., Parton, W. J., Wiens, J. A. 1981. *The True Prairie Ecosystem.* Stroudsburg, Penn: Hutchinson Ross. 557 pp.
71. Ritland, K., Ganders, F. R. 1985. Variation in the mating system of *Bidens menziesii* (Asteraceae) in relation to population substructure. *Heredity* 55:235–44
72. Ritland, K., Ganders, F. R. 1987. Covariation of selfing rates with parental gene fixation indices within populations of *Mimulus guttatus. Evolution* 41:760–71
73. Robertson, A. 1975. Gene frequency distributions as a test of selective neutrality. *Genetics* 81:775–85
74. Schaal, B. A. 1974. Isolation by distance in *Liatris cylindracea. Nature* 252:703
75. Schaal, B. A. 1975. Population structure and local differentiation in *Liatris cylindracea. Am. Nat.* 109:511–28
76. Schaal, B. A. 1980. Measurement of gene flow in *Lupinus texensis. Nature* 284:450–51
77. Schaal, B. A., Smith, W. G. 1980. The apportionment of genetic variation within and among populations of *Desmodium nudiflorum. Evolution* 34:214–21
78. Schnabel, A., Hamrick, J. L. 1990. Organization of genetic diversity within and among populations of *Gleditsia triacanthos* (Leguminosae). *Am. J. Bot.* 77:1060–69
79. Schnabel, A., Lauschman, R. H., Hamrick, J. L. 1991. Comparative genetic structure of two co-occurring tree species, *Maclura pomifera* (Moraceae) and *Gleditsia triacanthos* (Leguminosae). *Heredity.* In press
80. Schoen, D. J., Latta, R. G. 1989. Spatial autocorrelation of genotypes in populations of *Impatiens pallida* and *Impatiens capensis. Heredity* 63:181–90
81. Slatkin, M. 1977. Gene flow and genetic drift in a species subject to frequent local extinction. *Theor. Popul. Biol.* 12:253–62
82. Slatkin, M. 1985. Gene flow in natural populations. *Annu. Rev. Ecol. Syst.* 16:393–430
83. Slatkin, M. 1987. Gene flow and the geographic structure of natural populations. *Science* 236:787–92
84. Slatkin, M., Arter, H. E. 1991. Spatial autocorrelation methods in population genetics. *Am. Nat.* In press
85. Slatkin, M., Barton, N. H. 1989. A comparison of three indirect methods for estimating average levels of gene flow. *Evolution* 43:1349–68
86. Smouse, P. E., Long, J. C., Sokal, R. R. 1986. Multiple regression and correlation extensions of the Mantel test of matrix correspondence. *Syst. Zool.* 35:627–32
87. Smouse, P. E., Wood, J. W. 1987. The genetic demography of the Gainj of

Papua New Guinea: Functional models of migration and their genetic implications. In *Mammalian Dispersal Patterns,* ed. B. D. Chepko-Sade, Z. T. Halpin, pp. 211–24. Chicago: Univ. Chicago Press

88. Snaydon, R. W., Davies, M. S. 1976. Rapid population differentiation in a mosaic environment. IV. Populations of *Anthoxanthum odoratum* at sharp boundaries. *Heredity* 37:9–25

89. Sobrevila, C. 1988. Effects of distance between pollen donor and pollen recipient on fitness components in *Espeletia schultzii*. *Am. J. Bot.* 75:701–24

89a. Sokal, R. R., Jacquez, G. M. 1991. Testing inferences about microevolutionary processes by means of spatial autocorrelation analysis. *Evolution* 45:152–68

90. Sokal, R. R., Jacquez, G. M., Wooten, M. C. 1989. Spatial autocorrelation analysis of migration and selection. *Genetics* 121:845–55

91. Sokal, R. R., Oden, N. L. 1978. Spatial autocorrelation in biology. I. Methodology. *Biol. J. Linnaean Soc.* 10:199–228

92. Sokal, R. R., Wartenberg, D. E. 1983. A test of spatial autocorrelation analysis using an isolation-by-distance model. *Genetics* 105:219–37

93. Turner, M. E., Stephens, C., Anderson, W. W. 1982. Homozygosity and patch structure in plant populations as a result of nearest-neighbor pollination. *Proc. Natl. Acad. Sci. USA* 79:203–7

94. Uyenoyama, M. K. 1986. Inbreeding and the cost of meiosis: The evolution of selfing in populations practicing biparental inbreeding. *Evolution* 40: 388–404

94a. Van Dijk, H., Wolff, K., De Vries, A. 1988. Genetic variability in *Plantago* species in relation to their ecology. 3. Genetic structure of populations of *P. major, P. lanceolata* and *P. coronopus*. *Theor. Appl. Genet.* 75:518–28

95. Waller, D. M., Knight, S. E. 1989. Genetic consequences of outcrossing in the cleistogamous annual, *Impatiens capensis*. II. Outcrossing rates and genotypic correlations. *Evolution* 43:860–69

96. Waser, N. M. 1987. Spatial genetic heterogeneity in a population of the montane perennial plant, *Delphinium nelsoni*. *Heredity* 58:249–56

97. Waser, N. M., Price, M. V. 1983. Optimal and actual outcrossing in plants, and the nature of plant-pollinator interaction. In *Handbook of Experimental Pollination Biology,* ed. C. E. Jones, R. J. Little, pp. 341–59. New York: Van Nostrand Reinhold

98. Waser, N. M., Price, M. V. 1989. Optimal outcrossing in *Ipomopsis aggregata:* seed set and offspring fitness. *Evolution* 43:1097–1109

99. Watkins, L., Levin, D. A. 1990. Outcrossing rates as related to plant density in *Phlox drummondii*. *Heredity* 65:81–89

100. Weir, B. S. 1990. Sampling properties of gene diversity. See Ref. 6, pp. 23–42

101. Weir, B. S., Cockerham, C. C. 1984. Estimating F-statistics for the analysis of population structure. *Evolution* 38: 1358–70

102. Whitlock, M. C., McCauley, D. E. 1990. Some population genetic consequences of colony formation and extinction: Genetic correlations within founding groups. *Evolution* 44:1717–24

103. Wright, S. 1922. Coefficients of inbreeding and relationship. *Am. Nat.* 56:330–38

104. Wright, S. 1943. Isolation by distance. *Genetics* 28:114–38

105. Wright, S. 1943. An analysis of local variability of flower color in *Linanthus parryae*. *Genetics* 28:139–56

106. Wright, S. 1946. Isolation by distance under diverse systems of mating. *Genetics* 31:39–59

107. Wright, S. 1951. The genetical structure of populations. *Ann. Eugen.* 15:323–54

108. Wright, S. 1969. *Evolution and the Genetics of Populations,* Vol. 2: *The Theory of Gene Frequencies*. Chicago: Univ. Chicago Press

109. Wright, S. 1970. Random drift and the shifting balance theory of evolution. In *Mathematical Topics in Population Genetics,* ed. K. Kojima, pp. 1–31. New York: Springer-Verlag

110. Wright, S. 1977. *Evolution and the Genetics of Populations,* Vol. 3: *Experimental Results and Evolutionary Deductions*. Chicago: Univ. Chicago Press

111. Wright, S. 1978. *Evolution and the Genetics of Populations,* Vol. 4: *Variability Within and Among Natural Populations*. Chicago: Univ. Chicago Press

Annu. Rev. Ecol. Syst. 1991. 22:357–78

RECENT ADVANCES IN STUDIES OF BIRD MIGRATION

P. Berthold[1]

Max-Planck-Institute for Behavioral Physiology, Vogelwarte, Schloss Moeggingen, D-7760 Radolfzell, Germany

S. B. Terrill

Harvey and Associates, Ecological Consultants, 906 Elizabeth Street, P.O. Box 1180, Alviso, CA 95002

KEY WORDS: migratory strategies, animal migration, behavioral genetics, migratory orientation, bird-population declines, migratory ecology

---

INTRODUCTION

During the past decade, studies on avian migration have made substantial progress through both field work and experimental research. After 90 years of extensive bird banding (introduced by the Vogelwarte Rossitten in 1901—116), the descriptive study of avian migration recently acquired an important new dimension: virtually permanent recording of migratory flights by satellite tracking (52, 72, 92). Due to the novelty of this approach, a review devoted to this subject would be premature; however, several other current topics require an overview. We have selected six themes which, in our opinion, are currently developing rapidly and stimulating great interest.

First, the task of selectively breeding migrant passerines has been perfected to the degree that large numbers of migrants have now been bred in aviaries. This achievement has resulted in substantial advances in our understanding of

[1]With support by the Deutsche Forschungsgemeinschaft in studies on genetics, annual rhythms and desert ecology

0066-4162/91/1120-0357$02.00

the links between genetic information and migratory behavior. Second, some important recent advances have occurred in research on adaptive aspects of interactions between endogenous annual rhythms and the photoperiod in determining overt migratory behavior. Third, from the extensive field of migratory orientation and navigation, we have chosen to review the ontogeny of multiple cue usage. Fourth, with respect to field studies, an actual stepwise approach employed by several research groups attempts to answer the question of how migrants overcome major ecological barriers to migration such as deserts, mountains, and oceans; this technique is producing some interesting answers. Fifth, much current interest focuses on ecophysiology and migratory strategies at stopover sites and on the wintering grounds. Finally, the worldwide decline of many migratory species, now well documented in both the Eurasian-African and the American bird migration systems, requires our full attention.

We have to emphasize that there are a number of other important topics on which enormous progress has also been recently made, including the timing of bird migration in relation to weather (95), endocrine mechanisms (122), energetics (49), the development of optimal models (7), the winter ecology of migrants (77) and others. These areas also deserve to be reviewed, but space limitation imposes a major constraint. Thus, these topics will be covered in a further, forthcoming review.

GENETIC CONTROL MECHANISMS

As recently as 10 years ago, the degree to which genetic factors are involved in the immediate control of avian migratory behavior was entirely unknown. We even lacked data concerning the genetic regulation of such basic migrant morphological and physiological traits as relatively long, pointed wings and migratory fat deposition. Development of the ability to breed songbirds, primarily old-world *Sylvia* warblers, on a large scale in captivity (27) resulted in a breakthrough and initiated the systematic study of the role played by genetic factors in the control of avian migration.

The Urge to Migrate

Since 1702 (117) it has been assumed that, in typical migrants, departure is triggered not by environmental factors but rather by internal stimuli (instinct). This view was essentially supported by the discovery of endogenous annual cycles (57, see below) which include migratory events. A series of recent experiments by Berthold and his colleagues have demonstrated that migratory behavior, including the urge to migrate, is highly heritable. When exclusively migratory blackcaps, *Sylvia atricapilla* (from southern Germany) and resident conspecifics (from the Cape Verde Islands) were cross-bred by Berthold and coworkers, migratory activity was transmitted genetically into 30% of the

F_1-hybrids (35). A detailed comparison of the population-specific onsets both of the endogenously controlled migratory activity in captive individuals and of actual migration in free-living conspecifics strongly supports the view that the inherited urge to move is, in fact, the fundamental releaser for departure in typical migrants (22).

Time Course of Migration

The experimental demonstration that both species- and population-specific patterns of endogenous migratory activity (Zugunruhe) in captive migrants correlate significantly with the actual migratory distance covered by these populations has long led to the assumption that migratory activity might be the result of endogenous time-programs for migration (103). Again, a cross-breeding experiment, this time using long-distance migrants (blackcaps, from southern Germany) and short-distance migrants (from the Canary Islands), demonstrated that migratory activity is a population-specific, quantitatively inherited characteristic. Combined with the results of other experimental studies on the structure of migratory activity (30) these findings support the view that inborn, temporal patterns of migratory activity serve as time-programs which enable inexperienced, first-time migrants to 'automatically' cover the distance between the breeding grounds and their specific wintering areas (vector-navigation hypothesis; 16).

Orientation

Based on the often repeated result that naive captive migrants show seasonally appropriate, oriented migratory activity in orientation cages (as well as on the results of a few displacement experiments in the field), it appeared likely that typical migrants inherit a program for migratory movement in a preferred direction (67). Two recent studies specifically demonstrated the heritability of migratory direction. In the cross-breeding experiment with migratory and resident blackcaps mentioned above, migratory activity along the directional axis which the migratory parents normally use was genetically transmitted into the migratory hybrids (35). When blackcaps from both sides of the Central European migration divide that migrate to Mediterranean and African winter quarters to the southwest and southeast, respectively, were cross-bred, F_1-hybrids behaved in a phenotypically intermediate fashion (67). Investigations with F_2-hybrids, currently in progress, indicate that only a few genes may be involved in the control of migratory directional preference. This is in contrast to patterns of migratory activity (above) and partial migration (below) that appear to involve complex genetic effects.

Partial Migration

With respect to the control of facultative, partial migration (i.e. in populations performing irregular, irruptive movements—112), environmental factors such as population density and food supply are likely of prime importance (21). On the other hand, in the case of obligate partial migration, i.e. in populations with regularly migrating fractions, both a behavioral-constitutional and a genetic hypothesis have been proposed to explain the control of the two behavioral traits, migratoriness and sedentariness (15). In blackcaps (29) and in European robins, *Erithacus rubecula* (37), selective breeding experiments with migratory and nonmigratory individuals have shown that both traits have a strong genetic basis. Similar evidence is provided by field and experimental studies involving three other species (19). These experimental studies suggest that both behavioral traits are threshold characters determined by multiple loci (16). When selection for migratory and resident behavior in the blackcap was continued beyond the F_1-generation, a two-way selective breeding experiment demonstrated that a partially migratory population (comprising a majority of birds showing migratory tendencies) can become fully migratory in only three generations or largely nonmigratory in only four to six generations. Thus, the experiment revealed an extremely rapid selection response and high heritability values, and it implies a strikingly high evolutionary potential with respect to strong selection pressures (26).

Novel Migratory Habits

Beginning about 25 years ago, blackcaps from Central Europe have been migrating, in increasing numbers, in a novel northwesterly direction, establishing a new wintering area on the British Isles. The present view of this recent microevolutionary process is that the widespread practice of feeding wild birds in the British Isles has provided the permissive base for the establishment and growth of this novel population. The new behavior appears to be controlled by the inheritance of the novel migratory direction and to be reinforced by successful wintering in the new area. In addition, individuals that winter in Britain and Ireland may benefit from lower intraspecific competition, shorter migratory distances relative to traditional migrating populations, shorter photoperiods that accelerate vernal migration back to the breeding areas and reproductive condition; thus, they may experience enhanced fitness (34). This extremely interesting case of recent evolution of migration patterns is currently being analyzed in a series of experimental and field studies (113).

Other Features

Cross-breeding experiments have documented that morphological correlates of migratory behavior such as body weight and wing length are under direct genetic control (28). The same holds true for the adaptive course of juvenile moult (65). Present investigations indicate that there is also high genetic variability in the amount of migratory activity within populations. This may provide a base for fairly rapid microevolutionary changes of programmed migratory distances if selection pressures change, e.g. with respect to the global warming, the hypothesized result of anthropogenic "greenhouse" effects (18).

Final Consideration

In view of the progress achieved in the field of genetics of avian migration, two important points have to be kept in mind: (*a*) most of the information obtained thus far originates from the experimental study of a single species, and (*b*) virtually nothing is known about genetic-environmental interactions in the development of migratory behavior. In addition, other species should be investigated in more detail. In general, preliminary studies favor the opinion that genetic factors play an important role in the control of bird migration, but this remains to be tested on a broader scale.

INTERPLAY BETWEEN ENDOGENOUS ANNUAL RHYTHMS AND PHOTOPERIOD

Ever since Rowan's pioneering experiments in avian ecophysiology in 1925 (97), the significance of photoperiod in the control of annual cycles in birds and for proper timing of annual events within the framework of the annual cycle has been increasingly demonstrated. Only 25 years ago, the first convincing evidence of endogenous annual cycles (circannual rhythms) in birds first surfaced (56). Currently, endogenous circannual rhythms have been experimentally demonstrated in approximately 15 bird species, including resident and migratory forms (long-, middle-, short-distance, and partial migrants) comprising temperate as well as tropical species (31, 57, 62, 64). Finally, a concept of exclusive photoperiodic control of avian annual cycles has been developed (51). Whether such "pure photoperiodic" species exist is an open question. Recent studies show, however, a complex interplay between circannual rhythms and photoperiod in which photoperiod acts as a synchronizer as well as a permissive factor for the expression of circannual rhythms. Examples concerning the control of bird migration are outlined.

Timing of Homeward Migration in African Wintering Areas

During the past decade, Gwinner and his coworkers have initiated a series of experiments to elucidate the role of different photoperiodic schedules on the temporal organization of events in the annual cycle of garden warblers, *Sylvia borin,* wintering at vastly different latitudes on the African wintering grounds (58–60). This western Palaearctic warbler winters in central and southern Africa between about 15°N and 25°S; populations that breed in northern Europe tend to winter more toward the southern part of the wintering area. The species shows pronounced circannual rhythms that rigidly control a variety of annual events including migratory activity and fat deposition. These endogenous rhythms are expressed over many years under a wide range of constant photoperiodic conditions (from periods of at least 10 to 16 hr of daily light period). Under various constant photoperiods, the period length of the annual cycle is approximately constant. The most important synchronizer (Zeitgeber) for these circannual rhythms is photoperiod (57). Individuals that winter close to the equator live up to six months under largely constant photoperiods. In these birds, it is reasonable to conclude that vernal migration is stimulated largely by internal factors (60). When experimental groups were kept in simulated winter photoperiodic regimes either of the equator, or of 20°S, most of the typical annual events occurred earlier in the second group. Individuals under 20°S conditions started and ended winter moult earlier, showed a more rapid course of moult, initiated earlier spring migratory activity, and experienced accelerated gonadal development. All these results are in agreement with observations in the field. The experiment indicates that longer photoperiods experienced at a critical time by individuals wintering far south in the wintering area phase-advance spring migration and the preparation for the reproductive period. Because these birds have farther to travel to reach the breeding grounds, an earlier start would be adaptive. Consequently, "endogenous circannual components and Zeitgeber stimuli constitute a functional entity that provides as a whole for adaptive temporal programming" (61).

Photorefractoriness and Photoperiod as a Permissive Factor

It has long been speculated that photorefractoriness (the inability to react to long photoperiods) must be different between resident, temperate-zone species and long-distance (especially transequatorial) migrants. If this were not the case, long-distance migrants should come into reproductive condition a second time in the wintering area (102). Studies of garden warblers and muscicapid flycatchers (60) showed that in birds that winter in regions near, or beyond, the equator: (*a*) photorefractoriness is terminated under very long days and (*b*) the termination of refractoriness is a more gradual process than in temperate-zone species. In experiments with garden warblers, the responsiveness of individuals (kept in constant equatorial conditions) to long days increased steadily between November and April. As shown in the previous section, responsiveness can be accelerated by longer day lengths (of 20°S), but this acceleration is so small that breeding condition does not develop before homeward migration (60).

A comparative study of two old-world flycatchers—the pied flycatcher, *Ficedula hypoleuca,* and the collared flycatcher, *F. albicollis*—produced some interesting results. In *hypoleuca,* a circannual rhythmicity persists *only* if the winter photoperiod is at least as short as that normally experienced by the species in its wintering grounds which lie slightly north of the equator. In *albicollis* rhythmicity is expressed under a much wider range of photoperiod. This species winters over a substantially greater range of latitudes and thus is programmed to respond under a diversity of winter photoperiodic conditions (61). These results, in combination with data obtained on other species (57), indicate that the overt expression of circannual rhythms is somehow dependent on the presence of particular permissive environmental conditions. To better understand the extremely complex control mechanisms of timing of bird migration, a more detailed analysis of these endogenous-exogenous interactions would be necessary.

ORIENTATION SYSTEMS USED BY MIGRANT BIRDS

Compass Systems, Ontogeny, Calibration

One of the most fascinating pursuits in bird migration research involves unraveling the mystery of how birds orient their migratory flights. As we have mentioned earlier, naive birds performing their first migration possess innate information providing the appropriate migratory distance and direction for their first migration. But migrants need external references to choose the correct direction to migrate and to remain on course once aloft. How migrants perform this feat has generated a great deal of research, especially in the past 20 years. Although only a few species of migrants have been intensively studied thus far, it appears that migrants use a variety of cues available to them for orientation and that these various compass systems interact in complex ways during ontogeny and migration itself. Since the ability of individual migrants to utilize directional cues from a variety of sources is now well established, most current research has focussed on the ontogenetic and hierarchical relationships between multiple compass mechanisms. Although not a primary objective of studying the ontogeny of orientation behavior (which was to elucidate these relationships), this research has illuminated some interesting details of the ways in which experience with specific stimuli interacts with innate rules in the development of complex behavior (2). Migrant birds are now known to possess the ability to use the sun, the earth's geomagnetic field, polarized light, and landmarks to orient (2, 20, 21, 99, 108, 121).

Heritable information concerning migratory direction is represented twice in those species studied in detail; it is encoded within the framework of a magnetic compass and within a compass mechanism based on celestial rotation (2, 4, 120, 121). These two orientation systems will develop fully in hand-reared birds in the absence of information from any other system. Thus, birds raised in the earth's magnetic field in the absence of visual cues (e.g. no view of the sky, etc) orient along an axis appropriate for migration using cues only from the geomagnetic field. Birds raised in a null magnetic field that are exposed to rotational star patterns orient in the direction appropriate for migration using the axis of rotation as north.

During ontogeny, the development of the stellar compass appears to be free of the influence from the magnetic compass. Later in life, however, during migration, the stellar compass is calibrated by information derived from the magnetic compass, which by that time seems to be rigidly fixed and independent of the axis of celestial rotation. This relationship may have evolved because familiar star patterns descend towards the horizon and finally disappear to be replaced by new stars as birds migrate south. Thus, migrants may use the magnetic field to recalibrate the stellar compass as patterns of stars change with migratory distance.

On the other hand, the magnetic compass is modifiable during early development by directional information derived from celestial cues. This relationship has been demonstrated by raising birds in situations with conflicting compass information provided by celestial cues (either daytime or nighttime skies) versus magnetic cues. First, birds will allow directional information derived from celestial rotation to override conflicting directional information provided by an experimentally shifted magnetic field. Secondly, birds will recalibrate their magnetic compass so it conforms with information provided by celestial cues. The adaptive significance of this relationship is unclear, but it may have something to do with corrections for declination (2).

Sunset Cues

A number of experiments with several North American migrants indicate that information obtained around sunset seems to be of primary importance in choosing the direction for that night's migratory flight (1, 2, 85). In European migrants studied thus far, the importance of cues at sunset is less clear (68, 121). Whether these cues actually provide directional information or whether they act only as prominent visual marks that aid in confirming orientation based in other primary cues in European species has yet to be established (121).

Both the sun itself (migrant birds have been shown to possess a sun compass, but the degree to which it is used in migration probably varies greatly between migrants depending on diel timing of migratory flights, weather conditions, and other factors) and skylight polarization seem to be important, although some evidence seems to indicate that polarized light provides the primary orientation information at sunset, at least in savannah sparrows, *Passerculus sandwichensis* (2).

Further, recent results with savannah sparrows have indicated that these birds calibrate their polarized light compass using information provided by their magnetic compass during the first three months of life (2, 3).

Unanswered Questions

Several cautionary notes should be emphasized with respect to migrant orientation research with migrant birds. First, although basic results appear to more-or-less agree for the several different species studied in detail, only a few species have been examined and many of the experiments have not been repeated. Second, apparent interrelationships among cues may be, in part, a function of experimental design rather than demonstrating that particular reference systems impose dominant directional information that overrides information provided by other systems. Migrants may utilize redundant compass mechanisms to arrive at a 'weighted average' of input from multiple systems when performing migratory orientation in nature (2).

One very interesting question that currently remains unaddressed is how transequatorial migrants deal with the fundamental changes in orientation cues associated with crossing the equator (99, 121). With respect to the magnetic field, birds have a magnetic compass that does not use polarity, but rather uses the axial course of the field lines and their inclination in space (120). When transequatorial migrants cross the equator, information from the magnetic compass becomes ambiguous. In addition, south of the equator, new star patterns become visible, and the axis of rotation indicates south rather than north. Thus, decision rules based on the primary innate compasses (magnetic and celestial rotation of the night sky) must somehow be reversed after crossing the equator. This apparent remarkable attribute requires experimental investigation.

MIGRATORY STRATEGIES TO OVERCOME ECOLOGICAL BARRIERS

Many migratory birds are regularly faced with major ecological barriers which may occasionally prevent successful migration (6) but which are frequently overcome by a series of remarkable adaptations. Major geographical barriers are oceans (especially for species that cannot land on water), high mountains, and deserts. Less severe barriers include tropical rain forests, arctic icefields, and temporary meteorological events. The crossing of mountains and oceans by migratory birds has been studied for a fairly long time (48, 119), while the question of how migrants cross major deserts has only recently begun to be extensively investigated; the strategies migrants use to overcome desert barriers are currently under debate (10, 11, 81, 82). We thus concentrate on problems related to trans-desert migration and only mention a few pertinent points concerning the crossing of mountains and oceans.

Intermittent Strategy Vs Nonstop Flight Across Deserts

It has been known for a long time (43) that even small songbirds are able to perform extended nonstop flights and to cross, nonstop, distances of over

1000 km (e.g. the Gulf of Mexico between North and South America). Based on this information, the distribution of ringing recoveries, and the small number of migrants formerly observed within the desert during migration, Moreau (89, 90) developed a concept of nonstop migration across the entire Sahara Desert and the Mediterranean Sea via an extended flight of about 30–60 hr. In 1973, an initial study by Lavee & Safriel (76) at a small oasis in the Sinai Desert found considerable landfall by numerous individuals representing a number of species. The great majority of the grounded individuals were fairly heavy, due to considerable fat depots, and were in good health. Many of those birds stayed for just one day and thus were not considered as being grounded by energetic constraints.

Subsequently, three groups of investigators started independent studies at approximately the same time in the western Sahara (13), the eastern Sahara (38), and in central Asian deserts (46), and all have recently compiled their results (10, 39, 47). These results can be summarized as follows: Birds representing nearly all of the species known to cross deserts are found resting during the day in oases and even in open desert. Large numbers are recorded from some groups including warblers, wagtails, and flycatchers. Individuals with sufficient fuel to overcome the remaining part of the desert are found resting during the day in shaded hiding-places. Lean individuals appear to search for oases where they have been shown to be able to replenish fat depots during an extended stopover period of several days to several weeks. These findings resulted in the hypothesis that deserts are crossed by an intermittent strategy: Nocturnal migrants approach deserts with fat reserves as high as possible, and these allow for quiet resting periods in microhabitats that meet thermoregulatory requirements. After resting during the day they take off for the next stage of the migratory journey the following evening. In this manner migrants cross the inhospitable desert in a few nocturnal flights. It is currently unclear whether this strategy represents the rule and the nonstop flight hypothesis is largely incorrect, or to what extent both strategies might possibly be used. Because so many grounded warblers, wagtails, etc are seen, the intermittent strategy is clearly followed by certain species. For more aerial species such as swallows, the data support more a nonstop strategy (10).

The concept of endogenously programmed courses of migration (see below) is in better accordance with the idea of an intermittent strategy (17). But even for species regularly found resting in the desert, nonstop crossing by a portion of the population cannot be excluded. Data are not yet available that would address the question of whether most trans-Sahara migrants fly nonstop or use an intermittent migration, i.e. numbers of birds remaining aloft versus numbers of birds grounded (39). Therefore, a future multimethodological approach including radar studies, satellite-tracking of individuals equipped with transmitters, moon-watching and a variety of other field studies and field experiments will be necessary to clarify this fascinating problem. Other efforts will have to concentrate on the question of why obviously healthy migrants rest during the day when their fat depots would allow an uninterrupted nonstop crossing. Initial ecophysiological analyses indicate that permissive environmental factors are important. These factors include specific wind conditions, reduced risk of hypothermy and dehydration during the night (39), as well as specific stress factors involved in the physiology of migrants which could necessitate rests after extended flights in extreme situations (12).

Aspects of Crossing Mountains and Oceans

It has been well known for many years that migratory birds cross even extremely high mountains such as the Himalayas (101). Recently, the unusual hypoxic tolerance of birds recorded in altitudes of up to 11,300 m has been elucidated. The answer to the question of how birds cope with high altitudes (at least for a few species) turned out to be a two- or three-stage cascade of hemoglobins of graded oxygen affinities which guarantee oxygen supply over a wide range of altitudes (69).

A detailed analysis of how migrants deal with crossing the Alps (42) has recently shown that they employ a variety of complex strategies to cross or circumvent mountains. A considerable number of migrants apparently possess innate programs that enable them to circumvent the Alps. A tendency to cross the Alps is characteristic of birds with strong north-south migration routes; long-distance migrants with long distances remaining on their migration; birds that typically migrate at relatively high altitudes (which are often drifted towards mountains by prevailing winds at high altitudes); larger species with higher flight speeds; individuals with relatively high fat depots; experienced adult birds; and individuals approaching the Alps early in the night. Soaring species and wetland birds appear more likely to circumvent the mountains.

Extensive world-wide studies of transoceanic bird migration have shown that overwater flight per se appears to pose no barrier to migrants, including land birds. Even land birds fly over almost all the major oceans of the world, and some of their overwater flights may reach 7500 km. Observations and estimates indicate that some limicoline species are capable of making nonstop flights of this magnitude (119). There is now a fairly complete knowledge of how these extreme migratory journeys are completed (7, 71, 95, 119). Transoceanic migrants show three different phases of flight behavior. At takeoff, they are sensitive to local wind and weather conditions. Once transoceanic flight is initiated, they appear to be largely insensitive to these conditions as well as to land marks and appear to be guided by internal programs. In this way they migrate nonstop for up to 100 hours (often at altitudes of up to 6 km) without any apparent reference to local conditions or topography en route. During the final phase of their journey, transoceanic migrants descend considerably and again become sensitive to local conditions, and this facilitates landfall. A recent simulation of flight from the Aleutian Islands to Hawaii under the observed range of wind conditions en route showed that constant compass orientation was sufficient to bring migrants to the chain of Hawaiian Islands even under adverse weather conditions in the Gulf of Alaska (75, 119). This result strongly favors the vector-navigation hypothesis (above).

ECOPHYSIOLOGY AND MIGRATORY STRATEGIES AT STOPOVER SITES AND ON THE WINTERING GROUNDS

Stopover Sites

The ecology and physiology of birds at migratory stopover sites have recently emerged as areas of relatively intense research. This is important research because it contributes substantially toward a better understanding of the selection pressures and the complex behavior and physiology associated with migration. Of course, a more complete understanding of the ecological role of stopover sites in migration is critical in designing and implementing migrant conservation programs (8, 33, 86, 88, 111).

Selection pressures on migrants during migration are complex and intense (5, 6, 79, 106). In a broad sense, these pressures result from: (*a*) the long distances that are frequently involved (79, 109), (*b*) the fact that migrants continually encounter new areas and unfamiliar habitats as they migrate, especially during the first migration (79), and (*c*) the relatively severe predation and competition pressures encountered during migration (78, 88).

Selection pressures that occur during migration should result in behavioral and physiological attributes that maximize the probability of survival and minimize the costs associated with migration. Stopover sites are probably very important in achieving this goal in that migrants use them for refueling, resting, and shelter. The energetic costs of migration are extremely high, and migrants must achieve energy intake rates that enable them to mobilize energy reserves to fuel migration, survive in unfamiliar stopover sites until they find food, and to hedge against unpredictable weather conditions and resource availability along the migratory route. The relationship between rate of migration, probability of survival, rate of deposition of energy stores, and length of stopover have recently received theoretical and empirical attention. Alerstam & Lindström (7) have introduced models that deal with these relationships. These optimal migration models generate predictions that will, we hope, make it possible to distinguish which selective forces—time, energy, or risk of predation—are primarily responsible for the migratory characteristics of different species under different seasonal and environmental conditions. The empirical data gathered so far appears generally to support the importance of maximizing the speed of migration (the time-minimizer model) relative to the energy minimization (55, 79).

Birds maximizing the rate of migration should be sensitive to variation in rate of fat deposition at stopover sites. This sensitivity should translate into decisions concerning the length of stopover at a particular site. The higher the rate of fat deposition, the greater the fat load at migratory departure. Their time-minimizer model predicts that a migrant should depart a stopover site when the instantaneous speed of migration has dropped to the expected speed of migration at stopover sites within the range of the bird's fat load (79). Migrants should not remain at stopover sites that do not allow them to achieve a sufficient rate of fat deposition.

So far, evidence generally supports the time minimization hypothesis. For example, the length of stopover by migrants at oases in the Sahara is correlated with fat deposition rate (9, 38, 40, 76, 98, and above). Biebach, Gwinner, and their colleagues were able to simulate this situation in the lab using spotted flycatchers, *Muscicapa striata* (38), and garden warblers (63, 66). If migrants were deprived of food, their migratory activity increased. If the birds were again given access to food, their migratory activity stopped until they had accumulated fat stores that were comparable to those held when the birds had previously had access to ad lib food and were showing normal levels of migratory activity. A. Lindström and T. Alerstam (personal communication) and Gudmundsson et al (55) found that the behavior of migrating bluethroats, *Luscinia svecica,* rufous hummingbirds, *Selasphorus rufus,* and shorebirds qualitatively fits the behaviors predicted by the time-minimization model: the higher the fat deposition rates, the fatter the birds upon departure.

Another aspect of stopover ecology that has received recent interest is the manner in which migrants respond to the extreme energetic demands associated with migration in unfamiliar, and often unpredictable, environments associated with stopover sites. Migrants should be selected to maximize the probability of survival and the rate of energy intake to improve the chance of a successful migration (7). Genetic programs facilitate this undertaking. For example, migrant habitat selection, and possibly associated diet choice, appears to be largely the result of endogenous influences (8, 21, 41). This endogenous influence most likely minimizes search time at stopover sites. Nevertheless, migrants must cope with the extreme demands of maximizing energy intake rates in novel, often hostile, environments. Thus, we should expect selection pressure to diversify migrant foraging behavior at stopover sites to meet these demands (86).

In a series of studies, Moore and his colleagues have found that migrants do indeed diversify their foraging behavior (80, 86, 87). Loria & Moore (80) found that fat-depleted red-eyed vireos, *Vireo olivaceus,* moved at higher velocities, increased their degree of turning following a feeding attempt, broadened their use of microhabitat space, and expanded their feeding repertoire. Moore & Simm (87) found that yellow-rumped warblers, *Dendroica coronata,* in migratory disposition choose a variable reward (risk-acceptance), but warblers that were not in migratory condition consistently preferred a constant reward (risk aversion). The increased use of space, expanded foraging repertoire, and risk-prone behavior all indicate that extreme demands of migration can result in behavioral plasticity, even in light of costs associated with that plasticity.

Competition for resources can be important in stopover ecology. Moore & Yong (88) recently used a predator-exclosure experiment to demonstrate that passerine migrants depress food abundance at coastal stopover sites following migration across the gulf of Mexico. They further showed that, in general, birds were more likely to gain mass when the density of migrants at stopover sites was low. With respect to social interactions, these results, coupled with the results concerning fat deposition rate and length of stopover reviewed above, indicate that subordinate individuals that cannot achieve sufficient rates of food intake (due to dominant individuals monopolizing resources) should move on, while dominant birds remain until fat levels are regained (see also 105–109).

Wintering Grounds

In general, breeding and wintering sites have been considered relatively fixed sites, with migrants traversing the distance between them as rapidly as possible (45, 107, 115). This description often seems generally appropriate, but in at least some situations the system appears more complex. This complexity is brought about by migratory movements that occur during the winter, within the winter quarters, in some species. In some cases, these extended migrations appear to be the result of endogenous programs that produce spontaneous migratory behavior throughout the winter—such as in the marsh warbler, *Acrocephalus palustris* (25). In other species such as the yellow-rumped warblers (114, 115), and the yellow wagtails, *Motacilla flava* (123), these movements are apparently the result of facultatively continued migration in response to reduced food availability.

As migrants proceed into the wintering grounds during migration, at some point a stopover site may become an overwinter site. What factors determine whether an area will be a stopover site or become a wintering site? The evidence presented above (see under Genetic Control Mechanisms) indicates that an endogenous program enables migrants to reach the wintering grounds. During this endogenous period (the *obligate phase*—see 112), as we have seen above, the impetus to continue migration from a stopover site is basically

(*a*) expressed when birds are in an energetic state that enables them to meet the energetic demands of migration, even if the stopover sites contain abundant resource availability, (*b*) expressed when birds are confronted with poor prospects for maintaining or improving energy budgets, and (*c*) inhibited when migrants are low on energy reserves, but arrive at a stopover site that contains sufficient resources to improve their condition. If birds are able to renew their fat reserves at a sufficient rate, they once again express migratory behavior in accordance with state (*a*). Evidence from several species indicates that as the autumnal migration proceeds in time and distance, state (*a*) is dropped from the migratory program, leaving states (*b*) and (*c*) functional (66, 105, 107–110). Thus, if migrants end up at a stopover site favorable for overwinter survival, they remain at that location because they are no longer motivated by endogenous mechanisms to continue to migrate. Rather, the birds now utilize resources and fat reserves to survive the winter period at a particular site. If, however, they wind up at a stopover site that contains insufficient resources, states (*b*) and (*c*) continue to operate, and the birds continue on their migration until they reach a favorable site (the *facultative phase*—see 112).

Evidence for such a system is based on both field work (see 107) and laboratory experiments (66, 105). With respect to an experimental approach, Gwinner et al (66) have demonstrated the existence of a system such as that described above in the garden warbler. During the autumn migratory period, well-fed garden warblers put on substantial fat reserves and show spontaneous migratory activity in response to their endogenous migratory program. If the birds are deprived of food, their migratory activity increases significantly. When the warblers are again allowed access to food, their migratory behavior ceases until they have regained their former weights. Migratory behavior is then expressed once again. If food continues to be available, the birds will eventually stop showing spontaneous migratory activity for the winter. After termination of spontaneous migratory activity, however, migratory behavior can be reactivated by subjecting the birds to food deprivation.

The behavior, ecology, and physiology of migrants at stopover sites and on the wintering grounds remains largely unexplored for the vast majority of species (17, 107). We are only just beginning to appreciate the importance of these areas in the complex life histories of these birds.

MIGRANT POPULATION DECLINES, CONSERVATION, AND BASIC RESEARCH

Recent Declines

As a group, long-distance migrants in both the New World and the Old World appear to be in trouble (e.g. 23, 24, 44, 91, 96, 104). Perhaps the most urgent challenge for basic research in avian migration involves illuminating the factors responsible for widespread declines in these populations. The spatially and ecologically complex life histories of migrants make elucidation of the underlying causes of these declines elusive. The apparently rapid pace at which some migrant populations are declining, coupled with a lack of basic knowledge of population dynamics and ecology, especially during migration and on the wintering grounds, makes large-scale research efforts imperative.

In the Old World, substantial decreases in many migrant populations have been detected for some time now. Europe has a long tradition of well-organized, extensive banding and population monitoring programs, and the data generated by these programs indicate that, at present, roughly one third to one half of the central European bird fauna are declining in number, (14, 24, 36, 83). Of the 15 species experiencing the most significant, widespread declines over large areas (according to the MRI monitoring program), 14 species are long-distance migrants that winter in Africa (17). In the New World, decreases in long-distance migrant populations were first noted in the late 1970s (104); these decreases included species that were thought to have stable, or increasing, populations up to this time (53, 96).

Causes of Declines

Migrants are vulnerable on the breeding grounds, along the migratory route, and on the wintering grounds. Habitat loss and degradation have been, and are currently, occurring on a massive scale in all three of these (14, 104). In addition, other factors negatively affect migrant populations. These factors include the use of pesticides and other environmentally dangerous chemicals, pollution, altered agricultural and forestry practices, nest parasitism, increased predation due to habitat fragmentation and high numbers of predators, and probably increased numbers of more resident avian species that compete for nest sites and food (21, 50, 93, 100, 104). In addition, changes on a global scale (e.g. global warming due to the greenhouse effect) may have started to affect migrant populations negatively (18, 23). Teasing apart relative roles and evaluating cumulative impacts of these factors on the world's migrants represents a substantial challenge.

Habitat destruction and degradation may represent the most general, widespread effect on a global scale. Although the large-scale destruction of tropical habitats is generally much more recent than the clearing of temperate forests, tropical forests are being lost at an estimated rate of 11 million hectares per year, and 40% of the world's closed tropical forests are already gone (124). If this clearing continues, a minimum of 2225 million hectares of tropical forests will be cleared by the year 2000. Satellite images show that tropical habitat loss is particularly advanced in areas where large numbers of migrants winter (104). Several lines of evidence suggest that neotropical migrants are being affected to a greater degree by habitat loss on the wintering grounds. There is a great deal more landmass and habitat in the breeding area relative to the wintering grounds (104). Thus, each acre of habitat loss on the wintering grounds affects more birds per unit area than on the breeding grounds. This indirect evidence is supported by the recent results of Robbins et al (96) which indicate that neotropical migrants using primarily forested habitats in either the wintering or the breeding areas are declining. Robbins et al interpret this result as indicating that tropical deforestation is having a more direct impact on neotropical migrants than is the loss and fragmentation of forest habitat in North America. On the other hand, clearings in forests only sparsely used by wintering migrants can enhance the use of an area by providing more secondary growth (e.g. 70), and the relative importance of primary versus secondary growth in the wintering ecology of many migrants is still unclear (77).

The impact of pesticides on migrants is well-known in some cases, such as the effect of DDT on avian metabolism (84). The effects of a multitude of other potentially harmful chemicals are much less well-known. Pesticides that might not harm birds directly may negatively affect avian populations by reducing food availability. For example, locust populations which provide a major food source for wintering white storks, *Ciconia ciconia,* in Africa are declining substantially due to pesticide use. Although the use of DDT and other environmentally persistent chemicals used in pesticides has declined in the northern hemisphere, such chemicals are still widely used in developing countries. The impact of the continued use of these pesticides on wintering migrants appears to be considerable, yet direct effects remain obscure (e.g. 54, 94). Long-term monitoring programs in central Europe indicate that the fat stores of migrating passerines have, in recent years, possibly been below optimum levels. This generalized condition may be due to an overall decrease in food availability due to habitat loss, extensive use of herbicides, or both (73).

Artificial sources of food are apparently causing increases in populations of avian competitors, predators, and parasites of migrants. In North America, bird feeding is not only promoting increases in numbers of resident competitors, it is enhancing predator populations as well. Jays, which are significant predators of the eggs and nestlings of migrants, benefit widely from artificial sources of food (104). Finally, in North America, cowbirds, *Molothrus,* which are the major nest-parasites of neotropical migrants, are increasing greatly due to bird feeders and other human-derived sources of food such as cattle feedlots and agriculture. These birds have played a major role in the near extinction of one neotropical migrant, the Kirtland's warbler, *Dendroica kirtlandii* (118), and appear to be having major negative impact on a number of other migrant species (104). In Europe, some short-distance migrants, such as the blackcap, may be out-competing some long-distance migrant populations. European blackcap populations are increasing. These populations are reducing their migratory distance to some extent and increasingly using feeders during the winter; both of these may be facilitating population increases. This species is a well-known competitor of the long-distance migrant, the garden warbler (32).

Urgent Information Needed

It would appear that the most urgent information necessary for conservation of the world's migrants is data on habitat use and the full spectrum of ecological requirements on the breeding grounds, during migration, and on the wintering areas. Although a fair amount of information has been collected on the breeding grounds, we lack necessary information concerning these aspects of migrant natural history away from the breeding areas. The difficulties of the situation were emphasized at a recent symposium on the white stork. This species was used in the initial pioneering banding studies of nearly a century ago, and documentation of population declines began about the same time. In spite of the fact that virtually every possible cause for population declines in this species has been considered, and despite long-term population monitoring programs, it still remains open whether this species can be saved from extinction (74, 94, 100).

In both the Old World and the New World, a systematic, standardized approach for monitoring migrant populations and their ecological requirements during migration and on the wintering grounds needs to be further developed. Successful standardized monitoring programs have already been initiated by the British Trust for Ornithology in Britain and Ireland and by the Vogelwarte Radolfzell in Central Europe (24, 83). Only through more intensified internationally standardized research can we obtain accurate data concerning migrant population trends and ecology.

Literature Cited

1. Able, K. P. 1982. Skylight polarization patterns at dusk influence the migratory orientation of birds. *Nature* 299:550–51
2. Able, K. P. 1990. Experimental studies of the development of migratory orientation mechanisms. In *Orient. Birds, Experientia* 46:388–94
3. Able, K. P., Able, M. A. 1990. Ontogeny of migratory orientation in the savannah sparrow *(Passerculus sandwichensis):* Calibration of the magnetic compass. *Anim. Behav*. 39:1179–98
4. Able, K. P., Bingman, V. P. 1987. The development of orientation and navigation behavior in birds. *Q. Rev. Biol.* 62:1–29
5. Alerstam, T. 1981. The course and timing of bird migration. In *Animal Migration,* ed. D. I. Aidley, pp. 9–54. Cambridge: Cambridge Univ. Press. 264 pp.
6. Alerstam, T. 1990. Ecological causes and consequences of bird orientation. In *Orient. Birds, Experientia* 46:405–15
7. Alerstam, T., Lindström, A. 1990. Optimal bird migration: the relative importance of time, energy, and safety. See Ref. 61a, pp. 331–51
8. Bairlein, F. 1981. Ökosystemanalyse der Rastplätze von Zugvögeln. *Ökol. Vögel* 3:7–137
9. Bairlein, F. 1985. Body weights and fat deposition of Palaearctic passerine mi-

grants in the central Sahara. *Oecologia* 66:141–46
10. Bairlein, F. 1988. How do migratory songbirds cross the Sahara? *Trends Ecol. Evol.* 3:191–94
11. Bairlein, F. 1989. Reply. *Trends Ecol. Evol.* 4:23
12. Bairlein, F. 1990. Physiology of trans-desert migration. *Rep. 4 Ornithol. Expedition to study physiology of trans-Sahara flights in passerine migrants*. Köln. 24 pp.
13. Bairlein, F., Beck, P., Feiler, W., Querner, U. 1983. Autumn weights of some Palaearctic passerine migrants in the Sahara. *Ibis* 125:404–7
14. Bauer, S., Thielcke, G. 1982. Gefährdete Brutvogelarten in der Bundesrepublik Deutschland und im Land Berlin: Bestandsentwicklung, Gefährdungsursachen und Schutzmaβnahmen. *Vogelwarte* 31:183–391
15. Berthold, P. 1984. The control of partial migration in birds: a review. *Ring* 10:253–65
16. Berthold, P. 1988. Evolutionary aspects of migratory behavior in European warblers. *J. Evol. Biol.* 1:195–209
17. Berthold, P. 1988. The biology of the genus Sylvia—a model and a challenge for Afro-European co-operation. *Tauraco* 1:3–28
18. Berthold, P. 1990. Die Vogelwelt Mitteleuropas: Entstehung der Diversität, gegenwärtige Veränderungen und Aspekte der zukünftigen Entwicklung. *Verh. Dt. Zool. Ges.* 83:227–44
19. Berthold, P. 1990. Genetics of migration. See Ref. 61a, pp. 269–80
20. Berthold, P. 1990. Orientation in birds: concluding remarks; a brief definition of the research field's position. In *Orient. Birds, Experientia* 46:426–28
21. Berthold, P. 1990. *Vogelzug*. Darmstadt: Wissenschaftliche Buchgesellschaft. 252 pp.
22. Berthold, P. 1990. Wegzugbeginn und Einsetzen der Zugunruhe bei 19 Vogelpopulationen—eine vergleichende Untersuchung. *Proc. Int. 100. DO-G Meeting, Curr. Top. Avian Biol., Bonn, 1988. J. Ornithol. 131, Sonderh.*, pp. 217–22
23. Berthold, P. 1991. Patterns of avian migration in light of current global 'greenhouse' effects: a central European perspective. *Acta 20th Congr. Int. Ornithol., Christchurch, 1990*. In press
24. Berthold, P., Fliege, G., Querner, U., Winkler, H. 1986. Bestandsentwicklung von Kleinvögeln in Mitteleuropa: Analyse von Fangzahlen. *J. Ornithol.* 127:397–437
25. Berthold, P., Leisler, B. 1980. Migratory restlessness of the marsh warbler, *Acrocephalus palustris. Naturwissenschaften* 67:472
26. Berthold, P., Mohr, G., Querner, U. 1990. Steuerung und potentielle Evolutionsgeschwindigkeit des obligaten Teilzieherverhaltens: Ergebnisse eines Zweiweg-Selektionsexperiments mit der Mönchsgrasmücke *(Sylvia atricapilla). J. Ornithol.* 131:33–45
27. Berthold, P., Querner, U. 1981. Genetic basis of migratory behavior in European warblers. *Science* 212:77–79
28. Berthold, P., Querner, U. 1982. Genetic basis of moult, wing length, and body weight in a migratory bird species, *Sylvia atricapilla. Experientia* 38:801
29. Berthold, P., Querner, U. 1982. Partial migration in birds: experimental proof of polymorphism as a controlling system. *Experientia* 38:805
30. Berthold, P., Querner, U. 1988. Was Zugunruhe wirklich ist—eine quantitative Bestimmung mit Hilfe von Video-Aufnahmen bei Infrarotlichtbeleuchtung. *J. Ornithol.* 129:372–75
31. Berthold, P., Querner, U. 1991. The annual cycle of the blackcap *(Sylvia atricapilla)* on the Cape Verde Islands: characteristics, environmental and genetic control. *Proc. 7th Panafrican Ornithol. Congr., Nairobi, 1988*. In press
32. Berthold, P., Querner, U., Schlenker, R. 1990. *Die Mönchsgrasmücke, Sylvia atricapilla*. Die Neue Brehm-Bücherei. Wittenberg Lutherstadt: Ziemsen. 180 pp.
33. Berthold, P., Schlenker, R. 1975. Das "Mettnau-Reit-Illmitz-Programm"—ein langfristiges Vogelfangprogramm der Vogelwarte Radolfzell mit vielfältiger Fragestellung. *Vogelwarte* 28:97–123
34. Berthold, P., Terrill, S. B. 1988. Migratory behaviour and population growth of blackcaps wintering in Britain and Ireland: some hypotheses. *Ringing Migration* 9:153–59
35. Berthold, P., Wiltschko, W., Miltenberger, H., Querner, U. 1990. Genetic transmission of migratory behavior into a nonmigratory bird population. *Experientia* 46:107–8
36. Bezzel, E. 1982. *Vögel in der Kulturlandschaft*. Stuttgart: Ulmer. 350 pp.
37. Biebach, H. 1983. Genetic determination of partial migration in the European robin, *Erithacus rubecula. Auk* 100: 601–6
38. Biebach, H. 1985. Sahara stopover in migratory flycatchers: fat and food affect the time program. *Experientia* 41:695–97

39. Biebach, H. 1990. Strategies of trans-Sahara migrants. See Ref. 61a, pp. 352–67
40. Biebach, H., Friedrich, W., Heine, G. 1986. Interaction of bodymass, fat, foraging and stopover period in trans-Sahara migrating passerine birds. *Oecologia* 69:370–79
41. Brensing, D. 1989. Ökophysiologische Untersuchungen der Tagesperiodik von Kleinvögeln. *Ökol. Vögel* 11:1–148
42. Bruderer, B., Jenni, L. 1990. Migration across the Alps. See Ref. 61a, pp. 60–77
43. Cooke, W. W. 1905. Routes of bird migration. *Auk* 22:1–11
44. Cramp, S. 1978. *Schicksal und Zukunft der Vögel Europas*. Greven: Kilda. 71 pp.
45. Curry-Lindahl, K. 1981. *Bird Migration in Africa*. Vols. 1,2. London: Academic. 444 pp. 251 pp.
46. Dolnik, V. R. 1985. Nocturnal bird migration over arid and mountainous regions of middle Asia and Kazakhstan. *Proc. Zool. Inst. USSR Acad. Sci. Leningrad* 169:1–147
47. Dolnik, V. R. 1990. Bird migration across arid and mountainous regions of middle Asia and Kazakhstan. See Ref. 61a, pp. 368–86
48. Dorka, V. 1966. Das jahres- und tageszeitliche Zugmuster von Kurz- und Langstreckenziehern nach Beobachtungen auf den Alpenpässen Cou/Bretolet. *Ornithol. Beob.* 63:165–223
49. Drent, R., Piersma, T. 1990. An exploration of the energetics of leap-frog migration in arctic breeding waders. See Ref. 61a, pp. 399–412
50. Ehrlich, P. R., Dobkin, D. S., Wheye, D. 1988. *The Birders Handbook: A Field Guide To the Natural History of North American Birds*. New York: Simon & Schuster. 785 pp.
51. Farner, D. S. 1966. Über die photoperiodische Steuerung der Jahreszyklen bei Zugvögeln. *Biol. Rundsch.* 4:228–41
52. Fuller, H. R., Levanon, N., Strikwerda, T. E., Seegar, W. S., Wall, J., et al. 1984. Feasibility of a bird-borne transmitter for tracking via satellite. *Biotelemetry* 8:375–78
53. Gauthreaux, S. A. 1990. Long-term patterns of trans-Gulf migration in spring: a weather radar analysis. *Acta 20th Congr. Int. Ornithol., Christchurch, 1990, Suppl., Programme Abstr.*, p. 337
54. Grimmett, R. 1987. *A Review of the Problems Affecting Palaearctic Migratory Birds in Africa*. Cambridge: Int. Council Bird Preservation Migratory Birds Prog. 25 pp.
55. Gudmundsson, G. A., Lindström, A., Alerstam, T. 1991. Optimal fat loads and long-distance flights by migrating knots, sanderlings and turnstones. *Ibis*. In press
56. Gwinner, E. 1967. Circannuale Periodik der Mauser und der Zugunruhe bei einem Vogel. *Naturwissenschaften* 54:447
57. Gwinner, E. 1986. *Circannual Rhythms*. Berlin/Heidelberg: Springer. 154 pp.
58. Gwinner, E. 1987. Annual rhythms of gonadal disposition and molt in garden warblers *Sylvia borin* exposed in winter to an equatorial or a southern hemisphere photoperiod. *Ornis Scand.* 18:251–56
59. Gwinner, E. 1987. Photoperiodic synchronization of circannual rhythms in gonadal activity, migratory restlessness, body weight, and molt in the garden warbler *(Sylvia borin)*. *Comp. Physiol. Environm. Adapt.* 3:30–44
60. Gwinner, E. 1988. Photorefractoriness in equatorial migrants. *Acta 19th Int. Ornithol. Congr., Ottawa, 1986,* pp. 626–58. Ottawa: Univ. Ottawa Press. 2815 pp.
61. Gwinner, E. 1989. Photoperiod as a modifying and limiting factor in the expression of avian circannual rhythms. *J. Biol. Rhythms* 4:237–50
61a. Gwinner, E., ed. 1990. *Bird Migration: Physiology and Ecophysiology*. Berlin: Springer
62. Gwinner, E. 1990. Endogenous-exogenous interactions in circannual rhythms. In *Progress in Comparative Endocrinology*, ed. A. Epple, C. G. Scanes, M. H. Stetson, pp. 632–38. New York: Wiley-Liss. 752 pp.
63. Gwinner, E., Biebach, H., von Kries, I. 1985. Food availability affects migratory restlessness in garden warblers *(Sylvia borin)*. *Naturwissenschaften* 72:51–52
64. Gwinner, E., Dittami, J. 1990. Endogenous reproductive rhythms in a tropical bird. *Science* 249:906–8
65. Gwinner, E., Neusser, V. 1985. Die Jugendmauser europäischer und afrikanischer Schwarzkehlchen *(Saxicola torquata rubicula und axillaris)* sowie von F1-Hybriden. *J. Ornithol.* 126:219–20
66. Gwinner, E., Schwabl, H., Schwabl-Benzinger, I. 1988. Effects of food deprivation on migratory restlessness and diurnal activity in the garden warbler, *Sylvia borin*. *Oecologia* 77:321–26
67. Helbig, A. J. 1989. Angeborene Zugrichtungen nachts ziehender Singvögel: Orientierungsmechanismen, geographische Variation und Vererbung. *Dissertation, Univ. Frankfurt/Main*. 148 pp.

68. Helbig, A., Wiltschko, W. 1989. The skylight polarization pattern at dusk affects the orientation behavior of blackcaps, *Sylvia atricapilla*. *Naturwissenschaften* 76:227–29
69. Hiebl, I., Braunitzer, G. 1988. Anpassungen der Hämoglobine von Streifengans *(Anser indicus)*, Andengans *(Chloephaga melanoptera)* und Sperbergeier *(Gyps rueppellii)* an hypoxische Bedingungen. *J. Ornithol.* 129:217–26
70. Hutto, R. L. 1989. The effect of habitat alteration on migratory land birds in a west Mexican tropical deciduous forest: a conservation perspective. *Conserv. Biol.* 3:138–48
71. Johnson, S. R., Herter, D. R. 1990. Bird migration in the arctic: a review. See Ref. 61a, pp. 22–43
72. Jouventin, P., Weimerskirch, H. 1990. Satellite tracking of wandering albatrosses. *Nature* 343:746–48
73. Kaiser, A. 1989. Körpergewicht, Fettdepots und theoretische Zugstreckenleistung wegziehender Kleinvögel am Bodensee (SW-Deutschland). *Diplomarbeit, Univ. Mainz*. 113 pp.
74. Kanyamibwa, S., Schierer, A., Pradel, R., Lebreton, J. D. 1990. Changes in adult annual survival rates in a western European population of the white stork *(Ciconia ciconia)*. *Ibis* 132:27–35
75. Kloeckner, P. D., Williams, J. M., Williams, T. C. 1982. Radar and visual observations of transpacific migrants. *Elepaio* 42:77–80
76. Lavee, D., Safriel, U. 1973. Utilization of an oasis by desert-crossing migrant birds. *Israel J. Zool.* 22:219
77. Leisler, B. 1990. Selection and use of habitat of wintering migrants. See Ref. 61a, pp. 156–74
78. Lindström, A. 1989. Finch flock size and risk of hawk predation at a migratory stopover site. *Auk* 106:225–32
79. Lindström, A. 1990. *Stopover ecology of migrating birds*. PhD thesis. University of Lund. 133 pp.
80. Loria, D. L., Moore, F. R. 1990. Energy demands of migration on red-eyed-vireos, *Vireo olivaceus*. *Behav. Ecol.* 1:24–35
81. Lövei, G. L. 1989. Passerine migration between the Palaearctic and Africa. In *Current Ornithology*, ed. D. M. Power, 6:143–74. New York/London: Plenum. 332 pp.
82. Lövei, G. L. 1989. Trans-Sahara bird migration. *Trends Ecol. Evol.* 4:72
83. Marchant, J. H., Hudson, R., Carter, S. P., Whittington, P. A. 1990. *Population Trends in British Breeding Birds*. Tring: British Trust Ornithol. 300 pp.
84. Miller, G. T. 1988. *Living in the Environment. An Introduction to Environmental Science*. Belmont: Wadsworth. 603 pp. 5th ed.
85. Moore, F. R. 1987. Sunset and the orientation behavior of migrating birds. *Biol. Rev.* 62:65–86
86. Moore, F. R. 1991. Ecophysiological and behavioral response to energy demand during migration. *Acta 20th Congr. Int. Ornithol., Christchurch, 1990*. In press
87. Moore, F. R., Simm, P. A. 1985. Migratory disposition and choice of diet by the yellow-rumped warbler *(Dendroica coronata)*. *Auk* 105:820–26
88. Moore, F. R., Yong, W. 1990. Evidence of food-based competition among passerine migrants during stopover. *Behav. Ecol. Sociobiol.* In press
89. Moreau, R. E. 1961. Problems of Mediterranean-Sahara migration. *Ibis* 103:580–623
90. Moreau, R. E. 1972. *The Palaearctic-African Bird Migration Systems*. London/New York: Academic. 384 pp.
91. Nowak, E. 1988. Internationaler Biotopschutz für wandernde Tierarten. In *Biotopverbund in der Landschaft*, ed. Akademie für Naturschutz und Landschaftspflege. Laufener Seminarbeitr. 10/86:116–128
92. Nowak, E., Berthold, P., Querner, U. 1990. Satellite tracking of migrating Bewick's swans. *Naturwissenschaften* 77:549–50
93. Rappole, J. H., Morton, E. S., Lovejoy, T. E. III, Ruos, J. L. 1983. *Nearctic Avian Migrants in the Neotropics*. Washington, DC: US Dept. Interior, Fish and Wildlife Service. 646 pp.
94. Rheinwald, G., Ogden, J., Schulz, H. 1989. *Weißstorch–White stork*. Proc. First Int. Stork Conserv. Symp. Walsrode, 1967. Dachverbd. Dt. Avifaunisten Nr. 10. 741 pp.
95. Richardson, W. J. 1990. Timing of bird migration in relation to weather: updated review. See Ref. 61a, pp. 78–101
96. Robbins, C. S., Sauer, J. R., Greenberg, R. S., Droege, S. 1989. Population declines in North American birds that migrate to the neotropics. *Proc. Natl. Acad. Sci. USA* 86:7658–62
97. Rowan, W. 1925. Relation of light to bird migration and developmental changes. *Nature* 115:494–95
98. Safriel, U. N., Lavee, D. 1988. Weight changes of cross-desert migrants at an oasis—do energetic considerations alone

determine the length of stopover? *Oecologia* 67:611–19
99. Schmidt-Koenig, K. 1990. The sun compass. *Orient. Birds, Experientia* 46:336–42
100. Schulz, H. 1988. *Weißstorchzug—Ökologie, Gefährdung und Schutz des Weißstorchs in Afrika und Nahost.* Weikersheim: Margraf. 459 pp.
101. Schüz, E. 1971. *Grundriß der Vogelzugskunde*. Berlin/Hamburg: Parey. 390 pp. 2nd ed.
102. Schüz, E., Berthold, P. 1966. Über das Brüten fernziehender Vogelarten an beiden Polen der Jahresverbreitung. *Vogelwarte* 23:233–34
103. Stresemann, E. 1934. Aves. In *Handbuch der Zoologie* 7, ed. W. Kükenthal, T. Krumbach. Berlin/Leipzig: de Gruyter. 899 pp.
104. Terborgh, J. 1989. *Where Have All the Birds Gone?* Princeton, NJ: Princeton Univ. Press. 207 pp.
105. Terrill, S. B. 1987. Social dominance and migratory restlessness in the dark-eyed junco *(Junco hyemalis). Behav. Ecol. Sociobiol.* 21:1–11
106. Terrill, S. B. 1988. The relative importance of ecological factors in bird migration. *Acta 19th Congr. Int. Ornithol., Ottawa, 1986*, pp. 2180–90. Ottawa: Univ. Ottawa Press. 2815 pp.
107. Terrill, S. B. 1990. Ecophysiological aspects of movements by migrants in the wintering quarters. See Ref. 61a, pp. 130–43
108. Terrill, S. B. 1990. Evolutionary aspects of orientation and migration in birds. *Orient. Birds, Experientia* 46:395–404
109. Terrill, S. B. 1990. Food availability, migratory behavior, and population dynamics of terrestrial birds during the nonreproductive season. *Stud. Avian Biol.* 13:438–43
110. Terrill, S. B. 1990. The regulation of migratory behavior: interactions between exogenous and endogenous factors. *Proc. Int. 100. DO-G Meeting, Curr. Top. Avian Biol., Bonn, 1988. J. Ornithol. 131, Sonderh.*, pp. 211–15
111. Terrill, S. B. 1991. Opening remarks: new aspects of avian migration systems. *Acta 20th Congr. Int. Ornithol., Christchurch, 1990.* In press
112. Terrill, S. B., Able, K. P. 1988. Bird migration terminology. *Auk* 108:205–06
113. Terrill, S. B., Berthold, P. 1990. Ecophysiological aspects of rapid population growth in a novel migratory blackcap *(Sylvia atricapilla)* population: an experimental approach. *Oecologia*. In press
114. Terrill, S. B., Crawford, R. L. 1988. Additional evidence of nocturnal migration by yellow-rumped warblers *(Dendroica coronata)* in winter. *Condor* 90:261–63
115. Terrill, S. B., Ohmart, R. D. 1984. Facultative extension of fall migration by yellow-rumped warblers *(Dendroica coronata). Auk* 101:427–38
116. Thienemann, J. 1927. *Rossitten*. Neudamm: Neumann. 327 pp.
117. von Pernau, A. F. 1702. *Unterricht. Was mit dem lieblichen Geschöpff, denen Vögeln, auch ausser dem Fang, nur durch die Ergründung deren Eigenschafften und Zahmmachung oder anderer Abrichtung man sich vor Lust und Zeitvertreib machen könne*. Nürnberg
118. Walkinshaw, L. H. 1983. *The Kirtland's Warbler*. Cranbrook Inst. Sci. 58:1–207
119. Williams, T. C., Williams, J. M. 1990. The orientation of transoceanic migrants. See Ref. 61a, pp. 7–21
120. Wiltschko, W., Wiltschko, R. 1988. Magnetic versus celestial orientation in migrating birds. *Trends Ecol. Evol.* 3:13–15
121. Wiltschko, W., Wiltschko, R. 1990. Magnetic orientation and celestial cues in migratory orientation. In *Orient. Birds, Experientia* 46:342–52
122. Wingfield, J. C., Schwabl, H., Mattocks, P. M. 1990. Endocrine mechanisms of migration. See Ref. 61a, pp. 232–56
123. Wood, B. 1979. Changes in numbers of over-wintering yellow wagtails *Motacilla flava* and their food supplies in a west African savanna. *Ibis* 121:228–31
124. World Resources Institute 1988. *Tropical Forests: A Call for Action*. Part 1, the Plan. Washington DC: World Resources Inst. 49 pp.

Annu. Rev. Ecol. Syst. 1991. 22:379–407

SOCIAL AND POPULATION DYNAMICS OF YELLOW-BELLIED MARMOTS: Results from Long-Term Research

Kenneth B. Armitage

Department of Systematics and Ecology, The University of Kansas, Lawrence, Kansas 66045-2106

KEY WORDS: kinship, reproductive success, individuality, recruitment, dispersal

---

INTRODUCTION

Population dynamics is the consequence of the fitness strategies of individuals. Fitness is measured as reproductive success, i.e. the production of descendents. Individuals may forego short-term reproduction to increase survival (51), thereby increasing the incidence of iteroparity and lifetime reproductive success (LRS). Measured as production of offspring, LRS is not a panacea for answering evolutionary questions (64). However, the value of LRS is improved if the reproductive success of descendents is determined. Measuring LRS requires detailed, longitudinal studies of individuals. Such studies allow us to partition and evaluate the significance of various components of LRS, such as age of first reproduction, degree of iteroparity, production and survival of offspring, and recruitment (43).

A major component of the fitness strategy of an individual is its social behavior with conspecifics. Social behavior can affect various components of LRS. Aggressive behavior may reduce fecundity by inhibiting reproduction (112). Social behavior affects spacing behavior (82, reviewed in 45) that, in

0066-4162/91/1120-0379$02.00

turn, may limit immigration and group size and cause some individuals to disperse (35, 61, 70, 83). Dispersal may be density-independent (62) and is related to mating systems (65, 93). Among mammals in general, juvenile males are the predominant dispersers, especially in those species with promiscuous or polygynous mating systems (48). Generally, male mammals compete for mates, and females compete for resources for rearing young (e.g. 34). Hence, both males and females play key roles in the processes of dispersal and recruitment, especially with individuals of the same sex. Although dispersal should be viewed as a process to increase an individual's fitness (4, 52, 90), dispersal has demographic consequences (84, 102). Finally, social dynamics may be kin-biased (19); therefore, the population consequences of social behavior may depend on the relatedness of the individuals involved.

The information needed to assess the factors governing LRS is that needed to describe population dynamics and the mechanisms of population regulation. One advantage of focusing on individual fitness and deriving population dynamics from the activities of individuals is that we are freed from focusing on density as the most important agent acting on populations (45). Furthermore, life-history traits of individuals of a behavioral phenotype (41, 47) or of members of a kin group may be combined so that the demographic characteristics of alternative phenotypes or of kin groups of different sizes may be compared. Because survival is an exercise in probability (113), individuals should act to increase the probability of survival of their offspring and/or to reduce the probability of survival of the offspring of conspecifics (e.g. 33, 38, 75). Thus, the behavior of individuals can provide insights into population processes (68).

Detailed, longitudinal studies of individuals are readily conducted on relatively large, diurnal mammals. We initiated research on the yellow-bellied marmot *(Marmota flaviventris)* at the Rocky Mountain Biological Laboratory, Gothic, Gunnison County, Colorado, in 1962, at 2900 m. Additional work occurred in North Pole Basin at 3400 m. The semi-fossorial rodent is a member of the subfamily Marmotinae of the family Sciuridae (squirrels) and is closely allied to the prairie dogs and spermophiles (58, 67). This species occurs at high elevations, usually above 2000 m, in western United States, southcentral British Columbia, and southern Alberta (58).

In the Gothic area, the distribution of marmots is clumped and conforms closely to the mosaic formed by meadow and forest vegetation (103). Marmots occupy the open area relatively free of trees and shrubs, but which contains talus, rock outcrops, or scattered boulders under which burrows are constructed. Grasses *(Festuca, Bromus, Poa)* and large showy perennials characterize the meadows. The grasses contribute 9% to 41% of the total dry biomass; the variation occurs among sites and seasonally within a site (57, 80).

Marmot habitat patches are classified as colonial or satellite (= isolate; 49). A colony consists of one or more males, resident females, usually yearlings (animals one year old), and young (animals <4 months old) (5, 25). Satellite sites rarely have more than one female, a male may not be present, yearlings rarely are present, but young usually are present (25). Colonial sites have larger openings ($\bar{x}$ = 58 ha vs 6.6 ha), more residents ($\bar{x}$ = 3.2 vs 1.4), more resident females ($\bar{x}$ = 2.1 vs 1.0), and more burrows in use ($\bar{x}$ = 14 vs 2.3). Vegetative characteristics do not differ between the two types of sites (103). In effect, satellites are minihabitat patches, and there is a continuum of habitat sizes ranging from about 0.01 ha to 70 ha or more (15, 20). Area is correlated with the mean number of resident females (r = 0.86) and with the total number of resident adults (r = 0.83) (103).

ACTIVITY PATTERNS

Density-independent weather factors may affect survival and reproduction. Animals may adjust their activity patterns to minimize their exposure to environmental stress. The alpine and subalpine environments where marmots live are characterized by large diurnal fluctuations in temperature, intense sunshine during the summer, and low temperatures and several months of snow cover during the winter (42, 63).

Annual Cycle of Reproduction and Hibernation

PATTERN AND TIMING The annual cycle is a circannual rhythm (46, 110) with two phases—heterothermal and homeothermal (89). Immergence, hibernation, emergence, reproduction, and growth and preparation for immergence are sequentially linked. Emergence and immergence patterns follow an age-sex sequence (6, 78, 81). Adult males emerge first in late April or early May and are soon followed by adult females, yearling males, and yearling females. The same sequence occurs during immergence except that females that produced litters immerge later than males and nonreproductive females, and young immerge last, with female young immerging before males (24). Immergence begins in late August and usually is completed by mid September.

PHYSIOLOGICAL CHARACTERISTICS The physiology of the heterothermal phase is characteristic of hibernators (53, 59, 60, 71). Adults spontaneously terminate hibernation, but young do not terminate hibernation until fed or emaciated (60). Spontaneous termination of hibernation apparently is related to energy resources. The larger adults can withstand a prolonged fast, whereas the young cannot (59), and emerge when vegetation is available (78, 103). Also, adults must reproduce as early as possible; survival of young weaned late in the homeothermal phase is less than 10% (26).

All age classes gain mass at the rate of about 12 to 14 g/day, in the Gothic area, but young grow at a significantly greater rate at 3400 m (2, 26). Reproductive females initiate mass gain about three weeks later than yearlings, adult males, and nonreproductive females. The larger the young are at immergence the more likely they are to survive (26); the critical factor may be the amount of fat accumulated because young that fail to fatten do not survive hibernation (87). In the laboratory, metabolic rate, change in body mass, and food consumption follow a circannual rhythm with the maximum and minimum values of metabolic rate preceding the maximum and minimum value of food consumption by at least one month and those of body mass by at least two months. The rhythm may be a critical mechanism that shifts energy expenditures from maintenance to production in preparation for immergence (29, 115). Adult marmots decrease the time spent foraging in late summer (55, 78, 80). One intuitively expects an increase in foraging time as marmots fatten for hibernation; however, foraging time may be adjusted to an individual's energy balance, which is affected by the phase of the annual cycle. Thus, ground squirrels deprived of food for several weeks and returned to ad libitum feeding increase consumption sufficiently to return body mass to the level expected for that phase of the annual cycle (46). The relationship between the annual cycle of metabolism and change in mass and foraging needs further investigation in free-ranging marmots and other ground-dwelling sciurids.

VARIATION IN WEATHER PATTERNS Survivorship and reproduction could be affected by the length of the active season, which varies from year to year as a consequence of variation in the onset and/or termination of snow cover. When data from all colonies are lumped, reproductive females survive better the later the onset of winter, and young survive better the earlier winter terminates (25). Both mean litter size and percent total survival are positively correlated with the length of the growing season (99). The length of the growing season varies among marmot localities over a distance of 4.8 km in the East River Valley where the greatest difference in elevation between colonies is 165 m. Mean date of 50% snow cover differs among localities by as much as 21 days; the latest date was May 30. The number of litters per female and mean litter size are negatively correlated with time of 50% snow cover (109). This relationship probably is a phenotypically plastic response to environmental variation and not a heritable trait. The East River populations lie within the distances readily transversed by dispersers (108), and considerable gene flow occurs among these populations (97). The significance of the length of the growing season for reproduction is dramatically emphasized by the pattern in North Pole Basin where the mean date of 50% snow cover is June 30. About one fourth of the females weaned a litter each year, compared to one half in the East River Valley (15), and no female produced litters in consecutive years during six years of study (78). Annual reproductive success of females is related to food resources (2). Females lost mass early in the active season and probably relied on body fat as a major source of energy. The failure to breed in successive years also appears to be phenotypic; about one fourth of the females each year initiated reproduction (including some that weaned litters the previous year) but failed to wean a litter, which suggests that their energy reserves were insufficient to sustain both maintenance and reproduction.

The annual cycle is a major constraint on population dynamics. The need to satisfy the energy requirements for hibernation limits reproduction to a single annual event occurring immediately after emergence. The short active season combined with large body size delays reproductive maturity until two years of age. Post-natal reproductive investment is extended into the second summer of a juvenile's life and probably is the major factor leading to sociality in marmots and other ground-dwelling sciurids (12). The need to mobilize energy for reproduction and then prepare for hibernation in a short time period undoubtedly accounts for the energy conservative physiology of this species and probably of other hibernating sciurids (29, 80, 81, 87, 88).

Daily Cycles

SEASONAL PATTERN During the first two weeks after emergence the daily activity cycle is unimodal with peak activity in the early afternoon. As daily temperatures increase, the activity cycle becomes bimodal with peaks at 1000 and 1700 hrs (6). As temperatures continue to warm, the peak morning activity shifts to 0800 hrs and the afternoon peak to 1800 hrs, with some variation due to the direction of the slope on which a colony is located (5, 80, 87, 88, 106). As temperatures and photoperiod decrease in late summer, peak activity shifts to a later time in the morning and an earlier time in the afternoon and becomes unimodal just before immergence (80, 86).

TIME BUDGET On average, yellow-bellied marmots spend 72 to 84% of a 24-hr day in their burrows, 8 to 12% sitting above ground, and 7 to 16% in moderate activity and foraging (81, 87, 106). During lactation, females spend an average of 29% of their above ground time foraging. Foraging time decreases to 22% ($0.1 > p > 0.05$) during the first four weeks post-emergence of the litter and to 18% in late summer about when the molt is completed (55). The seasonal change in the time budget of yearling females is similar except the decrease in foraging time occurs during the first four weeks post-emergence of the young and coincides with the completion of the molt. Oxygen consumption decreases markedly after the annual molt (96). The pattern of seasonal change in foraging is similar to the pattern of change in metabolism and food consumption of laboratory animals. This similarity suggests that foraging is adjusted to current metabolic needs. The precise nature of the metabolic needs (e.g. lactation, growth, mate seeking) and its relationship to metabolic rate and the time budget remains to be determined.

PHYSIOLOGICAL CONSTRAINTS The thermoneutral zone extends approximately from 15 to 20°C for adults and 20 to 25°C for young (29, 86). Adult activity decreases when ambient temperature exceeds 20°C (5), and foraging occurs five times as frequently on cloudy as on sunny days (106). Air temperatures usually are lower than the lower critical temperature during the morning activity period. However, marmot activity is affected primarily by the standard operative temperature (T_{es}), a measure that integrates all factors affecting thermal energy exchange (88). Foraging time decreases hyperbolically when T_{es} exceeds 25°C, frequently between 0900 and 1000 hrs. Marmots extend foraging time by becoming hyperthermic, but they cease foraging when body temperature approaches 40°C (88). The high T_{es} persists until late afternoon; marmots must either forage at T_{es}'s below the thermoneutral zone or engage in short foraging bouts during midday. Marmots do the former primarily but pay a cost of increased metabolic rate. This cost is reduced by minimizing activity at stressful T_{es}'s and by metabolic rates and thermal conductances that are much lower than predicted for a mammal of this body size (29, 80). Thus, thermoregulatory costs constitute only 1% to 6% of daily energy expenditure (87).

The activity pattern of marmots suggests that they are lazy (69). If, as suggested, foraging time is adjusted to metabolic rates, marmot foraging may be an example of Herber's case 2. At least part of the effect of being lazy is that it conserves energy (87), but it is unclear why much of the lazy time is spent sitting and sunning instead of in the burrow. Sunning does not affect body temperature but could possibly affect metabolic rate (106), especially when T_{es} is below the lower critical temperature. High T_{es} limits adult foraging time such that a marmot attempting to escape aggression by shifting its activity period to midday would probably reduce its effective foraging time by one half and would encounter a considerable energy deficit because it could fill its gut once a day instead of twice (88). Young can forage more extensively during midday because the T_{es} of young is lower; thus young encounter less heat stress and avoid much of the cold stress by foraging at higher T_{es}'s than adults (88).

Water balance Although marmots shift foraging areas away from water-stressed plants (57) and reject dried leaves in feeding trials (11), water metabolism only slightly affects diet choice. Water turnover rates in field animals may be up to five times those of laboratory animals provided water ad libitum (87, 111). However, several features of water balance indicate that yellow-bellied marmots conserve water and that this conservation is related to metabolic rate and hibernation. The rate of evaporative water loss (mg $H_2O{\cdot}ml\ O_2^{-1}$) is similar to that of desert dwellers and at the lower end of the range of other sciurids (32). Marmots on a restricted water regimen concentrate urine more than similarly stressed ground-dwelling sciurids (32, 111) and significantly reduce food consumption and metabolic rate (29, 111). The inability of marmots to mobilize water for evaporative cooling may be related to meeting water requirements during hibernation; at low temperatures characteristic of hibernation only about one half of metabolic water production is lost through evaporation, thereby providing water for urine formation (29). Marmots probably have no other water input during hibernation. Although low metabolic rates and low conductances (including low evaporative water loss) are essential for conserving energy, yellow-bellied marmots living in lowland, semixeric areas have higher metabolic rates at low temperatures associated with higher conductances and a greater ability to dissipate heat evaporatively (29).

RESOURCE UTILIZATION

The two major resources of marmots are food and burrows. Young are weaned (emerge from the natal burrow) in late June to mid July when the standing crop of vegetation reaches 75% to 100% of its maximum (57, 81). All age classes gain mass; food does not appear to limit growth and preparation for hibernation (2, 26, 81).

Annual Energy Budget

Marmots consume 0.8 to 3.1% of the aboveground primary production. Although the assimilation efficiency of 71% to 75% falls in the range of mammalian herbivores (66), the production efficiency of 22.8% is about seven times greater than that of other mammals (76, 81). The consequent high production/maintenance ratio of yellow-bellied marmots seems to be a consequence of the energy-conserving physiology.

Marmots are generalist herbivores (57), but their choice of plants is restricted by plant secondary compounds (11). Food selectivity is based primarily on relative abundance, phenology, nutritional quality, and energy requirements (57). Unfortunately plant epidermal cells do not survive digestive processes equally well, thus precluding the use of fecal analysis to assess diet. Diet choice requires more study; e.g. there is some indication that plants may be selected on the basis of their lipid composition (54). Access to foraging areas is strongly influenced by kinship; only closely related adult females share foraging patches (56). Mothers and juveniles and littermates (both young and yearlings) have nearly identical foraging areas.

Burrows

Burrows are of three types: home or nest, flight, and hibernating. These are not mutually exclusive; the same burrow may be used for all three functions. When alarmed, marmots run to the nearest flight burrow; if possible, they return to the nest burrow (5). Burrows are widely distributed in marmot habitats; a marmot seldom is far from this place of refuge, typically less than 20 m (5, 56, 103). Flight burrows often have only one entrance whereas a nest burrow typically has several entrances. Most nest burrows and hibernacula occur in rocks or are dug in soil under rocks, logs, or bushes. Rocks and tree roots support the structure (104). In contrast to the alpine *(M. marmota)* and steppe *(M. bobac)* marmots, yellow-bellied marmots almost never construct a nest burrow in open meadow because of their vulnerability to badgers *(Taxidea taxus)* (3).

Quality burrows, especially those that function as a hibernaculum, may be the critical resource that determines breeding ratios (73), but the presence of several hibernacula in the territory of *M. marmota* (31) suggests other factors are also important. Group hibernation may be critical for survival and lifetime reproductive success where low temperatures increase the use of fat reserves (30, 31). At North Pole Basin, group hibernation was common and we knew of no instance in which less than two animals occupied a hibernaculum (78). We do not know the extent of group hibernation in the East River Valley. Dispersers fitted with radiotransmitters hibernated singly and about 12% died during hibernation, a figure not statistically different from the 10% hibernation mortality of residents (108). Much more needs to be learned about the energetics of hibernation of field marmots and its relationship to social structure and lifetime reproductive success.

MATING SYSTEMS AND SOCIAL DYNAMICS

Male Reproductive Tactics

SEX RATIOS Although sex ratios at weaning are 1:1 (18, 25), adult sex ratios are biased toward females (5, 6, 25, 78, 92). The female bias, readily apparent at colonial sites (Figures 1, 2), is a consequence of differential mortality; more males, especially at two years of age when they move around seeking females, fall to predators (108). Males may be monogamous in small habitat patches or may defend several females, each living alone, over an area twice the size of the territory of a colonial male (8, 108). Polygyny characterizes the mating system where females are clumped on colonial sites (5, 6, 8, 49); the average breeding sex ratio at these sites is 1:2.3 (15).

TERRITORIALITY Adult males are territorial and defend their territory against incursions by peripheral or transient males (8, 25). Living peripherally to clumped females may represent a waiting tactic; 30 of 52 colonial males

lived peripherally for one or more years before succeeding to the territory (15). Adult males generally behave agonistically to yearling males, including their sons (8). Rates of amicable behavior among males of all age groups are lower and rates of agonistic behavior are higher than expected based on the class composition of the population (27). Although rates of social behavior are highly variable, they are independent of the density of yearling and adult males (8). One source of the variability stems from the individual behavioral phenotypes that characterize marmots (105), but the precise contribution of individual variability has not been quantified. The consequence of adult male agonism to yearling males is that most yearling males disperse (8, 25). Dispersal may be delayed until age two where habitat structure and the dispersion of adult males results in local refugia where the yearlings can avoid the adults (8). Because all males disperse, virtually all males associated with females were born elsewhere and immigrated into their breeding population (8, 14, 25, 97). Fights among males are rare and we never observed a territorial male displaced by an intruder. Male turnover seems to occur when a resident male dies overwinter and a new male becomes resident after spring emergence (15).

REPRODUCTIVE SUCCESS The most important factor determining reproductive success is the ability to associate with adult females. Residence with females typically occurs at age three or older. The probability that a young male will survive to become a yearling is 0.47 (25) and the probability that a yearling male will live to age three is 0.36 (108). Thus, only about 17% of the males born reach an age that makes reproduction likely, but we have yet to determine what proportion do so. The second critical factor is the number of females that comprise the male's harem. Although the number of young or yearlings per female decreases as harem size increases, the number of young or yearlings per male increases (15). For each additional harem female, a male gains about 1.5 young and 0.6 yearlings. The number of females in a harem is determined by adult females; there is no evidence that adult males recruit females or exclude potential female immigrants (7, 9, 13, 25). The third factor affecting male reproductive success is length of residency. Mean length of residency is 2.24 years (15), but some males are resident for from 4 to 6 years (Figure 1; 14, 18). Considerable variance in the reproductive success of males is indicated, but this variance has not been calculated. The average male can expect a lifetime reproductive success of 11.1 young; polygyny is an evolutionarily stable strategy (15, 85).

Female Reproductive Tactics

NATURE OF GROUPS The size of groups of adult females ranges from 1 to 5; mean size varies among colonies and ranges from 1.05 to 1.92; the grand mean is 1.47 (15, 20). Initially harems were treated as female social groups

(49), but now the harem ($\bar{x}$ size 2.27 adult females) is considered to be a male reproductive unit that consists of one or more female groups (15, 20). A female group may also include young of the year and yearlings from reproduction the previous year (Figures 1, 2).

Female groups consist of closely related kin, primarily mother:daughter or sister:sister pairs (14, 15, 20). Groups continue through time as matrilines that may increase in size through the recruitment of daughters (Figure 1). Recruitment may decrease average relatedness as aunt:niece, grandmother:granddaughter, and other kin relationships develop. Thus, matrilines divide to form independent groups. For example, when average relatedness of four females within a matriline decreased to 0.25, the matriline divided to form two groups of two sisters each; each matriline had an average relatedness of 0.5 (14). Another group of five female kin had an average r of 0.0625, but the females were organized into two kin groups, each with an average r of 0.5. Matrilines are readily identified by patterns of space-use; members of a matriline share common home ranges and often live in the same burrow system, whereas members of different matrilines never share burrows and their home ranges overlap little or not at all (14, 21, 56). The mean size of matrilines is negatively related ($p = 0.1$) with habitat area, but the mean number of matrilines is positively correlated ($p < 0.05$) with habitat area (20). Thus, the increase in the number of resident females that is positively correlated with habitat area ($p < 0.05$) does not increase the size of matrilines but increases the number of matrilines. The failure to form larger matrilines is the consequence of females forming stable associations only with individuals related by 0.5.

MATRILINE RECRUITMENT Replacement, that is, the addition of an adult female aged two or older to a population, is significantly correlated with recruitment ($R_s = 0.94$, $p = 0.01$). Recruitment is defined as either the retention of yearling daughters (14, 16) or of two-year-old daughters (20) in their natal colonies. The latter definition discounts mortality during hibernation following the yearling summer and is a more realistic index of reproductive success. Percent recruitment (the percentage of replacement that is recruitment) is unrelated to replacement ($p > 0.1$) but is significantly related with the mean number of females ($p = 0.1$) (20). Hence, replacement is more likely to be recruitment when more adult females are present. This relationship is interpreted to mean that a matrilineal group can exclude potential immigrants (an animal born elsewhere) and increase the probability that replacement involves a related recruit. Thus, population replacement is strongly affected by the social system.

About 53% of the yearling females become recruits; about one third of the recruitment events involve two or more yearlings from the same litter.

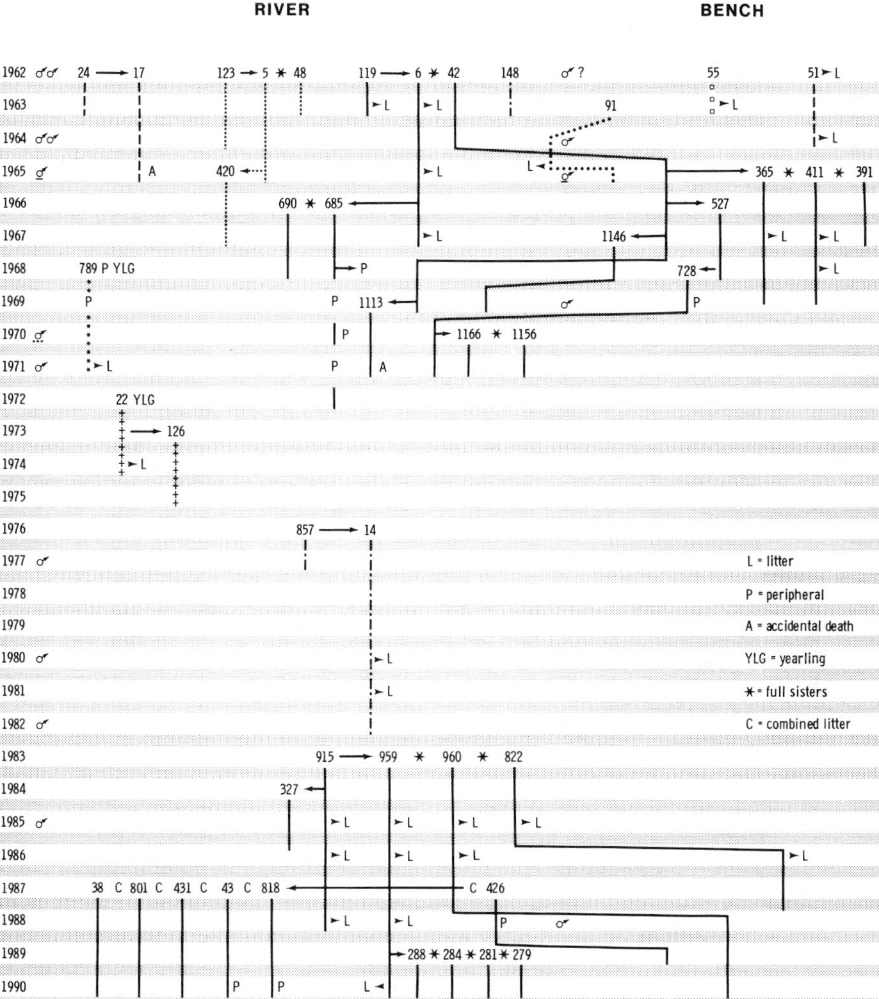

Figure 1 Patterns of residency, recruitment, and immigration. Each animal is identified by ear-tag number in the year of immigration or birth. Vertical lines show years of residency. Matrilines are represented by vertical lines of the same pattern. Recruits are indicated by a short arrow in the year of birth; litters from which there were no recruits are indicated by an L. A male symbol indicates the year in which a given male became resident; male symbols with the same underlining indicate that the same male defended both River and Bench. The resident male of 1970 was born to female 6 in 1965. Bench colony was trapped out in 1969, and only a few animals have resided in that area since. A horizontal line crossing vertical lines indicates that young intermingled and maternity could not be determined. Updated from (14).

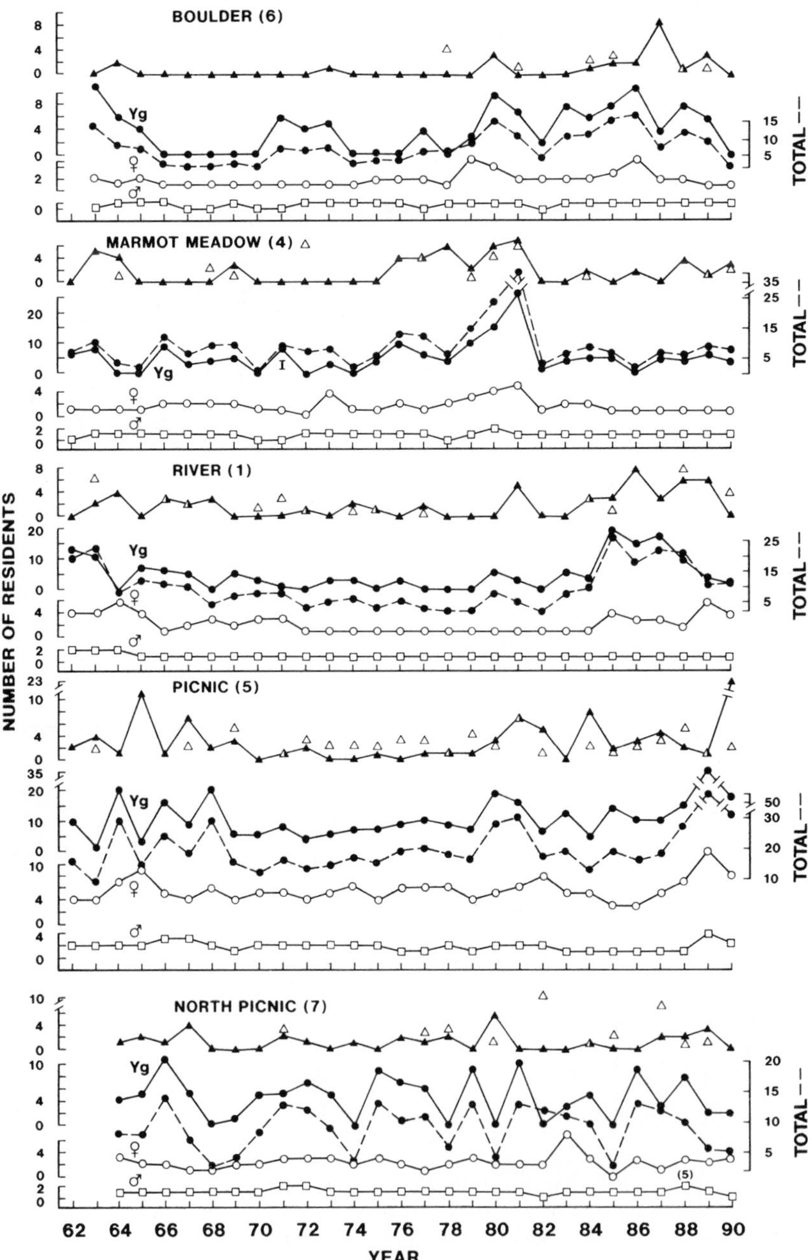

Figure 2 Population levels at five marmot colonies arranged in sequence with the smallest habitat area at the top and the largest habitat area at the bottom. The numbers in parentheses are the locality designations (8); number 1 is furthest down valley and number 7, furthest up valley. Male and female symbols refer to male and female adults, respectively. Yg = young. Solid triangles are yearlings that disappeared during the summer, and open triangles are yearlings that remained throughout the summer. Modified and updated from (25).

Recruitment is twice as likely to occur when the recruit's mother is present; the mother's reproductive status does not significantly affect recruitment (14). Recruitment occurs significantly more often than not when another adult female is present and when she is reproductive. Whether recruitment occurs is not affected by the kinship of the other females resident in the colony (14). We continue to build up sample sizes that will permit a more detailed analysis of recruitment. The structure of matrilines strongly implicates relatedness as critical to the recruitment process (14, 20, 21). For example, when immigration occurs, recruitment is highly unlikely; immigrants and potential recruits are unrelated. Recruitment is unaffected by males; newly resident males do not inhibit recruitment, and males are as likely to tolerate their daughters as not (14).

Of the adult females 40% successfully recruit one or more daughters. In comparison to nonrecruiters, recruiters are resident for a longer time (4.4 vs 2.6 years) and produce more litters (2.3 vs 0.7), more total young (9.8 vs 2.8), more female young (4.9 vs 1.4), and more female yearlings (2.9 vs 0.3). Mean litter size does not differ between the two groups (14). Three variables entered a stepwise discriminant analysis model of recruitment: number of female yearlings, number of female young, and number of litters. The number of female yearlings was the most important variable by far. The model correctly classified 89.6% of the 77 females used in the analysis. The significance of the number of yearlings agrees with life table analysis; the life expectancy of a young one is 1.7 years whereas that of a yearling is 2.0 years. An adult female that produces infants that grow into yearlings has a much greater probability of leaving reproductive descendents (14, 25).

SEX RATIO MANIPULATION Although the population sex ratio is 1:1, several lines of evidence suggest that sex-ratio is manipulated. The number of females per litter is higher for recruiters than for nonrecruiters (14). Sociable females over their lifetime produce more female young than other behavioral phenotypes whereas females classified as submissive-avoider produce more male young (16). Sex ratio does not vary with litter size, measurements of stress, density of adult females, or social environment, but does vary with age; young females produce significantly more daughters than sons (18). Social structure interacts with age; young females living with at least one additional female in the only matriline present on a habitat patch produce almost twice as many daughters as sons. Solitary young females, young females living alone or in a matriline when another matriline is present, and old females in all social organizations do not produce sex-biased litters. The young females living in a matriline with no other matriline present recruit more daughters per litter and more daughters per female young weaned than expected. By contrast, old females living alone with or without another matriline present recruit significantly fewer daughters per litter or per female young weaned (18). These results do not support local resource competition (40) or female condition (107) models. Rather the results are consistent with the hypothesis that females produce the sex conveying the greater fitness gain. Female-biased litters are produced by those yellow-bellied marmots with a high probability of recruiting their daughters into the local population (18).

REPRODUCTIVE SUCCESS: GROUP EFFECTS Living in groups entails benefits and costs (1). A major benefit of matrilineal organization in yellow-bellied marmots is that a dominant group can exclude competing conspecifics from resources (15, 21). For example, immigration almost never occurs into an established group (Figure 1, also Figure 1 in 19). Also, a dominant matrilineal group may acquire the best burrow sites and foraging areas by causing a subordinate matriline to move to an area of lower quality (56). Additionally, a dominant group may suppress reproduction of a female from a different matriline (15).

The major cost of living in a matrilineal group is reproductive inhibition. Note that none of five females born at River in 1987 (thus recruited as yearlings in 1988) weaned litters in the presence of the older female 959, who recruited her four daughters in 1990 (Figure 1). Two-year-old females reproduce less often than expected when adult females, including their mothers, are present (21); the likelihood that three-year-old or older pregnant females successfully wean a litter is significantly less for females living in proximity to other adult females than for females living solitarily (15).

Per capita reproductive output decreases as group size increases in yellow-bellied marmots (49), black-tailed prairie dogs (74), and red deer (44). This relationship could be the result of the reproductive inhibition described above. Reanalysis of the marmot data disclosed that group effects on reproduction depend on the nature of the group. The production of young and yearlings increases as the size of the matriline increases, but per capita production of young and yearlings does not decrease. Although the number of young and yearlings produced increases with increased size of harems, the per capita production of both young and yearlings significantly decreases with increased harem size (15). The reduced per capita reproductive output of harems occurs because the number of litters per female decreases significantly as harem size increases, but mean litter size is unaffected. The difference between the effects of group size on the per capita reproductive output of matrilines and harems is attributed to competition that occurs between matrilines when harems consist of two or more matrilines. For example, if two matrilines each consist of one female, reproduction is unaffected. But if one matriline consists of two or more females and the other matriline one female, the female living singly produces significantly fewer young and yearlings (15). The group effect may occur because only closely related females share foraging patches; the female living singly may be forced to forage in inferior patches during critical times, such as early post-emergence when little vegetation is available (2, 56, 104).

REPRODUCTIVE SUCCESS: HABITAT EFFECTS Habitat differences clearly exist. For example, at North Picnic, the largest habitat, the number of resident adult females consistently is lower than the number at Picnic, the second largest habitat (Figure 2). At Picnic, yearlings were recruited in 23 of 29 years, but only in 10 of 27 years at North Picnic. A critical question is whether any difference in habitat affects the reproductive output of individual females. One test for possible differences in habitat quality analyzes the variation in the reproductive output of individual females per year and in the reproductive output per female per matriline per year. The first analysis considers the lifetime reproductive output of individual females, expressed as average output per year. The second analysis examines the yearly per capita production of matrilines. In these analyses, variance is partitioned among colonies, matrilines, and individuals.

The production of young per female per year differs significantly among colonies, but not among matrilines. However, only 5% of the variance is explained by differences among colonies; 95% of the variance occurs among individual females. Similarly, the yearly per capita production of young does not differ among colonies nor among matrilines; 99% of the variance is attributable to differences among years (15). The production of yearlings per reproductive female and per matriline differ significantly among colonies but not among matrilines. Again, most of the variance is explained by differences among individuals (60%) or differences among years (93%). The difference among years is best interpreted as differences among females. These differences are expressed by either weaning or not weaning a litter; mean litter size is independent of the size of matrilines and of colonies. I conclude that variation in reproductive success is primarily a consequence of differences among individuals that are essentially independent of habitat quality and size of matrilines.

This is not to say that habitat quality is unimportant; it could affect either annual or lifetime reproductive success by its effect on survivorship. Juvenile survivorship was calculated as the ratio of the number of yearlings captured to the number of young weaned. Juvenile survivorship differs among colonies ($0.1 > p > 0.05$); lowest juvenile survivorship occurred in the smallest habitat and second lowest in the largest habitat (20). Adult survivorship was calculated as the ratio of the number of adult females present to the number of resident adult females the previous year. Adult survivorship does not differ among habitats, nor are adult survivorship and juvenile survivorship correlated among habitats. Adult survivorship was significantly correlated (p = 0.05) with juvenile survivorship only within one habitat and nearly so within two others (p = 0.1) (20). Furthermore, survivorship of adult females is unrelated to mean harem size in which each female lived (15).

The lack of any significant habitat effect on adult survivorship suggests that females settle where resources are at least adequate for survival, but where reproductive success may be uncertain. The low contribution of habitat differences to reproductive success measured as production of young indicates that most females have resources adequate for reproduction. The high percentage of variance (40%) in the per capita production of yearlings that is explained by differences among colonies (15) and the significant effect of colony on juvenile survivorship suggest that the key to understanding variation in success of reproductive females is to determine the factors affecting juvenile growth and survival. Although some of the year-to-year variation within a colony can sometimes be attributed to predation (25), unexplained mass mortality (but probably predation, 15), or inadequate preparation for hibernation in late-weaned litters (25, 87), there is no evidence that any of these accounts for differences among colonies. Furthermore, the differences in juvenile survivorship cannot be accounted for by differences in the density of adult females, although density probably has minor effects (20).

One of the costs of living in poor quality habitat is that a trade-off between survivorship and per capita production of young may exist. These two characters are negatively related ($r_s = -0.71$, $p \approx 0.1$) among colonies. For the two colonies with the highest mean rates of juvenile survivorship, adult survivorship is significantly positively correlated with the per capita production of young. Thus, good habitat supports both reproduction and survivorship (20).

Although reproductive success and survivorship are partially attributable to habitat differences and population density, much of the variation remains unexplained. The demonstration of reproductive inhibition suggests that this variation could be a consequence of social behavior.

SOCIAL DYNAMICS Rates of amicable and agonistic behavior are related to population density, the age-sex structure of the population, individual behavioral phenotypes, and length of shared residency. Year-to-year changes in amicable and agonistic behavior are not correlated within or between colonies (10).

Within matrilines, amicable behavior predominates among adult females (14, 16, 21, 27, 78) because females related by 0.5 engage in more amicable behaviors than would be expected by their relative abundance in the population (15, 21). When relatedness is 0.25 or less, amicable behavior may be much less than expected. Thus aunt:niece behaviors are characterized as either much less amicable or much more agonistic than expected based on the frequency of aunt: niece dyads in the population and their degree of sharing space (15, 21). When behavior among matrilineal females is primarily agonistic, relatedness is low (27).

Behavior among nonburrowmates living within a male's territory is predominantly agonistic but may be slightly biased toward amicable if the nonburrowmates are related by 0.25 or more. Behavior among marmots living in different harems is nearly always agonistic (27, 78). Behavior between yearlings and adults is highly variable; in general, female yearlings interact amicably with both parents, but agonistically with other females, whereas male yearlings interact agonistically with adult males and females but may behave amicably with their mothers (8, 27). Adult females may be highly agonistic to yearlings that are nonlittermate full sibs and to nieces; some females are agonistic toward their daughters (15).

Despite the complexity of individual variability, social relationships are strongly kin related. Virtually all of the significant differences in the frequency analysis of social behaviors among age-sex classes can be interpreted in terms of kinship (27). The kin-biased behavior suggests that marmots can discriminate kin from nonkin. Such discrimination has not been demonstrated, but kin discrimination based on familiarity is likely. For example, when two or three adult females wean their litters in the same or adjacent burrows so that young intermingle, there is no behavioral evidence for kin discrimination among the young or between adults and young (21). Young marmots interact only amicably with members of the matriline and share space with their mothers; space-use overlap between young and unrelated adult females is rare and slight (94). Also, young rarely share space with nonlittermates. Thus, young animals normally associate only with close kin, and this association may suffice as a basis for kin-biased behaviors.

INDIVIDUAL DIFFERENCES Much of the variability in social behaviors is related to individual behavioral phenotypes (105). Several factor analyses of behavior measured during mirror image stimulation (MIS) indicated that marmots could be classified as social or asocial; the asocial animals were either aggressive or submissive-avoiders (14, 105). These behavioral phenotypes apparently are stable among adults (103, 105) but may change during ontogeny (17). The social interactions among eight female young were significantly and inversely related to their ranking on the sociability factor determined from MIS and not to kinship (13). Six of the females were tested as yearlings. Social behaviors were positively correlated with rankings on the avoidance factor (17). When rankings on the MIS factors were compared between the young and yearlings, there was no significant correlation. Some animals had the same ranking each year; only a minority changed their behavioral phenotype. It is not surprising that ontogenetic experiences affect the development of behavioral phenotypes. Social play occurs among young and yearlings but declines to very low levels about midsummer of the yearling year and is not a part of adult behavior (77, 91). Social play was postulated to facilitate social dominance and the coordination of agonistic behavior; this interpretation is consistent with ontogenetic influences on the expression of behavioral phenotypes. The manner and the degree by which experience molds the expression of behavioral phenotypes is unknown.

Behavioral phenotype affects the expression of social behavior which, in turn, influences reproductive success. Over a two-year period, submissive females produced few young whereas social females produced about one third of the young (103). When females were classified as either social or asocial, social females recruited more yearling daughters (14). The relationship between behavioral phenotype and lifetime reproductive success was examined for 19 females. Lifetime reproductive success was unrelated to the rank order of the females on the three MIS factors; this fact indicates that individual differences are not continuous. When females were placed in one of three groups according to the MIS factor on which each had her highest score, rankings for number of female yearlings, number of recruits, and number of two-year-old daughters varied significantly among groups. Mean values of the measures were highest for females classified as sociable (16). Social behaviors were not correlated with the three MIS groups; however, several measures of lifetime amicable behavior were correlated with recruitment.

Social measures of behavioral individuality converge on one key index of reproductive success: Sociable females are by far the most successful recruiters. The question remains: Why is there so much behavioral variability? Possibly fitness may be related to heterozygosity or to developmental processes that by their nature produce individuals of different fitnesses. Alternatively, phenotypic plasticity could be an evolutionarily stable strategy. Social and ecological environments of marmots vary; perhaps each behavioral phenotype is most fit under a particular set of conditions. Thus, a female that produces variable offspring is more fit than a female that produces only one behavioral phenotype (for a more extensive discussion, see 16). The significance and evolutionary maintenance of individuality; i.e. variable behavioral phenotypes, remain to be determined.

POPULATION CONSEQUENCES OF REPRODUCTIVE TACTICS

Population Structure

Analysis of population structure focused on clumped breeding units consisting of 1 to 3 males, 1 to 12 adult females, up to 8 resident yearlings (mostly females), and up to 36 young (Figures 1, 2). Peripheral animals, those living adjacent to but not interacting with the resident population, may be present. Transients, those that move through a colony but do not remain for more than a few days, occur irregularly (25). These breeding populations are not isolated; nearly all resident males are born elsewhere and immigrate into the colonial habitat, and some female adults are immigrants (Figure 1; 14, 18, 20). Thus, the effective population size is larger but is unknown (100). The difficulties of measuring effective population size are discussed elsewhere (39). Given that marmot breeding units are clumped and that mating is nonrandom, one questions to what degree a calculation of effective population size is realistic. However, the subdivision of the marmot population into semi-isolated breeding units suggests that effective population size may be relatively small.

Population Dynamics

CHANGES IN DENSITY Clearly, adult males have little effect on the density of marmot colonies (Figure 2). Except for North Picnic in 1988 when five two-year-old males remained for all or part of the summer, typically one or two males reside at each colony. In some years no resident male was observed or trapped, even though females produced litters. This situation occurred primarily at the two smallest colonies. Sometimes males at those areas are wide-ranging and live elsewhere during the summer. Some absence of males probably results from predation, but specific information is lacking. The major change in density results from the annual production of young; the curves for total number and number of young are very similar for all colonies (Figure 2). The influence of yearling residents on total number varies widely among colonies. In only 6 of 28 years do yearling residents affect the total residents at Boulder, whereas resident yearlings affect population density in 22 of 29 years at Picnic. The number of adult females varies among colonies, and its effect on population density also varies. Thus, the number of adult females varied: from 3 to 11 at Picnic, reached a maximum of 4 twice at Boulder, and exceeded 3 only once at North Picnic, three times at Marmot Meadow, and seven times at River. Changes in density are not correlated among colonies (9, 10, 25). Each colony has its density characteristics; although mean density is affected by the size of the colony's habitat, the factors that determine what the mean density is for each colony are not clearly understood.

RECRUITMENT AND IMMIGRATION An increase in the number of adult residents or replacement of deceased adults occurs through immigration or recruitment. Immigration plus recruitment equals replacement, the addition of an adult female to the population. Replacement was calculated as the number of new adults added divided by the number of residents the previous year (20). Replacement was related to survivorship in only one colony; replacement decreased as survivorship increased. Because survivorship was calculated as a rate and the rate could be the same over a wide range of densities, survivorship was calculated as the number of adult females returning. The number of females returning was negatively correlated with replacement in three of the five colonies (20). However, the correlations explained only 12 to 22% of the variation in replacement; although density is a factor, other factors, especially social, are important. When rates of immigration and recruitment within colonies were analyzed separately, only two of ten correlations were significant. The lack of simple density-dependence was also supported by the low and insignificant correlations between survivorship and recruitment within colonies and recruitment and the mean number of adult females among colonies. Immigration varies from 0.15 to 0.46 females per year; by contrast recruitment varies from 0.23 to 1.07 females per year (22). Immigration or recruitment occurred in 73 of 141 colony-years (a colony-year is one colony studied for one year; a calendar year could have a maximum of six colony-years). In only five colony-years did immigration and recruitment occur in the same colony. Twice as many females became residents as recruits than did as immigrants (22).

Immigration occurs primarily when deceased residents are not replaced by recruits (Figure 1). For example, at Picnic Colony, only one female immigrant became a resident between 1974 and 1990. The immigrant occupied space that was vacant because two female residents died over winter (see Figure 1 in 18). During the same period, 27 females were recruited.

DENSITY-DEPENDENCE No simple density-dependence is evident in any marmot colony (Figures 1, 2). One reason for the lack of density-dependence is that virtually all male yearlings emigrate regardless of population density. Recruitment or immigration often does not occur at low densities; recruitment often occurs at high densities (Figure 1, also Armitage 14, 18). Demographic factors such as number of young per female, number of litters per female, number of female yearlings, and percentage of female yearlings becoming resident were not significantly related to number of female residents, number of female recruits, or percentage of female yearlings becoming resident, except that the recruitment of female yearlings was positively correlated with the number of litters per female (25). This relationship may mean simply that successful females both reproduce and recruit (see discussion of matriline recruitment) and that recruitment is a function of fitness strategies rather than of density-dependence (20, 21). Although social behavior may mediate replacement, mean interaction rate and mean number of residents were unrelated. Rates of amicable behavior were not related to population density in any colony; the rate of agonistic behavior was related to density in one colony (7, 9). Rates of amicable and of agonistic behavior between yearling females and either adult males or adult females, and among adult females, were independent of measures of population density whether colonies were analyzed separately or lumped (9). An examination of possible stress effects failed to find any relationship between blood corticosteroid concentrations and two measures of population density (23). Social status and social behavior were the major factors affecting corticosteroid concentrations. A stress response seems to be more strongly related to behavioral phenotype than to population density. Stressed females, those ranking high on the MIS submissive-avoider factor, produced litters at about one fourth the rate of unstressed females (18). A different measure of stress indicated that stressed females produced litters at about one third the rate of unstressed females (18). Densities at which both stressed and unstressed females lived varied; thus there was no evidence that density was the direct cause of the stress.

Dispersal

AGE AND PATTERN Dispersal occurs when a marmot moves away from its natal colony. Most dispersal occurs when marmots are yearlings, but some dispersal occurs as late as age three (25, 101, 108). Age of dispersal is unaffected by sex. Although dispersal occurs throughout the active season, it is concentrated from May to July. Males tend to disperse earlier in the season (May) than females (June). All males and slightly less than one half of the females disperse (8, 14). Dispersal occurs in all directions but is weakly correlated with topography, especially with the direction of the East River Valley (101, 108). Dispersal distance, measured as a straight line from the natal burrow to the first hibernaculum, is highly skewed; most dispersers moved 4 km or less, but distances were as great as 15.5 km for males and 6.4 km for females. Median dispersal distance was less than 400 m for females and greater than 1500 m for males (108). These differences may reflect different requirements for the two sexes; females need find only an unoccupied site with a suitable burrow and food whereas a male must locate both a suitable burrow and food and undefended females.

Three different dispersal patterns occur. Forty-one percent of dispersers abandoned their natal home range in a single, one-way, abrupt move to a new locality. Thirty-three percent emigrated by a gradual process of incremental home range extension until a new home range was established. Females dispersed by this process more than did males. This gradual process may enable females to find the closest adequate resources. Twenty-seven percent of dispersers emigrated in two stages. In the first stage, dispersers left their natal home range but established a new home range nearby, on average 265 m distant. After a mean of 41 days, the animals moved again. Males dispersed more often in two stages and often doubled body mass between the two movements (108). Apparently, the first move enabled animals to escape social stress, and the second move involved finding unoccupied habitat with a suitable hibernaculum.

SURVIVAL AND REPRODUCTION The survival rate during the first summer was about 20% less for dispersers than for residents (only females). Survival during the first winter was about the same for both dispersers and residents. During the second summer, the survival of female residents was unchanged, but that of male and female dispersers decreased. Survival of dispersers and residents was similar during the second winter. Mortality during the summer was caused by predators; 60% of the kills were by coyotes *(Canis latrans)* (108).

About the same proportion of female residents and of female dispersers first bred at age two. Thus, any loss of reproduction by residents because of reproductive inhibition by resident adults appears not to be compensated by dispersal. Fecundity of female dispersers older than two years was lower than that of resident females. Sample sizes are small but suggest that dispersers may suffer a reduction in the frequency of reproduction (108).

PROXIMAL CAUSES OF DISPERSAL The causes of dispersal likely differ between the sexes. Yearling males cannot compete with adult males; adult males do not tolerate potential reproductive competition. The dispersal of males thus seems inevitable (8, 50). However, the timing of dispersal may be affected by social or physiological factors. In general, dispersal of ground-dwelling sciurids is not correlated with rates of social behavior, but evidence suggests that social mechanisms are important (72). Yearling male yellow-bellied marmots remain longer at their natal site when rates of amicable behavior are high or when June body mass is low (50).

Because half of the females become recruits, the factors determining if females disperse are likely more complex than those for males. Social tolerance by adults seems to be critical; yearlings are more likely to become recruits if their home ranges overlap those of adult females, especially those of their mothers, by more than 50% (9, 14). Recruitment may be stressful; yearling recruits had higher concentrations of corticosteroids than those that disappeared ($p = 0.066$) (23). The timing of dispersal is independent of the number of yearling females or of adult females. Females disperse earlier when rates of aggression are high and remain longer when rates of amicable behavior are high (50). As discussed earlier, recruitment of yearling females is strongly associated with amicable behaviors between the yearlings and their mothers (16).

An alternative hypothesis is that dispersal is genetically determined (see 79 for discussion of genetic models). However, when adults were removed from North Picnic colony, none of six yearling females dispersed. The number of recruits in that year was greater than the total number for the previous 17 years (36). In another experimental population at Marmot Meadow, no yearling females dispersed when an adult male but not an adult female was present (17). These results and those described above suggest that dispersal is not genetically determined but is socially mediated by adults interacting with same-sex yearlings (50). Yearlings may assess the probability of future reproductive success in their natal home ranges and decide to remain or disperse. Amicable and/or agonistic behavior may be one but not the only cue to the decision-making process. Further research on this complex problem is necessary and must consider kinship, individual behavioral phenotypes, social behavior, population density, and patterns of space-use.

Genetic Structure

Neither gene frequencies nor heterozygosity at eight loci was associated with altitude, habitat, age, sex, survivorship, litter size, or a suite of behavioral variables (98). The formation of matrilines through yearling recruitment in discrete colonies, the low exchange rate of individuals between groups, and the restriction of mate selection to those in the group promote genetic heterogeneity among social groups (97). The genotype frequencies drift among colonies but do not go to fixation because of the high rate of male immigration (e.g. Figure 1). The social system thus supports the theory of gradual evolution, not of acclerated evolution in closed societies.

POPULATION DYNAMICS: THE ROLE OF DIRECT FITNESS

When this research began, the intent was to determine if social behavior played a critical role in a density-dependent population regulation (114). It soon became evident that density did not drive the system. In general, marmot biology is consistent with the model that self-regulation of population size may occur when resources are unequally partitioned among iteroparous individuals of different behavioral rank (84). Differential reproductive success coupled with either scramble or contest competition may initiate dispersal. An individual of low rank may not disperse because its reproductive success improves if others disperse, but it should disperse if its fitness as a resident is lower than its expected fitness as a disperser (84). This variability in the timing and incidence of dispersal is consistent with the frequent lack of density-dependence in dispersal and the lack of a consistent relationship between social behavior and dispersal (9, 14, 22, 25, 36, 45, 50, 62, 72, 82). In general, individual variation, spatial heterogeneity, behavioral interactions among individuals, and dispersal behavior determine the dynamics and stability of ecological systems (84). To this general model should be added the role of kinship.

As information on kinship accumulated, I formulated the hypothesis that the population system is driven by direct fitness strategies. Major increases in resident adult females occur when daughters are recruited (Figure 1, Figure 2: River 1964, 1983–1990; Marmot Meadow 1977, 1981; Boulder 1979, 1986; Picnic 1974, 1982, 1989; also see 7, 14, 15, 20, 21). Whether a female chooses to disperse or remain in her natal colony affects both her fitness and the fitness of her mother. Because fitness depends on producing reproductive descendents, a female should act to increase the probability that her daughters will survive to reproduce. Survivorship, and probably reproductive success, of daughters is higher for recruits than dispersers. The daughter should decide whether to disperse based on how her fitness is affected. Thus, arguments about whether dispersal benefits residents or dispersers are misdirected (4); the fitness of both is affected. Although a female recruits daughters, her fitness is increased by producing more daughters rather than granddaughters. However, the daughter's fitness is enhanced by reproducing rather than helping her mother. Gains in indirect fitness do not offset losses in direct fitness (19, 20, 21, 95). Kin selection (= indirect fitness) is a minor component in this system. The major importance of direct fitness is supported by the following: preponderance of mother:daughter matrilines, small size of matrilines, space-sharing by close kin, fission of matrilines as average r decreases, amicable behavior directed primarily to kin related by 0.5, and agonistic behavior to all related by 0.25 or less; reproductive suppression of daughters, and lack of any linear relationship between fitness benefits and degree of relatedness (21).

Reproductive competition develops both within and between matrilines as expressed through infanticide (28, 37) and reproductive suppression. Recruitment declines or ceases; presumably yearling females decide that reproductive opportunities lie elsewhere and disperse. The population of adult females declines as older females die; eventually the process is repeated. Competition seems to be for reproductive success, not resources per se. Animals may disperse to escape reproductive inhibition and not because burrows and food are limited. However, the possibility that the quality of a key resource limits reproduction, if not density (2), and causes dispersal cannot be excluded. Thus, population density may be a consequence of access to one or more critical resources or of behavioral/physiological competition for reproductive success. These issues can be clarified by examining blood hormones for evidence of reproductive suppression and by a more critical assessment of resource availability and use.

ACKNOWLEDGMENTS

Our research on the yellow-bellied marmot was supported by National Science Foundation grants G16354, GB-1980, GB-6123, GB-8526, GB-32494, BMS74-21193, DEB78-07327, BSR-8121231, BSR8614690, and BSR-9006772 and by grants from The University of Kansas General Research Fund. I thank Dirk Van Vuren for use of data from his unpublished doctoral dissertation. My thanks and deepest appreciation to those graduate students, trappers, and field assistants without whose energetic and committed assistance this work could not have been performed. The research was conducted at the Rocky Mountain Biological Laboratory.

Literature Cited

1. Alexander, R. D. 1974. The evolution of social behavior. *Annu. Rev. Ecol. Syst.* 5:325–83
2. Andersen, D. C., Armitage, K. B., Hoffmann, R. S. 1976. Socioecology of marmots: female reproductive strategies. *Ecology* 57:552–60
3. Andersen, D. C., Johns, D. W. 1977. Predation by badger on yellow-bellied marmot in Colorado. *Southwest. Nat.* 22:283–84
4. Anderson, P. K. 1989. *Dispersal in Rodents: A Resident Fitness Hypothesis. Am. Soc. Mammalogists Sp. Publ.* 9:1–141
5. Armitage, K. B. 1962. Social behaviour of a colony of the yellow-bellied marmot *(Marmota flaviventris). Anim. Behav.* 10:319–31
6. Armitage, K. B. 1965. Vernal behaviour of the yellow-bellied marmot *(Marmota flaviventris). Anim. Behav.* 13:59–68
7. Armitage, K. B. 1973. Population changes and social behavior following colonization by the yellow-bellied marmot. *J. Mammal.* 54:842–54
8. Armitage, K. B. 1974. Male behaviour and territoriality in the yellow-bellied marmot. *J. Zool., Lond.* 172:233–65
9. Armitage, K. B. 1975. Social behavior and population dynamics of marmots. *Oikos* 26:341–54
10. Armitage, K. B. 1977. Social variety in the yellow-bellied marmot: a population-behavioural system. *Anim. Behav.* 25: 585–93
11. Armitage, K. B. 1979. Food selectivity by yellow-bellied marmots. *J. Mammal.* 60:626–29
12. Armitage, K. B. 1981. Sociality as a life history tactic of ground squirrels. *Oecologia* 48:36–49
13. Armitage, K. B. 1982. Social dynamics of juvenile marmots: role of kinship and individual variability. *Behav. Ecol. Sociobiol.* 11:33–36
14. Armitage, K. B. 1984. Recruitment in yellow-bellied marmot populations: kinship, philopatry, and individual variability. In *Biology of Ground-Dwelling Squirrels,* ed. J. O. Murie, G. R. Michener, pp. 377–403. Lincoln: Univ. Nebraska Press
15. Armitage, K. B. 1986. Marmot polygyny revisited: determinants of male and female reproductive strategies. In *Ecological Aspects of Social Evolution,* ed. D. S. Rubenstein, R. W. Wrangham, pp. 303–31. Princeton: Princeton Univ. Press
16. Armitage, K. B. 1986. Individuality, social behavior, and reproductive success in yellow-bellied marmots. *Ecology* 67:1186–93
17. Armitage, K. B. 1986. Individual differences in the behavior of juvenile yellow-bellied marmots. *Behav. Ecol. Sociobiol.* 18:419–24
18. Armitage, K. B. 1987. Do female yellow-bellied marmots adjust the sex ratios of their offspring? *Am. Nat.* 129:501–19
19. Armitage, K. B. 1987. Social dynamics of mammals: reproductive success, kinship, and individual fitness. *Trends Ecol. Evol.* 2:279–84
20. Armitage, K. B. 1988. Resources and social organization of ground-dwelling squirrels. In *The Ecology of Social Behavior,* ed. C. N. Slobodchikoff, pp. 131–55. New York: Academic
21. Armitage, K. B. 1989. The function of kin discrimination. *Ethol. Ecol. Evol.* 1:111–21
22. Armitage, K. B. 1989. Dynamics of im-

migration into yellow-bellied marmot colonies. In *Abstracts of Papers and Posters,* p. 218. Rome: Fifth Int. Theriological Congr.

23. Armitage, K. B. 1991. Factors affecting corticosteroid concentrations in yellow-bellied marmots. *Comp. Biochem. Physiol.* 98A:47–54
24. Armitage, K. B. 1991. Unpublished data
25. Armitage, K. B., Downhower, J. F. 1974. Demography of yellow-bellied marmot populations. *Ecology* 55:1233–45
26. Armitage, K. B., Downhower, J. F., Svendsen, G. E. 1976. Seasonal changes in weights of marmots. *Am. Midl. Nat.* 96:36–51
27. Armitage, K. B., Johns, D. W. 1982. Kinship, reproductive strategies and social dynamics of yellow-bellied marmots. *Behav. Ecol. Sociobiol.* 11:55–63
28. Armitage, K. B., Johns, D., Andersen, D. C. 1979. Cannibalism among yellow-bellied marmots. *J. Mammal.* 60:205–7
29. Armitage, K. B., Melcher, J. C., Ward, J. M. Jr. 1990. Oxygen consumption and body temperature in yellow-bellied marmot populations from montane-mesic and lowland-xeric environments. *J. Comp. Physiol.* B 160:491–502
30. Arnold, W. 1988. Social thermoregulation during hibernation in alpine marmots *(Marmota marmota). J. Comp. Physiol.* B 158:151–56
31. Arnold, W. 1990. The evolution of marmot sociality: II. Costs and benefits of joint hibernation. *Behav. Ecol. Sociobiol.* 27:239–46
32. Bintz, G. L. 1984. Water balance, water stress, and the evolution of seasonal torpor in ground-dwelling sciurids. In *Biology of Ground-Dwelling Squirrels,* ed. J. O. Murie, G. R. Michener, pp. 142–65. Lincoln: Univ. Nebraska Press
33. Boonstra, R. 1984. Aggressive behavior of adult meadow voles *(Microtus pennsylvanicus)* towards young. *Oecologia* 62:126–31
34. Boonstra, R., Krebs, C. J., Gaines, M. S., Johnson, M. L., Crane, I. T. M. 1987. Natal philopatry and breeding systems in voles (*Microtus* spp.). *J. Anim. Ecol.* 56:655–73
35. Boyce, C. C. K., Boyce, J. L. III. 1988. Population biology of *Microtus arvalis.* III. Regulation of numbers and breeding dispersion of females. *J. Anim. Ecol.* 57:737–54
36. Brody, A. K., Armitage, K. B. 1985. The effects of adult removal on dispersal of yearling yellow-bellied marmots. *Can. J. Zool.* 63:2560–64
37. Brody, A. K., Melcher, J. C. 1984. Infanticide in yellow-bellied marmots. *Anim. Behav.* 33:673–74
38. Brooks, R. J. 1984. Causes and consequences of infanticide in populations of rodents. In *Infanticide Comparative and Evolutionary Perspectives,* ed. G. Hausfater, S. B. Hrdy, pp. 331–48. New York: Aldine
39. Chepko-Sade, B. D., Shields, W. M. 1987. The effects of dispersal and social structure on effective population size. In *Mammalian Dispersal Patterns,* ed. B. D. Chepko-Sade, Z. T. Halpin, pp. 287–321. Chicago: Univ. Chicago Press
40. Clark, A. B. 1978. Sex ratio and local resource competition in a prosimian primate. *Science* 201:163–65
41. Clark, A. B., Ehlinger, T. J. 1987. Patterns and adaptation in individual behavioral differences. In *Perspectives in Ethology,* Vol. 7, ed. P. G. Bateson, P. H. Klopfer, pp. 1–47. New York: Plenum
42. Climatological data. 1962–1986. *Colorado.* Vol. 67–91. Washington: U.S. Govt. Printing Off.
43. Clutton-Brock, T. H. 1988. Reproductive success. In *Reproductive Success,* ed. T. H. Clutton-Brock, pp. 472–85. Chicago: Univ. Chicago Press
44. Clutton-Brock, T. H., Guinness, F. E., Albon, S. D. 1982. *Red Deer Behavior and Ecology of Two Sexes.* Edinburgh: Edinburgh Univ. Press
45. Cockburn, A. 1988. *Social Behaviour in Fluctuating Populations.* London: Croom Helm
46. Davis, D. E. 1976. Hibernation and circannual cycles of food consumption in marmots and ground squirrels. *Q. Rev. Biol.* 51:477–514
47. Dijk, T. S. van. 1982. Individual variability and its significance for the survival of animal populations. In *Environmental Adaptation and Evolution,* ed. D. Mosakowski, G. Roth, pp. 233–51. New York: Gustav Fischer
48. Dobson, F. S. 1982. Competition for mates and predominant juvenile male dispersal in mammals. *Anim. Behav.* 30:1183–92
49. Downhower, J. F., Armitage, K. B. 1971. The yellow-bellied marmot and the evolution of polygamy. *Am. Nat.* 105:355–70
50. Downhower, J. F., Armitage, K. B. 1981. Dispersal of yearling yellow-bellied marmots *(Marmota flaviventris). Anim. Behav.* 29:1064–69
51. Elliott, P. F. 1975. Longevity and the evolution of polygamy. *Am. Nat.* 109:281–87

52. Fairbairn, D. J. 1978. Dispersal of deermice, *Peromyscus maniculatus:* Proximal causes and effects on fitness. *Oecologia* 32:171–93
53. Florant, G. L., Heller, H. C. 1977. CNS regulation of body temperature in euthermic and hibernating marmots *(Marmota flaviventris). Am. J. Physiol.* 232:203–8
54. Florant, G. L., Nuttle, L. C., Mullinex, D. E., Rintoul, D. A. 1990. Plasma and white adipose tissue lipid composition in marmots. *Am. J. Physiol.* 258:R1123–R1131
55. Frase, B. A. 1983. Spatial and behavioral foraging patterns and diet selectivity in the social yellow-bellied marmot. PhD thesis. Univ. Kansas, Lawrence
56. Frase, B. A., Armitage, K. B. 1984. Foraging patterns of yellow-bellied marmots: role of kinship and individual variability. *Behav. Ecol. Sociobiol.* 16:1–10
57. Frase, B. A., Armitage, K. B. 1989. Yellow-bellied marmots are generalist herbivores. *Ethol. Ecol. Evol.* 1:353–66
58. Frase, B. A., Hoffmann, R. S. 1980. *Marmota flaviventris. Mammalian Species* 135:1–8
59. French, A. R. 1986. Patterns of thermoregulation during hibernation. In *Living in the Cold: Physiological and Biochemical Adaptations,* ed. H. C. Heller, X. J. Musacchia, L. C. H. Wang, pp. 393–402. New York: Elsevier
60. French, A. R. 1990. Age-class differences in the pattern of hibernation in yellow-bellied marmots, *Marmota flaviventris. Oecologia* 82:93–96
61. Gaines, M. S., Johnson, M. L. 1987. Phenotypic and genotypic mechanisms for dispersal in *Microtus* populations and the role of dispersal in population regulation. In *Mammalian Dispersal Patterns,* ed. B. D. Chepko-Sade, Z. T. Halpin, pp. 162–79. Chicago: Univ. Chicago Press
62. Gaines, M. S., McClenaghan, L. R. Jr. 1980. Dispersal in small mammals. *Annu. Rev. Ecol. Syst.* 11:163–96
63. Gates, D. 1980. *Biophysical Ecology.* New York: Springer-Verlag
64. Grafen, A. 1988. On the uses of data on lifetime reproductive success. In *Reproductive Success,* ed. T. H. Clutton-Brock, pp. 454–71. Chicago: Univ. Chicago Press
65. Greenwood, P. J. 1980. Mating systems, philopatry and dispersal in birds and mammals. *Anim. Behav.* 28:1140–62
66. Grodzinski, W., Wunder, B. A. 1975. Ecological energetics of small mammals. In *Small Mammals: Their Productivity and Population Dynamics,* ed. F. B. Golley, K. Petrusewicz, L. Ryszkowski, pp. 173–204. New York: Cambridge Univ. Press
67. Hafner, D. J. 1984. Evolutionary relationships of the Nearctic Sciuridae. In *Biology of Ground-Dwelling Squirrels,* ed. J. O. Murie, G. R. Michener, pp. 3–23. Lincoln: Univ. Nebraska Press
68. Hassell, M. P., May, R. M. 1985. From individual behavior to population dynamics. In *Behavioural Ecology,* ed. R. M. Sibly, R. H. Smith, pp. 3–32. London: Blackwell Sci.
69. Herbers, J. 1981. Time resources and laziness in animals. *Oecologia* 49:252–62
70. Hestbeck, J. B. 1982. Population regulation of cyclic mammals: the social fence hypothesis. *Oikos* 39:157–63
71. Hock, R. J. 1969. Thermoregulatory variations of high-altitude hibernators in relation to ambient temperature, season, and hibernation. *Fed. Proc.* 28:1047–52
72. Holekamp, K. E. 1984. Dispersal in ground-dwelling sciurids. In *Biology of Ground-Dwelling Squirrels,* ed. J. O. Murie, G. R. Michener, pp. 297–320. Lincoln: Univ. Nebr. Press
73. Holmes, W. G. 1984. The ecological basis of monogamy in Alaskan hoary marmots. In *Biology of Ground-Dwelling Squirrels,* ed. J. O. Murie, G. R. Michener, pp. 250–74. Lincoln: Univ. Nebraska Press
74. Hoogland, J. L. 1981. The evolution of coloniality in white-tailed and black-tailed prairie dogs (Sciuridae: *Cynomys leucurus* and *C. ludovicianus*). *Ecology* 62:252–72
75. Hoogland, J. L. 1985. Infanticide in prairie dogs: lactating females kill offspring of close kin. *Science* 230:1037–40
76. Humphreys, W. F. 1979. Production and respiration in animal populations. *J. Anim. Ecol.* 48:427–51
77. Jamieson, S. H., Armitage, K. B. 1987.; Sex differences in the play behavior of yearling yellow-bellied marmots. *Ethology* 74:237–53
78. Johns, D., Armitage, K. B. 1979. Behavioral ecology of alpine yellow-bellied marmots. *Behav. Ecol. Sociobiol.* 5:133–57
79. Johnson, M. L., Gaines, M. S. 1990. Evolution of dispersal: theoretical models and empirical tests using birds and mammals. *Annu. Rev. Ecol. Syst.* 21: 449–80

80. Kilgore, D. L. Jr. 1972. Energy dynamics of the yellow-bellied marmot *(Marmota flaviventris)*: a hibernator. PhD thesis. Univ. Kansas, Lawrence
81. Kilgore, D. L. Jr., Armitage, K. B. 1978. Energetics of yellow-bellied marmot populations. *Ecology* 59:78–88
82. Krebs, C. J. 1985. Do changes in spacing behavior drive population cycles in small mammals? In *Behavioural Ecology*, ed. R. M. Sibly, R. H. Smith, pp. 295–312. London: Blackwell Sci.
83. Lidicker, W. Z. Jr. 1985. Dispersal. In *Biology of New World Microtus*, ed. R. H. Tamarin, pp. 420–54. *Am. Soc. Mammal. Sp. Publ. No. 8*
84. Łomnicki, R. 1988. *Population Ecology of Individuals*. Princeton: Princeton Univ. Press
85. Maynard Smith, J., Price, G. R. 1973. The logic of animal conflict. *Nature* 246:15–18
86. Melcher, J. C. 1987. The influence of thermal energy exchange on the activity and energetics of yellow-bellied marmots. PhD thesis. Univ. Kans., Lawrence
87. Melcher, J. C., Armitage, K. B., Porter, W. P. 1989. Energy allocation by yellow-bellied marmots. *Physiol. Zool.* 62:429–48
88. Melcher, J. C., Armitage, K. B., Porter, W. P. 1990. Thermal influences on the activity and energetics of yellow-bellied marmots *(Marmota flaviventris)*. *Physiol. Zool.* 63:803–20
89. Morrison, P., Galster, W. 1975. Patterns of hibernation in the arctic ground squirrel. *Can. J. Zool.* 53:1345–55
90. Murray, B. G. Jr. 1967. Dispersal in vertebrates. *Ecology* 48:975–78
91. Nowicki, S., Armitage, K. B. 1979. Behavior of juvenile yellow-bellied marmots: play and social integration. *Z. Tierpsychol.* 51:85–105
92. Pattie, D. L. 1967. Observations on an alpine population of yellow-bellied marmots *(Marmota flaviventris)*. *Northwest Sci.* 41:96–102
93. Pusey, A. E. 1987. Sex-biased dispersal and inbreeding avoidance in birds and mammals. *Trends Ecol. Evol.* 2:295–99
94. Rayor, L. S., Armitage, K. B. 1991. Social behavior and space-use of young of ground-dwelling squirrel species with different levels of sociality. *Ethol. Ecol. Evol.* 3. In press
95. Rubenstein, D. I., Wrangham, R. W. 1980. Why is altruism toward kin so rare? *J. Tierpsychol.* 54:381–87
96. Salsbury, C. M., Armitage, K. B. 1990. Factors affecting metabolism of field-trapped marmots. *Bull. Ecol. Soc. Am.* 71(2):312
97. Schwartz, O. A., Armitage, K. B. 1980. Genetic variation in social mammals: the marmot model. *Science* 207:665–67
98. Schwartz, O. A., Armitage, K. B. 1981. Social substructure and dispersion of genetic variation in the yellow-bellied marmot *(Marmota flaviventris)*. In *Mammalian Population Genetics*, ed. M. H. Smith, J. Joule, pp. 139–59. Athens: Univ. Georgia Press
99. Schwartz, O. A., Armitage, K. B. 1989. Density-independent correlates of life history characteristics of yellow-bellied marmots. Sixty-ninth Ann. Meet. Am. Soc. of Mammal. Abstract
100. Shields, W. M. 1987. Dispersal and mating systems: investigating their causal connections. In *Mammalian Dispersal Patterns*, ed. B. D. Chepko-Sade, Z. T. Halpin, pp. 3–24. Chicago: Univ. Chicago Press
101. Shirer, H. W., Downhower, J. F. 1968. Radio tracking of dispersing yellow-bellied marmots. *Trans. Kans. Acad. Sci.* 71:463–79
102. Stenseth, N. C. 1983. Causes and consequences of dispersal in small mammals. In *The Ecology of Animal Movement*, ed. I. R. Swingland, P. J. Greenwood, pp. 63–101. Oxford: Clarendon Press
103. Svendsen, G. E. 1974. Behavioral and environmental factors in the spatial distribution and population dynamics of a yellow-bellied marmot population. *Ecology* 55:760–71
104. Svendsen, G. E. 1976. Structure and location of burrows of yellow-bellied marmot. *Southwest. Nat.* 20:487–94
105. Svendsen, G. E., Armitage, K. B. 1973. An application of mirror-image stimulation to field behavioral studies. *Ecology* 54:623–27
106. Travis, S. E., Armitage, K. B. 1972. Some quantitative aspects of the behavior of marmots. *Trans. Kans. Acad. Sci.* 75:308–21
107. Trivers, R. L., Willard, D. E. 1973. Natural selection of parental ability to vary the sex ratio of offspring. *Science* 179:90–91
108. Van Vuren, D. 1990. Dispersal of yellow-bellied marmots. PhD thesis. Univ. Kans., Lawrence
109. Van Vuren, D., Armitage, K. B. 1991. Duration of snow cover and its influence on life history variation in yellow-bellied marmots. *Can. J. Zool.* 69. In press
110. Ward, J. M. Jr., Armitage, K. B. 1981. Circannual rhythms of food consump-

tion, body mass, and metabolism in yellow-bellied marmots. *Comp. Biochem. Physiol.* 69A:621–26

111. Ward, J. M. Jr., Armitage, K. B. 1981. Water budgets of montane-mesic and lowland-xeric populations of yellow-bellied marmots. *Comp. Biochem. Physiol.* 69A:627–30
112. Wasser, S. K., Barash, D. P. 1983. Reproductive suppression among female mammals: implications for biomedicine and sexual selection theory. *Q. Rev. Biol.* 58:513–38
113. White, T. C. R. 1978. The importance of a selective shortage of food in animal ecology. *Oecologia* 33:71–86
114. Wynne-Edwards, V. C. 1962. *Animal Dispersion in Relation to Social Behaviour,* Edinburgh: Oliver & Boyd
115. Zatzman, M. L., Thornhill, G. V., Ray, W. J., Ellersiek, M. R. 1984. Seasonal changes of food and water consumption and urine production of the marmot, *Marmota flaviventris. Comp. Biochem. Physiol.* 77A:735–43

Annu. Rev. Ecol. 1991. 22:409–429

CLUTCH SIZE

H. C. J. Godfray

Department of Biology and Centre for Population Biology, Imperial College at Silwood Park, Ascot, Berkshire SL5 7PY, United Kingdom

L. Partridge

Division of Biological Science, University of Edinburgh, Ashworth Building, West Mains Road, Edinburgh EH9 3JT, United Kingdom

P. H. Harvey

Department of Zoology, University of Oxford, South Parks Rd., Oxford, OX1 3PS, United Kingdom

KEY WORDS: Clutch size, life history theory, behavioral ecology

---

INTRODUCTION

In recent years, problems associated with the evolution of clutch size have evoked continued interest. Although it can claim to be one of the oldest topics in life history theory, a variety of new mechanisms have been proposed that may influence selection on clutch size. Notable advances have also occurred in the sophistication of the experimental (79, 104) and comparative techniques available for testing theory (60). In addition, the taxonomic scope of clutch size studies has widened appreciably: Clutch size theory was invented by ornithologists to help explain the reproductive behavior of birds and was largely developed by vertebrate biologists. Today, the theory is widely applied to invertebrates and even to plants (84, 151).

Many animals lay eggs or produce young in discrete groups or clutches. The relationship between per capita offspring fitness and clutch size (we shall refer to this as the offspring fitness curve) is of crucial importance in de-

0066-4162/91/1120-0409$02.00

termining the evolution of clutch size. Here, we review the selective processes that have been proposed to influence clutch size in animals and how these hypotheses may be tested. Our intention is not to provide an exhaustive review of experimental studies on clutch size (recent reviews for birds include 79, 83, 103, and for invertebrates 49), but to concentrate on conceptual issues and controversies. We also adopt a wide taxonomic perspective and discuss how theory may be applied to disparate animal groups—in particular, we make no special distinction between groups with and without parental care.

Two broad approaches to the study of adaptation can be distinguished. The behavioral ecological approach uses optimality models to predict phenotypes and tests the models using a combination of observation, manipulative experiments, and comparative studies. The assumption is made that the population will evolve to the optimum phenotype unhindered by genetic constraints. Here, we concentrate on the application of this approach to clutch size evolution. However, an important and complementary method of studying clutch size evolution is through the use of quantitative genetics which we briefly outline here. Additive genetic variation is usually detectable for both litter size in mammals (39) and for clutch size in birds (17). Clutch size is an important determinant of the fecundity component of fitness, and a number of workers have estimated selection differentials for clutch size in birds. Most studies have found large, consistent and positive selection differentials (18, 44, 116; but see 153) which at first sight suggest the presence of directional selection for larger clutch size. If the environment has recently changed, then it is possible that the species is not at evolutionary equilibrium (116) or that it lacks sufficient heritable variation (44, 116). However, the frequency with which positive selection differentials have been found has prompted the search for other explanations. One possibility is that there is negative genetic covariance between clutch size and other components of fitness (see the discussion on trade-offs below). Alternatively, Price & Liou (113) have shown that positive selection differentials can be maintained by selection acting on nonheritable traits which covary with clutch size, such a nutritional state, while Cooke et al (27) have shown that the same result can arise through a gene-environment interaction caused by antagonistic selection through competitive abilities—a Red Queen effect. A more prosaic explanation is that ornithologists tend to study birds in good habitats where gene flow from poor habitats may lead to the appearance of selection differentials.

We begin by discussing clutch size in circumstances in which the female is selected to maximize her gain in fitness from a single clutch and where what has become known as Lack's hypothesis applies. We then examine trade-offs between clutch size and future reproductive success and go on to discuss the influence of different forms of environmental variability. Finally, we explore conflicts of interest over clutch size, both between relatives and between nonrelatives.

LACK'S HYPOTHESIS

The modern study of the evolution of clutch size began with the work of the ornithologist David Lack in the 1940s (74–76). Lack provided an explanation of clutch size in terms of individual selection at a time when many (perhaps most) biologists considered that clutch size evolved to allow a population or species to persist without overexploiting limiting resources (e.g. 162, see also 158, 159). Lack saw that selection acting on the individual would normally outweigh selection acting on the population and argued that a bird would lay the number of eggs that resulted in the maximum number of fledged young. He argued further that, in altricial birds, the main factor limiting the number of fledged young was the ability of the parents to feed their offspring.

Lack's hypothesis applies when maximizing the fitness gain per clutch is equivalent to maximizing lifetime fitness. This is true when the animal produces a single clutch in a lifetime, but it will also be true if reproductive success is severely limited by opportunities to reproduce. A number of biological mechanisms can lead to limited opportunities for reproduction. In birds and mammals, as well as in other groups with extensive parental care, the physiological requirements for reproduction limit the opportunities to breed. Many invertebrates lay eggs in resource patches that are rare or difficult to find in the environment. Lack's hypothesis will not apply if there is a trade-off between the number of offspring in a clutch and future reproductive success; we discuss such trade-offs in the following section.

In a sightly modified form, Lack's hypothesis is at the core of modern clutch size theory, and the clutch size that maximizes the gain in parental fitness is sometimes called the "Lack clutch size" (22). Lack measured parental fitness using the number of fledged young as a currency. However, as Lack was aware, clutch size may affect the fitness of young after fledging; for example, poorly fed birds from large clutches may have low overwinter survival (83). Today it is normal to use a better measure of parental fitness, for example, the number of young surviving to breed or the number of grandchildren (54). In discussing clutch size in altricial birds, Lack suggested that offspring fitness would decline monotonically with increasing clutch size because of increased competition for food. However, other factors may cause a decreasing offspring fitness curve, and it may be domed or even monotonically increasing. We refer to the Lack clutch size as the clutch size that leads to the greatest gain in parental fitness, irrespective of the biological mechanisms relating clutch size to offspring fitness.

Other factors apart from food may influence clutch size in birds. Skutch (128) argued that the fitness of offspring in large clutches would decline because of an increased incidence of predation. Comparative work on New World passerine species has supported this idea: There is an association between the size of clutch laid and the probability of loss of offspring to predators (73). The risk of predation may increase with clutch size because large broods make more noise and attract predators (106, 128, 129, 134) or because they take longer to fledge (this assumes chicks in the nest are more susceptible to predation than recent fledglings) (9, 81). There is also evidence that the costs of incubation are related to clutch size (15, 56). Increasing clutch size may thus increase the risk of improper incubation or so weaken the parent that the quality of parental care is reduced.

In species where there is no parental care after the eggs hatch and where individuals from the same clutch do not compete, offspring fitness may still decline with clutch size if the resources used to provision eggs are limiting. In other words there is a trade-off between egg size and clutch size. The first analysis of this problem was by Smith & Fretwell (131) who assumed the parent had a fixed pool of resources (perhaps yolk) that is distributed among an indeterminate number of young. Offspring fitness is assumed to increase with greater parental investment, though with diminishing returns. Smith & Fretwell calculated the optimum egg size that maximized the parental fitness returns from the clutch. Two rather separate literatures have developed, one devoted to calculating optimum clutch size and the other to calculating optimum egg size (or equivalently optimum parental investment) (e.g. 26, 84, 98, 161). In fact, when the parent is selected to maximize the gain in fitness from a single clutch, the two approaches are often identical: The Smith & Fretwell result is obtained when the variable chosen for optimization is parental investment, while the Lack result is obtained when clutch size is the chosen variable. Complications arise when there are trade-offs between parental investment early in the life of the offspring (for example, investment in egg size) and subsequent competition between offspring for parental care or other resources (98).

A domed offspring fitness curve may be found in cases where small clutches are penalized. Again, this may be caused by a variety of biological processes. In birds, single chicks in a nest may be unable to thermoregulate efficiently (90). It has been suggested that large clutches of mollusc eggs gain protection from dessication (11) and that large clutches of lepidopteran caterpillars are able to protect themselves from environmental extremes by constructing shelters (135, 150). Individual larvae of some herbivorous insects are unable to initiate feeding in the absence of conspecifics (43). Small clutches of some gregarious parasitoid species fail as, unless all the host tissue is consumed, the larvae drown when they attempt to pupate (146). A variety of other mechanisms that may give rise to domed offspring fitness curves in invertebrates are discussed in Ref. 49.

While predation tends to penalize large clutches of birds, the reverse is often true for invertebrate clutches. If a predator becomes satiated before it consumes a whole clutch, or a parasitoid runs out of eggs, predation or parasitism will be inversely density dependent and the probability of an individual escaping attack will increase with clutch size. The effect of predator satiation will be accentuated if predator efficiency declines when attacking large broods. There is evidence that swarms of prey may confuse predators, lessening their attack rate (89). Large clutches may also lead to more efficient predator defense, by either physical or passive means. For example, in the presence of predators, clutches of newly hatched ascalaphid larvae *(Acaloptynx furciger)* aggregate, turn to the attacker, and rapidly and repeatedly snap their ferocious jaws (62). Individual females of some moths lay irregular piles of eggs, the interior eggs protected from parasitism by the eggs on the outside (36). The egg masses of other moths are arranged to mimic the tendrils of plants (25). Large clutches of lepidopteran larvae are able physically to conceal themselves from parasitism by building large communal webs (135). Large clutch size may also improve the efficiency of aposematic coloration or lead to its evolution because predators cease attack after experiencing one or few brightly colored, distasteful food items (45). In butterflies, at least in cross-species comparisons, there appears to be a clear correlation between aposematism, distasefulness, and gregarious oviposition (25, 28).

The Lack clutch size can still be calculated if per capita offspring fitness initially increases with clutch size, as long as offspring fitness ultimately declines. However, in some organisms, the advantages of producing large clutches may be so great that the Lack clutch size exceeds the egg capacity of the organism. The animal will then be selected to lay its complete complement of eggs as a single clutch. Observed clutch sizes in such species will be determined by the factors that limit total reproductive effort and by the dynamics of egg production. The evolution of reproductive effort is a classic problem of life history theory (41, 63, 148) though the evolutionary dynamics of egg production have been little studied. Begon & Parker (12) investigated clutch size theoretically in an organism that matured eggs while foraging and where parental fitness increased monotonically with clutch size. The optimum clutch size was determined by the trade-off between the advantages of laying a large clutch and the risk of dying before reproduction. In a number of species, several females tend to deposit their eggs in the same place despite an abundance of potential oviposition sites. For example, butterflies that lay large clutches of eggs are not infrequently observed to lay next to, or even on top of, clutches laid by other females (49). This provides strong though circumstantial evidence that the optimum clutch size exceeds that that can be laid by an individual. Brood fusion in precocial birds may also have a similar explanation (37).

The majority of clutch size models do not distinguish between the sex of the offspring. However, in circumstances where sibling competition is affected by sex, sex ratio and clutch size evolve together. Depending on the exact form of the competition, all clutches may be of the same size and uniformly biased toward the sex with least competition between siblings (40, 47, 144). Alternatively, where intersexual competition is greater than intrasexual competition, single sex clutches may be produced. The size of the separate clutches will be the "within-sex" Lack clutch size, and the overall sex ratio will be determined by the ratio of the two expected clutch sizes. Single sex clutches of different sizes are known in some parasitoid wasps (53). In lions an association exists between large litter size and a male-biased sex ratio (97). The functional explanation may be that the fitness of male, but not female, offspring increases sharply with the size of the same-sex cohort of relatives in which they are reared.

In some cases, the clutch size strategies available to a female may be severely constrained by the organism's breeding biology. Long-tailed skuas (jaegers) never lay a clutch size greater than two, evidently because they do not build a nest and instead incubate their eggs on their feet. This habit is apparently an evolutionary response to high rates of egg-predation; experimentally introduced nests greatly increased predation on eggs (4). A rather different constraint operates in the greater rhea. The environment deteriorates strongly during the breeding season, favoring the immediate incubation of eggs. Synchronous hatching is also selectively favored so that all the young in a clutch can be kept together to protect them from predators. However, females cannot lay all their eggs at once because it takes time to produce such large eggs. The evolutionary resolution to this conflict is for each female in a local geographical area to mate with and lay their eggs in the nests of different males, starting with the top dominant and working down the hierarchy. As a consequence each female lays a clutch of one egg that contributes to a much larger clutch of male eggs (19). Some butterflies lay their eggs in crevices, and clutch size is simply the number of eggs needed to fill the crevice (28).

The standard experimental technique for investigating the Lack clutch size is to manipulate brood size and then measure offspring and parental fitness. Such experiments have now been carried out on a wide variety of birds (surveys in 78, 79). Overall, most studies have shown that the observed clutch size is either equal to or a little smaller than the most productive clutch size. Relatively few manipulative studies have been performed on clutch size in invertebrates, but the available evidence (35, 142, 143) suggests that clutch size in these groups is considerably smaller than the Lack clutch size. While manipulation is a powerful experimental technique, care must be taken to avoid several potential problems. The female must respond to the manipulation which must not result in very high or very low clutch sizes, outside the range of a realistic phenotypic response. Costs of reproduction associated with the production of eggs are unaffected by the manipulation of brood size. Lindén & Møller (83) discuss some particularly ornithological problems: Manipulations are frequently carried out on hole-nesting birds in nest boxes where unnaturally high population densities, reduced predation, and reduced ectoparasitoid load may all influence clutch size dynamics.

Until recently, it has not been practical to manipulate egg size to investigate the relationship between egg size and fitness. However, Sinervo & Huey (125, 126) have demonstrated that it is possible to remove yolk from lizard eggs which go on to produce small, but entirely viable hatchlings. Small hatchlings had lower sprint speeds which probably increases their vulnerability to predation.

Comparisons among taxa in clutch size have often been reported, but making sense of them has led to several controversies. The main problem is that species differences in clutch size may be correlated with several ecological, morphological, and life-history variables, any number of which may be causal (60). The task is to distinguish confounding from causal associations. For example, Lack (76) suggested that individual birds might allocate finite resources into producing larger clutches of smaller eggs or smaller clutches of larger eggs. He considered that the idea was likely to apply most forcefully to precocial taxa in which the mother feeds the offspring little, if at all. Accordingly he compared waterfowl (Anatidae) species and found the predicted negative correlation between egg size and clutch size. Rohwer (117) argued that females from larger species, which are known to have smaller clutches, are likely to have more resources to invest in their offspring and would therefore produce larger eggs. He thus attempted to control statistically for body size in his analysis and found that the residual variance in clutch size accounted for only 13% of the variance in egg weight. However, Rohwer's study did not properly control for body size (he used deviations from a major axis rather than a Model 1 regression line), for phylogenetic association, or for species differences in life history and habitat utilization. Blackburn's (16) reanalysis of Rohwer's data controlled for these factors and found that clutch size accounted for at least 29% of the variance in egg weight after controlling body weight. Similar negative relationships have been revealed, using varying degrees of statistical sophistication, in several other taxa, including salamanders (122), chelonians (38), other birds (14) and mammals (114).

The same problems in inferring mechanisms from the cross-species comparison of clutch sizes also occur in cross-population studies within a species (105), although in such cases differences in clutch size correlated with environmental variation are much more likely to indicate causality (e.g. 71).

TRADE-OFFS BETWEEN CLUTCH SIZE AND FUTURE REPRODUCTIVE SUCCESS

A number of reasons may suggest why the production of a Lack clutch size might lead to a reduction in future reproductive success. The presence of trade-offs will lead to selection for clutches smaller than the Lack clutch size.

Much recent work on clutch size has involved the theoretical investigation and experimental measurement of trade-offs.

The first trade-off implicated in clutch size evolution was that between clutch size and the probability of surviving to breed again (21, 158). It was suggested that birds that *reared* large clutches would be so physically exhausted by the end of the breeding season that their probability of surviving to breed again would be reduced. This trade-off is an example of a cost of reproduction (13, 105) and will lead to selection for reduced clutch size. Large clutches may not only increase the risks of overwinter mortality but may also lead to loss of condition and reduced fecundity the following year. A related trade-off occurs if the production and rearing of a large clutch precludes a bird from breeding a second time in the same breeding season.

A slightly different trade-off has been discussed by Sibly & Calow (123). Suppose that laying a large clutch has no effect on the instantaneous risk of mortality but delays the next breeding attempt. This may have two effects on lifetime reproductive success. First, if the instantaneous risk of mortality is constant, delaying breeding increases the chances of dying prior to breeding again. Second, in an age-structured population with overlapping generations, there may be an advantage to breeding early so that the young reach maturity and themselves reproduce as quickly as possible. This advantage only occurs in increasing populations and is analogous to compound interest on a financial investment. It is therefore most likely to operate in organisms that reproduce several times within a breeding season or that are multivoltine. Delayed reproduction may also be important in birds that breed once a year if the production of a large clutch results in a delayed start to breeding in the next year because there is frequently a strong correlation between reproductive success and the date that nesting commences (e.g. 112).

A number of workers have attempted to detect these trade-offs in birds by manipulating clutch size (surveys in 79, 83, 103). Some studies have demonstrated impaired overwinter survival of birds with experimentally enlarged broods (34, 115), while other studies have found evidence for reduced fecundity in the subsequent breeding season (54, 55, 119). A delay in the time of breeding may also occur in the following year (78). Brood enlargement has been shown to reduce the probability of producing a second brood in the same season (82, 132) as well as the size and success of the second brood (61, 129). Overall, manipulative studies have usually found evidence of trade-offs, while nonmanipulative ones generally have not (103).

In species without parental care, other trade-offs will be more important. Perhaps the most fundamental trade-off involves egg limitation. Consider a species with a monotonic decreasing offspring fitness curve. Note that as each egg is added to make up the Lack clutch size, the fitness gain per egg for the parent decreases. Now compare an animal whose lifetime reproductive success is limited by opportunities to reproduce with an animal limited by the number of eggs it has to lay. The first animal should maximize its fitness gain per clutch while the second animal should maximize its fitness gain per egg. While the first animal should produce a Lack clutch size, the second animal should, in this case, lay a single egg in each clutch. Often, reproductive success will be partially limited both by the opportunities to produce clutches and by egg supply. The predicted clutch size will then be determined by the balance between the immediate gain in fitness from increasing the size of the current clutch and the potential loss in fitness through running out of eggs (22, 99, 155). Any factor that increases the likelihood of running out of eggs, for example, an increase in the opportunities to produce clutches or a decrease in the risk of mortality, will lead to selection for smaller clutch sizes. When the risk of mortality is independent of age, the optimum clutch size clearly should decrease through life as egg reserves are used up and the risk of egg exhaustion increases (68, 85). If, however, mortality increases with age, the optimum clutch size may increase in old animals with a short life expectancy, an example of terminal investment (109, 145). A final complication is found in animals which mature eggs throughout their life. In these animals, the optimum clutch size will be inversely related to the current egg reserves (85, 86).

Where the offspring fitness curve is domed, the parent achieves the maximum fitness gain per egg with a clutch size greater than one. Domed offspring fitness curves thus lead to selection for clutches greater than one, even when the parent is egg limited but has abundant opportunities to produce clutches. This may explain why animals such as folivorous insects, which often appear to have unlimited sites to lay eggs, choose to lay their eggs in clutches (46, 156).

Few manipulative experiments have attempted to test the idea that clutch size is related to egg supply. Rosenheim & Rosen (118) recently studied the influence of the size of egg reserves on the clutch of a parasitoid wasp. Egg reserves were manipulated by storing wasps at different temperatures prior to experimentation; egg reserves also varied in wasps of different sizes. Clutch size was found to be inversely related to egg supply, which, in a multiple regression analysis, was the most important factor influencing clutch size.

A third way in which the production of large clutches may reduce future reproductive success is by wasting time. Consider again the case of a monotonic decreasing offspring fitness curve and suppose that the oviposition of each egg takes an appreciable period of time. As each egg is added to the clutch, the fitness increment per egg declines, and a point may be reached when the female is selected not to "waste" time adding more eggs to create the Lack clutch size, but to begin searching for a site to lay a new clutch. This assumes that the female is under selection to maximize her rate of gain of fitness over time, just as classical foraging theory assumes an animal is selected to maximize its rate of gain of food. In fact, there is a very close parallel between foraging in a patchy environment and clutch size with time limitation. For example, the marginal value theorem (20) is used to calculate the optimum behavior in both cases. In foraging theory, longer patch residence is predicted in high quality patches or when travel time between patches is high (136). In contrast, larger clutch sizes are predicted on good oviposition sites or when travel time between oviposition sites is high (22, 23, 68, 99, 127). If the production of clutches involves time-consuming activities that do not depend on clutch size (for example, nest construction), then these will also lead to selection for increased clutch size (127). Models of time limitation are relatively simple and popular with theoreticians. In particular, complexities such as state-dependent decisions arise less readily (24, 86). However, there are probably relatively few cases when the time spent in oviposition is a substantial fraction of time available for reproduction.

THE EFFECT OF A VARIABLE ENVIRONMENT

The Lack clutch size is not constant for a species or a population. In species with parental care, different individuals will vary in their ability to feed their young and will thus have personal Lack clutch sizes. For all species, there is likely to be both temporal and spatial variation in the factors that affect offspring fitness leading to temporal and spatial variation in the Lack clutch size. This variation is likely to contain both predictable and unpredictable components. Another possible source of variability is in the production of the clutch itself: Some organisms may not be able to lay a precise number of eggs. Finally, there may be variation among the young themselves, for example, the youngest offspring is frequently a runt in birds and mammals. All these sources of variability may influence the evolution of clutch size.

The study of individual variation in optimum clutch size has largely concerned birds. There is normally considerable within-population variation in clutch size in birds. An individual's clutch size is likely to be influenced by proximate factors such as the physiological condition of the bird (either directly—7, 152, 160—or because birds in better condition obtain better territories—64). A bird's condition may act simply as a constraint, preventing the individual from producing its optimum clutch size (70), or natural selection may mould the behavior of the bird so that it produces the optimum clutch size appropriate to its condition. Proximate and ultimate factors thus interact in a complex manner, and physiological constraints and individual optima should not be treated as alternative hypotheses. A number of workers have suggested that individual adjustment in clutch size might explain the observation that the most productive clutch size is higher than the mean clutch size

(76, 107). This idea can be tested by manipulating brood size, both up and down, to see whether parents are able to rear more young than they lay and whether their survival to breed again is affected by the manipulation. The results from such experiments are mixed: Nur (94), reviewing a number of studies, concluded that while some results are consistent with the individual optimization hypothesis, there is no firm evidence in its support. However, more recent studies (55, 108) do provide better evidence for individual optimization though there is still much controversy. For example, Nur (94) and Pettifor et al (108) analyzed the same data on Great Tits and concluded respectively that individual optimization is absent and present. The difference of opinion arises from the use of different statistical techniques, and the question cannot be resolved without further analysis. Finally, it should be noted that experiments to manipulate brood size do not exclude all aspects of individual optimization. For example, birds may adjust their clutch size with respect to their ability to incubate eggs or to the ability of their mate to feed them while incubating (94).

Individual optimization may also be important when there is a cost to reproduction. Nur & Hansson (95) have modelled optimal reproductive effort when birds vary in their abilities to overwinter. They predict a positive correlation between clutch size and female weight (after controlling for body size, a measure of condition) *after* breeding. There is evidence of this from a number of bird species (94).

In the face of environmental variability, either spatial or temporal, an animal may evolve a single response to an "average" environment or may evolve phenotypic plasticity. Obviously, for plasticity to evolve, the environmental variability must be measurable or predictable. The average response will be a genetically determined compromise and subject to temporal and spatial variation in natural selection (33). Plastic responses, if perfect, will lead to optimal behavior at all times.

Consider first unpredictable spatial variability, for example, an insect ovipositing clutches of eggs in patches of unknown quality or a bird nesting in a heterogeneous habitat in which food quality varies. The optimum clutch size is determined simply by calculating the arithmetic mean fitness of different strategies over the range of environmental variability. It has been suggested (121) that when there is a high, clutch size–independent probability that some clutches will be completely destroyed, an organism will be selected to lay many small clutches instead of few large clutches. Although the mean fitness of the two strategies may not differ, producing many small clutches increases the likelihood of at least some young surviving and thus reduces the variance in fitness. Prima facie, this argument is incorrect as selection will not distinguish among strategies with the same within-generation arithmetic fitness. However, in finite populations with density-dependent mortality, within-generation variance in reproductive success may lead to variation between generations, in which case the strategy with the highest temporal geometric mean fitness will be selected.

Temporal variability occurs if, for example, conditions vary unpredictably between breeding seasons. In species with non-overlapping generations, the appropriate measure of the fitness of a trait is its geometric mean fitness across generations. The geometric mean is more sensitive to low values of fitness than the arithmetic mean and is zero overall if the fitness in any one year is zero. Temporal variability thus tends to select for more conservative traits than does spatial variability. Boyce & Perrins (18) have suggested that temporal variability may explain why the mean clutch size of birds is frequently less than the most productive. If large clutches are more productive but have a greater variance in fitness than small clutches, then the clutch size that results in the highest geometric mean fitness will be less than the clutch size that results in the highest arithmetic mean fitness. Their results can explain the difference between the mean and most productive clutch sizes in Great Tits. However, the problem of defining fitness in an age-structured population with temporal variability is poorly explored, and their definition of fitness and application of the geometric mean needs further study. Note that different patterns have also been claimed for the same data (108).

Many examples of phenotypic plasticity appear in conditions where environmental variability can be measured or predicted. There is evidence that birds lay small clutches in years where food supply is poor (8, 66, 106) or in areas where food is scarce (69). Supplying birds with additional food can lead to both an advance in laying date and larger clutch size (65, 72). However, it is often difficult to distinguish between whether the bird lays a small clutch because it is weakened by lack of food or as a strategy to anticipate future food shortages while rearing young. Phenotypic plasticity is easier to demonstrate in species without parental care. Several herbivores are known to lay clutches of different sizes on different host plants (46, 110). In both cases small clutches are laid on the plant species providing fewer resources for the developing young. Some of the best examples of phenotypic plasticity come from parasitoid wasps that lay clutches of eggs on the bodies of parasitized insects (hosts). The host provides the only sustenance for the developing young and thus the size of the host is crucial in determining the fitness of the clutch. It has been repeatedly shown that female wasps modulate their clutch size in response to both interspecific and intraspecific size variation in the host (reviews in 49, 154).

One strategy of coping with uncertain environmental conditions is always to produce the same size clutch but subsequently to reduce the size of the brood in poor years or in poor localities. In species where the parent remains with the brood, this could be done through direct intervention. Burying

beetles *(Nicrophorus)* lay clutches of eggs on dead mice and remain with their brood while they develop. They lay a fixed clutch size over a range of mouse sizes and then reduce their first instar brood by cannibalism to match the food reserves provided by the mouse (10). Direct intervention by the parent may not be necessary if the parent controls the food supply, as in altricial birds. An alternative strategy is thus to create a hierarchy in offspring size and allow smaller offspring to die when conditions are bad (75). Hierarchies in offspring size are common in birds and are produced by the parent commencing incubation before completing the clutch. Brood reduction occurs either because the smallest young are outcompeted by their nest mates or through sibling aggression. The evolution of asynchronous brooding has been a major preoccupation of modern experimental ornithology, and recent reviews list no less than eight competing explanations (80, 130). A number of studies have examined the effects of artificial synchrony on the fate of nestlings under different circumstances. Many of these found that synchrony did not reduce nestling survival or growth, and that the effect of asynchrony was to impose a cost in the form of the death of the youngest nestling without any clear compensating advantage (138). One study (87) did find a reduction in nestling survival and growth with imposed synchrony, but only under poor feeding conditions; the study did not demonstrate that this would compensate the costs of asynchrony under good feeding conditions.

The ability of a female's mate to provide resources for the growing brood provides another source of environmental variability. There is evidence that secondary female Pied Flycatchers that share a male's territory with a primary female and have to rear their clutch alone, lay comparatively small clutches (1). In contrast, facultatively polygamous female dunnocks lay larger clutches if they are tended by two males than do birds tended by a single male (30).

The optimum clutch size may also be affected by the inability of a female to produce a clutch of a precise size (18, 51, 92). The effect of this imprecision can be seen by weighing the penalties of overshooting and undershooting the deterministic optimum clutch size. To take an extreme example, if the most productive clutch size is x but if all members of a clutch of $x + 1$ died, the female may be selected to lay a clutch smaller than x to avoid the risk of a costly overshoot. Imprecise clutch sizes may arise if the parent is unable to gauge exactly the size of the clutch: This is very likely in invertebrates with large clutches—for example, some parasitoids oviposit 400 eggs in under ten seconds. A second cause of imprecision is egg infertility. Offspring hierarchies leading to brood reduction may also be an adaptation to cope with infertile eggs (3). For example, several eagle species almost invariably lay a clutch of two eggs but never, or very seldom, rear both young (42, 88). Where both eggs hatch, the larger chick kills its sibling in its first few days of life. In eagles, egg infertility can be as high as 10%, and laying an extra egg substantially lessens the risk of wasting a whole breeding season.

CONFLICT OVER CLUTCH SIZE

At least three potential sources of conflict exist over clutch size. First, natural selection operating on genes expressed in the young may have different results than natural selection operating on genes expressed in the parent: In other words there may be parent-offspring conflict in the sense of Hamilton (58) & Trivers (149). Second, the females of many organisms, especially invertebrates, lay clutches together on a single resource patch. The offspring compete together for resources, and the optimum clutch size for an individual depends on the strategy adopted by the rest of the population—an example of a classical evolutionary game. Finally, although it is not discussed here, there may also be conflicts of interest between the mother and the father over clutch size.

The most obvious way for an offspring to influence clutch size is by killing one or more siblings. While not common, siblicide is widespread in birds, especially in large raptors, herons and allies, gannets and skuas (3, 90, 91, 93, 124). An offspring may be selected to destroy a sibling if the fitness advantages to itself and to its surviving siblings compensate for the death of a relative (50, 58, 96, 101, 137). Conditions such as food shortage that favor parental brood reduction also favor brood reduction by siblings, though theory suggests that siblings will be more willing to reduce the size of the brood than is optimal for the parent. As mentioned above, where brood reduction does occur, it is normally the young and not the parents that cause the death of the runt. This suggests that the extent of brood reduction may be determined by selection acting on the offspring rather than the parent, but as yet no experimental study has succeeded in disentangling the two processes.

Brood reduction is found in a variety of invertebrate groups where it is often associated with cannibalism. It has been suggested (2) that some beetles lay large clutches of eggs that hatch over several days so that the larvae that hatch first are able to obtain their first meal by feeding on their unhatched siblings. A number of spiders and marine invertebrates lay large clutches of eggs, many of which are infertile (nurse eggs), and their only apparent function is to provide nourishment for the eggs that hatch (111, 147). The production of nurse eggs is an alternative strategy to increasing the yolk content of the egg. The larvae of many parasitoid wasp species kill all conspecifics in the same host, and this has a major influence on the possible clutch size that can be produced by the female (48).

Offspring may also have a more subtle, indirect influence on clutch size. In sexual species, offspring differ genetically and will compete among themselves for resources. For example, nestling birds compete with each other when begging food from their parents, while gregarious insect larvae may compete for food resources at the oviposition site. Sibling competition will

entail costs to the participants; for example, begging may use up energy and attract predators while a caterpillar might sacrifice assimilation efficiency for speed of ingestion. The fitness gain to the parent may thus be reduced by sibling competition, and the parent may be selected to reduce clutch size to lessen sibling conflict (52). It has been suggested that the production of an offspring hierarchy is a mechanism to reduce sibling competition (57, 58) though it is unclear how this may evolve (53, 102). Where the parent remains with the brood, sibling conflict may be reduced by direct parental intervention, an imposition of the parental optimum (2), though it is unlikely that the parent always "wins" (32, 59, 100). Direct intervention is, of course, not possible in species where the parent abandons her eggs.

In some species of parasitoid wasp, clutch size is under the control of the offspring. Polyembryonic wasps inject a single egg into the body of their host. The egg then divides asexually to produce a large number, sometimes several thousand, of genetically identical offspring (139). Because of the genetic identity, there is no sibling conflict among the larvae, and some species even have fighting larval morphs that protect their siblings but fail to mature sexually (29). As the clutch size is determined by the mass of larvae and there is no question of parental trade-offs, the species should evolve to the Lack clutch size. It is interesting that this group of parasitoids, where a strong a priori argument can be made for the Lack clutch size, has the largest clutch size of any parasitoid.

Conflict over clutch size among nonrelatives occurs if more than one female lays clutches of eggs in the same site. In species that feed their young, this phenomenon is known as brood parasitism. Intraspecific brood parasitism occurs when one female clandestinely lays an egg in another female's nest, and this is known to occur in many bird species (5, 120). With the occurrence of intraspecific brood parasitism, selection may occur for hosts to drop their clutch size, and some evidence suggests that this occurs in birds (6, though see 77). The well-known interspecific brood parasite, the European cuckoo, has a simple strategy for manipulating host clutch size: The young cuckoo ejects all the host offspring from the nest (31).

Many invertebrates lay several clutches of eggs in the same place. This behavior has been particularly studied in parasitoid wasps (where it is termed superparasitism) and more recently in other groups such as herbivorous and carrion insects (superoviposition) (49). If a resource patch already contains eggs laid by another female, then the value of the resource will normally be reduced and a second female may be expected to ignore the patch or to lay a reduced clutch size (22, 127). In parasitoid wasps, the clutch of eggs laid by a superparasitizing female is normally smaller than a primary clutch (154, 157). The calculation of optimum clutch sizes is complicated as, if superparasitism is common, females laying eggs on unexploited patches may experience selection to lay smaller clutches because of the high risk that the patch will be subsequently discovered by another female. This question has been investigated using game theoretic models (47, 67, 99, 133, 141, 157). Other complications arise when superparasitizing females destroy the eggs of previous females (133, 140) or adopt different sex ratio strategies (141, 157).

CONCLUSION

The behavioral ecological theory of the evolution of clutch size is well developed and rich in predictions. The pioneering work of field ornithologists in testing these predictions using field manipulations continues to provide remarkable insights into life history theory. The experimental approach to the study of clutch size is now being applied to novel taxonomic groups while recent developments in manipulating egg size allow the experimental investigation of new sets of questions. Recent advances in the comparative method offer the prospect of a far more rigorous approach to cross-species comparisons. Though only briefly considered here, the renaissance of quantitative genetics has provided a variety of new techniques for the study of clutch size, and while they complement the methods of behavioral ecology, greater integration of the two approaches is desirable. It thus seems likely that the modern study of clutch size, initiated by David Lack over 40 years ago, will continue at the forefront of the investigation of adaptation.

Literature Cited

1. Alatalo, R., Carlson, A., Lundberg, A., Ulfstran, S. 1981. The conflict between male polygamy and female monogamy: the case of the pied flycatcher *Ficedula hypoleuca*. *Am. Nat.* 117:738–53
2. Alexander, R. D. 1974. The evolution of social behavior. *Ann. Rev. Ecol. Syst.* 5:325–83
3. Anderson, D. J. 1990. Evolution of obligate siblicide in boobies. 1. A test of the insurance-egg hypothesis. *Am. Nat.* 135:334–50
4. Andersson, M. 1976. Clutch size in the Long-tailed Skua *Stercorarius longicaudus:* some field experiments. *Ibis* 118:586–88
5. Andersson, M. 1984. Brood parasitism within species. In *Producers and Scroungers,* ed. C. Barnard, pp. 195–228. London: Croom Helm
6. Andersson, M., Eriksson, M. O. G. 1982. Nest parasitism in goldeneyes *Bucephala clangula:* some evolutionary aspects. *Am. Nat.* 120:1–16
7. Ankney, C. D., MacInnes, C. D. 1978. Nutrient reserves and reproductive performance of female lesser snow geese. *Auk* 95:459–71
8. Arcese, O., Smith J. N. M. 1988. Effects of population density and supplemental food on reproduction in song sparrows. *J. Anim. Ecol.* 57:119–36
9. Arnold, T. W., Rohwer, F. C., Armstrong, T. 1987. Egg viability, nest predation, and the adaptive significance of clutch size in prairie ducks. *Am. Nat.* 130:643–53
10. Bartlett, J. 1987. Filial cannibalism in burying beetles. *Behav. Ecol. Sociobiol.* 21:179–83
11. Bayne, C. J. 1969. Survival of the embryos of the grey field slug *Agriolimax reticulatus* following dessication of the egg. *Malacologia* 9:391–401
12. Begon, M., Parker, G. A. 1986. Should egg size and clutch size decrease with age? *Oikos* 47:293–302
13. Bell, G., Koufopanou, V. 1986. The costs of reproduction. In *Oxford Surveys of Evolutionary Biology,* ed. R. Dawkins, M. Ridley, 3:83–131. Oxford: Oxford Univ. Press

14. Bennett, P. M. 1986. *Comparative studies of morphology, life history and ecology among birds*. PhD thesis. Univ. Sussex, England
15. Biebach, H. 1984. Effect of clutch size and time of day on the energy expenditure of incubating starlings *(Sternus vulgaris)*. *Phys. Zool.* 57:26–31
16. Blackburn, T. 1991. The interspecific relationship between egg size and clutch size in waterfowl—a reply to Rohwer. *Auk*. In press
17. Boag, P. T., van Noordwijk, A. J. 1987. Quantitative genetics. In *Avian Genetics: A Population and Ecological Approach,* pp. 45–78. London: Academic
18. Boyce, M. S., Perrins, C. M. 1988. Optimizing great tit clutch size in a fluctuating environment. *Ecology* 68: 142–53
19. Bruning, D. F. 1973. The greater rhea chick and egg delivery route. *Nat. Hist.* 82:68–75
20. Charnov, E. L. 1976. Optimal foraging, the marginal value theorem. *Theor. Popul. Biol.* 9:129–36
21. Charnov, E. L., Krebs, J. R. 1974. On clutch size and fitness. *Ibis* 116:217–19
22. Charnov, E. L., Skinner, S. W. 1984. Evolution of host selection and clutch size in parasitoid wasps. *Fla. Entomol.* 67:5–21
23. Charnov, E. L., Skinner, S. W. 1985. Complementary approaches to the understanding of parasitoid oviposition decisions. *Environ. Entomol.* 14:383–91
24. Charnov, E. L., Stephens, D. W. 1988. On the evolution of host selection in solitary parasitoids. *Am. Nat.* 132:707–22
25. Chew, F. S., Robbins, R. K. 1984. Egg-laying in butterflies. In *The Biology of Butterflies,* ed. R. I. Vane-Wright, P. R. Ackery, pp. 65–80. London: Academic
26. Clutton-Brock, T. H. 1991. *The Evolution of Parental Care*. Princeton: Princeton Univ. Press
27. Cooke, F., Taylor, P. D., Francis, C. M., Rockwell, R. F. 1990. Directional selection and clutch size in birds. *Am. Nat.* 136:261–67
28. Courtney, S. P. 1984. The evolution of batch oviposition by Lepidoptera and other insects. *Am. Nat.* 123:276–81
29. Cruz, Y. P. 1981. A sterile defender morph in a polyembryonic hymenopterous parasite. *Nature* 289:27–33
30. Davies, N. B. 1985. Cooperation and conflict among dunnocks, *Prunella modularis,* in a variable mating system. *Anim. Behav.* 33:628–48
31. Davies, N. B. 1988. Cuckoos versus reed warblers: adaptations and counteradaptations. *Anim. Behav.* 36:262–84
32. Dawkins, R. 1976. *The Selfish Gene*. Oxford: Oxford Univ. Press
33. Dhondt, A. A., Adriansen, F., Matthysen, E., Kempenaers, B. 1990. Non-adaptive clutch size in tits. *Nature* 348:723–25
34. Dijkstra, C. 1988. *Reproductive tactics in the Kestral*. PhD thesis. Univ. Groningen, The Netherlands
35. Dijkstra, L. J. 1986. Optimal selection and exploitation of hosts in the parasitic wasp *Colpoclypeus florus* (Hym., Eulophidae). PhD Thesis. Univ. Leiden, The Netherlands
36. Dowden, P. B. 1961. The gypsy moth egg parasite, *Ooencyrtus kuwanai,* in Southern Connecticut in 1960. *J. Econ. Entomol.* 54:876–78
37. Eadie, J. McA., Kehoe, F. P., Nudds, T. D. 1988. Pre-hatch and post-hatch brood amalgamation in North American Anatidae: A review of hypotheses. *Can. J. Zool.* 66:1709–21
38. Elgar, M. A., Heaphy, L. J. 1989. Covariation between clutch size, egg weight and egg shape: comparative evidence from chelonians. *J. Zool. Lond.* 219:137–52
39. Falconer, D. S. 1981. *Introduction to Quantitative Genetics*. London: Longman
40. Frank, S. A. 1990, Sex allocation theory for birds and mammals. *Annu. Rev. Ecol. Syst.* 21:13–55
41. Gadgil, M., Bossert, W. H. 1970. Life historical consequences of natural selection. *Am. Nat.* 104:1–24
42. Gargett, V. 1978. Sibling aggression in the Black Eagle in the Matapos, Rhodesia. *Ostrich* 49:57–63
43. Ghent, A. W. 1960. A study of the group feeding behavior of larvae of the jack-pine sawfly, *Neodiprion pratti banksianae* Rao. *Behaviour* 16:110–48
44. Gibbs, H. L. 1988. Heritability and selection on clutch size in Darwin's medium ground finches *(Geospiza fortis)*. *Evolution* 42:750–62
45. Gittleman, J. L., Harvey, P. H. 1980. Why are distasteful prey not cryptic? *Nature* 286:149–50
46. Godfray, H. C. J. 1986. Clutch size in a leaf-mining fly (*Pegomya nigritarsis:* Anthomyiidae). *Ecol. Entomol.* 11:75–81
47. Godfray, H. C. J. 1986. Models for clutch size and sex ratio with sibling interaction. *Theor. Popul. Biol.* 30:215–31
48. Godfray, H. C. J. 1987. The evolution

of clutch size in parasitic wasps. *Am. Nat.* 129:221–33
49. Godfray, H. C. J. 1987. The evolution of invertebrate clutch size. In *Oxford Surveys of Evolutionary Biology*, ed. P. H. Harvey, L. Partridge, 4:117–54. Oxford: Oxford Univ. Press
50. Godfray, H. C. J., Harper, A. B. 1990. The evolution of brood reduction by siblicide in birds. *J. Theor. Biol.* 145:163–75
51. Godfray, H. C. J., Ives, A. R. 1987. Stochastic models of invertebrate clutch size. *Theor. Popul. Biol.* 33:79–101
52. Godfray, H. C. J., Parker, G. A. 1991. Sibling competition, parent-offspring conflict and clutch size. *Anim. Behav.* In press
53. Godfray, H. C. J., Parker, G. A. 1991. Clutch size, fecundity and parent-offspring conflict. *Philos. Trans. R. Soc. Lond. B* 332:67–79
54. Gustaffson, L., Part, T. 1990. Acceleration of senescence in the collared flycatcher *Ficedula albicollis* by reproductive costs. *Nature* 347:279–81
55. Gustaffson, L., Sutherland, W. J. 1988. The costs of reproduction in the collared flycatcher *Ficedula albicollis*. *Nature* 335:813–15
56. Haftorn, S., Reinertsen, R. E. 1985. The effect of temperature and clutch size on the energetic cost of incubation in a free-living blue tit *(Parus caeruleus)*. *Auk* 102:470–78
57. Hahn, D. C. 1981. Asynchronous hatching in the laughing gull: cutting losses and reducing rivalry. *Anim. Behav.* 29:421–27
58. Hamilton, W. D. 1964. The genetical theory of social behaviour. I & II. *J. Theor. Biol.* 7:1–16, 17–51
59. Harper, A. B. 1986. The evolution of begging: sibling competition and parent-offspring conflict. *Am. Nat.* 128:99–114
60. Harvey, P. H., Pagel, M. D. 1991. *The Comparative Method*. Oxford: Oxford Univ. Press
61. Hegner, R. E., Wingfield, J. C. 1987. Effects of brood size manipulations on parental investment, breeding success, and reproductive endocrinology. *Auk* 104:470–80
62. Henry, C. S. 1972. Eggs and repagula of *Ululodes* and *Ascaloptynx* (Neuroptera: Ascalaphidae): a comparative study. *Psyche (Cambridge)* 79:1–22
63. Hirshfield, M. F., Tinkle, D. W. 1975. Natural selection and the evolution of reproductive effort. *Proc. Natl. Acad. Sci. USA* 72:2227–31
64. Hogstedt, G. 1980. Evolution of clutch size in birds: adaptive variation in relation to territory quality. *Science* 210: 1148–50
65. Hogstedt, G. 1981. Effect of additional food on reproductive success in the Magpie *Pica pica*. *J. Anim. Ecol.* 50:219–90
66. Hornfeldt, B., Eklund, U. 1990. The effect of food on laying date and clutch-size in Tengmalm's Owl *Aegiolus funereus*. *Ibis* 132:395–406
67. Ives, A. R. 1989. The optimal clutch size of insects when many females oviposit per patch. *Am. Nat.* 133:671–87
68. Iwasa, Y., Suzuki, Y., Matsuda, H. 1984. Theory of oviposition of parasitoids. I. Effect of mortality and limited egg number. *Theor. Popul. Biol.* 26: 205–27
69. Jarvinen, A. 1989. Clutch-size variation in the pied flycatcher *Ficedula hypoleuca*. *Ibis* 131:572–77
70. Jones, P. J., Ward, P. 1976. The level of reserve protein as the proximate factor controlling the timing of clutch size in the red-billed quelea. *J. Zool. (Lond.)* 189:1–19
71. Klomp, H. 1970. The determination of clutch size in birds, a review. *Ardea* 58:1–124
72. Korpimaki, E. 1989. Breeding performance of Tengmalm's owl *Aegiolus funereus:* effects of supplementary feeding in a peak vole year. *Ibis* 131:51–56
73. Kuleza, G. 1990. An analysis of clutch size in New World passerine birds. *Ibis* 132:407–22
74. Lack, D. 1947. The significance of clutch size. *Ibis* 89:309–52, 90:25–45
75. Lack, D. 1954. *Natural Regulation of Animal Numbers*. Oxford: Clarendon
76. Lack, D. 1966. *Population Studies of Birds*. Oxford: Oxford Univ. Press
77. Lank, D. B., Rockwell, R. F., Cooke, F. 1990. Frequency-dependent fitness consequences of intraspecific nest parasitism in snow geese. *Evolution* 44:1436–53
78. Lessells, C. M. 1986. Brood size in Canada geese: a manipulation experiment. *J. Anim. Ecol.* 55:669–89
79. Lessells, C. M. 1991. The evolution of life histories. In *Behavioral Ecology: an Evolutionary Approach*, ed. J. R. Krebs, N. B. Davies pp. 32–68. Oxford: Blackwell Sci. 3rd ed.
80. Lessells, C. M., Avery, M. I. 1989. Hatching asynchrony in European Bee-eaters *Merops apiastor*. *J. Anim. Ecol.* 58:815–35
81. Lima, S. L. 1987. Clutch size in birds: a predation perspective. *Ecology* 68: 1062–70

82. Lindén, M. 1988. Reproductive tradeoffs between first and second clutches in the Great Tit *Parus major:* an experimental study. *Oikos* 51:285–90
83. Lindén, M., Møller, A. P. 1989. Cost of reproduction and covariation of life history traits in birds. *Trends Ecol. Evol.* 4:367–71
84. Lloyd, D. 1987. Selection of offspring size at independence and other size versus number stategies. *Am. Nat.* 129: 800–17
85. Mangel, M. 1987. Oviposition site selection and clutch size in insects. *J. Meth. Biol.* 25:1–22
86. Mangel, M. 1989. Evolution of host selection in parasitoids: does the state of the parasitoid matter. *Am. Nat.* 133: 688–705
87. Magrath, R. D. 1989. Hatching asynchrony and reproductive success in the Blackbird. *Nature* 339:536–38
88. Meyburg, B.-U. 1974. Sibling aggression and mortality among nestling eagles. *Ibis* 116:224–28
89. Milinski, M. 1977. Do all members of a swarm suffer the same predation? *Z. Tierpsychol.* 45:373–88
90. Mock, D. W., Parker, G. A. 1986. Advantages and disadvantages of brood reduction in egrets and herons. *Evolution* 40:459–70
91. Mock, D. W., Ploger, B. J. 1987. Parental manipulation of optimal hatch asynchrony in cattle egrets: an experimental study. *Anim. Behav.* 35:150–60
92. Mountford, M. D. 1968. The significance of litter size. *J. Anim. Ecol.* 37:363–67
93. Nelson, B. 1989. Cainism in the Sulidae. *Ibis* 131:609
94. Nur, N. 1986. Is clutch size variation in the blue tit *(Parus caeruleus)* adaptive? An experimental study. *J. Anim. Ecol.* 55:983–99
95. Nur, N., Hansson, O. 1984. Phenotypic plasticity and the handicap principle. *J. Theor. Biol.* 110:275–97
96. O'Connor, R. J. 1978. Brood reduction in birds: selection for fratricide, infanticide and suicide? *Anim. Behav.* 26:79–96
97. Packer, C., Pusey, A. E. 1987. Intrasexual cooperation and the sex ratio in African lions. *Am. Nat.* 130:636–42
98. Parker, G. A., Begon, M. 1986. Optimal egg size and clutch size: effects of environment and maternal phenotype. *Am. Nat.* 128:573–92
99. Parker, G. A., Courtney, S. P. 1984. Models of clutch size in insect oviposition. *Theor. Popul. Biol.* 26:27–48
100. Parker, G. A., Macnair, M. R. 1979. Models of parent-offspring conflict. IV. Suppression: Evolutionary retaliation by the parent. *Anim. Behav.* 27:1210–35
101. Parker, G. A., Mock, D. W. 1987. Parent-offspring conflict over clutch size. *Evol. Ecol.* 1:161–74
102. Parker, G. A., Mock, D. W., Lamey, T. C. 1989. How selfish should stronger sibs be? *Am. Nat.* 133:846–68
103. Partridge, L. 1989. Lifetime reproductive success and life history evolution. In *Lifetime Reproduction in Birds,* ed. I. Newton, pp. 421–40. London: Academic
104. Partridge, L. 1989. An experimentalist's approach to the role of costs of reproduction in the evolution of life histories. In *Toward a More Exact Ecology,* ed. P. J. Grubb, J. B. Whittaker, pp. 231–46. Oxford: Blackwell Sci.
105. Partridge, L., Harvey, P. H. 1988. The ecological context of life history evolution. *Science* 241:1449–55
106. Perrins, C. M. 1965. Population fluctuations and clutch size in the Great Tit *Parus major* L. *J. Anim. Ecol.* 34:601–47
107. Perrins, C. M., Moss, D. 1975. Reproductive rates in the great tit. *J. Anim. Ecol.* 44:695–706
108. Pettifor, R. A., Perrins, C. M., McCleery, R. H. 1988. Individual optimization of clutch size in great tits. *Nature* 336:160–62
109. Pianka, E. R., Parker, W. S. 1975. Age-specific reproductive tactics. *Am. Nat.* 109:453–64
110. Pilson, D., Rausher, M. D. 1988. Clutch size adjustment by a swallowtail butterfly. *Nature* 333:361–63
111. Polis, G. A. 1981. The evolution and dynamics of intraspecific predation. *Annu. Rev. Ecol. Syst.* 12:225–51
112. Price, T., Kirkpatrick, M., Arnold, S. J. 1988. Directional selection and the evolution of breeding date in birds. *Science* 240:798–800
113. Price, T., Liou, L. 1989. Selection of clutch size in birds. *Am. Nat.* 134:950–59
114. Read, A. F., Harvey, P. H. 1989. Life history differences among the eutherian radiations. *J. Zool. (Lond.)* 219:329–53
115. Reid, W. V. 1987. The cost of reproduction in the glaucous-winged gull. *Oecologia* 74:458–67
116. Rockwell, R. F., Findlay, C. S., Cooke, F. 1987. Is there an optimal clutch size in snow geese. *Am. Nat.* 130:839–63
117. Rohwer, F. C. 1988. Inter- and intraspecific relationships between egg

size and clutch size in waterfowl. *Auk* 105:161–76
118. Rosenheim, J. A., Rosen, D. 1991. Foraging and oviposition decisions in the parasitoid *Aphytis lingnanensis:* distinguishing the influences of egg load and experience. *J. Anim. Ecol.* In press
119. Røskaft, E. 1985. The effect of enlarged brood size on the future reproductive potential of the rook. *J. Anim. Ecol.* 54:255–60
120. Rothstein, S. I. 1990. Brood parasitism and clutch-size determination in birds. *Trends Ecol. Evol.* 5:101–2
121. Rubenstein, D. I. 1982. Risk, uncertainty, and evolutionary strategies. In *Current Problems in Sociobiology,* ed. King's College Sociobiol. Group, pp. 91–111. Cambridge: Cambridge Univ. Press
122. Salthe, S. N. 1969. Reproductive modes and the number and sizes of ova in urodeles. *Am. Midl. Nat.* 81:467–90
123. Sibly, R., Calow, P. 1983. An integrated approach to life-cycle evolution using selective landscapes. *J. Theor Biol.* 102:527–47
124. Simmons, R. 1988. Offspring quality and the evolution of cainism. *Ibis* 130:339–57
125. Sinervo, B. 1990. The evolution of maternal investment in lizards: an experimental and comparative analysis of egg size and its effects on offspring performance. *Evolution* 44:279–94
126. Sinervo, B., Huey, R. B. 1990. Allometric engineering: an experimental test of the causes of interpopulational differences in performance. *Science* 248:1106–09
127. Skinner, S. W. 1985. Clutch size as an optimal foraging problem for insects. *Behav. Ecol. Sociobiol.* 17:231–38
128. Skutch, A. F. 1949. Do tropical birds rear as many young as they can nourish? *Ibis* 91:430–55
129. Slagsvold, T. 1984. Clutch size variation of birds in relation to nest predation: on the cost of reproduction. *J. Anim. Ecol.* 53:945–53
130. Slagsvold, T., Lijfield, J. T. 1989. Constraints on hatching asychrony and egg size in pied flycatchers. *J. Anim. Ecol.* 58:837–50
131. Smith, C. C., Fretwell, S. D. 1974. The optimal balance between the size and number of offspring. *Am. Nat.* 108:499–506
132. Smith, H., Källänder, H., Nilsson, J-Å. 1989. The trade-off between offspring number and quality in the great tit *Parus major*. *J. Anim. Ecol.* 58:383–401
133. Smith, R. H., Lessells, C. M. 1985. Oviposition, ovicide and larval competition in granivorous insects. In *Behavioural Ecology,* ed. R. M. Sibly, R. H. Smith, pp. 423–48. Oxford: Blackwell Sci.
134. Snow, B. K. 1970. A field study of the bearded bellbird in Trinidad. *Ibis* 112: 299–329
135. Stamp, N. E. 1980. Egg deposition patterns in butterflies: why do some species cluster their eggs rather than deposit them singly. *Am. Nat.* 115:367–80
136. Stephens, D. W., Krebs, J. R. 1986. *Foraging Theory*. Princeton: Princeton Univ. Press
137. Stinson, C. H. 1979. On the selective advantage of fratricide in raptors. *Evolution* 33:1219–25
138. Stouffer, P. C., Power, H. W. 1990. Density effects on asynchronous hatching and brood reduction in European Starlings. *Auk* 107:359–66
139. Strand, M. R. 1989. Clutch size, sex ratio and mating by the polyembryonic encyrtid *Copidosoma floridanum* (Hymenoptera: Encyrtidae). *Fla Entomol.* 72:32–42
140. Strand, M. R., Godfray, H. C. J. 1989. Superparasitism and ovicide in parasitic Hymenoptera: theory and a case study of the ectoparasitoid *Bracon hebetor*. *Behav. Ecol. Sociobiol.* 24:421–32
141. Suzuki, Y. & Iwasa, Y. 1980. A sex ratio theory of gregarious parasitoids. *Res. Popul. Ecol.* 22:366–82
142. Takagi, M. 1985. The reproductive strategy of the gregarious parasitoid *Pteromalus puparum* (Hymenoptera: Pteromalidae). I. Optimal number of eggs in a single host. *Oecologia* 68:1–6
143. Taylor, A. D. 1988. Host effects on larval competition in the gregarious parasitoid *Bracon hebetor*. *J. Anim. Ecol.* 57:163–72
144. Taylor, P. D. 1981. Intra-sex and inter-sex sibling interaction as sex ratio determinants. *Nature* 291:64–66
145. Taylor, P. D. 1990. Optimal life histories with age-dependent trade-off curves. *J. Theor. Biol.* 148:33–48
146. Taylor, T. H. C. 1937. *The Biological Control of an Insect in Fiji*. London: Imperial Inst. Entomol.
147. Thorson, G. 1950. Reproductive and larval ecology of marine bottom invertebrates. *Biol. Rev.* 25:1–45
148. Tinkle, D. W. 1969. The concept of reproductive effort and its relation to the evolution of life histories of lizards. *Am. Nat.* 103:501–16
149. Trivers, R. L. 1974. Parent-offspring conflict. *Am. Zool.* 14:249–64
150. Tsubaki, Y. 1981. Some beneficial

effects of aggregation in young larvae of *Pryeria sinica* Moore (Lepidoptera: Zygaenidae). *Res. Pop. Ecol.* 23:156–67

151. Uma Shaanker, R. U., Ganeshaiah, K. N., Bawa, K. S. 1988. Parent-offspring conflict, sibling rivalry, and brood-size patterns in plants. *Annu. Rev. Ecol. Syst.* 19:177–205
152. van Noordwijk, A. J., de Jong, G. 1986. Acquisition and allocation of resources: their influence on variation in life history tactics. *Am. Nat.* 128:137–42
153. van Noordwijk, A. J., van Balen, J. H., Scharloo, W. 1981. Genetic and environmental variation in clutch size of the great tit *Parus major*. *Neth. J. Zool.* 31:342–72
154. Waage, J. K. 1986. Family planning in parasitoids: adaptive patterns of progeny and sex allocation. In *Insect Parasitoids*, ed. J. K. Waage, D. J. Greathead, pp. 449–70. London: Academic
155. Waage, J. K., Godfray, H. C. J. 1985. Reproductive strategies and population ecology of insect parasitoids. See Ref. 133, pp. 449–70
156. Weis, A. E., Price, P. W., Lynch, M. 1983. Selective pressures on the clutch size of the gall maker *Asteromyia carbonifera*. *Ecology* 64:688–95
157. Werren, J. H. 1980. Sex ratio adaptations to local mate competition in a parasitic wasp. *Science* 208:1157–59
158. Williams, G. C. 1966. *Adaptation and Natural Selection*. Princeton: Princeton Univ. Press
159. Williams, G. C. 1971. *Group Selection*. Chicago: Aldine Atherton
160. Winkler, D. W. 1985. Factors determining a clutch size reduction in california gulls *(Larus californicus):* a multi-hypothesis approach. *Evolution* 39:667–677
161. Winkler, D. W., Wallin, K. 1987. Offspring size and number: a life history model linking effort per offspring and total effort. *Am. Nat.* 129:708–20
162. Wynne-Edwards, V. C. 1962. *Animal Dispersion in Relation to Social Behaviour*, Edinburgh: Oliver & Boyd

Annu. Rev. Ecol. Syst. 1991. 22:431–46

INTERACTIONS BETWEEN WOODY PLANTS AND BROWSING MAMMALS MEDIATED BY SECONDARY METABOLITES

John P. Bryant

Institute of Arctic Biology, University of Alaska, Fairbanks, Alaska 99775-0180

Frederick D. Provenza

Department of Range Science, Utah State University, Logan, Utah 84322-5230

John Pastor

Natural Resource Research Institute, University of Minnesota, Duluth, Minnesota 55811

Paul B. Reichardt and Thomas P. Clausen

Department of Chemistry, University of Alaska, Fairbanks, Alaska 99775-0180

Johan T. du Toit

Department of Biological Sciences, University of Zimbabwe, Harare, Zimbabwe

KEY WORDS: secondary metabolite, mammals, chemical defense, toxicity, community, ecosystem, learning

---

INTRODUCTION

Mammals must overcome several challenges to exploit woody plants; these include variation among plant species, individuals, growth stages, and parts in their nutritional value and mechanical and chemical defenses (91). The latter are especially significant because woody plants produce a variety of secondary metabolites (56), many of which are chemical defenses against browsing by mammals and some of which appear to be an evolutionary

0066-4162/91/1120-0431$02.00

response to browsing by mammals (12a). Not all secondary metabolites are equally effective as defenses against browsing, and none provides complete protection (98), because mammals have evolved anatomical, physiological, and behavioral counters to plant defenses (27, 61, 91). As a result, some woody plants are browsed more than others.

The effects of secondary metabolites occur at different hierarchical scales, ranging from individual plants to ecosystems. There are recent reviews of the evolutionary and environmental controls of chemical defenses against browsing, and of the physiological and biochemical mechanisms mammals use to counter these defenses (61, 81). This review concerns three aspects of chemically mediated interactions between woody plants and browsing mammals: (i) chemical specificity and toxicity, (ii) learning as a counter to chemical defense, and (iii) the effects of chemical defense on plant communities and ecosystems.

MODES OF CHEMICAL DEFENSE AGAINST MAMMALS

Specificity of Chemical Defenses Against Mammals

Until about a decade ago, knowledge of the chemical defenses of plants against mammals was limited to an understanding that the concentrations of several classes of secondary metabolites (e.g. resins, phenolics, tannins, alkaloids) were often inversely correlated with the use of plants by herbivores (11, 34, 102, 119). The potential role of toxins in the defense of plants against mammals was also recognized (34), and correlative evidence existed for the defensive roles of some individual plant metabolites (55, 103).

During the last decade, chemists have furthered our understanding of phytochemical defenses against mammals in two ways (98). First, they have identified specific chemical substances, rather than general classes of substances, that are responsible for a mammal's response. Second, they have linked avoidance or rejection of a plant to: (i) some fundamental nutritional deficiency in the plant, or (ii) phytochemicals that are unpalatable and adversely affect the mammal's physiology. This chemically oriented approach has increased interest in individual substances that mediate plant-mammal interactions.

One of the most intensively studied chemically mediated interactions involves snowshoe hares *(Lepus americanus)* and winter-dormant woody plants (9, 11, 12, 61, 110). In winter, snowshoe hares discriminate among woody plants based on species, growth forms (evergreen vs deciduous), developmental stages (juvenile vs adult), and plant parts. Much of this discrimination is related to secondary metabolites, and boreal woody plants contain a diversity of low molecular weight metabolites that serve as "antifeedents." Individual monoterpenes (97, 110), triterpenes (96, 98), and phenols (19, 52) deter feeding by hares, as do other substances of unknown biosynthetic origins (95). Three generalities arise from examining the relationship between hares and secondary metabolites in woody plants.

The first generalization is that phytochemicals belonging to similar biosynthetic classes do not necessarily have similar activities. For example, camphor contributes to the defense of white spruce *(Picea glauca),* but the structurally related monoterpene bornyl acetate does not (110). Similarly, Reichardt et al (97) identified six monoterpenes in the buds of balsam poplar *(Populus balsamifera),* of which only cineol was significantly unpalatable to hares. Analogue studies confirm the relationship between chemical structure and activity for defensive chemicals. For example, pinosylvin is a strong feeding deterrent, pinosylvin methyl ether is effective but less potent, and pinosylvin dimethyl ether is virtually inactive (19). Similarly, 2,4,6-trihydroxydihydrochalcone in juvenile balsam popular deters feeding by hares (52), but the structurally similar pinostrobin in green alder *(Alnus crispa)* does not (19).

Another generalization concerning the chemical defenses of woody plants is that deterrence varies with the concentrations and potencies of individual metabolites, but the situation within a plant can be quite complex. For example, Jogia et al (52) consider differences in the concentrations of 2,4,6-trihydroxydihydrochalcone to be primarily responsible for the low palatability to hares of juvenile as compared to mature balsam poplar. Reichardt et al (97), on the other hand, argue that the difference in palatability is largely due to different concentrations of salicaldehyde and 6-hydroxycyclohexenone in the two growth stages. These two studies suggest that the chemical defense of juvenile poplar against hares involves different phytochemicals and the additive and (or) synergistic effects of different metabolites.

The final generalization is that the chemical defenses of woody plants vary by growth stage and by plant parts within growth stages. For example, the internodes of the juvenile and adult stages of Alaska paper birch *(Betula resinifera)* differ qualitatively in chemical defenses, as do the chemical defenses of buds and internodes (96). Hares reject balsam poplar buds because of the presence of high levels of metabolites that are not even detectable in the internodes (97). Similarly, hares reject green alder buds because of high concentrations of metabolites that are found only in low concentrations in the internodes (19).

Dynamic Aspects of Chemical Defense

Although it has long been recognized that disruption of plant tissue by herbivores during ingestion may release harmful substances (e.g. cyanogenic glycosides; 23), the concept of "dynamic defenses" has not been extensively applied to mammals. However, recent reports indicate that dynamic defenses against mammals may be common (21, 97). For example, phenol glycosides are transformed enzymatically to substances (trichcoparpogenin and 6-hydroxycyclohexenone) that deter feeding by hares on quaking aspen *(Populus tremuloides)* and balsam poplar (97).

Relationship of Feeding Deterrence to Antibiosis

Chemical defenses have been classified as either toxins or generalized digestion inhibitors, although some overlap between the categories has been recognized (32, 100). There is growing evidence, however, that food selection and ingestion are regulated by toxins rather than by inhibition of protein or carbohydrate digestion. Studies of domestic and wild mammals fed unpalatable browse normally available to them, or artificial diets treated with extracts from this browse, confirm the importance of toxicity. For instance, condensed tannins supposedly limit intake by inhibiting ruminant digestion (102, 119), but deterrence of goat browsing is associated with toxicity, not digestion inhibition (27, 93). Indeed, most phytotoxins deter feeding by domestic mammals (93, 114). Moreover, snowshoe hares (95–97), microtine rodents (3, 53), and bushy tailed woodrats *(Neotoma lepida)* (74) voluntarily reduce food intake to well below maintenance when fed browse containing high concentrations of secondary metabolites or when fed artificial diets treated with extracts from this browse. Creosote brush *(Larrea tridentata)* resin that deters feeding by woodrats forms a complex with protein in vitro (100), but it does not affect protein digestion by woodrats that normally eat creosote brush (75). Hares that eat birch and evergreen conifers containing high concentrations of feeding deterrents lose more nitrogen in feces (96, 109), but the associated large losses of sodium and nitrogen in urine indicate that detoxification increased and that renal function was disrupted (86, 87, 96). Finally, essentially all known feeding deterrents extracted from food that could be browsed by snowshoe hares are lipid soluble low molecular weight substances that are toxic to mammals, insects, and microbes (Table 1).

Tannins as Defenses Against Mammals

Tannins are generally defined as water-soluble, high molecular weight (>500 amu) polyphenols capable of precipitating proteins. Only condensed and hydrolyzable tannins were considered in ecological studies at the beginning of the last decade, because tanning quantity was considered more important than tannin quality (32, 100). However, the importance of low molecular weight metabolites as chemical defenses against mammals is changing the view of tannins.

It is questionable that tannins deter browsing primarily by inhibiting digestion of proteins or carbohydrates. Instead, mammals may reject tannin-containing plants because they cause internal malaise (27, 92). Condensed

Table 1 Toxicity to mammals, insects, and microbes of substances that deter feeding by snowshoe hares (Numbers in parentheses are references)

Compound	Deterrent to mammals	Toxic to mammals	Toxic to insects	Toxic to microbes
Cineole	Yes (97)	Yes (43)	Yes (25)	Yes (1)
Alpha Bisabolol	Yes (97)	Yes (41)	Yes (2)	Yes (29)
Pinosylvin	Yes (19)	Yes (35)	Yes (123)	Yes (35)
Pinosylvin Methyl Ether	Yes (19)	?	Yes (123)	Yes (35)
Germaerone	Yes (95)	?	?	?
Campher	Yes (110)	Yes (99)	?	Yes (24)
Papyriferic Acid	Yes (96)	Yes (19, 96)	?	?
Condensed Tannins	Yes (20, 92)	?	? (5)	Yes (122)

tannins, particularly procyanidins (46), readily depolymerize under acidic conditions similar to those in portions of the digestive tract of most mammalian herbivores (16, 21). Hence, some condensed tannins may be toxic (16, 58, 70). In addition, hydrolyzable tannins in some plant species cause severe necrosis and ulceration of the epithelium of the esophagus, stomach, intestines, and proximal renal tubules (28, 70).

It is also clear that the effectiveness of condensed tannins as deterrents to browsing depends on tannin structure (20). Tannin structures are now routinely elucidated, and tannins are readily isolated even on a large scale (92). As a result, the use of commercially available tannins for bioassays, or as standards for analytical procedures, is being replaced by partially characterized tannins isolated from plants of interest (42).

LEARNING AS A MAMMALIAN COUNTER TO CHEMICAL DEFENSE

Evolutionary Considerations

Some aspects of plant defense theory assume that mammals have the innate ability to detect secondary metabolites and to avoid intoxification, but there is increasing evidence that mammals learn to avoid poisoning. Mammalian herbivores are generally long lived, many are social, their home ranges frequently encompass entire landscapes, and their diets are commonly catholic (108). Such conditions should confer a selective advantage to individuals whose diet selection is flexible (6, 90, 91). Learning provides the flexibility necessary for mammalian herbivores to maintain homeostasis in environments where the nutrient content and toxicity of potential foods change temporally and spatially. While some nutritious foods do not contain toxins, nearly all woody species contain potentially toxic secondary metabolites whose concentrations vary with environment, season, plant developmental phase, plant physiological age, and plant part (9, 12, 69).

In what follows, we must refer to the results of studies involving domestic herbivores such as sheep and goats because the role of learning in diet selection for most mammalian species is not known. While learning in diet selection has not been studied in many mammals, the mechanisms that different species of mammals use to learn about foods should be qualitatively similar (37, 90, 91) and are likely important in the feeding behavior of wild mammalian herbivores (61). From a theoretical standpoint, evolutionary processes create a match between environmental exigencies and behavior such as foraging (94, 111). Even in domestic herbivores the match should still exist because the behavioral changes brought about through domestication are more quantitative than qualitative (89).

Learning from Postingestive Consequences

One way animals can learn about foods is through postingestive feedback from nutrients and toxins. Animals learn about the postingestive consequences of foods through two interrelated systems, affective and cognitive (36). Taste plays an essential role in both systems. The affective system integrates the taste of food and its postingestive consequences, and changes the amount of food the animal will ingest, depending on whether the postingestive consequences are aversive or positive. Thus, the affective system provides feedback so mammals can learn to ingest nutritious foods and to avoid intoxification. The cognitive system integrates the taste of food with its odor and sight. Mammals use the senses of smell and sight to select or to avoid particular foods.

Research on conditioned food preferences and aversions shows that voluntary intake depends on postingestive consequences. If the food is nutritious (positive postingestive consequences), intake of the food increases. If toxicity ensues (aversive postingestive consequences), intake of the food is limited. For example, bitter and sweet are often considered feeding deterrents and attractants, respectively. However, intake of bitter substances increases when they are paired with positive postingestive consequences, and intake of sweet substances decreases when they are paired with aversive postingestive consequences (38).

Mammalian herbivores apparently learn to prefer foods with positive postingestive consequences, as demonstrated by lambs' strong preference for nonnutritive flavors paired with glucose over nonnutritive flavors paired with the nonnutritive sweetener saccharine (15). The amount ingested by other species of mammals is also increased by pairing nonnutritive foods or flavors with calories, recovery from nutritional deficiencies, and recovery from postingestive distress (93).

Animals also regulate their intake of potentially toxic foods by associating the taste of the foods with aversive postingestive consequences (36, 93). For example, current season's growth (CSG) of blackbrush *(Coleogyne ramosissima)* is rich in condensed tannin, while older growth (OG) is not. In less than four hours, goats switch from a diet largely composed of CSG to a diet largely composed of OG. Aversive feedback from condensed tannin in CSG causes this rapid dietary shift (27, 92). When purified condensed tannin from CSG is incorporated into OG pellets, goats eat until they experience aversive feedback, which sets an upper limit to the mass of OG that goats will eat (27, 92). In these studies, the taste of condensed tannin per se did not regulate goat feeding. Rather, goats associate the taste of the food (i.e. CSG or OG) with the degree of postingestive malaise caused by condensed tannin to determine how much to eat.

Odor is an important signal mammals use to avoid foods that elicit negative feedback (93), and many of the secondary metabolites that deter feeding by snowshoe hares in winter are volatile even at low temperatures (95–97, 110; Table 1). However, mammals require taste as well as odor to assess the quality of food (36). For example, animals that become ill after only smelling a novel food will later ingest that food, unless they have also tasted the food. Persistent sampling of foods, which is characteristic of mammalian herbivores (34), lets mammals determine when the association between odor, taste, and food quality have changed (60).

Mammalian herbivores have physiological and biochemical mechanisms to detoxify secondary metabolites (61, 67), and if the capacity of these detoxification systems is exceeded mammals becomes ill and may die. However, mammals usually adjust intake to avoid intoxification. To do so, they must sample foods to determine when the concentrations of nutrients and toxins change as a result of growth processes and previous herbivory (9). Sheep (13, 31, 60, 116), goats (27, 92), and cattle (88) sample foods and regulate their intake of nutritious plants that contain toxins. If toxicity decreases, the taste of the plant is no longer paired with aversive postingestive consequences. Any nutritional value that plant provides will constitute positive feedback, and subsequently cause intake of the plant to increase. In contrast, intake decreases as the toxicity of the plant increases.

Learning from Mother

Learning by trial and error, based solely on postingestive feedback, is one way to determine which foods to eat and which foods to avoid, but it can be inefficient and risky (6, 90, 91, 93). This could provide additional selective pressure for herbivores that feed in mixed-generation groups to rely on social learning, where information is passed from experienced to inexperienced foragers, generally from a mother to her offspring.

Learning from mother increases learning efficiency. For example, if a ewe has learned to avoid a food that causes postingestive distress, its lamb also learns to avoid that food much sooner than a lamb reared without its mother (13, 76–78). Likewise, a lamb learns to eat food preferred by its mother much sooner than a lamb not reared with its mother (64, 76–78, 117). Such socially mediated feeding behavior leads to foraging traditions (48, 54, 63, 101).

Learning from social models also decreases the risks of trial and error learning, provided young mammals remember what they ate and sample novel foods cautiously. Young mammalian herbivores can remember, for at least one to three years, specific foods with either aversive (14, 27, 59) or positive (27, 39, 113) postingestive consequences. They identify novel foods and sample them cautiously (13, 92, 116). Animals that experience either unpleasant or pronounced positive postingestive consequences attribute those consequences to the novel food, even when it has eaten several familiar foods and the consequences occur six to eight hours after ingestion (13, 15, 92).

EFFECTS OF SELECTIVE BROWSING ON COMMUNITIES AND ECOSYSTEMS

Differences in the chemical defenses of woody species, and in the abilities of mammalian herbivores to learn to distinguish among food items with different levels of defenses, have consequences for plant communities and ecosystems.

Consequences of Functional Responses to Toxin Satiation

Selective browsing can change the composition of woody plant communities (62). In biomes as diverse as boreal forests (8, 84, 85), arid shrublands of western North America (114), African savannas (49), and the tropical woodlands of Sri Lanka (Ceylon) (80), selective browsing increases the abundance of unpalatable species. These unpalatable species are often heavily defended chemically (8). For example, in the boreal forest of Isle Royale National Park (57, 68, 112) and the Alaskan taiga (65), browsing of poorly defended deciduous species such as willow *(Salix spp.)*, quaking aspen, and birch by snowshoe hare, beaver *(Castor canadensis)*, and moose *(Alces alces)* favors domination by more chemically defended evergreens such as white spruce and Labrador tea *(Ledum groenlandicum)* (11, 95, 109).

Selective browsing does not always favor dominance by chemically defended species, however, because browsing intensity depends on the relative abundance of the woody species in the vegetation. On Isle Royale (7) and in Newfoundland (4), browsing by moose removes balsam fir *(Abies balsamea)* from areas where it is not abundant (7), even though balsam fir is rich in terpenes that are potentially toxic to mammals (121). In contrast, in areas where fir density is high, enough young fir escape to ensure that it persists.

Similarly, although chemicals defend green alder against browsing by boreal mammals (11, 19), snowshoe hares browse green alder heavily when it grows in areas dominated by palatable species such as feltleaf willow *(Salix alaxensis)* (10); they browse it lightly in areas where it is moderately abundant. More intense browsing of a woody species when it is rare has been reported for other biomes as well (45).

Bryant & Fox (10) suggest that the inverse relationship between relative abundance and browsing intensity represents a functional response of a predator to its prey (47). A functional response assumes that at some prey biomass predators become satiated and allow some prey to escape. When applied to interactions between woody plants and mammals, the functional response relates the percent of the plant's biomass eaten (browsing intensity) to the relative biomass of the plant in the vegetation. Such a graph shows that a woody species is more likely to be browsed severely when it is rare than when it is abundant.

Intoxification apparently causes the functional response (10). When a woody species is so rare that its mass of toxins does not exceed the detoxification capacity of the browsing mammals, all of the individuals of the species can be eaten. If the biomass of the species exceeds the detoxification capacity of its predators, however, the average browsing intensity decreases. Thus, because the satiation biomass of the heavily defended species is less than that of poorly defended species, selective browsing is more likely to favor an increase in the abundance of a heavily defended species. This increase is less likely to occur if the heavily defended species is rare.

Effects of Selective Browsing on Nutrient Cycling in Ecosystems

The link between tissue secondary chemistry, selective foraging by mammals, and the abilities of woody species to survive selective browsing has important consequences for nutrient cycling in ecosystems. These consequences arise for two reasons. First, forage selection and litter decomposition are determined not only by nutrients, structural carbohydrates, and lignins, but also by secondary metabolites (8, 33, 72, 73, 82, 83, 85, 115); broad spectrum phytotoxins that affect mammals are also toxic to microbes (Table 1). Second, the chemical defenses of many woody species are directly correlated with their ability to tolerate nutrient stress (9, 22).

The selective foraging of mammals involves both positive and negative feedbacks in ecosystems. Positive feedbacks occur when selective foraging causes the rate of nutrient cycling to increase or decrease. This is especially true of nitrogen because its mineralization limits the net primary production in many terrestrial ecosystems (120), and the nitrogen content of plant tissues may determine herbivore feeding rates (66), as well as the kinds and amounts of chemical defenses (9). Negative feedbacks occur when physical disturbances (e.g. wildfires or outbreaks of pathogens or specialized insects) destroy woody plant species that are not browsed by mammals. Negative feedbacks appear to be exceptions to the correlation of feeding preferences, tissue chemistry, and litter decay rates.

Invasion and ultimate dominance by unpalatable, slowly growing species is apparently a more common response of woody vegetation to selective browsing (8, 82–85). The limited evidence for this hypothesis comes primarily from boreal forests and south African savanna-woodlands. On Isle Royale in Michigan and in interior Alaska, browsing by moose on willow, aspen, birch, and other palatable hardwoods increases the abundance of white spruce (57, 65, 68, 112). White spruce is a poor quality food for moose and other boreal mammals (11); it is also a low quality litter (33, 79, 82, 83, 85). The same chemical properties that reduce its food quality also reduce its value for soil microbes; those properties include high concentrations of toxic secondary metabolites, lignins, waxes, and cutins that are indigestible by mammals (102) and slow to decompose (115), and a low concentration of nitrogen (33). Likewise, in the dystrophic savanna woodlands of southern Africa, browsing by mammals apparently influences vegetation composition in the same way and may have similar effects on nutrient cycling (30).

Palatable species are browsed more intensively as chemically defended species become dominant, primarily because they represent a declining supply of good food (8, 83), and secondarily because of the functional response described above. Thus begins a decrease in nutrient cycling that leads to the establishment of pure stands of chemically defended species beneath which nitrogen availability is low (8, 30, 82–85). This sequence of events has been confirmed in exclosures on Isle Royale National Park in Minnesota (83, 85) and is suggested by the reduction in soil nutrients outside exclosures in African savanna ecosystems with high populations of mammalian browsers (44). It also occurs in exclosures in central Washington (118).

Soil fertility declines as the biomass of unpalatable species such as spruce comes to contain increasing amounts of nutrients, and as decomposition declines because of low litter quality (8, 82–85). Slowly growing species such as spruce, however, can persist in nutrient-deficient soils (17, 40), and a decline in soil fertility associated with increasing abundance of such species may actually enhance their dominance relative to species that require more nutrients (18).

In contrast to ecosystems dominated by woody plants, selective foraging can enhance nutrient cycles in grassland ecosystems such as the Serengeti of tropical East Africa and the Great Plains of temperate North America. In these ecosystems, grazing by mammals can stimulate grass regrowth and the rapid recycling of nutrients in urine, feces, and carrion (26, 71, 104, 106, 107). In the Hudson Bay lowlands, grazing by lesser snow geese also accelerates nutrient cycling and the growth of preferred graminoids through fecal deposition (50, 51, 105).

In grasslands, the enhancement of nutrient cycles through deposition of feces and urine depends on continued grazing of preferred species. In contrast, reduced nutrient cycling and decreased ecosystem productivity following the invasion of unpalatable woody species can continue even after browsers abandon an area, if the invading woody species have low nutrient demands (8, 18, 83). Moreover, selective browsing of less palatable species is unlikely to change the dominance of species that are heavily defended against browsing. Palatable species will proliferate, and ecosystem productivity will increase, only following negative feedback from physical disturbances (e.g. wildfires or outbreaks of pathogens or specialized insects) that destroy species that are not browsed.

CONCLUSIONS

Woody plants apparently rely on a chemically diverse array of specific secondary metabolites (rather than classes of secondary metabolites) as defenses against browsing by mammals. The effectiveness of these substances is primarily due to toxicity rather than digestion inhibition.

A mammal's most flexible counter to phytotoxins is to learn to avoid intoxification through trial and error and from social models like mother. Through learning, mammals can determine which foods to eat and which to avoid, and how much food they can safely ingest. The amount of each food they can safely ingest is usually lower than the mammal's maintenance food requirement, so browsing mammals usually starve before they eat enough browse to be poisoned.

The limits of food intake set by toxins suggest that interactions between woody plants and mammals can be analyzed as a functional response of a predator to its prey; the mammal is satiated when it is intoxified by the prey's chemical defenses. According to this hypothesis, a woody species should be most intensively browsed when it is rare. Heavily defended species should be more likely than poorly defended species to increase, because the detoxification capacity of mammals would be satiated sooner by heavily defended species. However, browsing does not always result in dominance by heavily defended species. If the biomass of a heavily defended species is too low to satiate the detoxification systems of mammals, browsing can eliminate the species.

Selective browsing by mammals can greatly affect the composition of plant communities. Browsing usually favors an increase in unpalatable species that are heavily defended chemically. Moreover, once selective browsing encourages invasion by chemically defended species, invasion is likely to continue because the low nutrient requirements of these woody species let them persist in a nutrient-deficient environment.

Selective browsing can slow nutrient cycling in ecosystems dominated by woody plants, but it can accelerate nutrient cycling in ecosystems dominated by graminoids. A major reason for this difference is that woody plants have more chemical defenses than grasses do against mammalian herbivores. The chemical defenses that decrease the quality of woody species as food also make their litter poor quality as a substrate for decomposers.

ACKNOWLEDGMENTS

We thank the National Science Foundation for grants to J. P. Bryant (BSR-870262 for long-term ecological research in the Alaskan tiaga), F. D. Provenza (BSR 8614856), and J. Pastor (BSR 8817665, BSR 8906843).

Literature Cited

1. Agawal, I., Mathela, C. S. 1979. Study of antifungal activity of some terpenoids. *Indian Drugs Pharm. Ind.* 14:19–21
2. Bar-Zeev, M. 1980. Studies of repellents against Panstrogylus megistus (Hemiptera: Reduviidae) in Brazil. *J. Med. Entomol.* 17:70–74
3. Batzli, G. O., Jung, H. G. 1980. Nutritional ecology of microtine rodents: Resource utilization near Atkasook, Alaska, *Arctic Alpine Res.* 12:483–99
4. Bergerud, A. T., Manuel, F. 1968. Moose damage to balsam fir-white birch forests in central Newfoundland. *J. Wildl. Manage.* 32:729–46
5. Bernays, E. A., Cooper-Driver, G., Bilgener, M. 1989. Herbivores and plant tannins. *Adv. Ecol. Res.* 19:263–302
6. Boyd, R., Richardson, P. J. 1985. *Culture and the Evolutionary Process*. Chicago: Univ. Chicago Press
7. Brandner, T. A., Peterson, R. O., Risenhoover, K. L. 1990. Balsam fir on Isle Royale: effects of moose herbivory and population density. *Ecology* 71: 155–64
8. Bryant, J. P., Chapin, F. S. III. 1986. Browsing-woody plant interactions during boreal forest plant succession. In *Forest Ecosystems in the Alaskan Taiga*, ed. K. Van Cleve, F. S. Chapin, III, P. W. Flanagan, L. A. Viereck, and C. T. Dyrness, pp. 313–25. New York: Springer-Verlag
9. Bryant, J. P., Chapin, F. S. III, Klein, D. R. 1983. Carbon/nutrient balance of boreal plants in relation to vertebrate herbivory. *Oikos* 40:357–68
10. Bryant, J. P., Fox, J. F. 1991. *Effects of browsing by mammals on the species composition of woody vegetation: importance of chemical defenses to functional responses by browsing mammals*. Manuscript
11. Bryant, J. P., Kuropat, P. J. 1980. Selection of winter forage by subarctic browsing vertebrates: The role of plant chemistry, *Annu. Rev. Ecol. Syst.* 11:261–85
12. Bryant, J. P., Kuropat, P. J., Reichardt, P. B., Clausen, T. P. 1991. Controls over allocation of resources by woody plants to chemical antiherbivore defense. In *Plant Defenses Against Mammalian Herbivory,* ed. T. Palo, C. Robbins. Boca Raton: CRC. In press

12a. Bryant, J. P., Reichardt, P. B., Clausen, T. P., Provenza, F. D., Kuropat, P. J. 1991. Woody plant-mammal interactions. In *Herbivores: Their Interaction with Plant Metabolites,* vol. 2, ed. G. A. Rosenthal and M. R. Berenbaum. New York: Academic. In press

13. Burritt, E. A., Provenza, F. D. 1989. Food aversion learning: Ability of lambs to distinguish safe from harmful foods. *J. Anim. Sci.* 67:1732–39
14. Burritt, E. A., Provenza, F. D. 1990. Food aversion learning in sheep: persistence of conditioned taste aversions to palatable shrubs (Cercocarpus montanus and Amelanchier alnifolia). *J. Anim. Sci.* 68:1003–7
15. Burritt, E. A., Provenza, F. D. 1991. Ability of lambs to learn with a delay between food ingestion and con-

sequences given meals containing novel and familiar foods. *Appl. Anim. Behav. Sci.* In press

16. Butler, L. G., Rogler, J. C., Mehansho, H., Carlson, D. M. 1986. Dietary effects of tannins. In *Plant Flavonoids in Biology and Medicine: Biochemical, Pharmacological and Structure-Activity Relationships,* ed. V. Cody, E. Middleton, J. B. Harborne, pp. 141–157. New York: Liss
17. Chapin, F. S. III. 1980. The mineral nutrition of wild plants, *Annu. Rev. Ecol. Syst.* 11:233–60
18. Chapin, F. S. III, Vitousek, P. M., Van Cleve, K. 1986. The nature of nutrient limitation in plant communities. *Am. Nat.* 127:48–58
19. Clausen, T. P., Bryant, J. P., Reichardt, P. B. 1986. Defense of winter-dormant green alder against snowshoe hares. *J. Chem. Ecol.* 12:2117–31
20. Clausen, T. P., Provensa, F. D., Burritt, E. A., Bryant, J. P., Reichardt, P. B. 1990. Ecological implications of condensed tannin structure: a case study. *J. Chem. Ecol.* 16:2381–92
21. Clausen, T. P., Reichardt, P. B., Bryant, J. P., Werner, R. A., Post, K. 1989. Chemical model for short-term induction in quaking aspen (Populus tremuloides) foliage against herbivores. *J. Chem. Ecol.* 15:2335–45
22. Coley, P. D., Bryant, J. P., Chapin, F. S. III. 1985. Resource availability and plant antiherbivore defense. *Science* 230:895–99
23. Conn, E. C. 1979. Cyanide and cyanogenic glycosides. In *Herbivores: Their Interaction With Secondary Plant Metabolites,* ed. G. A. Rosenthal, D. H. Janzen, pp. 387–412. New York: Academic. 718 pp.
24. Cox, P. H., Spanjers, F. 1970. Preparation of sterile implants by compression. *Pharm. Weekbl.* 105:681–84
25. Dassler, H. G., Dube, G. 1957. The action of terpene oxides on different classes of insects. *Anz. Schadlingskunde* 30:86–8
26. Day, T. A., Detling, J. K. 1990. Grassland patch dynamics and herbivore grazing preferences following urine deposition. *Ecology* 71:180–88
27. Distel, R. A., Provenza, F. D. 1990. Experience early in life affects voluntary intake of blackbrush by goats. *J. Chem. Ecol.* 17:431–50
28. Divers, T. J., Crowell, W. A., Duncan, J. R., Whitlock, R. H. 1982. Acute renal disorders in cattle: a retrospective study of 22 cases. *J. Am. Vet. Met. Assoc.* 181:694–99
29. Dull, G. G., Fairley, J. L. Jr., Gottshall, R. V., Lucus, E. H. 1957. Antibacterial substances in seed plants active against tubercle bacilli. IV. The antibacterial sesquiterpenes of Populus tacamahaca. *Antibiotics Annu.* 682:6
30. du Toit, J. T. 1991. Introduction of artificial waterpoints: potential impacts on nutrient cycling. In *Management of the Hwange Ecosystem,* ed. M. Jones, R. Martin. Harare, Zimbabwe: USAID/Zimbabwe Dept. Natl. Parks Wildl. Manage.
31. du Toit, J. T., Provenza, F. D., Nastis, A. S. 1991. Conditioned food aversions: How sick must a ruminant get before it detects toxicity in food? *Appl. Anim. Behav. Sci.* 30:35–46
32. Feeny, P. 1976. Plant apparency and chemical defense. In *Recent Advances in Phytochemistry,* Vol. 10. *Biochemical Interactions Between Plants and Insects,* ed. J. W. Wallace, R. L. Mansell, 10:1–40. New York: Plenum. 425 pp.
33. Flanagan, P. W., Van Cleve, K. 1983. Nutrient cycling in relation to decomposition of organic matter quality in taiga ecosystems. *Can. J. For. Res.* 13:795–817
34. Freeland, W. J., Janzen, D. H. 1974. Strategies in herbivory by mammals: the role of plant secondary compounds. *Am. Nat.* 108:269–89
35. Frykholm, K. O. 1945. Bacteriological studies of pinosylvin, its monomethyl ether and dimethyl ethers, and toxicological studies of pinosylvin. *Nature* 155:454–55
36. Garcia, J. 1989. Food for Tolman: cognition and cathexis in concert. In *Aversion, Avoidance and Anxiety,* ed. T. Archer and L. Nilsson, pp. 45–85. Hillsdale, NJ: Erlbaum
37. Garcia, J., Hankins, W. G., Coil, J. D. 1977. Koalas, men, and other conditioned gastronomes. In *Food Aversion Learning,* ed. N. W. Milgram, L. Krames, T. M. Alloway, pp. 195–92. New York: Plenum
38. Garcia, J., Holder, M. D. 1985. Time, space and value. *Hum. Neurobiol.* 4:81–89
39. Green, G. C., Elwin, R. L., Mottershead, B. E., Keogh, R. G., Lynch, J. J. 1984. Long-term effects of early experience to supplementary feeding in sheep. *Proc. Aust. Soc. Anim. Prod.* 15:373–75
40. Grime, J. P. 1977. Evidence for the existence of three primary strategies in plants and its relevance to ecological and evolutionary theory. *Am. Nat.* 111:1169–94
41. Habersang, S., Leuschner, F., Issac,

O., Thiemer, K. 1979. Pharmacological studies of chamomile constituents. IV. Studies on the toxicity of (–)-a-bisabolol. *Planta Medica* 37:115–23

42. Haggerman, A. E., Butler, L. G. 1989. Choosing appropriate methods and standards for assaying tannin. *J. Chem. Ecol.* 15:1795–1810
43. Haley, T. J. 1982. Cineole (1,8-cineole). *Dangerous Prop. Ind. Mater. Rep.* 2:10–12
44. Hatton, J. C., Smart, N. O. E. 1984. The effect of long-term exclusion of large herbivores on soil nutrient status in Murchison Falls National Park, Uganda. *Afr. J. Ecol.* 22:23–30
45. Heady, F. H. 1964. Palatability of herbage and animal preference. *J. Range. Manage.* 17:76–82
46. Hemingway, R. W., McGraw, G. W. 1983. Kinetics of acid-catalyzed cleavage of procyanidins. *J. Wood. Chem. Technol.* 3:421–35
47. Holling, C. S. 1959. The components of predation as revealed by a study of small-mammal predation of the European pine sawfly. *Can. Entomol.* 91: 293–320
48. Hunter, R. F., Milner, C. 1963. The behavior of individual, related and groups of south country Cheviot hill sheep. *Anim. Behav.* 11:507–513
49. Jachman, H., Croes, T. 1991. Effects of browsing by elephants on the Combretum/Terminalia woodland at the Nazinga Game Ranch, Burkina Faso. *Biol. Conserv.* In press
50. Jefferies, R. L. 1988. Vegetational mosaics, plant-animal interactions and resources for plant growth. In *Plant Evolutionary Biology,* ed. L. D. Gottlieb, S. K. Jain, pp. 341–69. London: Chapman & Hall
51. Jefferies, R. L. 1989. Pattern and process in arctic coastal vegetation in response to foraging by Lesser Snow Geese. In *Plant Form and Vegetation Structure, Adaptation, Plasticity and Relation to Herbivory,* ed. M. J. A. Werger, pp. 1–20. The Hague: SPB Academic Publ.
52. Jogia, M. K., Sinclair, A. R. E., Anderson, R. J. 1989. An antifeedent in balsam poplar inhibits browsing by snowshoe hares. *Oecologia* (Berl.) 79: 189–92
53. Jung, H. G., Batzli, G. O. 1981. Nutritional ecology of microtine rodents: Effects of plant extracts in the growth of Arctic microtines. *J. Mamm.* 62:286–92
54. Key, C., MacIver, R. M. 1980. The effects of maternal influences on sheep: breed differences in grazing, resting and courtship behavior. *Appl. Anim. Ethol.* 6:33–48
55. Kingsbury, J. M. 1964. *Poisonous Plants of the United States and Canada.* Englewood Cliffs, NJ: Prentice-Hall
56. Kramer, P. J., Kozlowski, T. T. 1979. *Physiology of Woody Plants.* New York: Academic. 811 pp.
57. Krefting, L. W. 1974. *The Ecology of Isle Royale Moose. Univ. Minn. Agric. Exp. Stn. Tech. Bull. 297, Forestry Series 15.* 75 pp.
58. Kumar, R., Singh, M. 1984. Tannins: their adverse role in ruminant nutrition. *J. Agric. Food Chem.* 32:447–53
59. Lane, M. A., Ralphs, M. H., Olsen, J. D., Provenza, F. D., Pfister, J. A. 1990. Conditioned taste aversion: Potential for reducing cattle loss to larkspur. *J. Range Manage.* 43:127–31
60. Launchbaugh, K. L., Provenza, F. D. 1991. Can plants practice mimicry to avoid mammalian herbivory? *Am. Nat.* Submitted
61. Lindroth, R. L. 1988. Adaptations of mammalian herbivores to plant chemical defenses. In *Chemical Mediation of Coevolution,* ed. K. C. Spencer, pp. 415–45. New York: Academic. 609 pp.
62. Lindroth, R. L. 1989. Mammalian herbivore-plant interactions. In *Plant-Animal Interactions,* ed. W. G. Abrahamson, pp. 163–204. New York: McGraw-Hill. 480 pp.
63. Lynch, J. J. 1987. The transmission from generation to generation in sheep of the learned behaviour for eating grain supplements. *Aust. Vet. J.* 64:291–292
64. Lynch, J. J., Keogh, R. G., Elwin, R. L., Green, G. C., Mottershead, B. E. 1983. Effects of early experience on the post-weaning acceptance of whole grain wheat by fine-wool Merino lambs. *Anim. Prod.* 36:175–183
65. MacAvinchey, R. J. P. 1991. *Winter herbivory by snowshoe hares and moose as a process affecting primary succession on an Alaskan floodplain.* Thesis. Univ. Alaska, Fairbanks. 69 pp.
66. Mattson, W. J. Jr. 1980. Herbivory in relation to plant nitrogen content. *Annu. Rev. Ecol. Syst.* 11:119–61
67. McArthur, C., Haggerman, A. E., Robbins, C. T. 1991. Physiological strategies of mammalian herbivores against plant defenses. In *Plant Defenses Against Mammalian Herbivory,* ed. T. Palo, C. Robbins. Boca Raton: CRC. In press
68. McInnes, P. F., Naimen, R. J., Pastor, J., Cohen, Y. 1991. *Effects of moose browsing on vegetation and litterfall of*

the boreal forest, Isle Royale, Michigan, USA. Manuscript

69. McKey, D. 1979. The distribution of secondary compounds within plants. In *Herbivores: Their Interaction With Secondary Plant Metabolites,* ed. G. A. Rosenthal, D. H. Janzen, pp. 55–133. New York: Academic
70. McLeod, M. N. 1974. Plant tannins—their role in forage quality. *Nutr. Abstr. Rev*. 44:803–15
71. McNaughton, S. J., Ruess, R. W., Seagle, S. W. 1988. Large mammals and process dynamics in African ecosystems. *BioScience* 38:794–800
72. Meetemeyer, V. 1978. Macroclimate and lignin control of litter decomposition. *Ecology* 59:465–72
73. Melillo, J. M., Aber, J. D., Muratore, J. F. 1982. Nitrogen and lignin control of hardwood leaf litter decomposition dynamics. *Ecology* 63:621–26
74. Meyer, M. W., Karazov, W. H. 1989. Antiherbivore chemistry of *Larrea tridentata:* Effects of woodrat *(Neotoma lepida)* feeding and nutrition. *Ecology* 70:953–61
75. Meyer, M. W., Karazov, W. H. 1991. Deserts. In *Plant Defenses Against Mammalian Herbivory,* ed. T. Palo, C. Robbins. Boca Raton: CRC. In press
76. Mirza, S. N., Provenza, F. D. 1990. Preference of the mother affects selection and avoidance of foods by lambs differing in age. *Appl. Anim. Behav. Sci*. 28:255–63
77. Mirza, S. N., Provenza, F. D. 1991. Effects of age and conditions of exposure on maternally mediated food selection in lambs. *Appl. Anim. Behav. Sci*. In press
78. Mirza, S. N., Provenza, F. D. 1991. Socially-induced food avoidance in lambs: Direct or indirect maternal influence? *Appl. Anim. Behav. Sci*. In press
79. Moore, T. R. 1984. Litter decomposition in a subarctic spruce-lichen woodland, eastern Canada. *Ecology* 65:299–308
80. Müller-Dombois, D. 1972. Crown distortion and elephant distribution in the woody vegetation of Ruhuna National Park, Ceylon. *Ecology* 53:208–26
81. Palo, T., Robbins, C. 1991. *Plant Defenses Against Mammalian Herbivory*. Boca Raton: CRC. In press
82. Pastor, J., Aber, J. D., McClaugherty, C. A., Melillo, J. M. 1984. Aboveground production and N and P cycling along a nitrogen mineralization gradient on Blackhawk Island, Wisconsin. *Ecology* 65:256–68
83. Pastor, J., Dewey, B., Naimen, R. J., McInnes, P. F., Cohen, Y. 1991. *Moose browsing and soil fertility in the boreal forests of Isle Royale National Park*. Manuscript
84. Pastor, J., Naimem, R. J. 1991. *Selective foraging and ecosystem processes in boreal forests*. Manuscript
85. Pastor, J., Naimen, R. J., Dewey, B., McInnes, P. 1988. Moose, microbes, and the boreal forest. *BioScience* 38:770–77
86. Pehrson, A. 1983. Digestibility and retention of food components in caged mountain hares (*Lepus timidus* L.) during the winter. *Holarctic Ecol*. 6:395–403
87. Pehrson, A. 1987. Maximal winter browse intake in captive mountain hares. *Finn. Game Res*. 41:47–55
88. Pfister, J. A., Provenza, F. D., Manners, G. D. 1990. Ingestion of tall larkspur by cattle: separating the effects of flavor from post-ingestive consequences. *J. Chem. Ecol*. 16:1697–1705
89. Price, E. O. 1984. Behavioral aspects of animal domestication. *Q. Rev. Biol*. 59:1–32
90. Provenza, F. D., Balph, D. F. 1987. Diet learning by domestic ruminants: theory, evidence and practical implications. *Appl. Anim. Behav. Sci*. 18:211–32
91. Provenza, F. D., Balph, D. F. 1990. Applicability of five diet-selection models to various foraging challenges ruminants encounter. In *Behavioral Mechanisms of Food Selection,* ed. R. N. Hughes, Vol. 20, NATO ASI Series G: Ecological Sciences, pp. 423–59. Heildelberg: Springer-Verlag
92. Provenza, F. D., Burritt, E. A., Clausen, T. P., Bryant, J. P., Reichardt, P. B. 1990. Conditioned flavor aversion: a mechanism for goats to avoid condensed tannins in blackbrush. *Am. Nat*. 136:810–28
93. Provenza, F. D., Pfister, J. A., Cheney, C. D. 1991. Mechanisms of learning in diet selection with reference to phytotoxicosis in herbivores. *J. Range Manage*. In press
94. Pyke, G. H. 1984. Optimal foraging theory: a critical review. *Annu. Rev. Ecol. Syst*. 15:523–75
95. Reichardt, P. B., Bryant, J. P., Anderson, B. J., Phillips, D., Clausen, T. P. 1990. Germacrone defends Labrador tea from browsing by snowshoe hares. *J. Chem. Ecol*. 16:1961–70
96. Reichardt, P. B., Bryant, J. P., Clausen, T. P., Wieland, G. D. 1984. Defense of

winter-dormant Alaska paper birch against snowshoe hares. *Oecologia* 65: 58–69

97. Reichardt, P. B., Bryant, J. P., Mattes, B. R., Clausen, T. P., Chapin, F. S. III. 1990. The winter chemical defense of balsam poplar against snowshoe hares. *J. Chem. Ecol.* 16:1941–60
98. Reichardt, P. B., Clausen, T. P., Bryant, J. P. 1987. Plant secondary metabolites as feeding deterrents to vertebrate herbivores. In *Proceedings-Symposium on Plant-Herbivore Interactions. USDA Forest Service Gen. Tech. Rept. INT-222,* ed. F. D. Provenza, J. T. Flinders, E. D. McArthur, pp. 37–42. Ogden/Intermontain Res. Sta. 179 pp.
99. Reut, N. A., Danvsevich, I. K., Zakharovskii, A. S., Kuzmitskii, B. B., Kevra, M. K. 1975. Toxicological properties of synthetic camphor prepared from pine tree oleoresins. *Farmakol. Toksikol. Nov. Prod. Khim. Sint., Mater. Resp. Korf.* 3:185–7
100. Rhoades, D. F., Cates, R. G. 1976. Toward a general theory of plant antiherbivore chemistry. In *Recent Advances in Phytochemistry,* Vol 10. *Biochemical Interactions Between Plants and Insects,* ed. J. W. Wallace, R. L. Mansell, 10:168–213. New York: Plenum. 425 pp.
101. Roath, L. R., Krueger, W. C. 1982. Cattle grazing and behavior on forested range. *J. Range Manage.* 35:332–38
102. Robbins, C. T. 1983. *Wildlife Nutrition.* New York: Academic. 343 pp.
103. Rosenthal, G. A., Janzen, D. H. 1979. *Herbivores: Their Interaction with Secondary Plant Metabolites.* New York: Academic. 718 pp.
104. Ruess, R. W. 1987. Grazing and the dynamics of nutrient and energy regulated microbial processes in the Serengeti grasslands. *Oikos* 49:101–10
105. Ruess, R. W., Hik, D. S., Jefferies, R. L. 1989. The role of lesser snow geese as nitrogen processors in a sub-arctic salt marsh. *Oecologia* 79:23–9
106. Ruess, R. W., McNaughton, S. J. 1984. Urea as a promotive coupler of plant-herbivore interactions. *Oecologia* 63: 331–37
107. Schimel, D. S., Parton, W. J., Adamsen, F. J., Woodmansee, R. G., Senft, R. L. 1986. The role of cattle in the volatile loss of nitrogen from a shortgrass steppe. *Biogeochemistry* 2:39–52
108. Senft, R. L., Coughenour, M. B., Bailey, D. W., Rittenhouse, L. R., Sala, O. E. 1987. Large herbivore foraging and ecological hierarchies. *Bioscience* 37:789–99
109. Sinclair, A. R. E., Jogia, M. K., Anderson, R. J. 1989. Camphor from juvenile white spruce as an antifeedent for snowshoe hares. *J. Chem. Ecol.* 14: 1505–14
110. Sinclair, A. R. E., Krebs, C. J., Smith, J. N. M., Boutin, S. 1988. Population biology of snowshoe hares. III. Nutrition, plant secondary compounds and food limitation. *J. Anim. Ecol.* 57:787–806
111. Skinner, B. F. 1981. Selection by consequences. *Science* 213:501–04
112. Snyder, J. D., Janke, R. A. 1976. Impact of moose browsing on boreal-type forests of Isle Royale National Park. *Am. Mid. Nat.* 95:79–92
113. Squibb, R. C., Provenza, F. D., Balph, D. F. 1990. Effect of age of exposure to a shrub on the subsequent feeding response of sheep. *J. Anim. Sci.* 68:987–97
114. Stoddart, L. A., Smith, A. D., Box, T. W. 1975. *Range Management.* New York: McGraw-Hill
115. Swift, M. J., Heal, O. W., Anderson, J. M. 1979. *Decomposition in Terrestrial Ecosystems.* Oxford: Blackwell
116. Thorhallsdottir, A. G., Provenza, F. D., Balph, D. F. 1987. Food aversion learning in lambs with or without a mother: discrimination, novelty and persistence. *Appl. Anim. Behav. Sci.* 18:327–40
117. Thorhallsdottir, A. G., Provenza, F. D., Balph, D. F. 1990. Ability of lambs to learn about novel foods while observing or participating with social models. *Appl. Anim. Behav. Sci.* 25:25–33
118. Tiedmann, A. R., Berndt, H. W. 1972. Vegetation and soils of a 30-year deer and elk exclosure in central Washington. *Northwest Sci.* 46:59–66
119. Van Soest, P. 1982. *Nutritional Ecology of the Ruminant.* Corvalis, Oregon: O & B Books. 374 pp.
120. Vitousek, P. 1982. Nutrient cycling and nutrient use efficiency. *Am. Nat.* 119:553–72
121. Von Rudloff, E. 1975. Volatile leaf oil analysis in chemosystematic studies of North American Conifers. *Biochem. Syst. Ecol.* 2:131–67
122. Waage, S. K., Hedin, P. A., Grimley, G. 1984. A biologically active procyanidin from Machsarium floribundum. *Phytochemistry* 23:2785–87
123. Wolcott, G. N. 1953. Stilbenes and comparable materials for dry-wood termite control. *J. Econ. Entomol.* 46:374–75

Annu. Rev. Ecol. Syst. 1991. 22:447–75

MORPHOLOGICAL AND MOLECULAR SYSTEMATICS OF THE DROSOPHILIDAE

Rob DeSalle

Department of Biology, Yale University, New Haven, Connecticut 06511*

David A. Grimaldi

Department of Entomology, American Museum of Natural History, New York, NY 10024

KEY WORDS: Drosophila, homoplasy, cladistics, DNA sequences, phylogenies

---

INTRODUCTION

The first response from other biologists to a statement of what we do as Drosophilidae systematists is, "Oh, you mean you study *Drosophila*." Most biologists will then add, "I thought that everything was already known about the evolution of *Drosophila*." They are referring, of course, to the tradition established in population, Mendelian, and quantitative genetics, speciation, studies of selection and natural populations, all by the study of just one of the approximately 65 genera in the family, the genus *Drosophila*. The Columbia Group (Morgan, Muller, Bridges, and Sturtevant) helped launch the field of modern genetics with their pioneering use of *Drosophila melanogaster*, by clarifying or discovering fundamental concepts such as mutation, linkage, crossing over, sex-linked inheritance, and the linear arrangements of genes on chromosomes. Their momentum carried them and their unprecedented discoveries into the 1940s. In 1944, Bridges & Brehme provided some of the first comprehensive treatments of *Drosophila* mutants. The 1941 publication of Dobzhansky's *Genetics and the Origin of Species* (36) and Mayr's *Sys-*

*Present address: Department of Entomology, American Museum of Natural History, New York, NY 10024

0066-4162/91/1120-0447$02.00

tematics and the Origin of Species (72) revolutionized biology. Ford, working in England, was very influential in showing polymorphisms and their genetic bases in natural populations of butterflies. From the joined forces of these scientists would come a rigorous formulation of the allopatric model of speciation and of genetic differences within and among populations. By this time, the University of Texas Genetics Foundation Group had been organized by Patterson and Stone, and a generation of students studying *Drosophila* arose from this group. At least two decades later a return to macroevolutionary questions and thought began with the translations of the works of Hennig (48).

It is the attention paid to mechanistic features of evolution, particularly speciation, that caused studies on the systematics of the Drosophilidae to languish. Obviously Mayr and Dobzhansky believed the problem of how species originate to be the central problem in evolution. This view was propagated as well by Huxley (56). Preoccupied with describing the phenotypic and genetic differences among geographical populations, drosophilists unquestioningly accepted the traditional classifications of *Drosophila* and the family, based largely on the 1921 monograph by Sturtevant (95), *The North American Species of Drosophila*. It was Sturtevant, a geneticist and entomologist, who revised the subgeneric classification of the genus *Drosophila* (96). This revision antedated the explicit cladistic method of Hennig (48). That subgeneric classification (and species groups within the subgenera *Sophophora* and *Drosophila*) by Patterson & Stone (78) is still in use today.

Although the mode and mechanism of speciation remains a perplexing problem, we believe that knowledge of phylogenetic relationships is the central issue in evolutionary biology. Without knowledge of species relationships, it is impossible to discern patterns in geographical distributions, such as the frequent allopatry of sister species. If phylogenetics is the central issue in evolutionary biology, and the concepts of homology and homoplasy (the latter, convergent similarity or shared pleisiomorphy) are central to phylogenetics, then the deciphering of homologous features is pivotal to biology in general. It is universally agreed that organisms can be grouped based upon the presence of shared derived characters. Unfortunately, character convergence makes this theoretically simple approach difficult in practice. Here we explore the use of morphological and molecular data used in the systematics of the Drosophilidae and evaluate their combined utility for the resolution of problems of convergence.

THE DROSOPHILIDAE

The Drosophilidae is a large family of muscomorphan Diptera, containing very nearly 3000 species around the world (110). The family is by far most speciose in the tropics. The range in habitats is the subarctic *(Drosophila subarctica)* and the windswept islands off the shore of Tristan da Cunha *(Scaptomyza)*, to the cloud forests and paramos of all tropical regions and deserts. The diversity of the habits of this family is astonishing: There are species whose larvae breed in sap fluxes of trees (various *Drosophila, Chymomyza, Neotanygastrella*), as parasites *(Cladochaeta)* and inquilines of spittle bugs (African *Leucophenga*); in various macrofungi *(Drosophila, Hirtodrosophila)*; in living (*D. flavipilosa* group, *Zapriothrica, Phloridosa*) and decaying flowers (many *Drosophila*); in decaying fruits, stems, leaves, cacti, etc (many *Drosophila*); in the nephric grooves of land crabs (1 *Lissocephala*, 2 *Drosophila*); as predators on blackfly larvae (*D. simulivora* group), scale insects *(Rhinoleucophenga, Pseudiastata, Mayagueza, Acletoxenus)*, bee larvae *(Gitona)*, spider embryos *(Titanochaeta)*, and even frog eggs (a *Hirtodrosophila*). The morphological diversity also probably exceeds that of most other muscomorphan groups. Included in this assemblage are 5 genera with species that have broad-headed males, several genera with elongate piercing oviscapts (ovipositor in the loose sense), 1 brachypterous *(Scaptomyza frustifera)* and 1 apterous *(Hypselothyrea)* species. Also contributing to the morphological diversity in the family are the myriad of species among the huge fauna endemic to Hawaii with bizarre male ornaments unlike that of any group of flies. It might be expected that such a rich natural history would attract systematic entomologists. In actuality, the generic-level taxonomy of the family has been neglected, probably because the community of *Drosophila* biologists and literature on *Drosophila* biology is daunting to the uninitiated. The drosophilid phenotype is still an unexplored realm. Perhaps a thousand species await description and collection. Detailed morphological studies for 99% of the described species do not exist, let alone the information on host use, distribution, mating behavior, and general natural history.

MORPHOLOGICAL CHARACTERS

Morphology is a very convenient aspect of the phenotype to work with. Morphological features are durable, in that dried insect cuticle endures for centuries. Morphology is easily observed, requiring optics that are relatively inexpensive compared to molecular instruments (as well as a penchant by the investigator for detailed, meticulous work). Examination of morphology facilitates comparisons of individuals, allowing species distributions to easily be made and homologous structures in different species to be quickly perused. Morphological characters allow explicit and unambiguous criteria for homologizing characters based on variation in shape, number, relative position, size, fine structure (e.g. surface microsculpturing). Most important is a rich source of characters that the analysis of morphology represents. One could easily survey the phenotypic products of thousands of structural and regulatory genes by a study of as few as one hundred morphological characters. Because the informative and convenient nature of morphology was realized centuries ago the study of systematics based on morphology is regarded as antiquated. Here we consider a short history of the characters used and review examples (Table 1) where the information in morphology has been especially revealing for systematists.

Traditional morphological characters useful at the genus level have included chaetotaxy (patterning of bristles and setae) of the head involving use of the positions of the frontal orbital setae and their relative sizes. Chaetotaxy of the thorax and a few characters on the wings were also used, such as the pointed wing apex of *Microdrosophila* and the costal lappet in *Microdrosophila* and several other genera. These features could be easily observed on a pinned specimen with a standard stereoscope so that no dissection or compound microscopy was involved. However, a German physician and ardent amateur taxonomist of acalyptrate flies, Duda, was a contemporary of Sturtevant and already well ahead of his time. He produced several monographic

Table 1 Morphological characters used prior to Grimaldi (1990)

Character	Possible States	Reference
CHAETOTAXY		
Rows of aerostichal scrotichal setulae	Number	38
Pairs of dorsocentral setae	Number	38
Prescutellar setae	Presence/absence	38
Katepisternal setae	Number	38
Katepisternal setae	Size	38
Rows of cuneiform setulae on leg	Presence/absence	75
CIBARIAL		
Cibarium	Shape	42
Cibarial sensilla	Pattern	42
GENITALIC		
Periphallic lobes	Shape, size, number	38, 75
Aedeagus	Shape, size	38, 75
Hypandrium	Shape, size	38, 75
Surstyli (Claspers)	Shape, size	38, 75
Folds in paragonia	Shape, size, number	78, 101
Testes	Pigmentation, coils	78, 101
Cecae of sperm pump	Presence/absence, shape	78, 101
Coiling of ventral receptacle	Length, shape	78, 101
WING		
Wing apex pegs	Presence/absence	75
Pointed wing apex	Presence/absence	38
PREADULT		
Filaments on egg	Number, position, shape	75, 78, 101
Anterior spiracles length	Ratio to length of pupae	75, 78, 101

works on Costa Rican, Palearctic, and Sumatran faunas in which he routinely described for the first time the complex male genitalia and the female genitalia (oviscapt and spermathecae), and included fine photographs of the wings. Frey (42) predated Duda (38) in dissecting the mouthparts and even in examining cibarial characters. Male genitalia have proved to be one of the richest sources of characters in distinguishing among closely related species, simply because of their complexity. It wasn't until the work of Okada (75) that the routine use of illustrating various internal features, including male and female genitalia, become a standard practice. Okada was also responsible for discovering several obscure but very important and interesting characters on the wings and legs.

The use of compound microscopy and the study of fine structure of male and female genitalia (97), mouthparts, and legs did not take hold in the United States. This may have been due to the emphasis placed on other character systems by Patterson & Stone (78). These were features of the soft internal reproductive organs and several egg and puparial characters (78). These character systems plus a general gestalt on the external habitus (coloration, size, and habits) provided the basis for Patterson & Stone's species groups of *Drosophila*.

Throckmorton was originally a graduate student at the University of Texas, and his morphological work exemplifies the bias of the characters considered to be the most important by Patterson & Stone. Throckmorton (101–103) prepared comprehensive comparative treatments of the internal reproductive organs, the ejaculatory apodeme, the anterior spiracles and eggs, for a great many species groups in *Drosophila* as well as several other genera. These data were the basis for his hypotheses on relationships among genera, subgenera, and species groups of *Drosophila* (105). This paper is one of the most frequently cited papers in biology. It is used today as a standard reference for phylogenetic relationships in the family and especially within *Drosophila*. Although the synapomorphies that support the groupings proposed in this paper are not shown or discussed, the hypotheses in the 1975 paper are based on extensive and carefully prepared morphological data sets, shown in Throckmorton's papers (101–103). The preoccupation with internal characters was probably a throwback to the view of Mayr et al (73) that internal characters are less subject to external vagaries and thus to selection pressures. Selection pressures were seen as the cause of adaptive convergence, the pitfall of morphologically based systematics. Despite the bias in Throckmorton's data it is useful for circumscribing some natural groups. For example, the development of long anterior and posterior caecae on the sperm (ejaculatory) pump is a highly derived feature linking many species in *Drosophila*. Similarly, the number and position of anterior filaments on the egg has proven to be a reliable feature for asserting the monophyly of several genera and subgenera of *Drosophila*. The subgenus *Sophophora* is well defined alone by the single pair of broad subapical egg filaments. Unfortunately, since the dissections must usually rely upon properly fixed living material, these characters can be used only for species that can be established in culture, and many *Drosophila* species cannot be cultured as yet. Also, for the purposes of taxonomic diagnoses and routine identification, these characters would be impossible to use, but they can be supplemented with more convenient external characters.

FOSSILS

Most organisms do not have a fossil record complete enough to support a reliable absolute date of origin of a clade. Ages based on vicariant distribution patterns and the geological timing of the areas are probably more useful. However, the data afforded by the scarce fossil records of most organisms, including the Drosophilidae, are still useful for several reasons. First, a fossil can lie outside the distribution of its living relatives and potentially force the revision of vicariant hypotheses based solely on the living species (37). Second, a fossil can be older than the geological events hypothesized to have caused a particular vicariant event and thus can revise (but not refute) the hypothesis. Third, the age and locality of the fossil can fall within the historical bounds hypothesized on the basis of living relatives and thus lend supporting data to the hypothesis. The minimum age of clades based on fossil data in itself is of importance in calibrating molecular studies where a clock-like rate of nucleotide substitution is assumed.

Prior to the study by Grimaldi (45), there were three species of Drosophilidae known in the fossil record, and all these are amber. Compression or mineral replacement fossils are generally lacking in so many fine details as to be useless for acalyptrate Diptera. The oldest drosophilid known, *Electrophortica succini* Hennig, is from Baltic amber. Unfortunately, much of the acalyptrate material studied and described by Hennig in the Konigsberg collection was discovered washed up on the Baltic shores, and so the exact depositional strata are unknown. Baltic amber can vary from Eocene to Oligocene in age, and it was an Eocene age that Hennig ascribed to *Electrophortica*. The first fossil drosophilid described was a specimen in Columbian amber (20), but the locality of this specimen has not been traced and it appears that the amber is probably subfossil or even recent in origin (45). Wheeler (109) gave an account of a third species, a *Neotanygastrella*, from the late Oligocene amber of Chiapas, Mexico.

The large deposits of Miocene amber and younger copal from the Dominican Republic have yielded a rich diversity of drosophilids, including nine new species (only one of them in the copal) and two new genera (45). The species include a *Chymomyza,* two *Drosophila* (subgenera uncertain), a *Hirtodrosophila,* a *Scaptomyza,* two Drosophilinae species (incertae sedis), and the newly named genera *Miomyia* and *Protochymomyza*. *Protochymomyza* is an extinct sister group to the cosmopolitan genus *Chymomyza*. *Miomyia,* having been reexamined, appears to be related to the Neotropical genera *Diathoneura* and *Cladochaeta*. *Protochymomyza* shows the fossilized basis for several character states of a hypothesized pleisiomorphic ancestor or sister group. None of the amber fossil drosophilids appears to force a reinterpretation of the history of drosophilids as based on living representatives, unlike some of the cases illustrated by Donoghue et al (37).

In summary, the amber fossil record indicates that many groups now present in the Caribbean and Central American region were present there, at least in southern Mexico and/or the Dominican Republic, during the Miocene. These include the subgenus *Drosophila, Hirtodrosophila, Chymomyza, Scaptomyza* and *Diathoneura*. Thus it is quite conceivable that these groups have been affected by plate tectonics in the Caribbean region and that they have a minimum age of at least 25 million years.

HIGHER LEVEL RELATIONSHIPS

There are at present three hypotheses of phylogenetic relationships among the Drosophilidae. These are by Throckmorton (105), Okada (76), and Grimaldi (46). Grimaldi (46) evaluated the hypotheses of Okada and Throckmorton based on a reanalysis of their own data, and he compared them with his own phylogenetic hypothesis based on additional data. In brief, for a survey of 205 adult morphological synapomorphies and a synoptic set of 160 representative species, Grimaldi (46) constructed a strict consensus cladogram of generic and subgeneric relationships (Figure 1). The results were in much better agreement with Throckmorton's original (105) tree and a revised tree based on his data than they were in comparison to Okada's phenetic tree. The subfamilies Steganinae and Drosophilinae remained intact as monophyletic groups, and each of the other subfamilies was hierarchically arranged into smaller sets of monophyletic groups. The relationships of some genera are definitive, as based on the types and numbers of apomorphies that they share. These are *Leucophenga* + *Stegana* and *Zapriothrica* + *Laccodrosophila* as sister groups, the monophyly of the *Amiota* and of a group of genera predacious as larvae on scale insects. The close relationships of *Mycodrosophila, Zygothrica, Hirtodrosophila,* and *Paramycodrosophila* reflect the conclusions of Throckmorton (105) and Grimaldi (45). Results that were somewhat surprising were that several "subgenera" such as *Hirtodrosophila, Scaptodrosophila,* and *Lordiphosa* do not belong in the genus *Drosophila*. Exclusion of these "subgenera" to other nodes of the generic cladogram results in a

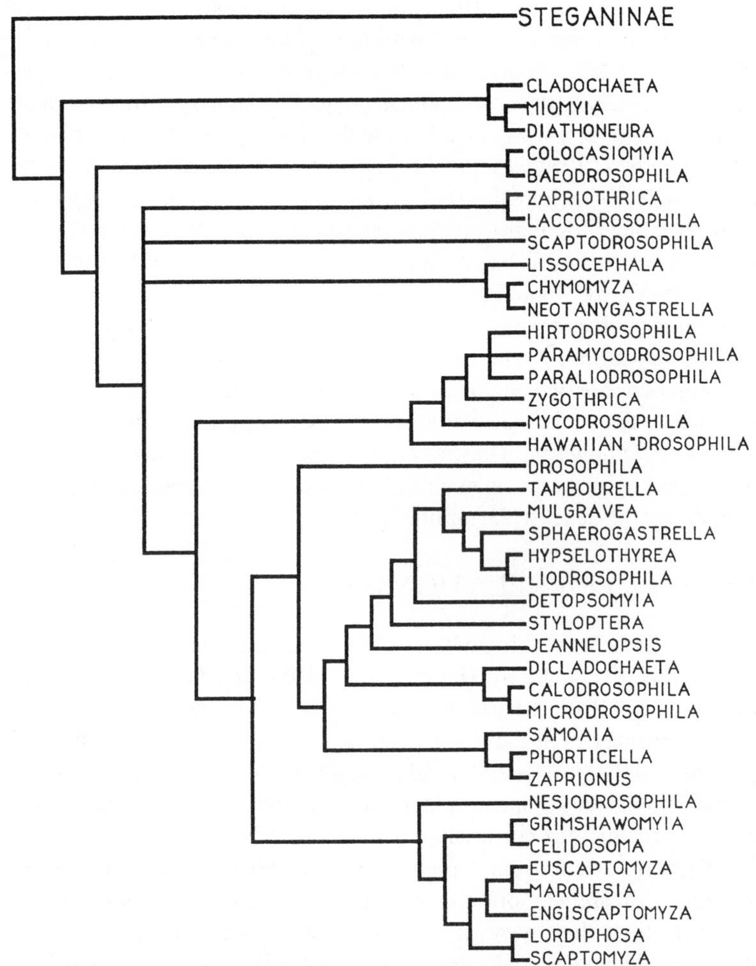

Figure 1 Morphological phylogeny of the genera in the subfamily Drosophilinae. The morphological characters and cladistic methods used to construct this phylogeny as well as the details of the relationships of the genera in the subfamily Steganinae are in Grimaldi (46).

monophyletic genus *Drosophila*. Traditionally, *Drosophila* was a paraphyletic catchall category, typical of many type genera in insects. Of particular importance is that the species enedemic to Hawaiii placed in the subgenus *Drosophila* (17) were found not to be *Drosophila* at all, because they lacked several of the distinctive morphological features diagnostic of a monophyletic genus *Drosophila*. For this reason, the old name given to some of the species *Idiomyia,* which was later synonomized by Kaneshiro (60) with *Drosophila,* has been resurrected by Grimaldi (46) for all Hawaiian *Drosophila*.

CHROMOSOMAL ANALYSIS

Because of the presence of giant polytene chromosomes in the salivary gland tissue of most Diptera, chromosomal phylogenies can be highly resolved. The degree of resolution for *Drosophila* at the level of comparisons of species within groups is paralleled in only a few insect groups, due to this technique, and many extensive studies have established detailed chromosomal phylogenies (2, 16, 67, 78, 108). Application of polytene chromosomal analysis to the comparison of distantly related taxa (species in different subgenera or even in different species groups) has been lacking due to an inability to homologize polytene banding patterns. This inability is due to the presence of multiply overlapping inversions that extensively scramble the polytene banding patterns so that homologous regions become difficult to recognize. Two studies have homologized short regions of polytene chromosomes in an attempt to establish phylogenetic relationships of distantly related *Drosophila* groups (92, 115).

Two techniques that could extend the applicability of polytene chromosome analysis to phylogenetic studies at greater taxonomic distances are the establishment of linkage groups and their assignments to particular chromosomal regions (H. L. Carson, personal communication) and the use of in situ hybridization of molecular probes to polytene chromosomes (26, 57, 94). Both techniques serve to redefine an inversion based on additional resolution of its fine structure (e.g. location of particular genes or linkage groups). The ability to homologize intergenerically compared inversions should be improved by these methods.

MOLECULAR ANALYSES

Throckmorton's Assessment

Throckmorton was instrumental in bringing the study of molecular comparisons into systematics. Hubby & Throckmorton's 1965 publication (52) was the paper that lit the systematics electrophoresis fire of the 1970s, by illustrating the extent of protein variation in natural populations and among species. Throckmorton's views on molecular systematics are traced to three main review papers (104, 106, 107). In 1966, Throckmorton (103) was concerned almost entirely with the concept and problem of homology at the moelcular level and was well ahead of most biologists in recognizing the dichotomy between phenetic (such as microcomplement fixation) and cladistic methodologies. He did arrive at a very eclectic conclusion, however, attempting to reconcile a need for both approaches in phylogenetic reconstruction. Throckmorton suggested that molecular systematics could be most useful in studying speciation, and he was clearly concerned about differences among very closely related species and populations. But here his conclusion was more cynical than optimistic, and quite vague. He did not formulate a method by which electrophoretic data could be analyzed but only cautioned that "the maximum parsimony methods are equally unsound [to phenetic analyses],. . . ." He refers to the "fact" that "there is nothing whatever in evolutionary theory to suggest that [evolution occurred] over the shortest possible pathway." By this thinking, he was allowing theories on the data to dictate the methodology and logic by which the data should be analyzed. Throckmorton's views reflected a concern with the problem of homology in molecular data and with the appropriate methodologies for analyzing electrophoretic data. His views are equally applicable to RFLP data, protein sequence data, and DNA sequence data.

The utility of the techniques described in the following sections in resolving the relationships of the Drosophilidae depends directly on the degree of homoplasy observed for the particular molecular probes discussed. Consequently in the following sections we assess the molecular techniques at several levels, one of which is the potential for establishing relationships such as those in Figure 1. It is surprising that with the large number of genera in the Drosophilidae only a few studies (6, 7, 29, 79) have attempted to examine the relationships of even a small number of these genera at the molecular level. The following discussion of molecular data reviews work on taxa within genera (the genera *Drosophila* and *Idiomyia* in particular), within subgenera *(Sophophora* and *Drosophila),* within species groups, and within species subgroups. The work pertinent to between-group, subgroup, and genera relationships is reviewed, but by far the largest amount of information exists for relationships within the genus *Drosophila*.

Proteins

Gene products (proteins and structural RNAs), gene regulators (assessed by patterns of gene product distribution), and the genes themselves (DNA sequenced directly from clones, PCR products, or from structural RNAs) have all been used in molecular systematics. Due to the presence of two recent extensive reviews of allozyme studies in *Drosophila* (69, 107) we omit discussion of these types of studies. After a brief discussion of two-dimensional gel electrophoresis and immunoprecipitation studies of proteins and gene regulatory studies, we concentrate on DNA-level studies and their usefulness in the systematics of Drosophilidae.

Most isozyme studies in *Drosophila* have concentrated on the relationships of closely related species. The reasons for this emphasis are discussed at length in MacIntyre & Collier (69) and Spicer (90). Two-dimensional gel electrophoresis involves the electrophoresis of protein extracts first in one direction, followed by a rotation of the electric field by 90 degrees, and has been used to examine closely related *Drosophila* species and populations (65, 66, 74) as well as distantly related species (90). At the level of closely related species and populations, the technique has a resolution similar to conventional allozyme analysis. MacIntyre & Collier (69) suggested that two-dimensional gel electrophoresis studies of slowly evolving less variable proteins should be useful in examining relationships of distantly related taxa. Spicer (90) confirmed this prediction by reconstructing the phylogeny of a small number of taxa in the genus *Drosophila* using two-dimensional gel electrophoresis. His phylogenetic results are in agreement with the conventional placement (105) of many of the *Drosophila* species in his study.

Immunoprecipitation studies have been used primarily to examine distantly related taxa in the genus *Drosophila* (Table 2). Immunoprecipitation studies for three proteins in *Drosophila* have been conducted [alpha GPDH (24), acid phosphatase-1 (68, 70) and larval serum protein (6, 7, 8)] and are reviewed in some detail in MacIntyre & Collier (69). Immunoprecipitation techniques result in a measure of immunological distance between taxa, so systematic relationships from these studies are necessarily obtained by phenetic analysis. The results of these studies showed a "striking resemblance to the phylogeny proposed by Throckmorton [101, 105] for these species" (69). One notable exception to this general agreement was the close apparent clustering of *D. immigrans* (a member of the subgenus *Drosophila*) with the *melanogaster* subgroup flies in the subgenus *Sophophora* in the acid phosphatase-1 study (70). This discrepancy was attributed to genuine convergence of the *D. immigrans* acid phosphatase-1 with sophophoran acid phosphatase-1 immunoprecipitation properties due to a drastic change in the net charge of the *D. immigrans* acid phosphatase-1 protein.

The larval hemolymph protein (LHP) studies (Bevereley & Wilson: 6–8) have been most useful in examining distantly related taxa in the genus *Drosophila* and in closely related genera such as *Chymomyza*. In addition, the relationships of the Hawaiian *Drosophila (Idiomyia)* were examined for this protein using immunoprecipitation techniques (8).

Patterns of gene regulation have also been suggested as useful phylogenetic tools (34, 35, 100) and have been used to examine phylogeny in the Hawaiian *Drosophila*. In general, these studies are in agreement with morphological and chromosomal phylogenies for the Hawaiian *Drosophila*. Some exceptions occur, and it should be noted that Dickinson (35) indicated such exceptions are probably the result of convergence of gene regulation patterns and are not a problem with morphological or chromosomal analysis.

Genes and DNA—Techniques

Three techniques have been used to obtain information from *Drosophila* DNA sequences. The first technique uses the single copy portion of the nuclear

Table 2 Species groups in the genus *Drosophila* examined using molecular techniques*

Species group	Number of taxa	Technique	Data analysis	References
nasuta	11 species 32 strains	RFLP mtDNA	cladistic	18
repleta	8 species	DNA-DNA hybrids	phenetic	86
virilis	12 species	2-D gels	cladistic and phenetic	89
planitibia	5 species	RFLP mtDNA	cladistic and phenetic	31, 33
	5 species	RFLP ADH	cladistic and phenetic	9, 55
	5 species	DNA-DNA hybrids	phenetic	53
	5 species	Sequence ADH	phenetic	55
obscura	8 species 12 strains	DNA-DNA hybrids	phenetic	43
	7 species 15 strains	RFLP mtDNA	cladistic and phenetic	63, 64
melanogaster	9 species	DNA-DNA hybrids	phenetic	14, 80
	8 species 13 strains	RFLP mtDNA	cladistic and phenetic	88
	5 species	Sequence ADH	cladistic and phenetic	10

*For discussion of allozyme studies see MacIntyre and Collier (69).

genome in soluble hybridization experiments (DNA-DNA hybridization or DNA reassociation). For recent critical appraisal of this technique, the reader is referred to Sarich et al (84) and Springer & Krajewski (91). The technique has been used in comparisons of *Drosophila* taxa for single copy genes from the nucleus (117), mtDNA (14, 80), and for the single copy component of the nuclear genome (14, 39, 43, 53, 54, 86). As in isozyme studies, the taxonomic range over which the technique appears to be useful is between closely related species (sometimes even within a species—12) and within species groups. Species group comparisons are impeded by low normalized percent hybridization extrapolation procedures (91) for T_{50H}, problems of nonadditivity for ΔT_m (14, 84) and the lack of observable ΔMode caused by extreme genetic distances. As with the immunoprecipitation approach, data from DNA-DNA hybridization studies give only distances between taxa and are therefore analyzed phenetically.

In general, RFLP techniques in *Drosophila* systematics are useful for intraspecific and within species subgroup comparisons. Most phylogenetic studies of nuclear genes using RFLPs have concentrated on within-species comparisons and the reconstruction of phylogenies within species groups such as the *melanogaster* subgroup and the *obscura* subgroup (both members of the subgenus *Sophophora*) and closely related Hawaiian *Drosophila* species. Repeated nuclear gene comparisons have concentrated mostly on *melanogaster* subgroup flies (21, 81). While problems with RFLP pattern convergence occur even at these levels, they can be dealt with statistically either using distances or character state analysis (28, 40, 98, 99). Over longer divergence times a problem similar to polytene chromosome band scrambling occurs. Patterns are hard to discern as homologous. In addition, the resolution of RFLP mapping techniques becomes problematic. Most restriction mapping studies are accurate to between 100 and 300 base pairs. It is probable that two restriction sites for the same enzyme could exist in such a short stretch of DNA and hence that one or the other would go undetected. The problems of convergent restriction sites or lack of homology are acute for rapidly evolving and moderately rapidly evolving genes such as ADH or parts of the mtDNA molecule.

Slowly evolving genes such as rDNA should be more appropriate genes for between subgroup and between species group comparisons using RFLP techniques. However, two technical problems exist with the *Drosophila* species examined to date for rDNA. The first is the presence of a heterogeneous class of insertions in one of the subunits of the genes in this cluster. This insertion makes restriction mapping for rDNA genes in these species extremely difficult. The second problem is the occurrence of a high degree of heterogeneity in these repeated genes even within individuals (22). Both of these problems have thwarted the detailed and extensive use of rDNA RFLPS as a phylogenetic tool in *Drosophila* as has been employed in plants (116) and vertebrates (50). The RFLP technique on rDNA could be used if particular groups of *Drosophila* lack this extensive heterogeneity or the troublesome insertions. For further discussion of the potential uses and problems with rDNA as a phylogenetic tool the reader is referred to Williams et al (114) and Hillis & Davis (51). RFLP techniques on other repeated, slowly evolving genes such as 5S RNA genes or histone genes may prove useful in phylogenetic analysis between species groups in *Drosophila* if the above requirements are met.

DNA sequencing is a technique that, if employed over a considerable part of the genome, yields the ultimate type of molecular data for systematic analysis. Recently, Pamilo & Nei (77) have cautioned that polymorphism in ancestral species can have a great effect on the equivalence of a gene tree and a species tree. However, both Pamilo & Nei (77) and Felsenstein (41) point out that if divergence times of the species are sufficiently large the probability that a gene tree and a species tree will be equal is high (Pamilo & Nei estimate $> 2Ne$). One obvious solution to the gene tree–species tree problem, with respect to DNA sequencing, would be to sequence many independently segregating genes. While this approach is time consuming and expensive, it can be reasonably accomplished when several independent laboratories generate sequences for different genes from the same organisms. Perhaps the best examples of this are the Hawaiian picture-winged species *D. heteroneura, D. silvestris, D. planitibia,* and *D. differens,* where gene trees have been constructed for several independently segregating genes—ADH (55), vitellogenin genes (M. Kambysellis, personal communication), chorion genes (71), and mtDNA (30).

The two DNA regions that have been most intensively studied at the sequence level in *Drosophila* systematics are the ADH gene region of the nuclear component and selected parts of the mtDNA molecule. The sequencing of the *Drosophila* ADH gene has been done in several labs. Other nuclear genes have been sequenced in *Drosophila* but only for a few interspecific comparisons. ADH sequences exist for several members of *Sophophora* (10, 25, 85), the *repleta* subgroup (1, 23), and the Hawaiian *Drosophila* (55, 82, 83). mtDNA sequences exist for large portions of the molecule, mostly in the *melanogaster* subgroup (19, 27), and for a smaller region of the molecule in several continental *Drosophila, Sophophora,* and Hawaiian *Drosophila* (14, 30, 32).

Drosophila *DNA Sequences and Homoplasy Problems*

Detailed molecular analysis of DNA sequences suggests that caution should be exercised when using these data in systematics. An appreciation of the consequences of several universal characteristics of DNA sequence change is essential for a complete understanding of the nature of homoplasy in certain DNA regions. For the protein coding DNA sequences, third positions in codons accrue change more rapidly than do second or first positions. Likewise intron sequences change more rapidly than exon regions. Related to this rapid third position change are two phenomena that complicate the patterns seen in third positions. In addition, structural RNA genes such as rDNA may be affected by character covariance.

SHIFT IN THIRD POSITION NUCLEOTIDE FREQUENCY In the nuclear gene ADH, a shift in the third position nucleotide composition is observed when members of the subgenus *Drosophila* are compared to *Sophophora*. In particular, there is an 11% increase in G + C nucleotides in third positions of the ADH gene in members of the subgenus *Sophophora* compared to the subgenus *Drosophila* (93).

TRANSITION TO TRANSVERSION BIAS The second phenomenon is the extreme transition to transversion bias observed for several genes in a wide variety of organisms. Following the suggestion of Brown et al (11) that a transition to transversion bias in vertebrate mtDNA existed, DeSalle et al (33) examined the frequency of transversions and transitions in the mtDNA of various *Drosophila* species. Using a geological calibration for divergence times of Hawaiian species based on insular vicariance and LHP calibration for more divergent comparisons (8), DeSalle et al (30) demonstrated an extreme transition to transversion bias for comparisons of taxa that have divergence times of 10 MY or less for a protein coding region in the mtDNA. No such bias existed for a 600-base pair sequence of the mitochondrial 16S rDNA gene for these same species. A similar attempt to detect transition to transversion bias in the *melanogaster* subgroup by comparing *D. yakuba* sequences to *D. melanogaster* sequences failed to detect a transition to transversion bias. This failure may be due either to a large divergence time between *D. yakuba* and *D. melanogaster* (62) or to an altered rate of mtDNA evolution in the *melanogaster* subgroup (13).

CHARACTER COVARIANCE IN SEQUENCES CODING FOR STRUCTURAL RNAS For structural RNA genes such as ribosomal RNAs or transfer RNAs, paired secondary structure is important (113), and so-called character covariance occurs between paired members in the secondary structure. Character covariance appears to be a major consideration affecting the determination of distant phylogenetic relationships such as comparisons between insect orders (113). However, for comparisons among Hawaiian *Drosophila, Sophophora,* and continental *Drosophila,* character covariance in the 16S rRNA doesn't appear to be a great problem (30). For comparisons of genera in the Drosophilidae, character covariance may pose a problem.

The occurrence of DNA sequence character covariance, third position composition shifts, and transition to transversion bias all could contribute to covergence or homoplasy problems in DNA sequences data sets. Two primary problems in *Drosophila* molecular systematics are due, in part, to a recent concentration of sequencing efforts of the ADH gene and mtDNA genes, and the problems involve the shift of third position base composition and transition to transversion bias. Although Starmer & Sullivan (93) point out that the G + C shift between the subgenus *Drosophila* and the subgenus *Sophophora* could produce problems with phylogenetic analysis, they offer no evidence that it does, nor do they suggest a solution to this potential problem. If phylogenetic comparisons are being made within either subgenus, the shift observed between the two subgenera is irrelevant and can be ignored. Comparisons between the two subgenera would have to take this shift into consideration. However, possible solutions are available. Third positions can be eliminated from the analysis, or the nucleotide sequence can be conceptually translated into amino acid sequence. More appropriately, character weighting of third position changes can be performed. A similar but more extreme problem arises with respect to the transition to transversion bias in mtDNA. DeSalle et al (30) have taken the approach of completely eliminating transitions from their phylogenetic analysis.

The Molecular Approach to Phylogeny in the Genus Drosophila

Phylogenetic questions in the Drosophilidae are localized to three major hierarchical levels. The first level is within-species group and subgroup relationships. The second level is the between-species group and subgroup relationships, while the third level involves the relationships of the genera and subgenera in the family Drosophilidae.

RELATIONSHIPS WITHIN AND BETWEEN SPECIES GROUPS Table 2 lists the species groups and subgroups that have been examined using molecular techniques. (Allozyme analyses have been omitted from the table and can be found in MacIntyre & Collier—69). Table 3 lists the species groups that have been examined using RFLPs and also summarizes these analyses. The *nasuta* subgroup has in fact been analyzed only using mtDNA RFLPs. The *obscura* group, and the *melanogaster* subgroup have all been examined using mtDNA RFLPs and DNA reassociation techniques on single copy nuclear DNA. The *melanogaster* subgroup has also been analyzed using DNA reassociation techniques on mtDNA. In addition, the *planitibia* subgroup and the *melanogaster* subgroup ADH genes have been sequenced and analyzed. Some interesting observations can be made on three of these species groups (*obscura, melanogaster,* and *planitibia*) with respect to congruency of data sets.

The obscura *group* Analysis of this species group is characterized by exact agreement of the molecular data sets (Figure 2). The DNA-DNA hybridiza-

Table 3 Summary of RFLP systematic studies on genus Drosophila species groups

Group	Number taxa	Map sites	Phylogenetically informative sites	Comment	Reference
nasuta mtDNA	16	33	12	46 parsimony trees RI = 0.41; CI = 0.50	18
planitibia mtDNA	9	115	61	1 parsimony tree RI = 0.76; CI = 0.68	30
planitibia ADH	5	34	6	1 parsimony tree RI = 0.67; CI = 0.75	9
obscura mtDNA	15	54	42	2 parsimony trees RI = 0.85; CI = 0.68	63
melanogaster mtDNA	11	87	31	1 parsimony tree RI = 0.60; CI = 0.60	88

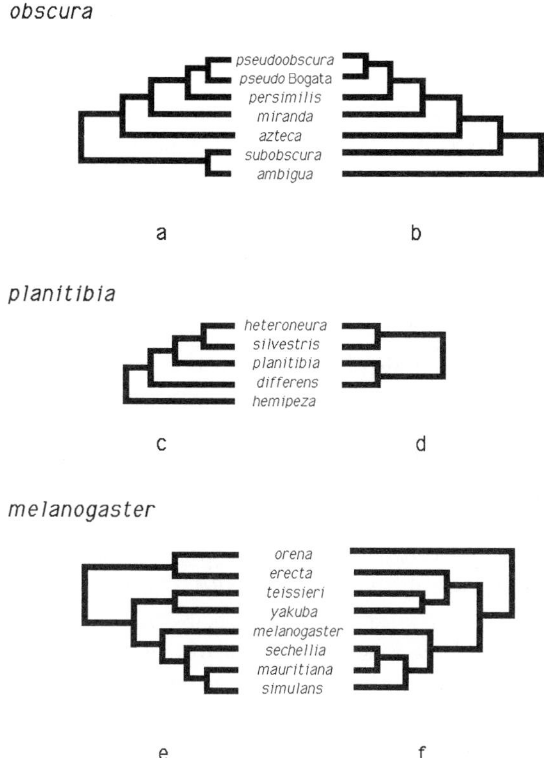

Figure 2 Comparison of mtDNA phylogenies (left side; a, c, and e) and DNA-DNA hybridization trees (right side; b, d, and f) for three species groups in the genus *Drosophila*.
a. mtDNA RFLP cladistic tree for the *obscura* group constructed from an exhaustive search using 47 phylogenetically informative restriction sites. For the placement of *D. subobscura* lines see Latorre et al (63, 64). *D. obscura* was used as an outgroup and a single most parsimonious tree was found (63).
b. DNA-DNA hybridization phenetic tree for the *obscura* group from the study of Goddard et al (43) with *D. algonquin* and *D. affinis* removed. The root of the tree was determined using midpoint rooting.
c. mtDNA RFLP cladistic tree for the *planitibia* subgroup of Hawaiian picture-winged *Drosophila* constructed from an exhaustive search using 61 phylogenetically informative restriction sites. *D. hemipeza* was used as an outgroup.
d. DNA-DNA hybridization phenetic tree for the *planitibia* subgroup of Hawaiian picture-winged *Drosophila* from the study of Hunt and Carson (53). The tree was rooted using *D. picticornis* (not shown) as an outside reference.
e. Consensus tree for the *melanogaster* group using a combination of several data sets (Lachaise et al—62).
f. DNA-DNA hybridization phenetic tree for the *melanogaster* group from the study of Caccone et al (13). The tree was rooted using midpoint rooting.

tion data (43) agree with the mtDNA RFLP data set (63, 64). Both of these data sets are also in good agreement with previous morphological and chromosomal data.

The planitibia *subgroup of Hawaiian* Drosophila DNA-DNA hybridization studies (53), and ADH RFLPs and sequences produce the same phylogenies. The mtDNA RFLP analysis produces a phylogeny that is at odds with the nuclear phylogenies (Figure 2). Interestingly, an analysis of gene regulation patterns (35) of several nuclear encoded genes agrees with the mtDNA phylogeny. Several cases of differences between mtDNA phylogenies and nuclear gene phylogenies exist, and this particular case could be the result of what Avise et al (3–5) call stochastic branching processes. The discrepancy in this particular case could also be due to recency of a common nuclear gene contract of *D. differens* and *D. planitibia* with asymmetric maternal gene introgression. In light of the strong mating asymmetries observed for the Hawaiian *Drosophila,* in general, and the *planitibia* subgroup in particular (61), this is a strong possibility.

The melanogaster *subgroup* This subgroup is by far the most intensely studied group of flies in the genus *Drosophila*. Much attention has been paid to the phylogenetic relationships of the species trio *D. seychellia, D. simulans,* and *D. mauritiana*. Resolution of this trio by mtDNA RFLP analysis is plagued by genomic introgression among these taxa, causing the mtDNA data to give phylogenies with polyphyletic origins for *D. mauritiana* and *D. simulans* (88). ADH sequences (10, 25) indicated uncertainty about the phylogenetic relationships of these species, although Lachaise et al (62) pointed out that the ADH sequence data support a more recent derivation of *D. sechellia* from *D. simulans* than does *D. mauritiana*. Unfortunately *D. sechellia* was the only taxon not analyzed for rDNA and histone genes (21) from this subgroup, and hence the rDNA/histone data are consistent with any arrangement of these three taxa.

Caccone et al (13), using DNA-DNA hybridization, claimed to have broken the trichotomy, placing *D. mauritiana* and *D. sechellia* as most closely related, with *D. simulans* next and then *D. melanogaster*. Lachaise et al (62) summarized much of these data (except for the DNA hybridization study) as well as the chromosomal, morphological, and behavioral data into a consensus phylogeny (Figure 2). There are several differences between these two trees; basically three of the five nodes defined in the consensus tree are supported in the DNA hybridization tree. The major difference is the placement of *D. erecta* in the two trees. In the consensus tree *D. erecta* is a sister taxon of *D. orena,* and in the DNA hybridization tree it is a sister taxon of *D. yakuba* and *D. tessieri*.

AMONG SPECIES GROUP RELATIONSHIPS As discussed in earlier sections only certain techniques are consistently useful in examining higher level questions. For instance, DNA reassociation and mtDNA RFLP analysis, although apparently extremely useful within species groups and subgroups, become unreliable above that level. DNA sequences, 2D protein electrophoresis, and immunoprecipitation have all claimed to be useful at higher levels. A combination of these three approaches has been used to examine the two following higher level questions in the genus *Drosophila*.

The groups and subgroups of Hawaiian Drosophila Bevereley & Wilson (8) used immunoprecipitation techniques on larval hemolymph protein to examine the phylogenetic relationships of the Hawaiian *Drosophila*. They were particularly interested in the relationship of species that were placed in the picture-winged species group of Hawaiian *Drosophila*, as based on chromosomal, morphological and behavioral data. Their results indicated that the picture-winged group, previously thought to be monophyletic, was polyphyletic by the LHP data, with two major lineages contributing to the group: the *adiastola* subgroup and a monophyletic assemblage of picture-winged species comprised of the *grimshawi* complex and the *planitibia* subgroup. These results, combined with the degree of immunological distance observed between the *adiastola* subgroup and the other Hawaiian *Drosophila*, prompted Bevereley & Wilson (8) to suggest an "ancient origin" for the Hawaiian *Drosophila*. DeSalle et al (30) used transversions in the 143 bases of the ND-1 gene from the mtDNA to examine the phylogenetic relationships of the picture wings. This study as well as DNA sequences of the ADH gene (J. A. Hunt, personal communication) confirmed the molecular hypothesis that the picture-winged *Drosophila* are indeed polyphyletic.

The subgenus Sophophora The relationships of the *obscura*, *melanogaster*, and *willistoni* species groups in the subgenus *Sophophora* have been examined using mtDNA sequences (29), nuclear rDNA sequences (79), LHP immunoprecipitation (7), and DNA reassociation experiments on the C-DNA portion of the nuclear genome (A. Caccone, personal communication). All of these analyses support the sister group relationship of the *melanogaster* and *obscura* species groups, with the *willistoni* species group branching first. The mtDNA sequences support this relationship less strongly than the other studies (29). An interesting hypothesis resulting from the nuclear rDNA sequence analysis is that the *Sophophora* are not monophyletic (79).

GENERA, SUBGENERA, AND SPECIES GROUPS IN THE DROSOPHILIDAE As mentioned in the previous sections of this review, Throckmorton (105) proposed the first comprehensive phylogeny for the Drosophilidae. The relationships that he proposed for the various genera and subgenera were not well defined or resolved, and he presented a classical figure of *Drosophila* phylogeny as a bush (105). Three molecular studies have directly attempted to examine in more detail the relationships that Throckmorton (105) proposed, and Grimaldi (46) revised. Bevereley & Wilson (6, 7) used immunoprecipitation of LHP, Spicer (90) used 2D-gel electrophoresis of soluble proteins, and DeSalle (29) used mtDNA sequences to examine several of these higher level questions. A more detailed examination of the molecular and morphological hypotheses of some of these relationships follows.

RECONCILING MOLECULAR AND MORPHOLOGICAL CHARACTER SCHEMES: TEST CASE OF THE ENIGMATIC HAWAIIAN DROSOPHILIDAE

The more than 400 species of Drosophilidae that are endemic to the Hawaiian archipelago provide an opportunity to evaluate the relative merits of molecular and morphological cladograms. The endemic species are undoubtedly the premier example of an adaptive radiation, and their highly modified morphology may have obscured attempts to decipher their relationships. Actually the study of Hawaiian drosophilid systematics has traditionally met with little controversy. Hardy (47) proposed several genera for some of the species now placed in the subgenus *Drosophila* by Kaneshiro (58–60). Other species are placed in the genera *Grimshawomyia, Celidisoma, Scaptomyza, Titanochaeta,* and *Engiscaptomyza* (the latter a subgenus of *Drosophila,* sensu Kaneshiro), and collectively these have been lumped into a group with the unfortunate name "scaptomyzoids" (103, 105). Essentially, Kaneshiro adopted Hardy's diagnosis of *Drosophila* and his placement of most of the Hawaiian "Drosophila" in this genus. Likewise, the view that the endemic Hawaiian species are indeed *Drosophila* has been widely accepted and used as evidence for an explosive radiation of species (17). Despite the wide acknowledgment that these species are *Drosophila* and *Scaptomyza,* efforts have so far proven unsuccessful in finding the sister group of the Hawaiian *Drosophila* within the genus or subgenus *Drosophila.*

One view of the Hawaiian drosophilids is that the entire fauna represents one monophyletic clade, with the large cosmopolitan genus *Scaptomyza* as one lineage in this clade. This view originated with Throckmorton (103) and has been actively espoused by most systematists investigating these taxa. In addition, this view is also corroborated by the molecular studies (32; J. A. Hunt, personal communication). On the basis of internal morphological characters, Throckmorton (103) maintained that *Scaptomyza* originated in Hawaii and colonized the world, because he felt that the Hawaiian *Scaptomyza* intergraded with the Hawaiian *Drosophila*, particularly the genus/subgenus *Engiscaptomyza* (60).

In the cladistic morphological study by Grimaldi (46), the Hawaiian "Drosophila" are not placed within or even near the genus *Drosophila* but are a sister group to a set of seven mycophagous genera. For this reason, the Hawaiian "Drosophila" were removed from the genus and their name replaced with the original one, *Idiomyia* (sensu lato) (46). Moreover, the genus *Scaptomyza* was resolved as monophyletic. The Hawaiian species, *Engiscaptomyza*. *Celidosoma* and *Grimshawomyia* are revealed as close relatives of *Scaptomyza* (Figure 1). If Grimaldi's (46) scheme is correct, then there are two and perhaps as many as five monophyletic groups of endemic Hawaiian drosophilids.

There are some major similarities in the morphological cladogram by Grimaldi (46) and the molecular one by DeSalle (29) and DeSalle & Kaneshiro (32) (Figure 3). Both studies found the subgenera *Scaptodrosophila* and *Hirtodrosophila* to be plesiomorphic with respect to *Drosophila* and some of the other drosophiline genera and that they should therefore be excluded from the genus *Drosophila*. The genus *Chymomyza* was in a similar plesiomorphic position. *Engiscaptomyza* is most closely related to the genus *Scaptomyza* in both the morphological and molecular cladograms. However, the molecular cladogram indicates that the Hawaiian species originally placed

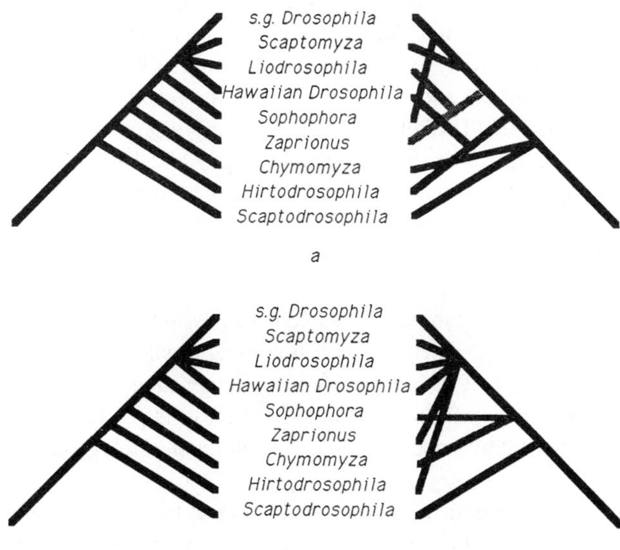

Figure 3 Cladograms showing the comparison of the molecular (mtDNA) hypothesis (on the left side) discussed in the text and two morphological hypotheses (on the right side); (*a*) the hypothesis advanced by Grimaldi (46) and (*b*) Throckmorton's hypothesis (105).

in the subgenus *Drosophila* should be the sister group to *Scaptomyza; Scaptomyza* + Hawaiian *Drosophila* is the sister group to subgenus *Drosophila.* We are at a point where the relationships of Hawaiian drosophilids are controversial, where molecular and morphological data sets differ in their results, and there is a clear need to examine the bases for the different hypotheses.

The hypothesis by Grimaldi (46) was reanalyzed and compared to the molecular hypotheses (29, 32) by first reducing the morphological and molecular data matrices down to the same nine terminal taxa (genera and subgenera), since the same representative species were not examined in the two studies. Characters were omitted from the global cladogram where they would be scored as autapomorphies on the local cladogram, or because they had no relevance to the nine terminal taxa. This resulted in a morphological matrix with 10 binary characters by 9 taxa, and a molecular matrix with 60 characters (with four possible character states A,G,T, and C) and 9 taxa. The most exhaustive parsimony algorithm produced a result of 1 tree with a length of 16 steps, CI = 0.62 and RI = 0.76 for the morphological data. The topology of this tree is the same as the relative relationships in the larger cladogram of Grimaldi (46) (Figure 1), so the hypothesis of relationships seems unaffected by additional taxa and characters. Exhaustive parsimony search of the molecular data gave two trees with a length of 139 steps, CI = 0.54 and RI = 0.41. When analyzed by a strict consensus procedure, the two parsimony trees indicate that the taxa *Liodrosophila, Drosophila,* Hawaiian *Drosophila,* and *Scaptomyza* are an unresolved polychotomy (Figure 3). In this case, removing taxa and characters affects the resolution of relationships in this part of the tree. The relationships obtained in this analysis for the rest of the tree are identical to those in DeSalle (29) and DeSalle & Kaneshiro (32). The molecular phylogeny is compared to the morphological hypotheses of Grimaldi (46) and Throckmorton (105) in Figure 3.

By imposing the topology of the molecular tree on Grimaldi's data matrix, a tree length of 23 steps, CI = 0.43, and RI = 0.48 was obtained. The molecular tree obviously imposes less consistency among the morphological characters. Conversely, when the morphological topology is imposed on the molecular data, a tree of 157 steps (18 steps longer than the parsimony tree), CI = 0.52, and RI = 0.38 is obtained. Interestingly, the CI and RI are reduced only slightly by imposing the morphological topology on the molecular data.

Is there any independent biological evidence to use in evaluating the molecular and morphological hypotheses, in conjunction with parsimony analysis? Morphological characters might be easier to use to assess the biological basis of these schemes, since homology decisions can be deliberated upon using a variety of structural and ontogenetic evidence. However, for molecular characters, factors such as length of the apomorphic nucleotide sequence, roles of third codon changes, and structural molecular constraints of particular coding regions can be taken into account in assessing homoplasious characters.

On the morphological tree, character 15 (as scored by Grimaldi—46) has the lowest CI and goes through three steps. That character is the presence of a large, keeled apodeme on the middle of the ventral floor of the cibarium of the mouthparts. When these same morphological characters are imposed on the molecular tree, characters 20, 102, and 201 each go through three steps. Character 20 is a lacinia (in the adult mouthparts) with long, thin dorsal and vental arms; character 102 is the presence of a row of setulae between the pair of large katepisternal setae; and character 201 is the presence of microtrichia on the paraphysis. As yet, there is no evidence from the fine structure of any of these characters [20, 102 and 201] to suggest whether they are independently derived or lost, whereas character 15 does show a great deal of variation in its shape and therefore may be truly homoplasious. The other characters [20, 102 and 201] might be good candidates for more precise examination at the developmental and morphological level.

Another approach to reconciling the morphological and molecular cladograms is to combine the data into one large data matrix. However, an overwhelming bias is created in favor of the molecular characters because the number of characters is six times larger for the sequence data. When the two data sets are combined and analyzed in an exhaustive search for parsimony trees, the same tree topology obtained for just the molecular data set is obtained. The topology of the tree is also quite stable as consensus trees constructed from parsimony trees and from the next nearly parsimonious trees that are two and three steps longer than the shortest tree retain the same topology seen for the molecular data. No appropriate weighting scheme is currently available for this system that would allow for compensation for numbers of characters in the two types of data.

CONCLUSIONS

While some molecular systematists and evolutionary biologists have claimed the primacy of the molecular approach in systematics (44, 87), some morphologists and theoretical systematists have claimed "molecules are bad for you" (Farris; described in 15). Some authors have attempted to reconcile the two approaches (49), while others (62) have suggested that the two approaches should be combined to produce consensus trees in systematic analysis. It would be nice if all studies using all approaches agreed, but very few systematics studies examined with the different approaches have shown agreement. What then can be done about data sets on the same taxa that contradict each other?

We recognize that there are several levels at which systematic studies may disagree. These levels include disagreement among morphological studies and disagreement among molecular studies. Comparison of Okada's and Throckmorton's morphological phylogenies (46) is a good example of the former, while the comparisons of Figure 2 are good examples of the latter. Discrepancies at this level are attributable in most cases to the method of analysis (phenetic or cladistic) or to the characters chosen. Assessment of how to reconcile such discrepancies will lie in our ability to discriminate which methodological technique and which characters are better suited to phylogenetic analysis. In some cases, we are constrained to use phenetic analysis (Tm, immunological distance, etc) and the problem then becomes one of accepting or rejecting a particular molecular technique as most appropriate.

Another level at which discrepancies occur is, of course, between molecular and morphological studies. When comparable types of analyses are used for both approaches [i.e. as in the morphological studies of Grimaldi (46) and the mtDNA studies (29, 32) on a subset of the same taxa], discrepancies can be attributed to the types of characters used. Reconciliation of such discrepancies requires reassessment of the types and numbers of characters that support the two different data sets.

On the molecular side, one must assess several aspects of the molecules examined. Some of these aspects are: mode of inheritance of the molecules, times of divergence that affect the gene tree–species tree argument, position in codon of characters, base composition, transition-transversion ratios, and character covariance. In addition it will be necessary to examine the nature of the characters that define nodes that are found to contradict morphological hypotheses and to discover if these characters have higher rates of convergence or base composition qualities that affect the cladistic analysis. One problem that plagues DNA sequencing studies is the choice of proper outgroups for polarizing the relationships of ingroups (112, 113). Because only four character states exist for any character in a nucleotide sequence, and these character states are highly reversible in some cases, outgroups can be completely random with respect to ingroup taxa. The ability to recognize this randomness should be an essential part of phylogenetic analysis of molecular characters.

On the morphological side, character types and quantities that define discrepant nodes could be reassessed in more detail. Comparison to molecular studies could be useful in determining which characters might be convergent or inappropriate for a particular analysis. In addition, molecular data could reveal that morphological characters that at first examination appear to be perfectly good characters could be placed under closer scrutiny. The developmental-molecular basis of many of the morphologies in the *Drosophila* bauplan is well understood, and it may be possible to show that particular morphological characters that appear to define monophyletic groups do not have the same genetic developmental basis in all taxa in those monophyletic groups. To our knowledge the family Drosophilidae is perhaps the first group of organisms where this level of examination will be consistently possible.

ACKNOWLEDGMENTS

We would like to thank Dr. John Hunt and Dr. Michel Solignac for sharing unpublished manuscripts with us. We would also like to thank Dr. Hamp Carson, Dr. Ken Kaneshiro, and Dr. Michael Kambysellis for stimulating discussion on Drosophilidae systematics. Finally we would like to thank Dr. Annie Williams for her critical reading of the manuscript.

Literature Cited

1. Atkinson, P. W., Mills, L. E., Starmer, W. T., Sullivan, D. T. 1988. Structure and evolution of the Adh genes of *Drosophila mojavensis*. *Genetics* 120:713–23
2. Ashburner, M., Lemeunier, F. 1976. Relationships within the *melanogaster* species subgroup of the genus *Drosophila (Sophophora)* I. Inversion polymorphisms in *Drosophila melanogaster* and *Drosophila simulans*. *Proc. R. Soc. Lond. B*. 193:137–57
3. Avise, J. C. 1986. Mitochondrial DNA and the evolutionary genetics of higher animals. *Philos. Trans. R. Soc. Lond. B* 312:325–42
4. Avise, J. C., Arnold, J., Ball, R. M., Bermingham, E., Lamb, T., Neigel, J. E., et al. 1988. Intraspecific phylogeography: mitochondrial DNA bridge between population genetics and systematics. *Annu. Rev. Ecol. Syst.* 18:489–522
5. Avise, J. C. Neigel, J. E., Arnold, J. 1984. Demographic influences mitochondrial DNA lineage survivorship in animal populations. *J. Mol. Evol.* 20:99–105
6. Bevereley, S. M., Wilson, A. C. 1982. Molecular evolution in the *Drosophila* and higher diptera. I. Micro-complement fixation studies of larval hemolymph protein. *J. Mol. Evol.* 18:251–64
7. Bevereley, S. M., Wilson, A. C. 1984. Molecular evolution in the *Drosophila* and higher diptera. II. A timescale for fly evolution. *J. Mol. Evol.* 21:1–13
8. Beverely, S. M., Wilson, A. C. 1985. Ancient origin for Hawaiian Drosophilinae inferred from protein comparisons. *Proc. Natl. Acad. Sci. USA* 82:4753–57
9. Bishop, J. G. III, Hunt, J. A. 1988. DNA divergence in and around the alcohol dehydrogenase locus in five closely related species of Hawaiian *Drosophila*. *Mol. Biol. Evol.* 5:415–31
10. Bodmer, M., Ashburner, M. 1984. Conservation and change in the DNA sequences coding for alcohol dehydrogenase in sibling species of *Drosophila*. *Nature* 309:425–17
11. Brown, W. M., Prager, E. M., Wang, A., Wilson, A. C. 1982. Mitochondrial DNA sequences of primates: Tempo and mode of evolution. *J. Mol. Evol.* 18:225–39
12. Caccone, A., Amato, G., Powell, J. R. 1987. Intraspecific DNA divergence in *Drosophila:* a study on parthenogenetic *D. mercatorum*. *Mol. Biol. Evol.* 4: 343–50
13. Caccone, A., Amato, G., Powell, J. R. 1988. Rates and patterns of scnDNA and mtDNA divergence within the *Drosophila melanogaster* subgroup. *Genetics* 118:671–83
14. Caccone, A., DeSalle, R., Powell, J. R. 1988. Calibration of the change in thermal stability of DNA duplexes and degree of base pair mismatch. *J. Mol. Evol.* 27:212–16
15. Carpenter, J. M. 1987. Cladistics of cladists. *Cladistics* 3:263–375
16. Carson, H. L. 1987. Tracing ancestry with chromosomal sequences. *Trends Ecol. Evol.* 2:203–7
17. Carson, H. L., Kaneshiro, K. Y. 1976. *Drosophila* of Hawaii: systematics and ecological genetics. *Annu. Rev. Ecol. Syst.* 7:11–45
18. Chang, H., Wang, D., Ayala, F. J. 1989. Mitochondrial DNA evolution in the *Drosophila nasuta* subgroup of species. *J. Mol. Evol.* 28:337–48
19. Clary, D. O., Wolstenholme, D. R.

1985. The mitochondrial DNA molecule of *Drosophila yakuba:* nucleotide sequence, gene organization and genetic code. *J. Mol. Evol.* 22:252–71

20. Cockerell, T. D. A. 1923. Insects in amber from South America. *Am. J. Sci.* 5:331–33
21. Coen, E., Strachan, T., Dover, G. 1982. Dynamics of concerted evolution of ribosomal DNA and histone gene families in the *melanogaster* species subgroup of *Drosophila. J. Mol. Biol.* 158:17–35
22. Coen, E. S., Thoday, J. N., Dover G. 1982. Rate of turnover of structural variants in the rDNA family of *Drosophila melanogaster. Nature* 295:564–68
23. Cohn, V. H., Moore, G. P. 1988. Organization and evolution of the alcohol dehydrogenase gene in *Drosophila. Mol. Biol. Evol.* 5:154–66
24. Collier, G. E., MacIntyre, R. J. 1977. Microcomplement fixation studies on the evolution of alpha-glycerophosphate dehydrogenase within the genus *Drosophila. Proc. Natl. Acad. Sci. USA* 74:684–88
25. Coyne, J. A., Kreitman, M. 1986. Evolutionary genetics of two sibling species, *Drosophila simulans* and *D. sechelia. Evolution* 40:673–91
26. Daniels, S., Strausbaugh, L. D. 1986. The distribution of P element sequences in *Drosophila:* the *Willistoni* and *saltans* species groups. *J. Mol. Evol.* 23:138–48
27. de Bruijn, M. H. L. 1983. *Drosophila melanogaster* mitochondrial DNA, a novel organization and genetic code. *Nature* 304:234–41
28. DeBry, R., Slade, N. A. 1985. Cladistic analysis of restriction endonuclease cleavage maps within a maximum likelihood framework. *Syst. Zool.* 34:21–34
29. DeSalle, R. 1991. The phylogenetic relationships of genera, subgenera and species groups in the family Drosophilidae from mtDNA sequences. *Mol. Syst. Evol.* Submitted
30. DeSalle, R., Friedman, T., Prager, E. M., Wilson, A. C. 1987. Tempo and mode of sequence evolution in mitochondrial DNA of Hawaiian *Drosophila. J. Mol. Evol.* 26:157–64
31. DeSalle, R., Giddings, L. V. 1986. Discordance of nuclear and mitochondrial DNA phylogenies in Hawaiian *Drosophila. Proc. Natl. Acad. Sci. USA* 83:6902–6
32. DeSalle, R., Kaneshiro, K. Y. 1991. Origin and possible time of divergence of the Hawaiian Drosophilidae from mtDNA sequences. *Mol. Biol. Evol.* Submitted
33. DeSalle, R., Templeton, A. R. 1988. Founder effects accelerate the rate of mtDNA evolution of Hawaiian *Drosophila. Evolution* 42:1076–85
34. Dickinson, W. J. 1980. Evolution of patterns of gene expression in Hawaiian picture-winged *Drosophila. J. Mol. Evol.* 16:73–94
35. Dickinson, W. J. 1989. Gene regulation and evolution. In *Genetics, Speciation and the Founder Principle,* ed. L. V. Giddings, K. Y. Kaneshiro, W. W. Anderson, pp. 181–204. Oxford: Oxford Univ. Press
36. Dobzhansky, T. 1941. *Genetics and the Origin of Species.* New York: Columbia Univ. Press. 2nd ed.
37. Donoghue, M. J., Doyle, J. A., Gauthier, J., Kluge, A. G., Rowe, T. 1989. The importance of fossils in phylogeny reconstruction. *Annu. Rev. Ecol. Syst.* 20:431–60
38. Duda, O. 1935. *Drosophilidae.* In *Die fliegen der Palearktischen Region,* ed. E. Linder, part 58g. Stuttgart: E. Schweizerbart'sche Verlagsbuchhandlung
39. Entigh, T. D. 1970. DNA hybridization in the genus *Drosophila. Genetics* 66:55–68
40. Felsenstein, J. 1985. Confidence limits on phylogenies with a molecular clock. *Syst. Zool.* 34:152–61
41. Felsenstein, J. 1988. Phylogenies from molecular sequences: Inferences and reliability. *Annu. Rev. Genet.* 22:521–65
42. Frey, R. 1921. Studien uber den Bau des Mundes Des Niederen Diptera Schizophora nebst bemerkungen uber dies systematik dieser Dipterengruppe. *Acta Soc. Fauna Flora Fenn.* 48:1–245
43. Goddard, K., Caccone, A., Powell, J. R. 1990. Evolutionary implications of DNA divergence in the *Drosophila obscura* group. *Evolution* 44:1656–70
44. Gould, S. J. 1985. A clock for evolution. *Nat. Hist.* 94:12–25
45. Grimaldi, D. A. 1987. Amber fossil Drosophilidae (Diptera), with particular reference to the Hispaniolan taxa. *Am, Mus. Nov.* 2880:1–23
46. Grimaldi, D. A. 1990. A phylogenetic, revised classification of the genera in the Drosophilidae (Diptera). *Bull. Am. Mus. Nat. Hist.* 197:1–139
47. Hardy, D. E. 1965. Diptera: Cyclorrapha. II. Series Schizophora, Section Acalypterae. I. Family Drosophilidae. In *Insects of Hawaii,* ed. E. C. Zimmerman, 12:1–814. Honolulu: Univ. Hawaii Press

48. Hennig, W. 1966. *Phylogenetic Systematics*. Urbana: Univ. Ill. Press
49. Hillis, D. M. 1987. Molecular versus morphological approaches to systematics. *Annu. Rev. Ecol. Syst.* 18:23–42
50. Hillis, D. M., Davis, S. K. 1986. Evolution of ribosomal DNA: fifty million years of recorded history in the frog genus *Rana*. *Evolution* 40:1275–88
51. Hillis, D. M., Davis, S. K. 1988. Ribosomal DNA: interspecific polymorphism, concerted evolution and phylogeny reconstruction. *Syst. Zool.* 32:63–66
52. Hubby, J. L., Throckmorton, L. 1965. Protein differences in *Drosophila*. II. Comparative species, genetics and evolutionary problems. *Genetics* 52:203–15
53. Hunt, J. A., Carson, H. L. 1983. Evolutionary relationships of four species of Hawaiian *Drosophila* as measured by DNA reassociation *Genetics* 104:353–64
54. Hunt, J. A., Hall, T. J., Britten, R. J. 1981. Evolutionary distances in Hawaiian *Drosophila* measured by DNA reassociation. *J. Mol. Biol.* 17:361–67
55. Hunt, J. A., Houtchens, K. A., Brezinsky, L., Shadravan, F., Bishop, J. G. 1988. Genomic DNA variation within and between closely related species of Hawaiian *Drosophila*. In *Genetics, Speciation and the Founder Principle*, ed. L. V. Giddings, K. Y. Kaneshiro, W. W. Anderson, pp. 168–80. Oxford: Oxford Univ. Press
56. Huxley, J. S. 1942. *Evolution, the Modern Synthesis*. New York: Harper
57. Jeffery, D. E., Farmer, J. L., Pliliy, M. D. 1988. Identification of Mullerian chromosomal elements in Hawaiian *Drosophila* by in situ hybridization. *Pac. Sci.* 42:48–50
58. Kaneshiro, K. Y. 1969. A study of the relationships of Hawaiian *Drosophila* species based on external male genitalia. *Univ. Texas Publ. 6918:55–70*
59. Kaneshiro, K. Y. 1974. Phylogenetic relationships of Hawaiian Drosophilidae based on morphology. In *Genetic Mechanisms of Speciation in Insects,* ed. M. J. D. White, pp. 102–10, Melbourne: Australia and New Zealand Book Co.
60. Kaneshiro, K. Y. 1976. A revision of generic concepts in the biosystematics of the Hawaiian Drosophilidae. *Proc. Hawaiian Entomol. Soc.* 22:255–78
61. Kaneshiro, K. Y. 1983. Sexual selection and direction of evolution in the biosystematics of the Hawaiian Drosophilidae. *Annu. Rev. Entomol.* 28:161–78
62. Lachaise, D., Cariou, M.-L., David, J. R., Lemunier, F., Tsacas, L., Ashburner, M. 1988. Historical biogeography of the *Drosophila melanogaster* species group. *Evol. Biol.* 22:159–225
63. Latorre, A., Barrio, E., Moya, A., Ayala, F. J. 1988. Mitochondrial DNA evolution in the *Drosophila obscura* group. *Mol. Biol. Evol.* 5:717–28
64. Latorre, A., Moya, A., Ayala, F. J. 1986. Evolution of mitochondrial DNA in *Drosophila subobscura*. *Proc. Natl. Acad. Sci. USA* 83:8649–53
65. Lee, T. J., Pak, J. H. 1986. Biochemical phylogeny of the *Drosophila aurata* complex. *Drosoph. Inf. Serv.* 63:81
66. Leigh Brown, A. J., Langley, C. H. 1979. Reevaluation of the level of genetic heterozygosity in natural populations of *Drosophila melanogaster* by two dimensional gel electrophoresis. *Proc. Natl. Acad. Sci. USA* 76:2381–84
67. Lemeunier, F., Ashburner, M. 1976. Relationships within the *melanogaster* species subgroup of the genus *Drosophila (Sophophora)* I. Phylogenetic relationships between six species based on polytene chromosome banding sequences. *Proc. R. Soc. Lond. B.* 193:275–94
68. MacIntyre, R. J. 1971. Evolution of acid phosphatase-I in the genus *Drosophila* as estimated by subunit hybridization. *Genetics* 68:483–508
69. MacIntyre, R. J., Collier, G. E. 1986. Protein evolution in the genus *Drosophila*. In *The Genetics and Biology of* Drosophila, ed. M. Ashburner, J. N. Thompson, H. L. Carson, 3e:39–146. London: Academic
70. MacIntyre, R. J., Dean, M. R. 1978. Evolution of acid phosphatase-1 in the genus *Drosophila* as estimated by subunit hybridization. Interspecific tests. *J. Mol. Evol.* 12:143–71
71. Martinez-Cruzado, J. C., Swimmer, C., Fenerjian, M. G., Kafatos, F. C. 1988. Evolution of the autosomal chorion locus in *Drosophila*. I. General organization of the locus and sequence comparisons of genes s15 and s19 in evolutionary distant species. *Genetics* 120: 663–677
72. Mayr, E. 1942. *Systematics and the Origin of Species*. New York: Columbia Univ.
73. Mayr, E., Linsley, E. G., Usinger, R. L. 1953. *Methods and Principles of Systematic Zoology*. New York: MacGraw-Hill
74. Ohnishi, S., Kawanishi, M., Watanabe, T. K. 1983. Biochemical phylogenies of *Drosophila:* protein differences detected by two-dimensional electrophoresis. *Genetica* 61:55–63
75. Okada, T. 1956. *Systematic Study of*

Drosophilidae and Allied Families of Japan. Tokyo: Gihodo
76. Okada, T. 1989. A proposal for establishing tribes for the family Drosophilidae with keys to tribes and general (Diptera). *Zool. Soc.* 6:391–99
77. Pamilo, P., Nei, M. 1988. Relationships between gene trees and species trees. *Mol. Biol. Evol.* 5:568–83
78. Patterson, J. T., Stone, W. S. 1952. *Evolution in the Genus* Drosophila. New York: MacMillan
79. Pelandakis, M., Higgins, D. G., Solignac, M. 1991. Molecular phylogeny of the subgenus *Sophophora* of *Drosophila* derived from large subunit of ribosomal RNA sequences. *Proc. Natl. Acad. Sci. USA* In press
80. Powell, J. R., Caccone, A. 1989. Intraspecific and interspecific genetic variation in *Drosophila*. *Genome* 31:233–38
81. Roiha, H., Christopher, A. R., Brown, M. J., Glover, D. M. 1983. Widely differing degrees of sequence conservation of the two types of rDNA insertion within the melanogaster species subgroup of *Drosophila*. *EMBO J.* 2:721–26
82. Rowan, R. G., Dickinson, W. J. 1988. Nucleotide sequence of the genomic region encoding alcohol dehydrogenase in *Drosophila affinidisjuncta*. *J. Mol. Evol.* 28:43–54
83. Rowan, R. G., Hunt, J. A. 1991. Rates of DNA change and phylogeny from the DNA sequences of the alcohol dehydrogenase gene for five closely related species of Hawaiian *Drosophila*. *Mol. Biol. Evol.* 8:49–70
84. Sarich, N., Schmid, C., Marks, J. 1989. DNA hybridization as a guide to phylogenies: a critical analysis. *Cladistics* 5:3–32
85. Schaefer, S. W., Aquadro, C. W. 1987. Nucleotide sequence of the ADH gene region of *Drosophila pseudoobscura:* evolutionary change and evidence for an ancient gene duplication. *Genetics* 117: 61–73
86. Schultze, D. H., Lee, C. S. 1986. DNA sequence comparisons among closely related *Drosophila* species of the *mulleri* complex. *Genetics* 113:287–303
87. Sibley, C. G., Alquist, J. E. 1987. Avian phylogeny reconstructed from comparisons of the genetic material DNA. In *Molecules and Morphology in Evolution: Conflict or Compromise*, ed. C. Patterson, pp. 95–121. Cambridge: Cambridge Univ. Press
88. Solignac, M., Monnerot, M., Mounolou, J.-C. 1986. Mitochondrial DNA evolution in the *melanogaster* species subgroup of *Drosophila*. *J. Mol. Evol.* 23:31–40
89. Spicer, G. S. 1985. Systematics of the *Drosophila virilis* species group as assessed by two dimensional electrophoresis. PhD thesis. Texas Tech Univ., Lubbock, Texas
90. Spicer, G. S. 1988. Molecular evolution among some *Drosophila* species groups as indicated by two dimensional electrophoresis. *J. Mol. Evol.* 27:250–60
91. Springer, M., Krajewski, C. 1989. DNA hybridization in animal taxonomy: a critique from first principles. *Q. Rev. Biol.* 64:291–318
92. Stalker, H. D. 1973. Intergroup phylogenies in *Drosophila* as determined by comparisons of salivary banding patterns. *Genetics* 70:457–74
93. Starmer, W. T., Sullivan, D. T. 1989. A shift in the third-codon-position nucleotide frequency in alcohol dehydrogenase genes in the genus *Drosophila*. *Mol. Biol. Evol.* 6:544–52
94. Steinemann, M., Pinsker, W., Sperlich, D. 1984. Chromosome homologies within the *Drosophila obscura* group probed by in situ hybridization. *Chromosoma* 91:46–53
95. Sturtevant, A. H. 1921. *The North American Species of* Drosophila. Washington: Carnegie Inst.
96. Sturtevant, A. H. 1942. The classification of the genus *Drosophila,* with descriptions of nine new species. *Univ. Texas Publ. 4213:*6–51
97. Takada, H. 1966. Male genitalia of some Hawaiian Drosophilidae. *Univ. Texas Publ. 6615:*315–33
98. Templeton, A. R. 1983. Phylogenetic inference from restriction endonuclease cleavage site maps with particular reference to the evolution of humans and apes. *Evolution* 37:221–44
99. Templeton, A. R. 1983. Convergent evolution and nonparametric inferences from restriction fragment and DNA sequence data. In *Statistical Analysis of DNA Sequence Data,* ed. B. Weir, pp. 151–79. New York: Marcel Dekker
100. Thorpe, P. A., Dickinson, W. J. 1988. The use of regulatory patterns in constructing phylogenies. *Syst. Zool.* 37: 97–105
101. Throckmorton, L. H. 1962. The problem of phylogeny in the genus *Drosophila*. *Univ. Texas Publ. 6205:* 207–343
102. Throckmorton, L. H. 1965. Similarity versus relationship in *Drosophila Syst. Zool.* 14:221–36
103. Throckmorton, L. H. 1966. The relationships of endemic Hawaiian Dro-

sophilidae. *Univ. Texas Publ. 6615:* 335–96

104. Throckmorton, L. H. 1968. Concordance and discordance of taxonomic characters in *Drosophila* classification. *Syst. Zool.* 17:355–87
105. Throckmorton, L. H. 1975. The phylogeny ecology and geography of *Drosophila*. In *Handbook of Genetics,* ed. R. C. King, 3:421–69. New York: Plenum
106. Throckmorton, L. H. 1977. *Drosophila* systematics and biochemical evolution. *Annu. Rev. Ecol. Syst.* 8:235–54
107. Throckmorton, L. H. 1978. Molecular phylogenetics. *Beltsville Symp. Agric. Res.* 2:221–39
108. Wasserman, M. 1976. Evolution of the *repleta* group. In *The Genetics and Biology of* Drosophila, ed. M. Ashburner, H. L. Carson, J. N. Thompson, 3b:61–139. London: Academic
109. Wheeler, M. R. 1963. A note on some fossil Drosophilidae (Diptera) from the amber of Chiapas, Mexico. *J. Paleontol.* 37:123–24
110. Wheeler, M. R. 1981. The Drosophilidae: a taxonomic overview. In *The Genetics and Biology of* Drosophila, ed. M. Ashburner, H. L. Carson, J. N. Thompson, 3a:1–97. New York: Academic
111. Wheeler, W. C. 1990. Combinatorial weights in phylogenetic analysis: A statistical parsimony procedure. *Cladistics* 6:269–75
112. Wheeler, W. C., 1991. Nucleic acid sequence phylogeny and random outgroups. *Cladistics* 6:363–68
113. Wheeler, W. C., Honeycutt, R. L. 1987. Paired sequence differences in ribosomal RNAs: evolutionary and phylogenetic implications. *Mol. Biol. Evol.* 5:90–96
114. Williams, S. M., DeBry, R. W., Feder, J. L. 1988. A commentary on the use of ribosomal DNA in systematic studies. *Syst. Zool.* 37:60–63
115. Yoon, Y. S. 1989. In *Genetics, Speciation and the Founder Principle*, ed. L. V. Giddings, K. Y. Kaneshiro, W. W. Anderson, pp. 129–48. Oxford: Oxford Univ. Press
116. Zimmer, E. A., Hambry, R. K., Arnold, M. L., Leblanc, D. A., Theriot, E. C. 1989. Ribosomal RNA phylogenies and flowering plant evolution. In *The Hierarchy of Life,* ed. B. Fernholm, K. Bremer, H. Jornvall, pp. 205–13. New York: Elsevier
117. Zweibel, L. J., Cohn, V. H., Wright, D. R., Moore, G. P. 1982. Evolution of single copy DNA and the ADH gene in seven drosophilids. *J. Mol. Evol.* 19:62–71

Annu. Rev. Ecol. Syst. 1991. 22:477–503

HERBIVORES AND THE DYNAMICS OF COMMUNITIES AND ECOSYSTEMS

Nancy Huntly

Department of Biological Sciences, Idaho State University, Pocatello, Idaho 83209-8007

KEY WORDS: biodiversity, competition, disturbance, vegetation, frequency-dependence

---

INTRODUCTION

Herbivores are taxonomically and ecologically diverse, ranging in size from microscopic zooplankton to the largest of land vertebrates. Aquatic grazers include zooplankton (28, 182), larger invertebrates such as snails, insects, and crayfish, and vertebrates such as waterfowl, tadpoles, fish, muskrats, and moose (11, 27, 73, 115, 162, 163). Insects and mammals are the most conspicuous terrestrial herbivores (2, 46–49, 125), but nematodes (20), crustaceans (152), molluscs (78), birds, and reptiles (69) can also be significant. Marine ecosystems are grazed primarily by crustaceans (57, 79, 112, 143), molluscs (15, 126), fish (83, 84, 92), echinoderms (24, 25, 65), and a few insects (170), reptiles and mammals (123, 185).

These herbivores affect plant communities in many ways. Feeding selectivity and feeding modes are highly varied; the terrestrial insect herbivores alone include phloem and xylem feeders, root grazers, gall formers, and folivores that mine, chew, roll, rasp, or pit leaves (194). Herbivores may select among plant parts, individuals, species, patches, and portions of landscapes (7, 55, 83, 95, 120, 126, 137, 138, 140). Many herbivores also clip or tear loose much plant biomass that is not consumed (5, 70, 104, 105, 122, 169). Herbivores change the environment by their trails, burrows, wallows, den building, foraging, social behavior, and other activities (1, 4, 15, 40, 50, 53, 94, 96, 98, 138, 148, 165, 167, 177). They convert plants into dung, frass,

0066-4162/91/1120-0477$02.00

feces, urine, and other excretions which can alter local nutrient availabilities (11, 24, 50, 52, 138, 140, 182). Additionally, herbivores may be vectors of plant disease (14, 145, 161).

The plants consumed by herbivores include single-celled and colonial phytoplankton, macroalgae, bryophytes, ferns, and angiosperms. There is a great variation in form and function within each of these groups. Marine plants include kelps, algal crusts, foliose and filamentous algae, seagrasses, phytoplankton, and periphyton (53, 54, 79, 128, 185); terrestrial angiosperms range from grasses and forbs to shrubs and trees (46); and aquatic ecosystems have plankton, periphyton, and emergent and submerged macrophytes (123). Even single-celled plants are diverse, including silica-impregnated diatoms and nitrogen-fixing blue-green algae, and they span five orders of magnitude in cell size (79, 112, 182, 190).

Given the vast variety and complexity of plants, herbivores, and their interactions, it is easy to conclude that a general understanding of how herbivores affect communities and ecosystems is not possible. Nevertheless, herbivory can be understood in a general way, and a variety of well-developed theory exists for doing so. Experiments have established that herbivores can significantly alter community structure and dynamics (e.g. 10, 24, 26, 48, 65, 73, 80, 83, 95, 97, 98, 115, 120, 126, 139, 140, 163–165), although they do not always do so (24, 48, 65, 70). Here, I attempt to put these results in context, asking *(a)* by what mechanisms do herbivores influence communities and ecosystems, and *(b)* whether the importance of herbivory to community dynamics varies systematically among ecosystems or habitats. I focus on effects of herbivores on plant diversity, species composition, standing crop, and productivity, and I compare results from terrestrial, aquatic, and marine ecosystems of various types. Effects of herbivores ramify to higher trophic levels (e.g. 96, 148), but these are not considered in detail here.

The literature on plant-herbivore interactions is extensive, and I have selectively cited more recent studies. To emphasize effects shown to occur in nature, I cite primarily field studies. Much literature deals with effects of herbivores on plant individuals or single populations; often these studies don't include information necessary to infer what patterns will be produced at higher levels of organization. Reviews that include more laboratory or behavioral and population studies are available for specific ecosystems and herbivore groups (26, 31, 46–49, 73, 85, 92, 111, 114, 115, 122, 125, 128, 174, 182).

BY WHAT MECHANISMS DO HERBIVORES AFFECT PLANT COMMUNITIES?

Herbivores can affect the numbers, kinds, or relative abundances of plant species in a community by several conceptually distinct mechanisms. These mechanisms aren't mutually exclusive: some can operate simultaneously, and one can sometimes produce another. For herbivory to change the species composition of a community, the net effects of herbivores must disproportionately affect the growth rates of populations of plants. For herbivores to alter the diversity of a community, they must create new limiting factors or reduce the effectiveness of a factor that is limiting in their absence (36). Herbivores may cause disproportionate mortality or tissue loss rates for certain plant species (41, 77, 100, 126) and may also cause the population dynamics of co-ocurring species to be linked, giving the appearance that they compete for resources (88). Additionally, herbivores may produce spatial or temporal heterogeneity in the environment or in plant demographic processes. They may affect community dynamics by causing disturbances to which plant species differentially respond (30, 53, 81, 97, 168, 179, 180), by creating distinct habitats or microhabitats (33, 83, 95, 118, 119, 138, 140), or by causing variation in plant demography through time (33–35, 67, 75, 192a). Although there are numerous biologically interesting ways for these processes to arise, there remain relatively few general ways in which herbivores can influence plant communities.

The question here is not whether herbivores increase or decrease diversity in plant communities. They can do either (77, 88, 89, 126, 149); they can do this at a variety of scales (7, 83, 97, 138, 143); and the patterns produced are dynamic, with trends often changing through time (26, 28, 97, 126). For example, pocket gophers influence diversity of successional old fields by creating initially barren mounds of soil. Plant diversity on mounds is not different from plant diversity on comparably sized plots away from mounds; however, whole field diversity tends to be increased. Whether field diversity increases depends on the age of a field. In older fields, mound soils are distinct from and retain a different flora from off-mound soils. In very young fields, these differences are minor or non-existent (97). On coral reefs, the presence of territorial fishes, roving fishes, and echinoderms can lead to patches of distinct algal composition and diversity. Algae within the grazing area of a fish, urchin, mollusc, or crab may be higher or lower than in adjacent ungrazed areas, and the relative diversity of grazed areas may change with successional stage, but overall diversity typically is higher than where herbivores are absent (24, 83, 92, 126, 143).

The question also is not whether particular plant species increase or decrease with herbivory. Under some conditions, herbivory results in a community dominated by highly productive species (11, 24, 29, 139, 164, 182); in others a community of slower-growing species results (126, 159, 178, 182). The question of interest here is how herbivory influences the dynamics of communities so that patterns of species composition and/or diversity emerge, being altered from or equal to what they would have been in the absence of an herbivore or herbivores.

Herbivores can affect plant populations either in density-independent or density-dependent ways. Density-independent herbivore effects can influence populations and communities in the same way that physical environmental conditions can, and herbivory can be a way in which effects of the physical environment are realized by plants (14, 125, 132, 187). Alternatively, herbivory can be correlated with plant density and add a density-dependent component to population dynamics (14, 38, 51). The rich variety of ways in which herbivores can contribute to population growth rates, and thus to community dynamics, makes herbivory particularly interesting. Even when the direct effects of herbivores are independent of plant density, they can cause systematic changes in species composition and diversity, through fluctuations in mortality or shifts in the relative importance of different limiting factors (34–36).

Compensatory Herbivory

Compensatory effects are those that disproportionately affect populations that are common or are capable of becoming more common in a community (143). Herbivores can affect plants in compensatory ways through density-dependent or frequency-dependent damage to plants. Although it has most often been sought at the level of the foraging individual, frequency dependence can be present at the level of a foraging population without individual herbivores showing frequency-dependent preferences (32). Frequency dependence may in fact be most likely to arise at the community level, because of the combined effects of herbivores with different preferences. Ultimately, long-term coexistence is equivalent to frequency dependence (34, 35, 37), which so many mechanisms indirectly generate over the long-term.

FREQUENCY DEPENDENCE AND DENSITY DEPENDENCE Frequency dependence and density dependence are closely related, and frequency dependence within a community can arise from herbivores that respond in a density-dependent way to their plant resources. The critical distinction is whether the response of an herbivore is to the relative frequencies of plants of two or more species or to absolute densities of a single species. Both could occur simultaneously, and some studies show behavioral responses of herbivores to both (109, 110, 172, 173). The net effect on plants of sessile herbivores that require time to kill plants depends not only on the responses of the herbivores, but also on the response of the herbivores' natural enemies (189). Kareiva (110) gives an example of herbivore density that is dependent on plant spacing and isolation, and in which the pattern is caused by search behavior of the coccinellid beetle predator of the herbivorous aphid.

Density-dependent herbivory can offset the tendency of a plant population to increase in abundance, monopolize resources, and exclude other species from a community. This is the situation described by Paine (157) in which predation by a starfish greatly reduced the abundance of a dominant consumer, allowing its competitors (which included algae) to increase. Lubchenco (126) reported a similar situation for tide pool algae and a grazing snail, *Littorina littorea*. Density-dependent behavioral or population responses of one or more herbivores can generate frequency dependence in a system of co-occurring plant populations. For instance, the antler moth, which feeds on the basal portions of grass tillers, increases in abundance as grasses increase. Ultimately, high moth densities result in sharp decreases in grass biomass, but forbs subsequently increase. As the moths are grass specialists, they fluctuate in abundance with their hosts. Moth populations are constrained also by weather, tending to reach high density only in cold springs (51). Frequency-dependent herbivory also can affect diversity. Herbivores that disproportionately damage the more common species in a patch or community tend to increase plant diversity (41, 100); those that disproportionately damage rare species (68, 158) will have the opposite effect.

Although few models have been developed specifically for herbivory, two that have invoke immediate frequency-dependent behavior of herbivores: The Janzen-Connell hypothesis of tropical forest diversity and associational resistance. The Janzen-Connell (JC: 41, 100) hypothesis describes a particular way that frequency dependence can be produced. Both Janzen and Connell hypothesized that the high diversity of tropical forests might result from disproportionate disease- or herbivore-caused mortality of seedlings growing near parent plants. Recent studies provide considerable support for the JC hypothesis. Connell et al (43) demonstrate compensatory frequency-dependent recruitment, growth, and survival of young age classes in rain forests of tropical and subtropical Australia, but the strength of the effect is not yet clear. For three of four species of tropical trees in Costa Rican primary evergreen forest, adults were more uniformly dispersed than were juveniles, and the adults were also more uniformly distributed than would be predicted by random mortality of the extant juveniles (184). Clark & Clark (38) and Connell et al (43) reviewed other data and found strong support for density- or distance-dependent mortality. The Janzen-Connell hypothesis has also been tested for temperate plants, including an annual grass, and found often to apply (135). At a more subtle level, Langenheim & Stubblebine (116) found that individual trees of *Hymenaea courbaril,* a tropical legume, produced seeds and seedlings of several biochemical phenotypes. Older surviving seedlings came to have only those phenotypes different from that of a nearby adult; thus, frequency-dependence occurs at the level of the biochemical phenotype.

It is clear that frequency-dependent regeneration of plants occurs. The strength of this effect, how much of it is contributed by natural enemies, and its overall contribution to plant dynamics and diversity remain to be determined. For organisms with lifespans of centuries, frequency-dependence could arise from occasional recruitment failure, the probability of which is frequency dependent. Fluctuating herbivore populations may often cause lasting effects, the production of which can only occasionally be observed (12, 150, 198).

ASSOCIATIONAL RESISTANCE Although the Janzen-Connell hypothesis implies frequency dependence, Root (172, 173) and Atsatt & O'Dowd (6) focused attention explicitly on relationships among co-occurring plant species, suggesting that the amount of herbivory a plant experiences depends on the local species composition. Individuals of one plant species, which might most obviously be regarded as competitors, may in fact have a net positive effect on another species by deterring the amount of herbivory the other experiences; associational resistance is an indirect mutualism. Associational resistance has been proposed to result from the tendency for herbivores to be attracted to and stay in patches of high density of a favored resource, the tendency for diverse stands to harbor more natural enemies of herbivores, chemical or structural interference with herbivore location or consumption of plants, and attraction of herbivores to alternative food plants (6, 172, 173).

Associational resistance has been demonstrated for marine algae (84) as well as for terrestrial plants (6, 16, 99). It also may be a mechanism contributing to differences in plankton composition between littoral and pelagic zones of some lakes (123), and to decreased loss rates of edible algae associated with large inedible ones (175, 182). Associational resistance may arise from events occurring during a single stage of a plant's life cycle, as when one plant species provides another with protection from herbivory during seedling or sporeling establishment (6, 133, 134, 187). A tendency for one plant species to establish only under another can lead to long-term cyclical replacements (192).

The effectiveness of associational resistance can depend on the foraging selectivity of an herbivore. Buffalo and wildebeest grazed a lower proportion of *Themeda triandra,* a palatable African savanna grass, from plots with higher proportions of less palatable species. However, this associational resistance was not effective for zebra and Thompson's gazelles, which are smaller and more selective feeders, and which fed without respect to the relative abundances of palatable or unpalatable plants (137). The tendency for insects that are specialized feeders to be deterred by associated plants suggests that mobility or host location factors also affect the tendency for associational resistance to be effective (147, 172, 173).

Associational resistance may or may not extend to entire ecosystems (16). Some postulated mechanisms could produce lower herbivory for an entire community (e.g. enemies), whereas others (e.g. attractant plants) would cause some species to realize increased herbivory. Proportional leaf area loss was lower in more diverse plots in successional neotropical communities. Some species experienced associational resistance and others associational susceptibility (110, 188), which may translate into apparent competition.

Clearly, associational resistance occurs in nature; its strength probably varies with herbivore and plant densities. Hay (84) demonstrated a competitive cost for palatable algae that grow in association with unpalatable algae; that cost, however, is lower than the benefit from reduced herbivory under some conditions. As a result, palatable algae are more abundant where there is at least a 20% cover of unpalatable algae when herbivorous fish are abundant. In aquatic and marine ecosystems, small herbivores such as snails or amphipods have been reported to improve the photosynthetic performance of macroalgae or macrophytes from which they graze periphytic algae (57, 123, 156). Whether periphyton or macrophytes are grazed may depend on periphyton density. More long-term and whole-system studies would greatly help to reveal the frequency of occurrence and overall significance of associational resistance.

Apparent Competition

Although predation, including herbivory, has often been considered to offset competition, predation can itself produce reciprocal negative interactions among prey (88–91). This relationship can arise from the joint contributions of two plant species to the population density of a shared herbivore, or from the behavior of a shared herbivore. In apparent competition, an increase in density of one plant species results in a decrease in the density of a second, not because they compete for the same resource(s), but because they are consumed by the same herbivore. Apparent competition is a sort of associational susceptibility; a plant experiences increased herbivore damage by virtue of the presence of another plant species. Apparent competition doesn't preclude the simultaneous existence of exploitative competition for resources.

Few data exist with which to evaluate the importance of apparent competition when the linkage between plant species is strictly as contributions to an herbivore's diet (44). However, Connell generalizes the idea of apparent competition to situations in which a shared predator is limited by the shelter provided by one plant species and then tends to eliminate other species from the vicinity of the shelter plant (9, 123, 158, 168, 175, 180). Apparent competition may contribute to habitat separation in plant communities (89, 90). Many data are compatible with this scenario, but direct tests of the contributions of apparent competition to habitat partitioning among plants have not been conducted. Apparent competition is likely to be common among plants that share herbivores, and further studies addressing this mechanism are needed.

Herbivores and Plant Competition

The effects of herbivores are realized in assemblages of plants that compete for resources or for space that allows access to resources. Herbivory is a source of loss of plant biomass or resources, and the differential effects of herbivory may alter patterns of plant performance and persistence. Traits that are positively associated with competitive ability may increase palatability to or vulnerability to herbivores. Most simply, losses to herbivores change the densities or biomasses of plant populations, but the interactions of the plants with their resources or with other plant species remain unaltered on a per capita or per biomass basis. Alternatively, herbivores may alter the density- or biomass-dependent demographic or resource depletion characteristics of plants (113). This may involve a change in plant form, phenology, or physiology, or may involve a change in the physical or chemical environment.

HERBIVORY AND PLANT BEHAVIOR The effects of predators on prey behavior have received much attention in systems of predators and their animal prey (111), but the potential for this idea to have value in understanding the dynamics of plant-herbivore systems has not been recognized. Although plants are most often conceived of as having specific per capita or per biomass competitive effects on each other, these interactions are dependent on particular organismal traits and environmental conditions (81, 190, 191). Herbivory may not only decrease plant biomass but may also make a plant of a given biomass behave differently, i.e. use resources or hold space differently.

Herbivores often cause changes in plant form or physiology. Secondary chemicals (85, 108) or morphological defenses may be induced by grazing (121, 199), and growth in the absence of herbivory may differ for morphs with different defense investment (39, 83, 85, 112). For terrestrial plants, removal of leaf tissue has frequently been reported to increase photosynthetic rates of remaining leaves; however, insects that don't remove tissue (e.g. gallers, stem borers, phloem feeders) are most often reported to decrease unit photosynthetic rates (194). Recent experiments show striking immediate carbon-flow alteration in grazed plants that presumably result in the allocation changes that emerge over longer time periods (59).

Herbivores can dramatically alter plant form. Terrestrial plants often allocate more carbon to above-ground vs below-ground tissues (23, 87, 139), or to foliage vs wood (114, 186) in response to herbivory by mammals or insects. Insect damage to terminal shoots of pinyon pines results in trees of shrubby growth form that produce only male cones (196). Similarly, removal of shoots by beavers, moose, giraffes, defoliating beetles, and a variety of other herbivores causes trees to develop a shrub, coppice, or irregular crown form (136, 139, 177, 193). Herbaceous plants also are altered in form

and phenology by herbivory (23, 124), with grazing often producing prostrate but rapidly growing grazing lawns (80, 139, 164, 165).

Among-species differences in allocation patterns correlate with differences in competitive abilities among plants (191). Plants of similar size but different form will most likely consume resources at different rates. Herbivores frequently have distinct patterns of tissue removal or damage, and these can result in differences in function between plants affected by different herbivores. Resource use and regrowth rates differed for grasses with biomass removed in two distinct spatial patterns that simulated those produced by large grazers vs more selective insects (71). Damage to roots may result in root proliferation that decreases susceptibility to disturbance and mortality rates (176). Plants of the strikingly different forms that herbivores cause must often differ in space-holding and light-reducing characteristics, and thus they would also differ in competitive abilities. This is a plausible explanation for the higher understory biomass of herbs that occurs below browsed trees of boreal and temperate forests and of savannas. The form of a plant can also influence the likelihood of further herbivore attack, as it may influence probabilities of location by herbivores or by their predators (3). Plant form itself may limit the ability of galling insects to manipulate plant physiology and control carbon flow to their own benefit (117).

Similar herbivore-caused changes in form are reported for algae. *Padina jamaicensis,* a tropical reef macroalga, exists in two morphological forms, a low branched turf morph and an erect foliose form. Phenotypic transformation of individuals from the low turf to the upright foliose form results when heavy grazing by fishes is reduced, and similar morphological changes occur in two other algal species (121). Additionally, only the erect forms of these algae were observed to produce heterospores. The freshwater filamentous alga *Cladophora* is altered in form, from long turfs to prostrate webs, by chironomid larvae (163). In addition to changing the form or physiology of grazed plants, herbivores may cause differential representation of morphs of heteromorphic algae (127) or of physiologically differing karyophases of isomorphic algae (129).

Changes in plant form and function in response to herbivory also could occur over a longer time scale. Evolutionary changes in plant function are suggested by differences among grazing ecotypes under uniform conditions (56, 59, 139) and by biogeographic variation associated with differing grazing pressures (21, 64, 116, 153).

HERBIVORY AND THE ENVIRONMENT Herbivores may cause changes in competitive interactions among plant species because they alter the physical environment of plants. Physical factors often can be understood as changing the favorability of an environment by changing the resource uptake, growth, or survival rates of plants (190). Thus, altering the environment can create habitats in which different sets of species persist. Differential responses of plants to the wide variety of habitat patches produced by herbivores is probably an important mechanism underlying the effects of herbivory on plant species composition and diversity.

Animals may have significant long-term effects on geomorphology (15, 45, 63, 92, 148) and geo- or hydro-chemistry (148, 159). The tree-cutting and damming activities of beaver result in long-term alteration of entire watersheds (148). Even such small animals as pocket gophers, mole rats, and ctenomyid rodents contribute to major long-term alterations of topography and soils (45). Some grazing marine fishes cause reef erosion and sedimentation by biting and breaking coral (92), and snails and fishes can alter sedimentation rates in streams, lakes, marshes, and subtidal marine habitats (15, 27, 63, 164).

Herbivores also have more immediate effects on the physical environment. Changes in abundance of freshwater macrophytes influence water flow, oxygenation, pH, and temperature of lakes and rivers (27). Gopher mounds often differ in nutrient and water contents from adjacent undisturbed areas (96). The removal of plant biomass from an area results in a decrease in the rate at which local resources are depleted, which can result in higher levels of resources for remaining or colonizing plants (96, 97, 182). Sterner (183) suggests that zooplankton change the competitive arena for phytoplankton by regenerating nitrogen and phosphorus at different rates. The responses of the algae to their resources presumably are not changed, but the resource characteristics of the environment are.

Effects of herbivory on plant behavior and on the environment intergrade. Holland & Detling (87) report evidence that the higher productivity and nutrient status of terrestrial plant communities subjected to long-term grazing result in part from alteration of nutrient cycling. They suggest that a decrease in C allocation to roots causes decreased C input to decomposers, resulting in less N immobilization and more net N mineralization. Thus, plant-available nitrogen is increased. Similarly, increases in plant allocation to secondary chemicals can affect soil microbes, decreasing plant available nitrogen (93, 159).

SECONDARY CHEMISTRY Plant secondary chemistry has been viewed as important to the dynamics of plant-herbivore systems for several decades, as secondary chemicals influence palatability of plants to herbivores (e.g. 55, 85, 122, 181). Secondary chemistry can influence plant-herbivore dynamics in two distinct ways. Levels of secondary chemicals that vary independently of the amount of herbivory a plant experiences contribute to determining environmental favorability for both plants and herbivores, and thus to setting their population levels (14). Levels of secondary chemicals that change in response to herbivory may increase or decrease susceptibility to herbivores, adding density-dependence to the dynamics of both plant and herbivore populations (14, 60, 61, 108). The effects of secondary chemistry on plant-herbivore dynamics need not be direct but can be realized via routes involving the natural enemies of herbivores. Secondary chemicals may decrease herbivory by increasing predation on herbivores (166) or may increase herbivory by reducing predation (84a, 106).

Secondary chemistry also can affect community and ecosystem dynamics via feedback effects on nutrient cycling. Many secondary chemicals inhibit endosymbiotic microbes that degrade cellulose and also may inhibit the decomposer microbes of soil ecosystems (93, 159). Thus, herbivory can affect nutrient cycling through effects on secondary chemistry. Pastor et al (159) suggest that herbivory causes or speeds succession from more palatable tree species, such as alder or birch, to less palatable species, such as hemlock and spruce, which may persist in part because of the effects of their secondary chemistry on nutrient availability.

GRADIENTS OF HERBIVORY Herbivory is sometimes envisioned as varying primarily in extent, being simply a loss rate of some magnitude (190). Although herbivores are almost always at least somewhat selective, many polyphagous herbivores have similar feeding preferences (M. A. Bowers, unpublished manuscript; 122, 128, 143), and selectivity may be relatively low in situations where herbivory is high (92, 95, 151).

The idea of gradients of herbivory has had utility in explaining patterns of plant response to herbivory in many systems (83, 95, 125, 143, 151; M. A. Bowers, unpublished manuscript). Several models of herbivory postulate correlations of herbivore pressure with productivity or food web structure (28, 144, 155). Phytoplankton succession is thought to result in part from changes in grazing pressure (182). Algal diversity in the Bay of Panama was well-correlated with the number of herbivore groups removed from small plots; additional variation in algal diversity was interpreted as reflecting unique effects of different kinds of herbivores (143). M. A. Bowers (unpublished manuscript) reports similar effects of mammalian herbivores in successional old-fields. Noy-Meir et al (151) interpret differences among pastures grazed by domestic herbivores as reflecting responses to a grazing intensity gradient; however, they suggest that the mechanism of the effect switches as grazing intensity increases. At low grazing pressure, disturbance results from patchy removal of plants, whereas with high grazing differential loss rates dominate, caused by higher vulnerability to grazing for taller species. Herbivore species may differ primarily in rate of herbivory (24, 143, 182). The difference in algae found within damselfish territories vs those browsed by parrotfish may reflect differing intensities of herbivory by the two fishes. Succession in cages that excluded all grazing fishes proceeded from green to thick filamentous red algae; succession within damselfish territories, which are grazed at moderate rates, was similar but slower, perhaps stopping at an intermediate high diversity stage. Succession outside of damselfish territories, where parrotfish graze at high rates, involved a totally different group of species, primarily algal crusts (M. A. Hixon, W. N. Brostoff unpublished manuscript).

Herbivores that forage around refuges that offer them protection from their own predators often exert extreme influence over plant communities, and the resulting patterns of vegetation can be well-explained as resulting from local gradients in amount of herbivory. Refuges cause both local grazing gradients and mosaics of patches that differ in amount of grazing. Algivorous fishes in Panamanian and North America rivers concentrate in areas with lower exposure to predators, resulting in higher algal biomass along pool edges and in pools containing piscivorous fishes (162). Reef fishes also forage near refuges (83), as do smaller reef herbivores such as urchins and crabs. These animals often shelter in crevices or holes and may cause local grazing lawns or barren areas to develop (24, 25, 31, 65, 143, 187). In lakes, crayfish and some fish may forage near shelter from predators (123). The nests, dens, burrows or other refuges of small mammals typically are surrounded by prostrate or unpalatable plants; a different community develops in the absence of these animals (4, 9, 95, 195). Gradients of herbivory also may be caused by proximity to other features of the environment, such as waterholes or streams (4, 50, 58, 105).

The outcome of moderate to high grazing often is a community of low-growing or prostrate plants with high nutrient content and productivity—a grazing lawn; however, in some cases unpalatable plants of low productivity result. These alternative outcomes seem to be the result of a balance between environmental limits on primary productivity and amount of herbivory. Rapidly growing, small or low plants result from the balance of high loss rates to herbivory against high growth rates (11, 24, 25, 27, 28, 73, 80, 115, 139, 164, 165). Grazing-resistant species predominate when plants deter herbivory by unpalatability (85, 159, 175). Similar processes occur in planktonic communities, where the outcome of manipulations of herbivory often depends on the initial species composition of plankton, particularly whether large or toxic blue-green algae are present (26, 175). In low-productivity environments, in which plant growth is limited by other nutrients (not carbon), high investment in carbon-based secondary chemicals that deter herbivory is likely, whereas in productive environments, rapid growth is the more likely response (39, 154). du Toit et al (58) suggest that grazing may sometimes increase productivity and nutrient concentration of shoots of woody plants but that these tend to be replaced by other vegetation. Presumably the difference stems from the inability of plants with high structural maintenance costs and slow growth rates to sustain high long-term losses to herbivores. The effects on community productivity of herbivores that don't remove plant tissue are as yet not well-studied, but appear to be much more generally negative (194).

THE COMPETITION HERBIVORY TRADE-OFF It often is assumed that compensatory herbivory arises from a trade-off such that plant traits that result in superior competitive ability for resources or space entail higher losses to herbivores. Most often, a trade-off between rapid growth or erect form and susceptibility to herbivores is identified (83, 95, 120, 126, 156, 165). However, growth rates and maximum height and/or lifespan tend to be negatively correlated for plants (190), macroalgae (83, 120), and phytoplankton (28, 79, 112). Rapid growth requires low investment in structural tissues and high concentrations of free sugars and amino acids; thus, fast-growing plants often are preferred by herbivores (83, 130, 178). Alternatively, erect form often makes plants differentially vulnerable to a variety of grazers, because they lose more mass to herbivores that graze down through a canopy (80, 83, 139, 151, 165; M. A. Bowers, unpublished manuscript). Thus, both fugitive and resource- or space-holding strategies may engender higher herbivory. This probably explains why herbivores are sometimes reported to speed and sometimes to slow, stop, or reverse succession (19, 46, 77, 78, 126, 170, 178).

Many results are consistent with the operation of a competition/herbivory trade-off. In desert ecosystems, selective consumption of large-seeded annual plants can offset the tendency for these plants to become dominant, thus increasing plant diversity (17). Marine algae that are competitively dominant often are limited by herbivores to habitats of low herbivory (83, 120), and plants that persist as fugitives often are limited in abundance by herbivores (19, 126, 178). Recent experimental work shows that insect herbivores can prevent community dominance by goldenrod during old-field succession (W. P. Carson, R. B. Root, personal communication). Burdon & Chilvers (22) suggest that specialized insect herbivores contribute to the high diversity of eucalypt forests, with each tree species limited by its own herbivore. Nevertheless, a relationship between palatability to herbivores and dominance in competition clearly does not always occur. Lubchenco (126) reports habitat-specific patterns of competitive dominance and invariant herbivore preferences such that a trade-off occurs for algae grazed by the snail *Littorina littorea* in tidepools but not on emergent substrata. Noy-Meir's (151) results suggest that there is not a simple competition/herbivory trade-off because the mechanisms by which plants are made differentially vulnerable to herbivores also vary with herbivore type or herbivore pressure.

Heterogeneity and the Effects of Herbivory

Communities are dynamic, showing fluctuations, successions, and cyclical changes, and they are spatially heterogeneous (79, 146, 179, 182). This high variability both affects and is affected by herbivores. Herbivore populations and the effects of herbivores on plants vary greatly both from place to place and over time (e.g. 25, 26, 42, 50, 51, 55, 66, 83, 92, 96, 97, 101, 102, 120, 137, 139, 177, 182, 187, 197, 198). This high variation probably contributes significantly to the effects of herbivores. Environmental variability caused by herbivores can either decrease or increase diversity, but biological details suggest that increasing diversity will be common (33–36).

Most simply, heterogeneity produces structure in populations inhabiting a heterogeneous environment. Structure can reflect age or life history stage (which may simply involve distinguishing juveniles from long-lived adult or resting stages), or phenotypic differentiation of organisms in different habitats or microhabitats. When subpopulations make distinct contributions to population growth and persistence, that structure must be taken into account to understand population dynamics (34, 35, 81, 82, 118, 119). There are several ways in which herbivores can influence plant communities by causing or responding to population structure; they may cause disturbance, may cause or respond to purely spatial environmental patchiness, and may have time-variant effects on plants.

BIOLOGICAL DISTURBANCE Both herbivory and disturbance are sources of loss for plants. Conceptual models of herbivory as biological disturbance emphasize spatial patchiness in plant loss rates, whereas those of herbivory as consumption of plants emphasize selective losses among plant populations. A model of an herbivore that clears patches is basically a model of disturbance. If the herbivore has preferences, then it is a model of nonrandom disturbance. Caswell's (30) and Hastings's (81) models of non-equilibrium predator-mediated coexistence are not formally different from a disturbance model; predators are simply invoked to remove plants from patches. The mechanism by which community dynamics and coexistence are affected is one of succession in discrete patches, with diversity resulting from trade-offs in colonization vs competitive ability or persistence; regional diversity tends to be promoted by the presence of patches at a variety of successional stages (36). Much of the diversity of ecosystems is realized through successional sequences which herbivores may contribute to initiating.

Herbivores often have been recognized as influencing community dynamics and diversity by producing disturbance (53, 86, 96, 97, 128, 174). In marine hard-bottom ecosystems, many sorts of herbivores scrape or otherwise clear hard surfaces or disrupt soft bottom sediments (15, 53). Terrestrial ecosystems are affected by a variety of herbivore-generated disturbances, including trails, burrows, diggings, earth mounds, and wallows (40, 45, 50, 96, 138). Equivalent disturbances may have different effects depending on precise spatial location or form or on temporally varying environmental factors (119). Intertidal algae and a grassland forb were both less successful at colonizing small disturbances because these were grazed by herbivores that apparently gained shelter from vegetation at the disturbances' edges (168, 180). Strong recruitment of an annual grass required the coincidence of unusually wet weather with the presence of gopher mounds (86).

SPATIAL VARIATION Plant community composition, herbivore behavior, and herbivore population density are spatially heterogeneous (1, 8, 20, 66, 79, 83, 95, 96, 137, 138, 140, 143, 162, 187). Herbivores may cause differences in plant demography among sites or patches or along gradients. Pattern in the physical environment affects many herbivores (83, 95, 99, 109, 110, 137–141, 143, 150), and herbivores may themselves impose pattern on a landscape (7, 45, 83, 92, 95, 96, 120, 138–140, 195). Below-ground herbivores typically are aggregated on many scales (1, 20, 107), as are African ungulates (138, 140), many group-living small mammals (78, 95, 195), and zooplankton (111). Herbivores' activities cause pattern in the environment and in plant growth and recruitment across spatial scales ranging from the microscopic to hundreds of km^2.

Sites may differ consistently over the long term in level of herbivory (8, 83, 95, 120, 125, 162). Dry and low-nutrient sites consistently have higher densities of European pine shoot moth and western pine-shoot borer (14), and the forb *Cardamine cordifolia* is consistently more damaged by herbivores in very wet or very dry sites, resulting in habitat restriction (125). Portions of reef habitats offering shelter from predators have more herbivores (93), and differential habitat use by many other herbivores is imposed by their own risks of predation (9, 62, 95, 99, 143, 162, 187). Subtle site differences such as micronutrient levels can influence levels of herbivory (98, 141).

Herbivores can alter the physical environment experienced by plants, and plant communities frequently differ between animal-generated landforms and adjacent unaffected areas (45, 148). Gophers that move low organic matter subsoil to the soil surface produce patches that remain distinct for long periods of time (97). Herbivores defecate, urinate, or produce other excretions, which can cause net nutrient transport from one habitat to another (50, 92, 123, 138, 140) or can produce small-scale local patchiness (52, 98, 182). Urine patches produced distinct plant patches in short-grass prairie, and these patches had higher productivity and supplied a disproportionately high amount of biomass and nitrogen to herbivores (52).

Structuring of the environment by herbivores may be subtle. For instance, the vegetation over and adjacent to the underground tunnels of pocket gophers is altered (165), and the influence of tunnels may extend 0.5 m each side of a tunnel (O. J. Reichman, J. Benedix, T. R. Seastedt, unpublished manuscript). Pocket gophers also alter subsurface soil structure and nutrient distribution by backfilling old tunnels (1) and by caching vegetation or concentrating urine and feces in dens (K. C. Zinnel, J. R. Tester, unpublished data). Even hoofprints can function as distinct microenvironments for germination (78). The microscale and transient nutrient patches that grazing zooplankton regenerate are believed to be very important to productivity, coexistence, and spatial patterning of planktonic algae (79, 182, 183).

TEMPORAL VARIATION The populations of virtually all herbivores fluctuate through time, often being influenced by weather or disease (8, 10, 12, 14, 41, 84, 94, 01, 102, 103, 110, 150, 159, 175, 182, 197, 198), and thus the effects of herbivory on plant growth, establishment, and mortality rates vary. Time-varying demographic rates can have major influences on the coexistence of species (33–35), and this may be an important way in which herbivory works. Fluctuations in herbivory occur on diurnal, seasonal, and multiannual scales. Occasional high densities can cause extensive damage to plants, resulting in greatly decreased growth or reproductive rates or in mortality (8, 177, 187). Noy-Meir (150) documents a case of voles increasing dramatically over a large region of Israel and causing increased abundance of forbs in pasture vegetation. This impact also varied spatially, due to the existing grazing regimes of pastures. Similarly, Berdowski & Zeilinga (12) document the replacement of heath (*Calluna vulgaris*) by grassland following a high density of heather beetles.

Juvenile plants may be particularly sensitive to herbivory (41, 42, 47–49, 125), so herbivores can cause fluctuations in plant recruitment. Chesson (34, 35) notes the potential for the storage effect to arise and promote diversity in systems with fluctuating recruitment. The storage effect occurs when populations recruit strongly under occasional good conditions (e.g. years in which herbivores are rare and weather is good) and subsequently retain (store) the results of that successful recruitment as individuals that are relatively invulnerable to such things as herbivores and weather. For instance, mature trees of a species that is unrepresented in the understory often are reported to occur as isolated cohorts that established during periods of low herbivore pressure (42). Pastor (159) notes adult birch persisting in an area of Isle Royale heavily browsed by moose and attributes their presence to a period of low moose numbers some 50 years previous. Tegner & Dayton (187) record an instance of establishment of canopy kelps (*Macrocystis*) that depended on a catastrophic die-off of sea urchins. The resting stages of planktonic organisms can act similarly as stores of recruitment bursts (112).

More subtle variation in growth periodicity of plants also is documented in response to fluctuating herbivory. Periodic cicadas, which can affect all deciduous trees of the eastern United States, cause periodic growth reductions in their host trees (107), and tree ring data identify long-term growth periodicity for forest species influenced by outbreaks of spruce budworm and other forest insects (67, 171, 192a). Temporal variation in growth of individuals that can store accumulated biomass or nutrients also can contribute to species coexistence, as species can specialize on transiently occurring conditions.

These arguments suggest that effects of herbivory on ecosystems are often best understood in the context of relatively long-term dynamics. Mattson & Addy (131) argued that herbivory functioned over the long-term as a cybernetic mechanism. Although their suggestion of cybernetic regulation generally has not been supported (131, 171, 186), the point that effects of herbivores are best understood as part of a long-term process remains well-taken. How long a process depends on both the period over which environmental conditions, including herbivory, vary, and the life spans of the plants. Terrestrial ecosystems may often be understood only as systems that change over at least decades and perhaps centuries, whereas marine systems may often be understandable at the level of years (spanning the time involved in successions, the life-spans of algae, and repeated disturbances) or decades (covering the period of larger-scale climatic influences such as the El Niño-Southern Oscillation). Planktonic algae operate in part on the fastest time scales, having rapid generation times and often showing marked specialization on seasonal conditions, but algae also show long-term time trends (79). Ecologically meaningful variation may be expected in all plant communities on the scale of the life span of a top predator, as numerical responses of these feed back on density and composition of lower trophic levels (26).

VARIATION IN THE IMPORTANCE OF HERBIVORY

It is of interest to know whether herbivory may be more important under some sorts of ecological or environmental conditions than under others. At the coarsest level, this resolves to a question of regulation of trophic levels: Does herbivory significantly limit the biomass of vegetation, and thus pose a strong selective force on plant populations, under some predictable set of conditions but not under others?

Hairston, Smith & Slobodkin (HSS: 76) first took the approach of analyzing ecosystems as food chains, representing trophic levels as if they were populations. They proposed that herbivores had little impact on the dominant species within nonsuccessional (equilibrium) terrestrial communities, because the dominant herbivore populations ordinarily were limited by their predators, not their food supply. Their result assumes a food chain structured so that productivity differences flow through to the top consumer level, but populations at lower trophic levels may have their abundance determined by their predators. Several alternatives to the HSS hypothesis have been developed, each of which posits increased importance of herbivores under some conditions. These models highlight the influences of productivity and disturbance, which are recognized among plant ecologists to have strong effects on community dynamics (74, 77, 190), and of particular species composition.

Oksanen & Fretwell (OF: 155) proposed that trophic structure varies with primary productivity; more productive communities have longer food chains. Thus, the HSS hypothesis becomes a special case, a three-trophic-level (plants, herbivores, and carnivores) ecosystem. The OF model predicts that the importance of herbivory varies with primary productivity because of this change in food chain length: In very unproductive habitats, resident herbivore populations are not supported by the low phytomass, and plants are resource limited. As productivity increases, herbivore populations can be supported and the vegetation becomes limited by herbivory. Further increases in productivity allow carnivores to persist, producing the HSS scenario. Higher yet productivity might support a fourth trophic level, resulting in herbivores again limiting vegetational biomass. Data from some tundra, island, and stream ecosystems are in agreement with the predictions of OF, as are data from many lakes (153–155, 160, 163). In terrestrial ecosystems, the OF hypothesis seems supported for very unproductive and moderately unproductive environments. However, it is not yet clear whether herbivory is less limiting to vegetation significant in more productive habitats or whether primary productivity typically is the major determinant of food chain length. Although the OF model seems to have been particularly successful in predicting dynamics of aquatic ecosystems, their food chain lengths seem often to be less than productivity could support, due to such things as winter fish kills (175).

The OF model has been challenged because of its assumption that the dynamics of predator-prey interactions are laissez-faire, i.e. that the functional responses of predators are influenced only by the density of their prey, not by predator density. Arditi et al (5) propose a ratio-dependent model, which makes dynamics dependent equally on predator and prey densities. This model predicts that, at equilibrium, all trophic levels increase proportionately with productivity, and the authors view the commonly observed positive correlations of plant and herbivore biomasses, nutrient supply, and productivity as evidence that ratio-dependent dynamics are the rule. However, linear regressions probably provide little power to distinguish between two hypotheses that both predict a generally positive association of these parameters over many productivity ranges, and the nonequilibrium dynamics of the two models are similar.

Carpenter & Kitchell have emphasized dynamics within trophic levels in food chain models such as OF, and they have proposed the cascading trophic interactions hypothesis (CK: 27). CK predict that predators, including herbivores, regulate the productivity of lower trophic levels by determining species composition of their prey. Thus, potential productivity is set by the physical environment, including inherent fertility and climatic effects, but actual primary productivity reflects also the particular species that become dominant, their growth rates, and the nutrient cycling regime. A small plankton biomass with a high productivity is predicted to result from moderately high grazing, which favors algae with high growth rates and keeps the pool of nutrients circulating more rapidly. Cascading trophic interactions have been demonstrated in many lake ecosystems (26, 28, 175), and Pastor et al (159) apply this idea to boreal forest ecosystems. Cascading trophic interactions are one way in which herbivory can alter plant community productivity.

Ecologists working in marine ecosystems have emphasized the role of physical or physiological stresses, such as wave disturbance or desiccation associated with tidal flow (42, 144). The most recent synthesis of these ideas is that of Menge & Sutherland (MS: 144). They postulate that organisms at higher trophic levels are differentially vulnerable to stresses, with sessile organisms, including plants, being least susceptible; thus environmental stress is predicted to reduce the importance of herbivory. Menge & Farrell (142) recently reviewed results of a wide variety of experimental studies in marine ecosystems and concluded that the MS model does not apply to terrestrial, planktonic, or marine subtidal habitats, but that it may be useful in understanding marine intertidal habitats.

CONCLUSIONS

Herbivores can be understood as influencing plant communities via a few general sorts of processes. Herbivores influence growth, recruitment, and mortality rates of plants and may do so in ways correlated with plant density, frequency, or other neighborhood traits, or with competitive abilities. Herbivores may increase, offset, or generate reciprocal negative interactions (competition) among plants. Herbivory is highly variable in space and in time, and these spatial and temporal patterns of herbivory can generate structure in plant populations, the existence of which strongly influences community dynamics. Effects of herbivores are most interpretable when measured in terms of rates of plant growth, reproduction, dispersal, recruitment, and mortality for subpopulations that reflect the range of conditions plants experience. Field studies should be designed to evaluate the contributions of particular mechanisms to herbivore effects, not simply to determine whether communities differ when herbivores are present or absent.

The studies reviewed here suggest that the mechanisms by which herbivores influence plant communities do not differ fundamentally for terrestrial, aquatic, and marine ecosystems. Herbivory can significantly alter community composition and productivity in all ecosystems. Secondary chemistry is an important influence on herbivory. It may be a cost to plants and herbivores, thus setting limits to population density. Also, levels of secondary chemicals may covary with herbivory, producing density-dependent population dynamics. Herbivores often are observed to be constrained in their foraging to specific areas or microhabitats; the result is strong spatial gradients and mosaics in amount of herbivory and thus in plant composition. Keystone herbivores, which cause development of a fundamentally different biological community, often with strongly altered ecosystem-level characteristics, have been demonstrated in marine (63; M. A. Hixon, W. N. Brostoff, unpublished manuscript), terrestrial (18; W. P. Carson, R. B. Root, unpublished data) and freshwater (26, 175) ecosystems. Examples also exist of herbivores that cause changes of the keystone sort in ecosystems but that do so primarily by changing the physical environment rather than by consuming plants (96, 148). Data are as yet inadequate to evaluate the relative contributions particular mechanisms make to the overall dynamics of plant communities.

There are more direct experimental demonstrations of the effects of herbivores on plants in aquatic and particularly marine ecosystems than in terrestrial. This largely reflects differences in the temporal and spatial scales at which these ecosystems vary. Marine ecologists frequently observe full successional sequences in short time periods (142). For an extreme example, a significant 25% increase in algal diversity and 36% decrease in algal productivity occurred within 5 days of the die-off of a major grazing urchin (*Diadema*) in a Caribbean reef ecosystem (25). Successions to and in forest, tundra, or perennial grassland may take decades or centuries. Planktonic algae have doubling times ranging from 1 day to about a week (79); this time-scale introduces its own complications; it is so rapid as to be difficult to study. These differences in time-scale of community dynamics, and thus of response to naturally occurring and experimental manipulations, reflect the different investments in structural tissues made by terrestrial plants, aquatic macrophytes and macroalgae, and planktonic algae. Terrestrial ecosystems may also be more heterogeneous than are marine systems.

Since temporal and spatial variation in herbivore populations often is high and may be of major significance, studies need to be designed to include measures of variability in herbivore impact over the range of environmental and biotic conditions that occur. We need long-term and large-scale studies, but other approaches are necessary as well. Creatively designed shorter-term and smaller-scale studies are needed, in which the environments are manipulated so the full range of conditions that occur can be observed.

ACKNOWLEDGMENTS

Preparation of this review was supported by National Science Foundation grants BSR-8706278 and BSR-8811884 and by the Idaho State University Faculty Development Fund. I thank W. P. Carson, M. I. Dyer, M. A. Hixon, D. M. Lodge, O. J. Reichman, and R. B. Root for sharing unpublished data and manuscripts, and T. Angradi, P. Chesson, and R. Inouye for reading and commenting on the manuscript.

Literature Cited

1. Andersen, D. D. 1987. Below-ground herbivory in natural communities: a review emphasizing fossorial animals. *Q. Rev. Biol.* 62:261–86
2. Anderson, A. N., Lonsdale, W. M. 1990. Herbivory by insects in Australian savannas: a review. *J. Biogeogr.* 17: 433–44
3. Andow, D. A., Prokrym, D. R. 1990. Plant structural complexity and host-finding by a parasitoid. *Oecologia* 82: 162–65
4. Andrew, M. H. 1988. Grazing impact in relation to livestock watering points. *TREE* 3:336–39
5. Arditi, R., Ginzburg, L. R., Akcakaya, H. R. 1991. Variation in plankton densities among lakes: a case for ratio-dependent predation models. *Am. Nat.* In press
6. Atsatt, P. R., O'Dowd, D. J. 1976. Plant defense guilds. *Science* 193:24–29
7. Bakker, J. P., de Leeuw, J., van Wieren, S. E. 1983. Micropatterns in grassland vegetation created and sustained by sheep-grazing. *Vegetatio* 55:153–61
8. Barbosa, P., Schultz, J. C., eds. 1987. *Insect Outbreaks*. San Diego: Academic
9. Bartholomew, B. 1970. Bare zone between California shrub and grassland communities: the role of animals. *Science* 170:1210–12
10. Batzli, G. O., Pitelka, F. A. 1970. Influence of meadow mouse populations on California grassland. *Ecology* 51: 1027–39
11. Bazely, D. R., Jeffries, R. L. 1986. Changes in composition and standing crop of salt marsh communities in response to removal of a grazer. *J. Ecol.* 74:693–706
12. Berdowsky, J. J. M., Zeilinga, R. 1987. Transition from heathland to grassland: damaging effects of the heather beetle. *J. Ecol.* 75:159–75
13. Bernays, E. A., ed. 1989. *Insect-Plant Interactions, vol. 1*. Boca Raton: CRC
14. Berryman, A. A., Stenseth, N. C., Isaev, A. S. 1987. Natural regulation of herbivorous forest insect populations. *Oecologia* 71:174–84
15. Bertness, M. D. 1984. Habitat and community modification by an introduced herbivorous snail. *Ecology* 65:370–81
16. Brown, B. J., Ewel, J. J. 1987. Herbivory in complex and tropical successional ecosystems. *Ecology* 68:108–16
17. Brown, J. H., Davidson, D. W., Munger, J. C., Inouye, R. S. 1986. Experimental community ecology: the desert granivore system. In *Community Ecology*, ed. J. Diamond, T. J. Case, pp. 41–61. New York: Harper & Row
18. Brown, J. H., Heske, E. J. 1990. Control of a desert-grassland transition by a keystone rodent guild. *Science* 250: 1705–7
19. Brown, V. K., Gange, A. C. 1989. Differential effects of above- and below-ground insect herbivory during early plant succession. *Oikos* 54:67–76
20. Brown, V. K., Gange, A. C. 1990. Insect herbivory below ground. *Adv. Ecol. Res.* 20:1–58
21. Bryant, J. P., Tahvanainen, J., Sulkinoja, M., Julkinen-Thtto, R., Reichardt, P., Green, T. 1989. Biogeographic evidence for the evolution of chemical defense by boreal birch and willow against mammalian browsing. *Am. Nat.* 134: 20–34
22. Burdon, J. J., Chilvers, G. S. 1974. Fungal and insect parasites contributing to niche differentiation in mixed species stands of eucalypt saplings. *Aust. J. Bot.* 22:103–14
23. Cain, M. L., Carson, W. P., Root, R. B. 1991. Long-term suppression of insect herbivores increases the production and growth of *Solidago altissima* rhizomes. *Oecologia*. In press
24. Carpenter, R. C. 1986. Partitioning herbivory and its effects on coral reef algae. *Ecol. Monogr.* 56:345–63
25. Carpenter, R. C. 1988. Mass mortality of a Caribbean sea urchin: Immediate effects on community metabolism and other herbivores. *Proc. Natl. Acad. Sci.* 85:511–14
26. Carpenter, S. R., ed. 1988. *Complex Interactions in Lake Ecosystems*. New York: Springer-Verlag
27. Carpenter, S. R., Kitchell, J. F. 1988. Consumer control of lake productivity. *BioScience* 38:764–69
28. Carpenter, S. R., Kitchell, J. F., Hodgson, J. R., Cochran, P. A., Elser, J. J., et al. 1987. Regulation of lake ecosystem primary productivity by food web structure in whole-lake experiments. *Ecology* 68:1863–76
29. Carpenter, S. R., Lodge, D. M. 1986. Effects of submersed macrophytes on ecosystem processes. *Aquat. Bot.* 341–70
30. Caswell, H. 1978. Predator-mediated coexistence. *Am. Nat.* 112:127–54
31. Chapman, A. R. O., Underwood, A. J. eds. 1990. Determinants of structure in intertidal and subtidal macroalgal assemblages. *Hydrobiologia* 192:1–121
32. Chesson, P. L. 1984. Variable predators

and switching behavior. *Theor. Popul. Biol.* 26:1–26

33. Chesson, P. 1985. Coexistence of competitors in spatially and temporally varying environments: a look at the combined effects of different sorts of variability. *Theor. Popul. Biol.* 28:263–87
34. Chesson, P., Huntly, N. 1988. Community consequences of life-history traits in a variable environment. *Ann. Zool. Fenn.* 25:5–16
35. Chesson, P., Huntly, N. 1989. Short-term instabilities and long-term community dynamics. *TREE* 4:293–98
36. Chesson, P., Huntly, N. 1991. Roles of mortality, disturbance, and stress in the dynamics of ecological communities. *Science*. Submitted
37. Chesson, P., Rosenzweig, M. 1991. Behavior, heterogeneity and the dynamics of interacting species. *Ecology*. In press
38. Clark, D. A., Clark, D. B. 1984. Spacing dynamics of a tropical rain forest tree: evaluation of the Janzen-Connell model. *Am. Nat.* 124:769–88
39. Coley, P. D., Bryant, J. P., Chapin, F. S. III. 1985. Resource availability and plant antiherbivore defense. *Science* 230:895–99
40. Collins, S. L., Barber, S. C. 1985. Effects of disturbance on diversity in mixed grass prairie. *Vegetatio* 64:87–94
41. Connell, J. H. 1971. On the role of natural enemies in preventing competitive exclusion in some marine animals and in rain forest trees. In *Dynamics of Populations,* ed. P. J. den Boer, G. R. Gradwell, pp. 298–310. Wageningen: Ctr. Agric. Publ. Docu.
42. Connell, J. H. 1975. Some mechanisms producing structure in natural communities: a model and evidence from field experiments. In *Ecology and Evolution of Communities,* ed. M. L. Cody, J. M. Diamond, pp. 460–90. Cambridge: Belknap
43. Connell, J. H., Tracey, J. G., Webb, L. J. 1984. Compensatory recruitment, growth, and mortality as factors maintaining rain forest diversity. *Ecol. Monogr.* 54:14–64
44. Connell, J. H. 1990. Apparent versus "real" competition in plants. See Ref. 72, pp. 445–74
45. Cox, G. W., Gakahu, C. G. 1985. Mima mound topography and vegetational pattern in Kenyan savannas. *J. Trop. Ecol.* 1:23–36
46. Crawley, M. J. 1983. *Herbivory: the Dynamics of Animal-Plant Interactions.* Oxford: Blackwell Sci.
47. Crawley, M. J. 1988. Herbivores and plant dynamics. In *Plant Population Ecology,* ed. A. J. Davy, M. J. Hutchings, A. R. Watkinson, pp. 367–92. London: Blackwell Sci.
48. Crawley, M. J. 1989. The relative importance of vertebrate and invertebrate herbivores. See Ref. 13, pp. 45–71
49. Crawley, M. J. 1989. Insect herbivores and plant population dynamics. *Annu. Rev. Entomol.* 34:531–64
50. Cumming, D. H. M. 1982. The influence of large herbivores on savanna structure in Africa. See Ref. 94, pp. 217–45
51. Danell, K., Ericson, L. 1990. Dynamic relations between the antler moth and meadow vegetation in northern Sweden. *Ecology* 71:1068–77
52. Day, T. A., Detling, J. K. 1990. Grassland patch dynamics and herbivore grazing preference following urine deposition. *Ecology* 71:180–88
53. Dayton, P. K. 1971. Competition, disturbance, and community organization: the provision and subsequent utilization of space in a rocky intertidal community. *Ecol. Monogr.* 41:351–89
54. Dayton, P. K. 1985. Ecology of kelp communities. *Annu. Rev. Ecol. Syst.* 16:215–45
55. Denno, R. F., McClure, M. S., ed. 1983. *Variable Plants and Herbivores in Natural and Managed Systems.* New York: Academic
56. Detling, J. K., Painter, E. A. 1983. Defoliation responses of western wheatgrass populations with diverse grazing histories. *Oecologia* 57:65–71
57. Duffy, J. E. 1990. Amphipods on seaweeds: partners or pests. *Oecologia* 83:267–76
58. du Toit, J., Bryant, J. P., Frisby, K. 1990. Regrowth and palatability of *Acacia* shoots following pruning by African savanna browsers. *Ecology* 71:149–54
59. Dyer, M. I., Acra, M. A., Wang, G. M., Coleman, D. C., Freckman, D. W., et al. 1991. Source-sink carbon relations in two *Panicum coloratum* ecotypes in response to herbivory. *Ecology*. In press
60. Edelstein-Keshet, L. 1986. Mathematical theory for plant-herbivore systems. *J. Math. Biol.* 24:25–58
61. Edelstein-Keshet, L., Rauscher, M. D. 1989. The effects of inducible plant defenses on herbivore populations. I. Mobile herbivores in continuous time. *Am. Nat.* 133:787–810
62. Edwards, J. 1983. Diet shifts in moose due to predator avoidance. *Oecologia* 60:185–89
63. Estes, J. A., Palmisano, J. F. 1974. Sea otters: their role in structuring nearshore communities. *Science* 185:1058–60

64. Estes, J. A., Steinberg, P. D. 1988. Predation, herbivory, and kelp evolution. *Paleobiology* 14:19–36
65. Fletcher, W. J. 1987. Interactions among subtidal Australian sea urchins, gastropods and algae: effects of experimental removals. *Ecol. Monogr.* 57:89–109
66. Fox, L. R., Morrow, P. A. 1986. On comparing herbivore damage in Australian and north temperate systems. *Austr. J. Ecol.* 11:387–93
67. Fritts, H. C., Swetnam, T. W. 1989. Dendroecology: a tool for evaluating variations in past and present forest environments. *Adv. Ecol. Res.* 19:111–88
68. Futuyma, D. J., Wasserman, S. S. 1980. Resource concentration and herbivory in oak forests. *Science* 210:920–22
69. Gibson, C. W. D., Hamilton, J. 1983. Feeding ecology and seasonal movements of giant tortoises on Aldabra Atoll. *Oecologia* 56:84–92
70. Gibson, D. J., Freeman, C. C., Hulbert, L. C. 1990. Effects of small mammal and invertebrate herbivory on plant species richness and abundance in tallgrass prairie. *Oecologia* 84:169–75
71. Gold, W. G., Caldwell, M. M. 1990. The effects of the spatial pattern of defoliation on regrowth of a tussock grass. III. Photosynthesis, canopy structure and light interception. *Oecologia* 82:12–17
72. Grace, J. B., Tilman, D., ed. 1990. *Perspectives on Plant Competition*. San Diego: Academic
73. Gregory, S. V. 1983. Plant-herbivore interactions in stream ecosystems. In *Stream Ecology.*, ed. J. R. Barnes, G. W. Minshall, pp. 157–89. New York: Plenum
74. Grime, J. P. 1977. Evidence for the existence of three primary strategies in plants and its relevance to ecological and evolutionary theory. *Am. Nat.* 111: 1169–94
75. Grubb, P. J. 1977. The maintenance of species richness in plant communities: the regeneration niche. *Biol. Rev.* 52:107–45
76. Hairston, N. G., Smith, F. E., Slobodkin, L. B. 1960. Community structure, population control, and competition. *Am. Nat.* 94:421–25
77. Harper, J. L. 1969. The role of predation in vegetational diversity. *Brookhaven Symp. in Biology* 22:48–62
78. Harper, J. L. 1977. *Population Biology of Plants*. New York: Academic
79. Harris, G. P. 1986. *Phytoplankton Ecology: Structure, Function and Fluctuation*. Cambridge: Chapman & Hall
80. Hart, D. D. 1985. Grazing insects mediate algal interactions in a stream benthic community. *Oikos* 44:40–46
81. Hastings, A. 1978. Spatial heterogeneity and the stability of predator-prey systems: predator-mediated coexistence. *Theor. Popul. Biol.* 14:380–95
82. Hastings, A. 1988. Food web theory and stability. *Ecology* 69:1665–68
83. Hay, M. E. 1985. Spatial patterns of herbivore impact and their importance in maintaining algal species richness. *Proc. Fifth Intl. Coral Reef Cong.* 4:29–34
84. Hay, M. E. 1986. Associational plant defenses and the maintenance of species diversity: turning competitors into accomplices. *Am. Nat.* 128:617–41
84a. Hay, M. E., Duffy, J. E., Fenical, W. 1990. Host-plant specialization decreases predation on a marine amphipod: an herbivore in plant's clothing. *Ecology* 71:733–43
85. Hay, M. E., Fenical, W. 1988. Marine plant-herbivore interactions: the ecology of chemical defense. *Annu. Rev. Ecol. Syst.* 19:111–45
86. Hobbs, R. J., Mooney, H. A. 1991. Effects of rainfall variability and gopher disturbance on serpentine annual grassland dynamics. *Ecology* 72:59–68
87. Holland, E. A., Detling, J. K. 1990. Plant response to herbivory and belowground nitrogen cycling. *Ecology* 71: 1040–49
88. Holt, R. D. 1977. Predation, apparent competition, and the structure of prey communities. *Theor. Popul. Biol.* 12: 197–229
89. Holt, R. D. 1984. Spatial heterogeneity, indirect interactions, and the coexistence of prey species. *Am. Nat.* 124:377–406
90. Holt, R. D., 1987. Prey communities in patchy environments. *Oikos* 50:276–90
91. Holt, R. D., Kotler, B. P. 1987. Short-term apparent competition. *Am. Nat.* 130:412–30
92. Horn, M. H. 1989. The biology of marine herbivorous fishes. *Mar. Biol. Oceanog. Annu. Rev.* 27:167–272
93. Horner, J. D., Gosz, J. R., Cates, R. G. 1988. The role of carbon-based plant secondary metabolites in decomposition in terrestrial ecosystems. *Am. Nat.* 132:869–83
94. Huntley, B. J., Walker, B. H., ed. 1982. *Ecology of Tropical Savannas*. Berlin: Springer-Verlag
95. Huntly, N. J. 1987. Effects of refuging consumers (pikas: *Ochotona princeps*)

on subalpine vegetation. *Ecology* 68: 274–83
96. Huntly, N., Inouye, R. 1988. Pocket gophers in ecosystems: patterns and mechanisms. *BioScience* 38:786–93
97. Inouye, R. S., Huntly, N. J., Tilman, D., Tester, J. R. 1987. Pocket gophers *(Geomys bursarius),* vegetation, and soil nitrogen along a successional sere in east-central Minnesota. *Oecologia* 72: 178–84
98. Inouye, R. S., Huntly, N. J., Tilman, D. 1987. Responses of *Microtus pennsylvanicus* to fertilization of plants with various nutrients, with particular emphasis on sodium and nitrogen concentrations in plant tissues. *Holarc. Ecol.* 10:110–13
99. Jaksic, F. M., Fuentes, E. R. 1980. Why are native herbs in the Chilean matorral more abundant beneath bushes: microclimate or grazing? *J. Ecol.* 68: 665–69
100. Janzen, D. H. 1970. Herbivores and the number of tree species in tropical forests. *Am. Nat.* 104:501–27
101. Janzen, D. H. 1981. Patterns of herbivory in a tropical deciduous forest. *Biotropica* 13:271–82
102. Janzen, D. H. 1988. The lepidopteran fauna of a Costa Rican dry forest. *Biotropica* 20:120–35
103. Jassby, A. D., Powell, T. M., Goldman, C. R. 1990. Interannual fluctuations in primary productivity: direct physical effects and the trophic cascade at Castle Lake, California. *Limnol. Oceanog.* 35:1021–38
104. Joern, A. 1989. Insect herbivory in the transition to California annual grasslands: did grasshoppers deliver the coup de gras? In *Grassland Structure and Function: California Annual Grasslands,* ed. L. F. Huenneke, H. A. Mooney, pp. 117–34. Dordrecht: Kluwer Academic
105. Johnston, C. A., Naiman, R. J. 1990. Browse selection by beaver: effects on riparian forest composition. *Can. J. For. Res.* 20:1036–43
106. Jones, C. G., Whitman, D. W., Silk, P. J., Blum, M. S. 1988. Diet breadth and insect chemical defenses: a generalist grasshopper and general hypotheses. See Ref. 182, pp. 477–512
107. Karban, R. 1980. Periodical cicada nymphs impose periodical oak tree wood accumulation. *Nature* 287:326–27
108. Karban, R., Myers, J. H. 1989. Induced plant responses to herbivory. *Annu. Rev. Ecol. Syst.* 20:331–48
109. Kareiva, P. 1983. Influence of vegetation texture on herbivore populations: resource concentration and herbivore movement. See Ref. 55, pp. 259–89
110. Kareiva, P. 1986. Patchiness, disperal, and species interactions: consequences for communities of herbivorous insects. In *Community Ecology,* ed. J. Diamond, T. J. Case, 192–206. New York: Harper & Row
111. Kerfoot, W. C., Sih, A., eds. 1987. *Predation: Direct and Indirect Impacts on Aquatic Communities.* Hanover: Univ. Press New Engl.
112. Kilham, P., Hecky, R. E. 1988. Comparative ecology of marine and freshwater phytoplankton. *Limnol. Oceanogr.* 33:776–95
113. Kotler, B. P., Holt, R. D. 1989. Predation and competition: the interaction of two types of species interactions. *Oikos* 54:256–60
114. Kulman, H. M. 1971. Effects of insect defoliation on growth and mortality of trees. *Annu. Rev. Entomol.* 16:289–324
115. Lamberti, G. A., Moore, J. W. 1984. Aquatic insects as primary consumers. In *The Ecology of Aquatic Insects,* ed. V. H. Resh, D. M. Rosenberg, pp. 164–95. New York: Praeger
116. Langenheim, J. H., Stubblebine, W. H. 1983. Variation in leaf resin composition between parent tree and progeny in *Hymenaea:* implications for herbivory in the humid tropics. *Biochem. Syst. Ecol.* 11:97–106
117. Larson, K. C., Whitham, T. G. 1991. Sink-source relationships as determinants of aphid performance: an alternative to defensive chemistry. *Oecologia.* In press
118. Levin, S. A. 1974. Dispersion and population interactions. *Am. Nat.* 114:103–14
119. Levin, S. A. 1976. Population models and community structure in heterogeneous environments. *Annu. Rev. Ecol. Syst.* 7:287–310
120. Lewis, S. M. 1986. The role of herbivorous fishes in the organization of a Caribbean reef community. *Ecol. Monogr.* 56:183–200
121. Lewis, S. M., Norris, J. N., Searles, R. B. 1987. The regulation of morphological plasticity in tropical reef algae by herbivores. *Ecology* 68:636–41
122. Lodge, D. M. 1991. Herbivory on freshwater macrophytes. *Aquatic Bot.* In press
123. Lodge, D. M., Barko, J. W., Strayer, D., Melack, J. M., Mittelbach, G. G., et al. 1988. Spatial heterogeneity and habitat interactions in lake communities. See Ref. 26, pp. 181–208
124. Louda, S. M. 1984. Herbivore effect on

stature, fruiting and leaf dynamics of a native crucifer. *Ecology* 65:1379–86
125. Louda, S. M., Keeler, K. H., Holt, R. D. 1990. Herbivore influences on plant performance and competitive interactions. See Ref. 72, pp. 445–74
126. Lubchenco, J. 1978. Plant species diversity in a marine intertidal community: importance of herbivore food preference and algal competitive abilities. *Am. Nat.* 64:1116–23
127. Lubchenco, J., Cubit, J. 1980. Heteromorphic life histories of certain marine algae as adaptations to variations in herbivory. *Ecology* 61:676–87
128. Lubchenco, J., Gaines, S. D. 1981. A unified approach to marine plant-herbivore interactions. I. Populations and communities. *Annu. Rev. Ecol. Syst.* 12:405–37
129. Luxoro, C., Santelices, B. 1989. Additional evidence for ecological differences among isomorphic reproductive phases of *Iridaea laminarioides* (Rhodophyta: Gigartinales). *J. Phycol.* 25:206–212
130. Mattson, W. J. 1980. Herbivory in relation to plant nitrogen content. *Annu. Rev. Ecol. Syst.* 11:119–61
131. Mattson, W. J., Addy, N. D. 1975. Phytophagous insects as regulators of forest primary productivity. *Science* 190:515–22
132. Mattson, W. J., Haack, R. A. 1987. The role of drought stress in provoking outbreaks of phytophagous insects. See Ref. 8, pp. 365–407
133. McAuliffe, J. R. 1984. Prey refugia and the distributions of two Sonoran Desert cacti. *Oecologia* 65:82–85
134. McAuliffe, J. R. 1986. Herbivore-limited establishment of a Sonoran desert tree, *Cercidium microphyllum*. *Ecology* 67:276–80
135. McCanny, S. J., Cavers, P. B. 1987. The escape hypothesis: a test involving a temperate, annual grass. *Oikos* 49:67–76
136. McGinley, M. A., Whitham, T. G. 1985. Central place foraging by beavers *(Castor canadensis):* a test of foraging predictions and the impact of selective feeding on the growth form of cottonwoods *(Populus fremontii)*. *Oecologia* 66:558–62
137. McNaughton, S. J. 1978. Serengeti ungulates: feeding selectivity influences the effectiveness of plant defense guilds. *Science* 199:806–7
138. McNaughton, S. J. 1983. Serengeti grassland ecology: the role of composite environmental factors and contingency in community organization. *Ecol. Monogr.* 53:291–320
139. McNaughton, S. J. 1984. Grazing lawns: animals in herds, plant form, and coevolution. *Am. Nat.* 124:863–86
140. McNaughton, S. J. 1985. Ecology of a grazing ecosystem: the Serengeti. *Ecol. Monogr.* 55:259–94
141. McNaughton, S. J. 1988. Mineral nutrition and spatial concentrations of African ungulates. *Nature* 334:343–45
142. Menge, B. A., Farrell, T. M. 1989. Community structure and interaction webs in shallow marine hard-bottom communities: tests of an environmental stress model. *Adv. Ecol. Res.* 19:189–262
143. Menge, B. A., Lubchenco, J., Ashkenas, L. R. 1985. Diversity, heterogeneity and consumer pressure in a tropical rocky intertidal community. *Oecologia* 65:394–405
144. Menge, B. A., Sutherland, J. P. 1987. Community regulation: variation in disturbance, competition, and predation in relation to environmental stress and recruitment. *Am. Nat.* 130:730–57
145. Menges, E. S., Loucks, O. L. 1984. Modeling a disease-caused patch disturbance: oak wilt in the midwestern United States. *Ecology* 65:487–98
146. Miles, J. 1982. *Vegetation Dynamics*. New York: Chapman & Hall
147. Morrow, P. A., Tonkyn, D. W., Goldburg, R. J. 1989. Patch colonization by *Trirhabda canadensis* (Coleoptera: Chrysomelidae): effects of plant species composition and wind. *Oecologia* 81: 43–50
148. Naiman, R. J., Johnston, C. A., Kelley, J. C. 1988. Alteration of North American streams by beaver. *BioScience* 38:753–63
149. Noy-Meir, I. 1981. Theoretical dynamics of competitors under predation. *Oecologia* 50:277–84
150. Noy-Meir, I. 1988. Dominant grasses replaced by ruderal forbs in a vole year in undergrazed Mediterranean grasslands in Israel. *J. Biogeogr.* 15:579–87
151. Noy-Meir, I., Gutman, M., Kaplan, Y. 1989. Responses of Mediterranean grassland plants to grazing and protection. *J. Ecol.* 77:290–310
152. O'Dowd, D. J., Lake, P. S. 1990. Red crabs in rain forest, Christmas Island: differential herbivory of seedlings. *Oikos* 58:289–92
153. Oksanen, L. 1988. Ecosystem organization: mutualism and cybernetics or plain Darwinian struggle for existence? *Am. Nat.* 131:424–44
154. Oksanen, L. 1990. Predation, herbivory, and plant strategies along gradients

of primary productivity. See Ref. 72, pp. 445–74

155. Oksanen, L., Fretwell, S. D., Arruda, J., Niemela, P. 1981. Exploitation ecosystems in gradients of primary productivity. *Am. Nat.* 118:240–61
156. Olsen, A. M., Lubchenco, J. 1989. Competition in seaweeds: linking plant traits to competitive outcomes. *J. Phycol.* 26:1–6
157. Paine, R. T. 1966. Food web complexity and species diversity. *Am. Nat.* 100:65–75
158. Parker, M., Root, R. B. 1981. Insect herbivores limit habitat distribution of a native composite *Machaeranthera canescens*. *Ecology* 62:1390–92
159. Pastor, J., Naiman, R. J., Dewey, B., McInnes, P. 1988. Moose, microbes, and the boreal forest. *BioScience* 38: 770–77
160. Persson, L., Andersson, G., Hamrin, S. F., Johansson, L. 1988. Predator regulation and primary productivity along the productivity gradient of temperate lake ecosystems. See Ref. 26, pp. 45–65
161. Power, A. G. 1987. Plant community diversity, herbivore movement, and an insect-transmitted disease of maize. *Ecology* 68:1658–69
162. Power, M. E. 1987. Predator avoidance by grazing fishes in temperate and tropical streams: importance of stream depth and prey size. See Ref. 111, pp. 333–51
163. Power, M. E. 1990. Effects of fish in river food webs. *Science* 250:811–14
164. Power, M. E. 1990. Resource enhancement by indirect effects of grazers: armored catfish, algae, and sediment. *Ecology* 71:897–94
165. Power, M. E., Stewart, A. J., Matthews, W. J. 1988. Grazer control of algae in an Ozark mountain stream: effects of short-term exclusion. *Ecology* 69:1894–98
166. Price, P. W., Bouton, C. E., Gross, P., McPheron, B. A., Thompson, J. N., et al. 1980. Interactions among three trophic levels: influence of plants on interactions between insect herbivores and natural enemies. *Annu. Rev. Ecol. Syst.* 11:41–65
167. Reichman, O. J., Smith, S. C. 1985. Impact of pocket gopher burrows on overlying vegetation. *J. Mammal.* 66: 720–25
168. Rice, K. J. 1987. Interaction of disturbance, patch size and herbivory in *Erodium* colonization. *Ecology* 68: 1113–15
169. Risley, L. S., Crossley, D. A. Jr. 1988. Herbivore-caused greenfall in the southern Appalachians. *Ecology* 69:1118–27
170. Robles, C. D., Cubit, J. 1981. Influence of biotic factors in an upper intertidal community: dipteran larvae grazing on algae. *Ecology* 62:1536–47
171. Romme, W. H., Knight, D. H., Yavitt, J. B. 1986. Mountain pine beetle outbreaks in the Rocky Mountains: regulators of primary productivity? *Am. Nat.* 127:484–94
172. Root, R. B. 1973. Organization of a plant-arthropod association in simple and diverse habitats: the fauna of collards *(Brassica oleracea)*. *Ecol. Monogr.* 43:95–24
173. Root, R. B. 1975. Some consequences of ecosystem texture. In *Ecosystem Analysis and Prediction,* ed. S. A. Levin, pp. 83–92. Philadelphia: SIAM
174. Schowalter, T. D., Hargrove, W. W., Crossley, D. A. Jr. 1986. Herbivory in forested ecosystems. *Annu. Rev. Entomol.* 31:177–96
175. Shapiro, J. 1990. Biomanipulation: the next phase—making it stable. *Hydrobiologia* 200/201:13–27
176. Simberloff, D., Brown, B. J., Lowrie, S. 1978. Isopod and insect root borers may benefit Florida mangroves. *Science* 201:630–32
177. Sinclair, A. R. E., Norton-Griffiths, M., eds. 1979. *Serengeti: Dynamics of an Ecosystem*. Chicago: Univ. Chicago Press
178. Sousa, W. P. 1979. Experimental investigations of disturbance and ecological succession in a rocky intertidal algal community. *Ecol. Monogr.* 49:227–54
179. Sousa, W. P. 1984. The role of disturbance in natural communities. *Annu. Rev. Ecol. Syst.* 15:353–91
180. Sousa, W. P. 1984. Intertidal mosaics: Patch size, propagule availability, and spatially variable patterns of succession. *Ecology* 65:1918–35
181. Spencer, K. C., ed. 1988. *Chemical Mediation of Coevolution*. San Diego: Academic
182. Sterner, R. W. 1989. The role of grazers in phytoplankton succession. In *Plankton Ecology: Succession in Plankton Communities,* ed. U. Sommer, pp. 107–70. Berlin: Springer-Verlag
183. Sterner, R. W. 1990. The ratio of nitrogen to phosphorus resupplied by herbivores: zooploplankton and the algal competitive arena. *Am. Nat.* 136:209–29
184. Sterner, R. W., Ribic, C. A., Schatz, G. E. 1986. Testing for life historical changes in spatial patterns of four tropical tree species. *J. Ecol.* 74:621–33

185. Stevenson, J. C. 1988. Comparative ecology of submersed grass beds in freshwater, estuarine, and marine environments. *Limnol. Oceanogr.* 33:867–93
186. Swank, W. T., Waide, J. B., Crossley, D. A. Jr., Todd, R. L. 1981. Insect defoliation enhances nitrate export from forested ecosystems. *Oecologia* 51:297–99
187. Tegner, M. J., Dayton, P. K. 1987. El Niño effects on southern California kelp forest communities. *Adv. Ecol. Res.* 17:243–79
188. Thomas, C. D. 1986. Butterfly larvae decrease host plant survival in the vicinity of alternate host species. *Oecologia* 70:113–17
189. Thomas, C. D. 1990. Herbivore diets, herbivore colonization, and the escape hypothesis. *Ecology* 71:610–15
190. Tilman, D. 1988. *Plant Strategies and the Dynamics and Structure of Plant Communities*. Princeton: Princeton Univ. Press
191. Tilman, D., Kilham, S. S., Kilham, P. 1982. Phytoplankton community ecology: the role of limiting nutrients. *Annu. Rev. Ecol. Syst.* 13:49–72
192. Vandermeer, J. 1980. Saguaros and nurse trees: a new hypothesis to account for population fluctuations. *Southwest Nat.* 25:357–60
192a. Veblen, T. T., Hadley, K. S., Reid, M. S., Rebertus, A. J. 1991. The response of subalpine forests to spruce beetle outbreak in Colorado. *Ecology* 72:213–31
193. Vranjic, J. A., Gullam, P. J. 1990. The effect of a sap-sucking herbivore, *Eriococcus-Coriaceous* (Hompotera, Eriococcidae), on seedling growth and architecture in *Eucalyptus blakelyi*. *Oikos* 59:157–62
194. Welter, S. C. 1989. Arthropod impact on plant gas exchange. See Ref. 13, pp. 135–50
195. Whicker, A. D., Detling, J. K. 1988. Ecological consequences of prairie dog disturbances. *BioScience* 38:778–85
196. Whitham, T. G., Mopper, S. 1985. Chronic herbivory: impacts on tree architecture and sex expression of pinyon pine. *Science* 228:1089–91
197. Wolda, H. 1978. Fluctuations in abundance of tropical insects. *Am. Nat.* 112:1017–45
198. Wong, M., Wright, S. J., Hubbell, S. P., Foster, R. B. 1990. The spatial pattern and reproductive consequences of outbreak defoliation in *Quararibea asterolepis*, a tropical tree. *J. Ecol.* 78:579–88
199. Young, T. P. 1987. Increased thorn length in *Acacia depranolobium*—an induced response to browsing. *Oecologia* 71:436–38

Annu. Rev. Ecol. Syst. 1991. 22:505–23

APPLICATION OF ECOLOGICAL PRINCIPLES TO THE MANAGEMENT OF ENDANGERED SPECIES: The Case of the Red-Cockaded Woodpecker

Jeffrey R. Walters

Department of Zoology, North Carolina State University, Box 7617, Raleigh, North Carolina 27695-7617

KEY WORDS: conservation, population viability, cooperative breeding, cavity, management

---

INTRODUCTION

A major development in the biological sciences in the 1980s was the emergence of the discipline known as conservation biology. The rise of conservation biology was triggered by growing concern about environmental problems such as tropical deforestation and the loss of biodiversity (50). Historically, conservation biology may be viewed as a union between parts of the fields of wildlife biology and ecology (50) that reflects the maturation of both.

Wildlife biology has always focused on conservation and management of natural resources. Attention was originally concentrated on those few species that were hunted (45, p. 2) and, as in all disciplines, research was mostly descriptive. Management techniques developed largely through trial and error. As the field developed, its focus broadened to include nongame species, particularly endangered species. Interest in the basic biology of the species and ecosystems of concern grew, and management increasingly was based on deductions derived from fundamental understanding of the systems involved rather than from trial and error.

Within ecology, the relevant changes have been in theory. The theory developed in the 1950s, 1960s, and early 1970s (e.g. 37) can be characterized

0066-4162/91/1120-0505$02.00

as general and simple. The basic ecological principles contained within these theories, although of fundamental importance, had limited utility in natural resource management. Subsequently, ecological theory has become increasingly precise and realistic, and increasingly able to generate accurate predictions about the behavior of real systems. This has enabled ecologists more effectively to address natural resource management problems.

The union of ecology and wildlife biology in conservation biology thus is a natural one, but it is not without its difficulties. A gap between theory and practice, the width of which varies considerably among problems, remains. Collaboration between ecologists and wildlife biologists is inhibited by the separation of the two fields in the infrastructures of American universities and government agencies (50). The two disciplines have developed independently to a large extent, and their historic and current structural separation continues to hinder communication.

This paper illustrates both the potential and the problems characterizing the current contribution of conservation biology to management of endangered species, using as an example the red-cockaded woodpecker *(Picoides borealis),* an endangered species endemic to the southeastern United States. Use of artificial cavity construction illustrates the potential for deriving management techniques from an understanding of basic biology. Attempts to assess population viability illustrate the problems created when management is based on theory that is not yet sufficiently mature and when managers and researchers have difficulty communicating.

My thesis is that effective conservation of endangered species requires insightful research that incorporates theory, and that judicious use of research findings can result in designing successful management techniques. I choose to illustrate this by reviewing one particular case in detail, but other examples abound (e.g. 39). For instance Crouse et al (12) used sophisticated population modeling techniques to show that sea turtle population numbers are much more sensitive to mortality in the juvenile stage than at the egg or hatchling stage. In terms of effect on populations, saving one juvenile is equivalent to saving hundreds of hatchlings. This justifies conservation efforts to reduce juvenile mortality by ensuring that turtle-excluding devices are installed on shrimp boats, and it shifts emphasis away from more popular but less effective programs involving nest and hatchling protection (38).

THE BIOLOGY OF THE RED-COCKADED WOODPECKER

The red-cockaded woodpecker has in the last 20 years become one of the most studied avian species in the United States. Its biology has been the subject of two symposia (57, 67), two recovery plans (58, 59), and several recent reviews (25, 26, 36, 63). The red-cockaded woodpecker was once an abundant resident of the southeastern Piedmont and Coastal Plain, ranging from New Jersey to Texas, and inland to Kentucky, Tennessee, and Missouri (23). It is now virtually extirpated north of North Carolina and in all interior states but Arkansas. Most remaining populations are isolated, small, and fragmented (11, 24, 36, 59, 63), and only three number more than 300 breeding groups. Many populations continue to decline, some rapidly (8). Some populations are remarkably stable in their numbers (63), but none are increasing, a point to which I return.

The decline of the species is due primarily to habitat loss and alteration (25, 26, 36, 59). The species is highly adapted to pine savannahs, preferring especially longleaf pine *(Pinus palustris)*. These habitats were once abundant: Longleaf pine alone once covered 25 million ha in the Southeast (40). Vast expanses of pine savannah have been lost through conversion to agriculture and forestry plantations, development, and timbering (59, 62). Virtually all the remaining habitat has been altered in ways that adversely affect the species.

Habitat Requirements

Red-cockaded woodpeckers are nonmigratory and territorial throughout the year. Territories are large, ranging from 50–150 ha or more in size (14, 21, 41, 59, 63). In their foraging these woodpeckers are specialists on live pine (63), feeding on invertebrates which they obtain primarily by scaling bark and pecking. Although both sexes forage on the upper trunk, only females regularly forage low on the trunk, and only males forage regularly on the twigs and limbs.

The species is just as closely tied to live pine in its nesting. It invariably excavates cavities for roosting and nesting in live pines, a highly unusual habit that may be an adaptation to a low density of snags in the fire-maintained ecosystems in which it lives (25). The pines in which red-cockaded woodpeckers nest are highly resistant to fire. Excavating in live pine poses some special problems. The cavity chamber must be excavated in the tree's heartwood core, and cannot extend into the surrounding sapwood. Since the diameter of heartwood is largely a function of age, the birds can only excavate cavities in old trees. Cavity trees generally average 80–120 years of age (27, 59) and may be much older where many old trees are available (63). A major cause of the decline of the woodpecker species is the disappearance of trees of sufficient age for cavity excavation from much of the remaining pine savannahs, due primarily to timber harvest.

Second, the mechanics of excavating in live tissue are problematic. Heartwood is dense and difficult to work. However, the birds usually excavate in trees in which the heartwood has been softened by decay due to infestation by red-heart fungus *(Phellinus pini)* (7). Making an entrance tunnel through the sapwood to reach the heartwood is also difficult. This appears to limit the speed with which cavities can be constructed, presumably because sap leakage into the tunnel interrupts excavation. Cavities take at least 10 months to complete, typically several years (27), and the bulk of that time is spent excavating, intermittently, through the sapwood.

A final difficulty in using cavities in live pines is that the rough surface of the trunk enables predators, especially snakes, to climb to the cavity. Red-cockaded woodpeckers maintain resin wells, places where they chip into the sapwood, around their cavities. The resulting sap flow prevents snakes from reaching the cavity by smoothing the climbing surface and interfering with the action of the ventral scales used by the snakes in climbing (46).

Remaining habitat has also been altered through exclusion of fire. Historically the interval between fires in southeastern pine savannahs was only 1–5 years (26). In the absence of fire, a hardwood understory and midstory develops. There is a well-established correlation between development of this hardwood layer and abandonment of cavities by red-cockaded woodpeckers (8, 11, 24, 60). This may be related to the access to cavities provided to predators by encroaching hardwoods.

The Social System

Red-cockaded woodpeckers live in groups containing a breeding pair and 0–4 helpers, nearly all of which are male (63). There is no evidence that helpers participate in clutch production (63), but they assist in incubation and feeding of nestlings and fledglings (33, 35). Each member of the group has its own roost cavity. Hence a territory contains a cavity tree cluster, that is, a set of cavity trees, usually located in close proximity.

Group formation is best understood in terms of alternative life-history tactics practiced by young birds (63, 65, 66). Many fledglings disperse from their natal group during their first year to search for a breeding vacancy. This tactic is adopted by nearly all females (Figure 1) and many males (Figure 2). Although many early dispersers are breeders at age one (Figures 1 and 2), some are floaters, that is, individuals without a territory or mate (63), and some males are solitary, that is, they have a territory but no mate (Figure 2). Floaters and solitary males often become breeders in subsequent years, but their mortality rates are high (Figure 2) (63, 65). Individuals adopting this tactic may disperse long distances, although they do not always do so (Table 1).

Other individuals remain on the natal territory and act as helpers. Such individuals become breeders by inheriting breeding status on their natal territory or by dispersing at a later age to a nearby territory (Figure 2). Helpers rarely disperse long distances (Table 1) and may wait many years before

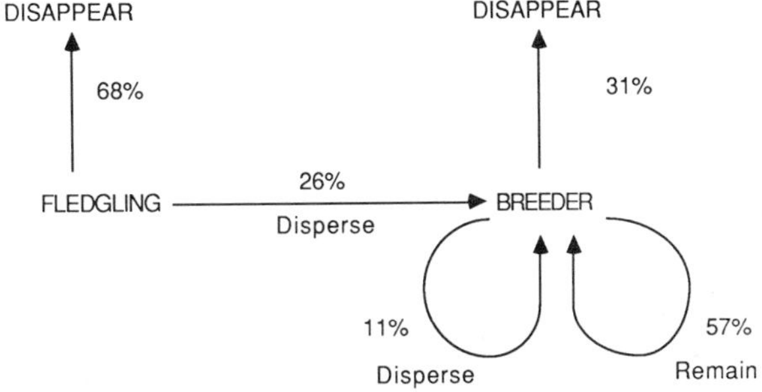

Figure 1 Annual transition probabilities for females, based on data from 765 fledgling and 1106 breeding females from the North Carolina Sandhills. Only transitions that account for at least 4% of transitions involving either class of individual are included. See Walters et al (65) for a complete accounting. Whether transitions involve remaining on the same territory or dispersing to another is indicated.

acquiring a breeding position. This tactic is adopted by many males, but very few females. Thus most helpers in this species are natal males that delay dispersal and reproduction. Once males acquire a breeding position they almost always hold it until they die (Figure 2). Breeding females, on the other hand, sometimes switch groups (Figure 1) (63, 65).

Whichever tactic individuals adopt, if they become breeders they almost always do so by replacing a deceased individual. In present populations red-cockaded woodpeckers compete for breeding vacancies in existing groups rather than form new groups. New groups might form by reoccupation of abandoned territories or creation of new territories. New territories may be created by pioneering, in which birds disperse into an area not previously occupied and construct a new cluster of cavities, or by budding, in which an existing territory and cavity tree cluster (and often the existing group) is split into two (20). In the population of over 200 groups I study in the North Carolina Sandhills, budding resulted in the formation of only 6 new groups in 8 years, and pioneering did not occur in that period (63).

Reoccupation of abandoned territories is more common, resulting in the formation of 22 new groups in 8 years in the Sandhills, for example (63). Still, in the Sandhills the rate of reoccupation of abandoned territories is only 8.7% annually (15). Once a territory is abandoned for longer than two years, it almost always stays abandoned.

That populations of red-cockaded woodpeckers are sometimes remarkably stable but seldom increase follows directly from the tendency of individuals to compete for breeding vacancies in existing groups rather than form new ones. Lack of new group formation has been a serious obstacle in recovery efforts. Research on the basic biology of the species' social structure suggests an ecological basis for the rarity of formation of new groups, and a solution to this management problem, cavity construction.

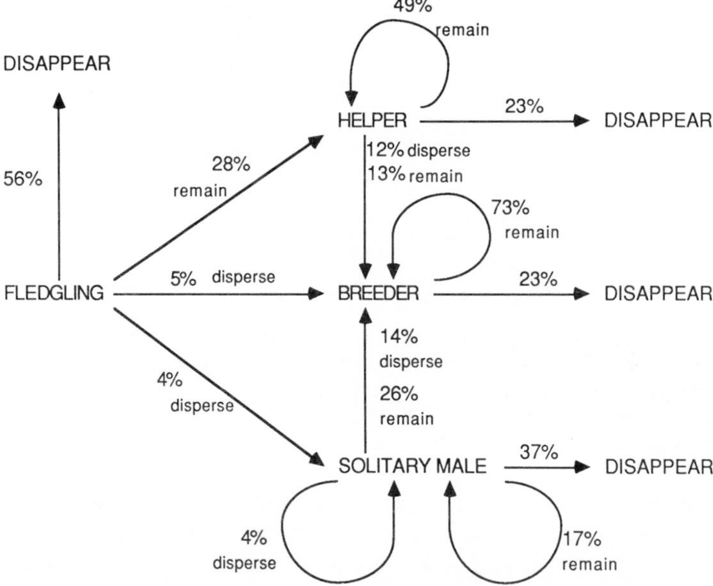

Figure 2 Annual transition probabilities for males, based on data from 775 fledglings, 354 helpers, 1033 breeders and 141 solitary males from the Sandhills of North Carolina. Only transitions involving at least 4% of all transitions for a particular class of individual are included. See Walters (63) for a complete accounting. Whether transitions involved remaining on the same territory or dispersing to another is indicated.

POPULATION DYNAMICS

The Ecology of Cooperative Breeding

The rarity with which red-cockaded woodpeckers form new groups can be related to the evolution of their cooperative breeding system. It is clear that there are several evolutionary pathways to cooperative breeding, and that no common set of ecological conditions applies to all the more than 200 species of birds characterized by this breeding system (18). Advantages inherent to group living may account for the evolution of cooperative breeding in some species, particularly those in which there are several breeders within the group, and in which per capita survival and reproductive success increases with group size (3, 53, 61). However, the most common pathway to cooperative breeding appears to be selection for delayed dispersal and reproduction, that is, retention of young within their natal group, as in red-cockaded woodpeckers.

The demographic conditions required for selection for retention of young have been described in progressively more explicit and detailed models (e.g. 17, 66, 68). These demographic models are simply an accounting of fitness,

Table 1 Dispersal distances of red-cockaded woodpeckers in the Sandhills of North Carolina (km)[a]

Category of disperser	Mean	Median	Maximum	Percentage to adjacent territory	N
Fledgling female	4.8	3.2	31.5	27%	217
Fledgling male	5.1	3.9	21.1	31%	88
Helper male	1.8	1.0	17.1	61%	41

[a]Modified from Walters (63).

using survival and reproductive schedules under alternative life-history tactics. Recent research provides evidence that the demographic conditions depicted in these models prevail in several species (e.g. 29, 68, 69).

What ecological factors produce the demographic conditions that favor retention of young? Emlen (17) outlined two ecological regimes under which remaining with the natal group may result in greater lifetime reproductive success than does attempting to disperse and breed early. One regime involves a harsh, unpredictable environment. In poor years, inexperienced birds reproduce poorly, so that living with the natal group is favored over independent reproduction.

The second, more common regime, traditionally termed habitat saturation because it is characterized by an apparent shortage of breeding vacancies (1, 2, 17, 52), is the one that applies to red-cockaded woodpeckers. An apparent lack of unoccupied territories has been noted in many cooperative breeders (18). The only attempt to conceptualize habitat saturation quantitatively is that of Koenig & Pitelka (28). They proposed that in species experiencing habitat saturation, reproductive success falls sharply between suitable and unsuitable habitat, and little marginal habitat exists between suitable and unsuitable habitat. Under these conditions, suitable habitat will be filled continuously.

Stacey & Ligon (54, 55) have proposed the benefits of philopatry model as an alternative to habitat saturation. According to this model, in species with unusually great variation in the quality of breeding positions, it pays individuals whose natal sites are of high quality to remain as helpers and potentially inherit breeding status, as this is an effective way to acquire a superior site. Variance in breeding positions may depend on effects of territory quality (42, 54, 55) or group size (43, 55) on fitness. If the unsuitable habitat in the Koenig & Pitelka model (28) is interpreted as incapable of supporting reproduction, then this model is distinct from the philopatry model. If unsuitable habitat is of low quality but is capable of supporting reproduction, then only high quality habitat, rather than all breeding habitat, is saturated. In this case, the two models are similar, differing primarily in the explicit role given to comparison between natal and nonnatal territories in the philopatry model (54).

The consensus is that individuals ignore vacant habitat in which they could breed successfully while competing intensely for other (higher quality) habitat. Delaying reproduction and altering dispersal behavior is viewed as an effective way of competing for high-quality breeding positions. Individuals remaining on the natal territory are supposed to have an advantage in competing for vacancies in their vicinity over those dispersing after fledging, a supposition supported by experimental evidence (70). Those adopting this tactic can be thought of as increasing their ability to compete for a restricted set of vacancies.

Testing Theory: the Cavity Construction Experiment

Research by my colleagues and me indicates that in the Sandhills population, fitness of males that delay dispersal and reproduction is comparable to that of males that disperse soon after fledging (66). Birds that disperse early do not have unusually low survival compared to other small landbirds, but they have an unusually low probability of acquiring a breeding position. Those that delay dispersal have a very high survival rate, both during their first year and during subsequent years spent helping. The extent to which reproductive performance improves with age among breeders is also unusually great. The average number of fledglings produced by males with both a territory and a mate increases steadily from 0.64 to 2.10 from age one to age six (66).

Thus, demography conducive to the evolution of delayed dispersal exists in this species, but this demography depends on individuals' competing for breeding positions within existing groups rather than their forming new groups. I have proposed that the basis of this selectivity, and the intense competition for breeding positions that results, is variation in territory quality, and further, that the basis of variation in territory quality is the cavity tree cluster (63, 65). Because of the time and energy required to construct a set of cavities, the worst territories with an existing set of cavities may be sufficiently better than the best ones without cavities that it pays to compete for territories with existing cavities and ignore those without them. This may be viewed as a version of the models described above in which territory quality has a discontinuous rather than continuous distribution (55).

To test this hypothesis, my colleagues and I constructed cavities in live pines in 20 sites in our Sandhills study area, using a drilling technique (9). In each site we drilled two complete cavities, plus three additional entrance tunnels. Half of these sites were abandoned territories that contained old cavities but had not been used by red-cockaded woodpeckers for at least 3 years, and the other half were vacant sites that contained neither cavities nor birds.

Each of the experimental abandoned sites was paired with a control site that was also abandoned. We cleared hardwood midstory and understory from the vicinity of cavity trees in both experimental and control sites. We used another set of abandoned sites in which we did no understory clearing as an additional control. We also delimited a set of vacant control sites. We cleared understory and midstory from the vicinity of trees in which we constructed cavities in vacant experimental sites, and from a set of comparable trees (in which no drilling was done) within vacant control sites. We constructed cavities from February 1988 to February 1989 and evaluated the response to the experiment in the breeding season (April–July) of 1989.

The results of the experiment were dramatic. Eighteen of 20 experimental sites were occupied by red-cockaded woodpeckers—nine abandoned sites and nine vacant sites—whereas no control sites were occupied. Although some experimental sites were used by previously existing groups of woodpeckers, the experiment resulted in a net addition of 12 groups to the population. This contrasts not only to the lack of response to controls, but the general rarity of new group formation in the population. Further details about the experiment are given elsewhere (10, 64).

A New Understanding of Population Dynamics

The results of the experiment suggest that red-cockaded woodpeckers compete intensely for some habitat while other potentially usable habitat goes unoccupied due to the absence of cavities. The population contains many nonbreeding adults (i.e. helpers), yet not all potential breeding habitat is filled. This necessitates taking an unconventional approach in modeling the population dynamics of the species.

Population dynamics are usually modeled using a life table approach in which changes in population size are a function of class- or age-specific mortality and survival parameters. Such models have proved valuable in conservation (e.g. 12, 48), but they cannot be applied to red-cockaded woodpeckers because they cannot incorporate the constraints imposed by the species' social structure. First, population reproduction is not a simple function of adult density. Because a pool of replacement breeders (helpers) exists, variation in breeder mortality has little effect on the future number of breeders. When an epidemic resulted in the loss of 45% of the breeders and all the fledglings in a population of another cooperative breeder, the Florida scrub jay *(Aphelocoma coerulescens),* the number of breeders was reduced by only 27% the next breeding season and returned to the pre-epidemic level the following year (68, p. 351). Because the probability of transition from helper to breeder is a function of breeder mortality, it cannot be treated as a constant parameter in a life table analysis.

Second, the dispersal behavior of helpers introduces spatial structure into the population dynamics. One cannot assume that a given nonbreeder can fill any breeding vacancy that occurs in the population. Less conventional models that incorporate social and spatial structure (e.g. 30, 56) are required for the red-cockaded woodpecker. One such model, which treats the population as a set of interacting groups, has been developed by L. A. Maguire and her students at Duke University (51). This and similar models could accurately predict changes in population size, the distribution of individuals among the helper, breeder, and floater classes, and population reproduction. However, they still cannot predict changes in the number of breeding groups accurately.

The number of groups in the population depends on a second process besides that which determines population size, namely that which determines distribution of cavity tree clusters and thus the number of acceptable territories. The domain of the models discussed above is the dynamics of individuals distributed among acceptable territories, not the dynamics of territory acceptability. A model of the population dynamics of red-cockaded woodpeckers must include a second level that treats territory acceptability.

Much of the observed behavior of red-cockaded woodpecker populations can readily be described in terms of population dynamics involving two levels, the first, the number of cavity tree clusters, and the second, the number of individuals distributed among those clusters. Changes in population size appear to involve primarily the first level. Territories can be lost due to hardwood encroachment on or destruction of cavities. In the absence of such losses of cavity tree clusters, populations will be extremely stable but will not grow appreciably. That is, population decline results from territory abandonment, not reductions in survival or reproduction. As populations decline, birds on occupied territories continue to do well, but fewer territories are occupied (65). Populations that have declined differ from those that have not in the proportion of territories abandoned rather than the survival and reproduction of birds on remaining territories (J. Walters, unpublished data).

The gradual nature of territory abandonment (15) can be explained in these terms. Breeders are often reluctant to abandon their territory, even if cavities are destroyed or lost to hardwood encroachment. However, nonbreeders may no longer find the territory acceptable and thus will not replace the breeders when the latter perish. A pair or a solitary male may linger on the territory for many years (63, 65), but when these individuals die, the territory is at last abandoned. It is abandoned not because replacements are unavailable, but because the territory is unacceptable.

Of course extreme variation in demography could cause some acceptable territories to be unoccupied or occupied by unpaired males. However, variation in demography appears to translate primarily into variation in group size. For example, high levels of reproduction are followed by increases in the size of the helper class and thus in group size rather than increases in the number of groups in a population (63, 65). The two levels of population dynamics are thus to a considerable degree independent.

Implications for Woodpecker Management

My interpretation of the population dynamics of red-cockaded woodpeckers has dramatic implications for their management. First, it suggests that management efforts that improve reproduction and survival have little potential to promote population recovery or to arrest decline. Such efforts would primarily affect the average group size in a population rather than the number of groups. Management should instead focus on factors that affect the number of acceptable cavity tree clusters to reduce loss of existing clusters and promote addition of new clusters.

Using this perspective, I evaluate the major techniques used in management of red-cockaded woodpeckers. These include control of hardwood understory and midstory by prescribed burning and mechanical and chemical removal, increasing the quantity and quality of foraging habitat, increasing old growth available for cavity excavation, and reducing usurpation of red-cockaded cavities by other species (11, 36, 59).

Southern flying squirrels *(Glaucomys volans)* are one of the primary users of active red-cockaded woodpecker cavities (4). Flying squirrels may force individuals to roost in the open, destroy eggs and young, and even prevent a group from nesting by temporarily usurping all of a group's cavities, but they do not destroy cavities. Thus flying squirrels have no effect on territory acceptability and do not cause territory abandonment. Controlling their interaction with red-cockaded woodpeckers may increase reproduction and survival, and thus group size, but not the number of groups in the population.

In contrast to flying squirrels, other woodpeckers that usurp cavities from red-cockaded woodpeckers destroy cavities by enlarging them. Thus these species can affect territory acceptability as well as reproduction and survival. The pileated woodpecker *(Dryocopus pileatus)* is particularly destructive. Pileated woodpeckers can enlarge all the cavities on a territory in a short period of time, thus rendering the territory unacceptable. Cavity restrictors, metal plates placed around cavity entrances, have proven effective in reducing loss of cavities to pileated woodpeckers and other species (4). Use of restrictors in areas where pileated woodpeckers and other species cause cavity loss should reduce loss of existing clusters and thus help prevent population decline.

Efforts to improve foraging habitat have the same limitations as efforts to control flying squirrels. Increased quality and quantity of foraging habitat is correlated with increased reproduction and group size (34). Foraging habitat may thus be thought of as a factor affecting quality among acceptable territories. Providing more and better foraging habitat may increase average group size but will not add groups to the population. It is likely that some amount of foraging habitat is necessary for a territory to be acceptable. The large size of territories suggests such a requirement, but there is no direct evidence that loss of foraging habitat is a factor in current population declines. The available evidence does not support a relationship between reduction in foraging habitat and territory abandonment (8).

Our experiment suggests that control of hardwood understory and midstory alone has limited potential to convert unacceptable territories into acceptable ones and thus to promote population recovery. Presumably this is because cavities deteriorate if they are long abandoned. However, hardwood control should be effective in reducing loss of existing clusters and thus preventing population decline.

Until recently, attempts to provide additional habitat in which new cavity tree clusters might be constructed consisted of establishing patches of old trees known as recruitment stands that the woodpeckers could colonize. That this approach was not immediately effective is consistent with the view of population dynamics presented here. One cannot expect this species to colonize habitat except at a very low rate. On the other hand, constructing cavities in recruitment stands should be an effective way to increase the number of groups in a population. This is the approach currently adopted by the US Forest Service. Using our technique and another they developed, since the fall of 1989 Forest Service personnel have been constructing cavities in abandoned territories, recruitment stands, and other vacant habitat in National Forests throughout the Southeast. Cavity construction is also beginning on other federal lands, state lands, and even private lands. The degree to which cavity construction can promote population expansion by increasing the number of acceptable territories will soon be apparent.

Cavity construction on already acceptable territories might reduce territory abandonment where cavities are lost faster than they can be replaced. In many areas, there is a scarcity of old trees in which the birds can construct replacement cavities (11). Cavity construction has been used to reduce territory abandonment on a large scale in one instance, again by the Forest Service. In September of 1989 Hurricane Hugo passed through the Francis Marion National Forest in South Carolina, devasting a population of nearly 500 groups of red-cockaded woodpeckers. About 60% of the birds perished, and 87% of the cavity trees were destroyed (22). That the number of acceptable territories was reduced even more than the number of birds created the potential for large-scale abandonment of territories among the surviving birds.

To avoid this possibility, the Forest Service constructed cavities on 222 territories (R. G. Hooper, personal communication). The birds used the cavities and even nested in them during the 1990 breeding season. Although 79 territories contained only single birds, the number of occupied territories in the 1990 breeding season was 65% of the number occupied before the storm. Thus, the number of groups was reduced by much less than the number of adults. By minimizing the loss of acceptable territories, cavity construction will allow the population to recover much more quickly than if the number of groups had been further reduced. Indeed, enough young were produced in 1990 to fill most of the remaining breeding vacancies on acceptable territories.

Implications for Conservation Biology

The use of artificial cavity construction in management of red-cockaded woodpeckers illustrates the potential of conservation biology to contribute to the management of endangered species. Where the theory is specific enough, ecological principles can contribute to our ability to deduce effective management principles. Application of theory in studies of basic biology has led to insights that provide a new perspective on population dynamics and suggest a solution to major management problems that had previously been intractable. This new perspective would probably not have developed except from a theory pertaining to the evolution of cooperative breeding. Further, this perspective differs sufficiently from the conventional one that it requires a reexamination of existing management techniques and even of the species' recovery plan (59). Finally, this case illustrates effective communication between researchers that enabled new research results to be quickly translated into productive management. For this, R. E. F. Escano and R. G. Hooper of the US Forest Service deserve much of the credit.

This case also illustrates the importance of considering the role of social structure in population dynamics. The principle illustrated is that it often is necessary to incorporate additional factors into general models to make them sufficiently accurate in specific cases. Incorporating social structure into population dynamics models may be important in a variety of contexts, for example, in examining effects of habitat fragmentation on bird populations. Social structure often affects dispersal behavior, and dispersal behavior affects the ability of birds to colonize habitat patches. Cooperative breeders are characterized by reduced long-distance dispersal capabilities and enhanced abilities to fill nearby breeding vacancies. Therefore, cooperative breeders might be unusually good at persisting in patches large enough to hold several neighboring groups, but unusually poor at persisting in smaller patches that require repeated colonization.

When populations fail to behave as expected in response to management, social structure may be a factor. The endangered Puerto Rican parrot *(Amazona vittata)* may be an example. Although many birds live in a particular area, there is only one breeding pair, or occasionally two (49). When a member of a breeding pair dies, it is immediately replaced by a formerly nonbreeding bird from the area. Releases of birds into an area may increase the population in that area, but not the number of nesting birds. Several explanations of the low recruitment rate of breeders have been offered (49). I offer another: Perhaps some complex social structure exists which limits the number of breeders in an area. There is no evidence of this, but there are enough unanswered questions to impel basic research on social structure.

MINIMUM VIABLE POPULATION SIZE

Just as application of theory about the evolution of cooperative breeding to management of red-cockaded woodpeckers illustrates the potential of conservation biology, application of theory about population viability illustrates its pitfalls. In a world of fragmented habitats, the importance of establishing self-sustaining populations of rare species is obvious. To be viable, a population must be large enough to resist a number of threats, including demographic stochasticity, environmental stochasticity, loss of genetic variability, and catastrophes (47).

Conservation biologists have attempted to determine how large red-cockaded woodpecker populations must be to be viable. Unfortunately, existing theory is not sufficiently precise to enable such a determination, and this attempt has been counterproductive. Hurricane Hugo has provided a textbook example of the vulnerability of even large populations to catastrophe, arguing for the necessity of maintaining several populations. This consideration is incorporated into the species' recovery plan (59), but here the enlightened treatment of minimum viable population size in red-cockaded woodpeckers ends. A major problem is our inability to model population dynamics (see above), a prerequisite of viability analyses involving demographic and environmental stochasticity (e.g. 48, 48a). Hence, viability has been assessed solely in terms of loss of genetic variability, even though it generally is demography and environmental stochasticity rather than genetics that limits population viability (31). For this reason alone, one can have no confidence that a population size indicated by a genetic model is really sufficient to ensure long-term survival.

Further, the genetic models are terribly imprecise. These models are based on idealized populations. Although the relationship between the size of these idealized populations (effective population size) and the rate of loss of genetic variability is well known, the relationship between loss of genetic variability and population viability is not (31, 48a). The effective size necessary for viability differs among species due to differences in demographic constraints, inherent variability, and evolutionary history (32, 48a). The capability to determine what the appropriate effective size is for any particular species does not exist. This has led some to argue against the use of genetic models of population viability in conservation (13).

Additional imprecision arises in computing a population's effective size. The effective size of a population is almost always much smaller than the

Table 2 Estimates of the number of breeders required for an effective size of 500 in the Sandhills population of red-cockaded woodpeckers, using different models and different estimates of demographic parameters.

Model	Demographic parameters	Estimate
Reed et al (44)	NCSU[a]	1018
USFWS (59)	none[b]	500
Emigh & Pollak (16)	combined[c]	664
Emigh & Pollak	NCSU[a]	794
Emigh & Pollak	NCSU, dispersal correction[d]	574
Emigh & Pollak	NCSU, +10% adult survival[e]	544
Emigh & Pollak	NCSU, −10% adult survival[f]	1012

[a] Data collected from Sandhills population by my colleagues and I. Assumes no dispersal out of study area.
[b] Assumes effective size equals the number of breeders
[c] From Heckel and Lennartz (19). They use a combination of our data on survival and their data from another population on reproduction
[d] NCSU data, corrected for dispersal out of study area (see 60)
[e] NCSU data without dispersal correction, adding 10% to our estimates of adult survival
[f] NCSU data without dispersal correction, substracting 10% from our estimates of adult survival

actual number of breeders (5, 48a). Effective size is estimated using mathematical models that make a number of simplifying assumptions about reproduction, mating structure, and variation among individuals (e.g. 44). These assumptions are surely violated in real populations, introducing error of unknown magnitude into calculations of effective size (48a). Two models have been used to calculate the effective size of red-cockaded woodpecker populations (19, 44). These two models give very different results when applied to the same population, and even the more complex, presumably more accurate, model gives widely varying results depending on the accuracy of estimates of demographic model parameters (Table 2). Further, these models assume that populations are closed, whereas many are not. If populations receive immigrants, estimates of the rate of loss of genetic variability must be corrected for addition of genetic variability due to gene flow from other populations.

Implications for Woodpecker Management

Clearly one cannot determine precisely how large a viable red-cockaded woodpecker population must be currently, especially if genetic models provide the only criteria. Yet that is exactly what is being done. The US Fish and Wildlife Service, in its recovery plan for the species (59) and its enforcement policies, has defined a minimum viable population size based on genetic models, and uses that definition in making decisions about whether habitat alterations jeopardize populations. The USFWS uses an effective size of 500 as its minimum standard. Originally effective size was equated to the number of breeders (59), but now various other values within the range given in Table 2 are used for the required number of breeders. I suggest that the genetic models are too imprecise to be useful in making management and policy decisions. The USFWS cannot be faulted for considering genetically based viability standards—the considerable literature on the topic essentially forces them to consider these standards. The fault lies with researchers involved in viability analysis, myself included, who have failed in communicating the limitations of these analyses to managers and policy-makers. This is an example of the premature application of ecological theory in management, resulting in an illusion of rigor that disguises the arbitrariness of individual decisions and inhibiting use of more reasonable approaches with other bases.

The red-cockaded woodpecker would be better served if the health of populations were assessed solely by criteria other than estimates of effective size. Other criteria are used by the USFWS in determining whether habitat alterations jeopardize populations, and these criteria could be given even more weight if genetically based viability assessments were excluded from consideration. Population goals are essential, but I suggest that the availability of genetically based standards inhibits the development of standards based on more reasonable approaches. Such approaches include demographic modeling and empirical studies (49). Even standards with little basis in theory, for example, deviation from maximum population density (6) may be more effective in achieving the goal of species preservation than use of genetically based viability assessments.

SUMMARY

The case of the red-cockaded woodpecker exemplifies how modern conservation biology can contribute to the management of endangered species. The value of artificial cavity construction illustrates how successful management can be deduced from ecological theory. Studies of basic biology have resulted in a new understanding of populations dynamics and a new management technique that, in combination with habitat preservation and restoration, has the potential to bring the species back from the brink of extinction.

Set against such promises are the problems involved in population viability analysis. In this case theory is not sufficiently developed, and it is best to resist the temptation to apply it. In such cases trial-and-error management, and use of criteria other than those derived from theory in decision making, are more appropriate. Improved communication between managers and researchers, based on increased understanding and appreciation of one another, is a necessity if we are to distinguish those situations in which we can use theory from those in which we cannot. The successes promise to increase and the problems to diminish as conservation biology matures, to the betterment of society and the natural world.

ACKNOWLEDGMENTS

My research on red-cockaded woodpeckers in the Sandhills has been supported by NSF grants BSR-8307090 and BSR-8717683, and the North Carolina Agricultural Research Service. I thank my collaborators, Drs. P. D. Doerr and J. H. Carter, III for their many contributions to the woodpecker research. I also thank the many graduate and undergraduate students who have contributed to data collection and analysis. M. Hassler assisted with the tables and E. Seaman with the figures. F. James provided a helpful review of the manuscript.

Literature Cited

1. Brown, J. L. 1974. Alternate routes to sociality in jays—with a theory for the evolution of altruism and communal breeding. *Am. Zool.* 14:63–80
2. Brown, J. L. 1978. Avian communal breeding systems. *Annu. Rev. Ecol. Syst.* 9:123–55
3. Brown, J. L. 1982. Optimal group size in territorial animals. *J. Theor. Biol.* 95:793–810
4. Carter, J. H. III, Walters, J. R., Everhart, S. H., Doerr, P. D. 1989. Restrictors for red-cockaded woodpecker cavities. *Wildl. Soc. Bull.* 17:68–72
5. Chessor, R. K. 1983. Isolation by distance: relationships to the management of genetic resources. In *Genetics and Conservation*, ed. C. M. Schonewald-Cox, S. M. Chambers, B. MacBryde, L. Thomas, pp. 66–95. Menlo Park, Calif: Benjamin/Cummings
6. Conner, R. N. 1988. Wildlife populations: minimally viable or ecologically functional? *Wildl. Soc. Bull.* 16:80–84
7. Conner, R. N., Locke, B. A. 1982. Fungi and red-cockaded woodpecker cavity trees. *Wilson Bull.* 94:64–70
8. Conner, R. N., Rudolph, C. D. 1989. *Red-cockaded Woodpecker Colony Status and Trends on the Angelina, Davy Crockett, and Sabine National Forests. Res. Pap. SO-250*. New Orleans: US Dep. Agric., For. Serv., Southern For. Exp. Stn.
9. Copeyon, C. K. 1990. A technique for constructing cavities for the red-cockaded woodpecker. *Wildl. Soc. Bull.* 18:303–311
10. Copeyon, C. K., Walters, J. R., Carter, J. H. III. 1991. Induction of red-cockaded woodpecker group formation by artificial cavity construction. *J. Wildl. Manage.* 55: In press
11. Costa, R., Escano, E. F. 1989. *Red-cockaded Woodpecker. Status and Management in the Southern Region in 1986*. Atlanta: US Dep. Agric., For. Serv.
12. Crouse, D. T., Crowder, L. B., Caswell, H. 1987. A stage-based model for loggerhead sea turtles and implications for conservation. *Ecology* 68:1412–23
13. Dawson, W. R., Ligon, J. D., Murphy, J. R., Myers, J. P., Simberloff, D., et al. 1987. Report of the scientific advisory panel on the spotted owl. *Condor* 89:205–29
14. DeLotelle, R. S., Epting, R. J., Newman, J. R. 1987. Habitat use and home range characteristics of red-cockaded woodpeckers in central Florida. *Wilson Bull.* 99:202–17
15. Doerr, P. D., Walters, J. R., Carter, J. H. III. 1989. Reoccupation of abandoned clusters of cavity trees (colonies) by red-cockaded woodpeckers. *Proc. Annu. Conf. Southeast. Assoc. Fish and Wildl. Agencies* 43:326–36
16. Emigh, T. H., Pollak, E. 1979. Fixation probabilities and effective population numbers in diploid populations with overlapping generations. *Theor. Pop. Biol.* 15:86–107
17. Emlen, S. T. 1982. The evolution of helping. I. An ecological constraints model. *Am. Nat.* 119:29–39
18. Emlen, S. T. 1991. The evolution of cooperative breeding in birds and mammals. In *Behavioral Ecology: an Evolu*

ionary Approach, ed. J. R. Krebs, N. B. Davies. Oxford: Blackwell. 3rd ed. In press

19. Heckel, D. G., Lennartz, M. R. 1992. Estimation of effective population size for the red-cockaded woodpecker. *Conserv. Biol.* In review
20. Hooper, R. G. 1983. Colony formation by red-cockaded woodpeckers: hypotheses and management implications. See Ref. 67, pp. 72–77
21. Hooper, R. G., Niles, L. J., Harlow, R. F., Wood, G. W. 1982. Home ranges of red-cockaded woodpeckers in coastal South Carolina. *Auk* 99:675–82
22. Hooper, R. G., Watson, J. C., Escano, R. E. F. 1990. Hurricane Hugo's initial effects on red-cockaded woodpeckers in the Francis Marion National Forest. *N. Am. Wildl. Natur. Resource Conf.* 55: 220–24
23. Jackson, J. A. 1971. The evolution, taxonomy, distribution, past populations and current status of the red-cockaded woodpecker. See Ref. 57, pp. 4–29
24. Jackson, J. A. 1978. Analysis of the distribution and population status of the red-cockaded woodpecker. In *Proceedings of Rare and Endangered Wildlife Symposium,* ed. R. R. Odum, L. Landers, pp. 101–110. Atlanta: Ga. Dep. Natural Resource, Game and Fish Div., Tech. Bull. WL4
25. Jackson, J. A. 1986. Biopolitics, management of federal lands, and the conservation of the red-cockaded woodpecker. *Am. Birds* 40:1162–68
26. Jackson, J. A. 1987. The red-cockaded woodpecker. In *Audubon Wildlife Report 1987,* ed. R. L. Sivestro, pp. 479–91. New York: Natl. Audubon Soc.
27. Jackson, J. A., Lennartz, M. R., Hooper, R. G. 1979. Tree age and cavity initiation by red-cockaded woodpeckers. *J. For.* 77:102–103
28. Koenig, W. D., Pitelka, F. A. 1981. Ecological factors and kin selection in the evolution of cooperative breeding in birds. In *Natural Selection and Social Behavior,* ed. R. D. Alexander, D. W. Tinkle, pp. 261–80. New York: Chiron
29. Koford, R. R., Bowen, B. S., Vehrencamp, S. L. 1986. Habitat saturation in groove-billed anis *(Crotophaga sulcirostris). Am. Nat.* 127:317–37
30. Lande, R. 1987. Extinction thresholds in demographic models of territorial populations. *Am. Nat.* 130:624–35
31. Lande, R. 1988. Genetics and demography in biological conservation. *Science* 241:1455–60
32. Lande, R., Barrowclough, G. F. 1987. Effective population size, genetic variation, and their use in population management. In *Viable Populations for Conservation,* ed. M. Soule, pp. 87–123. Cambridge: Cambridge Univ. Press
33. Lennartz, M. R., Harlow, R. F. 1979. The role of parent and helper red-cockaded woodpeckers at the nest. *Wilson Bull.* 91:331–35
34. Lennartz, M. R., Hooper, R. G., Harlow, R. F. 1987. Sociality and cooperative breeding of red-cockaded woodpeckers, *Picoides borealis. Behav. Ecol. Sociobiol.* 20:77–88
35. Ligon, J. D. 1970. Behavior and breeding biology of the red-cockaded woodpecker. *Auk* 87:255–78
36. Ligon, J. D., Stacey, P. B., Conner, R. N., Bock, C. E., Adkisson, C. S. 1986. Report of the American Ornithologists' Union Committee for the Conservation of the Red-cockaded Woodpecker. *Auk* 103:848–55
37. MacArthur, R. H. 1972. *Geographical Ecology.* New York: Harper & Row
38. Magnuson, J. J., Bjorndal, K. A., DuPaul, W. D., Graham, G. L., Owens, D. W., et al. 1990. *Decline of the Sea Turtles.* Washington: Nat. Acad.
39. Orians, G. H., Buckley, J., Clark, W., Gilpin, M. E., Jordan, C. F., et al. 1986. *Ecological Knowledge and Environmental Problem-solving: Concepts and Case Studies.* Washington: Nat. Acad.
40. Platt, W. J., Evans, G. W., Rathburn, S. J. 1988. The population dynamics of a long-lived conifer *(Pinus palustris). Am. Nat.* 131:491–525
41. Porter, M. L., Labisky, R. F. 1986. Home range and foraging habitat of red-cockaded woodpeckers in northern Florida. *J. Wildl. Manage.* 50:239–47
42. Powell, R. A. 1989. Effects of resource productivity, patchiness and predictability on mating and dispersal strategies. In *Comparative Socioecology,* ed. V. Standen, R. A. Foley, pp. 101–123. Oxford: Blackwell Sci.
43. Rabenold, K. N. 1984. Cooperative enhancement of reproductive success in tropical wren societies. *Ecology* 65:871–85
44. Reed, J. M., Doerr, P. D., Walters, J. R. 1988. Minimum viable population size of the red-cockaded woodpecker. *J. Wildl. Manage.* 52:385–91
45. Robinson, W. L., Bolen, E. G. 1984. *Wildlife Ecology and Management.* New York: Macmillan
46. Rudolph, D. C., Kyle, H., Conner, R. N. 1990. Red-cockaded woodpeckers vs. rat snakes: the effectiveness of the resin barrier. *Wilson Bull.* 102:14–22

47. Shaffer, M. L. 1981. Minimum viable population sizes for species conservation. *Bioscience* 31:131–34
48. Shaffer, M. L. 1983. Determining minimum viable population sizes for the grizzly bear. *Int. Conf. Bear Res. Manage.* 5:133–39
48a. Simberloff, D. 1988. The contribution of population and community biology to conservation science. *Annu. Rev. Ecol. Syst.* 19:473–511
49. Snyder, N. F. R., Wiley, J. W., Kepler, C. B. 1987. *The Parrots of Luquillo: Natural History and Conservation of the Puerto Rican Parrot.* Los Angeles: Western Found. Vertebrate Zool.
50. Soule, M. E. 1986. Conservation biology and the real world. In *Conservation Biology: The Science of Scarcity and Diversity,* ed. M. E. Soule, pp. 1–12. Sunderland, Mass: Sinauer
51. Spellman, C. B. 1987. *Simulation modeling of red-cockaded woodpecker clan dynamics.* MS thesis. Duke Univ., Durham, North Carolina
52. Stacey, P. B. 1979. Habitat saturation and communal breeding in the acorn woodpecker. *Anim. Behav.* 27:1153–66
53. Stacey, P. B. 1982. Female promiscuity and male reproductive success in social birds and mammals. *Am. Nat.* 120:51–64
54. Stacey, P. B., Ligon, J. D. 1987. Territory quality and dispersal options in the acorn woodpecker, and a challenge to the habitat saturation model of cooperative breeding. *Am. Nat.* 130:654–76
55. Stacey, P. B., Ligon, J. D. 1991. The benefits of philopatry hypothesis for the evolution of cooperative breeding: variation in territory quality and group size effects. *Am. Nat.* In press
56. Stevens, E. E. 1988. *Kin selection and inbreeding in the cooperatively breeding stripe-backed wren.* PhD thesis. Univ. North Carolina, Chapel Hill
57. Thompson, R. L., ed. 1971. *The Ecology and Management of the Red-cockaded Woodpecker.* Tallahassee: US Bur. Sport Fish. Wildl., Tall Timbers Res. Stn.
58. US Fish and Wildlife Service. 1979. *Red-cockaded Woodpecker Recovery Plan.* Atlanta: US Fish & Wildl. Serv.
59. US Fish and Wildlife Service. 1985. *Red-cockaded Woodpecker Recovery Plan.* Atlanta: US Fish & Wildl. Serv.
60. Van Balen, J. B., Doerr, P. D. 1978. The relationship of understory vegetation to red-cockaded woodpecker activity. *Proc. Annu. Conf. Southeast Assoc. Fish and Wildl. Agencies* 32:82–92
61. Vehrencamp, S. L. 1978. The adaptive significance of communal nesting in groove-billed anis *(Crotophaga sulcirostris). Behav. Ecol. Sociobiol.* 4:1–33
62. Wahlenberg, W. G. 1946. *Longleaf Pine: Its Use, Ecology, Regeneration, Protection, Growth and Management.* Washington: Charles Lathrop Pack For. Found.
63. Walters, J. R. 1990. Red-cockaded woodpeckers: a "primitive" cooperative breeder. In *Cooperative Breeding in Birds: Long-term Studies of Ecology and Behavior,* ed. P. B. Stacey, W. D. Koenig. pp. 69–101. Cambridge: Cambridge Univ. Press
64. Walters, J. R., Copeyon, C. K., Carter, J. H. III. 1992. An experimental test of the ecological basis of cooperative breeding in red-cockaded woodpeckers. *Auk.* In press
65. Walters, J. R., Doerr, P. D., Carter, J. H. III. 1988. The cooperative breeding system of the red-cockaded woodpecker. *Ethology* 78:275–305
66. Walters, J. R., Doerr, P. D., Carter, J. H. III. 1992. Delayed dispersal and reproduction as a life history tactic in cooperative breeders: fitness calculations from red-cockaded woodpeckers. *Am. Nat.* In press
67. Wood, D. A., ed. 1983. *Red-cockaded Woodpecker Symposium II.* Atlanta: Fla. Game & Fresh Water Fish Comm., US Fish & Wildl. Serv.
68. Woolfenden, G. E., Fitzpatrick, J. W. 1984. *The Florida Scrub Jay.* Princeton: Princeton Univ. Press
69. Zack, S., Ligon, J. D. 1985. Cooperative breeding in *Lanius* shrikes. I. Habitat and demography of two sympatric species. *Auk* 102:754–65
70. Zack, S., Rabenold, K. N. 1989. Assessment, age and proximity in dispersal contests among cooperative wrens: field experiments. *Anim. Behav.* 38:235–47

Annu. Rev. Ecol. Syst. 1991. 22:525–64

FUNGAL MOLECULAR SYSTEMATICS

Thomas D. Bruns

Department of Plant Pathology, University of California, Berkeley, California 94720

Thomas J. White

Hoffmann-La Roche, 1145 Atlantic Ave., Alameda, California 94501

John W. Taylor

Department of Plant Biology, University of California, Berkeley, California 94720

KEY WORDS: evolution, phylogenetic analysis, taxonomy, sequence, fingerprint

---

MOLECULAR EVOLUTION OF FUNGI

The fungi comprise both members of the kingdom Fungi as we now recognize it (Ascomycota, Basidiomycota, Zygomycota, and Chytridiomycota) and fungal-like protists such as the Oomycota and the cellular and acellular slime molds (Myxomycota and Acrasiomycota). Treating this admittedly polyphyletic assemblage as a group is useful because these organisms often fill rather similar roles within ecosystems, and they have traditionally been studied almost exclusively by mycologists and plant pathologists. Throughout this review, *Fungi* will refer to the Kingdom, *fungi* to the organisms studied by mycologists. The fungi, as thus defined, are of great importance for the following reasons: (*a*) They are the primary decomposers in all terrestrial ecosystems; (*b*) they are important symbiotic associates of vascular plants both in mutualistic and parasitic relationships; (*c*) they constitute the overwhelming majority of plant pathogens and as such have a tremendous eco-

0066-4162/91/1120-0525$02.00

nomic impact (Several significant human pathogens are also fungi); (*d*) they offer several well-developed genetic systems for molecular biologists *(Saccharomyces cerevisiae, Neurospora crassa, Aspergillus nidulans)*, and (*e*) they are crucial to the fermentation and biotechnology industries.

In spite of their importance, very little is known about evolutionary relationships within the fungi. Their simple and frequently convergent morphology, their lack of a useful fossil record, and their diversity have been major impediments to progress in this field. With the development of molecular techniques, many new avenues are now available that allow us to circumvent these obstacles.

Many techniques have been used to study evolution within the fungi. In this review we only consider those that compare nucleic acids, and we have made no attempt to cite all papers on this topic. Instead we have chosen those that enable us to illustrate specific points about different methods, types of analysis, modes of molecular evolution, or evolutionary relationships. We attempt to answer the questions of "what is the appropriate technique for a given level of divergence?" and "what are valid ways to analyze various types of molecular data?" We summarize what is known of evolution within and among the major divisions of fungi. Because only a small number of fungi have been sampled and the preliminary data are often fragmentary or equivocal, however, we have not proposed or endorsed any global classification scheme for the fungi.

METHODS USED IN FUNGAL MOLECULAR SYSTEMATICS

DNA-DNA Hybridization

The small size of fungal genomes and the fact that the proportion of repetitive sequences is much smaller than in plants or animals would seem to make them ideal for hybridization studies, but in fact the technique has very restricted usefulness because of the manner in which fungal genomes evolve. Virtually all studies in fungi have focused on the percentage of cross-hybridization between total DNA extracts rather than the thermal stability of hybrids. Several variations of both isotopic and spectrophotometric assays have been used and are reported to yield reasonably similar results if the percentage of cross-hybridization is greater than 90%. At lower percentages the results of different methods vary dramatically (78, 90). Perhaps the most interesting result to emerge from such studies is the observation that the percentage of DNA that cross-hybridizes between even closely related species is exceedingly low, typically less than 20%, while the percentage cross-hybridization between individuals within a biological species is typically greater than 90%. Species pairs exhibiting intermediate values are rarely reported. Vilgalys & Johnson (157) point out that the low level of cross-hybridization between what appear to be closely related species stands in stark contrast to the gradual reduction in hybridization between distantly related animals. The mechanisms that underlie this apparently rapid genomic turnover in fungi remain unknown and are an area ripe for further research. Whatever the cause, the phenomenon effectively limits the method to questions of very close relationships (78, 89). Some have used DNA hybridization for questions beyond the species level by focusing solely on ribosomal RNA genes. These highly conserved genes provide sufficient cross-hybridization for comparisons of greater phylogenetic distances, but resolution from such studies remains low (91).

Even in cases where the resolution is sufficient, the options for analyzing the hybridization distance values produced are more limited than for character data, and statistical testing of the resulting trees is difficult (38). Another major drawback is the need for pairwise comparisons. For N taxa the number of experiments equals N(N-1)/2. Thus, for a study with 25 taxa, 300 experiments are necessary. This number assumes no replications or even reciprocal hybridization experiments, both of which are typically needed to verify and improve the estimated hybridization values. As a result, large quantities of DNA are necessary for hybridization studies; this requirement eliminates the possibility of examining fungi that do not grow well in culture or produce sufficient quantities of easily collected biomass in nature. A further drawback is that the results of two different studies of related organisms are not readily compared without additional experiments.

The sharp drop off in cross-hybridization has commonly been used to help define species. This practice has proven to be of great utility in yeast systematics where correlation with mating studies has generally supported the concept that biological species can be recognized by levels of cross-hybridizations greater than 80% (89, 125). In filamentous fungi, defining species with levels of cross-hybridization has been used less often, and at least a few exceptions to strict correlation with ability to mate are known (33, 157). Even with the exceptions, however, cross-hybridization measurements represent the only simple one-number comparison that seems to correlate fairly well with ability to mate; for this reason alone the technique will probably continue to be used.

Restriction Enzyme Analysis

RFLPS VERSUS MAPPING Restriction patterns are generated by cleavage of DNA with restriction enzymes, followed by size separation of the resulting fragments via gel electrophoresis. The fragment patterns may be compared in two ways, either: (*a*) an RFLP (restriction fragment length polymorphism) approach is used in which the pattern of fragments or the individual fragments themselves are analyzed, or (*b*) a mapping approach is employed by using the pattern of fragments to deduce a map of the enzyme sites, which then become the units of analysis along with mapped length mutations. Mapping is the only way to determine the physical relationship of restriction fragments to each other, and as a result it has been and will continue to be the tool of choice for investigating structural variation in mitochondrial DNA (mtDNA). Mapping, however, is a time consuming and relatively error-prone process. In theory, maps can be compared directly, but in practice, detailed alignment of maps from different studies is often not possible without at least some direct fragment comparisons. The resulting resolution of phylogenetic relationships is often marginal because unlike plants and animals, fungi lack a large contiguous genomic region that is relatively free of length differences. Length mutations (i.e. insertions and deletions) complicate mapping, and restriction sites are prone to convergent loss and saturate quickly. For these reasons mapping to make phylogenetic comparisons among fungi is of limited value.

RFLP analyses can be very fast, and although a physical comparison is required, this can be done among 10 to 40 isolates simultaneously rather than by pair-wise experiments. The limits depend only on the number of lanes in a single gel. Problems exist with RFLPs, however, if one tries to use the number of fragment differences to estimate the degree of difference or nucleotide divergence or if one treats the individual fragments as independent characters. The greatest problem exists if length mutations are present and if multiple enzymes are surveyed (Figure 1). In this case a single mutation will be counted as at least 2N differences where N equals the number of enzymes surveyed. For obvious reasons, length mutations also invalidate estimations of nucleotide divergence. Even if no length mutations exist, the relationship between the number of fragment differences and the number of mutations is not very precise (Figure 2). Furthermore, even without length differences, fragments are clearly not independent characters because single site or length changes always result in multiple fragment changes. This lack of independence should concern anyone employing a parsimony analysis, and it could lead to erroneously high levels of confidence from bootstrapping. In the absence of length mutations, nucleotide divergence can be calculated by the method of Nei & Li (118). However, their formula for fragments employs no correction for multiple hits and so only makes reasonable estimates if the level of divergence is low (118). The only tests conducted with this formula were on circular DNA molecules (both simulated and real) where unobserved flanking regions do not exist; it is unclear how the estimates derived from random probe analysis would compare.

How serious are these problems? Unfortunately, length mutations occur at a high frequency in mtDNA and probably in nuclear DNA of fungi. Many

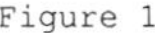

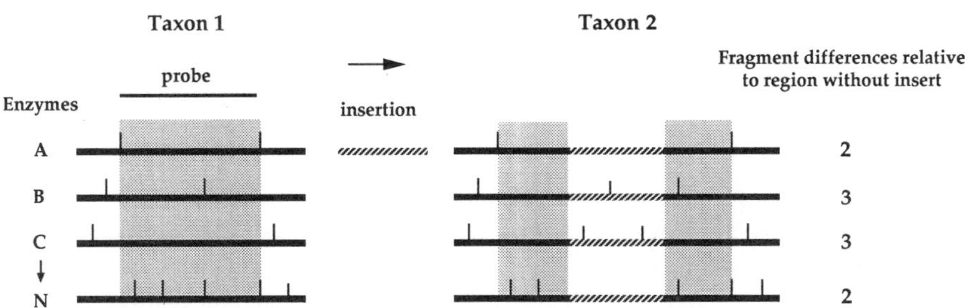

probe

taxa

Site and fragment differences

	A	B	C	D	E
A		1	2	3	4
B	0		1	2	3
C	2	2		1	2
D	3	3	3		1
E	3	3	3	2	

Figure 1 Problems with counting fragment differences when length differences exist. Restriction enzyme maps for a hypothetical region of taxon 1 are shown with sites for enzymes A to N given on separate lines. A probe based on a cloned fragment of enzyme A is used to visualize a portion of the area (shaded). The same region from taxon 2 contains a single insert. This single length difference would create RFLP differences in any enzyme surveyed. In the simplest case (enzymes A & N), this would result in the loss of one fragment from the taxon 1 pattern and the gain of a longer fragment in the taxon 2 pattern. An additional fragment difference appears if the insert contains sites for the given enzyme (B & C).

Figure 2 Problems with counting fragment differences when length differences are not a contributing problem. Maps are shown for the same enzyme in five different hypothetical taxa (A-E); each differs from the proceeding by a single site change. The area visualized by a probe is indicated. The matrix shows the number of site differences separating each taxon (top) and the number of observable fragment differences (bottom). Note that single site changes always result in 0, 2, or 3 changes in the number of fragments seen but additional site changes add either 0 or 1 additional fragments.

researchers have recognized the problem and tried to circumvent it by: (*a*) making course-maps to check for length mutations (117), (*b*) scanning probe regions with only a single enzyme to eliminate the multiple counting problem (47, 144), or (*c*) not analyzing the fragments as individual characters but instead cataloging the fragment patterns from a single probe region as different allelic forms of a given genetic locus (74, 108, 111). This latter approach is ideally suited to the investigation of population structure. The use of RFLPs as fingerprints for strain identification is another major strength (see below). For these purposes the use of RFLP analyses is likely to flourish.

TARGET REGIONS FOR RESTRICTION ENZYME ANALYSIS Mitochondrial DNA has been widely used for evolutionary studies in fungi. Several features contribute to its popularity: (*a*) It has a useful size (176-17 Kb), small enough so that it can be mapped or compared by RFLPs in its entirety, but large enough to supply many characters. (*b*) No evidence exists for methylation of bases—thus a potentially confounding factor of nuclear DNA is avoided. (*c*) The strongly A + T biased composition makes isolation of purified mtDNA relatively easy (73). (*d*) The high copy number of the mitochondrial genome makes it possible to visualize restriction fragments easily by hybridization of total DNA to mtDNA probes. Mitochondrial restriction fragments larger than 2Kb can often be visualized directly from stained gels of total DNA digested with restriction enzymes with G-C four-base recognition sites (e.g. HaeIII, CfoI, HpaII). These approaches are ideal for large comparative studies where crude DNA extracts from many isolates are the rule. (*e*) MtDNA is rich in RFLPs at the intraspecific level (17, 42–44, 47, 144, 152), and length mutations are the major cause of the variation in all cases where these polymorphisms have been mapped (17, 137, 152). Length mutations cause some analytical problems, but one major advantage is that, unlike site changes which are unique to a specific enzyme, length difference can be detected by virtually any restriction enzyme (Figure 1). (*f*) MtDNA is the best studied genomic element in fungi. Five ascomycetous mitochondrial genomes have been extensively sequenced: *Aspergillus nidulans, Neurospora crassa, Podospora anserina, Saccharomyces cerevisae,* and *Schizosaccharomyces pombe* (13, 27, 165). Maps of many other genomes are published, and most are summarized along with basic information about the structure and function of these genomes (72, 165).

Plasmids are extremely common in fungi (136) and potentially offer another multicopy target region for mapping or RFLP studies, but the disadvantages of using plasmids generally will outweigh any advantages. They are neither universally present nor conserved in sequence, and they may be horizontally transmitted (26, 107).

The nuclear ribosomal DNA repeat unit (rDNA) has also been used extensively for restriction enzyme studies in fungi, and many of these maps have been recently summarized by Garber et al (45). In most eukaryotes, including all true fungi, rDNA exists as a tandomly repeated array of the three largest rRNA genes separated by transcribed and nontranscribed spacers. In *Physarum* and *Dictyostelium,* however, these genes are arrayed in an extrachromosomal palindromic repeat (20, 25). Gene order appears to be un-

iversally conserved with the exception of the 5S gene, which may or may not be within the repeat and is reported to exist in the opposite orientation in at least some Basidiomycetes (23). The multiple copies of the repeat unit appear to homogenize quickly via concerted evolution (4a), and thus they generally behave like a single copy gene. A minor exception to this exists in some plants, animals, and Oomycota where different copies within the tandem repeat may vary in size owing to different copy numbers of small subrepeats within the intergenic spacer (IGS) (82, 131, 106).

The gene-spacer-gene arrangement of the multicopy array is part of the underlying reason that rDNA has been so popular. Portions of genic regions are so highly conserved that all heterologous probes hybridize strongly to them, yet other genic regions vary considerably at even moderate taxonomic levels. The spacers, particularly the intergenic spacer (IGS), often vary significantly even at the intraspecific level. Mapping studies have shown that both site and length differences occur in the repeat and that the level of RFLP variation often is similar to or slightly less than mtDNA (Table 1).

Universally conserved sequences within the ribosomal genes also make ideal priming targets for enzymatic amplification (see below); Vilgalys & Hester (156) have recently taken advantage of this fact to speed up the mapping of portions of the large subunit rRNA. Their method makes mapping only slightly more work than RFLP analysis, and for the additional effort one should gain considerable resolution and flexibility in analytical methods.

Single-copy and multi-copy anonymous clones have been used with increasing frequency in RFLP and mapping studies of fungi to examine dispersed genetic regions. Clones containing highly repeated DNA sequences can be quickly identified by colony hybridization using total genomic DNA as a probe (62, 111, 139). Single-copy regions are confirmed by probing southern blots of restricted DNA with individual clones (111).

Table 1 Relative levels of observed intraspecific RFLP variation[1]

Fungus	Random single-copy clones	MtDNA	rDNA	References
Agaricus bitorquis	nt	+	+	68
Armillaria spp.	nt	+++	+	3, 144
Coprinus cinereus	+++	+	+	166, 34
Fusarium oxysporum	nt	+++	+	81
Laccaria spp.	nt	+++	+	46, 47
Neurospora spp.	++	+	+	132, 117, 151, 152
Phytophthora spp.	+++	+	nt	44
Septoria tritici	+++	+	+	109, Bruce McDonald personal communication

[+]similar levels; ++, slightly greater; +++, much greater; nt, not tested
[1]comparisons are only made within species, not between

Studies employing random clones have revealed an amazing amount of the RFLP polymorphism both within and between species (Table 1); virtually every clone in some studies has been found to be polymorphic. (108). Many of the polymorphisms, particularly between species, involve apparent loss of the fragment or at least loss of ability to detect it via hybridization (44, 109, R. Vilgalys, personal communication). Mapping of the sites within six random clones in *Coprinus cinereus,* by Wu et al (166), demonstrated that many of the polymorphisms were caused by length mutations. They estimated that about 20% of random clones of 17 kb would contain unique inserts. These results are consistent with a model of frequent length mutation in the nuclear genome and may help explain the rapid drop off seen in DNA-DNA hybridization experiments discussed above.

Sequence Analysis

DIRECT SEQUENCING The use of DNA sequences for evolutionary studies can overcome many of the problems associated with restriction enzyme analysis. The large number of characters compared can substantially increase the resolving power. One can also observe the mode of sequence variation, i.e. whether a change is a transversion or transition, silent or selected, and can measure the degree of nucleotide bias; these observations may be incorporated into the phylogenetic analysis in various ways (see below). Results from different laboratories can be directly compared, and the publication of sequences and their deposition in electronic databases (GENBANK, EMBL) facilitate the confirmation of results and their application to other taxa without the need to obtain strains or clones or to repeat experiments.

Until recently, however, recombinant DNA techniques used to obtain sequence information were sufficiently difficult and laborious that the study of large numbers of species or individuals required exceptional efforts. Two methods—direct sequencing of ribosomal RNA and direct sequencing of amplified DNA fragments—have circumvented the need for cloning and thus dramatically reduced the time and effort required for comparative sequencing studies.

The first rRNA to be sequenced extensively was the 5S rRNA (10, 71, 158), but this gene is too small and evolves too rapidly to be suitable for the study of most fungal relationships (60, 148). Additionally, compensatory substitutions that maintain RNA secondary structure and the existence of multiple, independently evolving copies of the gene in filamentous *Ascomycota,* including *Neurospora* and *Aspergillus* (5, 140), present further problems.

Direct sequencing of the two largest rRNA by primer extension has broader applicability to systematic studies (61, 93). The rRNA template can be easily isolated in large quantities. This procedure has been used extensively for determining the 16S rRNA sequences of prokaryotes (164), and Lynn & Sogin (100a) have extended its use to the systematics of protists. A deficiency of the method is that the sequence of only one strand of the nucleic acid is obtained. This results in a relatively high frequency of errors (1–5%) because ambiguities cannot be resolved by comparison with the opposite strand. Another potential problem is that RNA sequences of some organisms have extensive posttranscriptional sequence modifications or "editing" and do not reflect the actual DNA sequence of the gene (149).

In contrast to rRNA sequencing, the polymerase chain reaction (PCR) developed by Mullis (116, 133) allows biologists to sequence DNA from many species or individuals in a few days. Several methods are available for direct sequencing of PCR products. These include methods that generate and sequence single strands (59, 67, 114) or that directly sequence double strands (22). Any of these approaches when working properly is ideally suited for large comparative studies. The accuracy of direct sequencing of PCR products is improved over rRNA sequencing because both strands can be determined. A second advantage is that regions other than nuclear rRNA genes are accessible for sequence analysis. "Universal" primer sequences for both nuclear and mitochondrial rRNA genes as well as the internal transcribed spacer have been described for fungi (156, 160), and in animal systems primers for cytochome **b** have also been developed (85). An underlying assumption of sequence analysis is that the phylogeny of the region is a good indicator of the phylogeny of the organisms. A good test of this assumption is to compare the results from regions that are physically and functionally unlinked; PCR makes such a test feasible with little additional effort. Another important advantage of PCR is that only minute amounts of DNA are needed. As a result, rare or obligately parasitic fungi are now accessible to molecular systematists.

One concern regarding direct sequencing of amplified DNA involves errors introduced during synthesis by the DNA polymerase. The initial observed error frequency for the *Taq* DNA polymerase was as high as one substitution per four hundred base pairs after 30 cycles of amplification (138). However, conditions can be optimized to reduce this error frequency to less than one substitution in 15,000 bp (49). Fidelity is usually of no concern in direct sequencing of amplified DNA (in contrast to sequencing single clones from amplified DNA) because the sequence obtained is a consensus of all molecules present in the reaction (59). Thus, each individual misincorporation will be represented only very infrequently in the population of DNA molecules to be sequenced. Even if the error is introduced in the first cycle, using a single molecule of target only 25% of the relevant band density on the sequencing gel will reflect the erroneous base. Although cloning is slower and potentially more error prone, individual clones may be worth sequencing when allelic variation is high, but several should be determined to confirm that the differences found are not artifactual.

SELECTION OF APPROPRIATE REGIONS TO SEQUENCE There are several factors to consider in selecting a region for sequence analysis (76). These include: (*a*) The region should be evolving at an appropriate rate for the comparison of interest. Ideally this means that the region supplies enough consistent differences to separate the taxa into statistically supported monophyletic groups. Regions that are too conserved will provide too few changes. Regions that are too variable will contain too many inconsistent characters due to multiple substitutions at single positions, and alignment may be an additional problem. (*b*) The region should be present ideally as a single copy or should at least evolve like a single copy region (e.g. rDNA, mtDNA). The danger of multicopy regions is that different copies might be compared in different species (i.e. paralogous comparisons). (*c*) The region should have the same function in all taxa. Evolution of a new function changes selective pressures and therefore the rate of sequence change. (*d*) The effect of base composition and codon bias should be examined. Both factors can distort estimates of divergence (36, 142).

Virtually all sequence studies in fungi have focused on the rRNA genes. Their popularity is caused primarily by the presence of universally conserved regions that serve as ideal primer sites. Different regions of the mitochondrial and nuclear rRNA genes diverge at different rates; therefore if one wants to get the maximum amount of information from a minimum amount of sequencing the question of which regions are most appropriate for a specific level of comparison is important. To this end a preliminary comparison among six different rRNA regions is relevant (Figure 3). The situation with rRNA genes is complicated because both base substitutions and length mutations make alignment of some regions ambiguous. Thus, as more distant comparisons are made the number of alignable nucleotides decreases. This problem is most acute in the mitochondrial genes (Figure 3); they are thus not particularly useful for distant phylogenetic comparisons. At intermediate taxonomic levels, however, the mitochondrial genes may have a higher proportion of variable sites. This point can be seen by comparing the MS1/2 region of the mitochondrial small subunit rRNA (SrRNA) gene and the NS3/4 region of the nuclear SrRNA gene (Figure 3). This comparison is reasonably sound because both regions overlap equivalent structural domains. Similar comparisons of LrRNA genes would be useful particularly for the regions of the nuclear gene that have been used for phylogenetic studies in fungi (56–58, 88, 167, 168).

Protein coding genes have some advantages over rRNA genes and spacers in that alignment of the sequence is less problematic. Protein sequences also lend themselves to differential weighting of bases by codon position, and third position sites can provide a relatively good estimate of the neutral

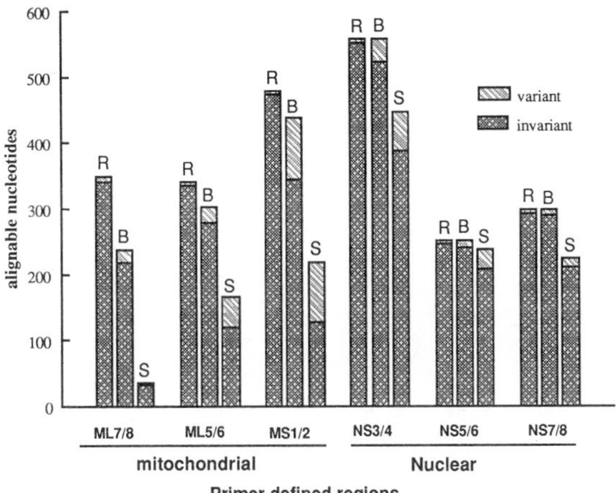

Figure 3 Size and variability of alignable sequences from six selected rRNA regions. Pairwise comparisons of sequences of *Suillus sinuspaulianus* versus *Rhizopogon subcaerulescens* (R), *Boletus satanas* (B), and *Saccharomyces cerevisae* (S) are shown. The regions are defined by primer pairs described by White et al (160) from the mitochondrial LrRNA (ML), mitochondrial SrRNA (MS), and the nuclear SrRNA (NS) genes. All regions were completely alignable between *Suillus* and *Rhizopogon* except within MS1/2 in which 59 bp of the *Suillus* gene could not be unambiguously aligned with Rhizopogon. Basidiomycete sequences are from Bruns et al 1989, 1990; T. D. Bruns, T. M. Szaro, unpublished; and the *Saccharomyces* sequences were from GENBANK.

substitution rate. Unfortunately, no "universal" primers have yet been developed for fungal protein genes. Part of the problem is that many or most of the well-characterized nuclear protein genes may be multicopy; for the reasons mentioned above this makes them unattractive. Mitochondrial protein genes have been favored in animal studies, but in fungi virtually all of the large mitochondrial proteins, except for cytochrome oxidase subunit III (121), are frequently interrupted by numerous large introns.

Electrophoretic Karyotyping

Recent advances in gel electrophoresis technology have enabled megabase-sized DNA molecules to be efficiently separated. These techniques have already been applied extensively to the separation of fungal chromosomes. Mills & McCluskey's (112) recent review lists 25 species of fungi whose electrophoretic karyotypes have been examined. One of the most surprising results of these studies is the high frequency of chromosomal length polymorphisms observed within species; this adds a new dimension to the emerging picture of fungal genome fluidity but essentially eliminates the method for phylogenetic studies above the population level.

ANALYTICAL OPTIONS FOR MOLECULAR DATA

To facilitate our later discussion of phylogenetic studies we briefly outline some of the methods currently used for analysis and recommend several reviews for a comprehensive discussion of the subject (38, 143a, 150). There are two main categories of methods for inferring phylogenetic trees: those based on distance, and those based on characters. Distance methods create trees by reducing all information on genetic characters to pairwise estimates of similarity or dissimilarity. Variations such as the UPGMA procedure assume or at least work best with equal rates of change on all lineages, while others such as the neighbor-joining method and maximum likelihood are less constrained (38, 135). Character-based methods such as parsimony search for the tree that minimizes the number of character state changes. Methods of both types such as maximum likelihood, evolutionary parsimony, and compositional statistics (142) are designed specifically for analysis of sequence data and are based on specific assumptions about the manner in which sequences evolve. Other more general methods can be adapted to sequence analysis by using a molecular evolution model to calculate the distances or by weighting certain types of characters or character state changes. An important point to keep in mind is that trees do not "reconstruct a phylogeny"—they are an inference of it. To paraphrase Felsenstein (38): The real question is not "what is the best tree?" but rather, "how much confidence can we place in a given phylogenetic estimate?" The latter question is a statistical one, and several methods can be used to address it. Bootstrapping (37) can be used to resample the data and assess the strength of internal branches. It has been used primarily with parsimony, but version 3.4 of PHYLIP allows it to be easily adapted to any method of analysis (39). A major advantage of bootstrapping is that it can be applied easily to moderately complex trees, although with large numbers of taxa the computational time becomes a major constraint. Several other methods—maximum likelihood, evolutionary parsimony, and the winning-sites method (38, 92, 124)—can be used to compare trees and ask if one is significantly better than another. With these methods a small number of taxa are used (typically four) in order to reduce the total number of trees compared to a manageable number.

When the number of variable characters is sufficiently large and when most characters have changed only once, all methods yield similar if not identical results (32, 80, 134). Problems arise as characters become saturated with mutations, and under these conditions methods often disagree and statistical evidence conflicts with different methods (54). Similar problems occur when multiple branches originate nearly simultaneously, such as during an adaptive radiation. The best documented case involves the phylogeny of the hominoid primates, which cannot be resolved with an acceptable level of statistical confidence despite large quantities of sequence data (69).

ANALYSIS AT THE POPULATION LEVEL

The population level is of great interest to mycologists and especially plant pathologists because it is a window into the process of speciation. The genetic structure of pathogens may indicate their potential for development of new races and fungicide-resistant strains. Sexual recombination, however, serves to shuffle genetic markers, and allelic differences often are not fixed. Therefore a dicotomously branching phylogenetic representation of relationships is often inappropriate. If the species are asexual, however, phylogenetic methods of analysis may be relevant.

RFLPs As Alleles

The use of RFLPs for genetic markers has now become a standard genetic tool, and their use in fungi was recently reviewed by Michelmore & Hulbert (111). Two later papers by these authors and their colleagues on *Bremia lactucae,* an obligately parasitic Oomycete, provide a model for the use of RFLPs as genetic markers in diploid fungal systems (74, 75). For haploid systems the recent work on *Septoria tritici* by McDonald & Martinez (108) provides a useful example. Their work with RFLPs is unique in that it focuses on the population structure within a single wheat field, and their results reveal a high level of genetic variation, including differences between different lesions on a single leaf. Their findings certainly demonstrate that for single-copy random clones, much of the variation may be within populations. As they point out this result raises questions about conclusions of other studies that employ relatively small intraspecific samples or assume that worldwide collections provide a good representation of population-level variation, but this concern is probably most relevant to the particular technique used: RFLPs of random clones.

A recently developed PCR-based method, termed RAPD (random amplified polymorphic DNA), avoids the need for restriction enzymes, blotting, probing, or cloning and produces fragment differences similar to RLFP analysis (162). Typically several fragments of varying intensity are produced by amplification with a single short primer of arbitrary sequence. Standardization of amplification conditions is very important because the fragment pattern observed is highly sensitive to concentrations of Mg^{+2}, DNA polymerase, primers and template DNAs, and cycling temperatures. Mapping of individual fragments has demonstrated that they behave as simple Mendelian loci (162). Several potential problems exist in the use of RAPDs. First, alternative alleles are usually seen as the absence of the fragment; this means heterozygotes and homozygotes are often not distinguishable. Second, the multiple fragments of a single RAPD sometimes map to the same or nearly the same locus and so may not be independent loci (162). Third, different band intensities may make scoring patterns ambiguous. When used as markers for constructing genetic maps these problems are all relatively minor because segregation analysis can be used to clear up most of the ambiguities. If RAPDs are used to analyze population genetics, however, these problems will need to be addressed; the speed and convenience of the method virtually ensures that they will.

Fingerprinting

For distinguishing strain or species differences a unique RFLP pattern can be used as a fingerprint. For this purpose, clones containing repeated elements have proven extremely useful. In *Candida albicans* and *Magnaporthe grisea,* closely related species or different pathogen-defined portions of the same species can be distinguished by their lack of dispersed copies of the repeat element (62, 139). In both cases, intraspecies RFLP variation was very high. In *Magnaporthe grisea* every rice pathogen had a distinct pattern; this finding is particularly impressive given that isozymes found little or no variation in these strains (62). In the case of *Candida albicans,* RFLP patterns actually changed at measurable rates from serial culturing on rich media and apparently also changed in isolates taken over the course of infection in a single individual (139). In *Colletotrichum lindemuthianum* a cloned repetitive element revealed a conserved RFLP pattern within the species (R. J. Rodriguez, personal communication), but a different pattern between species. Subsequent sequence analysis demonstrated that tandemly repeated copies of a degenerate trinucleotide, CAN, were responsible for the multilocus hybridization pattern (R. J. Rodriguez, unpublished).

Multilocus fingerprints can also be produced by hybridization of simple-sequence synthetic oligonucleotide probes to minisatellite DNAs. This approach avoids the necessity of cloning, and is beginning to be used widely in animal systems (2, 155). In fungal systems the approach has recently been used with *Penicillium, Aspergillus,* and *Trichoderma* (110a).

While multilocus probes are very powerful for fingerprinting, numerous problems exist if one tries to use similarity among fingerprints to discern genetic relationships (99, 100). Some of these problems would be less significant in haploid asexual fungi because recombination and the problem of ascertaining whether an individual is homo- or heterozygous would not be a factor. Nevertheless, one can be misled by nonhomologous fragments with similar electrophoretic migration and missing or unresolved alleles. All of the difficulties with respect to analyzing fragment differences as characters also apply to fingerprints. Furthermore, a recent study by McDonald & Martinez (109) has shown that little correlation exists between genetic distances calculated from allele frequencies of single copy RFLPs and from fragment differences visualized from multilocus probes. Even distances calculated from two different multilocus probes did not correlate well (109).

Strain and species differences frequently have been correlated with rDNA RFLPs. For example, seven species of *Sclerotina,* six species of *Canidia,* and 15 different intraspecific groups of *Rhizoctonia solani* can all be separated by rDNA RFLPS (86, 101, 157a). In other cases such as *Cenococcum geophilum* the amount of intraspecific variation appears to be considerably higher. This may be an artifact caused by a broadly defined morphological species that includes several distinct biological species (97), but the amount of intraspecific RFLP variation may also be highly variable among unrelated taxa. Differences in size and structure of the IGS spacer, the level of methylation, and the amount of sexual recombination clearly could affect the level of intraspecific rRNA variability.

Vilgalys & Hester (156) also used a portion of the rDNA repeat to distinguish the human pathogen *Cryptococcus neoformans* from three other species, but their RFLPs were derived from a PCR amplified portion of the repeat. This approach is not only faster, it also avoids the potential complication of RFLPs caused by methylation differences. In addition, it allows one to limit the survey to a portion of the rDNA that is evolving at the appropriate rate for the comparison of interest.

MtDNA RFLPs have served as fingerprints, but their use is often most straightforward at the strain rather than the species level because of the high RFLP variability of the genome. Different levels of variation are often observed among equivalent taxonomic levels of closely related fungi (44, 47, 151). The reasons for these differences in the level of intraspecific mtDNA variation are unclear and certainly are complicated by human agricultural practices, sample size differences, and the basis of the taxonomic categories. If one is planning to use mtDNA RFLPs for fingerprinting, the take home message is that the level of variation one might see below the species level cannot be anticipated from previous studies but must be determined empirically.

PHYLOGENETIC STUDIES

Owing to their economic importance, their laboratory tractability, or the hopeless condition of their classification, a few fungi have been subjected to intense molecular phylogenetic scrutiny by a variety of methods.

Yeasts

Clark-Walker and colleagues (24), have mapped mtDNA, determined its gene order, and sequenced cytochrome oxidase subunit II *(coxII)* genes in *Dekkera* species and species of the anamorphic genera *Brettanomyces* and *Eeniella.*

With these data, they have asked an important question about mtDNA evolution: Do phylogenetic relationships inferred from gene order or length mutations agree with those based on nucleotide substitution?" They conclude that, "neither length nor gene rearrangements are useful characteristics for establishing relationships between these molecules or the yeasts harboring them." It is worth noting, however, that both mtDNA size and gene order in these yeasts are at least consistent with their sequence data. Species with the same gene order also have similar sized mtDNAs, and branch together on the unrooted *cox II* tree. Two large rearrangements hold the large genome species together; *B. custersii* retains this order, while additional unique rearrangements occur in the two other lineages. What the sequence data have provided is much greater resolution and a likely root for the tree. Although mtDNA sizes make sense when reviewed on the *coxII* tree, size differences are great enough that length mutation mapping would be useful only between the two *Dekkera* species. Their similar mtDNAs are like those of outbreeding *Neurospora* species, among which mapped length mutations have been phylogenetically useful (151).

Clark-Walker et al (24) also recognize the problem of using phylogeny to define taxa. They ask, "Can species and genera be defined by DNA similarity coefficients?" and recommend lumping the two *Dekkera* species and putting *Eeniella* back in *Brettanomyces*. Lumping sexually reproducing species with very similar mtDNAs such as *Dekkera* may be premature because members of different *Neurospora* biological species do have identical mtDNAs (151). However, putting *Eeniella* in *Brettannomyces* seems logical because it branches within *Brettanomyces* and no mating-based species concept is possible with these asexually reproducing fungi.

The use of nuclear small subunit rRNA sequence to determine fungal phylogeny has been swiftly applied to yeast classification. Barns and colleagues (4b) used the gene to examine relationships among 10 of the most commonly pathogenic species of *Candida*. Their results confirmed the heterogeneous nature of the taxon and produced the first estimate of phylogenetic relationships within it. Their analysis included a filamentous ascomycete. *Aspergillus,* and three additional ascomyceteous yeasts. Addition of Basidiomycota to the analysis should prove interesting, because it has long been thought that *Candida* may include members of both divisions.

To obtain complete rRNA sequence for the many fungi required for systematics is both time consuming and expensive. Kurtzman & colleagues (57, 58, 88) have addressed this problem and argue that partial sequences of small and large subunit nuclear rRNAs can provide enough characters to classify ascomycete and basidiomycete yeasts. Determining the statistical support for the resulting phylogenies would seem essential to this approach. Guého et al (57) considered the question of significance in an intuitive way when they used 350 bp of large subunit and 338 bp of small subunit sequence to compare basidiomycete yeasts, including animal dermatophytes in the genus *Trichosporon*. They found that maximum likelihood analyses of each subunit alone produced different trees, and the deep branches in the phylogeny based on the combined sequence were marked to reflect this uncertainty. Alternative trees can be compared using a likelihood ratio test (38); if this test had been used, conclusions about the polyphyly of *Trichosporon* species might be stronger. Instead, branch rearrangements in the uncertain region can produce a monophyletic *Trichosporon* clade.

In a study of *Sterigmatomyces* sensu lato, Guého et al (58) showed that the four segregate taxa fell into two groups. Yamada and colleagues (167) took the same approach and problem further and claimed that one of the segregate genera, *Tsuchiyaea,* was distinct from another, *Fellomyces,* based on distance analysis of partial small subunit rRNA sequence They made this claim although their analysis of large subunit sequence contradicted it. In this case, statistical tests are important. Of the 160 bp of small subunit rRNA sequence given, only one site could be used to compare the taxa in question with the outgroup, *Sterigmatosporidium polymorphym*—clearly not enough to establish statistical support. The creation of new taxa is a matter of opinion, but if this approach is used to justify the new taxa (168), statistical testing cannot be ignored. Demonstrating statistical significance does not necessarily require long sequences. Mitchell & colleagues (115) used parsimony analysis and bootstrapping with the short 5.8S rRNA sequences of yeast-like basidiomycetes to support morphologically based classification of *Filobasidiella* and *Filobasidium* species.

Filamentous Ascomycetes

NEUROSPORA *Neurospora* has been the subject of a surprising number of evolutionary studies, considering its reputation as a laboratory denizen. When length mutations were used to compare four biological *Neurospora* species, most of the variability, and hence the oldest divergences, were found in *N. crassa* and *N. intermedia,* neither of which were monophyletic; *N. tetrasperma* and *N. sitophila* formed monophyletic groups with less variability (151). This result is consistent with at least three different but not mutually exclusive explanations: (i) the ancestral taxon is *N. crassa / N. intermedia,* and *N. tetrasperma* and *N. sitophila* have diverged recently (ii) *N. crassa* and *N. intermedia* became sympatric before reproductive isolation was complete and have exchanged mtDNAs; (iii) the conflict is illusory because too few mutations are seen to establish statistical significance. In support of the first explanation, Taylor & Natvig (151) argue that among closely related species, intraspecific diversity of clonally propagated molecules [i.e. mtDNA of *N. crassa* (103) and *N. tetrasperma* (95)] should be greatest in the ancestral species. This same argument has been made for mammals (163). By this argument, interpretation of mtDNA variation in closely related species must account for variation present in the ancestral population before species diverged. Only when extinctions have reduced intraspecific variation relative to interspecific variation will phylogenetic trees be reliably supported. Nuclear genomes in sexually reproducing fungi will also contain alleles present before divergence, but the frequencies of alleles should differ between populations. The second explanation, hybridization following allopatric speciation, is supported by the interfertility observed between *N. crassa* and *N. intermedia* (120). The third explanation, that too few data exist, is true and would be best addressed by sequencing studies, but to date no region with enough variation has been found. Anonymous clone nuclear DNA studies have not contradicted the mtDNA tree, but with the exception of *N. tetrasperma,* neither have they confirmed it because too few common isolates have been examined (117). Low resolution mapping indicated that length mutations were rare in the nuclear fragments, but sites outside the region of hybridization made interpretation somewhat more difficult.

Other genera of the Sordariaceae have been compared to *Neurospora,* first by DNA-DNA hybridization (33) which showed that they are unusually closely related ascomycete genera (87). Recently DNA sequence comparison of nuclear ITS and mitochondrial SrDNAs (153) showed that the sordariaceous genera are no more divergent in these DNA regions than are species of *Laccaria* (48) or *Suillus* (6). This disparity of generic and specific variation in the two groups cannot be ascribed solely to taxonomic opinion, as it holds for biologically determined species of *Neurospora* and *Laccaria.* Either ITS and mitochondrial SrDNA evolve more slowly in the Sordariaceae than the Agaricales, or speciation occurs more rapidly in the Sordariaceae. Those that favor a molecular clock with a constant rate, at least in higher fungi, might look to differences in life histories of saprobic ascomycetes and mycorrhizal basidiomycetes, or differences between monokaryon or dikaryon dominated life cycles, for an explanation of different rates of speciation.

FUSARIUM The appeal of this fungus for evolutionary studies is obvious: It is economically important, well suited to the laboratory, and no stranger to taxonomic controversy. Jacobson & Gordon (77) surveyed mtDNA variability within 78 strains of *Fusarium oxysporum f. sp. melonis.* This sample included all known vegetative compatibility groups (VCGs) and host-defined races within the taxon. Initial comparison by RFLPs revealed seven forms of the mtDNAs. Each of these forms was subsequently mapped, and the restriction site and length differences were then analyzed by both distance and parsimony methods. The results of the analysis showed that VCGs were strongly correlated with mtDNA type: six of the VCGs exhibited unique mtDNA patterns, while two shared a single mtDNA pattern. This strong correlation stems in part from the strictly asexual life cycle of this fungus, but it also suggests that either the rate of evolution of new VCGs, which is presumably based on nuclear genes, is similar to the rate of RFLP evolution in the mtDNA or that mtDNA recombination may serve to homogenize variants within portions of the population that retain the ability to fuse. The pattern of host-defined races, however, was not strongly correlated with the inferred phylogenetic relationships, suggesting that they may have evolved both more rapidly and convergently (77).

Guadet & colleagues used direct sequencing of two portions of the nuclear LrRNA gene to examine phylogenetic relationships within 52 strains of eight species of the genus *Fusarium* (56). The results of their phylogenetic analysis correlate well with traditional classification, based on both anamorphs (asexual stages) and teleomorphs (sexual). One species, *F. nivale,* was unrelated, but even this result is supported by prior taxonomic opinions. Their analysis, however, does not permit evaluation of the statistical strength of the conclusions. To examine this question we reanalyzed their data by a bootstrap parsimony method, which shows that most of their conclusions are strongly supported (Figure 4). The removal of *F. nivale* from the genus *Fusarium* and the monophylesis of two of the three teleomorphic genera and two of the five anamorphic sections are all supported above the 95% level. Section *Liseola* is the only supraspecific taxon not supported above the 80% level. The overall strength of these branches combined with the correlation with prior taxonomic opinions demonstrates that their analysis represents a very reasonable initial estimate of evolutionary relationships within *Fusarium*. The usefulness of partial sequences from nuclear LrRNA for phylogenetic analysis at intermediate to distant levels is also evident, and the database that they have started will make it easy to add additional taxa such as the related anamorphic genera *Cylindrocladium, Cylindrocarpon,* and *Tubercularia*.

Although the broad framework of relationships within *Fusarium* was clearly an appropriate taxonomic level for the approach of Guadet and colleagues, the minor sequence divergences revealed within and between closely related species were generally insufficient to resolve these lower-level relationships. Their analysis of multiple varieties and isolates, however, clearly shows that although variation exists at virtually all levels, intraspecific variation does not have a strong effect on tree topology at higher levels. This is an important assumption of molecular systematic analysis that has not often been tested thoroughly.

Questions about higher level ascomycete systematics have also been addressed, including the placement of enigmatic organisms such as *Pneumocystis* (35), as well as problems of fungi possessing what have been considered convergent, reduced morphologies (e.g. Plectomycetes with their simple asci)

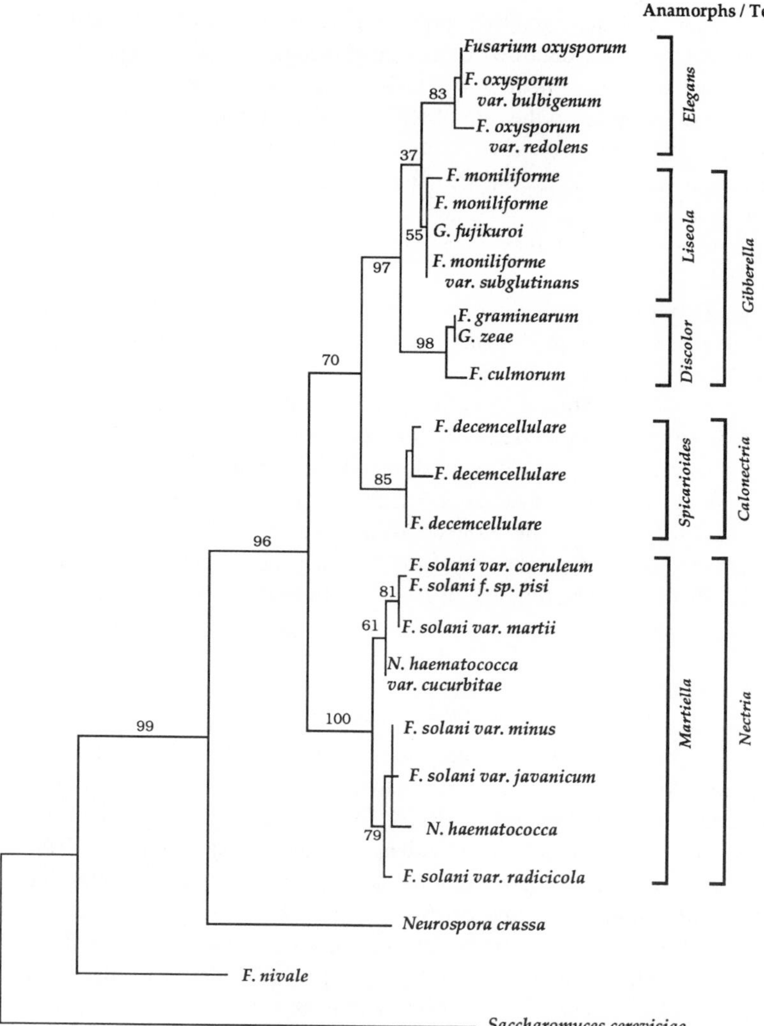

Figure 4 Bootstrap analysis of phylogenetic relationships within Fusarium (*Based on the data of Guadet et al*—56). Relationships depicted are based on Parsimony analysis. Horizontal distances correspond to the minimum number of inferred mutations; vertical distances are arbitrary. Numbers on subterminal branches correspond to the number of trees from a bootstrap sample size of 100 in which the clade existed as a monophyletic group.

or contradictory morphologies (e.g. *Ophiostoma* with its pyrenomycete perithecium and plectomycete asci). Berbee & Taylor (9) analyzed 18S rDNA sequences using parsimony analysis with bootstrapping to show that Plectomycetes are a monophyletic group compared to Pyrenomycetes and yeasts, and that, for *Ophiostoma* the ascocarp is a better indicator of phylogenetic relationships than the ascus. Bowman et al. (12) examined *Coccidioides immitis,* an anamorphic human pathogen claimed for the Ascomycetes and Zygomycetes (129), and showed that its 18 SrRNA sequence groups it with other ascomycete pathogens in the Plectomycetes.

Filamentous Basidiomycetes

Within the filamentous basidiomycetes, most phylogenetic studies have thus far been limited to mushroom-forming taxa: agarics, boletes, and their gasteroid relatives. Most of the published studies of agarics have focused on biological species complexes (i.e. mating-defined species that are difficult to separate morphologically) or closely related morphological species. At both levels lack of resolution has been a common problem with the techniques used.

In the *Armillaria mellea* complex, both mtDNA RFLPs and rDNA maps have been used to study relationships among the biological species of *Armillaria,* and both resulted in relatively limited resolution. For rDNA, ten variable sites and two length differences were identified; based on these differences the 13 North American and European taxa were placed into six groups, and relationships among these groups were analyzed by compatibility and parsimony analysis. The resulting unrooted network shows that differences between the six groups are limited to one or two site or length differences, and four of the nine informative changes occur convergently (3). In the case of mtDNA the patterns between species are so dissimilar that RFLP comparison is not possible (79, 144).

Relationships among *Laccaria laccata, L. bicolor,* and *L. amethystina* were also examined with both rDNA and mtDNA RFLPs, with results very similar to those of the *Armillaria* studies. Little variation was found in rDNA RFLPS, while mtDNA RFLPS were so variable that most intraspecies comparisons were out of range (46, 47). In both *Armillaria* and *Laccaria,* intraspecific mtDNA variation was so common that it proved useful for strain fingerprinting (47a, 145).

In the *Collybia dryophila* complex, DNA-DNA hybridization was used to study relationships (157). A virtual continuum of intermediate hybridization percentages enabled estimation of phylogenetic relationships. The tree produced correlated well with mating-compatible groups, and divergence within these potential breeding units provided evidence for allopatric speciation. Intermediate hybridization percentages have also been reported among European species of *Armillaria,* but these were not used for phylogenetic inference (79). In both the *Collybia* and the *Armillaria* studies, estimation error associated with interspecific comparisons appears to be fairly high; differences of greater than 10% existed, dependent on which isolates were chosen for intraspecific comparison. This variation is typical of DNA hybridization; whether it is biological or methodological in origin, it would certainly have an effect on the confidence of inferred interspecific relationships.

Morphological species of *Suillus* were examined by mtDNA mapping (16, 17). The extreme size differences among mtDNAs of *Suillus* species made detailed alignments of mtDNAs impossible. To overcome this problem fine-structure maps of the mitochondrial rRNA genes were constructed and aligned. This time-consuming approach yielded only 47 shared differences among the 19 taxa investigated. Parsimony analysis of these data clearly separated *Suillus* from the related genera and thereby resolved a long-standing controversy concerning the position of *Paragyrodon*. Reanalysis of these data by bootstrapping shows that this result is supported above the 99% level, and the result has since been confirmed within a subsample of the taxa by sequence analysis with the same level of confidence (15). Relationships within *Suillus,* however, were only defined by one to three site changes, and a reanalysis of these data with bootstrapping shows that none of the branches are supported at even the 95% confidence level. Failure to resolve the relationships within *Suillus* is caused primarily by the low sequence divergence (0.3–2.9%) found within the genus in the regions mapped. Direct sequencing of PCR products has now confirmed these low estimates in several portions of the mitochondrial rRNA genes (14, 15) but has failed to produce a robust estimate of evolutionary relationships within the genus (T. Bruns, unpublished results). Although the tree produced from the restriction site mapping study was not statistically significant, its correlation with mtDNA size and basidiocarp morphology suggests that it is at least a reasonable estimate. The close placement of *S. spraguei (pictus)* and *S. luteus* was the only really glaring conflict with morphology (14). In lieu of stronger molecular evidence the meaning of the apparent conflict remains uncertain.

The question of what technique is appropriate for analysis of closely related species of basidiomycetes remains largely unanswered. Direct sequencing of the nuclear ITS spacers, however, is one possible approach. The region differs by approximately 1–3% within a morphological species of *Suillus* and by about 10% or more between species examined, yet most of the region is alignable among what we now know to be the closest relatives of *Suillus: Rhizopogon, Gomphidius, Chroogomphus* (6, 15, 18). ITS variation among *Laccaria* species also ranges from about 1–3% (48) Within *Armillaria,* however, ITS sequence variation may be insufficient to resolve relationships within all species, but a portion of the IGS appears to be more useful (J. Anderson, personal communication).

Several ongoing studies based on rDNA mapping and on sequencing of nuclear or mitochondrial rRNA genes are focused on intra- and inter-family level relationships within the agarics and boletes (18, 126, 130). The results from these studies should soon provide a general framework of relationships, but many family-level relationships are likely to remain unresolved. Two published studies may portend the morphological surprises likely to arise from ongoing works. The first utilized partial sequence data from mitochondrial LrRNA gene to show that the false-truffle *Rhizopogon* is more closely related to *Suillus* than the latter is to three other members of the Boletaceae (15). This result was supported at the 99% confidence level through bootstrap analysis and was also corroborated by a shared mitochondrial gene order found in *Suillus* and *Rhizopogon*. This close relationship between *Suillus* and *Rhizopogon* can now also be supported with high confidence levels by data from two other portions of the mitochondrial rRNA genes as well from the nuclear SrRNA gene (14, 18; Figure 3). In this case the explanation for the dramatic morphological shift had been proposed 60 years earlier (102), but the closeness of the relationship between mushrooms and false-truffles revealed by the molecular data was still startling. Work in the genus *Coprinus* has now shown similar relationships between secotioid genera *Montagnea* and *Podaxis* and mushrooms in the genus *Coprinus* (70). Another morphological surprise appeared in a recent study by Hibbett & Vilgalys (66). In this case the results suggest that the genus *Lentinus* is polyphyletic and the closest relatives of *L. tigrinus* are polypores. No statistics were used to evaluate the tree, and their analysis was flawed by the use of fragments as characters and the likely presence of length mutations in the region analyzed, but their result is also reported to be strongly supported by sequence data (D. S. Hibbett, personal communication). The morphological implications of their conclusions are certainly significant: the switch between gilled and tubular hymenophores had to occur at least twice during the evolution of *Lentinus*. Similar results have also been reported for the Boletales (16). Taken together these results suggest that the underlying developmental difference between gills and tubes may not be very great, and evolutionary switching between these two types of hymenophores may be relatively easy, at least within lineages that have the developmental potential to make tubes.

Oomycota

Oomycota, particularly *Phytophthora* and *Pythium,* have been the focus of mtDNA and rDNA studies aimed at the species level and below. The most popular approach has involved mtDNA. In Oomycota, the molecule is circular (84) and evolves through a combination of nucleotide substitutions, length mutations (21, 42, 43, 106), and rearrangements (141). Most taxa have a long inverted repeat containing the rRNA genes (72, 110). Judging from the systematic distribution and sizes of the repeat, and the presence of very short inverted repeats in taxa that lack it (e.g. *Phytophthora, Leptomitus, Apodachlya;* 110, 141), the repeat is an ancestral feature of Oomycota, has been lost independently at least twice, and has changed size significantly in several lineages. Independent support for the convergent loss of the inverted repeat in *Apodachlya* and *Phytophthora* is provided by a mapping study of nuclear rDNA (83).

Inheritance of mtDNA has been claimed to be uniparental in *Pythium* (105a) and *Phytophthora* (40). In *Pythium,* however, no confirmation that mating actually occurred was offered, and in *Phytophthora,* recombination of mtDNAs could not be ruled out as many of the smaller mtDNA fragments were obscured by nuclear DNA fragments.

Frequent occurrence of both length mutations and rearrangements, coupled with uncertainty about the mode of inheritance, makes it difficult to expoit Oomycete mtDNA variation for phylogenetic purposes. One straightforward use of mtDNA RFLPs is for fingerprinting. An example is the correlation of asexual "hyphal swelling" *Pythium* isolates and their closest sexual relatives (105b). For phylogenetic studies, Förster and colleagues (42, 43) examined isolates of *Phytophthora megasperma, P. parasitica,* and *P. infestans* by comparison of restriction fragments. Because they ignored length mutation, their estimates of nucleotide substitution among isolates became meaningless. With mapped mutations, however, they show that *P. megasperma f.sp. glycinea* and *P. m f.sp. medicaginis* are as distant from each other as *P. infestans* is from *P. parasitica;* from this comparison they conclude that *P. megasperma* comprises more than one species. This conclusion assumes that interspecific variation should be constant throughout a genus. Yet where mtDNA variation has been studied within and among biological species, it was anything but constant (47, 151).

Förster and colleagues (44) have extended their studies of restriction fragment patterns to show that the amount of mtDNA diversity is variable within six *Phytophthora* species and to compare intraspecific phylogenies inferred from mtDNAs and nuclear DNAs. Although length mutations were not accounted for in using unmapped fragments to estimate diversity, and alternative branching patterns were not evaluated statistically, there was good agreement between phenograms based on mitochondrial and nuclear DNA for *P. capsici,* although less agreement for *P. parasitica.*

An examination of morphological, physiological, isozyme, and mtDNA variation in many isolates of *Phytophthora cryptogea* and *P. drechsleri* provided convincing evidence of morphological and physiological misidentification of these fungi (113) and again makes it clear that some species harbor more variation than others. As a result, Mills and colleagues proposed that groups of isolates with similar amounts of variability be called "molecular species." Here too, their phenograms are based on fragment comparison and do not account for length mutations; and trees with alternative branching patterns are not ruled out. In the absence of a clear understanding of the mode of mitochondrial inheritance, with variation exaggerated due to length mutations, with no statistical testing of alternative trees, and without a thorough comparison of known biological species, it seems premature to define clades as "molecular species."

Sequencing of PCR amplified elements of the rDNA repeat has been applied to the former "morphological forms" of *Phytophthora palmivora* (94). Use of neighbor-joining (135) and parsimony with bootstrapped samples (37) showed that *P. palmivora* and *P. megakarya* are significantly more closely related to each other than either is to *P. capsici* or *P. citrophthora* when *P. cinnamomi* is the outgroup. In line with the findings of Föster and colleagues, variation within and among species is itself variable; ITS regions that easily separated *P. palmivora* and *P. megakarya* were too similar to be used confidently to separate, *P. capsici* and *P. citrophthora*. Again, should we expect variation to be equal in all species?

Neglected Groups

Chytridiomycota and Zygomycota have received only a little attention. Based on 18S rDNA sequence, the rumen anaerobe *Neocallimastix* branches within the chytrids and is closest to *Spizellomyces* (11). Sequence analysis of 18S rDNA were also used to show that *Glomus, Gigaspora,* and *Endogone* are true fungi but not ascomycetes. Unfortunately, the lack of any zygomycete sequences for comparison prevented testing of their putative relationship (L. Simon, pers. comm.). In the Zygomycota, rDNA RFLPs from the animal dermatophyte *Basidiobolus* have been analyzed (119). Here, the human isolates show little variation compared to those isolated from other sources, but parsimony analysis of RFLPs used to understand *Basidiobolus* phylogeny suffered from the lack of an outgroup and the presence of length mutations. In Myxomycota and Acrasiomycota, the field is completely open.

Relationships Among the Major Groups

The big picture of the evolution of fungi and their place in biota has been a preoccupation of mycologists since the classics (19). Sogin and colleagues have obtained nuclear small subunit rRNA sequences from many organisms, including fungi (41, 147). They have shown with these sequences that the organisms traditionally claimed by mycologists fail to form one monophyletic branch on the eukaryotic evolutionary tree (Figure 5). The cellular and plasmodial slime molds diverge early, in the region populated by protists, and the Oomycota share a branch with chrysophyte algae. Farther up the tree, three familiar groups diverge, the land plants, animals, and Fungi. Included in the monophyletic kingdom Fungi are representatives of the divisions, Chytridiomycota, Ascomycota and Basidiomycota, and we anticipate that the Zygomycota can be included when a sequence is published.

The distance method used by Sogin and colleagues to infer phylogenetic

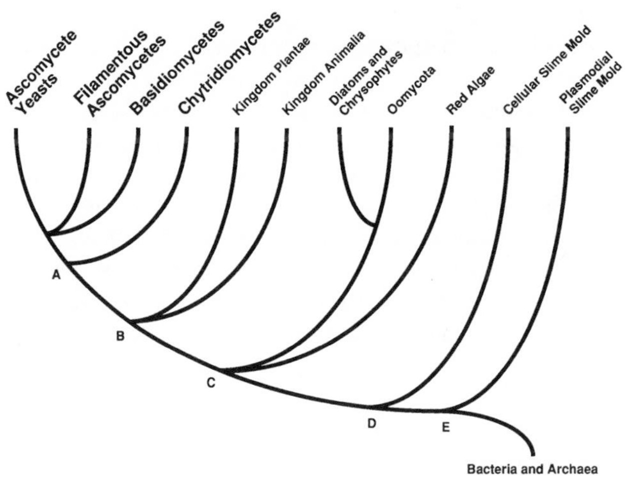

Is it in the Kingdom Fungi?	Answer	Significant Branch
Physarum, Plasmodial Slime Mold	No	B - E
Dictyostelium, Cellular Slime Mold	No	B - D
Achlya, Oomycota	No	A - C
Gracilaria, Red Alga	No	A - C
Chytridium, Chytridiomycota	Yes	A - B
Zygomycota	no data	

Figure 5 An evolutionary tree based on parsimony analysis of 18S rDNA sequence. The statistical support for internal branches is discussed in the section on "Relationships Among the Major Groups" and is summarized as a series of questions about membership in the kingdom Fungi. Branch lengths are arbitrary. GenBank 18S rDNA sequences from the following taxa were used in the analysis. Fungi: *Saccharomyces cerevisiae, Neurospora crassa, Spongipellis* (=*Tyromyces*) *unicolor* (11), *Chytridium confervae* (11 et al?). Plants: *Oryza sativa.* Animals: *Xenopus laevis. Achlya bisexualis, Gracilaria lemaneiformis, Dictyostelium discoideum,* and *Physarum polycephalum.*

relationships from small subunit rRNA does not lend itself to explicit statistical tests of the robustness of internal branches and branching order. Evidence that more than one evolutionary history can be obtained for the same organisms using the same DNA sequence is found by comparing the nuclear SrRNA trees of Förster et al (41) and Hendriks et al (64, 65), which conflict in their placement of ciliates, sporozoa, plants, animals, and fungi. The sensitivity of these trees to sequence alignment is made clear by Hendriks and colleagues (64), whose alignment of all eukaryote sequences fails to place Fungi on a single clade, while alignment of only the higher eukaryotes does.

Understanding the branching order of the three groups of higher eukaryotes is a puzzle of great general interest. Hasegawa et al (63) applied a maximum likelihood method to sequences from both the large and small rRNA subunits to show that plants diverged prior to Fungi and animals, but they noted that this result is not significant at the 95% confidence limit. Gouy & Li (55) used transformed distance and parsimony methods with an estimation of the standard error of branch lengths to compare rRNA sequence, tRNA sequence, and several protein sequences in hopes of solving this puzzle of higher eukaryote radiation. Their tRNA tests were not significant at the 95% level and favored either an early fungal divergence or an early plant divergence, depending on which prokaryote served as the outgroup. The amino acid sequences favored an early fungal divergence with 95% confidence but included GA3PDH, which may not be an appropriate protein to answer such ancient questions (see below). The combined small and large subunit rRNA sequences also favored an early fungal divergence with 95% confidence, but alternative trees in which animals or plants diverge first required only 7 or 9 more character state changes out of ca. 1550 total changes. Perhaps the clearest and most interesting fact about the radiation of higher eukaryotes is not the exact order but that the radiation appears to have occurred very quickly compared to subsequent evolution, so that each kingdom has essentially the same relationship to the others.

Statistical testing of branching order and internal branches has become a necessary feature of phylogenetic studies. As an illustration, we used three methods amenable to statistical evaluation—winning sites (124), evolutionary parsimony (92), and parsimony analysis with bootstrapped samples (37)—to examine the internal branches and branching order of small subunit rRNA trees composed of organisms studied by mycologists (41, 65). We have framed these tests as questions about the inclusion of various organisms in a monophyletic fungal kingdom.

Do the plasmodial slime molds belong in the Fungi? Hasegawa et al (63) used large and small subunit rRNA sequences with maximum likelihood to show that *Physarum polycephalum* was outside the Fungi at well above the 95% confidence level. To include *Physarum* in the fungi would mean including many protists, the groups of eukaryotic algae, the animals and the plants. Clearly, plasmodial slime molds should not be included among true fungi.

Do the cellular slime molds belong in the Fungi? Again, Hasegawa et al (63) used large and small subunit rRNA sequences and maximum likelihood to show that *Dictyostelium discoideum* was not a member of the Fungi. Using small subunit rRNA data, we made a winning sites test of *Dictyostelium* against *Neurospora* and rice, with *Physarum* as the outgroup. There is significant support for the branch uniting *Neurospora* and rice to the exclusion of *Physarum* and *Dictyostelium*. To include *Dictyostelium* among the Fungi would mean including the plants. Parsimony analysis of 100 bootstrapped samples of the same data supported the same branch at the 100% level. The cellular slime molds are outside the Fungi.

Do the Oomycota belong in the Fungi? A winning site test of small subunit rRNA sequence from *Achlya bisexualis* against *Neurospora* and rice, with *Dictyostelium* as the outgroup did not support the branch uniting *Achlya* and *Dictyostelium* at the 95% level. Parsimony analysis with 100 bootstrapped samples of the same molecule for the same taxa plus *Chytridium confervae,* yeast, and the basidiomycete *Spongipellis (=Tyromyces) unicolor,* supported the branch uniting *Achlya,* rice and *Dictyostelium* to the exclusion of others at the 100% level. However, the branch uniting *Achlya* and *Dictyostelium* to the exclusion of rice and the Fungi was not supported at the 95% confidence level. The possibility that *Achlya* could form a monophyletic branch with the Fungi cannot be ruled out, but if the Oomycota were included in a fungal branch they would diverge earlier than the Chytridiomycota. This broadened fungal kingdom would necessarily include the Chrysophyte algae and diatoms.

Do Chytridiomycota belong in the Fungi? Although mycologists have long claimed these flagellated organisms, some have allied them with the ciliates (104). Förster & colleagues (41) used distance methods and 18S rRNA sequence to show that a monophyletic fungal branch could include chytrids. Bowman & colleagues (11) used neighbor joining and winning sites tests on 18S rDNA sequence to show that the branch uniting chytrids with the rest of the Fungi, to the exclusion of a ciliate, was supported at the 95% level. Chytrids are Fungi.

Do the red algae belong in the Fungi? This question has intrigued several generations of mycologists with adherents on both sides of a debate over phenotypic attributes (28). Proponents see the red algae as ancestors to the Ascomycota. The relationship between the red algae and the Fungi based on small subunit rRNA is similar to that of the Oomycota. A winning sites test of sequence from *Gracilaria lemaneiformis* against *Neurospora* and rice with *Dictyostelium* as the outgroup supported the branch uniting *Dictyostelium* and *Gracilaria,* but not at the 95% level. If *Chytridium* was substituted for rice in this test, the branch uniting *Dictyostelium* and *Gracilaria* was supported at the 99% level. Parsimony analysis with 100 bootstrapped samples gave the same result. The branch uniting the Fungi was supported 100%, but the branch uniting plants with the Fungi to the exclusion of *Gracilaria* and *Dictyostelium* was not supported at the 95% level. The possibility of red algae forming a monophyletic group with the Fungi cannot be ruled out, but if they do they must branch earlier than the Chytridiomycota and cannot be direct ancestors to the Ascomycota. This conclusion supports that of a prior DNA hybridization study of red algal affinities using cloned basidiomycete rDNA, which showed that red algae were no closer to Ascomycota than were green algae (91).

Phylogenetic comparisons of nucleotide sequences of 5S rRNA (71) or GA3PDH (146) have produced trees that would exclude from the Fungi the entire Ascomycota or just the yeasts, or require broadening the Fungi to include plants and animals. Winning sites tests of small subunit rRNA sequences of yeast against *Neurospora* and either *Xenopus* or rice, with *Achlya* (Oomycota) as the outgroup, support the branch uniting yeast and *Neurospora* at the 99% level. Bowman & colleagues (11) used the winning sites test and neighbor joining to show that *Neurospora* and the basidiomycete *Spongipellis (=Tyromyces)* were closest relatives compared to a plant with a ciliate as an outgroup. In addition, parsimony analysis of 100 bootstrapped samples of Chytridiomycota, Ascomycota, and Basidiomycota with rice and *Xenopus* supports the notion that the branch unites the Fungi against rice and *Xenopus* at 99%. Therefore, with small subunit rRNA sequence, neither the Ascomycota nor the Fungi are polyphyletic, compared to animals or plants.

This result is unsettling because the trees based on GA3PDH and 5S rRNA contradict those based on morphological and physiological phenotype as well as nuclear small subunit rRNA sequence. The difficulty of using 5S rRNA to analyze older evolutionary divergences has already been discussed. The apparent conflict between the 18S rDNA and GA3PDH sequence data is worth examining further because, at our present level of sophistication, analysis of protein coding genes is more complex than that of rDNA genes.

With GA3PDH, Smith (146) used evolutionary parsimony to analyze nucleotide sequences from *E. coli,* yeasts, filamentous Ascomycota and Basidiomycota, plants, and animals. In the tree resulting from this analysis, the branch grouping *E. coli* and yeasts with plants against filamentous fungi and animals was supported at the 99% level.

E. coli has been shown to have two genes for GA3PDH (1), and it is postulated that one of them, the one used by Smith, has come to *E. coli* by horizontal transfer from a eukaryote (31). If this hypothesis is true, it could explain why yeast appeared to branch close to *E. coli*. It does not explain Smith's demonstration that a branch uniting yeasts and plants to the exclusion of filamentous ascomycetes and animals is supported by evolutionary parsimony analysis.

The evolutionary rates of molecules must match the antiquity of the divergences in question. This may not be the case for GA3PDH and higher eukaryotes. Pairwise comparison of amounts of nucleotide substitution in GA3PDH between *Saccharomyces cerevisiae* and *Cochliobolus heterostrophus* shows percentages of nucleotide substitution for the first, second, and third codon positions of this protein coding gene to be 32%, 20%, and 55% respectively. Given that total substitution in GA3PDH between two isolates of *S. cerevisiae* can be as high as 11.3%, nucleotide replacement may be near saturation and is well above the point where multiple mutations at single sites should confound analysis and produce an anomalous result.

In comparing animals, plants, yeasts, and filamentous ascomycetes, one would expect that the same result would be obtained regardless of which representative species were used. This is not the case with GA3PDH; 59% of the possible four-taxa evolutionary parsimony tests were not significant. Omitting the extremely variable third colon position resulted in 55% insignificant results. An assumption of evolutionary parsimony is that nucleotide substitution is unbiased, but GA3PDH in yeast is known to have extreme codon bias (8), a feature it shares with *E. coli*. Highly expressed genes in fermentative organisms may be under selective pressures unknown to their homologues in organisms capable of growing into new food sources. Nucleotide composition is also biased, e.g. the percentage of cytosine is 22% in *Saccharomyces,* 31% in filamentous ascomycetes, and 24% in *E. coli*. Compositional statistics (142) can be applied to evolutionary parsimony when nucleotide substitution bias is known to exist. In this case, it did not help; analysis of all possible four-taxa groups with compositional statistics returned 69% insignificant results with all three codon positions, and 57% with the third position omitted.

The high levels of nucleotide substitution, the codon and nucleotide substitution bias, and the dependence of significant evolutionary parsimony results on specific groups of four taxa—all indicate that GA3PDH is evolving too rapidly to be useful in determining the divergence of plants, animals, and fungi. Loomis & Smith (98) used a similar approach with eight different protein coding genes to study the phylogeny of *Dictyostelium*. In a comparison of these genes in *Dictyostelium* and *S. cerevisiae,* all except the actin gene had at least 40% of their amino acids substituted. The unusual relationships proposed for *Dictyostelium* by these authors were based on parsimony analysis and distance analyses uncorrected for multiple hits. It is very likely that these genes are not appropriate for studies of divergences as old as that of *Dictyostelium* and Fungi or mammals.

SUMMARY

Figure 6 provides a visual summary of the way we perceive the match between methods available and taxonomic levels for which they are appropriate. The necessary caveat is that equivalent taxonomic categories of different lineages often vary in terms of the level of molecular divergence; therefore, choosing an approach for a given study will almost always require an empirical survey. Throughout this review we have followed the lead of Felsenstein (38) in stressing the importance of the statistical testing of phylogenetic hypotheses, but as indicated in Figure 6 we believe that the usefulness of a method is not always limited to levels that yield a statistically significant result. Correlation with prior phylogenetic hypotheses based on morphology,

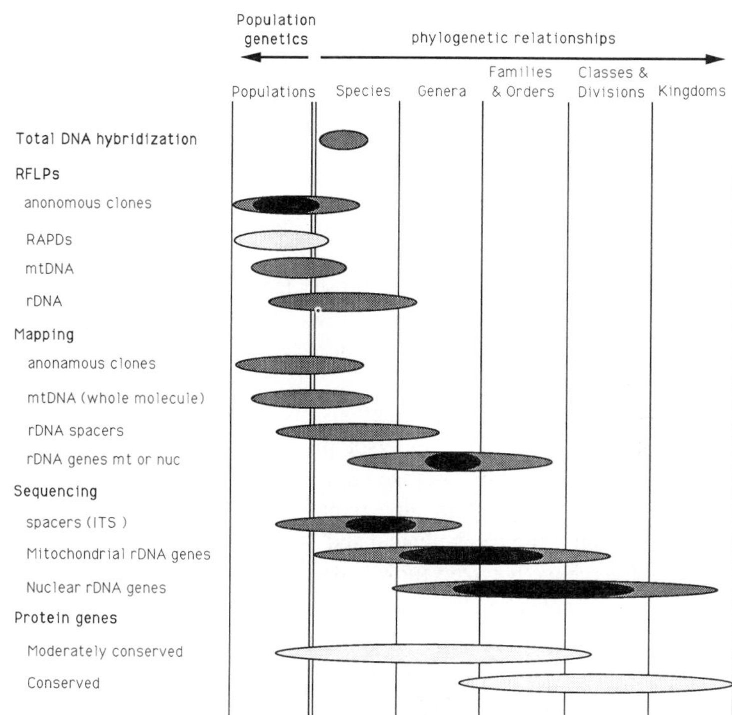

Figure 6 Qualitative summary of molecular methods and their useful ranges in fungal systematics. Dark stippling indicates the useful ranges of currently tested methods, no stippling indicates the projected range of RAPDs and direct protein sequencing if the appropriate genes can be identified. Black indicates ranges in which sufficient numbers of consistent characters might be expected to yield statistically significant results.

ecology, physiology, or other molecular data may provide additional support for molecular phylogenies. Indeed, it is in cases where molecular phylogenies conflict with other available evidence that it is most important to address the question of statistical significance.

Figure 6 also illustrates two important gaps in our current methodology: relationships among closely related species and among Divisions and Kingdoms. At the former level many methods are available, but to date none have yielded statistically significant results. Sequence analysis of IGS or perhaps third position sites of a variable protein may be useful at this level, but currently no studies are available in fungi to evaluate these options. At the level of Kingdoms and Divisions most attention has been focused on nuclear ribosomal RNA genes, but analysis of conserved proteins is starting to be used more frequently (51).

FUTURE PROSPECTS

Beyond establishing the evolutionary relationships among fungi, perhaps the most important unanswered question involves establishing a timeframe for it. The fungal fossil record is scanty, but minimum divergence dates for at least a few modern groups can be obtained (30, 122, 154). Indirect dating based on specific associations could be used to obtain maximum dates of divergence for other groups. For example, diversification within genera of rumen fungi is unlikely to have preceded the divergence of the ruminants. Similarly, taxa of rusts or other obligate parasites that are restricted to a given plant taxon might be assumed to be no older than their hosts. Thus, rough divergence dates could be coupled with sequence divergence observed in extant lineages to get initial calibration points for fungal molecular clocks. A second and more direct method of clock calibration may also be possible. This would involve obtaining both sequence data and isotopic age estimates from specimens preserved in amber, or semi-fossilized mud (52, 123). If such specimens can be unambiguously assigned to an extant group, the amount of sequence divergence per unit time can be calculated. A similar approach could be used to examine divergence between fungal strains recovered from buried periglacial soil of known dates and their closest relatives from extant arctic soils (7), or perhaps from frozen Antarctic soils of greater antiquity.

Another area for future research involves estimates of fungal biodiversity for those species that cannot be cultured in the laboratory. For example, the diversity of soil fungi is clearly underestimated by current soil dilution plate methods. The molecular approaches used for estimating diversity for thermophilic bacteria and plankton should also prove useful for fungi (50, 159).

The expanding capability to type strains and species of fungi molecularly will have a profound effect on the analysis of natural populations. Inheritance of nuclear and mitochondrial genes will be easier to study in the field. The work of Smith et al (145) provides a good example. The persistence, succession, and dispersal of individual strains, whether symbionts or pathogens, will be accessible to study using hyphae or spores instead of cultured isolates. This approach should be particularly useful for epidemiological investigations of the spread of plant and human pathogens, e.g., in determining whether an infection is the result of expansion of a virulent clone or importation of a new strain. The application to fungi of PCR-mediated identification systems developed for medical purposes (161) has begun with a study of *Phytophthora, Laccaria,* and *Cryptococcus* species (48, 96, 156). Realizing the full potential of this method will require its testing on many isolates.

At the level of experimental genetics, it should be possible to estimate recombination frequencies of linked genes and construct genetic maps via analysis of individual meiotic spores. By analogy to studies on individual sperm from mammals, this may be the only way to obtain genetic maps for some species where genetic crosses are impractical. This may also improve our ideas on the relationship of physical to genetic distance in the fungal genome. In the area of developmental biology, understanding the evolutionary relatedness of other fungal species to those that have been extensively analyzed, i.e. *Saccharomyces* and *Neurospora,* may make it easier to identify a gene's homolog in a novel organism.

A final area involves nomenclature and classification. If all fungi can be compared through their nucleic acids and placed on a single phylogenetic tree (56), do we need to maintain the Deuteromycota (127, 128)?

ACKNOWLEDGMENTS

We thank all the people who sent us prepublication data and manuscripts, also Barbara Bowman for editorial help, Tim Szaro for assistance with graphics and the data analysis, and Eric Swan for helpful editorial comments. This review was prepared with support from NSF (BSR-8700391, BSR-8918454) and NIH (RO1 AI28545).

Literature Cited

1. Alefounder, P. R., Perham, R. N. 1989. Identification, molecular cloning and sequence analysis of a gene cluster encoding the class II fructose 1,6-bisphosphate aldolase, 3-phosphoglycerate kinase and a putative second glyceraldehyde 3-phosphate dehydrogenase of *Escherichia coli*. *Mol. Microbiol.* 3:723–32
2. Ali, S., Müller, C. R., Epplen J. T. 1986. DNA finger printing by oligonucleotide probes specific for simple repeats. *Hum. Genet.* 74:239–43
3. Anderson, J. B., Bailey, S. S., Pukkila, P. J. 1989. Variation in ribosomal DNA among biological species of *Armillaria,* a genus of root infecting fungi. *Evolution* 43:1652–62

4a. Arnheim, N., Krystal, M., Schmickel, R., Wilson, G., Ryder, O., Zimmer, E. 1980. Molecular evidence for genetic exchange among ribosomal genes on nonhomologous chromosomes in man and apes. *Proc. Natl. Acad. Sci. USA* 77:7323–27

4b. Barnes, S. M., Lane, D. J., Sogin, M. L., Bibeau, C., Weisburg, W. G. 1990. Evolutionary relationships among pathogenetic *Candida* species and relatives. *J. Bacteriol.* 173:2250–55

5. Bartoszewski, S., Borsuk, P., Kern, I., Bartnick, E. 1987. Microheterogeneity in *Aspergillus nidulans* 5S rRNA genes. *Curr. Genet.* 11:571–73
6. Baura, G., Szaro, T. M., Bruns, T. D. 1991. *Gastroboletus laricinus* is a recent derivative of *Suillus grevillei:* molecular evidence. *Mycol. Soc. Am. Newsl.* (Abstr.) 42:5
7. Beiswenger, J. M., Christensen, M. 1989. Fungi as indicators of past environments. *Curr. Res. Pleistocene* 6: 54–56
8. Bennetzen, J. L., Hall, B. D. 1982. Codon selection in yeast. *J. Biol. Chem.* 257:3026–31
9. Berbee, M. L., Taylor, J. W. 1991. Ribosomal DNA evidence for two classes of Ascomycetes based on fruiting body type. *Mycol. Soc. Am. Newsl.* (Abstr.) 42:6
10. Blanz, P. A., Gottschalk, M. 1986. Systematic position of *Septobasidium, Graphiola* and other Basidiomycetes as deduced on the basis of their 5S ribosomal RNA nucleotide sequences. *System. Appl. Microbiol.* 8:121–27
11. Bowman, B., Taylor, J. W., Brownlee, A. G., Lee, J., Lu, S.-D., White, T. J. 1991. Molecular evolution of the fungi: relationship of the Basidiomycetes, Ascomycetes and Chytridiomycetes. *Mol. Biol. Evol.* In press

12. Bowman, B., Taylor, J. W., White, T. J. 1991. Molecular evolution and detection of human pathogenic fungi. *Mycol. Soc. Am. Newsl.* (Abstr.) 42:7
13. Brown, T. A., Waring, R. B., Scazzocchio, C., Davies, R. W. 1985. The *Aspergillus nidulans* mitochondrial genome. *Curr. Genet.* 9:113–17
14. Bruns, T. D., Fogel, R., Taylor, J. W. 1990. Amplification and sequencing of DNA from fungal herbarium specimens. *Mycologia* 82:175–84
15. Bruns, T. D., Fogel, R., White, T. J., Palmer, J. D. 1989. Accelerated evolution of a false-truffle from a mushroom ancestor. *Nature* 339:140–42
16. Bruns, T. D., Palmer, J. D. 1989. Evolution of mushroom mitochondrial DNA: *Suillus* and related genera. *J. Mol. Evol.* 28:349–62
17. Bruns, T. D., Palmer, J. D., Shumard, D. S., Grossman, L. I., Hudspeth, M. E. S. 1988. Mitochondrial DNAs of *Suillus:* three fold size change in molecules that share a common gene order. *Curr. Genet.* 13:49–56
18. Bruns, T. D., Szaro, T. M. 1990. Phylogenetic relationships within the Boletales: molecular perspectives. In *Fourth International Mycological Congress Abstracts,* ed. A. Reisinger, A. Bresinsky, p. 12. (Abstr.) Regensburg Germany: Bot. Inst. Univ. Regensburg
19. Buller, A. H. R. 1914. The fungus lore of the Greeks and Romans. *Trans. Br. Mycol. Soc.* 5:21–66
20. Cambell, G. R., Littau, V. C., Melera, P. W., Allfrey, V. G., Johnson, E. M. 1979. Unique sequence arrangement of ribosomal genes in the palindromic rDNA molecule of *Physarum polycephalum. Nucl. Acids Res.* 6:1433–47
21. Carter, D. A., Archer, S. A., Buck, K. W., Shaw, D. S., Shattock, R. C. 1990. Restriction fragment length polymorphisms of mitochondrial DNA of *Phytophthora infestans. Mycol. Res.* 94:1123–28
22. Casanova, J.-L., Pannetier, C., Jaulin, C., Kourilsky, P. 1990. Optimal condition for directly sequencing double-stranded PCR products with sequenase. *Nucl. Acids Res.* 18:4028
23. Cassidy, J. R., Pukkila, P. J., 1987. Inversion of 5S ribosomal RNA genes within the genus *Coprinus. Curr. Genet.* 12:33–36
24. Clark-Walker, G. D., Hoeben, P., Plazinska, A., Smith, D. K., Wimmer, E. H. 1987. Application of mitochondrial DNA analysis to yeast systematics. *Stud. Mycol.* 30:259–66
25. Cockburn, A. F., Tayler, W. C., Firtel, R. A. 1978. *Dictyostelium* rDNA consists of non-chromosomal palindromic dimers containing 5s and 36s coding regions. *Chromosoma* 70:19–29
26. Collins, R. A., Saville, B. J., 1990. Independent transfer of mitochondrial chromosomes and plasmids during unstable vegetative fusion in *Neurospora. Nature* 345:177–79
27. Cummings, D. J., McNally, K. L., Domenico, J. M., Matsuura, E. T. 1990. The complete DNA sequence of the mitochondrial genome of *Podospora anserina. Curr. Genet.* 17:375–402
28. Demoulin, V. 1985. The red algal-higher fungi phylogenetic link: the last ten years. *BioSystems* 18:347–56
29. Deleted in proof
30. Dilcher, D. L. 1965. Epiphyllous fungi from Eocene deposits in western Tennessee, USA. *Palaeontographica* 116: 1–54
31. Doolittle, R. F., Feng, D. F., Anderson, K. L., Alberro, M. R. 1990. A naturally occurring horizontal gene transfer from a eukaryote to a prokaryote. *J. Mol. Evol.* 31:383–88
32. Duncan, T., Phillips, R. B., Wagner, W. H. Jr. 1980. A comparison of branching diagrams derived by various phenetic and cladistic methods. *Syst. Bot.* 5:264–93
33. Dutta, S. K. 1976. DNA homologies among heterothallic species of *Neurospora. Mycologia* 68:388–401
34. Economou, A., Castleton, L. A., 1989. Polymorphism of the mitochondrial L-RNA gene of the basidiomycete *Coprinus cinereus. Curr. Genet.* 16:41–46
35. Edman, J. C., Kovacs, J. A., Masur, H., Santi, D. V., Elwood, H. J., Sogin, M. L. 1988. Ribosomal RNA sequence shows *Pneumocystis carinii* to be a member of the fungi. *Nature* 334:519–22
36. Edwards, S. V., Arctander, P., Wilson, A. C. 1991. Mitochondrial resolution of a deep branch in the genealogical tree for perching birds. *Proc. R. Soc. Lond.* 243:99–107
37. Felsenstein, J. 1985. Confidence intervals on phylogenies: an approach using the bootstrap. *Evolution* 39:783–91
38. Felsenstein, J. 1988. Phylogenies from molecular sequences: inference and reliability. *Annu. Rev. Genet.* 22:521–65
39. Felsenstein, J. 1991. PHYLIP 3.4. Department of Genetics SK-50, Univ. Wash., Seattle, Wash. 98195
40. Förster, H., Coffey, M. D. 1990. Mating behavior of *Phytophthora parasitica:* evidence for sexual recombination in

oospores using DNA restriction fragment length polymorphisms as genetic markers. *Exp. Mycol.* 14:351–59

41. Förster, H., Coffey, M. D., Elwood, H., Sogin, M. L. 1990. Sequence analysis of the small subunit ribosomal RNAs of three zoosporic fungi and implications for fungal evolution. *Mycologia* 82:306–12
42. Förster, H., Kinscherf, T. G., Leong, S. A., Maxwell, D. P. 1988. Estimation of relatedness between *Phytophthora* species by analysis of mitochondrial DNA. *Mycologia* 80:466–78
43. Förster, H., Kinscherf, T. G., Leong, S. A., Maxwell, D. P. 1989. Restriction fragment length polymorphisms of the mitochondrial DNA of *Phytophthora megasperma* isolated from soybean, alfalfa and fruit trees. *Can. J. Bot.* 67:529–37
44. Förster, H., Oudemans, P., Coffey, M. D. 1990. Mitochondrial and nuclear DNA diversity within six species of *Phytophthora*. *Exp. Mycol.* 14:18–31
45. Garber, R. C., Turgeon, B. G., Selker, E. U., Yoder, O. C. 1988. Organization of ribosomal RNA genes in the fungus *Cochliobolus heterostrophus*. *Curr. Genet.* 14:573–82
46. Gardes, M., Fortin, J. A., Mueller, G. M., Kropp, B. R. 1990. Restriction fragment length polymorphisms in the nuclear ribosomal DNA of four *Laccaria spp: L. bicolor, L. laccata, L. proxima,* and *L. amethystina*. *Phytopathology* 80:1312–17
47. Gardes, M., Mueller, G. M., Fortin, J. A., Kropp, B. R. 1991. Mitochondrial DNA polymorphisms in *Laccaria bicolor, L. laccata, L. proxima,* and *L. amethystina*. *Mycol. Res.* 95:206–16
48. Gardes, M., White, T. J., Fortin, J. A., Bruns, T. D., Taylor, J. W. 1991. Identification of indigenous and introduced symbiotic fungi in ectomycorrhizae by amplification of nuclear and mitochondrial ribosomal DNA. *Can. J. Bot.* 69:180–90
49. Gelfand, D. H., White, T. J. 1990. Thermostable DNA polymerases. In *PCR Protocols: A Guide To Methods and Applications,* ed. M. A. Innis, D. H. Gelfand, J. S. Sninsky, T. J. White. 16:129–141. New York: Academic. 482 pp.
50. Giovannoni, S. J., Britschgi, T. B., Moyer, C. L., Field, K. G. 1990. Genetic diversity in Sargasso sea bacterioplankton. *Nature* 345:60–63
51. Gogarten, J. P., Kibak, H., Dittrich, P., Taiz, L., Bowman, E. J., et al. 1989. Evolution of the vaculolar H^+ – ATPase: implications for the origin of eukaryotes. *Proc. Natl. Acad. Sci. USA* 86:6661–65
52. Golenberg, M. E., Giannasi, D. E., Clegg, M. T., Smiley, C. J., Durbin, M., et al. 1990. Chloroplast DNA sequences from a Miocene *Magnolia* species. *Nature* 334:656–58
53. Deleted in proof
54. Gouy, M., Li, W.-H. 1989. Phylogenetic analysis based on rRNA sequences supports the archaebacterial rather than the eocyte tree. *Nature* 339:145–47
55. Gouy, M., Li, W-H. 1989. Molecular phylogeny of the kingdoms Animalia, Plantae, and Fungi. *Mol. Biol. Evol.* 6:109–22
56. Guadet, J., Julien J., Lafay, J. F., Brygoo, Y. 1989. Phylogeny of some *Fusarium* species, as determined by large-subunit rRNA sequence comparison. *Mol. Biol. Evol.* 6:227–42
57. Guého, E., Kurtzman, C. P., Peterson, S. W., 1989. Evolutionary affinities of heterobasidiomycetous yeast estimated from 18S and 25S ribosomal RNA sequence divergence. *System. Appl. Microbiol.* 12:230–36
58. Guého, E., Kurtzman, C. P., Peterson, S. W. 1990. Phylogenetic relationships among species of *Sterigmatomyces* and *Fellomyces* as determined from partial rRNA sequences. *Int. J. Syst. Bact.* 40:60–65
59. Gyllensten, U. B., Erlich, H. A. 1988. Generation of single-stranded DNA by the polymerase chain reaction and its application to direct sequencing of the HLA-DQα locus. *Proc. Natl. Acad. Sci. USA* 85:7652–56
60. Halanych, K. M. 1991. 5S ribosomal RNA sequences inappropriate for phylogenetic reconstruction. *Mol. Biol. Evol.* 8:249–53
61. Hamby, R. K., Sims, L. E., Issel, L. E., Zimmer, E. A. 1988. Direct ribosomal RNA sequencing: optimization of extraction and sequencing methods for work with higher plants. *Plant Biol. Rep.* 6:175–92
62. Hamer, J. E., Farrall, L., Orbach, M. J., Valent, B., Chumley, F. G. 1989. Host species-specific conservation of a family of repeated DNA sequences in the genome of a fungal plant pathogen. *Proc. Natl. Acad. Sci. USA* 86:9981–85
63. Hasegawa, M., Iida, Y., Yano, T-A., Takaiwa, F., Iwabuchi, M. 1985. Phylogenetic relationships among eukaryotic kingdoms inferred from ribosomal RNA sequences. *J. Mol. Evol.* 22:32–38
64. Hendriks, L., De Baere, R., Van de Peer, Y., Neefs, J., Goris, A., De

Wachter, R. 1991. The evolutionary position of the rhodophyte *Porphyra umbilicalis* and the basidiomycete *Leucosporidium scottii* among other eukaryotes as deduced from complete sequences of small ribosomal subunit RNA. *J. Mol. Evol.* 32:167–77

65. Hendriks, L., Goris, A., Neefs, J.-M., Van de Peer, Y., Hennebert, G., De Wachter, R. 1989. The nucleotide sequence of the small ribosomal subunit RNA of the yeast *Candida albicans* and the evolutionary position of the fungi among the Eukaryotes. *Syst. Appl. Microbiol.* 12:223–29
66. Hibbett, D. S., Vilgalys, R. 1991. Evolutionary relationships of *Lentinus* to the Polyporaceae: Evidence from restriction analysis of enzymatically amplified Ribosomal DNA. *Mycologia* 83:In press
67. Higuchi, G. R., Ochman, H. 1989. Production of single-stranded DNA templates by exonuclease digestion following the polymerase chain reaction. *Nucl. Acids Res.* 17:5865
68. Hintz, W. E. A., Anderson, J. B., Horgen, P. A. 1988. Relatedness of the three species of *Agaricus* inferred from restriction fragment length polymorphism analysis of the ribosomal DNA repeat and mitochondrial DNA. *Genome* 32:173–78
69. Holmquist, R., Miyamoto, M. M., Goodman, M. 1988. Higher-primate phylogeny—why can't we decide? *Mol. Biol. Evol.* 5:201–16
70. Hopple, J. S. Jr. 1990. Phylogenetic relationships within the genus *Coprinus* based on molecular and morphological evidence. See ref. 18, p 25
71. Hori, H., Osawa, S. 1987. Origin and evolution of organisms as deduced from 5S ribosomal RNA sequences. *Mol. Biol. Evol.* 4:445–72
72. Hudspeth, M. E. S. 1991. The fungal mitochondrial genome—a broader perspective. In *Handbook of Applied Mycology,* Vol. IV, *Biotechnology*. pp 213–41. New York: Marcel Dekker
73. Hudspeth, M. E. S., Shumard, D. S., Tatti, K. M., Grossman, L. I. 1980. Rapid purification of yeast mitochondrial DNA in high yield. *Biochim. Biophys. Acta* 610:221–28
74. Hulbert, S. H., Ilott, T. W., Legg, E. J., Lincoln, S. E., Lander, E. S., Michelmore, R. W. 1988. Genetic analysis of the fungus, *Bremia lactucae,* using restriction fragment length polymorphisms. *Genetics,* 120:947–58
75. Hulbert, S. H., Michelmore, R. W. 1988. DNA restriction fragment length polymorphism and somatic variation in the lettuce downy mildew fungus, *Bremia lactucae. Mol. Pl. Microbe Interact.* 1:17–24
76. Irwin, D. M., Wilson, A. C. 1991. Limitations of molecular methods for establishing the phylogeny of mammals, with special reference to the position of elephants. In *American Museum of Natural History Symposium on Mammalian Phylogeny,* ed. F. S. Szalay, M. J. Novacek, M. C. McKenna. New York: Springer Verlag. In press
77. Jacobson, D. J., Gordon, T. R. 1990. The variability of mitochondrial DNA as an indicator of the relationships between populations of *Fusarium oxysporum f.sp. melonis. Mycol. Res.* 94:734–44
78. Jahnke, K.-D., Bahnweg, G. 1986. Assessing natural relationships in basidiomycetes by DNA analysis. *Trans. Br. Mycol. Soc.* 87:175–91
79. Jahnke, K.-D., Bahnweg, G., Worrall, J. J. 1987. Species delimitation in the *Armillaria mellea* complex by analysis of nuclear and mitochondrial DNAs. *Trans. Br. Mycol. Soc.* 88:572–75
80. Kim. J., Burgman, M. A. 1988. Accuracy of phylogenetic-estimation methods under unequal evolutionary rates. *Evolution* 42:596–602
81. Kistler, H. C., Bosland, P. W., Benny, U., Leong, S., Williams, P. H. 1987. Relatedness of strains of *Fusarium oxysporum* from Crucifers measured by examination of mitochondrial and ribosomal DNA. *Phytopathology* 77:1289–93
82. Klassen, G. R., Buchko, J. 1990. Subrepeat structure of the intergenic region in the ribosomal DNA of the oomycetous fungus *Pythium ultimum. Curr. Genet.* 17:125–27
83. Klassen, G. R., McNabb, S. A., Dick, M. N. 1987. Comparison of physical maps of ribosomal DNA repeating units in *Pythium, Phytophthora* and *Apodachlya. J. Gen. Microbiol.* 133:2953–59
84. Klimczak, L. J., Prell, H. H. 1984. Isolation and characterization of mitochondrial DNA of the oomycetous fungus *Phytophthora infestans. Curr. Genet.* 8:323–26
85. Kocher, T. D., Thomas, W. K., Meyer, A., Edwards, S. V., Pääbo, S., et al. 1989. Dynamics of mitochondrial DNA evolution in animals: amplification and sequencing with conserved primers. *Proc. Natl. Acad. Sci. USA* 86:6196–6200
86. Kohn, L. M., Petsche, D. M., Bailey, S. R., Novak, L. A., Anderson, J. B. 1988. Restriction fragment length polymorphisms in nuclear and mitochondrial

DNA of *Sclerotinia* species. *Phytopathology* 78:1047–51

87. Kurtzman, C. P. 1985. Molecular taxonomy of the fungi. In *Gene Manipulations in Fungi,* ed. J. W. Bennett, L. L. Lasure, pp. 35–63. Orlando, Fla: Academic
88. Kurtzman, C. P. 1989. Estimation of phylogenetic distances among ascomycetous yeasts from partial sequencing of ribosomal RNA. *Yeast* 5:S351–S354
89. Kurtzman, C. P., Phaff, H. J., Meyer, S. A. 1983. Nucleic acid relatedness among yeasts. In *Yeast Genetics Fundamental and Applied Aspects,* ed. J. F. T. Spencer, D. M. Spencer, A. R. W. Smith, pp 139–66. New York: Springer-Verlag
90. Kurtzman, C. P., Smiley, M. J., Johnson, C. J., Wickerham, L. J., Fuscon, G. B. 1980. Two new and closely related heterothallic species, *Pichia amylophila* and *Pichia mississippiensis:* Characterization by hybridization and deoxyribonucleic acid reassociation. *Int. J. Syst. Bacteriol.* 30:208–16
91. Kwok, S., White, T. J., Taylor, J. T. 1986. Evolutionary relationships between fungi, red algae, and other simple eucaryotes inferred from total DNA hybridizations to a cloned Basidiomycete ribosomal DNA. *Exp. Mycol.* 10:196–204
92. Lake, J. A. 1987. A rate-independent technique for analysis of nucleic acid sequences: evolutionary parsimony. *Mol. Biol. Evol.* 4:167–91
93. Lane, D. J., Pace, B., Olsen, G. J., Stahl, D. A., Sogin, M. L., Pace, N. R. 1985. Rapid determination of 16S ribosomal RNA sequences for phylogenetic analyses. *Proc. Natl. Acad. Sci. USA* 82:6955–59
94. Lee, S. B., Taylor, J. W. 1991. Phylogeny of five *Phytophthora* species as determined by ribosomal DNA internal transcribed spacer sequence comparison *Mol. Biol. Evol.* In press
95. Lee, S. B., Taylor, J. W. 1991. Uniparental inheritance and replacement of mitochondrial DNA in *Neurospora tetrasperma. Genetics.* Submitted
96. Lee, S. B., White, T. J., Taylor, J. W. 1991. Detection of species-specific DNA variability in *Phytophthora* by oligonucleotide hybridization to amplified ribosomal DNA. *Phytopathology.* Submitted
97. LoBuglio, K. F., Rogers, S. O., Wang, C. J. K. 1990. Variation in ribosomal DNA among isolates of the mycorrhizal fungus *Cenoccum geophilum* Fr. See Ref. 18, p. 330
98. Loomis, W. F., Smith, D. W. 1990. Molecular phylogeny of *Dictyostelium discoideum* by protein sequence comparison. *Proc. Natl. Acad. Sci. USA.* 87:9093–97
99. Lynch, M. 1988. Estimation of relatedness by DNA fingerprinting. *Mol. Biol. Evol.* 5:584–99
100. Lynch, M. 1990. The similarity index and DNA fingerprinting *Mol. Biol. Evol.* 7(5):478–84

100a. Lynn, D. H., Sogin, M. L. 1988. Assessment of phylogenetic relationships among ciliated protists using parial ribosomal RNA sequences derived from reverse transcriptase. *Biosystems* 21:249–54

101. Magee, B. B., D'Souza, T. M., Magee, P. T. 1987. Strain and species identification by restriction fragment length polymorphisms in the ribosomal DNA repeat of *Candida* species. *J. Bact.* 169:1639–43
102. Malençon, G. 1931. La série des Astérosporés. *Trav. crypto. déd a L. Mangin.* 1:337–96
103. Mannella, C. A., Pittenger, T. H., Lambowitz, A. M. 1979. Transmission of mitochondrial deoxyribonucleic acid in *Neurospora crassa* sexual crosses. *J. Bacteriol.* 137:1449–51
104. Margulis, L., Schwartz, K. V. 1988. *Five Kingdoms: An Illustrated Guide To the Phyla of Life on Earth.* New York: W. H. Freeman 2nd ed.

105a. Martin, F. N. 1989. Maternal inheritance of mitochondrial DNA in sexual crosses of *Pythium sylvaticum. Curr. Genet.* 16:373–74

105b. Martin, F. N. 1989. Taxonomic classification of asexual isolates of *Pythium ultimum* based on cultural characteristics and mitochondrial DNA restriction patterns. *Exp. Mycol.* 14:47–56

106. Martin, F. N. 1991. Variation in the ribosomal DNA repeat unit within single oospore isolates of the genus *Pythium. Genome* 33:585–91
107. May, G., Taylor, J. W. 1989. Independent transfer of mitochondrial plasmids in *Neurospora crassa. Nature* 339:320–22
108. McDonald, B. A., Martinez, J. P. 1990. DNA restriction fragment length polymorphisms among *Mycosphaerella graminicola* (Anamorph *Septoria tritici*) isolates collected from a single wheat field. *Phytopathology* 80:1368–73
109. McDonald, B. A., Martinez, J. P. 1991. DNA fingerprinting of the plant pathogenetic fungus *Mycosphaerella graminicola* (Anamorph *Septora tritica*). *Exp. Mycol.* 15:146–58

110. McNabb, S. A., Klassen, G. R. 1988. Uniformity of mitochondrial DNA complexity in Oomycetes and the evolution of the inverted repeat. *Exp. Mycol.* 12:233–42

110a. Meyer, W., Kock, A., Beyermann, B., Epplen, J. T., Börner, ?. 1991. Differentiation of species and strains among filamentous fungi by DNA printing. *Curr. Genet.* 19:239–42

111. Michelmore, R. W., Hulbert, S. H. 1987. Molecular markers for genetic analysis of phytopathological fungi. *Annu. Rev. Phytopathol.* 25:383–404

112. Mills, D., McCluskey, K. 1990. Electrophoretic karyotypes of fungi: the new cytology. *Mol. Pl. Microb. Interact* 3:351–57

113. Mills, S. D., Förster, H., Coffey, M. D. 1991. Taxonomic structure of *Phytophthora cryptogea* and *P. drechsleri* based on isozyme and mitochondrial DNA analysis. *Mycol. Res.* 95:31–48

114. Mitchell, L. G., Merril, C. R., 1989. Affinity generation of single-stranded DNA for dideoxy sequencing following the polymerase chain reaction. *Anal. Biochem.* 178:239–42

115. Mitchell, T. G., White, T. J., Taylor, J. W. 1991. Comparison of ribosomal DNA sequences among the basidiomycetous yeast genera, *Cystofilobasidium, Filobasidium* and *Filobasidiella. J. Bacteriol.* Submitted

116. Mullis, K. B., Faloona, F. A. 1987. Specific synthesis of DNA *in vitro* via a polymerase-catalysed chain reaction. *Methods Enzymol.* 155:335–50

117. Natvig, D. O., Jackson, D. A., Taylor, J. W. 1987. Random-fragment hybridization analysis of evolution in the genus *Neurospora:* the status of four-spored strains. *Evolution* 41:1003–21

118. Nei, M., Li, W-H. 1979. Mathematical model for studying genetic variation in terms of restriction endonucleases. *Proc. Natl. Acad. Sci. USA* 76:5269–73

119. Nelson, R. T., Yangco, B. G., Te Strake, D., Cochrane B. J. 1990. Genetic studies in the genus *Basidiobolus.* II. Phylogenetic relationships inferred from ribosomal DNA analysis. *Exp. Mycol.* 14:197–206

120. Perkins, D. D., Turner, B. C. 1988. *Neurospora* from natural populations: toward the population biology of a haploid eukaryote. *Exp. Mycol.* 12:91–131

121. Phelps, L. G., Burke, J. M., Ullrich, R. C., Novotny, C. P. 1988. Nucleotide base sequence of the mitochondrial COIII gene of *Schizophyllum commune. Curr. Genet.* 14:401–403

122. Pirozynski, K. A., Dalpé, Y. 1989. Geological history of the Glomaceae with particular reference to mycorrhizal symbiosis. *Symbiosis 7:*1–36

123. Poinar, G. O., Singer, R. 1990. Upper Eocene gilled mushroom from the Dominican Republic. *Science* 248:1099–1101

124. Prager, E. M., Wilson, A. C. 1988. Ancient origin of lactalbumin from lysozyme: analysis of DNA and amino acid sequences. *J. Mol. Evol.* 27:326–35

125. Price, C. W., Fuson, G. B., Phaff, H. J. 1978. Genome comparison in yeast systematics: delimitation of species within the genera *Schwanniomyces, Saccharomyces, Debaryomyces,* and *Pichia. Microbiol. Rev.* 42:161–93.

126. Rehner, S. A., Bruns, T. D., Vilgalys, R. 1990. Molecular systematic investigations in the Agaricales. See Ref. 16, p. 339 (Abstr).

127. Reynolds, D. R., Taylor, J. W. 1991. DNA specimens and the international code of botanical nomenclature. *Taxon* 40:311–15

128. Reynolds, D. R., Taylor, J. W. 1992. Article 59: Reinterpretation or revision? *Taxon* 42: In press

129. Rippon, J. W. 1988. *Medical Mycology.* Philadelphia: W. B. Saunders. 3rd. ed.

130. Rogers, S. O., Ammirati, J. F., LoBuglio, K., Rehner, S. A., Bledsoe, C. 1990. Molecular taxonomy and phylogenetics of Basidiomycetes. (Abstr.) See Ref. 18, p. 339

131. Rogers, S. O., Honda, S., Bendich, A. J. 1986. Variation in the ribosomal RNA genes among individuals of *Vicia faba. Plant Mol. Biol.* 6:339–45

132. Russell, P. J., Wagner, S., Rodland, K. D., Feinbaum, R. L., Russel, J. P., et al. 1984. Organization of the ribosomal ribonucleic acid genes in various wild-type strains and wild-collected strains of *Neurospora. Mol. Gen. Genet.* 196: 275–82

133. Saiki, R. K., Gelfand, D. H., Stoffel, S., Scharf, S. J., Higuchi, R. G., Horn, G. T., Mullis, K. B., Erlich, H. A. 1988. Primer-directed enzymatic amplification of DNA with a thermostable DNA polymerase. *Science* 239:487–91

134. Saitou, N., Imanishi, T. 1989. Relative efficiencies of the Fitch-Margoliash, maximum-parsimony, maximum-likelihood, minimum-evolution, and neighbor-joining methods of phylogenetic tree construction in obtaining the correct tree. *Mol. Biol. Evol.* 6:514–25

135. Saitou, N., Nei, M. 1987. The neighbor-joining method: a new method for reconstructing phylogenetic trees. *Mol. Biol. Evol.* 4:406–25

136. Samac, D. A., Leong, S. A. 1989. Mitochondrial plasmids of filamentous fungi: characteristics and use in transformation vectors. *Mol. Plant-Microbe Interact.* 2:155–59

137. Sanders, J. P. M., Heyting, C., Verbett, M. P., Meijlink, F. C. P. W., Borst, P. 1977. The organization of genes in yeast mitochondrial DNA. III. Comparison of the physical maps of the mitochondrial DNAs from three wild-type *Saccharomyces* strains. *Mol. Gen. Genet.* 157: 239–61

138. Scharf, S. J., Horn, G. T., Erlich, H. A. 1986. Direct cloning and sequence analysis of enzymatically amplified genomic sequences. *Science* 233:1076–78

139. Scherer, S., Stevens, D. A. 1988. A *Candida albicans* dispersed, repeated gene family and its epidemiologic applications. *Proc. Natl. Acad. Sci. USA* 85:1425–56

140. Selker, E. U., Stevens, J. N., Metzenberg, R. L. 1985. Heterogeneity of 5S RNA in fungal ribosomes. *Science* 227:1340–43

141. Shumard-Hudspeth, D. S., Hudspeth, M. E. S. 1990. Genic rearrangements in *Phytophthora* mitochondrial DNA. *Curr. Genet.* 17:413–15

142. Sidow, A., Wilson, A. C. 1990. Compositional statistics: An improvement of evolutionary parsimony and its application to deep branches in the tree of life. *J. Mol. Evol.* 31:51–68

143a. Simon, C. 1991. Molecular systematics at the species boundary: exploiting conserved and variable regions of the mitochondrial genome of animals via direct sequencing from amplified DNA. In *Molecular Taxonomy,* ed. G. M. Hewitt, A. W. B. Johnston. Berlin: Springer Verlag. In press

144. Smith, M. L., Anderson, J. B. 1989. Restriction fragment length polymorphisms in mitochondrial DNAs of *Armillaria:* identification of North American biological species. *Mycol. Res.* 93:247–56

145. Smith, M. L., Duchesne, L. C., Bruhn, J. N., Anderson, J. B. 1990. Mitochondrial genetics in a natural population of the plant pathogen *Armillaria. Genetics* 126:575–82

146. Smith, T. L. 1989. Disparate evolution of yeasts and filamentous fungi indicated by phylogenetic analysis of glyceraldehyde-3-phosphate dehydrogenase genes. *Proc. Natl. Acad. Sci. USA* 86: 7063–66

147. Sogin, M. L. 1989. Evolution of eukaryotic microorganisms and their small subunit ribosomal RNAs. *Am. Zool.* 29:487–99

148. Steele, K. P., Holsinger, K. E., Jansen, R. K., Taylor, D. W. 1991. Assessing the reliability of 5S rRNA sequence data for phylogenetic analysis in green plants. *Mol. Biol. Evol.* 8:240–48

149. Stuart, K., Feagin, J. E., Abraham, J. M. 1989. RNA editing: the creation of nucleotide sequences in messenger RNA—a minireview. *Gene* 82(1):155–60

150. Swofford, D. L., Olsen, G. J. 1990. Phylogeny reconstruction. In *Molecular Systematics,* ed. D. M. Hillis, C. Moritz, pp. 411–501. Sunderland, Mass: Sinauer

151. Taylor, J. W., Natvig, D. O. 1989. Mitochondrial DNA and evolution of heterothallic and pseudohomothallic *Neurospora* species. *Mycol. Res.* 93: 257–72

152. Taylor, J. W., Smolich, B. D., May, G. 1986. Evolution and mitochondrial DNA in *Neurospora crassa. Evolution* 40:716–39

153. Taylor, J. W., White, T. J. 1991. Molecular evolution of nuclear and mitochondrial ribosomal DNA region in *Neurospora* and other Sordariaceae and comparison with the mushroom, *Laccaria Fung. Genet. Newsl.* 38:26

154. Taylor, T. N. 1990. Fungal associations in the terrestrial paleoecosystem. *TREE 5*:21–25

155. Turner, B. J., Elders, J. F., Laughlin, T. F., Davis, W. P. 1990. Genetic variation in clonal vertebrates detected by simple-sequence DNA fingerprinting. *Proc. Natl. Acad. Sci. USA* 87:5653–57

156. Vilgalys, R., Hester, M. 1990. Rapid genetic identification and mapping of enzymatically amplified ribosomal DNA from several *Cryptococcus* species. *J. Bacteriol.* 172(8):4238–46

157. Vilgalys, R. J., Johnson, J. L. 1987. Extensive genetic divergence associated with speciation in filamentous fungi. *Proc. Natl. Acad. Sci. USA* 84:2355–58

157a. Vilgalys, R., Gonzalez, D. 1990. Ribosomal DNA restriction fragment length polymorphisms in *Rhizoctonia solani. Phytopathology* 80(2):151–62

158. Walker, W. F., Doolittle, W. F. 1982. Redividing the basidiomycetes on the basis of 5S rRNA sequences. *Nature* 299:723–24

159. Ward, D. M., Weller, R., Bateson, M. M. 1990. 16S rRNA sequences reveal numerous uncultured microorganisms in a natural community. *Nature* 345:63–65

160. White, T. J., Bruns, T. D., Lee, S.,

Taylor, J. W. 1990. Amplification and direct sequencing of fungal ribosomal RNA genes for phylogenetics. See Ref. 49, 38:315–22

161. White, T. J., Madej, R., Persing, D. H. 1991. The polymerase chain reaction: clinical applications. In *Advances in Clinical Chemistry* 29, ed. H. Spiegel. New York: Academic Press. In press

162. Williams, J. G. K., Kubelik, A. R., Livak, K. J., Rafalski, J. A., Tingey, S. V. 1990. DNA polymorphisms amplified by arbitrary primers are useful as genetic markers. *Nucl. Acids Res.* 18: 6531–35

163. Wilson, A. C., Cann, R. L., Carr, S. M., George, M., Gyllensten, U. B., Hel m-Bychowski, K. M. Higuchi, R. G., Palumbi, S. R., Prager, E. M., Sage, R. D., Stoneking, M. 1985. Mitochondrial DNA and two perspectives on evolutionary genetics. *Biol. J. Linnean Soc.* 26:375–400

164. Woese, C. R., Kandler, O., Wheelis, M. L. 1990. Towards a natural system of organisms: Proposal for the domains Archaea, Bacteria, and Eucarya. *Proc. Natl. Acad. Sci. USA* 87:4576–79

165. Wolf, K., Del Giudice L. 1988. The variable mitochondrial genome of Ascomycetes: organization, mutational alterations, and expression. *Adv. Genet.* 25:185–308

166. Wu, M. M. J., Cassidy, J. R., Pukkila, P. J. 1983. Polymorphisms in DNA of *Coprinus cinereus*. *Curr. Genet.* 7:385–92

167. Yamada, Y., Kawasaki, H., Nakase, T., Banno, I. 1989. The phylogenetic relationship of the conidium-forming anamorphic yeast genera *Sterigmatomyces, Kurtzmanomyces, Tsuchiyaea* and *Fellomyces,* and the teleomorphic yeast genus *Sterigmatosporidium* on the basis of the partial sequences of 18S and 26S ribosomal ribonucleic acids. *Agric. Biol. Chem.* 53:2993–3001

168. Yamada, Y., Nakagawa, Y., Banno, I. 1989. The phylogenetic relationship of Q9-equipped species of the heterobasidiomycetous yeast genera *Rhodosporidium* and *Leucosporidium* based on the partial sequences of 18S and 26S ribosomal ribonucleic acids: the proposal of a new genus *Kondoa*. *J. Gen. Appl. Microbiol.* 35:377–85

Annu. Rev. Ecol. Syst. 1991. 22:565–92

SYSTEMATICS AND EVOLUTION OF SPIDERS (ARANEAE)*

Jonathan A. Coddington

Department of Entomology, National Museum of Natural History, Smithsonian Institution, Washington, DC 20560

Herbert W. Levi

Museum of Comparative Zoology, Harvard University, Cambridge, Massachusetts 02138

KEY WORDS: taxonomy, phylogeny, cladistics, biology, diversity

---

INTRODUCTION

In the last 15 years understanding of the higher systematics of Araneae has changed greatly. Large classical superfamilies and families have turned out to be poly- or paraphyletic; posited relationships were often based on symplesiomorphies. In this brief review we summarize current taxonomic and phylogenetic knowledge and suggest where future efforts might profitably be concentrated. We lack space to discuss fully all the clades mentioned, and the cited numbers of described taxa are only approximate. Other aspects of spider biology have been summarized by Barth (7), Eberhard (47), Jackson & Parks (72), Nentwig (105), Nyffeler & Benz (106), Riechert & Lockley (134), Shear (149) and Turnbull (160).

Diversity, Paleontology, Descriptive Work, Importance

The order Araneae ranks seventh in global diversity after the five largest insect orders (Coleoptera, Hymenoptera, Lepidoptera, Diptera, Hemiptera) and Acari among the arachnids (111) in terms of species described or anticipated. Spiders are among the most diverse groups on earth. Among these taxa, spiders are exceptional for their complete dependence on predation as a trophic strategy. In contrast, the diversity of insects and mites may result from their diversity in dietary strategies—notably phytophagy and parasitism (104).

Roughly 34,000 species of spiders had been named by 1988, placed in about 3000 genera and 105 families (117). A small percentage of those species names will turn out to be synonyms. Families with over 1000 species described are Salticidae (jumping spiders; ca. 490 genera, 4400 species); Linyphiidae (dwarf or money spiders, sheet web weavers; ca. 400 genera, 3700 species), Araneidae (common orb weavers; ca. 160 genera, 2600 species); Theridiidae (cob web weavers; ca. 50 genera, 2200 species); Lycosidae (wolf spiders; ca. 100 genera, 2200 species), Gnaphosidae (ground spiders; ca. 140 genera, 2200 species); and Thomisidae (crab spiders; ca. 160 genera, 2000 species). Although the aforementioned families are cosmopolitan, the linyphiids are most diverse in the north temperate regions, whereas the others are most diverse in the tropics or show no particular pattern. Fourteen spider families are monotypic at the generic level, and 15 are known from 10 or fewer species.

Because spiders are not thoroughly studied, estimates of total species diversity are difficult. The faunas of Western Europe (especially England) and Japan are most completely known (136, 137, 166). The Nearctic fauna is perhaps 80% described (33), New Zealand perhaps 60–70% (36, 51, 52, 54, 55, 60, 61), and Australia perhaps 20% (131). Other areas, especially Latin America, Africa, and the Pacific region are much more poorly known. In several recent revisions of Neotropical orb weavers, 60–70% of the species in available collections were new. But for each 50 previously known species about 75 names exist, as common species had been given different names in different countries (96–99). Recent revisions by Baehr & Baehr (4, 5) of Australian hersiliids had 93% new species. In a monograph on the poorly known south temperate family Orsolobidae (57), 85% of the species were new. Finally, available collections are biased toward medium- and larger-sized species from easily accessible habitats. There are very few places on earth where even desultory searching does not yield new species of spiders. About one third of all genera (1090 in 83 families) occur in the Neotropics. If the above statistics suggest that 20% of the world fauna is described, then about 170,000 species of spiders are extant.

PALEONTOLOGY The earliest spider fossil is *Attercopus fimbriunguis* from the Middle Devonian (380–374 million years BP) Gilboa site in New York State (146). The spinner spigots of this fossil resemble those of recent mesotheles but also share features with primitive opisthotheles (150). Other fossil spiders formerly attributed to the Paleozoic either are not spiders or else are too incomplete to permit certain identification.

Two species of orb weaving spiders are known from Early Cretaceous limestone in Spain (144). These animals can be placed in modern families or superfamilies—Tetragnathidae and Deinopoidea. Eskov (48, 49) described a new family of orb weavers, the Juraraneidae, and a new archaeid subfamily from the Lower-Middle Jurassic. Given the placement of Orbiculariae and Palpimanoidea in current phylogenetic systems (see below), Araneae may have originated in late Silurian or Early Devonian, with the major radiation of Araneomorphae in late Paleozoic or early Mesozoic times. Jeram et al (74) report trigonotarbids, sister to Pedipalpi plus Araneae, from the Upper Silurian of England. Selden (145) offers a brief but intriguing review of the fossil record of Arachnida and Araneae, evaluating its support for various cladistic hypotheses at the ordinal level.

Amber fossils of about 400 species of Eocene to Miocene age are known, mainly from Baltic or Dominican amber. The latter are mainly small species of the family Theridiidae and males of the tetragnathid genus *Nephila*. Altogether 45 families are represented in Baltic and Dominican amber, of which 2–3 are extinct and of which 30 are found in both ambers. Spiders from Dominican amber mostly belong in recent genera (165).

DESCRIPTIVE TAXONOMIC WORK Descriptive taxonomic work on spiders is not much different from that on any poorly known arthropod group not susceptible to automated or mass sampling techniques. Few specimens of most species are available. Accurate identification is only feasible with adult specimens. Perhaps half of all named species were originally described from a single specimen. Spiders are predators, and adults of many species are rare. Roughly half of all species taken from single tropical sites are singletons, even in large samples (32). Although characteristic of tropical arthropod communities, this rarity may also be due to spiders living in habitats difficult for humans to access, such as tree canopies (28, 89).

Lack of material affects the taxonomist in many ways. Collections often contain one sex but not the other, and associating isolated males and females can be difficult. In some species of the orb-weaving genera *Witica* and *Micrathena,* males and females had originally been placed in different genera because they looked extremely different (93, 94).

Variation is always apparent among even a few specimens of a species, especially if from widely separated localities. However, variation in tiny samples is intractable statistically. One must always question whether this variation indicates separate species or reflects geographical or individual variation. Variation in spiders also arises from their propensity to mature in any of several molts. Some *Nephila* adult males are twice the length of others

(34). Female spiders also undergo a variable number of molts and may mature at different sizes (73). Despite allometric growth and variable morphology, the dimensions of adult genitalia from the same population vary less than do their coloration, body dimensions, or proportions (37, 91). Many papers are concerned with variation (6, 37), but few deal with the genetics of this variation or heredity of color patterns (109, 132).

Early taxonomic work focussed on faunas and new species descriptions, often the bounty of travelers and explorers. Revisionary work was hindered in the past because holotypes (voucher specimens for names of new species) were rarely loaned through the mail. Now large collections and loans of valuable specimens are routine. At present the best taxonomic research is done in the context of revisionary studies. In contrast, isolated papers on "new" species in unrevised groups may result in a new crop of synonyms. High quality revisions still present all relevant comparative data on all species known in a genus, but they are also heavily illustrated to facilitate identification of species by nonspecialists.

Kaston (76, 77) and Roth (139) have made it possible to key to genus most North American spider families. Comparable literature is available for England, Japan, and Western Europe. The work of Forster and collaborators (36, 51–55, 60, 61) provides the only concerted treatment of an Australasian fauna. Although dealing only with New Zealand and still incomplete, it is the reference for the whole region. As noted above, it will be a huge task to revise the many genera of spiders in unstudied areas of the world.

First revisions of spider groups rely heavily on good illustrations of genitalia, the most accessible and likely mark of specific identity. It has been known since Lister in the seventeenth century (Philip H. Schwann, personal communication) that spider species differ in their genitalia: the epigynum (female copulatory pores) and the male palp. Clerck (24) illustrated genitalia in his 1757 treatise on Swedish spiders, but later authors often ignored them in their eagerness to name new species. The critical question of why genitalia reflect species differences has been discussed by Eberhard (44), who favors sexual selection on genitalic morphology by female choice during copulation.

It is as yet impractical to start a revision or to identify voucher specimens with molecular or biochemical methods. Such methods are excellent to test genetic or phylogenetic hypotheses among or within named species, or to resolve cases in which morphological comparison is insufficient. Similarly, morphometric treatments of somatic characters are usually unnecessary to identify species with complex genitalia but are useful in the infraorder Mygalomorphae (128). They also offer much promise in answering specific research questions, such as partitioning variation into heritable and phenotypic components.

Spider taxonomists have been lucky to have a series of up-to-date taxonomic catalogs. Bonnet (12) in France and Roewer (138) in Germany independently prepared catalogs that were complete up to 1939 and 1942, respectively. Brignoli (18) included species described from 1940 to 1981, and Platnick (117) those from 1981 through 1987.

ARANEOLOGISTS AND ARANEOLOGICAL COLLECTIONS The situation for systematic araneologists in North America is probably typical for the rest of the world. About seven araneologists did systematic work in the 1940s in North America, and the number of paid professional systematic araneologists is similar now (33). Those few are supplemented by about 25 professionals with largely nonresearch jobs and consequently limited time for systematic work. Paralleling the loss in taxonomic expertise world-wide, the job situation for systematic araneologists is poor enough that many have left the field and few are entering it. The age structure of systematic araneologists is therefore significantly skewed towards older workers, compared to nonsystematic araneologists in North America (33). The number of araneologists in nonsystematic disciplines has increased much more rapidly, and consequently the need for identifications and taxonomic advice has outstripped the ability of systematists to supply it (33, 135). About 24 arachnological societies exist around the world, 8 of which publish research journals (35). The Centre International de Documentation Arachnologique, with about 750 members, is the major international society for nonacarine arachnid researchers.

Major collections of spiders accumulated at many institutions in the past when natural history was more in vogue, but many of those institutions no longer employ systematic araneologists. Consequently, many collections have become nearly static and often are poorly maintained. The largest spider collections are at the American Museum of Natural History in New York and the Museum of Comparative Zoology at Harvard University.

ECOLOGICAL, ECONOMIC, AND MEDICAL IMPORTANCE Spiders are diverse and abundant terrestrial predators. New England has almost 700 species (77); Great Britain and Ireland about 600 (136, 137); the Berlin area of Germany about 500 (165). Larger areas for which estimates exist are Japan with about 1100 (166), North America with about 4000 (33), and Australia with about 9000 (131). No comparable estimates exist for tropical regions, but a few hectares of tropical wet forest have numbers comparable to those cited above for immensely larger temperate areas (32). Bristowe (21) found about five million spiders per hectare in an abandoned field in Sussex, England. Linyphiid densities reached 29,000 individuals per cubic meter among filter-beds of an English sewage treatment plant; they fed principally on enchytraeid worms and dipteran larvae (42). As generalist predators, spiders are abundant in all terrestrial ecosystems. Turnbull (160) reported abundances ranging from 0.64 to 842 per square meter.

Control by spiders of insect populations in agricultural and epidemiological settings is receiving more attention as integrated pest management replaces the use of chemical pesticides (13, 14, 106, 133, 134). Spider neurotoxins are much used in neurobiological research (72), and they may have potential as insecticides (157). Fiber scientists study silks (164).

At least four genera are consistently responsible for medically serious or life-threatening bites: the Australian funnel-web mygalomorph *Atrax* (Hexathelidae), the brown recluse *Loxosceles* (Loxoscelidae), the widow spiders *Latrodectus* (Theridiidae), and the tropical wolf spiders *Phoneutria* (Ctenidae). Other ctenid genera occurring in the tropics are probably also responsible for serious bites. Sao Paulo has 100 serious bites a year (103). At least 20 other genera have been responsible for bites requiring medical attention (142).

PRIOR PHYLOGENETIC WORK ON SPIDERS

Before the 1880s, spider classification was based on broad categories of lifestyles. Important and widely accepted suprafamilial categories were Tubitelae (tube-dwellers), Orbitelae (orb web weavers), Saltigradae (jumpers), and Citigradae (runners). The classification became distinctly more artificial in the latter nineteenth century. A consensus developed to construct monothetic classifications based on strict character dichotomies. For example, groups were defined by two *or* three claws, presence *or* absence of a cribellum, paraxial *or* diaxial chelicerae, one *or* two pairs of booklungs. Taxa based on plesiomorphies, "not," or "absent" characters came to exist at all levels of the taxonomy. Examples of higher level "taxa" were Trionycha (three claws is primitive for spiders), Cribellatae (the cribellum is primitive for the suborder Araneomorphae), Tetrapneumonae (four booklungs is primitive for spiders), Orthognatha (paraxial chelicerae is primitive for spiders), and Haplogynae (all female spiders primitively lack fertilization ducts). Consequently, about half of the major suprafamilial taxa were paraphyletic.

Unfortunately, these erroneous groups were adopted by Eugène Simon, the most knowledgeable araneologist to date. His *Histoire Naturelle des Araignées* (156) was encyclopedic, detailed, and widely accepted; it has not yet been equalled or even approached. Catalogers such as Roewer (138) preserved most of the hierarchy embodied in Simon's system, while amplifying it and changing ranks to accommodate increased knowledge. Although several authors criticized Simon's system in one respect or another (20, 113, 114), it remained the consensus view until a furor erupted over a monograph on cribellate spiders (84).

The Collapse of the Cribellatae

Lehtinen (84) focussed on the Cribellatae, one of the artificial taxa mentioned above. The cribellum is an anterior median spinneret homolog, a flat plate bearing hundreds of densely packed spigots that produce persistently sticky silk (43). Lehtinen argued that all araneomorph spiders were once cribellate and that any ecribellate araneomorph was so secondarily. At that time roughly a fourth of all spider families were exclusively cribellate and regarded as one monophyletic lineage. Even though close relationships had been suggested between cribellate and ecribellate lineages (113, 114), they were highly controversial because they struck at the fundamental taxonomic dichotomy in the infraorder Araneomorphae. Paraphyly of the Cribellatae implied a wholesale review of araneomorph classification and phylogeny.

Lehtinen's argument received empirical support from the discovery of many clear cribellate-ecribellate close relatives in the New Zealand fauna (52) and an objective consideration of the most obvious European example, Uroc-teidae-Oecobiidae (8, 82). Whereas there were relatively few other cribellate-ecribellate sister taxa in the north temperate fauna, austral faunas, even within genera, were evidently full of them.

During the same period, cladistic theory began to revolutionize systematics. None of the authors of this rather "fact-based" challenge to the old araneomorph systematics used a cladistic approach, but it has been commonly used since then to rework and justify many of those arguments. The validity of Orthognatha, Tetrapneumonae, Trionychae, Haplogynae, Ctenizidae, Argiopidae, Agelenidae, Amaurobiidae, and Clubionidae, to name a few of the larger taxa, was obviously questionable. Lehtinen's work was not consistently phylogenetic (he recognized many paraphyletic groups), and there were enough loose ends and mistaken details to obscure the fundamental insight with controversy. However, in retrospect his challenge to the old system was unmistakably mortal.

Monophyly and Cladistic Relations of Araneae

The monophyly of Araneae is supported by several complex and unique synapomorphies. The most important are abdominal appendages modified as spinnerets, silk glands and associated spigots, cheliceral venom glands, male pedipalpal tarsi modified as secondary genitalia (sperm transfer organs), and loss of abdominal segmentation (external traces are clear in Mesothelae and faint in a few Mygalomorphae; all possible sister taxa to Araneae are segmented). Spiders also lack the trochanterofemoral depressor muscle in the walking legs (152).

The traditional view (163) has placed spiders as sister to Amblypygi (tailless whip scorpions). Shultz (152) added many new characters to the cladistic analysis of arachnid orders, emphasizing especially muscles and joints.

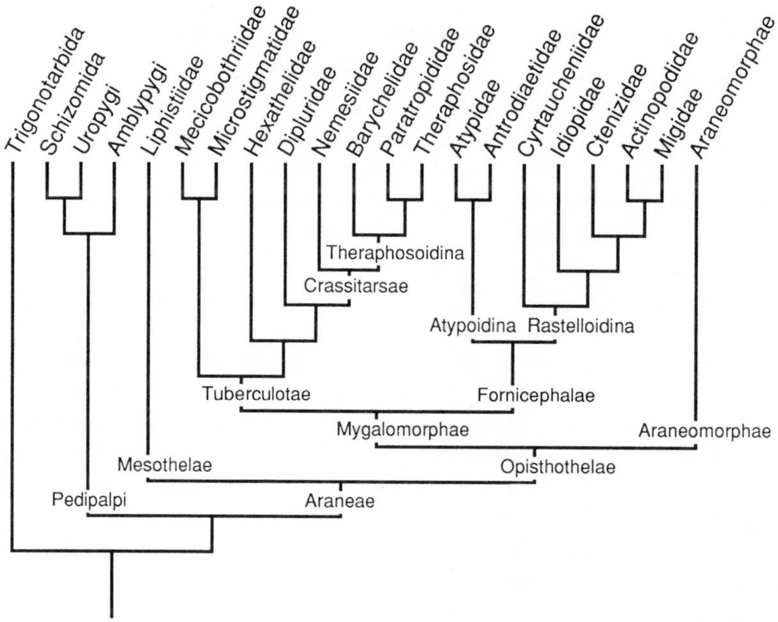

Figure 1 Cladistic hypothesis for Araneae and outgroups emphasizing cladistic structure of infraorders Mesothelae and Mygalomorphae. See Figure 2 for Infraorder Araneomorphae

Araneae emerged as sister to the Pedipalpi (Amblypygi, Schizomida, Uropygi) as a whole (Figure 1), based on six synapomorphies for Pedipalpi, and eight that linked Araneae to it. Shear et al (151) and Selden et al (146) agreed that Araneae was sister to Pedipalpi and further suggested that the extinct order Trigonotarbida was sister to the two together. Homann (71) argued that the plagula ventralis, a small sclerite associated with the cheliceral fang, was also a Pedipalpi-Araneae synapomorphy. Pedipalpi itself contains only about 200 species in three orders. The validity of the Pedipalpi-Araneae hypothesis is important for evolutionary studies on spiders. Reconstructing the ground plan for spiders now requires consideration of variation within all Pedipalpi rather than just Amblypygi.

CLADISTIC STRUCTURE OF ARANEAE

Three major monophyletic groups of spiders exist: Mesothelae, Mygalomorphae, and Araneomorphae. Although most workers recognized them as different groups, for many years their interrelationships were not so clear.

Many authors (62, 76) recognized the taxon Orthognatha as including Mesothelae and Mygalomorphae. However, in perhaps the first explicitly cladistic treatment of spiders, Platnick & Gertsch (121) showed that paraxial chelicerae (the "ortho" in Orthognatha) were primitive; Orthognatha was therefore paraphyletic. Instead Mygalomorphae and Araneomorphae are united as infraorders within Opisthothelae based on several characters such as terminally positioned spinnerets, coalesced neuromeres, and reduction of external evidence of abdominal segmentation. This work was an early example of the application of cladistic reasoning to a problem for which taxonomists did not lack data so much as they lacked any coherent rationale for preferring one solution over another.

Mesothelae

The suborder Mesothelae contains the single family Liphistiidae (2 genera; ca. 40 species), limited to China, Japan, southeast Asia, and Sumatra (123). Liphistiids confirm the metameric ancestry of spiders (tergites, sternites, dorso-ventral musculature, etc), dispersed (versus coalesced) ventral ganglia, and four distinct pairs of anteriorly placed spinnerets (versus fewer terminal spinnerets). One is left unsure whether many unique mesothele features, such as male genitalic morphology or pseudo-segmented spinnerets, are the plesiomorphic condition for all spiders or mesothele autapomorphies. This dilemma is particularly vexing in work on the evolution of spinnerets and spigots (110, 150).

Liphistiid monophyly is supported by five morphological synapomorphies, including a unique cuticular modification that apparently functions to detect leg flexion (122, 123). Because liphistiids are the sister group to the remaining spiders, their biology may give some indication of the ecological setting in which spiders evolved. Liphistiids are tube-dwelling sit-and-wait predators that construct rudimentary trap doors. Some make silk "trip-lines" radiating away from the burrow entrance that extend the sensory radius of the animal. They are active mainly at night, live for several years, have very low vagility, and consume a catholic diet of mainly walking prey. Females molt after sexual maturity. Because the internal spermathecal lining is shed during a molt, presumably they must mate again to continue to lay fertile eggs. While females rarely leave their burrows, adult males wander in search of the females. Their respiratory system consists only of booklungs, a possible obstacle to high activity levels (2, 3). This predisposition to forgo high activity levels and mobility (which characterize pterygote insects at least) for a low-cost, sit-and-wait strategy is a common, plesiomorphic, and perhaps constraining pattern in spider evolution.

Mygalomorphae

Mygalomorphs include the baboon spiders or tarantulas (Theraphosidae), trap-door spiders (Ctenizidae, Actinopodidae, Migidae, etc), purse-web spiders (Atypidae), funnel web spiders (Hexathelidae), and several other groups with no common name (Figure 1). With 15 families (ca. 260 genera, 2200 species) (117), Mygalomorphae are more diverse than Mesothelae, although they do not approach the diversity of araneomorphs. Mygalomorph monophyly rests mainly on spinneret and male genitalic characters. They lack any trace of the anterior median spinnerets present in mesotheles, whereas at least primitive araneomorphs retain the cribellum as a homolog of those spinnerets. In mygalomorphs the anterior lateral spinnerets are much reduced if not absent altogether; in araneomorphs these spinnerets are the largest and best developed. The male pedipalpal genital bulb is fused in most mygalomorphs, but primitive araneomorphs and mesotheles have two to three divisions. Homann (71) wrote that only mygalomorphs have a small ancillary dorsal sclerite near the fang tooth. Raven (129) reviews other possible mygalomorph synapomorphies as well.

The family Theraphosidae or "tarantulas" contains almost three times as many species as any other mygalomorph family (ca. 800 species). It is not clear why they are so speciose: the most obvious synapomorphy for the family is dense tarsal scopulae on the last two pairs of legs as well as the first. This feature may aid in locomotion or in prey-handling (see below under Dionycha). Theraphosids are famous for their large size (at 10 cm body length the South American *Theraphosa leblondi* is probably the largest spider), but some are only a centimeter, and the smallest mygalomorph is less than a millimeter. The Australian *Atrax* (Hexathelidae) is extremely venomous and dangerous to humans.

Like mesotheles, mygalomorphs usually live in tube retreats or burrows. The popular stereotype of tarantulas as vagabond predators is inaccurate (38). Instead many tube dwellers extend the range over which they can sense prey by constructing silk lines away from the retreat entrance or arranging debris in radial patterns. These elaborations rarely gain the animal more than a few centimeters in range, although the foraging area can equal that accessed by typical web spiders. Some diplurids, however, build extensive and elaborate capture webs that approach a meter in diameter (38). Mygalomorphs are also capable of spinning at least slightly adhesive silk (39, 159). Although use of silk by mygalomorphs is more diverse than commonly appreciated (38), it is not so developed as in some araneomorphs. Mygalomorphs display a limited diversity of silk glands and spigot types (78, 79, 110).

Raven (129) recently reviewed and revamped the systematics of mygalomorphs. For the first time, a cladogram for families, subfamilies, and many generic groupings was proposed. The prior classification included 9–11 families (18, 149) and, in general, lacked justification. Earlier cladistic analyses had contested the monophyly of the Atypoidea (63, 116) and linked Actinopodidae and Migidae (124). Perhaps the worst cladistic problem was the symplesiomorphic Ctenizidae, a large, amorphous "dumping ground" classically known as the trap-door spiders (130). At the family level, most of Raven's changes involved relimiting the Ctenizidae and recognizing groups formerly subsumed in it, although he did synonymize one family name. Figure 1 reproduces the cladistic structure among mygalomorphs presented by Raven.

His results show two major lineages, Fornicephalae (7 families, ca. 60 genera, 700 species), and the more speciose Tuberculotae (7 families, ca. 200 genera, 1500 species). Almost all the diversity in Tuberculotae is due to Theraphosidae (ca. 80 genera, 800 species). The Fornicephalae include two subsidiary branches, the atypoids and the rastelloids. It is interesting that one apomorphy of the atypoids is the great reduction or absence of tarsal trichobothria, mechanoreceptors sensitive to vibration and near-field sound. The rastelloids are united by the possession of a rastellum or digging rake on the chelicerae—they are all tube-dwellers and most make trap doors. A number of more subtle features unite the Fornicephalae as a whole, such as an arched head region, stout tarsi, and the first two pair of legs being more slender than the last. Tuberculotae, in contrast, have a sloping thoracic region, a serrula (saw-like row of teeth) on the maxillae that probably help to grasp and crush food items, and a distinct eye tubercle. Within these mygalomorphs, one well-defined group is the Crassitarsae (Nemesiidae and three families comprising Theraphosoidina), which share tarsal scopulae and a reduced median tarsal claw. Monophyly of Theraphosodina is based on presence of claw tufts (see also *Dionycha,* below) and the form of the tibial hook used by the male to catch the chelicerae of the female during mating.

The scope and results of Raven's work are impressive. He exposed many long-standing absurdities, and his work has become the point of departure for mygalomorph higher level systematics. It also substantiates the more general impression that uniquely derived and unreversed synapomorphies are not common in spiders. Inferring phylogeny is not so much a question of finding characters as it is of allocating homoplasy. Raven used 39 characters of which only 7 were fully consistent on his tree; three groupings at the family level were supported only by homoplasious characters.

Some of Raven's results are, of course, controversial. Speaking only of interfamilial relationships, perhaps the most significant controversy concerns the dismemberment of the Atypoidea (Atypidae, Antrodiaetidae, Mecicobothriidae). Araneologists often considered the atypoids as the most primitive mygalomorphs (23). Platnick (116) and Raven (129, 130) argued that atypoids were a symplesiomorphic group, although they accept the linkage of Atypidae and Antrodiaetidae. Eskov & Zonshtein (50) countered that mecicobothriids do form a monophyletic group with the atypoids and that Atypoidea in their sense are indeed sister to the remaining mygalomorphs, termed Theraphosoidea. However, Eskov & Zonshtein explicitly accepted grades in their scheme, and they excluded many apparently informative characters to arrive at a considerably less parsimonious explanation of mygalomorph relationships. The debate is productive and focused on characters, a direct benefit of competing explicit phylogenetic hypotheses.

Araneomorphae

The infraorder Araneomorphae, sometimes referred to as "true" spiders, includes all remaining taxa, some 90 families, 2700 genera, and 32,000 species described.

Diversification rates in spiders between sister taxa (by definition of equal age) tend to be unequal. Mesothelae has a few dozen known species, Opisthothelae has 34,000. Mygalomorphae has roughly 2,000 species, Araneomorphae has the rest. Hypochilidae (9 species) are sister to the Neocribellatae (32,000 species), and Austrochiloidea (23 species) are sister to the Araneoclada (59, Figure 2). Entelegynae includes roughly 30,000 species, whereas Haplogynae numbers only about 2500 (Figure 2). Diversification rates within "higher" Entelegynae have been somewhat more equal. Orbiculariae (orb weavers and their descendants; ca. 10,300 species) may be sister to all taxa that have a retrolateral tibial apophysis on the male palp ("RTA Clade," Figure 2, ca. 18,000 species). Within Araneoidea, Araneidae (ca. 2600 species) are apparently sister to the rest (ca. 7200 species) although the position of the large family Linyphiidae is still controversial. However, Deinopoidea (ca. 300 species) is putatively sister to Araneoidea (26, 30).

As noted above, plesiomorphic araneomorphs are unique in retaining the cribellum, a functional homolog of the anterior median spinnerets that produces extremely sticky silk. It is tempting to speculate that the diversity of araneomorph spiders is related to this important innovation. However, cribellate taxa are not very speciose, and for nearly all cribellate-ecribellate sister clades the cribellate lineage is less diverse. Examples are Filistatidae versus the remaining haplogynes (ca. 90 species versus 2400; Figure 2) and Deinopoidea versus Araneoidea (300 species versus 10,000; Figure 2). Only about 180 araneomorph genera in 22–23 families still contain cribellate members, although the diverse Australian cribellate fauna is mostly undescribed. However, that fauna may be another example of atypically high Australian diversity within an otherwise relict and depauperate clade (e.g. marsupials).

ARANEOCLADA This large group is defined by numerous synapomorphies. The most salient may be the transformation of the posterior booklungs into tracheae, the first appearance of a tracheal system in spiders. Here again, when a large group is defined by a synapomorphy with such an important function, it is tempting to suppose some linkage between diversification rates and evolutionary innovation. Although not as extensive as the tracheae of

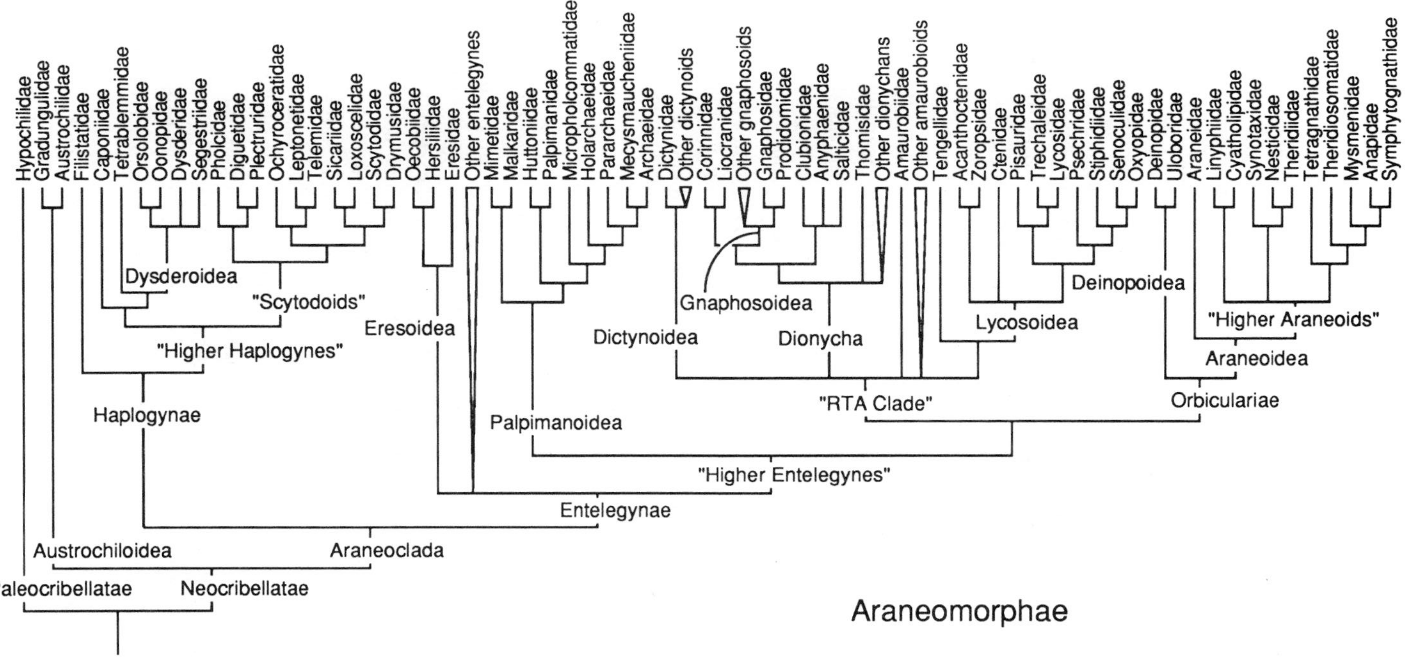

Figure 2 Cladistic hypothesis for Infraorder Araneomorphae. See text for synapomorphies defining formal superfamilies and labelled nodes. Valid families omitted from cladogram are lumped into categories labelled as "other" entelegynes, "other dictynoids," etc, because a more definite placement is unknown or controversial. Families in these categories would otherwise make a large multifurcation at that node, implying resolution at that node or any more distal than the one indicated. "Other" entelegynes include Nicodamidae, Cycloctenidae, Cryptothelidae, Zodariidae, Miturgidae, and Homalonychidae. "Other" dictynoids include at least Desidae, Cybaeidae, Argyronetidae, Hahniidae, and Neolanidae. "Other" amaurobioids include at least Titanoecidae, Agelenidae and Amphinectidae. "Other" gnaphosoids include Gallieniellidae, Ammoxenidae, Cithaeronidae, Trochanteriidae, and Lamponidae. "Other" dionychans include Zoridae, Selenopidae, Sparassidae, and Philodromidae

higher insects, the tracheal system in spiders does correlate with higher metabolic rates and better water conservation (2).

The most recent cladistic work in spiders has focused on araneocladan groups. The higher taxa Haplogynae (17 families), Orbiculariae (13 families), Dysderoidea (4 families), Palpimanoidea (10 families), Lycosoidea (10 families), Gnaphosoidea (7 families), Deinopoidea (2 families), and Araneoidea (11 families) have all received recent cladistic treatment (30, 31, 56, 57, 65, 88, 118, 119, 154). Important exceptions are the clubionoid assemblage, the dictynoid assemblage, dionychan subgroups, and the amaurobioid assemblage (but see 10, 64), all of which suffer from varying degrees of polyphyly or paraphyly. Generally concordant cladistic hypotheses relating most of these large groups have recently been presented (31, 119). Figure 2 summarizes these arguments.

HAPLOGYNAE Any female reproductive system that lacked fertilization ducts was called haplogyne, a "non" category that rightly received much criticism (15, 16, 19, 115). Although the Haplogynae were originally defined on the basis of this plesiomorphy, cladistic evidence suggests that the Filistatidae, Dysderoidea, and "scytodoids" (Figure 2) are a monophyletic group after all (30, 31, 119). Alberti & Weinmann (1) reported peculiar sperm morphology distributed among Filistatidae, Oonopidae, Dysderidae, and Scytodidae that may turn out to be phylogenetically informative when all relevant groups are studied. In addition, the fused and pyriform bulb in the male palp has been reinterpreted as derived (80, 143), although it was formerly considered primitive by analogy to the mygalomorph condition. The tripartite palps of Paleocribellatae, Austrochiloidea, and Mesothelae are critical to this inference (59, 81). Other characters supporting the monophyly of the Haplogynae are a basal fusion of the chelicerae, a lamina along the fang groove instead of teeth, and the enigmatic absence of "tartipores" (150), apparently the vestigial traces of spinneret spigots present in previous instars (167).

Within Haplogynae, the cribellate Filistatidae are apparently sister to the entirely ecribellate remainder. The monophyly of the Dysderoidea (four families, ca. 120 genera) was rigorously justified by Forster & Platnick (57) on the basis of two clear synapomorphies: a novel posterior bursa in the female genitalia to store sperm, and the anterior position of the abdominal tracheal trunk. Haplogynae exhibit many diverse morphologies related to sperm storage, obviously an important functional problem (53). The functional significance of a relatively anterior versus posterior spiracle is unclear. So far as is known, most dysderoids usually live in leaf litter or similar protected habitats. Some such as segestriids spin relatively simple webs with trip lines radiating from the tubular retreat; most presumably are vagabonds.

Platnick et al (119) found several features that placed two enigmatic families, Tetrablemmidae ("armored" spiders; 87, 147) and Caponiidae (no common name) close to the dysderoids. Tetrablemmids are related to the classical dysderoids by the loss of the anterior median eyes. Caponiidae are related to this clade by the advancement of the posterior tracheal spiracle to just behind the gonopore.

The remaining haplogynes, informally labelled "scytodoids" in Figure 2, are defined by the loss of the AME, a parallelism with the situation just noted for Tetrablemmidae-Dysderoidea. It includes the cellar spiders (Pholcidae) and diguetids, both of which spin elaborate webs, the spitting spiders (Scytodidae) and recluse spiders (Loxoscelidae), as well as lesser known groups. It differs from Scytodoidea as defined by Brignoli (16) by the transfer of Caponiidae and Tetrablemmidae to the dysderoids, and the placement of several small groups about which Brignoli was uncertain. Other hypotheses concerning haplogyne relationships (15, 88, 158) considered fewer taxa and characters, lacked data matrices, and were nonquantitative and less parsimonious. Subtle alternative explanations for character distributions (and implied groupings) are difficult to detect unless a quantitative approach is used.

ENTELEGYNAE Entelegynae includes all the remaining superfamilies and about 70 families. Entelegyne female genitalia have paired copulatory pores that usually open on a sclerotized plate, the epigynum. Grooves, projections, and cavities of the epigynum offer the male a complex structure to affix his palpal bulb during copulation. How copulating pairs that lack these structures solve the problem of orienting their complicated genitalia during copulation is unknown, and therefore the mechanical significance of the epigynum is difficult to assess. Separate fertilization ducts lead from the spermathecae to the gonoduct; thus the reproductive system opens twice to the outside. Presumably this "flow-through" sperm management system is also an important functional difference from the haplogyne condition, but precise comparative data on well-chosen groups are lacking. Entelegyny is more uniform morphologically than haplogyny and is apparently only rarely lost among higher spiders. Reversal to a "haplogyne" condition occurs in a few genera of Uloboridae, Tetragnathidae, Anapidae, and at least three times within the Palpimanoidea (119). Interpretation of this character system is subject to lively debate (30, 31, 53, 57, 119). Eberhard (46) also discovered details of cribellate silk spinning behavior that seem to confirm entelegyne monophyly.

OTHER ENTELEGYNES Several entelegyne families remain unplaced in Figure 2 (see legend). The large family Zodariidae (ca. 50 genera, 420 species) recently received cladistic treatment (75), although its affinities remain elusive. Miturgidae are also a large family (ca. 20 genera, 150 species), but their monophyly is dubious (64, 65).

Eresoidea Oecobiidae and Hersiliidae share elongate posterior spinnerets used in a distinctive attack behavior—they rapidly circle around prey to wrap and immobilize it with silk. Eresidae are a distinct family, and it contains some odd genera (e.g. *Wajane*); its phylogenetic placement is controversial. Based on a selection of taxa and characters chosen to answer another question, the conclusions of Platnick et al (119) were that the oecobiid lineage and eresids were sister taxa based on secondary loss of the paracribellum, a distinctive set of silk spigots. Coddington (31) instead found that eresids and oecobiids (hersiliids were not considered) were adjacent outgroups to the remaining Entelegynae. Both studies agreed that they were basal within Entelegynae. Both Oecobiidae and Eresidae are cribellate. They are pivotal in Figure 2 because, although entelegyne, they lack derived eye tapeta (69), a retrolateral tibial apophysis, and the derived trichobothrial pattern (see below). Ethological studies of mating positions during copulation roughly confirm the positions of Oecobiidae and Eresidae as basal within the Entelegynae (161). These families appear cladistically intermediate between two great groups of "lower" and "higher" neocribellate spiders.

"HIGHER ENTELEGYNES" The "higher" entelegynes have specialized spigots in females that are used to make egg-sacs (29, 78, 119). Where studied, cylindrical glands serve these spigots, also only present in female spiders. Eggsac characteristics can deter parasitoid attacks (68) and decrease desiccation (108); perhaps cylindrical fibers contribute to these functions. The same group has either the canoe-shaped tapetum or the yet more derived grate-shaped tapetum in the indirect eyes, as opposed to the primitive tapetum (69, 83). By influencing how photons reflect within the eye, tapetal structures can strongly influence optical performance (11).

Palpimanoidea Classically the palpimanoids included only three closely related haplogyne families—Huttoniidae, Palpimanidae, and Stenochilidae. The latter is now included within Palpimanidae. Forster & Platnick (56) redefined the superfamily to include the haplogyne archaeids, mecysmaucheniids, the entelegyne Holarchaeidae, Pararchaeidae, Mimetidae, and the mostly entelegyne Micropholcommatidae. Two synapomorphies define the group—cheliceral glands on an elevated mound and cheliceral "peg" teeth (modified setae). These new additions were previously placed in Araneoidea. Malkaridae were recently recognized as the sister taxon of the Mimetidae (40, 120, 165), thus adding an additional entelegyne group to this superfamily.

Although these four groups (the classical palpimanoids, the four "archaeoid" families, Mimetidae-Malkaridae, Micropholcommatidae) are each clearly monophyletic, the resulting Palpimanoidea is heterogeneous. Mimetids and some archaeoid taxa specialize as predators of other spiders.

Araneophagy occurs in other palpimanoids and thus may corroborate monophyly (F. Murphy, personal communication). Palpimanids, huttoniids, and the archaeoid families live in leaf litter and probably don't spin webs; otherwise their habits are poorly known. The entelegyne mimetoid families are basal within Palpimanoidea. Parsimony thus suggests that entelegyny is plesiomorphic for Palpimanoidea and that haplogyny in this case is secondary (119). Forster (53) has argued that the morphological transformations required by the latter hypothesis are implausible compared with multiple origins of the entelegyne condition. Once again, all possible resolutions to this problem require homoplasy. The diverse genitalic morphologies subsumed under the term "haplogyne" (53, 56, 57) need to be cladistically analyzed. The male genitalia of Mimetidae are reminiscent of araneoids in some respects, though few potential synapomorphies have been proposed (30, 148). If further research shows mimetids, archaeoids, and micropholcommatids to be araneoids, then haplogyny might be the primitive state for the classical palpimanoids, thus contesting their placement within Entelegynae. Even if the former families are not palpimanoids, at least some Palpimanidae have canoe tapeta (69), thus still favoring placement within the "higher" entelegynes.

Within "higher" entelegynes, the "RTA Clade" and Orbiculariae are united by the presence of distinct pseudoflagelliform spigots on the posterior lateral spinnerets (Figure 2; 119). This spigot contributes the fibers that support the sticky cribellate silk in Deinopoidea and presumably in other cribellates as well. The homology of this spigot with distinctive spigots in Austrochiloidea and Filistatidae has not been established (119). Should homology be confirmed, the justification for this node will rest solely on special similarities between the pseudoflagelliform spigot in the "RTA Clade" and the cribellate orbicularians.

"RTA CLADE" The retrolateral tibial apophysis on the male palpal tibia is another higher level synapomorphy among spiders, ubiquitous among Dionycha, Lycosoidea, amaurobioids, and dictynoids (10, 30, 31, 64, 65, 154). Haplogynes, orbicularians (except linyphiid erigonines), Oecobiidae, and Eresidae (*Wajane* is an exception) lack the structure. In two cases the tibial apophysis stabilizes the highly expansible palp during copulation (9, 155). Heimer (66) suggests that the paracymbium, an araneoid synapomorphy, may have a similar function. How other spiders lacking either structure solve the same functional problem is unknown. Comparative studies of copulatory position roughly confirm the RTA clade (161).

The distribution of trichobothria (fine sensory hairs) on the metatarsi and tarsi is another important character (30, 31, 85, 86). The plesiomorphic araneomorph pattern seems to be absence or near absence on the metatarsi and tarsi (although present on mygalomorph and mesothele tarsi). The derived condition is single or multiple rows of trichobothria, often increasing in length toward the leg tip. Trichobothria detect slight air movements and may help to detect prey and predators (7). This feature apparently supports the monophyly of the RTA clade.

Amaurobioidea and Dictynoidea These superfamilies are among the largest cladistic problems at the family level. It is no coincidence that they are mainly cribellate groups. Although certainly entelegyne, no strong synapomorphies have yet been found to define them or to resolve their exact placement. This difficulty partly stems from heterogeneity within families. Agelenidae, Desidae, Dictynidae, Hahniidae, Miturgidae, and Amaurobiidae, to name the largest, probably will be more or less drastically redefined. Some spin large funnel webs, some small cribellate sticky webs, some are wanderers. Griswold (64) offered a succinct definition of Amaurobiidae but could not allocate the excluded genera to other families on objective grounds, thus leaving them effectively *incertae sedis*. In a revision of nearctic Cybaeidae, Bennett (10) discussed the composition of Dictynoidea and found several features (secondary gland pores in the female genitalia, a male palpal patellar apophysis, loss of male palpal tegular apophyses) to define all or most of the Dictynoidea. Aygyronetidae, a monotypic family including only the European water spider, are probably close to the Cybaeidae (84). Hahniids may share palpal apomorphies with dictynids and cybaeids, but this argument is still preliminary. Almost no recent work has concerned the very diverse amphinectids and desids. Lehtinen's Amaurobioidea lacked well-defined and defended synapomorphies (84). Forster and Wilton (52, 61) defined Amaurobioidea on the basis of unbranched abdominal median tracheae, and Dictynoidea on the basis of branched tracheae. Coddington (30) argued that by outgroup comparison the unbranched condition was plesiomorphic, but that the branched condition could indicate monophyly of Dictynoidea. The placement of many families in Amaurobioidea and Dictynoidea continues more by tradition than explicit justification (see legend to Figure 2).

Lycosoidea The classical lycosoid synapomorphy is the specialized structure of the tapetum of the lateral and posterior eyes—known as a "grate-shaped tapetum" (69). The feature is a transformation of the "canoe" tapetum, itself a transformation of the "primitive" type. The roughly ten lycosoid families (ca. 235 genera, 3700 species) include both cribellate and ecribellate taxa. As usual, this situation required a wholesale reevaluation of the higher level systematics (65). As for all true spiders, the plesiomorphic lifestyle apparently was as a sedentary web spinner. This still characterizes a few taxa (the cribellate Psechridae and Tengellidae). Most members of this superfamily, including nearly all of the speciose family Lycosidae, have forsaken the web habit for a more cursorial hunting style.

Lycosoid relationships have been recently studied (41, 64, 65, 153, 154). Griswold gives the most complete phylogenetic treatment of the superfamily, and his arrangement is used in Figure 2. He and Sierwald have shown that Lycosidae (wolf spiders), Pisauridae (nursery web spiders), rhoicinines (no common name), and Trechaleidae (no common name, see 22 and 154) share synapomorphies of male palp structure. Griswold (65) also proposed the small cribellate family Tengellidae as the sister group of Lycosoidea based on the presence of an oval rather than a linear calamistrum (the patch of hairs that functions with the cribellum to produce cribellate silk). In Figure 2 the families Acanthoctenidae, Zoropsidae, Ctenidae, Psechridae, and Stiphidiidae contain cribellate members. Despite its scattered distribution on the cladogram, the cribellum still emerges as primitive for this mixed cribellate-ecribellate lineage.

Larger problems in lycosoid systematics include the definition and possible polyphyly of the large family Ctenidae, or tropical wolf spiders. For example, the cribellate genus *Acanthoctenus,* often regarded as a ctenid, grouped with the cribellate family Zoropsidae in Griswold's analysis (on the basis of other characters than the retention of the cribellum). Similar questions about monophyly concern the Pisauridae, Bradystichidae, and Dolomedidae (153, 154).

DIONYCHA Primitively, all spiders were three-clawed. The derived two-clawed condition, in which the third claw is generally lost or very reduced and tufts of setae adorn the leg tip, characterizes the Dionycha. The monophyly of Dionycha has not yet been confirmed by a thorough analysis. These are primarily hunting animals that have forsaken webs as a foraging technique. The claw tuft, and/or the continuation of dense setae along the tarsus as a scopula, is known to improve traction on smooth surfaces (140). Another role of the scopula may be to improve prey manipulation during attacks. Some evidence suggests that removal of scopulae or claw tufts makes the animals less adept (141). The dionychan condition also occurs elsewhere in spiders, e.g. in Palpimanidae or Ctenidae, but sporadically at lower taxonomic levels. Several dionychan families remain unplaced in Figure 2 (see legend).

Gnaphosoidea and their relatives Three synapomorphies link the gnaphosoid families: flattened, irregularly shaped posterior median eyes, obliquely depressed endites, and heavily sclerotized anterior lateral spinnerets (118). The Gnaphosoidea currently includes seven families: Gnaphosidae ("ground" spiders;" ca. 110 genera); Prodidomidae (ca. 30 genera); Gallieniellidae (3 genera); Ammoxenidae (1 genus); Cithaeronidae (1 genus); Trochanteriidae (5 genera) and Lamponidae (1 genus) (118). Gnaphosidae and Prodidomidae are thought to be sister taxa, but other gnaphosoid interrelationships remain ambiguous (118). Gnaphosids don't spin catching webs; presumably the other families related to them are also wanderers.

Penniman (112) suggested tentatively that Liocranidae or Corinnidae, formerly parts of the old Clubionidae, might, based on eye features, separately or together be the sister group to the Gnaphosoidea. The classical Clubionidae (or Clubionoidea) is therefore paraphyletic. He placed Anyphaenidae and Clubionidae as outgroups to this clade based on morphology of the sternum. This evidence has not been assessed by other workers, but the hypothesis is reasonable. Anyphaenidae, Clubionidae, and Salticidae lack cylindrical glands and spigots, otherwise characteristic of "higher" entelegynes (see above), and may therefore be united by this secondary loss (79, 119).

Salticidae or "jumping spiders" are the largest family of spiders, quite specialized, and proportionately enigmatic in their relationships. The most salient family synapomorphy is the highly specialized pair of anterior median eyes. These ocelli have impressive optical properties that enable the visually based, stalking attack for which jumping spiders are famous (11, 83). The chief cladistic problem has always been the difficulty of estimating intrafamilial cladistic structure, although some progress has been made (162). Loerbroks (102) found evidence in palp structure and function to relate Salticidae and Thomisidae. Blest (11) sought the sister group of salticids among web-building spiders, thus arguing against Thomisidae. In a study of silk manipulation during the spinning of trail lines, Eberhard (45) found evidence to relate Salticidae, Gnaphosidae, and "Clubionidae," but that excluded Anyphaenidae and Thomisidae (more homoplasy).

The placement of Thomisidae within Dionycha is also unclear. Philodromidae, Sparassidae, and Selenopidae may eventually be placed near Thomisidae because they have laterigrade legs, which make their locomotion appear crab-like. The laterigrade condition makes hiding and maneuvering in narrow crevices easier. Based on eye structure, Homann (70) argued that Sparassidae and Philodromidae were sister taxa, but that Thomisidae were sister to Lycosoidea. In the light of Griswold's analysis of Lycosoidea, this hypothesis seems improbable, although Griswold did not include Thomisidae in his analysis. Once again, serious homoplasy is evident. Progress on dionychan relationships probably will be slow until all comparative data are considered simultaneously in one analysis.

ORBICULARIAE Reconstitution of the orb weavers, or Orbiculariae (25), also resulted from cladistic analysis of a classical cribellate-ecribellate dichotomy. Classically orb webs were thought to have evolved twice; once among the [paraphyletic] Cribellatae, and once among the [polyphyletic] Ecribellatae. The reputation of the orb as extremely adaptive confused the issue; some workers suggested as many as six separate origins of the orb geometry (67). Various authors considered a link between Uloboridae and

Araneidae or Araneoidea and explored its logical consequence; secondary loss of the orb among most of the remaining Araneoidea (17, 90, 107). Given the collapse of the Cribellatae and Ecribellatae as valid taxa, the orb web itself constituted initial evidence for monophyly. A series of detailed ethological and morphological investigations has failed to refute this hypothesis, thus corroborating that cribellate orb weavers (Deinopoidea) are the sister group of Araneoidea (26, 30, 31). If true, then the orb web evolved earlier than formerly believed and was subsequently lost in the large linyphiid and theridiid clades (31).

Orbiculariae are primitive spiders in many respects. Cribellate species preserve the plesiomorphic entire cribellum, and nearly all representatives preserve the apparently primitive pattern in trichobothrial distribution and tracheae. Although the fossil record of spiders is poor, orbicularians are among the earliest known entelegynes in the record (48, 50, 144, 145).

Deinopoidea The entirely cribellate Deinopoidea contains two families, about 25 genera, and some 300 species. All spin modified orbs. Some controversy existed in the past over placement of Deinopidae (ogre-faced spiders), but ethological work showed that they shared derived motor patterns unique to orb weavers, despite the derived web architecture (26). Uloborid genera have been revised recently (107), although recent work on their outgroups has suggested some changes in generic interrelationships (30). Uloborids are interesting phylogenetically because they may still retain many primitive aspects of orb weaver biology. In other aspects they are clearly derived, especially the complete loss of poison glands.

Araneoidea Araneoidea includes 11 families (ca. 740 genera, 10,000 species) or about a third of all described spiders. The most salient synapomorphies are the behavioral and morphological features that produce highly elastic viscid silk lines. Viscid silk is neither as sticky nor as durable as cribellate sticky silk, but it is faster and probably more economical to produce. Controversy over secondary loss of this ability as opposed to primary absence underlies most of the controversies about the composition of Araneoidea (e.g. the placement of various families here or in Palpimanoidea, see above). Araneoids are morphologically a rather compact group, despite their species-level diversity.

Taxonomic progress at the family level in Araneoidea has come mainly through relimitation and redefinition of the large families Araneidae and Theridiidae, the common orb weavers and cob web weavers (58, 100). The same fate may await Linyphiidae. Before the last decade, Araneidae usually included all ecribellate orb weavers and was thus defined by a plesiomorphy. Araneidae used to include the families Theridiosomatidae and Tetragnathidae.

The former araneid subfamilies Metinae and Nephilinae are now placed within Tetragnathidae (95, 101) based on apomorphies in male genitalia. Theridiidae used to include all or parts of the symphytognathoid families Anapidae, Mysmenidae, and Symphytognathidae (27, 125–127). Araneidae are now more compact and diagnosable (92), although still one of the largest spider families.

Unresolved problems in Araneoidea concern the affinities of the theridiid lineage (Synotaxidae, Nesticidae, Theridiidae, the "cob web" weavers) and the linyphiid lineage (Cyatholipidae and Linyphiidae, the "sheet web" spiders). Both groups are highly derived and thus difficult to place among the relatively more plesiomorphic araneoid groups. Somatic morphology and details of the spinning apparatus ally linyphiids with the "higher" araneoids, i.e. tetragnathids and the symphytognathoid families (31), but details in palp structure may place the linyphiids with the araneids (30). Despite much work, the placement of Theridiidae and its relatives within Araneoidea remains ambiguous.

CONCLUSIONS

Figures 1 and 2 compile the progress to date in proposing explicit cladograms that relate families of spiders. To what extent these often initial hypotheses will survive test by the addition of new taxa and evidence remains to be seen. Although strict phylogenetic reasoning has advanced spider systematics tremendously, conceptions of spider phylogeny in the past lacked more than a classical Hennigian basis. As quantitative analyses covering a broad range of character systems accumulate, it is apparent that character systems conflict. Phylogenies wholly consistent with supposed transformations in one character system are frequently inconsistent with others. Allocating this homoplasy is a serious problem and requires a quantitative approach, if only to establish an objective point of departure. Phylogenetic hypotheses based on single character systems that are oblivious to others are usually less useful. Likewise, comparative morphology is also less useful if authors fail to confront the phylogenetic implications of their hypotheses. In view of the weaknesses of these approaches, the most fruitful course will be for workers to consider carefully the phylogenetic implications of their own results in the context of other studies. Synthesis of comparative data is the core task of systematics; we hope that the cladograms compiled here will be useful to evolutionary biologists interested in spiders and in a common goal—the reconstruction and explication of evolutionary history.

ACKNOWLEDGMENTS

We first thank the many authors who made this review possible. Several were kind enough to share results in advance of publication. S. N. Austad, F. A.

Coyle, R. R. Forster, S. Glueck, P. A. Goloboff, C. E. Griswold, G. Hormiga, Y. Lubin, F. Murphy, B. D. Opell, W. F. Piel, N. I. Platnick, R. J. Raven, N. Scharff, W. A. Shear, and P. Sierwald provided helpful comments on an earlier version of the manuscript. L. Levi and L. Leibensperger rewrote some sections. Parts of the research reviewed in this article were supported by grants from the National Science Foundation, the Smithsonian Scholarly Studies Program, The Biological Diversity in Latin America Program (BIOLAT Project), and The Neotropical Lowlands Research Program. This is contribution number 22, BIOLAT Project, Smithsonian Institution.

Literature Cited

1. Alberti, G., Weinmann, C. 1985. Fine Structure of spermatozoa of some labidognath spiders (Filistatidae, Segestriidae, Dysderidae, Oonopidae, Scytodidae, Pholcidae; Araneae; Arachnida) with remarks on spermiogenesis. *J. Morphol.* 185(1):1–35
2. Anderson, J. F. 1970. Metabolic rates of spiders. *Comp. Biochem. Physiol.* 33 (1):51–72
3. Anderson, J. F., Prestwich, K. N. 1982. Respiratory gas exchange in spiders. *Physiol. Zool.* 55(1):72–90
4. Baehr, B., Baehr, M. 1987. The Australian (Hersiliidae: Arachnida: Araneae): taxonomy, phylogeny, zoogeography. *Invertebr. Taxon.* 1(4):351–438
5. Baehr, B., Baehr, M. 1988. On Australian Hersiliidae from the South Australian Museum (Arachnida, Araneae). Supplement to the revision of the Australian Hersiliidae. *Rec. S. Aust. Mus.* (Adelaide) 22(1):13–20
6. Barrientos, J. A., Ribera, C. 1988. Algunas reflexiones sobre las especies del grupo *atrica* en la Peninsula Iberica (Araneae, Agelenidae, *Tegenaria*). *Rev. Arachnol.* 7(4):141–62
7. Barth, F. G., Ed. 1985. *Neurobiology of Arachnids*. Berlin: Springer-Verlag
8. Baum, S. 1972. Zum "Cribellaten-Problem": die Genitalstrukturen der Oecobiinae und Urocteinae (Arachn: Araneae). *Verh. Naturw. Ver. Hamburg* (N.F.) 16:101–53
9. Bennett, R. G. 1988. The spider genus *Cybaeota* (Araneae: Agelenidae). *J. Arachnol.* 16(1):103–20
10. Bennett, R. G. 1991. *The systematics of the North American cybaeid spiders (Araneae: Dictynoidea: Cybaeidae)*. PhD thesis, Univ. Guelph, Ontario
11. Blest, A. D. 1985. The fine structure of spider photoreceptors in relation to function. See Ref. 7, pp. 79–102
12. Bonnet, P. 1945–1961. *Bibliographia Araneorum*. Toulouse: Vol. 1–3. 6481 pp
13. Breene, R. G., Sterling, W. L., Dean, D. A. 1988. Spider and ant predators of the cotton fleahopper on woolly croton. *Southwest Entomol.* 13(3):177–84
14. Breene, R. G., Sweet, M. H., Olson, J. K. 1988. Spider predators of mosquito larvae. *J. Arachnol.* 16(2):275–77
15. Brignoli, P. M. 1975. Über die Gruppe der Haplogynae (Araneae). *Proc. 6th Int. Arachnol. Congr.* (Amsterdam IV. 1974) 1974:33–38
16. Brignoli, P. M. 1978. Some remarks on the relationships between the Haplogynae, the semi-Entelegynae and the Cribellatae (Araneae). *Symp. Zool. Soc. London* 42:285–92
17. Brignoli, P. M. 1979. Contribution à la connaissance des Uloboridae paléarctiques (Araneae). *Rev. Arachnol.* 2:275–82
18. Brignoli, P. M. 1983. *Catalogue of the Araneae*. Manchester, England: Manchester Univ. Press
19. Brignoli, P. M. 1986. Phylogenèse et radiation adaptative des Araneae. *Boll. Zool.* 53:271–78
20. Bristowe, W. S. 1938. The classification of spiders. *Proc. Zool. Soc. London* (B) 108:285–322
21. Bristowe, W. S. 1958. *The World of Spiders*. London: Collins. 304 pp
22. Carico, J. E. 1986. Trechaleidae: A "new" American spider family. (Abstr.). See Ref. 4a, p. 305
23. Chamberlin, R. V., Ivie, W. 1945. On some Nearctic mygalomorph spiders. *Ann. Entomol. Soc. Am.* 38:549–58
24. Clerck, C. 1757. *Aranei Suecici*. Stockholm. 154 pp
25. Coddington, J. A. 1986. The monophyletic origin of the orb web. See Ref. 149, pp. 319–63
26. Coddington, J. A. 1986. Orb webs in non-orb-weaving ogre-faced spiders

(Araneae: Deinopidae): a question of genealogy. *Cladistics* 2(1):53–67
27. Coddington, J. A. 1986. The genera of the spider family Theridiosomatidae. *Smithson. Contrib. Zool.* 422:1–96
28. Coddington, J. A. 1987. Notes on spider natural history: the webs and habits of *Araneus niveus* and *A. cingulatus* (Araneae, Araneidae). *J. Arachnol.* 15(2):268–70
29. Coddington, J. A. 1989. Spinneret silk spigot morphology. Evidence for the monophyly of orb-weaving spiders, Cyrtophorinae (Araneidae), and the group Theridiidae-Nesticidae. *J. Arachnol.* 17(1):71–95
30. Coddington, J. A. 1990. Ontogeny and homology in the male palpus of orb weaving spiders and their relatives, with comments on phylogeny (Araneoclada: Araneoidea, Deinopoidea). *Smithson. Contrib. Zool.* 496:1–52
31. Coddington, J. A. 1990. Cladistics and spider classification: Araneomorph phylogeny and the monophyly of orb weavers (Araneae: Araneomorphae; Orbiculariae). *Acta Zool. Fennica* 190:75–87
32. Coddington, J. A., Griswold, C. E., Silva, D., Peñaranda, E., Larcher, S. F. 1991. Designing and testing sampling protocols to estimate biodiversity in tropical ecosystems. In *The Unity of Evolutionary Biology: Proc. Int. Congr. Syst. Evol. Biol.*, ed. E. C. Dudley, pp. 44–60. Portland, Ore: Dioscorides Press
33. Coddington, J. A., Larcher, S. F., Cokendolpher, J. C. 1990. The systematic status of Arachnida, exclusive of Acarina, in North America north of Mexico (Arachnida: Amblypygi, Araneae, Opiliones, Palpigradi, Pseudoscorpiones, Ricinulei, Schizomida, Scorpiones, Solifugae, Uropygi). In *Diversity and Dynamics of North American Insect and Arachnid Fauna,* ed. M. Koztarab, C. W. Schaeffer, pp. 5–20. Blacksburg, Va: Va Polytech. Inst.
34. Cohn, J. 1990. Is it size that counts? Palp morphology, sperm storage and egg-hatching frequency in *Nephila clavipes. J. Arachnol.* 18:59–71
35. Cokendolpher, J. C. 1988. Arachnological publications and societies of the world. *Am. Arachnol.* 38:10–12
36. Court, D. J., Forster, R. R. 1988. The spiders of New Zealand. Part VI. Family Araneidae. *Otago Mus. Bull.* 6:68–124
37. Coyle, F. A. 1985. Two year life cycle and low palpal character variance in a Great Smoky Mountain USA population of the lampshade spider (Araneae: Hypochilidae) *Hypochilus. J. Arachnol.* 13(2):211–18
38. Coyle, F. A. 1986. The role of silk in prey capture by non-araneomorph spiders. See Ref. 149, pp. 269–305
39. Coyle, F. A., Ketner, N. D. 1990. Observations on the prey and prey capture behaviour of the funnelweb mygalomorph spider genus *Ischnothele* (Araneae, Dipluridae). *Bull. Br. Arachnol. Soc.* 8(4):97–104
40. Davies, V. T. 1980. *Malkara loricata,* a new spider (Araneidae: Malkarinae) from Australia. See Ref. 65a, pp. 377–82
41. Dondale, C. D. 1986. The subfamilies of wolf spiders (Araneae: Lycosidae). In *Actas X Congr. Int. de Aracnol.*, Vol. 1, ed. J. A. Barrientos, pp. 327–32. Jaca, Spain: Inst. Pirenaico de Ecol.
42. Duffey, E., Green, M. B. 1975. A linyphiid spider biting workers on a sewage-treatment plant. *Bull. Br. Arachnol. Soc.* 3(5):130–31
43. Eberhard, W. G. 1980. Persistent stickiness of cribellum silk. *J. Arachnol.* 8(3):283
44. Eberhard, W. G. 1985. *Sexual Selection and Animal Genitalia.* Cambridge, Mass: Harvard Univ. Press
45. Eberhard, W. G. 1986. Trail line manipulation as a character for higher level spider taxonomy. See Ref. 47a, pp. 49–51
46. Eberhard, W. G. 1988. Combing and sticky silk attachment behaviour by cribellate spiders and its taxonomic implications. *Bull. Br. Arachnol. Soc.* 7(8):247–51
47. Eberhard, W. G. 1990. Function and phylogeny of spider webs. *Annu. Rev. Ecol. Syst.* 21:341–72
47a. Eberhard, W. G., Lubin, Y. D., Robinson, B. C. 1983. *Proceedings of the Ninth International Congress of Arachnology, Panama, 1983*. Washington: Smithson. Inst. Press
48. Eskov, K. Y. 1984. A new fossil spider family from the Jurassic of Transbaikalia (Araneae: Chelicerata). *Neues Jb. Geol. Palaont. Mh.* 1984(11):645–53
49. Eskov, K. Y. 1987. A new archaeid spider (Chelicerata: Araneae) from the Jurassic of Kazakhstan, with notes on the so-called "Gondwanan" ranges of recent taxa. *Neues Jb. Geol. Paleont. Abh.* 175:81–106
50. Eskov, K. Y., Zonshtein, S. 1990. First Mesozoic mygalomorph spiders from the lower Cretaceous of Siberia and Mongolia, with notes on the system and evolution of the infraorder Mygalomorphae (Chelicerata: Araneae). *Neues Jb. Geol. Paleont. Abh.* 178(3):325–68
51. Forster, R. R. 1967. The spiders of New

Zealand. Part 1. *Otago Mus. Bull.* 1:1–124

52. Forster, R. R. 1970. The spiders of New Zealand. Part 3 (Desidae, Dictynidae, Hahniidae, Amaurobioididae, Nicodamidae). *Otago Mus. Bull.* 3:1–184
53. Forster, R. R. 1980. Evolution of the tarsal organ, the respiratory system, and the female genitalia in spiders. See Ref. 65a, pp. 269–84
54. Forster, R. R. 1988. The spiders of New Zealand. Part 6. Family Cyatholipidae. *Otago Mus. Bull.* 6:7–34
55. Forster, R. R., Blest, A. D. 1979. The spiders of New Zealand. Part 5. Cycloctenidae, Gnaphosidae, Clubionidae. Linyphiidae—Mynogleninae. *Otago Mus. Bull.* 5:1–173
56. Forster, R. R., Platnick, N. I. 1984. A review of the archaeid spiders and their relatives, with notes on the limits of the superfamily Palpimanoidea (Arachnida, Araneae). *Bull. Am. Mus. Nat. Hist.* 178(1):1–106
57. Forster, R. R., Platnick, N. I. 1985. A review of the austral spider family Orsolobidae (Arachnida, Araneae), with notes on the superfamily Dysderoidea. *Bull. Am. Mus. Nat. Hist.* 181:1–230
58. Forster, R. R., Platnick, N. I., Coddington, J. A. 1990. A proposal and review of the spider family Synotaxidae (Araneae, Araneoidea), with notes on theridiid interrelationships. *Bull. Am. Mus. Nat. Hist.* 189:1–116
59. Forster, R. R., Platnick, N. I., Gray, M. R. 1987. A review of the spider superfamilies Hypochiloidea and Austrochiloidea (Araneae; Araneomorphae). *Bull. Am. Mus. Nat. Hist.* 185(1):1–116
60. Forster, R. R., Wilton, C. L. 1968. The spiders of New Zealand. Part 2. Ctenizidae, Dipluridae, Migidae. *Otago Mus. Bull.* 2:1–180
61. Forster, R. R., Wilton, C. L. 1973. The spiders of New Zealand. Part IV. Agelenidae, Stiphiidae, Amphinectidae, Amaurobiidae, Neolanidae, Ctenidae, Psechridae. *Otago Mus. Bull.* 4:1–309
62. Gertsch, W. J. 1979. *American Spiders,* New York: Van Nostrand Reinhold. 2nd ed.
63. Gertsch, W. J., Platnick, N. I. 1980. A revision of the American spiders of the family Atypidae (Araneae, Mygalomorphae). *Am. Mus. Novitates* 2704:1–39
64. Griswold, C. E. 1990. A revision and phylogenetic analysis of the spider subfamily Phyxelidinae (Araneae, Amaurobiidae). *Bull. Am. Mus. Nat. Hist.* 196:1–206
65. Griswold, C. E. 1992. Investigations into the phylogeny of the lycosoid spiders and their kin (Arachnida: Araneae: Lycosoidea). *Smith. Contr. Zool.* In press

65a. Gruber, J. 1980. *Proc. 8th Internationaler Arachnologen—Kongress abgehalten an der Universität für Bodenkultur Wien, 7–12 Juli, 1980.* Vienna: H. Egermann

66. Heimer, S. 1982. Interne Arretierungsmechanismen an den Kopulationsorganen männlichen Spinnen (Arachnida: Araneae), ein Beitrag zur Phylogenie der Araneoidea. *Entomol. Abh. Mus. Tierk. Dresden* 45:35–64
67. Heimer, S., Nentwig, W. 1982. Thoughts on the phylogeny of the Araneoidea Latreille, 1806 (Arachnida, Araneae). *Z. Zool. Syst. Evolut.-forsch.* 20(4):284–95
68. Hieber, C. S. 1984. Egg predators of the cocoons of the spider *Mecynogea lemniscata* (Araneae: Araneidae): rearing and population data. *Florida Entomol.* 67(1):176–78
69. Homann, H. 1971. Die Augen der Araneae. Anatomie, Ontogenie und Bedeutung für die Systematik (Chelicerata, Arachnida). *Z. Morphol. Tiere* 69:201–72
70. Homann H. 1975. Die Stellung der Thomisidae und der Philodromidae im System der Araneae (Chelicerata, Arachnida). *Z. Morphol. Tiere* 80:181–202
71. Homann, H. 1985. Die Cheliceren der Araneae, Amblypygi, und Uropygi mit den Skleriten, den Plagulae (Chelicerata, Arachnomorpha). *Zoomorphologie* 105:69–75
72. Jackson, H., Parks, T. N. 1989. Spider toxins: recent applications in neurobiology. *Annu. Rev. Neurosci.* 12:405–14
73. Jakob, E. M., Dingle, H. 1990. Food level and life history characteristics in a pholcid spider *Holocnemus pluchei. Psyche* 97:95–110
74. Jeram, A. J., P. A. Selden, D. E. Edwards, 1990. Land animals in the Silurian: arachnids and myriapods from Shropshire, England. *Science* 250:658–61
75. Jocqué, R. 1991. A generic revision of the spider family Zodariidae (Araneae). *Bull. Am. Mus. Nat. Hist.* 201:1–160
76. Kaston, B. J. 1978. *How To Know the Spiders*. Dubuque, Iowa: Wm. C. Brown. 272 pp. 3rd ed.
77. Kaston, B. J. 1981. Spiders of Connecticut. *Bull. Conn. Geol. Nat. Hist. Surv.* 70:1–1020
78. Kovoor, J. 1977. La soie et les glandes

séricigènes des Arachnides. *Ann. Biol.* 16:97–141

79. Kovoor, J. 1987. Comparative structure and histochemistry of silk-producing organs in arachnids. See Ref. 105, pp. 160–86
80. Kraus, O. 1978. *Liphistius* and the evolution of spider genitalia. *Symp. Zool. Soc. London* 42:235–54
81. Kraus, O. 1984. Male spider genitalia: evolutionary changes in structure and function. *Verh. Naturw. Ver. Hamburg* 27:373–82
82. Kullmann, E., Zimmermann, W. 1976. Ein neuer Beitrag zum Cribellaten-Ecribellaten-Problem: Beschreibung von *Uroecobius ecribellatus* n. gen. n. sp. und Diskussion seiner phylogenetischen Stellung (Arachnida, Araneae, Oecobiidae). *Entomol. Germ.* 3(1–2):29–40
83. Land, M. F. 1985. The morphology and optics of spider eyes. See Ref. 7, pp. 53–78
84. Lehtinen, P. T. 1967. Classification of the Cribellate spiders and some allied families, with notes on the evolution of the suborder Araneomorpha. *Ann. Zool. Fennici* 4:199–467
85. Lehtinen, P. T. 1978. Definition and limitation of supraspecific taxa in spiders. *Symp. Zool. Soc. London* 42:255–71
86. Lehtinen, P. T. 1980. Trichobothrial patterns in high level taxonomy of spiders. See Ref. 65a, pp. 493–98
87. Lehtinen, P. T. 1981. Spiders of the Oriental-Australian region. III: Tetrablemmidae, with a world revision. *Acta Zool. Fennici* 162:1–151
88. Lehtinen, P. T. 1986. Evolution of the Scytodoidea. See Ref. 47a, pp. 149–59
89. Levi, H. W. 1973. Small orb-weavers of the genus *Araneus* North of Mexico (Araneae, Araneidae). *Bull. Mus. Comp. Zool.* 145:473–552
90. Levi, H. W. 1980. Orb-webs: primitive or specialized. See Ref. 65a, pp. 367–70
91. Levi, H. W. 1981. The American orb-weaver genera *Dolichognatha* and *Tetragnatha* (Araneae: Araneidae, Tetragnathinae). *Bull. Mus. Comp. Zool.* 149:271–318
92. Levi, H. W. 1983. The orb-weaver genera *Argiope, Gea* and *Neogea* from the western Pacific region (Araneae: Argiopinae, Araneidae). *Bull. Mus. Comp. Zool.* 150(5):247–338
93. Levi, H. W. 1986. Ant-mimicking orb-weavers of the genus *Ildibaha*. See Ref. 47a, pp. 159–62
94. Levi, H. W. 1986. The orb-weaver genus *Witica* (Araneae: Araneidae). *Psyche* 93:35–46
95. Levi, H. W. 1986. The neotropical orb-weaver genera *Chrysometa* and *Homalometa*. *Bull. Mus. Comp. Zool.* 151: 91–215
96. Levi, H. W. 1988. The neotropical orb-weaving spiders of the genus *Alpaida* (Araneae: Araneidae). *Bull. Mus. Comp. Zool.* 151:365–487
97. Levi, H. W. 1989. The neotropical orb-weaving genera *Epeiroides, Bertrana,* and *Amazonepeira* (Araneae: Araneidae). *Psyche* 96:75–99
98. Levi, H. W. 1991. The neotropical and Mexican species of the orb-weaver genera *Araneus, Dubiepeira* and *Aculepeira* (Araneae: Araneidae). *Bull. Mus. Comp. Zool.* 152(5):167–315
99. Levi, H. W. 1991. The Neotropical Orb-weaver genera *Edricus* and *Wagneriana* (Araneae: Araneidae). *Bull. Mus. Comp. Zool.* 152:363–417
100. Levi, H. W., Coddington, J. A. 1983. Progress report on the phylogeny of the orb-weaving family Araneidae and the superfamily Araneoidea. In *Taxonomy, Biology and Ecology of the Araneae,* ed. O. Kraus, pp. 151–54. Hamburg: Naturwiss Verein
101. Levi, H. W., von Eickstedt, R. D. 1989. The Nephilinae spiders of the Neotropics. *Mem. Inst. Butantan* 51:43–56
102. Loerbroks, A. 1984. Mechanik der Kopulationsorgane von *Misumena vatia* (24) (Arachnida: Araneae: Thomisinae). *Verh. Naturw. Ver. Hamburg* 27:383–403
103. Maretic, Z., Lebez, D. 1979. Araneism with special reference to Europe. Belgrade: Nolit Publ.
104. Mitter, C., Farrell, B., Wiegmann, B. 1988. The phylogenetic study of adaptive zones: has phytophagy promoted insect diversification? *Am. Nat.* 132(1): 107–28
105. Nentwig, W., ed. 1987. *Ecophysiology of Spiders*. Berlin: Springer-Verlag
106. Nyffeler, M., Benz, G. 1987. Spiders in natural pest control: a review. *J. Appl. Entomol.* 103(4):321–39
107. Opell, B. D. 1979. Revision of the genera and tropical American species of the spider family Uloboridae. *Bull. Mus. Comp. Zool.* 148(10):443–549
108. Opell, B. D. 1984. A simple method for measuring desiccation resistance of spider egg sacs. *J. Arachnol.* 12(2):245–47
109. Oxford, G. S. 1989. Genetics and distribution of black spotting in *Enoplognatha ovata* (Araneae: Theridiidae), and the role of intermittent drift in population differentiation. *Biol. J. Linnean Soc.* 36:111–28
110. Palmer, J. M. 1991. *Comparative*

morphology of the external silk production apparatus of primitive spiders. PhD thesis. Harvard University, Cambridge, Mass.

111. Parker, S. P., ed. 1982. *Synopsis and Classification of Living Organisms*. Vol. 2. New York: McGraw-Hill
112. Penniman, A. J. 1985. *Revision of the* britcheri *and* pugnata *groups of* Scotinella *(Araneae: Corinnidae, Phrurolithinae) with a reclassification of phrurolithine spiders*. PhD thesis. Ohio State Univ., Columbus
113. Petrunkevitch, A. 1928. Systema Aranearum. *Trans. Conn. Acad. Arts Sci.* 29:1–270
114. Petrunkevitch, A. 1933. An inquiry into the natural classification of spiders, based on a study of their internal anatomy. *Trans. Conn. Acad. Arts Sci.* 31:303–89
115. Platnick, N. I. 1975. On the validity of Haplogynae as a taxonomic grouping in spiders. *Proc. 6th Int. Arachnol. Congr.* (Amsterdam IV.1974):30–32
116. Platnick, N. I. 1977. The hypochiloid spiders: a cladistic analysis, with notes on the Atypoidea (Arachnida: Araneae). *Am. Mus. Novitates* 2627:1–23
117. Platnick, N. I. 1989. *Advances in Spider Taxonomy, 1981–1987*. Manchester, UK: Manchester Univ. Press
118. Platnick, N. I. 1990. Spinneret morphology and the phylogeny of ground spiders (Araneae, Gnaphosidae). *Am. Mus. Novitates* 2978:1–42
119. Platnick, N. I., Coddington, J. A., Forster, R. R., Griswold, C. E. 1991. Spinneret morphology and the phylogeny of haplogyne spiders. *Am. Mus. Novitates*. 3016:1–73
120. Platnick, N. I., Forster, R. R. 1987. On the first American spiders of the subfamily Sternodinae (Araneae: Malkaridae). *Am. Mus. Novitates* 2894:1–12
121. Platnick, N. I., Gertsch, W. J. 1976. The suborders of spiders: a cladistic analysis (Arachnida, Araneae). *Am. Mus. Novitates* 2807:1–15
122. Platnick, N. I., Goloboff, P. A. 1985. On the monophyly of the spider suborder Mesothelae (Arachnida: Araneae). *J. N.Y. Entomol. Soc.* 93(4):1265–70
123. Platnick, N. I., Sedgwick, W. C. 1984. A revision of the spider genus *Liphistius* (Araneae, Mesothelae). *Am. Mus. Novitates* 2781:1–31
124. Platnick, N. I., Shadab, M. U. 1976. A revision of the Mygalomorph spider genus *Neocteniza* (Araneae, Actinopodidae). *Am. Mus. Novitates* 2603:1–19
125. Platnick, N. I., Shadab, M. U. 1978. A review of the spider genus *Mysmenopsis* (Araneae, Mysmenidae). *Am. Mus. Novitates* 2661:1–22
126. Platnick, N. I., Shadab, M. U. 1978. A review of the spider genus *Anapis* (Araneae, Anapidae), with a dual cladistic analysis. *Am. Mus. Novitates* 2663: 1–23
127. Platnick, N. I., Shadab, M. U. 1979. A review of the spider genera *Anapisona* and *Pseudanapis* (Araneae, Anapidae). *Am. Mus. Novitates* 2672:1–20
128. Raven, R. J. 1978. Systematics of the spider subfamily Hexathelinae (Dipluridae: Mygalomorphae: Arachnida). *Aust. J. Zool. (Suppl.)* 65:1–75
129. Raven, R. J. 1985. The spider infraorder Mygalomorphae: Cladistics and systematics. *Bull. Am. Mus. Nat. Hist.* 182:1–180
130. Raven, R. J. 1986. A cladistic reassessment of mygalomorph spider families (Araneae). See Ref. 47a, pp. 223–27
131. Raven, R. J. 1988. The current status of Australian spider systematics. In *Australian Arachnology,* ed. A. D. Austin, N. W. Heather, pp. 1–137. *Aust. Entomol. Soc., Misc. Publ. No. 5*
132. Reillo, P. R., Wise, D. 1988. Genetics of color expression in the spider *Enoplognatha ovata* (Araneae: Theridiidae) from coastal Maine. *Am. Midland Natural* 119:318–26
133. Riechert, S. E., Bishop, L. 1990. Prey control by an assemblage of generalist predators: spiders in garden text systems. *Ecology* 71:1441–50
134. Riechert, S. E., Lockley, T. C. 1984. Spiders as biological control agents. *Annu. Rev. Entomol.* 29:299–320
135. Riechert, S. E., Uetz, G. W., Abrams, B. 1985. The state of arachnid systematics. *Bull. Entomol. Soc. Amer.* 31:4–5
136. Roberts, M. J. 1985. *The Spiders of Great Britain and Ireland*. Vols. 1, 3. Colchester: Harley. 229 pp. 256 pp.
137. Roberts, M. J. 1987. *The Spiders of Great Britain and Ireland*. Vol. 2. *Linyphiidae and Checklist. Martins, Essex:* Harley. 204 pp.
138. Roewer, C. F. 1942. *Katalog der Araneae von 1758 bis 1940, bzw. 1954*. Vol. 1. Bremen. [Publ. not given] 1040 pp.
139. Roth, V. 1985. *Spider Genera of North America*. Gainesville, Fla: Am. Arachnol. Soc.
140. Rovner, J. S. 1978. Adhesive hairs in spiders: behavioral functions and hydraulically mediated movement. *Symp. Zool. Soc. London* 42:99–108
141. Rovner, J. S. 1980. Morphological and ethological adaptations for prey capture

in wolf spiders (Araneae, Lycosidae). *J. Arachnol.* 8(3):201–15
142. Russell, F. E., Gertsch, W. J. 1983. Letter to the Editor. *Toxicon* 21(3):337–39
143. Schult, J. 1983. Taster haplogyner spinnen unter phylogenetischem aspekt (Arachnida: Araneae). *Verh. Naturw. Ver. Hamburg* 26:69–84
144. Selden, P. A. 1990a. Lower Cretaceous spiders from Sierra de Montsech, northeast Spain. *Paleontology* 33(2):257–85
145. Selden, P. A. 1990b. Fossil history of the arachnids. *Br. Arachnol. Soc. Newsl.* 58:4
146. Selden, P. A., Shear, W. A., Bonamo, P. M. 1991. A spider and other arachnids from the Devonian of New York, and a reinterpretation of Devonian fossils previously assigned to the Araneae. *Paleontology* 34(2): In press
147. Shear, W. A. 1978. Taxonomic notes on the armored spiders of the families Tetrablemmidae and Pacullidae. *Am. Mus. Novitates* 2650:1–46
148. Shear, W. A. 1981. Structure of the male palpal organ in *Mimetus, Ero* and *Gelanor* (Araneoidea, Mimetidae). *Bull. Am. Mus. Nat. Hist.* 170(1):257–62
149. Shear, W. A., ed. 1986. *Spiders: Webs, Behavior and Evolution*. Stanford: Stanford Univ. Press
150. Shear, W. A., Palmer, J. M., Coddington, J. A., Bonamo, P. M. 1989. A Devonian spinneret: early evidence of spiders and silk use. *Science* 246:479–81
151. Shear, W. A., Selden, P. A., Rolfe, W. D. I., Bonamo, P. M., Grierson, J. D. 1987. New fossil arachnids from the Devonian of Gilboa, New York. *Am. Mus. Novitates* 2901:1–74
152. Shultz, J. W. 1990. Evolutionary morphology and phylogeny of Arachnida. *Cladistics* 6:1–38
153. Sierwald, P. 1989. Morphology and ontogeny of female copulatory organs in American Pisauridae, with special reference to homologous features. *Smithson. Contrib. Zool.* 484:1–24
154. Sierwald, P. 1990. Morphology and homologous features in the male palpal organ in Pisauridae and other spider families, with notes on the taxonomy of Pisauridae. *Nemouria* 35:1–59
155. Sierwald, P., Coddington, J. A. 1988. Functional aspects of the copulatory organs in *Dolomedes tenebrosus*, with notes on the mating behavior (Araneae: Pisauridae). *J. Arachnol.* 16(2):262–65
156. Simon, E. 1892–1903. *Histoire naturelle des araignées*. Paris: Roret. Vol. 1:1–1084, Vol. 2:1–1080
157. Stapleton, A., Blankenship, D. T., Ackermann, B. L., Chen, T. M., Gorder, G. W., et al 1990. *Curtatoxins*. Neurotoxic insecticidal polypeptides isolated from the funnel-web spider Hololena curta. *J. Biol. Chem.* 265(4):1990; 2054–59
158. Starabogatov, Y. I. 1985. Taxonomic position and the system of the order of spiders (Araneiformes). *Proc. Zool. Inst. Acad. Sci. USSR* 139:4–16
159. Strohmenger, T., Nentwig, W. 1987. Adhesive and trapping properties of silk from different spider species. *Zool. Anz. (Leipzig)* 218(1–2):9–16
160. Turnbull, A. L. 1973. Ecology of the true spiders (Araneomorphae). *Annu. Rev. Entomol.* 18:305–48
161. von Helverson, O. 1976. Gedanken zur Evolution der Paarungsstellung bei den Spinnen (Arachnida, Araneae). *Entomol. Germ.* 3(1–2):13–28
162. Wanless, F. R. 1984. A review of the spider subfamily Spartaeinae nom. n. (Araneae: Salticidae), with descriptions of six new genera. *Bull. Br. Mus. Nat. Hist. (Zool.)* 46(2):135–205
163. Weygoldt, P., Paulus, H. F. 1979. Untersuchungen zur Morphologie, Taxonomie, und Phylogenie der Chelicerata. II. Cladogramme und die Entfaltung der Chelicerata. *Z. Zool. Syst. Evolut.-forsch.* 17:177–200
164. Work, R. 1976. The force elongation behavior of web fibers and silks forcibly obtained from orb-web-spinning spiders. *Texile Res. J.* 46:485–92
165. Wunderlich, J. 1986. *Spinnenfauna gestern und heute. I. Fossile Spinnen in Bernstein und ihre heute lebenden Verwandten.* Wiesbaden: Erich Bauer Verlag bei Quelle & Meyer
166. Yaginuma, T. 1977. A list of Japanese spiders (revised in 1977). *Acta Arachnol.* 27 (special No.):367–406
167. Yu, L., Coddington, J. A. 1990. Ontogenetic changes in the spinning fields of *Nuctenea cornuta* and *Neoscona theisi* (Araneae; Araneidae). *J. Arachnol.* 18(3):331–45

SUBJECT INDEX

B

C

H

Q

R

CUMULATIVE INDEXES

CONTRIBUTING AUTHORS, VOLUMES 18–22

CHAPTER TITLES, VOLUMES 18–22